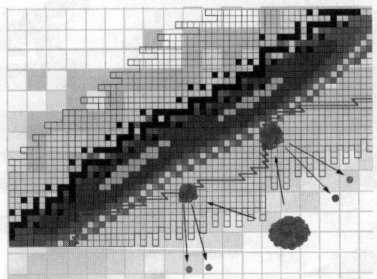

Proceedings of the Third International Conference

Fission and Properties of Neutron-Rich Nuclei

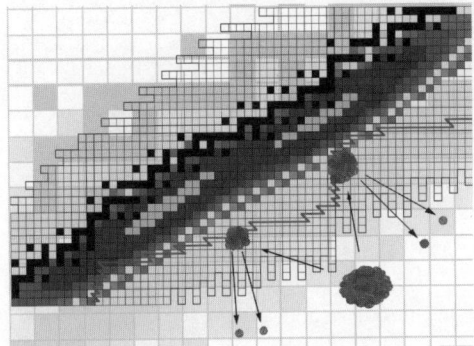

Proceedings of the Third International Conference

Fission and Properties of Neutron-Rich Nuclei

Sanibel Island, Florida, USA   3 – 9 November 2002

edited by

J H Hamilton
Vanderbilt University, USA

A V Ramayya
Vanderbilt University, USA

H K Carter
Oak Ridge Associated Universities, USA

 **World Scientific**

NEW JERSEY • LONDON • SINGAPORE • SHANGHAI • HONG KONG • TAIPEI • BANGALORE

Published by

World Scientific Publishing Co. Pte. Ltd.

5 Toh Tuck Link, Singapore 596224

USA office: Suite 202, 1060 Main Street, River Edge, NJ 07661

UK office: 57 Shelton Street, Covent Garden, London WC2H 9HE

British Library Cataloguing-in-Publication Data
A catalogue record for this book is available from the British Library.

Cover design by Susan Jacques

FISSION AND PROPERTIES OF NEUTRON-RICH NUCLEI
Proceedings of the Third International Conference

ISBN 981-238-386-7

Printed by FuIsland Offset Printing (S) Pte Ltd, Singapore

PROGRAM COMMITTEE AND ADVISORY COMMITTEE

International Advisory Committee

Joseph H. Hamilton (Chair), Vanderbilt University
Juha Äystö, EP-ISOLDE, CERN
Jon Batchelder, UNIRIB/Oak Ridge Associated Universities
Rafael J. Broda, Institute of Nuclear Physics, Poland
Jerry D. Cole, Lockheed Martin Idaho Technologies
A. Covello, Dipartimento di Scienze Fisiche
George D. Dracoulis, Australian National University
Amand Faessler, Universität Tübingen
Friedrich Goennenwein, Universität Tübingen
Walter Greiner, Johann Wolfgang von-Goethe Universität
Konstantin A. Gridnev, St. Petersburg University
Robert V.F. Janssens, Argonne National Laboratory
I. Y. Lee, Lawrence Berkeley National Laboratory
Yixiao Luo, Lawrence Berkeley National Laboratory
W.C. Ma, Mississippi State University
Witold Nazarewicz, University of Tennessee
Yuri Oganessian, Flerov Laboratory of Nuclear Reactions
W. R. Phillips, University of Manchester
D. Poenariu, National Institute of Physics
John O. Rasmussen, Lawrence Berkeley National Laboratory
Lee L. Riedinger, Oak Ridge National Laboratory
Ernst Roeckl, Gesellschaft für Schwerionenforschung
K. Rykacewski, Oak Ridge National Laboratory
Aurel Sandulescu, Institute of Atomic Physics
Bradley M. Sherrill, NSCL, Michigan State University
Michael Scott Smith, Oak Ridge National Laboratory
Isao Tanihata, RIKEN
G.M. Ter-Akopian, Flerov Laboratory of Nuclear Reactions
Piet Van Duppen, University of Leuven
C. Wagemans, University of Gent
Michael Wiescher, University of Notre Dame
Shengjiang Zhu, Tsinghua University

Local Organizing Committee

C.R. Bingham, University of Tennessee
H.K. Carter, Oak Ridge Associated Universities (Chair)
Joseph H. Hamilton, Vanderbilt University (Co-chair)
Sherry Lamb, University of Tennessee (Co-secretary)
A.V. Ramayya, Vanderbilt University
Carol Soren, Vanderbilt University (Secretary)
Carlene Stewart, Oak Ridge Associated Universities (Secretary)
E.F. Zganjar, Louisiana State University

Sponsors

Vanderbilt University
UNIRIB
Oak Ridge Associated Universities
Oak Ridge National Laboratory
Joint Institute for Heavy Ion Research

PREFACE

This Third International Conference on Fission and Properties of Neutron-Rich Nuclei was held November 3–9, 2002 at Sanibel Island, Florida. The conference followed very closely the spirit set by the previous two held in 1997 at Sanibel and 1999 at St. Andrews, Scotland. These conferences demonstrate the strong interest worldwide in these fields. This third conference documents very well the continued growth of new data, developments, and interest in this field as shown by over 100 papers presented over five and one-half days of sessions.

The conference assembled many of the world's experts in the fields of fission, properties of neutron-rich nuclei, astrophysics, and super heavy elements, both theoretical and experimental. Speakers presented overviews and the most recent data that provide an up-to-the-minute picture of these fields. Evolutions of these fields were quite evident. For example, in 1997 radioactive ion beam facilities were discussed as proposals in one talk. In this third conference an entire session was devoted to radioactive beam facilities, and data from some of these facilities were presented in other sessions. The interest in properties of neutron-rich nuclei and astrophysics has grown to occupy approximately half of the presentations. The future of these fields appears to be very bright as evidenced by the results presented and by the presence of many young researchers and graduate students not only as attendees but also as speakers.

These proceedings are an attempt by the editors to capture the conference in a pedagogical format. The papers have been arranged according to the principle topics and within each section they are arranged in order of presentation in the program with the short poster session papers at the end of the section.

We want to especially thank the international advisory committee and the local organizing committee listed next for all their diligent work. The scientific success was assured by the many thoughtful suggestions for speakers and topics by the international advisory committee. As planning proceeded, several members of this committee took very active roles in helping to ensure that the most current and important results from all laboratories were presented and to help balance speakers and topics. Unfortunately, we did not learn early enough of new regulations that significantly extended the visa application processing time for scientists from a number of countries. Therefore, several speakers did not receive their visas in time. In some cases, other co-authors on the paper presented their results. A few other papers that were not presented orally are included in these proceedings. The conference setting, in addition to its natural beauty, provided convenient arrangements for informal discussions and for enjoyable outings. The Sundial Beach Resort's entire staff went beyond their job descriptions in providing excellent surroundings, food, and excursion, and so is due special thanks.

We want to thank each of our sponsors: Vanderbilt University, UNIRIB, Oak Ridge Associated Universities, Oak Ridge National Laboratory, and the Joint Institute for Heavy Ion Research for their financial support of the conference. In particular, Oak Ridge Associated Universities is recognized for providing the conference chair with significant funds to organize a conference of his choosing in conjunction with presenting him their first Outstanding Service Award for the year 2000. The underpinning of any conference is the secretarial staff that sees to the endless details from long before to long after the conference including preparation of the proceedings. We are most appreciative to Carlene Stewart, UNIRIB/ORAU; Carol Soren, Vanderbilt University; and Sherry Lamb, the Joint Institute for Heavy Ion Research/University of Tennessee for their tireless efforts and very pleasant demeanor and calmness in the face of many demands in carrying out the conference. The editors and local organizing committee appreciate the skill and artistic abilities of Susan Jacques in providing pleasant and meaningful logos, book covers, and illustrations and Jill Krase in attending to the details in preparing the proceedings. We also want to thank J. K. Hwang for his help in getting the submitted manuscripts ready for the proceedings.

J. H. Hamilton
Vanderbilt University

A.V. Ramayya
Vanderbilt University

H. K. Carter
UNIRIB
Oak Ridge Associated Universities

Contents

x

Nuclear Astrophysics

Radioactive Ion Beam Facilities

Nuclear Structure

STRUCTURE OF NEUTRON-RICH NUCLEI

W. NAZAREWICZ[1-3] AND M. PŁOSZAJCZAK[4]

[1] *Department of Physics, University of Tennessee, Knoxville, Tennessee 37996*
[2] *Physics Division, Oak Ridge National Laboratory, Oak Ridge, Tennessee 37831*
[3] *Institute of Theoretical Physics, Warsaw University*
ul. Hoża 69, PL-00-681 Warsaw, Poland
[4] *Grand Accélérateur National d'Ions Lourds (GANIL), CEA/DSM –*
CNRS/IN2P3, F-14076 Caen, France

Structure of exotic radioactive nuclei having extreme neutron-to-proton ratios is different from that around the stability line. This short review discusses the progress in modeling of exotic nuclei in the neutron-rich "Terra Incognita". The consistent theoretical description of weakly bound systems requires a synergy between nuclear structure and nuclear reaction methods.

1. Introduction

Low-energy nuclear physics is undergoing a renaissance. Experimentally, there has been a technological revolution in the radioactive nuclear beam (RNB) experimentation. The next-generation tools invite us on the journey to the vast territory of nuclear landscape which has never been explored by science. Hand in hand with experimental developments, a qualitative change in theoretical modeling is taking place. Due to the progress in computer technologies and numerical algorithms, it has became exceedingly clear that the unified microscopic understanding of the nuclear many-body system is no longer a dream.

During recent years, we have witnessed substantial progress in many areas of theoretical nuclear structure. Effective field theory offers hope for a link between QCD and nucleon-nucleon forces. New interactions have been developed which, together with a powerful suite of *ab-initio* approaches, provide a quantitative description of light nuclei. For heavy systems, *global* modern shell-model approaches and self-consistent mean-field methods offer a level of accuracy typical of phenomenological approaches based on parameters *locally* fitted to the data. By exploring connections between models in various regions of the chart of the nuclides, nuclear theory aims

4

to develop a comprehensive, unified theory of the nucleus across the entire nuclear landscape.

From a theoretical point of view, short-lived exotic nuclei far from stability with "abnormal" neutron-to-proton ratios offer a unique test of those aspects of the many-body theory that depend on the isospin degrees of freedom [1]. The challenge to microscopic theory is to develop methodologies to reliably calculate and understand the origins of unknown properties of new physical systems, physical systems with the same ingredients as familiar ones but with totally new and different properties. The hope is that after probing the limits of extreme isospin, we can later go back to the valley of stability and improve the description of normal nuclei.

2. Nuclear structure theory: questions and challenges

Theoretical nuclear structure deals with the nuclear many-body problem in the very finite limit of particle number. In the non-relativistic limit, the goal is to solve the many-body Schrödinger equation with the nuclear Hamiltonian $\hat{H}$:

$$\hat{H}\Psi = E\Psi. \tag{1}$$

Unlike other areas of the many-body problem (atomic physics, condensed matter physics), nuclear physics is still struggling to understand the origin of the inter-nucleonic force which produces nuclear binding. Although it is clear that the nucleon-nucleon (NN) interaction has its roots in quark-gluon dynamics, the microscopic derivation is not yet in place. In addition, due to strong in-medium effects, additional complications arise when one tries to derive the *effective* interaction in the heavy nucleus. This brings us to the first major scientific question pertaining to Eq. (1): *What is the effective nuclear Hamiltonian?* In this context, some specific issues related to the RNB experimentation are: What is the $(N - Z)$ and A dependence (i.e., isovector and isoscalar density dependence) of the effective NN interaction? What is the NN interaction dependence on spin degrees of freedom? What is the nuclear matter equation of state?

In this context, significant progress in the area of the bare nucleon-nucleon force [2] is worth noting. In addition to several excellent phenomenological NN forces (both non-local and local) fitted to the two-body data, new interactions have been obtained in the framework of chiral perturbation theory (or low-momentum expansion) [3,4]. In addition, three-nucleon forces have been derived in the chiral effective field theory [5]. The chiral forces are highly nonlocal; hence it is difficult to use them in ab-initio quantum Monte Carlo calculations [6].

The second major challenge pertaining to Eq. (1) – *What is the nature of the nucleonic matter?* – concerns the properties of the many-body wave function Ψ. Here, the specific fundamental questions are: What is the microscopic mechanism of nuclear binding? Which combinations of protons and neutrons make up a nucleus? What is the single-nucleonic motion in a very neutron-rich environment? What are the collective phases of nucleonic matter? What is the nature of the collective modes of the nucleus (a finite fermion system having a pronounced surface)? What are the relevant collective degrees of freedom? How to understand microscopically the large-amplitude nuclear collective motion (fusion, fission, coexistence phenomena)? Most of these questions are not new. Still, the microscopic answer is missing.

3. The territory of nucleonic matter

Figure 1 shows the vast territory of various domains of nuclear matter characterized by the neutron excess, $(N - Z)/A$, and the isoscalar nucleonic density ($\rho = \rho_n + \rho_p$). In this diagram, the region of finite (i.e., particle-bound) nuclei extends from the neutron excess of about -0.2 (proton drip line) to 0.5 (neutron drip line). The next-generation RNB facilities will provide a unique capability for accessing the very asymmetric nuclear matter and for compressing neutron-rich matter approaching density regimes important for supernova and neutron star physics that are indicated in Fig. 1.

Measurements of neutron skin and radii at RNB facilities will enable us to build an intellectual bridge between finite nuclei and bulk nucleonic matter. Indeed, the thickness of the skin in a heavy nucleus depends on the pressure of neutron-rich matter. The same pressure supports a neutron star against gravity. Thus, models with thicker neutron skins often produce neutron stars with larger radii [8] (see also Ref. [9]). This suggests an inverse relationship: the thicker the neutron-rich skin of a heavy nucleus, the thinner the solid crust of a neutron star. It is an extrapolation of 18 orders of magnitude from the neutron radius of a heavy nucleus (several fm) to the approximately 10 km radius of a neutron star. Yet both radii depend on our incomplete knowledge of the density functional of the neutron-rich matter.

The nuclear equation of state (EOS) describes the possibility of com-

6

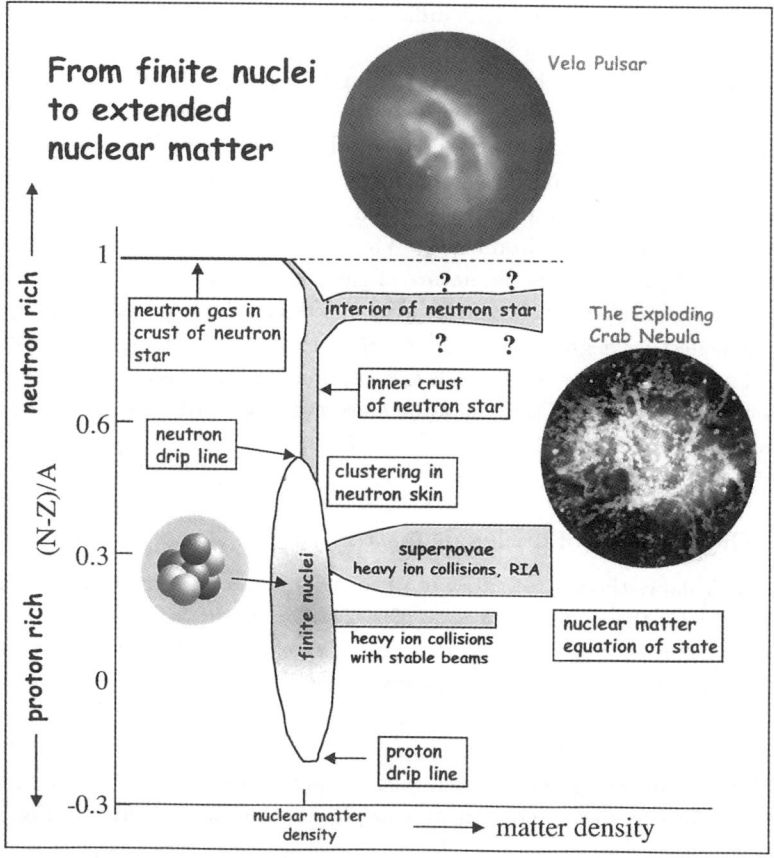

Figure 1. Diagram illustrating the range of nucleonic densities and neutron excess of importance in various contexts of the low- and intermediate-energy nuclear many-body problem. The territory of various domains of nucleonic matter is characterized by the neutron excess and the nucleonic density. The full panoply of bound nuclei comprises the vertical ellipse. Densities accessible with different reactions, and the properties of neutron star layers, are indicated. The new-generation RNB facilities will provide a unique capability for accessing very neutron-rich nuclei – our best experimentally accessible proxies for the bulk neutron-rich matter in the neutron star crust. They will also enable us to compress neutron-rich matter in order to explore the nuclear matter equation of state – essential for the understanding of supernovae and neutron stars. (Based on Ref. [7].)

pressing nuclear matter. It plays a central role in nuclear structure and in heavy ion collisions. It also determines the static and dynamical behavior of stars, especially in supernova explosions and in neutron star stability and evolution. Unfortunately, our knowledge of the EOS, especially at high densities and/or temperatures, is very poor. In nuclear collisions at RIA induced by neutron-rich nuclei, a transient state of nuclear matter with an appreciable neutron-to-proton asymmetry, as well as large density, can be created. This will offer the unique opportunity to study the N/Z-dependence of the EOS, crucial for the supernova problem.

3.1. *How to extrapolate to neutron-rich matter*

Unfortunately, the theoretical knowledge of the equation of state of pure neutron matter is poor; the commonly used energy-density functionals give different predictions for neutron matter. Figure 2 illustrates difficulties with

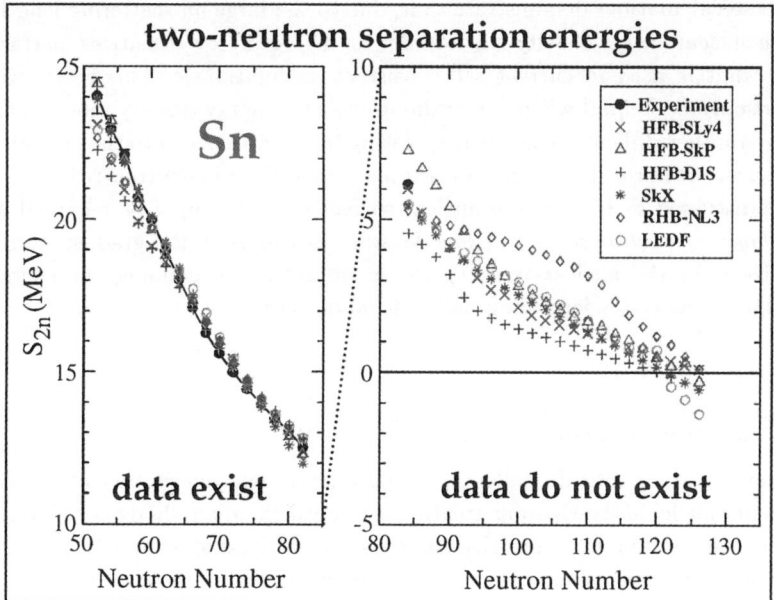

Figure 2. Predicted two-neutron separation energies for the even-even Sn isotopes using several microscopic models based on effective nucleon-nucleon interactions and obtained with phenomenological mass formulas (shown in the inset at top right). (Taken from Ref. [10].)

making theoretical extrapolations into neutron-rich territory. It shows the two-neutron separation energies for the even-even Sn isotopes calculated in several microscopic models based on different effective interactions. Clearly, the differences between forces are greater in the neutron-rich region than in the region where masses are known. Therefore, the uncertainty due to the largely unknown isospin dependence of the effective force (in both particle-hole and particle-particle channels) gives an appreciable theoretical "error bar" for the position of the drip line. Unfortunately, the results presented in Fig. 2 do not tell us much about which of the forces discussed should be preferred since one is dealing with dramatic extrapolations far beyond the region known experimentally. However, a detailed analysis of the force dependence of results may give us valuable information on the relative importance of various force parameters.

Many insights can be obtained from microscopic calculations of neutron matter using realistic nucleon-nucleon two-body and three-body forces [11,12]. These calculations demonstrate that, due to the large nn scattering length, the nuclear energy density functional must diverge at low densities (contrary to what is used in current self-consistent calculations). This result will certainly be helpful when constraining realistic energy density functionals.

Another difficulty when extrapolating from finite nuclei to the extended nuclear matter is due to the diffused neutron surface in neutron-rich nuclei. As discussed in Ref. [13], the nuclear surface cannot simply be regarded as a *layer of nuclear matter at low density*. In this zone the gradient terms (absent in the nuclear matter) are as important in defining the energy relations as those depending on the local density.

4. Continuum shell-model

The major thoretical challenge in the microscopic description of weakly bound nuclei is the rigorous treatment of both the many-body correlations and the continuum of positive-energy states and decay channels. Weakly bound states or resonances cannot be described within the closed quantum system formalism. For bound states, there appears a virtual scattering into the continuum phase space involving intermediate scattering states. Continuum coupling of this kind affects also the effective nucleon-nucleon interaction. For unbound states, the continuum structure appears explicitly in the properties of those states. The consistent treatment of continuum in multi-configuration mixing calculations is the domain of the continuum shell model (CSM) (see Ref. [14] for a review).

4.1. *Gamow Shell Model*

Recently, the multiconfigurational CSM in the complete Berggren basis, the so-called Gamow Shell Model (GSM), has been formulated [15,16]. The s.p. basis of GSM is given by the Berggren ensemble [17] which contains Gamow states (or resonant states and the non-resonant continuum).

The resonant states are the generalized eigenstates of the time-independent Schrödinger equation which are regular at the origin and satisfy purely outgoing boundary conditions. They correspond to the poles of the S matrix in the complex energy plane lying on or below the positive real axis. In the GSM framework, the number of particles in the scattering continuum is not predetermined, but it results from a variational calculation. GSM is a natural generalization of the SM concept for the open quantum systems. And, as such, it is a tool *par excellence* for nuclear structure studies.

4.1.1. *Completness relation involving Gamow states*

There exist several completeness relations involving resonant states. In the heart of GSM is the Berggren completeness relation [17] :

$$\sum_n |u_n\rangle\langle\tilde{u}_n| + \int_{L_+} |u_k\rangle\langle\tilde{u}_k| dk = 1, \qquad (2)$$

where $|u_n\rangle$ are the Gamow states (both bound states and the decaying resonant states lying between the real k-axis and the complex contour L_+) and $|u_k\rangle$ are the scattering states on L_+. The resonant states are normalized according to the squared radial wave function and not to the modulus of the squared radial wave function. This is a consequence of the analytical continuation which is used to introduce the normalization of Gamow states. In practical applications, one has to discretize the integral in (2). Such a discretized Berggren relation is formally analogous to the standard completness relation in a discrete basis of L^2-functions and, in the same way, leads to the eigenvalue problem $H|\Psi\rangle = E|\Psi\rangle$. However, as the formalism of Gamow states is non-hermitian, the matrix H is complex symmetric. The discretized Berggren basis can be a starting point for establishing the completeness relation in the many-body case in full analogy with the standard SM in a complete (discrete) basis of L^2-functions.

4.1.2. *Determination of many-body bound and resonance states*

In a standard SM, one often uses the Lanczos method to find the low-energy eigenstates (bound states) in very large configuration spaces. This popular method is unfortunately useless for the determination of many-body resonances because of a huge number (continuum) of surrounding many-body scattering states, many of them having lower energy than the resonances. A practical solution to this problem is the procedure proposed in Ref. [15]. In the first step, one performs the pole approximation; i.e., the Hamiltonian is diagonalized in a smaller basis consisting of s.p. resonant states only. Here, some variant of the Lanczos method can be applied. In the second step, one includes couplings to non-resonant continuum states. Finally, one searches among the solutions for the eigenvector which has the largest overlap with the unperturbed state.

This procedure allows for an efficient determination of physical states within the set of all eigenvectors of a given Lanczos subspace. Figure 3 shows the GSM eigenvalue spectrum in the complex energy plane for the 0^+ states of ^{20}O. While the two lowest (bound) states can be simply identified by inspection, for the higher-lying states it is practically impossible to separate the resonances from the non-resonant continuum. However, the procedure outlined above makes it possible to identify unambiguously the many-body resonance states.

4.1.3. *GSM Study of Helium Isotopes*

A description of neutron-rich helium isotopes, including Borromean nuclei ^{6,8}He, is a challenging theoretical problem. The nucleus ^{4}He is a well-bound system with the one-neutron emission threshold at 20.58 MeV. On the contrary, the nucleus ^{5}He is a broad resonance. The nucleus ^{6}He, which consists of two neutrons outside ^{4}He, is bound with the two-neutron emission threshold at 1.87 MeV. The first excited 2_1^+ state in ^{6}He at 1.8 MeV is neutron-unstable with a width $\Gamma = 113$ keV.

In our GSM calculations, the s.p. configuration space includes both resonances $0p_{3/2}$, $0p_{1/2}$ and the two associated complex continua $p_{3/2}$ and $p_{1/2}$ which are discretized with 5 points each. Figure 4 shows the lowest energy states of helium isotopes calculated with the surface delta interaction with the strength $V_{SDI} = 1670$ MeV·fm^3. The $0p_{3/2}$, $0p_{1/2}$ s.p. resonances are generated by a Woods-Saxon potential with the parameters chosen to reproduce experimental energies and widths of the $3/2_1^-$ and $1/2_1^-$ resonances of ^{5}He.

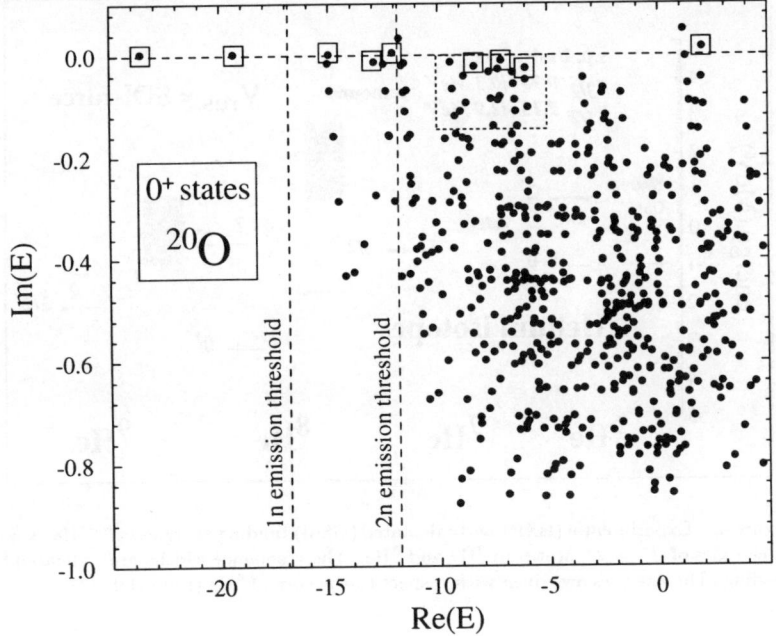

Figure 3. Complex energies of the 0^+ states in ^{20}O obtained by the diagonalization of the GSM Hamiltonian. One- (1n) and two-neutron (2n) emission thresholds are indicated. The physical bound and resonance states are matched by squares. The remaining eigenstates represent the non-resonant continuum (from Ref. [16]).

It is found that the non-resonant continuum contributions are *always essential* and, in some cases (e.g., ^{8,9}He), they dominate the structure of the g.s. wave function. Moreover, the wave function components having many neutrons in the non-resonant continuum give an essential contribution to the binding energy. Without the non-resonant (contour) states, the predicted g.s. energy of ^{8}He is $+2.08$ MeV. The inclusion of scattering states lowers the binding energy to -1.6 MeV. GSM calculations reproduce the most important feature of ^{6,8}He: *the ground state is particle bound, despite the fact that all the basis states lie in the continuum.* The odd-N isotopes of ^{7,9}He are calculated to be wide neutron resonances. The neutron separation energy anomaly, i.e., the *increase* of one-neutron separation energy when going from ^{6}He to ^{8}He, is reproduced. This anomaly is explained in GSM by a large contribution from non-resonant continuum states. This generic mechanism, expected to be present in loosely bound systems, may give rise

12

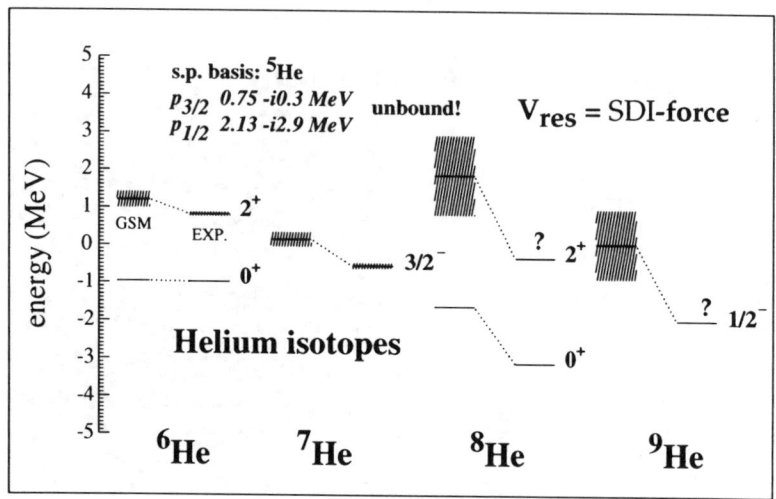

Figure 4. Experimental (EXP) and calculated (GSM) binding energies of $^{6-9}$He as well as energies of $J^\pi = 2^+$ states in ^{6}He and ^{8}He. The resonance widths are indicated by shading. The energies are given with respect to the core of ^{4}He (from Ref. [16]).

to the formation of multineutron Borromean systems, changing the *drip line* into a porous *drip zone*.

5. Conclusions

In years to come, we shall see substantial progress in our understanding of nuclear structure – a rich and interdisciplinary field. An important element in this task will be to extend the study of nuclei into new domains. New radioactive beam facilities, together with advanced multi-detector arrays and mass/charge separators, will be essential in probing nuclei in new domains where new phenomena, likely to be different from anything we have observed to date, will occur. The new data are expected to bring qualitatively new information about the fundamental properties of the nucleonic many-body system and will be crucial for developing a unified description of the nucleus.

The material contained in this paper was obtained in collaboration with J. Dobaczewski, N. Michel, and J. Okołowicz. This work was supported in part by the U.S. Department of Energy under Contract Nos. DE-FG02-96ER40963 (University of Tennessee) and DE-AC05-00OR22725 with UT-Battelle, LLC (Oak Ridge National Laboratory).

References

1. J. Dobaczewski and W. Nazarewicz, Phil. Trans. R. Soc. Lond. A **356**, 2007 (1998).
2. R. Machleidt, Nucl. Phys. A **689**, 11c (2001).
3. D.R. Entem and R. Machleidt, Phys. Lett. **524B**, 93 (2002).
4. S. Bogner, T.T.S. Kuo, L. Coraggio, A. Covello, and N. Itaco, Phys. Rev. C **65**, 051301 (2002).
5. E. Epelbaum, A. Nogga, W. Glockle, H.Kamada, Ulf.-G.Meissner, and H. Witała, Phys. Rev. C **66**, 064001 (2002).
6. S.C. Pieper and R.B. Wiringa, Ann. Rev. Nucl. Part. Sci. **51**, 53 (2001).
7. C.J. Pethick and D.G. Ravenhall, Annu. Rev. Nucl. Part. Sci. **45**, 429 (1995).
8. C. Horowitz and J. Piekarewicz, Phys. Rev. Lett. **86**, 5647 (2001).
9. R.J. Furnstahl, Nucl. Phys. A **706**, 85 (2002).
10. *Scientific Opportunities With an Advanced ISOL Facility*, Report, November 1997; http://www.er.doe.gov/production/henp/isolpaper.pdf.
11. J. Morales, V.R. Pandharipande, and D.G. Ravenhall, Phys. Rev. C **66**, 054308 (2002).
12. J. Carlson, J. Morales Jr., V.R. Pandharipande, and D.G. Ravenhall, nucl-th/0302041 (2003).
13. J. Dobaczewski, W. Nazarewicz, and M. V. Stoitsov, Eur. Phys. J. A **15**, 21 (2002).
14. J. Okołowicz, M. Płoszajczak, and I. Rotter, Phys. Rep. **374**, 271 (2003).
15. N. Michel, W. Nazarewicz, M. Płoszajczak, and K. Bennaceur, Phys. Rev. Lett. **89**, 042502 (2002).
16. N. Michel, W. Nazarewicz, M. Płoszajczak, and J. Okołowicz, Phys. Rev. C. in press; nucl-th/0302060.
17. T. Berggren, Nucl. Phys. A **109**, 265 (1968).

STRUCTURE OF NEUTRON-RICH NUCLEI IN THE ^{132}Sn REGION

A. COVELLO, L. CORAGGIO, A. GARGANO, N. ITACO

*Dipartimento di Scienze Fisiche, Università di Napoli Federico II,
and Istituto Nazionale di Fisica Nucleare,
Complesso Universitario di Monte S. Angelo, Via Cintia, I-80126 Napoli, Italy
E-mail: covello@na.infn.it*

We report on a study of neutron-rich nuclei around ^{132}Sn in terms of the shell model employing a realistic effective interaction derived from the CD-Bonn nucleon-nucleon potential. We present results for some Sb and Te isotopes. Comparison shows that our results are in very good agreement with the available experimental data supporting confidence in the predictions of our calculations. This may stimulate experimental efforts to gain more information on these nuclei lying well away from the valley of stability.

1. Introduction

The study of neutron-rich nuclei in the ^{132}Sn region is a subject of great interest. This is related to the fact that ^{132}Sn is a very good doubly magic nucleus whose neighbors provide the opportunity for testing the basic ingredients of shell-model calculations, especially the matrix elements of the effective interaction, well away from the valley of stability.

From the experimental point of view, it is a very hard task to obtain information on these nuclei. In the last few years, however, substantial progress has been made to access the limits of nuclear stability, which has paved the way to spectroscopic studies in the ^{132}Sn region. A summary of recent experimental efforts in this area including references through 2000 is given in Ref. 1.

Motivated by these experimental achievements, in recent years we have studied[2,3] several nuclei around ^{132}Sn in terms of the shell model employing realistic effective interactions derived from modern nucleon-nucleon (NN) potentials.

The main aim of this paper is to report on some selected results of our current work in this region, which have been obtained starting from the CD-

Bonn free NN potential.[4] In particular, we shall consider two odd-odd antimony isotopes, ^{130,132}Sb, and two even-odd tellurium isotopes, ^{135,137}Te.

As regards the former, we shall focus attention on the proton particle-neutron hole multiplets which play a special role for the understanding of the proton-neutron interaction around closed shells. More than thirty years ago the study of these multiplets in the Pb region was the subject of great experimental and theoretical interest.[5,6,7,8] In this region, through pick-up and stripping reactions, several particle-hole multiplets were identified[5,6] in ^{208}Bi. In this context, it is worth mentioning that a very good agreement with experiment was obtained in Ref. 7 using particle-hole matrix elements deduced from the Hamada-Johnston potential.[9]

Despite these early achievements in the study of the neutron-proton interaction in the vicinity of doubly magic ^{208}Pb, little work has been done ever since. By considering the new data which are becoming available in the ^{132}Sn region, it is high time to revive theoretical interest in this subject and perform shell-model calculations making use of a modern NN potential and improved many-body methods for deriving the effective interaction.

In Sec. 2 we give a bare outline of the theoretical framework in which our realistic shell-model calculations have been performed. In Sec. 3 we present and discuss our results comparing them with the available experimental data. Sec. 4 presents some concluding remarks.

2. Theoretical framework

We assume that ^{132}Sn is a closed core and let the valence protons and neutron holes occupy the five single-particle levels $0g_{7/2}$, $1d_{5/2}$, $1d_{3/2}$, $2s_{1/2}$, and $0h_{11/2}$ of the 50-82 shell. Similarly, for the valence neutrons in ^{135,137}Te the model space includes all the six single-particle levels $0h_{9/2}$, $1f_{7/2}$, $1f_{5/2}$, $2p_{3/2}$, $2p_{1/2}$, and $0i_{13/2}$ of the 82-126 shell. The single-proton and single-hole energies have been taken from the experimental spectra[10,11,12] of ^{133}Sb and ^{131}Sn, respectively. The only exception is the proton $\epsilon_{s_{1/2}}$ which was taken from Ref. 13, since the corresponding single-particle level is still missing in ^{133}Sb. As regards the single-neutron energies, they have been taken from the experimental spectrum[14] of ^{133}Sn , except that relative to the $i_{13/2}$ level which has not been observed. The latter has been taken from Ref. 15.

As already mentioned in the Introduction, in our shell-model calculations we have made use of a realistic effective interaction derived from the CD-Bonn free nucleon-nucleon potential.[4] This high-quality NN potential, which is based upon meson exchange, fits very accurately (χ^2/datum ≈ 1) the world NN data below 350 MeV available in the year 2000.

The shell-model effective interaction V_{eff} is defined, as usual, in the following way. In principle, one should solve a nuclear many-body Schrödinger equation of the form

$$H\Psi_i = E_i\Psi_i \tag{1}$$

with $H = T + V_{NN}$, where T denotes the kinetic energy. This full-space many-body problem is reduced to a smaller model-space problem of the form

$$PH_{\text{eff}}P\Psi_i = P(H_0 + V_{\text{eff}})P\Psi_i = E_iP\Psi_i. \tag{2}$$

Here $H_0 = T + U$ is the unperturbed Hamiltonian, U being an auxiliary potential introduced to define a convenient single-particle basis, and P denotes the projection operator onto the chosen model space.

A main difficulty one is confronted with in the derivation of V_{eff} from a modern NN potential, such as CD-Bonn, is the existence of a strong repulsive core which prevents its direct use in nuclear structure calculations. This difficulty is usually overcome by resorting to the time-honored Brueckner G-matrix method. Here, we have made use of a new approach[16] which provides an advantageous alternative to the use of the above method. It consists in constructing a low-momentum NN potential, V_{low-k}, that preserves the physics of the original potential V_{NN} up to a certain cut-off momentum Λ. In particular, the scattering phase shifts and deuteron binding energy calculated by V_{NN} are reproduced by V_{low-k}. The latter is a smooth potential that can be used directly as input for the calculation of shell-model effective interactions. A detailed description of our derivation of V_{low-k} can be found in Ref. 16, where a criterion for the choice of the cut-off parameter Λ is also given. We have used here the value $\Lambda = 2.1$ fm^{-1}.

Once the V_{low-k} is obtained, the calculation of the matrix elements of the particle-particle and hole-hole interaction is carried out within the framework of a folded-diagram method, as described, for instance, in Refs. 17 and 18. It should be pointed out that for ^{130,132}Sb the proton-neutron effective interaction has been explicitly derived in the particle-hole formalism. A description of the derivation of the particle-hole effective interaction is given in Ref. 3.

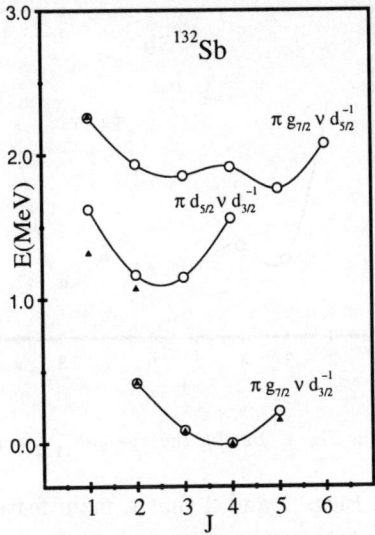

Figure 1. Proton particle-neutron hole multiplets in ^{132}Sb. The theoretical results are represented by open circles while the experimental data by solid triangles. The lines are drawn to connect the points.

3. Results

We report here some selected results of our study of neutron-rich nuclei in the ^{132}Sn region. All calculations have been performed using the OXBASH shell-model code.[19]

We start by considering the Sb isotopes. Some calculated multiplets for ^{132}Sb are reported in Fig. 1 and compared with the existing experimental data.[20,21] We see that the calculated energies are in very good agreement with the observed ones. In fact, the discrepancies are all in the order of tens of keV, except for the 1^+ state of the $\pi d_{5/2} \nu d_{3/2}^{-1}$ multiplet, which lies 300 keV above the experimental counterpart. Some other calculated multiplets having the neutron hole in the $h_{11/2}$ level are reported in Ref. 3.

As regards ^{130}Sb, we report in Fig. 2 the $\pi g_{7/2} \nu h_{11/2}^{-1}$ multiplet, for which several members have been experimentally identified.[22,23] We see that the agreement between experiment and theory is remarkably good. It should be noted that both the experimental and calculated energies are relative to the 8^- state, which has been experimentally observed to be the ground state. Our calculations predict[3] for the ground state $J^\pi = 4^+$, the 8^- state lying at 92 keV excitation energy.

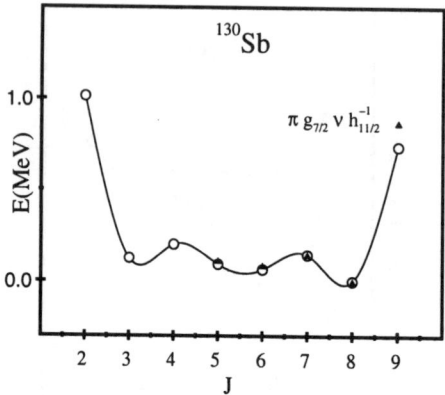

Figure 2. Same as Fig. 1 but for the $\pi g_{7/2}\nu h_{11/2}^{-1}$ multiplet in ^{130}Sb.

It is evident from Figs. 1 and 2 that a main feature of the calculated multiplets is that the states with minimum and maximum J have the highest excitation energy and are well separated from the other states, for which the splitting is relatively small. This pattern is in agreement with the experimental one for the $\pi g_{7/2}\nu d_{3/2}^{-1}$ multiplet in ^{132}Sb and the experimental data available for the other multiplets also go in the same direction. It should be pointed out that this behavior is quite similar to that exhibited[7,8] by the multiplets in the heavier particle-hole nucleus ^{208}Bi. Also, it is worth noting that in all of our calculated multiplets (including those reported in Ref. 3), the state of spin $(j_\pi + j_\nu - 1)$ is the lowest, in agreement with the early predictions of the Brennan-Bernstein coupling rule.[24]

We turn now to the tellurium isotopes, ^{135,137}Te. These nuclei have been the subject of recent experimental studies,[25,26] where high-spin states of a particularly simple structure were identified. In this context, it should be mentioned that also in the lighter odd isotope ^{133}Te several high-spin states have been recently identified.[27,28] The results of a shell-model calculation performed for this nucleus in the same way as described in Sec. 2 are reported in Ref. 3, where it can be seen that they are in very good agreement with experiment.

The theoretical spectrum of ^{135}Te is compared with the experimental one in Fig. 3, where all the observed[29] and calculated states up to 1.3 MeV are included. In the higher energy region up to about 2.2 MeV we compare the four observed high-spin states[25] with those predicted by the theory. As regards ^{137}Te, in Fig. 4 we compare the experimental levels[26] below 2 MeV with those predicted by our calculations.

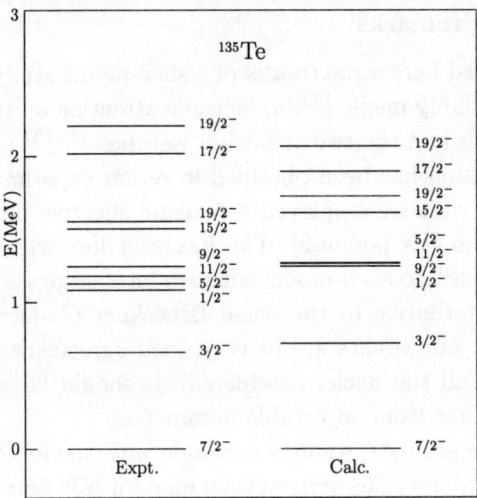

Figure 3.   Experimental and calculated spectra of ^{135}Te.

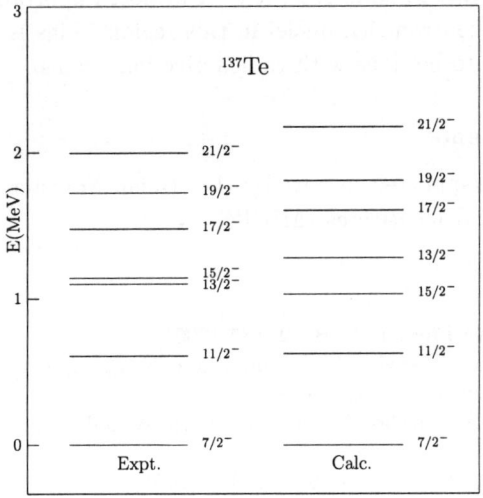

Figure 4.   Experimental and calculated spectra of ^{137}Te.

From Figs. 3 and 4 we see that the calculated spectra are in quite good agreement with the experimental ones, the largest discrepancy being about 170 keV in both cases.

4. Concluding remarks

We have presented here some results of a shell-model study of neutron-rich nuclei close to doubly magic ^{132}Sn, focusing attention on the two antimony isotopes ^{130,132}Sb and the two tellurium isotopes ^{135,137}Te, for which new relevant information has been obtained in recent experimental studies. In our calculations we have employed a realistic effective interaction derived from the CD-Bonn NN potential. This has been done within the framework of a new approach[16] to shell-model effective interactions which provides an advantageous alternative to the usual Brueckner G-matrix method. We have shown that our results are in very good agreement with the experimental data for all the nuclei considered. It should be stressed that our calculations are free from adjustable parameters.

On the above grounds, we may conclude with the following remarks.

(i) Effective interactions derived from modern NN potentials are able to describe with quantitative accuracy the spectroscopic properties of nuclei in the ^{132}Sn region far from stability. This gives confidence in their predictive power.

(ii) On the experimental side, it is of utmost importance to gain more information on neutron-rich nuclei in this region. This is certainly a very exciting physics to be done with radioactive ion beams.

Acknowledgments

This work was supported in part by the Italian Ministero dell'Istruzione, dell'Università e della Ricerca (MIUR).

References

1. H. Mach, *Acta Phys. Pol.* B **32**, 887 (2001).
2. A. Covello, L. Coraggio, A. Gargano, and N. Itaco, *Acta Phys. Pol.* B **32**, 871 (2001), and references therein.
3. L. Coraggio, A. Covello, A. Gargano, N. Itaco, and T. T. S. Kuo, *Phys. Rev.* C **66** (2002), and references therein.
4. R. Machleidt, *Phys. Rev.* C **63**, 024001 (2001).
5. J. R. Erskine, *Phys. Rev.* **135**, B110 (1964).
6. W. P. Alford, J. P. Schiffer, and J. J. Schwartz, *Phys. Rev. Lett.* **21**, 156 (1968).
7. T. T. S. Kuo, *Nucl. Phys.* A **122**, 325 (1968).
8. M. Moinester, J. P. Schiffer, and W. P. Alford, *Phys. Rev.* **179**, 984 (1969).
9. T. Hamada and I. D. Johnston, *Nucl. Phys.* **34**, 382 (1962).
10. M. Sanchez-Vega, B. Fogelberg, H. Mach, R. B. E. Taylor, A. Lindroth, J. Blomqvist, A. Covello, and A. Gargano, *Phys. Rev.* C **60**, 024303 (1999).

11. B. Fogelberg and J. Blomqvist, *Phys. Lett.* B **137**, 20 (1984).
12. B. Fogelberg and J. Blomqvist, *Nucl. Phys.* A **429**, 205 (1984).
13. F. Andreozzi, L. Coraggio, A. Covello, A. Gargano, T. T. S. Kuo, and A. Porrino, *Phys. Rev.* C **56**, R16 (1997).
14. P. Hoff *et al.*, *Phys. Rev. Lett.* **77**, 1020 (1996).
15. L. Coraggio, A. Covello, A. Gargano, and N. Itaco, *Phys. Rev.* C **65**, 051306(R) (2002).
16. S. Bogner, T. T. S. Kuo, L. Coraggio, A. Covello, and N. Itaco, *Phys. Rev.* C **65**, 051301 (2002).
17. A. Covello, F. Andreozzi, L. Coraggio, A. Gargano, T. T. S. Kuo, and A. Porrino, *Prog. Part. Nucl. Phys.* **38**, 165 (1997).
18. L. Coraggio, A. Covello, A. Gargano, N. Itaco, and T. T. S Kuo, *J. Phys.* G **26**, 1697 (2000).
19. B. A. Brown, A. Etchegoyen, and W. D. M. Rae, *The computer code OXBASH*, MSU-NSCL, Report No. 524.
20. C. A. Stone, S. H. Faller, and W. B. Walters, *Phys. Rev.* C **39**, 1963 (1989).
21. H. Mach, D. Jerrestam, B. Fogelberg, M. Hellström, J. P. Omtvedt, K. I. Erokhina, and V. I. Isakov, *Phys. Rev.* C **51**, 500 (1995).
22. W. B. Walters and C. A. Stone, in *Proceedings of International Workshop on Nuclear Fission and Fission Product Spectroscopy*, Seyssins, France, edited by H. Faust and G. Fioni, ILL Report N* 94FA05T, 1994, p.182.
23. J. Genevey, J. A. Pinston, C. Foin, M. Rejmund, H. Faust, and B. Weiss, *Phys. Rev.* C **65**, 034322 (2002).
24. M. H. Brennan and A. M. Bernstein, *Phys. Rev.* **120**, 927 (1960).
25. B. Fornal *et al.*, *Phys. Rev.* C **63**, 024322 (2001).
26. W. Urban *et al.*, *Phys. Rev.* C **61**, 041301(R) (2000).
27. P. Bhattacharyya *et al.*, *Phys. Rev.* C **64**, 054312 (2001).
28. J. K. Hwang *et al.*, *Phys. Rev.* C **65**, 034319 (2002).
29. Data extracted using the NNDC On-line Data Service from the ENSDF database, file revised as of October 18, 2002.

ISOMER SPECTROSCOPY OF IN-FLIGHT FISSION FRAGMENTS NEAR ^{132}SN AT THE GSI FRAGMENT SEPARATOR

M. HELLSTRÖM[1,2,a], M.N. MINEVA[2], A. BLAZHEV[1], H.J. BOARDMAN[3],
J. EKMAN[2], K. GLADNISHKI[4], H. GRAWE[1], J. GERL[1], R. PAGE[3] Z. PODOLYÁK[4]
AND D. RUDOLPH[2] FOR THE GSI-FRS ISOMER COLLABORATION

[1] *Gesellschaft für Schwerionenforschung mbH, D-64291 Darmstadt, Germany*
[2] *Dept. of Physics, Lund University, P.O. Box 118, SE-221 00 Lund, Sweden*
[3] *Oliver Lodge Laboratory, University of Liverpool, Liverpool, L69 7ZE, UK*
[4] *Department of Physics, University of Surrey, Guildford, GU2 7XH, UK*

At GSI-Darmstadt, a variety of neutron-rich isotopes in the vicinity of ^{132}Sn have been produced in relativistic projectile fission of ^{238}U. The fragments of interest were mass separated and identified using the fragment separator FRS before being implanted inside a high-resolution Ge-detector array. Delayed heavy-ion tagged γ-ray spectroscopy was applied to study the decay of relatively long-lived ($T_{1/2}$ in the 100ns - 100 µs range) isomeric states in the implanted fission fragments, including new isomers in $^{126-128}$In and ^{125}Cd.

1. Introduction

The doubly magic nucleus ^{132}Sn (Z=50, N=82) marks one of only two double shell closures now accessible in the very neutron rich (N/Z>1.6) region of the nuclear chart. The shell structure and residual interaction at these fix-points are intima-tely related to many important issues, such as the proposed quenching of shell-closure strength for extremely neutron rich nuclei, the development of low-lying deformed intruder configurations, and the astrophysical r-process. That the detailed understanding of neutron rich systems far from stability is far from complete is illustrated by the different predictions of e.g. shell evolution obtained by various theoretical approaches. More experimental data on both global nuclear properties as well as spin-dependent ones are thus urgently needed for neutron rich systems.

Recently, ion-tagged delayed γ-coincidence spectroscopy of isomeric states produced directly in nuclear reactions [1] has emerged as a successful alternative to β-decay studies for studying excited states in very neutron rich nuclei far from stability. We have applied this method to produce and investigate µs-range isomers in the region around ^{132}Sn using projectile fission

---

[a] E-mail: m.hellstroem@gsi.de

induced by a Pb target. In this reaction, the projectile is electromagnetically excited and then fissions. The resultant relatively cold fission products subsequently deexcite, mainly via γ-ray emission. In this latter process, isomeric states, if present in the final nucleus, may be populated directly or "from above" by acting as a trap for γ-ray cascades from states of higher spin.

The spectroscopy of isomeric decays provide a unique and quite selective method to probe not only transition rates but also level energies in the nuclei in which they occur. The method is an ideal complement to β-decay and in-beam experiments, which often require higher count rates. From the isomer decay pattern as well as the properties of intermediate states populated in the decay, entities such as single-particle energies, residual interaction strength, shell occupation and development of collectivity can be extracted.

Previous investigations of the tin region isomers include studies of fission product ß-decay after ISOL mass separation (limited by the availability of high-spin ß-decay parents), measurements of prompt γ-decay of fission products with large Ge-detector arrays (limited by low production yields to the Z>50, N>82 region), and experiments at the GSI fragment separator [2].

This paper reports on the results of a recent experiment performed at GSI-Darmstadt, where we searched for and studied relatively long-lived (100ns - 1ms) isomeric states in neutron rich isotopes in the region of ^{132}Sn. The study hopes to provide deeper insight into nuclear structure far from stability as well as the role of angular momentum in the production reaction.

2. Experiment

The present experiment was performed in July-August 2002 at the FRagment Separator [3] (FRS) at GSI, Darmstadt. Neutron-rich nuclei in the vicinity of ^{132}Sn were produced by projectile fission of ^{238}U at the relativistic energy of 732 MeV/nucleon impinging on a 1 g/cm^2 Pb target. The average beam intensity from the SIS heavy ion synchrotron was $4 \cdot 10^8$ ions per cycle, with each cycle consisting of 5 s acceleration and 5 seconds extraction. The data presented here were obtained during an effective measurement time of 8 hours with the FRS optimized for the transmission of ^{130}Sn.

The ions of interest were separated by combining magnetic analysis with energy loss in matter [3]. The FRS was operated in a standard achromatic mode, with a wedge-shaped aluminium degrader with thickness selected to 50% of the range of the fragments of interest placed at the intermediate dispersive focal plane of the spectrometer. Figure 1 shows a schematic view of the FRS and the associated detectors setup.

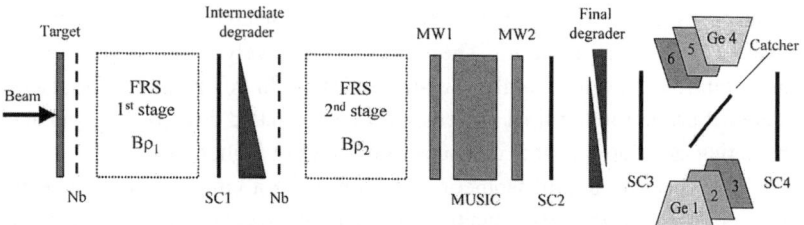

Figure 1. Schematic view of the FRS and the experimental setup.

Under the present conditions, typically about 20 different fragment species distributed over 3-4 elements were transmitted at each setting of the separator. The fragment identification was performed according to the so-called $B\rho$-ΔE-TOF method, which combines information on the particle trajectories (from the magnetic rigidity $B\rho$ and position information) with measurements of velocity (through time-of-flight (TOF)) in the second half of the FRS and energy loss (ΔE) at the final focus to obtain the proton number Z and mass-to-charge ratio A/q for all transmitted fragments. At the relativistic energies used, the percentage of non-fully stripped ions in the tin region is small, leading to q=Z. Figure 2 illustrates the good resolution that was obtained, resulting in unambiguous particle identification.

At the focal plane of the FRS, the transmitted fragments were slowed down in a variable-thickness aluminium degrader before being implanted in a plastic catcher foil. The implantation process was controlled by means of scintillator detectors. The catcher was surrounded by six segmented Clover-type detectors in which delayed γ-rays emitted by the implanted ions were detected.

The energy and time of all "first hits" in the Ge detectors within a 80 µs interval following the implantation of an ion were recorded together with the particle identification information for the respective ion. This allowed the construction of heavy ion-gated γ-ray energy versus time matrices for both prompt and delayed ion-gamma coincidences.

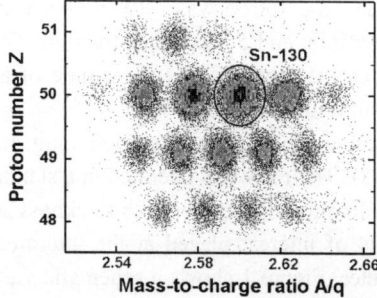

Figure 2. Particle identification spectrum showing proton number Z versus the mass-to-charge ratio A/q for fragments reaching the implantation setup with the FRS optimized for ^{130}Sn

3. Analysis and preliminary results

As the analysis is still in progress, the results presented here should be considered *preliminary*. Nevertheless, it seems clear that the present study may add considerably to the experimental data for selected ions "southwest" of ^{132}Sn.

Figures 3 and 4 show delayed γ-ray energy spectra of selected tin, cadmium and indium fragments exhibiting isomeric decays. The spectra were obtained by projecting the above mentioned γ-ray energy versus time matrices for the time interval 1-35 μs after implantation. This suppresses most of the "prompt" events which mainly result from radiation produced during the slowing-down process and/or fragments breaking up in the degrader or catcher, but especially at lower energies the background remains high, as illustrated e.g. by the ^{130}Sn spectrum.

By conversely defining gates on γ-ray energies, the decay time information of the transitions of interest could be obtained by matrix projection. To reduce the influence of prompt radiation, time spectra from background gates were subtracted before fitting the resultant distributions with a single exponential decay on top of a flat background. Table 1 summarizes the properties of the isomers observed in the present study, while Figure 5 illustrates such time spectra with fits for the isomers observed in the present experiment.

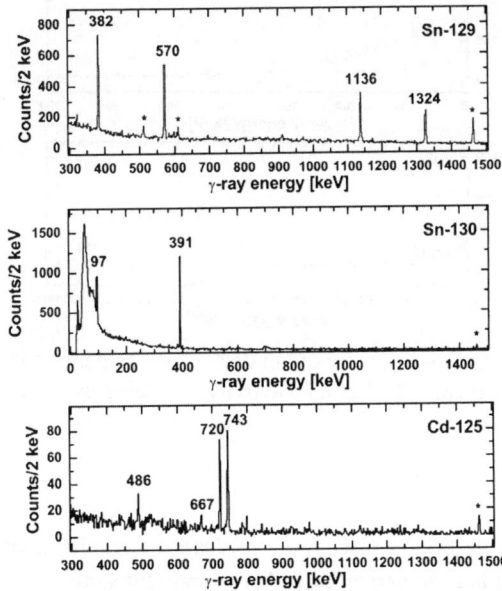

Figure 3. Gamma-ray spectra recorded in delayed coincidence (1.0-35.0 μs) with Sn and Cd fission fragments. Background activity is labeled by asterisks.

26

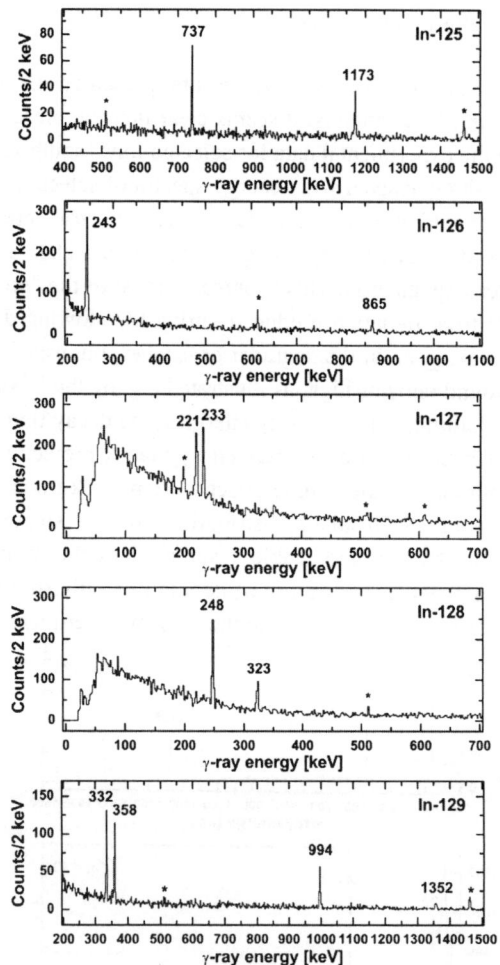

Figure 4. Gamma-ray spectra recorded in delayed coincidence (Δt=1.0-35.0 μs) with In fission fragments. Background activity is labeled by asterisks.

4. Discussion

In the following, we briefly discuss the *preliminary* results obtained so far for some of the individual isomers observed in the present study:

^{129}Sn: The half-life we obtain, 3.9(4) μs, is in good agreement with Genevey *et al.*'s value of 3.6(2) μs [4] for a proposed 19/2$^+$ isomer. This could indicate that in-flight Pb-target induced fission does not strongly populate the higher-lying 2.4(2) μs 23/2$^+$ isomer [4].

^{130}Sn: The good agreement between our half-life value of 1.5(2) μs with the 1.61(15) μs measured by Fogelberg *et al.* for the 10$^+$ isomeric state [5] gives us confidence in the experimental technique and the evaluation procedure.

^{125}Cd: This previously unknown isomer shows two strong γ-transitions with similar energy, intensity and decay behavior. Comparing with the A=120-130 tin isotopes, where the lowest lying state is alternately 2$^+$ or 15/2$^-$, a possible interpretation of this decay would be a cascade starting with a hindered M2 transition deexciting a 19/2$^+$ isomer via an 15/2$^-$ level down to a known (11/2$^-$) state, which could be the ground state (Ref. [8] quotes its energy as 50(70) keV.) Two much weaker delayed γ-transitions at 486 and 667 keV are also observed.

^{125}In: We observe an isomer as first observed by Fogelberg *et al.* [6], decaying by a strong cascade of 737 and 1173 keV transitions.

^{126}In: This previously unreported isomer exhibits a strong γ-ray at 244 keV. In analogy with ^{128}In (see below), we tentatively interpret this as the primary isomeric transition connecting the 1$^-$ member of the $\pi g_{9/2}^{-1} \nu h 11_{/2}^{-1}$ multiplet with the 3$^+$ ground state. The origin of the much weaker 614 and 865 keV transitions also present in the delayed γ-ray spectrum is presently not understood.

^{127}In: We observe an isomer decaying with only two γ-rays of similar intensity, energy and half-life. The decay pattern and half-life, which show very little resemblance to the neighbors ^{125,129}In (see Fig. 4 and Table 1), could belong to an isomer on top of the presumably long-lived 23/2$^-$ state [6], the decay of which escapes the observation time window of our experiment. We

Table 1. Some properties of isomers observed in the present study.

Nucleus	Spin	Observed delayed γ-rays* [keV]	Half-life* [μs] This work	Previous
^{129}Sn	19/2$^+$	382, 570, 1136, 1324	3.9(4)	3.6(2) [4]
^{130}Sn	10$^+$	97, 391	1.5(2)	1.61(15) [5]
^{125}Cd	(19/2$^+$)	720, 743	14(2)	
^{125}Cd		486, 667		
^{125}In	19/2$^+$	737, 1173	13(2)	9.4(6) [6]
^{126}In	(1$^-$)	244	30(3)	
^{126}In		614, 865		
^{127}In	(29/2$^+$)	221, 233	13(2)	
^{128}In	(1$^-$)	248	170(80)	>10μs [7]
^{128}In		323		
^{129}In	(17/2$^-$)	332, 358, 994, 1352	11(2)	2.0(5) [4]

* Values are preliminary and may change as the analysis progresses.

28

tentatively assign a spin of $(29/2^+)$ from the $\pi g_{9/2}^{-1} \nu h_{11/2}^{-2}$ configuration to this proposed high-lying state, which is likely present also in the neighbors but may have escaped detection up to now due to its half-life being too short or long [6]. ^{128}In: Our measurement confirms the isomeric (1^-) state observed by Fogelberg [7] to deexcite to the $(3)^+$ ground state by a 248 keV γ-ray with a quite long half-life. The weaker transition at 323 keV present in our spectrum could connect a second, higher-lying (1^-) or (5^+) isomeric level to the $(3)^+$ ground state. Such states are expected from the $\pi p_{1/2} \nu d_{3/2}$ and $\pi p_{1/2} \nu h_{11/2}$ configurations.

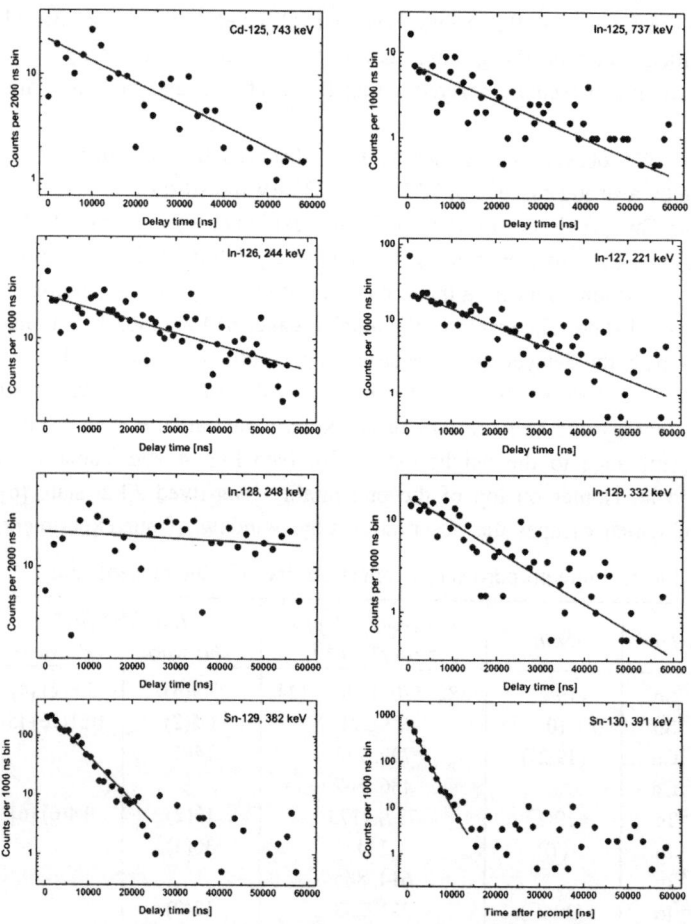

Figure 5. Background-subtracted decay time distributions with one-exponential fits for selected γ-transitions. See Table 1 for deduced half-lives.

^{129}In: We observe an isomeric decay with the same γ-transitions as reported by Genevey *et al.* [4] but exhibiting a much longer half-life of 11(2) μs. This decay can probably be interpreted as that of a (17/2⁻) state, as any higher spin would not allow for the know 700 ms (23/2⁻) isomer at ~1900 keV observed by Fogelberg *et al.* [6], and positive parity would enable a fast E2 transition to the (13/2⁺) state at 1352 keV. The existence of an undedected low-energy E1 primary isomeric transition cannot, however, be excluded from our data.

5. Outlook

Although this and other recent studies directed at nuclei close to doubly magic ^{132}Sn have provided a wealth of new data, their interpretation and the details of nuclear structure in this region is not yet clear. A number of outstanding issues remain, including the possibility of shell quenching. At GSI, the program of isomer spectroscopy of neutron rich exotic nuclei is planned to be continued as part of the stopped-beam phase of the RISING project [9].

In addition to this and further experimental efforts elsewhere, we would like to encourage an active and continuous involvement of the nuclear structure shell model theory community, whose help to develop and perform calculations using realistic interactions, especially for nuclei requiring large model spaces is much appreciated.

References

[1]   R. Grzywacz et al., *Phys. Rev.* **C55**, 1136 (1997)
[2]   M.N. Mineva et al., *Eur. Phys.J.* **A11**, 9 (2001)
[3]   H. Geissel et al., *Nucl. Instr. Meth. in Phys. Res.* **B70**, 286 (1992)
[4]   J. Genevey et al., *Phys. Rev.* **C65**, 034322 (2002)
[5]   B. Fogelberg et al., *Nucl. Phys.* **A352**, 157 (1981)
[6]   B. Fogelberg et al., *AIP Conf. Proc.* **447**, 191 (1998)
[7]   B. Fogelberg, *Proc. Int. Conf. Nuclear Data in Science and Technology,* Mito, Japan, p. 837 (1988)
[8]   J. Katakura, Nucl. Data Sheets **86**, 855 (1999)
[9]   See e.g. http://www-aix.gsi.de/~wolle/EB_at_GSI/main.html and references therein.

NUCLEAR STRUCTURE STUDIES WITH ISOL BEAMS AT THE HRIBF*

CYRUS BAKTASH

Physics Division, Oak Ridge National Laboratory, Oak Ridge, Tennessee, 37831

The Holifield Radioactive Ion Beam Facility (HRIBF) is currently the only facility in the world that is capable of delivering accelerated proton- and neutron-rich ISOL beams up to the Coulomb barrier for nuclear structure, reaction and astrophysics research. To take advantage of these unique beams, we have developed several state-of-the-art detector systems and new experimental techniques that have allowed us to cope with the difficulties of using low-intensity radioactive ion beams for nuclear physics research in general, and gamma-ray spectroscopy in particular.

Since the mid-nineties, Coulomb excitations of radioactive ion beams (RIBs) at projectile-fragmentation facilities have provided unique insight into the structures of the neutron-rich nuclei below $Z=20$, including early evidence regarding possible modification of the $N=20$ magic number in neutron-rich nuclei. Since Coulomb excitation of ISOL beams scattered off high-Z targets allows studies of multi-step processes, it would provide an even deeper insight into the nature of quadrupole collectivity in exotic nuclei. With this prospect in mind, we have initiated a Coulomb excitation program with the accelerated RIBs to systematically study the B($E2$) transition strengths in neutron-rich nuclei near magic number $N=50$, and those neighboring ^{132}Sn. The first results for several isotopes of Sn and Te indicate that while nearly all of the deduced B($E2$; $0 \rightarrow 2$) values follow the expected trends, the value for ^{136}Te is unexpectedly small: It is nearly half of that in ^{132}Te in spite of its much smaller excitation energy. This anomalous result could not be reproduced within the framework of shell model calculations. However, recent QRPA and QMCD calculations have predicted smaller transition rates that are close to the experimental value. The differences in these results primarily reflect the different degrees of contributions of neutrons to the wave function of this state. Within the QRPA model, the very large neutron component of the wave function is attributed to the small neutron pairing strength in this nucleus. We plan to further pursue these studies to provide some constrains on the choice of input parameters for microscopic models.

31

We have also begun to explore spectroscopy of the single-particle states in the vicinity of ^{132}Sn using selective heavy-ion transfer reactions, such as (^{9}Be, ^{8}Be), (^{13}C, ^{12}C), (^{7}Li, ^{8}Be), and (^{11}B, ^{10}B), in inverse kinematics. Early results from these reactions and for incomplete fusion reactions with ^{6,7}Li targets are very encouraging. We are also exploring the possibility of using light-ion transfer reactions, such as (d, p) and (^{3}He, ^{4}He), to probe spectroscopic factors of single particle states near magic numbers.

* Oak Ridge National Laboratory is managed by UT-Battelle, LLC for the U.S. DOE under contract DE-AC05-00OR22725.

ANOMALOUS BEHAVIOR OF 2⁺ EXCITATIONS OF TELLURIUM ISOTOPES AROUND N = 82 *

J. TERASAKI,[1-3] J. ENGEL,[4] W. NAZAREWICZ,[1,2,5] AND
M. STOITSOV[1-3,6]

[1] *Department of Physics, University of Tennessee, Knoxville, Tennessee 37996, USA*

[2] *Physics Division, Oak Ridge National Laboratory, P.O. Box 2008, Oak Ridge, Tennessee 37831, USA*

[3] *Joint Institute for Heavy Ion Research, P.O. Box 2008, Building 6008, MS 6374 Oak Ridge, Tennessee 37831, USA*

[4] *Department of Physics and Astronomy, University of North Carolina, CB 3255, Phillips Hall, Chapel Hill, North Carolina 27599, USA*

[5] *Institute of Theoretical Physics, University of Warsaw, ul. Hoża 69, PL-00-681 Warsaw, Poland*

[6] *Institute of Nuclear Research and Nuclear Energy, Bulgarian Academy of Science, Sofia-1784, Bulgaria*

In certain neutron-rich Te isotopes, a decrease in the energy of the first excited 2^+ state is accompanied by a decrease in the $E2$ strength to that state from the ground state, contradicting simple systematics and general intuition about quadrupole collectivity. We use a separable quadrupole-plus-pairing Hamiltonian and the quasiparticle random phase approximation to calculate energies, $B(E2, 0^+ \to 2^+)$ strengths, and g factors for the lowest 2^+ states near ^{132}Sn ($Z \geq 50$). We trace the anomalous behavior in the Te isotopes to a reduced neutron pairing above the $N = 82$ magic gap. In addition, we briefly discuss the magicity of ^{68}Ni.

*This work is supported in part by the U.S. Department of Energy under Contract Nos. DE-FG02-96ER40963 (University of Tennessee), DE-AC05-00OR22725 with UT-Battelle, LLC (Oak Ridge National Laboratory), and DE-FG02-97ER41019 (University of North Carolina), and by the National Science Foundation Contract No. 0124053 (U.S.-Japan Cooperative Science Award).

1. Introduction

As experiment pushes towards the nuclear drip line, it is becoming possible to examine isotopic chains over increasingly large ranges of N and Z. We have new opportunities to test systematics and the ideas that underlie them. One region in which experimental progress has been made recently surrounds the neutron-rich doubly magic isotope ^{132}Sn. In particular, recent Ref. [1] reports measurements of the transition rates $B(E2; 0^+ \to 2^+)$ $(B(E2)\uparrow)$ from the ground state to the lowest 2^+ state for ^{132}Te, ^{134}Te, and ^{136}Te. The authors discovered that $B(E2)\uparrow$'s and the energies of the lowest 2^+ states (E_{2+}) behave differently in the $N = 80$, 82, and 84 Te isotopes than in isotones of Xe, Ba, and Ce which have more protons. In most isotopic chains, a decrease in E_{2+} is accompanied by an increase in $B(E2)\uparrow$ as the states become collective. This is not the case in ^{132,136}Te, where the $B(E2)\uparrow$ decreases as E_{2+} decreases (see Fig. 1). The purpose of our work is to explain this unusual behavior. Our tool is the quasiparticle random phase approximation (QRPA), in conjunction with a simple schematic interaction, which we apply to even-even nuclei in the mass region $50 \leq Z \leq 58$ and $80 \leq N \leq 84$ (and a much larger range of N for the Sn chain).

2. QRPA calculation

The Hamiltonian we use in our QRPA calculation consists of a single-particle (s.p.) Hamiltonian, a monopole pairing field, and a residual two-body interaction. For the latter, we use an isoscalar QQ force, an isovector QQ force, and the quadrupole pairing force. For the details of the calculation, we refer the reader to Ref. [2]. Figure 1 shows the calculated E_{2+} energies and $B(E2)\uparrow$ values, along with the experimental data. The neutron number dependence in Ba and Ce is similar to that of Xe. The calculations reproduce the experimental trend quite well, in particular the asymmetry around $N = 82$ of the $B(E2)\uparrow$'s in the Te isotopes. We also predict an inverted, and more symmetric, curve for the $B(E2)\uparrow$'s in the Sn isotopes with $N = 80$–84. This kind of inversion is well known to occur in the Pb region[3] around $N = 126$.

What is the reason for the unusual behavior of the Te isotopes around $N = 82$, i.e. the fact that *both* E_{2+} and $B(E2)\uparrow$ are smaller in ^{136}Te than in ^{132}Te? The ingredient in our calculations that displays the most asymmetry around $N = 82$ is the neutron pairing gap: 0.574 MeV and 0.376 MeV for ^{132}Te and ^{136}Te, respectively (cf. Ref. [2] for derivation

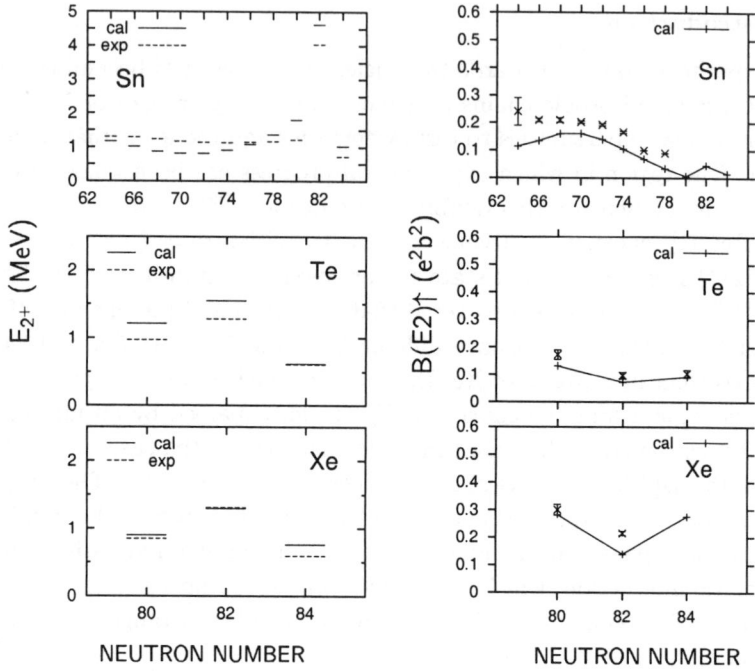

Figure 1. E_{2^+}'s (left) and $B(E2)\uparrow$'s (right) from the QRPA calculation and the experimental data[1,3] The bare charges were used for the calculations of $B(E2)\uparrow$'s.

of those values). To understand how it affects the results, we performed QRPA calculations for ^{136}Te for different values of Δ_n, and it turned out that as Δ_n decreases in a certain region, both E_{2^+} and $B(E2)\uparrow$ values decrease, indeed suggesting that Δ_n plays the key role in producing the unusual trend. To get more insight, we consider the forward ($\psi_{\mu\nu}$) and backward ($\varphi_{\mu\nu}$) QRPA amplitudes in the lowest-energy 2^+ excited states. The QRPA amplitudes $\psi_{\mu\nu}$ and $\varphi_{\mu\nu}$ depend on the ratios

$$\frac{\langle\mu||Q^\tau||\nu\rangle}{E_\mu + E_\nu - E_{2^+}} \quad \text{and} \quad \frac{\langle\mu||Q^\tau||\nu\rangle}{E_\mu + E_\nu + E_{2^+}}, \tag{1}$$

respectively, where E_μ is the BCS quasiparticle energy, and $\tau = $ p or n. We found that with decreasing Δ_n the neutron amplitude increases, while the proton one decreases.

The reason for the unusual behavior can be surmised from the behavior of E_{2^+}, $\psi_{\mu\nu}$, and $\varphi_{\mu\nu}$. The decreased neutron pairing gap in ^{136}Te implies that the lowest neutron quasiparticle energies are lower than those in ^{132}Te (0.792 MeV for ^{132}Te and 0.460 MeV for ^{136}Te). As a result, the energy of the lowest 2^+ state decreases when one crosses $N = 82$. However, the low-lying neutron quasiparticle energies also cause the neutron amplitudes in the wave function to increase, hence the proton amplitudes to decrease, cf. Eq. (1). Since $B(E2)\uparrow$ is determined solely by protons (we use the bare charges), it is found to decrease as well. In other words, the behavior of the lowest 2^+ states reflects the asymmetry of the pairing gap around ^{134}Te, and this asymmetry can be attributed to the fact that the density of s.p. levels below $N = 82$ is greater than above the gap[4]. This mechanism plays more important role in Te isotopes, than collective quadrupole enhancement induced by the residual interaction. This does not come as a surprise given that both ^{132,136}Te have only 2 valence neutrons (or neutron holes).

In the Xe, Ba, and Ce, isotopes, the increased number of valence protons makes proton pairing and the neutron-proton quadrupole-quadrupole interaction more important and reduces the effectiveness of the s.p. mechanism just described.

The value of $B(E2)\uparrow$ in ^{134}Te is smaller than that of ^{132}Te, in spite of the large proton amplitude (0.99). However, the 2^+ state in ^{134}Te corresponds to a two-quasiparticle configuration $(g_{7/2})^2$, while the strength in ^{132}Te and ^{136}Te is more fragmented, indicating the collective character of the 2^+ state.

We close this section by discussing the behavior of $B(E2)\uparrow$ of ^{130}Sn–^{134}Sn (see Fig. 1). The 2^+ states of ^{130,134}Sn consist almost completely of neutron excitation. However, in ^{132}Sn, *both* proton and neutron low-energy excitations are hindered; therefore the neutron amplitude decreases and the proton contribution increases, compared to the other isotopes. This change causes a local increase in $B(E2)\uparrow$ at ^{132}Sn. (When the collectivity is small, $B(E2)\uparrow$ reflects the magnitude of the proton amplitudes directly.) Since the nucleus is in a neutron-rich region, however, matrix elements of the quadrupole operators of the neutrons are larger, on average, near the Fermi surface than those of the protons. Thus, excitations of the neutrons are still much larger than that of the proton in the 2^+ state of ^{132}Sn.

3. g factors of Xe, Te, and Sn isotopes

The abnormal behavior of the E_{2^+}'s and $B(E2)\uparrow$'s around ^{132}Sn reflects the variations of proton and neutron amplitudes in the wave function of the lowest 2^+ state. Therefore, we analyze the g factor in neighboring nuclei; they are very sensitive to relative proton/neutron contributions. The calculated g factors are: $g_{\text{cal}}(^{134}\text{Xe}) = 0.585$, $g_{\text{cal}}(^{136}\text{Xe}) = 0.716$, and

$g_{cal}(^{138}Xe) = 0.291$, while experimentally they are: $g_{exp}(^{134}Xe) = 0.354(7)$ and $g_{exp}(^{136}Xe) = 0.766(45)$. As usual, we multiplied the bare spin g_s factors by 0.7, and took bare g_l factors, see, e.g., Ref. [5]. The predicted g factor in ^{136}Xe is larger than that in ^{134}Xe, though not by as much as in the data (see also Ref. [6]). We found by analyzing the wave functions that the main component of the 2^+ states of ^{134}Xe and ^{136}Xe is $\pi(1g_{7/2})^2$, while those of ^{138}Xe are $\pi(1g_{7/2})^2$ and $\nu(2f_{7/2})^2$. Our calculations for ^{134}Xe and ^{136}Xe support the idea proposed in Ref. [7] that the states of these nuclei consist mainly of proton excitations.

Calculated g factors of Te isotopes are $g_{cal}(^{132}Te) = 0.491$, $g_{cal}(^{134}Te) = 0.695$, and $g_{cal}(^{136}Te) = -0.174$. The neutron dominance in the QRPA wave function of ^{136}Te gives rise to low g factor. It would be interesting to test this prediction experimentally.

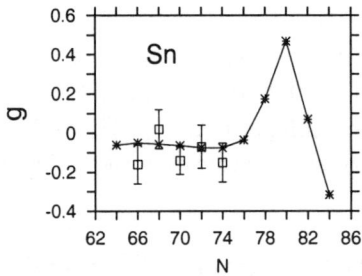

Figure 2. Calculated (asterisks) and experimental[8] (open squares with error bars) g factors of the lowest 2^+ states for Sn isotopes.

Figure 2 shows calculated g factors of the lowest 2^+ states in even-even Sn isotopes compared to the experimental data. The behavior of the g factors up to $N = 74$ can be understood in terms of the negative single-neutron g-factors of the $1h_{11/2}$, $2d_{5/2}$, and $3s_{1/2}$ shells (see Ref. [8]). Around $N=78$, however, the $2d_{3/2}$ orbital carrying a positive g factor becomes occupied, and this gives rise to positive g-factors in $^{128,130,132}Sn$. Above $N=82$, the structure of the lowest 2^+ state is dominated by the $2f_{7/2}$ orbit, and g-factors drop.

4. Is ^{68}Ni a doubly magic nucleus?

We briefly mention QRPA calculation for ^{68}Ni, for which a question has been asked if it is a doubly magic nucleus[9,10,11]. It has been argued that the energy and $B(E2)\uparrow$ of the first 2^+ state of ^{68}Ni are much closer to those of ^{56}Ni than to the midshell Ni isotopes. We performed systematic QRPA calculations for the Ni isotopes and found that the $B(E2)$ strength distribution of ^{68}Ni is fragmented, while that of ^{56}Ni is concentrated in the first 2^+ state. Thus judging from the behavior of the 2^+ state above, it is impossible to conclude that ^{68}Ni is a doubly magic nucleus. This question will be discussed in detail elsewhere[12], also using Shell Model Monte Carlo and a large-scale diagonalization shell model.

5. Summary

In this paper, we have investigated the irregular behavior of E_{2^+}'s and $B(E2)\uparrow$'s in ^{132}Te–^{136}Te through the QRPA with a simple separable interaction. Our QRPA calculations reproduce the behavior seen experimentally, and we trace it back to the difference in neutron pairing. A related finding is that the $B(E2)\uparrow$ in ^{132}Sn should be larger than in the neighboring even-even Sn isotopes, as is the case around ^{208}Pb. We hope that this prediction will stimulate further measurements in the neutron-rich region around ^{132}Sn.

To strengthen our argument, we also calculated g factors of the Xe, Te, and Sn isotopes. We reproduced the experimental trends seen in Xe and found that while protons dominate the excitation amplitudes in ^{134}Xe and ^{136}Xe, the g factor of the 2^+ state of ^{136}Te is dramatically reduced. The experimental confirmation of this effect as well as rapid change of g factor in ^{128}Sn–^{134}Sn would validate our understanding of the structure of nuclei around ^{132}Sn.

References

1. D.C. Radford *et al. Phys. Rev. Lett.* **88**, 222501 (2002).
2. J. Terasaki, J. Engel, W. Nazarewicz, and M. Stoitsov, to be published in *Phys. Rev. C*.
3. S. Raman, C.W. Nestor, Jr., and P. Tikkanen, *Atom. Data Nucl. Data Tab.* **78**, 1 (2001).
4. V.I. Isakov *et al.*, preprint, arXiv:nucl-th/0202044 .
5. K. Heyde, *The Nuclear Shell Model* (Springer-Verlag, Berlin, 1990).
6. L.S. Kisslinger and R.A. Sorensen, *Rev. Mod. Phys.* **35**, 853 (1963).
7. G. Jakob *et al. Phys. Rev.* **C65**, 024316 (2002).
8. M. Hass, C. Broude, Y. Niv, and A. Zemel, *Phys. Rev.* **C22**, 97 (1980).
9. M. Bernas *et al., Phys. Lett.* **B113**, 279 (1982).
10. O. Sorlin *et al., Phys. Rev. Lett.* **88**, 092501 (2002).
11. H. Grawe and M. Lewitowicz, *Nucl. Phys.* **A693**, 116 (2001).
12. K. Langanke *et al.*, in preparation.

CHARGE RADII AND NUCLEAR MOMENTS AROUND ^{132}SN

F. LE BLANC, E. COTTEREAU S. ESSABAA, J. OBERT, J.OMS, A.
OUCHRIF, B. ROUSSIÈRE, J. SAUVAGE AND D. VERNEY

Institut de Physique Nucléaire,
91406 Orsay cedex, France

L. CABARET AND J. PINARD

Laboratoire Aimé Cotton,
91405 Orsay cedex, France

R. HORN, G. HUBER AND J. LASSEN

Institut für Physik der Universität Mainz,
55099 Mainz, Germany

J.E. CRAWFORD AND J.K.P. LEE

Foster Radiation Laboratory,
Mc Gill University,
H3A2T8 Montréal, Canada

J. GENEVEY

Institut des Sciences Nucléaire,
CNRS-Université Joseph Fourier,
38026 Grenoble cedex, France

G. LE SCORNET AND ISOLDE COLLABORATION

CERN,
1211 Genève 23, Switzerland

Laser spectroscopy measurements have been carried out on the very neutron-rich tin isotopes with the COMPLIS experimental setup. Using the $5s^2 5p^2\ ^3P_0 \rightarrow 5s^2 5p6s\ ^3P_1$ optical transition, hyperfine spectra of $^{126-132}$Sn and $^{125m,127m,129m-131m}$Sn where recorded for the first time. The variation of the mean square charge radius ($\delta < r^2 >$) between these nuclei and nuclear moments of the isomers and the odd isotopes were thus measured. An odd even staggering which inverts at A=130 is observed showing the appearance of a plateau on the

$\delta < r^2 >$. The magnetic moments measured are compared with the Schmidt values calculations. Moreover, from the quadrupole moments values, all these nuclei appear to be spherical.

1. Introduction

The doubly-magic nuclei are of great interest in nuclear physics because their properties (binding energy, radius...) are the basis of the parametrization of the effective interactions used for mean-field calculations [1,2]. For the last two decades, these calculations [3,4,5,6] as well as the relativistic mean-field theory [7] successfully described the global properties of the nuclear ground states [6,8,9]. In the same way, a lot of new and accurate results were obtained providing systematic data available along isotopic series from light to heavy nuclei. This motivated many theoretical works in particular to improve the parameters of the effective interactions currently used. The goal of these theoretical studies is to define an effective interaction valid not only along the stability line but also for exotic nuclei. To constrain the nuclear forces we have studied the effect of the shell closure at $N=82$ on the mean square charge radius variation ($\delta < r^2 >$) far from stability. At the crossing of such a magic shell, a shell effect provokes a kink on the $\delta < r^2 >$ as it can be seen on the stability valley for Cs^{10}, Ba^{11} and Xe^{12}. If we now compare tin with another Z-magic element like lead ($Z=82$), one also observe a kink on the $\delta < r^2 >$ at $N=126$ [13]. This has been successfully reproduced by relativistic mean field Hartree-Fock (NL-SH and NL1) [14] and non relativistic Hartree-Fock calculations using the Sly10 force [15]. So, the most important question to address is: will the $\delta < r^2 >$ exhibit a slope change at ^{132}Sn as it does for ^{208}Pb ? The observation of a plateau at $A = 132$ should be a strong indication of a neutron skin for such neutron rich nuclei. $\delta < r^2 >$ curves have been calculated for these isotopes [16,17] and the predictions depend on the type of calculations.

From the measurement of the isotope shift we have a direct access to the $\delta < r^2 >$ along isotopes series. To perform such measurements on tin isotopes, we have successfully used a technique of ion-beam implantation followed by Resonant Ionisation Spectroscopy (RIS) studies of the laser desorbed radioactive element. Such a system (COMPLIS) is installed at the ISOLDE-Booster facility.

In this contribution, we report on recent laser spectroscopy measurements performed on the heavy tin isotopes up to A=132. From the $5s^2 5p^2 \; ^3P_0 \rightarrow 5s^2 5p6s \; ^3P_1$ optical transition, the hyperfine spectra of

$^{126-132}$Sn as well of these of $^{125m,127m,129m-131m}$Sn where recorded for the first time. The variation of the mean square charge radius $\delta < r^2 >$ between these nuclei and the nuclear moments of the isomers and the odd isotopes were thus measured. These results are discussed and compared with different theoretical predictions.

2. Experimental set-up

The COMPLIS experimental setup has already been described in Ref. [18]. For the tin studies, the experiment is performed as follows. The radioactive tin isotopes are produced via fission reactions on the ISOLDE UC_2 target with the 1GeV CERN PS-Booster proton beam. The ions are extracted at 60 kV and mass separated by the Ground Purpose Separator (GPS) of ISOLDE. The ions enter the COMPLIS beam line, are slowed to 1 kV and are thus deposited on the first atomic layers of a rotating graphite substrate. Once the amount of the collected atoms is optimum (the collection time depending on the half-life of the isotope to be studied) they are desorbed by a Nd:YAG laser and selectively ionized by a set of two pulsed, tunable dye lasers where the first excitation step at 286.3 nm $(5s^2 5p^2\,^3P_0 \to 5s^2 5p6s\,^3P_1)$ is obtained from frequency doubling. The ions are finally detected with time-of-flight identification using a microchannel plate detector. This experimental set up is shown on Fig. 1.

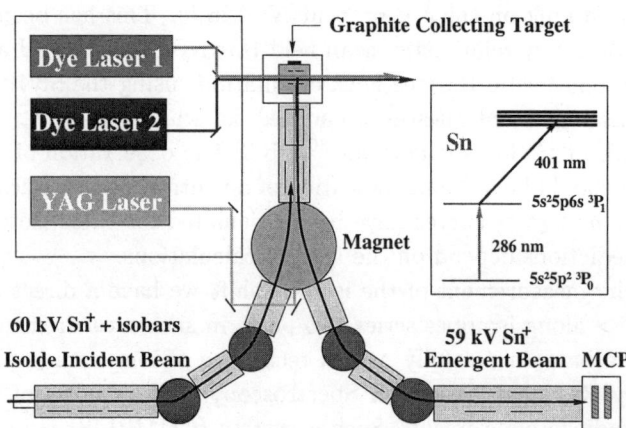

Figure 1. COMPLIS experimental set-up. Insert shows the RIS scheme used for tin

The frequency scan over the hyperfine structure of a given isotope (and eventually isomer) is made as follows : after a sufficient collection time, the desorption of the tin atoms is made over the entire collection spot on the slowly rotating target at a given frequency step. After the desorption is complete, a new cycle of implantation desorption is run at an advanced frequency step. Whenever the laser frequency corresponds to a hyperfine transition, the desorbed atoms are excited and ionized by the other fixed laser frequency. The number of counted ions at the detector is directly proportional to the intensity of the hyperfine transition. With this apparatus, the efficiency we measured is of about 10^{-6} with a resolution of 170 MHz. It is worth noting that, with the hot-plasma source we used, many other elements like In, Cd, Sb, Te, I and Cs are ionized and for example at mass unit 132, tin represent only 0.24 % of the produced nuclei. For the first experiment on tin, we measured for the first time all the isotopes and isomers from A=125 to A=132. From the displacement of the centers of gravity of the hyperfine spectra, we were able to extract the isotope shift. From the relative position of the three lines of each isomer and odd isotope we can extract the magnetic and quadrupole moments.

3. Experimental results

The magnetic moments of the isomers and the odd isotopes are obtained from A_i of the atomic excited state and the precisely known value of μ_I of ^{121}Sn [19]:

$$\mu_I(^x Sn) = 0.4191(5) \cdot A_i(^x Sn) \cdot I(^x Sn)$$

The μ_I values are presented in Table 1. The value measured by Anselment et al. [20] for ^{125}Sn^g is consistent with our measurements.

Also, for the isomers and the odd isotopes, the B_i factors of the $^3P_1$ atomic excited state have been extracted from the hyperfine spectra. Q_s is related to B_i via:

$$Q_s(^x Sn) = -7.25 \cdot B_i(^x Sn)$$

which is obtained from the calculated electric field gradient as described in Ref. [19]. The measured Q_s values are presented in Table 1.

The experimental isotope shift consists of a mass shift $\delta\nu_{MS}$ and a field shift $\delta\nu_{FS}$; it is from this last contribution that $\delta < r_c^2 >$ beween two nuclei A and A' can be extracted. Between two masses A and A', the isotope shift can thus be expressed as:

42

$$\delta\nu^{A,A'} = K.F_\lambda.\delta < r_c^2 >^{A,A'} + M\left(\frac{A - A'}{AA'}\right),$$

where $K = 0.975$ [21]. To evaluate F_λ and M we have performed a King Plot taking into account the error bars in both axis using a linear fit program as described in [22]. This King Plot is made with the isotope shift measured in [20] versus the muonic $\delta < r_c^2 >$ measured in [23] taking including all isotope pair combinations as described in [24]. This gives :

$$\delta\nu^{A,A'}.\left(\frac{AA'}{A' - A}\right) = K.F_\lambda.\delta < r_c^2 >^{A,A'}.\left(\frac{AA'}{A' - A}\right) + M$$

The result of the fit gives $F_\lambda = 3.30(27)\ GHz/fm^2$ and $M = -761(200)\ GHz$. Knowing the value of $< r_c > (^{120}Sn)$ from [23] we have calculated the values of all the absolute radii. The corresponding updated curve is presented Fig. 2.

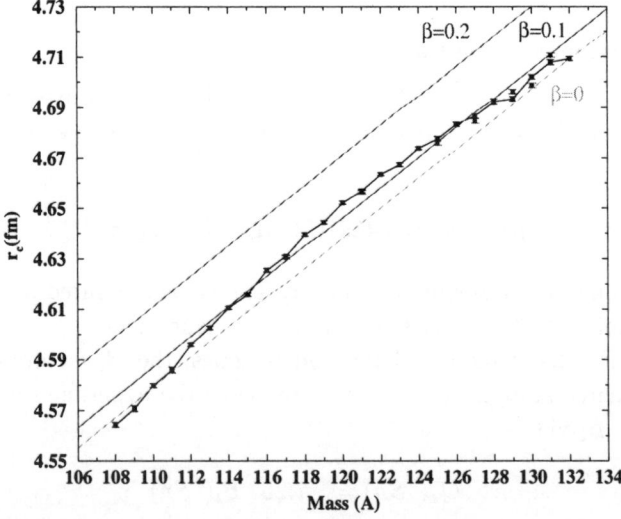

Figure 2. Absolute radii of the tin isotopes up to A=132. The different β deformation lines using the droplet model have also be put for comparison. The errors bars only include statistical errors.

4. Discussion

4.1. *Nuclear Moments*

If one assume axial symmetry, one can easely extract the intrinsic quadrupole moment and then the β deformation (see Table 1). Except for two nuclei, they all lead to a deformation close to zero which confirms a spherical shape for all these nearly doubly-magic nuclei. The only exceptions are for the two isomers ^{125}Sn^m and ^{127}Sn^m which have $\beta = 0.27$ and $\beta = 0.18$ respectively. This quite high for such Z-magic nuclei. This means that ^{125}Sn^m and ^{127}Sn^m are probably triaxial.

For all the deduced spherical nuclei, we have calculated the magnetic moments assuming spherical states in order to use the Schmidt model. For a mean value of $g_s = 0.7g_{s,free}$ the calculations very well reproduce the experimental values which means that the states labelled for each nucleus are well defined (see Table 1). This is in full agreement with the spin values of all these states that were uncertain up to now.

Table 1. Nuclear moments of the heavy tin isotopes. The β deformation is extracted assuming axial symmetry. For the magnetic moments, the states have only been labelled when the deformation is found to be close to zero. The calculated magnetic moments have been calculated with $g_s = 0.7g_{s,free}$.

A	I^π	Q_s[b]	β	$\mu_I[\mu_N]$	State	$\mu_{Icalc.}[\mu_N]$
125^g	11/2$^-$	+0.14(21)	+0.02(3)	-1.348(6)	$1h_{11/2}$	-1.34
125^m	3/2$^+$	+0.79(7)	+0.27(3)	+0.764(3)		
127^g	11/2$^-$	+0.30(13)	+0.04(2)	-1.329(7)	$1h_{11/2}$	-1.34
127^m	3/2$^+$	+0.60(6)	+0.20(2)	+0.757(4)		
129^g	3/2$^+$	+0.05(11)	+0.02(4)	+0.754(6)	$2d_{3/2}$	+0.80
129^m	11/2$^-$	-0.18(17)	-0.02(2)	-1.297(5)	$1h_{11/2}$	-1.34
130^m	7$^-$	-0.36(11)	-0.04(2)	-0.381(3)		
131^g	3/2$^+$	-0.04(8)	-0.01(2)	+0.747(4)	$2d_{3/2}$	+0.80
131^m	11/2$^-$	+0.02(20)	+0.002(20)	-1.276(5)	$1h_{11/2}$	-1.34

Moreover the μ_i of ^{130}Sn has not been calculated but the measured value is not as far of these of the other first 7$^-$ states in ^{118}Sn and in ^{114}Sn (see [25]). More specicifically, the μ_I of ^{131}Sn^m is compared with the value deduced from g-factor new shell model calculations in [26]. The agreement is very good since the theoretical values gives μ_I=-1.227 instead of -1.276(5) that we measured. The systematics of the 11/2$^-$ magnetic moment in the odd tin nuclei can be compared with that measured for the Te isotones (Z=52).

The values are quite the same but the evolution is somewhat different; in the tin, the magnetic moments increases as one approaches of N=82 which is the opposite in Te (see [26]). Contrary to the Te isotopes as described in [26], the trend of the $\mu_I(11/2^-)$ in the tin isotopes is in agreement with that predicted by the models.

4.2. *Charge radii*

The charge radius variation above mass 125 follows the same trend as between A=108 and A=125. One can compare the experimental values for the even isotopes with data extracted from fully self-consistent calculation with suitable relativistic or nonrelativistic effective interactions. The comparison is made with relativistic NL3 force [27] and the non relativistic Gogny[17], SLy4 and SLy7[28] non relativistic forces. NL3 perfectly reproduces the radius of ^{132}Sn but not the general trend. Gogny reproduces the general trend quite good but not the ^{132}Sn radius and the best agreement is obtained for the two Syrme forces (see Fig. 3).

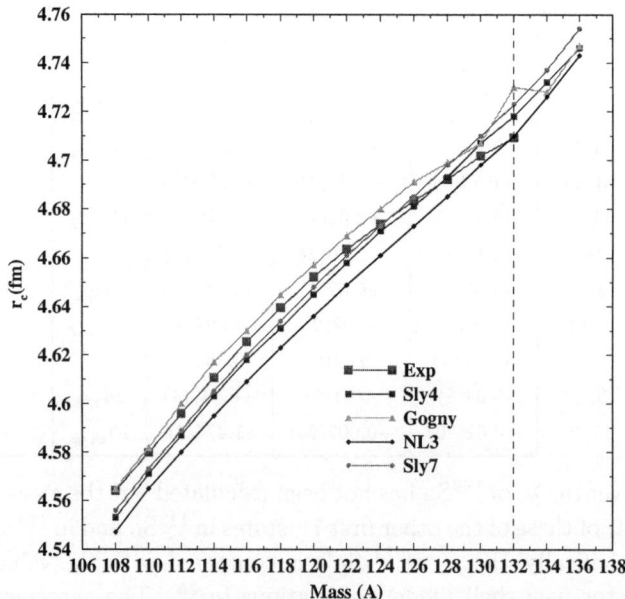

Figure 3. Comparison between the experimental $< r_c >$ of the even-even isotopes and the theoretical values using some microscopic mean-field models.

If we now make a zoom on the $\delta < r_c^2 >$ at the N=82 region and compare with the neighboring elements (see Fig. 4) one can see that the odd-even staggering also appearing for Cs and Ba inverts in the case of tin at N=81 and that a plateau is appearing then, which is not the case for the neighbors. This is a good indication of a neutron skin for this element but $\delta < r_c^2 >$ measurements up to A=134 should confirm this prediction.

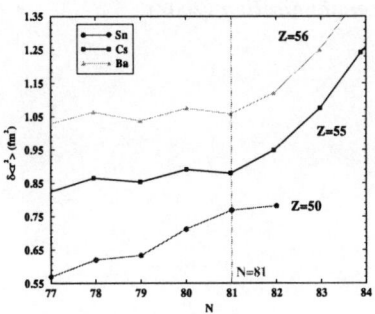

Figure 4. Zoom of the $\delta < r^2 >$ curve at N=81 to compare tin with two neighboring elements: Ba and Cs.

References

1. D. Vautherin and D.M. Brink, *Phys. Rev.* **C5**, (1972) 626.
2. J. Dechargé and D. Gogny, *Phys. Rev.* **C21** (1980) 1568.
3. P. Quentin and H. Flocard, *Annu. Rev. Nucl. Sci.* **28** (1978) 523.
4. M. Girod and B. Grammaticos, *Phys. Rev.* **C27** (1983)2317.
5. M. Girod et al., *Phys. Rev. Lett.* **62** (1989) 2452.
6. Z. Patyk et al., *Phys. Rev.* **C59** (1999) 704.
7. J.D. Walecka, *Annu. Phys. N.Y.* **83** (1974) 491.
8. P.G. Reinhard, *Rep. Prog. Phys.* **52** (1989) 439.
9. V. Blum et al.,*Phys. Rev.* **B323** (1994) 262.
10. C. Thibaud et al., *Nucl. Phys.* **A367** (1981) 1.
11. K. Wendt et al., *Z. Phys.***A318** (1984) 125.
12. W. Borchers et al., *Phys. Lett.* **216B** (1989) 7.
13. R.C. Thomson et al., *J. Phys.* **G9** (1983) 443.
14. M.M. Sharma, G.A. Lalazissis and P. Ring, *Phys. Lett.* **317B** (1993) 9.
15. E. Chabanat et al., *Nucl. Phys.* **A635** (1998) 231.
16. P.G. Reinhardt, *private communication* (1997).
17. M. Girod and S. Peru, *private communication* (1998).
18. J. Sauvage et al., *Hyp. Int.* **129** (2000) 303.
19. J. Eberz et al., *Z. Phys.* **A326** (1987) 121.

46

20. M. Anselment et al., *Nucl. Phys.* **A451** (1986) 471.
21. G. Torbohm et al., *Phys. Rev* **A31** (1985) 2038.
22. P.H. Borcherds and C.V. Smith, *Eur. Jour. of Physics* **16** (1995) 204.
23. C. Piller et al., *Phys. Rev.* **C42** (1990) 182.
24. W. Neu et al., *Z. Phys.* **D7** (1987) 193.
25. P. Raghavan, *At. Data Nucl. Data Tables* **42** (1989) 189.
26. G. Jakob et al., *Phys. Rev.* **C65** (2002) 024316.
27. G.A. Lalazissis et al., *At. Data Nucl. Data Tables* **71** (1999) 1.
28. J. Meyer, *private communication* (2002).

DECAY OF HIGH SPIN MICROSECOND ISOMERS IN ^{129}IN, ^{129}SN AND ^{130}SB NUCLEI

J. GENEVEY AND J. A. PINSTON

Institut des Sciences Nucléaires IN2P3-CNRS/ Université Joseph Fourier,
F-38026 Grenoble Cedex, France

H. FAUST, A. SCHERILLO, G. SIMPSON AND I. TSEKHANOVICH

Institut Laue-Langevin, F-38042 Grenoble Cedex, France

In this work microsecond isomers of the three nuclei ^{129}In, ^{129}Sn, and ^{130}Sb were investigated. These nuclei were produced by the thermal-neutron-induced fission of ^{239}Pu and ^{241}Pu. The detection is based on time correlation between fission fragments selected by the LOHENGRIN spectrometer at the ILL (Grenoble) and the γ rays or conversion electrons from the isomers. The interpretations of the level schemes are mainly based on a truncated shell model calculation using empirical two body interactions. Several new B(E2) strengths of isomeric transitions were measured and are discussed in the paper.

1. Introduction

During the last few years there has been significant progress in the experimental study of the high-spin yrast excitations in neutron rich nuclei with few particles or holes outside the doubly magic ^{132}Sn and for a neutron number N$\leq$82 [1,2,3]. The attention has been focussed on these nuclei because their yrast levels are expected to have a particularly simple structure for testing the basic ingredients of the nuclear shell model, such as the the two-body matrix elements of the residual interactions. In this paper we have extended these previous investigations to the three neutron hole $^{129}_{49}$In, $^{129}_{50}$Sn, and $^{130}_{51}$Sb nuclei.

2. Experimental procedure

The nuclei of the A=129 and 130 mass chains were produced by thermal-neutron-induced fission of ^{239}Pu and ^{241}Pu targets. The LOHENGRIN spectrometer at the ILL has been used to separate the fission fragments (FF) recoiling from thin targets of about 400 μg/cm^2, according to their

A/q ratios. The FF are detected by a ΔE gas detector, and subsequently stopped in a Mylar foil 12 μm thick. The γ-rays deexciting the isomeric states are detected by two large-volume Ge detectors and the conversion electrons are detected by two cooled adjacent Si(Li) detectors covering a total area of 2×6 cm^2 and located at 7 mm behind the Mylar window. The electron detection effficiency is very high, about 30%. The gas pressure of the ionization chamber was tuned to stop the FF at about 2 μm from the outer surface of the Mylar window to minimize electron absorption and to have good energy resolution. With this set-up, it is possible to detect conversion electrons down to about 15 keV. Note that a very low energy-detection threshold and a very high detection efficiency are absolutely necessary to observe the very low energy isomeric transitions expected in nuclei close to doubly-magic systems.

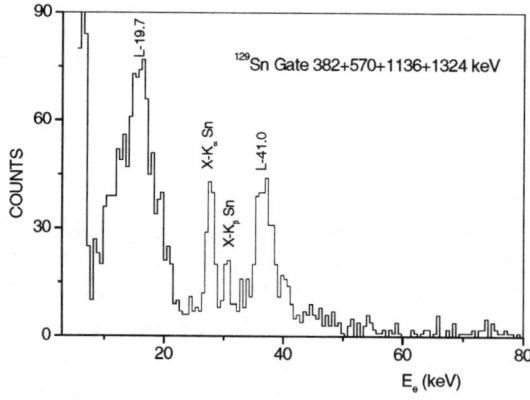

Figure 1. Si(Li) spectrum of ^{129m}Sn gated by the sum of the 382, 570, 1136 and 1324 keV γ-rays.

3. Theoretical calculations

A semi-empirical shell model calculation was performed using the OXBASH [4] code. A truncated configuration was used including only the orbitals $\pi g_{9/2}$ or $\pi g_{7/2}$ for the proton and $\nu d_{3/2}$ and $\nu h_{11/2}$ for the neutrons. The details concerning the calculations of the energy levels are given in [3]. The two-body matrix elements of the residual interaction were extracted,

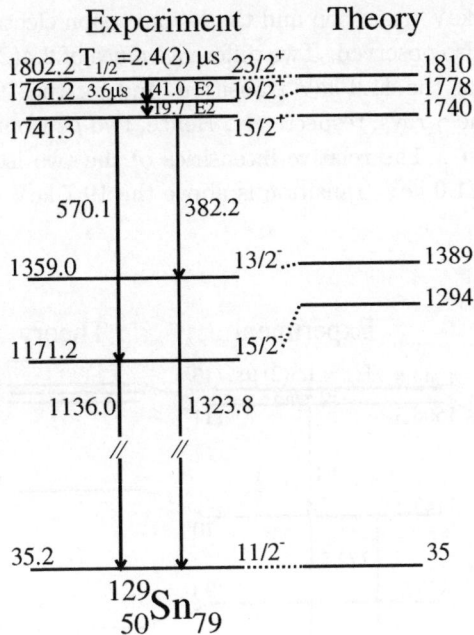

Figure 2. Decay scheme of the ^{129}Sn isomers. The calculated energies are given relative to the low-lying $11/2^-$ state at 35 keV.

wherever possible, from experimental data. However, the complete experimental data are not available for the n-p interactions and we have used the calculated values of Andreozzi et al. [5] for the $\pi g_{7/2} \nu h_{11/2}$ configuration and the values of Van Maldeghem et al. [6] for the $\pi g_{9/2} \nu h_{11/2}$ configuration. Only yrast or near yrast states having rather pure configurations are considered in the calculation. A comparison between the experimental levels of ^{129}Sn, ^{130}Sb and ^{129}In and the calculations is shown in Fig. 2, 3 and 5 respectively.

4. Results

4.1. $^{129}_{50}Sn$

The Si(Li) spectrum in coincidence with the four γ rays of ^{129}Sn is shown in Fig. 1. In addition to the Sn K_α and K_β X-rays, the L-conversion electrons of a 41.0 keV transition and the L-conversion electrons of a 19.7 keV transition are also observed. Two different values of 2.4(2) μs and 3.6(2) μs are measured for the 41.0 keV transition, and for the 19.7 keV transition as well as for the γ rays, respectively. Hence, two μs isomers in cascade are present in ^{129}Sn . The relative intensities of the two isomeric transitions show that the 41.0 keV transition is above the 19.7 keV one.

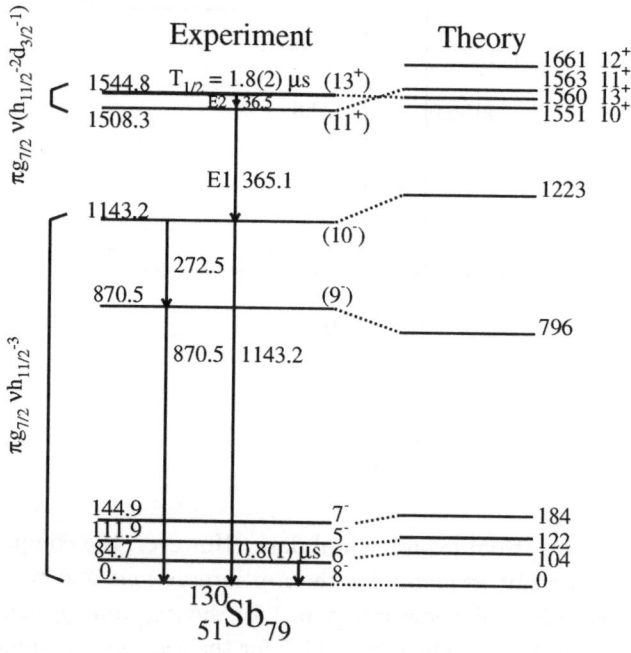

Figure 3. Decay scheme of the ^{130}Sb isomer. The theoretical energies are determined relative to the 8^- ground state. The two negative parity states at 111.9 and 144.9 keV are from Ref.[7].

The level scheme of ^{129}Sn is shown in Fig. 2. The two $19/2^+$ and $23/2^+$ isomers in cascade in ^{129}Sn are well reproduced by the model which shows that the $\nu(h_{11/2}^{-2}d_{3/2}^{-1})$ is the leading configuration of these states. The negative parity states belong to the $\nu h_{11/2}^{-3}$ configuration. However, the fully aligned $27/2^-$ state of that configuration was not observed in this experiment. May be its half-life is too short as discussed in Ref. [2] and it decays during its 2.2 μs flight time through the LOHENGRIN spectrometer.

4.2. $^{130}_{50}Sb$

The odd-odd ^{130}Sb nucleus has the same number of neutrons and one more proton than ^{129}Sn. A new isomer of 1.8 μs half-life, decaying by a low-energy E2 transition of 36.5 keV energy, was observed. The conversion electrons of another isomeric transition were also observed in the Si(Li) spectrum; they correspond to the already known 6^- → 8^- E2 transition [7]. A value of $T_{1/2}=0.8(1)$ μs was measured for its half-life for the first time.

The decay scheme of the new isomer of ^{130}Sb is shown in Fig. 3. The first excited state at 870.5 keV is the unique overlap between the levels fed by the microsecond isomer and the levels fed by β-decay. However, this level which feeds exclusively the 8^- ground state and not the 6^- state at 84.7 keV has very likely a spin and parity value $I^\pi=9^-$ and not the value $I^\pi=7^-$ previously proposed by Walters et al. [7]. The level scheme proposed supposes that the levels are close to the yrast line, as in the other microsecond isomers in this mass region fed by fission.

In ^{130}Sb, the calculation shows that the 10^+, 11^+ and 13^+ states of the $\pi g_{7/2}\nu(h_{11/2}^{-2}d_{3/2}^{-1})$ configuration are all in an energy range of only 12 keV. This result provides strong support for an isomeric 13^+ → 11^+ transition of very low energy and E2 multipolarity, although the precision of the calculation is not sufficient to reproduce the observed order of the levels. This 13^+ isomer in ^{130}Sb is the analogous to the $19/2^+$ isomer in ^{129}Sn and in fact, they have comparable excitation energies 1545 and 1761 keV respectively, in these two nuclei.

The negative parity states belonging to the $\pi g_{7/2}\nu(h_{11/2}^{-3})$ configuration are well reproduced by the model. In the absence of more complete experimental data in ^{132}Sb, this feature shows *a posterori* that the realistic effective interaction derived from the Bonn potential and used by Andreozzi et al. reproduces correctly the n-p $\pi g_{7/2}\nu h_{11/2}^{-1}$ interaction.

4.3. $^{129}_{49}In$

The nuclear structure information on the heavy In isotopes is very scarce. The most important results were the possible evidence by Fogelberg et al. [8] of the high-spin yrast-traps $23/2^-$ and $29/2^+$ in ^{129}In, belonging to the aligned configurations $(\pi g_{9/2}^{-1}\nu(h_{11/2}^{-1}d_{3/2}^{-1}))23/2^-$ and $(\pi g_{9/2}^{-1}\nu(h_{11/2}^{-2})29/2^+$ respectively.

In this work, we have observed a new 7.3 (5) μs isomer in ^{129}In decaying by four γ-rays of 333.5, 358.9, 995.2 and 1354.0 keV energy. The K-shell conversion electrons of the 333.5 keV transition was also observed in the

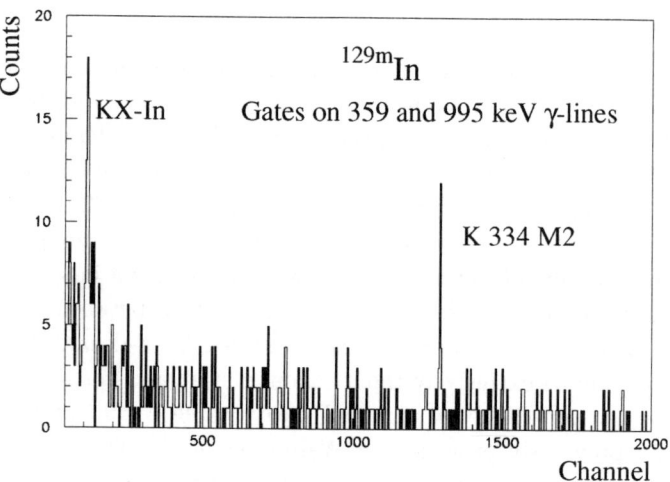

Figure 4. Si(Li) spectrum of ^{129m}In gated by the sum of 359 and 995 keV γ-rays.

Si(Li) spectrum and shown in Fig. 4. The computed conversion coefficient is compatible with either M2 or E3, but its lifetime is characteristic of M2 as suggested by its reduced transition probability B(M2)= 0.038 W. u.

In Fig. 5 the level scheme of the ^{129}In nucleus obtained in this work is compared with a theoretical calculation, where only the positive parity states belonging to the $\pi g_{9/2}^{-1} \nu h_{11/2}^{-2}$ were computed. Good agreement is obtained between the experiment data and theory, and the splitting between the $11/2^+$ and $13/2^+$ states is well reproduced. The isomeric state belongs to the $\pi g_{9/2}^{-1} \nu (h_{11/2}^{-1} d_{3/2}^{-1})$ configuration.

5. E2 transition rates in Sn isotopes

In Fig. 6 are plotted the $B(E2, 19/2^+ \rightarrow 15/2^+)$ and $B(E2, 23/2^+ \rightarrow 19/2^+)$ values for the odd Sn isotopes as well as the $B(E2, 10^+ \rightarrow 8^+)$ values for the even Sn isotopes against the mass number A; the first two correspond to the transitions involving states of the $\nu(h_{11/2}^2 d_{3/2})$ configuration, and the last corresponds to the transitions involving states of the $\nu h_{11/2}^2$ configuration. The data for ^{129}Sn are from the present work, while the other values have been taken from the literature [9,10,11,12]. The B(E2) values in the odd and even Sn nuclei show the same trend : a strong

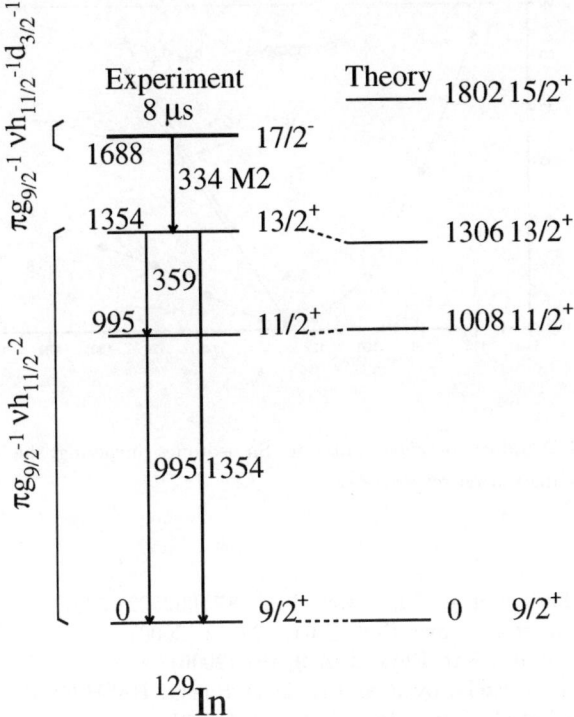

Figure 5. Decay scheme of the ^{129}In isomer. The theoretical energies are determined relative to the $9/2^+$ ground state.

decrease of the E2 strength when A decreases, with a deep minimum at A=123. This behavior reflects the filling of the $\nu h_{11/2}$ neutron subshell and the minimum corresponds to its half filling, which means that for the neutron number N=73 in the Sn isotopes, 6 neutrons occupy the $\nu h_{11/2}$ orbital. Note that the increase in the B(E2) strength above the deep minimum is much more dramatic for the levels of the $\nu(h^2_{11/2}d_{3/2})$ configuration than for the states belonging to the configurations involving the $\nu h_{11/2}$ orbital only. It is a challenge for the future to complete the data of the $\nu(h^2_{11/2}d_{3/2})$ configuration below A=123 where an increase in the B(E2) values is expected.

54

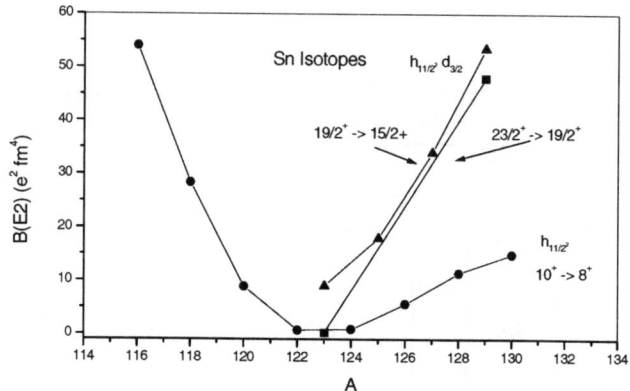

Figure 6. B(E2) values of even and odd Sn isotopes involving the $\nu(h^2_{11/2})$ and $\nu(h^2_{11/2}d_{3/2})$ configuration respectively.

References

1. P. Bhattacharyya et al., Phys. Rev. Lett. **87**, 062502 (2001).
2. J. A. Pinston et al., Phys. Rev. C **61**, 024312 (2000).
3. J. Genevey et al. , Eur. Phys. J. A **9**, 191 (2000).
4. B. A. Brown, A. Etchegoyen, and W. D. H. Rae OXBASH (1984) unpublished.
5. F. Andreozzi et al., Phys. Rev. C **59**, 746 (1999).
6. J. Van Maldeghem, Phys. Rev. C **32**, 1067 (1985).
7. W. B. Walters, and C. A. Stone, Proc. Intern. Workshop Nucl. Fission and Fission Product Spectroscopy, edited by H. Faust and G. Fioni, Seyssins, France, p. 182 (1994).
8. B. Fogelberg et al., Intern. Workshop Nucl. Fission and Fission Product Spectroscopy,2nd, Seyssins, France, edited by H. Faust and G. Fioni and F.-J. Hambsch, AIP Conf. Proc. 447, p. 191, (1998).
9. B. Fogelberg et al., Nucl. Phys. **A352**, 157 (1981).
10. R. Broda et al., Phys. Rev. Lett. **68**, 1671 (1992).
11. J. Genevey et al., AIP Conf. Proc. No. 455 (AIP, Woodburry, NY, 1998) p. 694.
12. C. T. Zhang et al., Phys. Rev. C **62**, 057305 (2000).

NEW EVIDENCE FOR NEUTRON MAGIC NUMBER N=16
AT BORDER LINE

Z. DLOUHÝ, D. BAIBORODIN, J. MRÁZEK, G. THIAMOVÁ

Nuclear Physics Institute, ASCR,
CZ 250 68 Řež, Czech Republic
for the GANIL - Orsay - Dubna - Řež - Bucharest collaboration

We present a survey of experimental results obtained at GANIL (Caen, France) resulting in an appearance of a new magic number, N=16, in very neutron-rich nuclei. Recent data on mass measurements of neutron-rich nuclei at GANIL and some characteristics of binding energies in this region are discussed. Nuclear binding energies are very sensitive to the existence of nuclear shells and together with the measurements of instability of the doubly magic nucleus ^{28}O they provide new information on changes in neutron shell closures of very neutron-rich isotopes from carbon up to calcium. The behavior of the two neutron separation energies S_{2n} gives a very clear evidence for the existence of the new shell closure N=16 for Z=9 and 10 appearing between $2s_{1/2}$ and $1d_{3/2}$ orbitals. This fact, strongly supported by the instability of C, N and O isotopes with N¿16, confirms the magic character of N=16 for the region $6 \leq Z \leq 10$, while the shell closure at N=20 tends to disappear for $Z \leq 13$.

The nuclear structure along the the beta-stability has been well studied both experimentally and theoretically. On the contrary, the yet not well known structure of extremely neutron- or proton-rich nuclei of light elements is currently of great interest and represents a very important topic in modern nuclear physics. In extreme configurations of the nuclear matter the higher sensitivity to the choice of the nuclear models offers a unique test of those components of effective interactions that depend on isospin degrees of freedom. Since the parameters of interactions used in the usual mean-field calculations are determined so as to reproduce the properties of beta-stable nuclei, these parameters may not always be optimal around drip lines due to the extrapolations involved.

Research in this field has revealed the "halo" and/or "skin" nuclear structure [1,2], phenomena quite different from the matter distribution of other particle stable nuclei. The halo nucleus is supposed to consist of a core that is surrounded by one or two loosely bound neutrons moving far away from the core. The core has the same size as the bare nucleus forming the core and having the radius which corresponds to the mass of the core nucleus. Thus adding one or two loosely bound valence neutrons to this core the radius of the next neutron-rich isotope (i.e. the halo nucleus) can suddenly increase due to relatively large distance of its valence neutrons from the core.

A major break through in nuclear physics was the discovery of shell structure in the nucleus and the introduction of the magic numbers in the frame of the shell model. Nuclei at shell closures have gaps of the order of 5-8 MeV and are classified as magic in the analogy with noble gas configurations in the atoms. The magic numbers expected for a harmonic oscillator mean field potential (2,8,20,40,70...) are perturbed in the nucleus by strong spin-orbit interaction, which is pushing the highest spin state out of a major oscillator shell towards the lower ones. As a result, new magic numbers like 28,50,82 do appear, while some other harmonic oscillator magic numbers completely disappear. The nuclear structure is experimentally confirmed for beta-stable nuclei, and thus the relevant part of the one-body potential can be approximated by the oscillator potential with the spin-orbit potential.

For the case of neutron-rich nuclei with many neutrons outside the closed shell in spherical potential the degeneracy in quantum spectra implies the possibility to gain energy by breaking away from spherical symmetry. This can be the reason why the nuclei near the drip line exhibit the type of deformation that originate from the new shell structure with new magic numbers. The breaking of magicity derived from the nuclei in the valley of stability has already been observed at the N=20 shell closure where an "island of inversion" in shell ordering has been shown to exist [3]. Such behaviour can result as a consequence of the instability of the both doubly magic neutron-rich nuclei ^{10}He [4,5] and ^{28}O [6].

During more than 30 years a number of experiments have been carried out to search for the doubly magic ^{10}He nucleus which, if bound, would possess the highest N/Z ratio among all known nuclei. The experiments supported by fragmentation studies led to conclusion, that ^{10}He is probably unbound. Only in 1994 Korsheninnikov et al.[4] investigated the ^{11}Li fragmentation using an invariant mass measurement and was successful in the observation of a peak in the ^{8}He+n+n coincidences which could be explained by the ^{10}He resonance at the energy 1.2 $\pm$ 0.3 MeV above the ^{8}He+n+n threshold.

At the same time, Ostrowski et al. [5] measured the mass and two excited states of ^{10}He with the ^{10}Be(^{14}C,^{14}O)^{10}He reaction at E/A=24 MeV and confirmed the instability of ^{10}He. Two excited states were observed at 3.24(20) and 6.80(7)MeV with preliminary spin assignment 2^+ and 3^-. From the existence of the relatively low-lying (2^+)-state, we can conclude that the shell closure probably does not take place at the neutron number N=8, but is shifted down to N=6. There are indications in neighboring nuclei that the $\nu\, 2s_{1/2}$-orbital is lowered for neutron-rich nuclei and crosses $1d_{5/2}$-orbital and moves even close to the $\nu\, 2p_{1/2}$-shell.

Several experiments [6,7,8] were performed on the search for the particle stability of nuclei near ^{28}O. In the last experiment performed on the LISE3 spectrometer at GANIL, we have used the fragmentation of the neutron-rich projectile ^{36}S to produce and study very neutron-rich nuclei in vicinity of a doubly magic nucleus ^{28}O. However, no events were observed corresponding to even ^{26}O and ^{28}O, odd oxygen isotopes ^{25,27}O and also ^{24,25}N. Till now, the heaviest experimentally found oxygen isotope remains ^{24}O. Our finding that ^{28}O is particle unstable [6] fairly supports the idea of the disappearance of magicity of the N=20 shell closure in ^{28}O and its influence to the Ne-Al region. The instability of ^{25,26,27,28}O and ^{24,25}N was also confirmed by the experiment performed by Sakurai et al. [9] at the fragment separator RIPS at RIKEN in which, however, a new isotope, ^{31}F, was observed for the first time. The calculated and observed yields of isotopes are in a good agreement and provide a strong evidence for the particle instability of ^{24,25}N, ^{25,26,27,28}O and ^{30}F.

Therefore, the heaviest experimentally found isotopes of carbon, nitrogen and oxygen [6,9] are ^{22}C, ^{23}N and ^{24}O, respectively, with the same neutron number, N=16, while the heaviest isotope of fluorine was found to be ^{31}F with N=22. It should be noted that it is a rather interesting behaviour among the light nuclei. Usually, in the region further from the shell closure the neutron numbers of heaviest isotopes of neighbor elements are gradually increasing with Z. Therefore, the sudden step in the largest neutron number from N=16 for carbon, nitrogen and oxygen to N=22 for fluorine may correspond to a substantial change in shell structure.

Till lately no experimental evidence about magic numbers, substituting N=20, was available, though some theoretical calculations predict the existence of new neutron magic numbers N=16, 34 [10,11]. Recently, two articles have appeared where the evidence on existence of neutron magic number, N=16, for nuclei near the drip line has been presented. In our article [12] the evidence has been obtained from dependence of two-neutron separation energy on Z and N, in the parallel experimental work of Ozawa et al. [13] the information has been obtained from interaction cross sections and one-neutron separation energies.

Moreover, the determination of the lifetime [14] and of the deformation of ^{44}S [15] has indicated the existence of a similar effect at N=28. This is the first shell closure which arises from the spin-orbit splitting, and is responsible for the gap between $1f_{7/2}$ and $2p_{3/2}$ orbitals. The study of shell closures N=20 and 28 is particularly interesting since the vanishing of the latter one could be the first evidence of the weakening of the spin-orbit force in neutron-rich nuclei. The determination of a neutron and proton drip line is also very important since it limits the particle stability region.

The particle stability of nuclei is directly related to the masses and nuclear binding energies, which are very sensitive to the existence of shells and may provide clear signatures of shell closures [16]. An experiment on mass measurement of neutron-rich nuclei using a direct time of flight technique was undertaken by Sarazin et al. [17] in order to investigate the N=20 and N=28 shell closures for nuclei from Ne (Z=10) to Ar (Z=18) and thus to bring some clarifications concerning the behaviour of magic numbers far from stability. The nuclei of interest were produced by the fragmentation of a 60 AMeV ^{48}Ca beam on a Ta target. The separation energies of two last neutrons, S_{2n} derived from the measured masses are displayed in Figure 1. The new data [17] are presented with error bars while the other, except of the data in circles, are taken from Audi et al. [18]. The Ca, K and Ar isotopes show a behaviour typical of the filling of shells, with the two shell closures at N=20 and N=28 being evidenced by the corresponding sharp decrease of S_{2n} for next two isotopes and a moderate decrease of S_{2n} for subsequent ones as the filling of the next shell starts to influence S_{2n}. The sharp drop at N=22, shown by the dashed vertical line and corresponding to the existence of the shell N_{sh}=20, is clearly visible through all the Si-Ca region while going to lower Z into the Al-Na region this drop seems to move towards lower N.

This was the reason why we made an attempt to clear up an evolution of two-neutron separation energies in this region. We used a fact that several particle stable nuclei [6,9,19] were found to exist in this region, however, their masses are not known yet. Nevertheless, their S_{2n} values must be positive and therefore, we included the estimated S_{2n} values of the heaviest ^{23}N, ^{22}C and ^{29,31}F particle stable isotopes as well as some others to the graph and marked them by circles. The most important has proved to be the inclusion of S_{2n} values for ^{29}F and ^{31}F that allowed us to observe the sharp

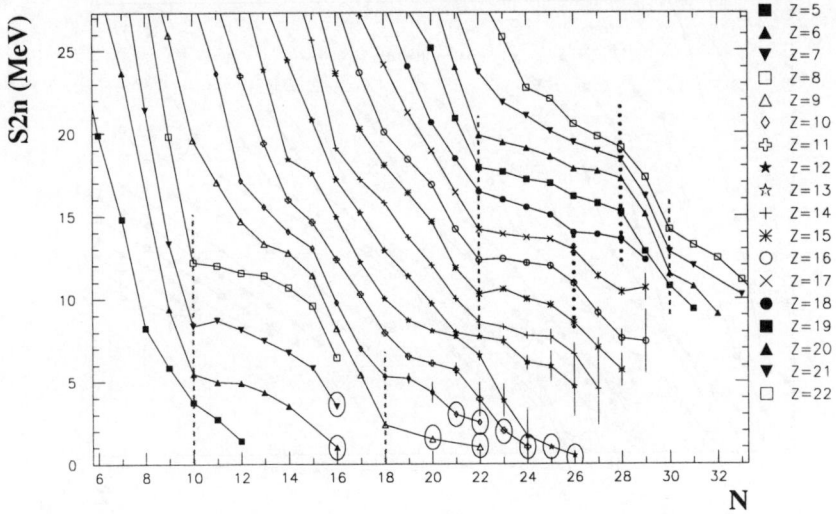

Figure 1. Two-neutron separation energy S_{2n} versus N. Expected S_{2n} values are encircled. Dotted and dashed vertical lines correspond to the neutron shells N_{sh} and a sharp drop at $N=N_{sh}+2$, respectively.

drop of ^{27}F value followed by a moderate decrease of S_{2n} values for ^{29}F and ^{31}F giving a very clear evidence for the existence of the new shell closure at N=16 for fluorine. A similar behaviour confirming the N=16 shell closure one can also see at neon isotopes that exhibit a moderate decrease of S_{2n} values for ^{29}Ne and ^{30}Ne. We have already mentioned that in the Al-Na region the sharp drop in S_{2n} values seems to move towards the lower N with decreasing Z. Now, we can make firm conclusion that it finally stabilizes at the vertical line at N=18 (new shell N_{sh}=16) for F and Ne.

It should be noted that the evidence for a new magic number N=16 follows also from Figure 2. where the S_{2n} values are plotted versus atomic number Z. The position of various possible shells or pseudo-shells are also shown. The shells N=20 and 28 appearing in Figure 1 are very clearly seen as large gaps in Figure 2. The dashed lines in Figure 2 symbolize the changes of neutron shell closures from 28 to 26 and from 20 to 16 in neutron-rich nuclei. However, both gaps are narrowing going to lower Z, till finally, at least the gap corresponding to N=20 completely disappears at Z=13, to emerge as the new N=16 gap at Z=10. This new gap governs over most of light Z neutron-rich nuclei and extends from carbon to neon.

In order to establish the firm evidence of existence the N=16 shell we

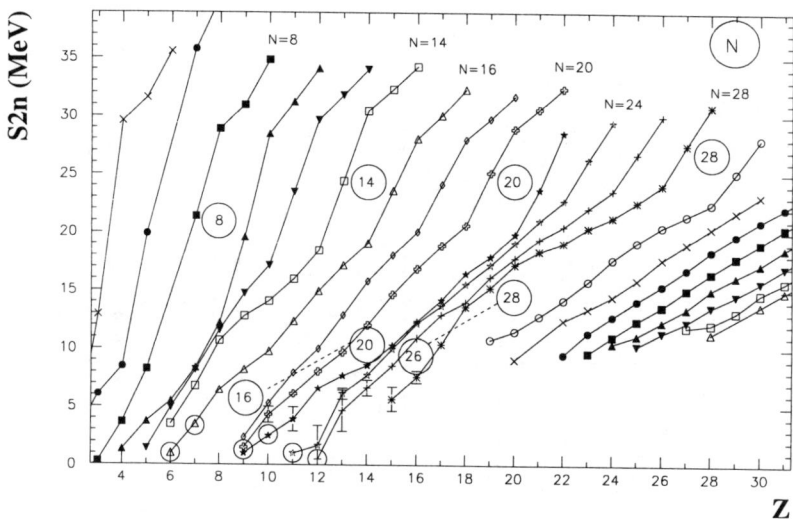

Figure 2. Experimental S_{2n} values versus proton number Z. Dashed lines symbolize the changes of neutron shell closures

have tried to analyze the form of the two neutron separation energy dependence on neutron number near the shell closures. Since the slope of S_{2n} for the nuclei located in the valley of stability changes very drastically, mainly at the drop points $(N_{sh}+2)$ corresponding to the shell closures at $N_{sh}=8$ and 20, we decided to make next step, and compared the angles between the slope for isotopes with $N<(N_{sh}+2)$ and the slope corresponding to $N>(N_{sh}+2)$, i.e. before the shell closure and during the filling the next shell. The dependence of the angle α (shown in the inset of Figure 3) on the proton number for $N_{sh}=8$ and $N_{sh}=20$ is shown in Figure 3 as open squares and circles, respectively. One can clearly see two peaks corresponding to the doubly magic nuclei ^{16}O (N=Z=8) and ^{40}Ca (N=Z=20). Moreover, the peak corresponding to the ^{36}S (N=20, Z=16) is also clearly visible, and probably points out the large region of the N=20 magicity inside the valley of stability. The neutron N=16 magicity in the region from carbon up to neon is confirmed by full triangles.

So we can state that a new shell closure at N=16 has appeared in neutron-rich nuclei for Z$\leq$10 between the $2s_{1/2}$ and $1d_{3/2}$ orbits in the good agreement with Monte Carlo shell model calculations of Utsuno [20] and Otsuka [21]. This fact, strongly supported by the instability of C, N and O isotopes with N > 16, confirms the magic character of N=16 for the

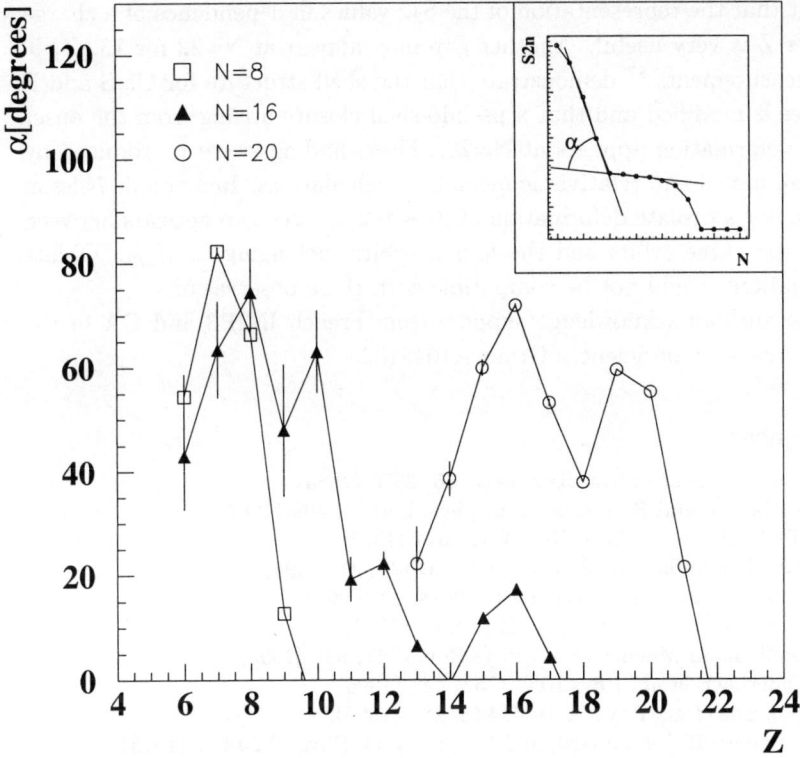

Figure 3. Dependence of the angle α on the proton number. The inset shows the derivation of the angle from the dependence of two-neutron separation energy on the neutron number at the drop points $N_{sh}+2$ corresponding to $N_{sh}=8$ (open squares), $N_{sh}=16$ (filled triangles), $N_{20}=20$ (open circles).

neutron-rich nuclei in the region $6\leq Z \leq 10$, while the shell closure at N=20 tends to disappear for Z <14.

This is in a good agreement with recently published work of Ozawa et al. [13] where, however, the information on the magicity of N=16 has been obtained from one neutron separation energies S_{1n} and the measurement of interaction cross section of radioactive beams σ_I.

As has been already pointed out in ref. [17] the Cl, S and P isotopes, however, exhibit a pronounced change of slope around N=26. Moreover, this change in the Cl and S isotopes is confirmed by the sharp drop at N=28. The discontinuity observed at N=26 in Figure 1 appears in Figure 2 as a sufficiently large gap at N=26 for Z=15,16 and 17. It is therefore evident that the representation of the S_{2n} values in dependence of a charge number Z is very useful. Another gap may appear at N=22 for $13 \geq Z \geq 9$. The measurements [17] demonstrate that the shell structure for Cl, S and P isotopes is modified and that a pseudo-shell closure arising from the onset of the deformation appears at N=26. These findings were reproduced by the shell model and relativistic mean field calculations. In a simple Nilsson picture, for a prolate deformation of $\beta_2 \sim 0.2$, a large gap appears between the lowest three orbits and the fourth orbital belonging to $1f_{7/2}$. Oblate deformations would not be compatible with these observations.

The authors acknowledge support from French IN2P3 and GA of the Czech Academy of Sciences Grant A1048102.

References

1. I. Tanihata et al., Phys. Rev. Lett. **55**, 2676 (1985).
2. P.G. Hansen and B. Jonson, Europhys. Lett. **4**, 409 (1987).
3. C. Thibault et al., Phys. Rev. **C12**, 644 (1975).
4. A. Korsheninnikov et al., Phys. Lett. **B326**, 31 (1994).
5. A.N.Ostrowski et al., Phys. Lett. **B338**, 13 (1994).
6. O. Tarasov et al., Phys. Lett. **B409**, 64 (1997).
7. D.Guillemaud-Mueller et al., Phys.Rev. **C41**, 937 (1990).
8. M.Fauerbach et al., Phys.Rev. **C53**, 647 (1996).
9. H. Sakurai et al., Phys. Lett. **B448**, 180 (1999).
10. M. Beiner, R.J. Lombard, and D.Mas, Nucl. Phys. **A249**, 1 (1975).
11. T. Otsuka et al., Nucl. Phys. **A685**, 100c (2001).
12. Z. Dlouhý et al., Contr. to RNB2000, 3-8 April 2000, Divonne (France); Nucl. Phys. **A701**, 189c (2002).
13. A.Ozawa et al., Contr. to RNB2000, 3-8 April 2000, Divonne (France); Phys. Rev. Lett. **84**, 5493 (2000).
14. O. Sorlin et al., Phys. Rev. **C47**, 2941 (1993).
15. T. Glasmacher et al., Phys. Lett. **B395**, 163 (1997).
16. W. Mittig, A. Lépine-Szily and N. Orr, Annu. Rev. Nucl. Sci. **47**, 27 (1997).
17. F. Sarazin et al., Phys. Rev. Lett. **84**, 5062 (2000).
18. G. Audi et al., Nucl. Phys. **A624**, 1 (1997).
19. H. Sakurai et al., Phys. Rev. **C54**, R2802 (1996).
20. Y. Otsuno et al., Phys. Rev. **C60**, 054315-1 (1999).
21. T. Otsuka et al., Nucl. Phys.**A682**, 155c (2001).

PROBING THE HALO STRUCTURE OF LIGHT NEUTRON-RICH NUCLEI BY INTERMEDIATE ENERGY ELASTIC PROTON SCATTERING*

P. EGELHOF

Gesellschaft für Schwerionenforschung (GSI), D-64291 Darmstadt , Germany

The investigation of light-ion induced direct reactions using radioactive beams in inverse kinematics gives access to a wide field of nuclear structure studies in the region far off stability. The present contribution is mainly focused on the investigation of light neutron-rich halo nuclei. The method of proton-nucleus elastic scattering at intermediate energies is demonstrated to be an effective means for obtaining accurate and detailed information on the size and radial shape of halo nuclei. Differential cross sections for small-angle scattering were measured at energies near 700 MeV/u for the neutron-rich lithium isotopes ^{8}Li, ^{9}Li and ^{11}Li. The experimental concept and the procedure of the data analysis in terms of the Glauber multiple scattering theory are discussed and the results on the nuclear matter radii, the matter distributions, and the significance of the data for a halo structure are presented. In a very recent experiment data on p^{6,8}He scattering at higher momentum transfer, covering the region of the first diffraction minimum, were taken. Such data are expected to give more detailed information on the structure of the α-like core in these halo nuclei. First results are presented. The experimental conditions at the next generation facilities for radioactive ion beams presently proposed or under construction will tremendously improve our research potential for nuclear structure studies on exotic nuclei. A feasibility study for the investigation of the nuclear matter distributions of very neutron rich Sn isotopes is given as an example.

1. Introduction

One of the most powerful classical methods for obtaining spectroscopic information on the structure of nuclei is the investigation of light-ion induced direct reactions. The study of such reactions, as for example elastic and inelastic proton-, deuteron-, alpha-scattering, or one- and few-nucleon transfer reactions, or charge exchange and knock-out reactions, etc. contributed in the past substantially to our present knowledge about stable nuclei. Of course, before the availability of radioactive ion beams, this method was limited to stable or very long-lived nuclei, which allow to produce targets. In the recent past the use of good-quality secondary radioactive ion beams, now available at several major accelerator facilities world wide, enabled to extend such studies on exotic nuclei by using the method of inverse kinematics. Despite the experimental challenge in performing such experiments with relatively low secondary beam intensities, investigations with radioactive beams have opened new territories of nuclei in the nuclear chart to be explored. Within the last two decades a number of first generation experiments on direct reactions with radioactive beams have been performed (see Ref. [1] for an overview). Such experiments were mainly, without a few exceptions, concentrated on light neutron-rich nuclei with mass number A < 20, but nevertheless, when reaching extreme neutron/proton ratios near the neutron drip line, allowed the discovery of new exciting phenomena of nuclear structure.

One of the most outstanding discoveries was the finding that the nuclear matter may appear under certain conditions with a qualitatively new type of nuclear structure, so-called "halo"-structure. Compared to the stable nuclei and those close to stability, in which all the protons and neutrons are essentially distributed uniformly over the nuclear volume, it was found that some light neutron-rich nuclei located at or near the neutron drip line exhibit a widely extended low-density distribution of loosely bound valence neutrons (the "halo") surrounding a compact distribution of the majority of nucleons (the "core"). The discovery of this phenomenon and its interpretation in terms of the halo picture was initiated by the pioneering work of Tanihata et. al.[2] where total cross sections σ_I were determined for the interaction of light neutron-rich isotopes with various targets. The surprisingly steep rise of σ_I for ^{11}Li in the sequence of the Li isotopes and also, but less pronounced, for the He isotopes ^{6}He and ^{8}He, was interpreted as being due to a pronounced increase of the nuclear matter radius. Later a similar behavior was found for a couple of other isotopic chains. In order to get a deeper insight into the structure of neutron-rich nuclei, the halo phenomenon has been the subject of numerous studies during the last decade applying various experimental methods[3]. The picture of nuclear halo structure established soon after the observation of the large interaction radii was qualitatively confirmed. Halo nuclei are thus assumed to be composed of a nuclear core and a number (one, two or four) of valence neutrons, and are characterized, besides the large interaction cross sections, by weak binding energies of the valence nucleon(s), and narrow momentum distributions of the reaction products following fragmentation.

It should be pointed out, that obtaining precise information on the radial structure and the size of halo nuclei from experiments on total interaction cross sections and momentum distributions of the fragments may inherently be limited by the incomplete knowledge of the rather complicated mechanisms and dynamics of the reactions involved. Hence, systematic errors in the determination of matter radii are likely to appear[3]. Therefore, accurate and detailed information on the matter radii and on the radial shapes of halo nuclei is desirable from other sources. The present contribution deals with the application of a (for the investigation of exotic nuclei) new and independent method, i.e. elastic proton scattering at intermediate energies.

2. Intermediate Energy Elastic Proton Scattering - a Tool to Study the Radial Shape of Halo Nuclei

The size of nuclei and the radial shape of the distribution of nuclear matter and charge are fundamental properties of nuclei, and therefore of high interest for various fields in nuclear physics. Over the years a large variety of experimental methods were developed[4], using leptonic probes (as electrons, muons, etc.) for investigating nuclear charge distributions, and hadronic probes (as protons, α-particles, etc.) for exploring the distributions of nuclear matter. While all these methods were applied successfully for many years for the study of stable nuclei, the investigation of the size and radial shape of exotic nuclei has become a new and exciting field of research.

Proton-nucleus elastic scattering at intermediate energies around 700 - 1000 MeV is known to be a method well established for obtaining accurate nuclear matter distributions[4,5]. This method was applied at GSI Darmstadt for the first time for the investigation of exotic nuclei by using the technique of inverse kinematics. Simulation calculations performed for pHe and pLi scattering[6,7] have clearly demonstrated that proton scattering in the region of small momentum transfer is particularly sensitive to the nuclear matter radius, and the halo structure of nuclei. Thus, besides determining precisely the nuclear matter radius, the shape of the nuclear matter distribution in halo nuclei can be explored already from data in the limited range of momentum transfer at small scattering angles. Moreover, simulation calculations performed for the region of higher momentum transfer[8] covering the region of the first diffraction minimum have shown, that such data are more sensitive to the internal structure of nuclei, and thus allow to get information on the central density and on the size and the structure of the core in halo nuclei.

Differential cross sections for elastic proton scattering at small momentum transfer were measured at GSI Darmstadt at energies around 700

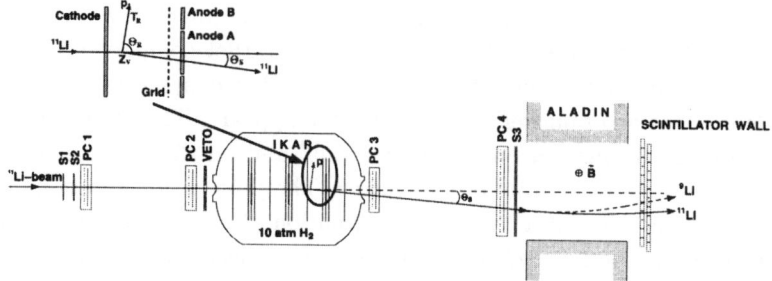

Figure 1. Schematic view of the experimental setup for small-angle elastic proton scattering on exotic nuclei in inverse kinematics. The central part shows the hydrogen-filled ionization chamber IKAR which serves simultaneously as a gas target and a detector system for recoil protons. In 6 independent sections of this detector the recoil angle θ_R, the recoil energy T_R, and the vertex point Z_V are determined (see insert). The forward spectrometer consisting of four multiwire proportional chambers PC1 - PC4 determines the scattering angle θ_S of the projectile. The scintillation counters S1 - S3, VETO are used for beam identification and trigger. The ALADIN magnet with a position sensitive scintillator wall behind allows to discriminate break-up channels.

MeV/u in inverse kinematics for the neutron-rich helium isotopes ^{6}He, ^{8}He[9], and more recently for the neutron-rich lithium isotopes ^{8}Li, ^{9}Li, and ^{11}Li[10]. The secondary He and Li beams were produced by fragmentation of ^{18}O ions delivered from the heavy-ion synchrotron SIS, and were isotopically separated by the fragment separator FRS. Beam intensities were of the order of 10^3 sec^{-1}. The experimental method (see Fig. 1) is based on the high-pressure hydrogen-filled ionization chamber IKAR, which serves simultaneously as gas target and recoil proton detector. The use of this dedicated active-target technique enabled us to detect the low energy recoil protons with sufficient angular and energy resolution covering the full solid angle of interest (for details see [9]). In addition projectile scattering angles were measured with multiwire tracking detectors and the elastically scattered particles were identified by a magnetic rigidity analysis to allow for a safe separation of breakup channels. In a very recent experiment[11] data on p^{6,8}He elastic scattering at higher momentum transfer were measured with a modified experimental setup (see below).

3. Results on Nuclear-Matter Density Distributions and Nuclear-Matter Radii

The differential cross sections deduced from the experiment which covered the low momentum transfer region are displayed for the Li isotopes investigated in Fig. 2. For establishing the nuclear matter density distributions from the measured cross sections, the Glauber multiple scattering theory was applied. Calculations were performed using the basic Glauber formalism[5] for proton-nucleus elastic scattering, and taking experimental data on the elementary proton-proton and proton-neutron scattering amplitudes as input. For modelling the nuclear matter distributions, various parametrical functions were used. The free parameters of these density distribution parametrizations were then determined by a least-square fit of the calculated cross sections to the experimental ones.

Within the present analysis, besides a one-parameter Gaussian (G) function, a symmetrized Fermi (SF) parametrization, a parametrization consisting of two Gaussian distributions (GG), and a parametrization consisting of a Gaussian and a 1p-shell harmonic oscillator-type density distribution (GO) were used, the latter ones having two free parameters each (for details see [6]). It should be noted, that only the parametrizations GG and GO allow to distinguish between a core and a halo distribution, whereas the parametrizations G and SF do not allow such a differentiation. As a result of the fitting procedure it turns out that for the ^{8}Li and ^{9}Li nuclei, supposed to have no halo structure, the cross sections calculated by using all four parametrizations for the density distribution fit equally well (with a reduced χ^2/N value around unity, see solid lines in Fig. 2 for the GG parametrization as an example) to the experimental data on $d\sigma/dt$. The corresponding four values of the nuclear matter radii R_m also closely resemble each other, within rather small errors (see [10] for a detailed discussion). In contrast, for the ^{11}Li nucleus the parametrizations G and SF fail to give

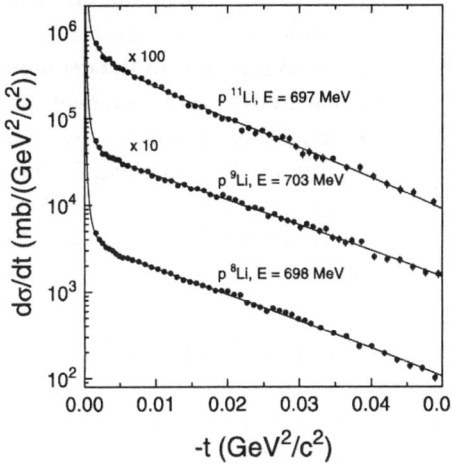

Figure 2. Differential cross sections $d\sigma/dt$ versus the four-momentum transfer squared t for $p^{8,9,11}$Li elastic scattering. Energies are the equivalent proton energies for normal kinematics. Lines are the result of fits to the data performed on the basis of the Glauber multiple scattering theory (for details see text).

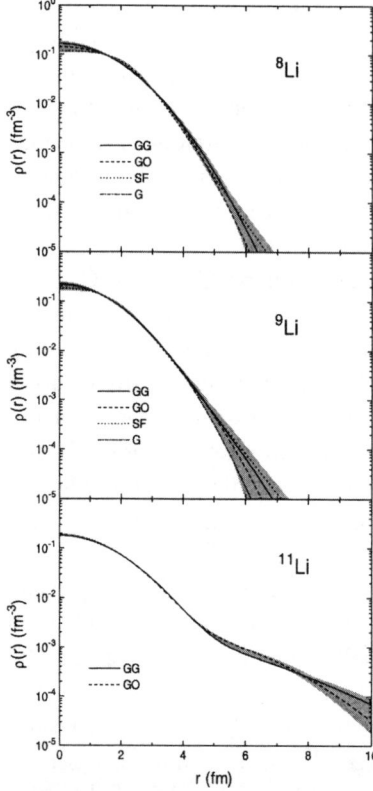

Figure 3. Nuclear matter density distributions $\rho(r)$ for the ^{8}Li, ^{9}Li and ^{11}Li nuclei deduced from the experimental cross section data displayed in Fig. 2. Labels denote different parametrizations of phenomenological density distributions used in the analysis (for details see text). The shaded areas confine the envelopes of the nuclear matter density variation within the model parametrizations applied, superimposed by the statistical errors calculated from the fit parameters and their mutual correlations.

a good description of the cross section data, whereas only the parametrizations GG and GO, which allow for different distributions for the core and halo nucleons, have permitted a good fit (with a reduced χ^2/N value around unity for both cases, see solid line in Fig. 2 for the GG parametrization).

The present results on the nuclear matter distributions obtained for the ^{8,9,11}Li isotopes (still preliminary) are summarized in Fig. 3. The matter distributions obtained for all parametrizations considered (see above) agree reasonably well within small errors. As compared to ^{8}Li and ^{9}Li a rather extended nuclear matter distribution is obtained for ^{11}Li, the matter density decreasing much slower with the radius parameter than the ones for ^{8,9}Li, and therefore representing a clear signature of a pronounced neutron halo in ^{11}Li. This behavior is also reflected in the deduced nuclear matter radii (average values resulting from the various parametrizations applied, including systematical errors), where the result for ^{11}Li: $R_m = 3.53$ (17) fm has to be compared with

$R_m = 2.49$ (7) fm for ^{8}Li and $R_m = 2.43$ (7) fm for ^{9}Li, respectively. For attaining a deeper insight into the nuclear structure, i.e. on the extension of the core and the halo inside the ^{11}Li nucleus, an assumption on the partition of the nucleus into halo neutrons and core nucleons has to be made. This

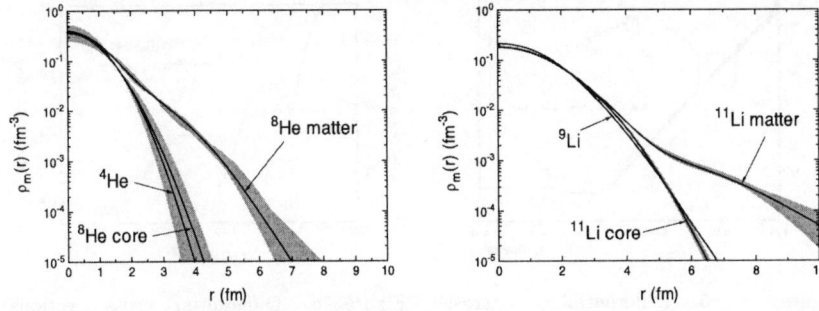

Figure 4. Right side: The nuclear matter and nuclear core density distributions of ^{11}Li, deduced from the present cross section data (average of the distributions obtained from the GG and GO analysis), are compared with the nuclear matter density distribution obtained for ^{9}Li (average of the results presented in Fig. 3). Left side: Corresponding results on the matter and core distributions for ^{8}He and the matter distribution for ^{4}He.

was done by using the GG and GO parametrizations, and assuming the ^{11}Li nucleus to consist of a ^{9}Li-like core and two halo neutrons. The resulting core and halo radii obtained by taking the average values deduced from the GG and GO analysis are $R_c = 2.54$ (7) fm and $R_h = 6.28$ (48) fm. The radial distributions of total nuclear matter and core matter deduced from the present data (again the average of the distributions obtained from the GG and GO analysis) are displayed in Fig. 4 (right side) for the ^{11}Li nucleus together with the deduced matter distribution for the ^{9}Li nucleus (average of the results presented in Fig. 3). The corresponding results deduced from a previous experiment on the ^{8}He nucleus[6] assuming an α-core + 4 neutron-halo structure are presented on the left side of Fig. 4. Obviously the results obtained represent clear evidence for significant neutron halos in both the ^{11}Li and ^{8}He nuclei. The direct comparison shows that ^{11}Li exhibits the much more pronounced halo structure as compared to ^{8}He. This is also reflected in the deduced halo radii, the one for ^{11}Li: $R_h = 6.28$ (48) fm being more as twice as large as the one for ^{8}He: $R_h = 3.08$ (10) fm. The close resemblance of the core distributions in ^{11}Li and ^{8}He with the matter distributions of the ^{9}Li and ^{4}He nuclei supports the picture of a two-neutron halo structure in ^{11}Li, and a four-neutron halo structure in ^{8}He.

In a very recent experiment[11] the previous data on p^{6,8}He scattering[9] have been extended to substantially higher momentum transfer using a modified experimental technique. The forward spectrometer for tracking and identifying the scattered projectile nuclei was similar to the one used in the previous experiments (see Fig. 1). The major difference was that,

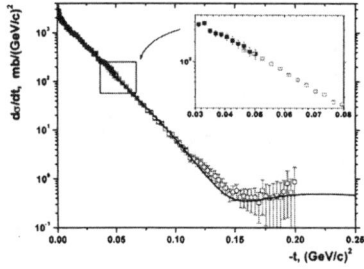

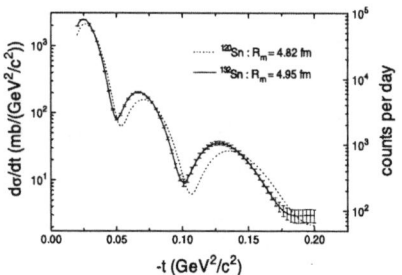

Figure 5. Differential cross section $d\sigma/dt$ versus the four-momentum transfer squared t for p^6He elastic scattering. Full symbols are the results of the low momentum transfer experiment[9], empty symbols represent still preliminary data from the recent high momentum transfer experiment[11]. The insert shows the area of overlap. The solid line is the result of a fit of the combined data set (for details see text).

Figure 6. Differential cross sections $d\sigma/dt$ versus the four-momentum transfer squared t for $p^{120}Sn$ (dotted line) and $p^{132}Sn$ (solid line) elastic scattering at E = 700 MeV resulting from a simulation calculation assuming nuclear matter distributions for the $^{120,132}Sn$ nuclei as predicted by [14]. The scale displayed on the right side of the figure and statistical error bars for $p^{132}Sn$ scattering correspond to typical experimental conditions as expected at the new generation radioactive beam facilities (for details see text).

due to the high recoil proton energies up to 160 MeV, instead of the active gaseous target IKAR a 600 mg/cm^2 liquid hydrogen cylinder could be used as the proton target and a position sensitive scintillator wall measured the recoil proton energies via time-of-flight. The experimental arrangement allowed for a very low-background data taking. At present, only preliminary results for the $p^{6,8}He$ differential cross sections at the larger t-range have been evaluated. An example of the measured cross sections is shown in Fig. 5 together with the previous data. Both data sets are in excellent agreement in the region of overlap. The solid line in Fig. 5 represents a fit of the combined data sets at low and high momentum transfer using the GG parametrization for modelling the nuclear matter distribution in core and halo. A reasonably good description of the data in the whole t-range is obtained. As a preliminary result it turned out, that the deduced nuclear matter distributions and the nuclear matter radii are within errors fully consistant with the previous results for the low-t region alone, and that the central density and the size of the core are for ^6He and ^8He determined with higher precision than from the previous low-t data alone (for details see [11]).

4. Future Perspectives at the Next Generation Radioactive Beam Facilities

The final part of this contribution is intended to focus on the "long range" perspectives and possibilities which will be opened by the next generation radioactive beam facilities presently under construction or planned, such as for example at RIKEN, Japan, GSI Darmstadt, Germany, or elsewhere. The expected performance of these facilities (the following discussion will concentrate on the GSI future project[12] as an example) will provide unique opportunities for nuclear structure studies on nuclei far off stability, and will enable to explore completely new and unknown regions in the chart of nuclei. In particular, the expected substantial increase in luminosities will open a large potential for reaction studies with radioactive beams, and will allow to extend previous investigations to the exploration of the structure and dynamics of medium heavy and heavy nuclides in the regions far off stability, including the pathways of stellar nucleosynthesis. For a detailed discussion of the tremendous research potential and the physics questions to be addressed see Refs. [12,13]. Among a number of observables to be investigated the radial shape of nuclei, namely their nuclear matter density distribution $\rho_m(r)$ including its first moment (the nuclear matter radius R_m), as well as the nuclear charge distribution $\sigma_{ch}(r)$, and the nuclear charge radius R_{ch}, will be of high interest. As has turned out from the previous investigations discussed above, it will be important to measure both, the matter and charge distributions, for the same nuclei in order to gain the full and model independent information.

Concerning future experimental concepts, secondary beams provided by a fragment separator may be injected, stored and cooled into an ion storage ring. For the investigation of the charge distributions, elastic electron scattering may be studied at the interaction zone of an electron-ion collider (for details see [12]). In contrast, the nuclear matter distributions could most favourably be studied by intermediate energy elastic proton scattering using an internal hydrogen target included into the ion storage ring. As compared to the use of an external target (which will be limited in thickness when low energy recoil protons are to be detected) this concept provides, because of the high repetition rates allowing many beam-target interactions, considerably higher luminosities (see [13] for a detailed discussion). For a feasibility study, the neutron-rich Sn isotopes around the doubly magic ^{132}Sn are considered as an example. Theoretical predictions for the charge and matter radii, and the difference of the neutron and proton radii (i.e. the neutron

72

skin thickness) are given in [14]. It turns out that the uncertainties in these predictions, dependent on different effective nucleon-nucleon interactions, different pairing forces, different parameters of the equation of state of nuclear matter, etc. chosen, amount to about $\Delta R = 0.2 - 0.3$ fm, and therefore demand for an experimental accuracy of the order of $\delta R \leq 0.05 - 0.1$ fm for providing a sensitive test of the theoretical predictions. In Fig. 6 the expected differential cross section for (p,p) scattering on ^{120}Sn and ^{132}Sn at an incident energy of 700 MeV is displayed. The simulation calculation was performed assuming nuclear matter distributions of ^{120,132}Sn as predicted in [14], and applying the Glauber multiple scattering theory as discussed in Section 3. On the right side of Fig. 6 the differential cross section is given for the case of p^{132}Sn scattering in units of "counts per day" for typical experimental conditions predicted[12,13]. A comparison of the cross sections for p^{120}Sn and p^{132}Sn scattering, taking into account the statistical error bars for ^{132}Sn after a measuring time of one day (^{120}Sn, as a stable nucleus, is assumed to be measured with much better accuracy), gives an impression on the sensitivity and accuracy for the determination of the nuclear matter radius when keeping in mind that the two nuclei considered differ in their matter radii only by $\Delta R_m = 0.13$ fm.

References

1. I. Tanihata, (ed.), Nucl. Phys. A 693 (2001).
2. I. Tanihata et al., Phys. Rev. Lett. 55 (1985) 2676.
3. P. Egelhof, Act. Phys. Pol. B30 (1999) 487 and refs. therein.
4. J.W. Negele et al., Adv. Nucl. Phys. 19 (1989) 1.
5. G.D. Alkhazov et al., Phys. Rep. C 42 (1978) 89.
6. G.D. Alkhazov et al., Nucl. Phys. A 712 (2002) 269.
7. P. Egelhof et al., Prog. Part. Nucl. Phys. 46 (2001) 307.
8. L.V. Chulkov et al., Nucl. Phys. A 587 (1995) 291.
9. S.R. Neumaier et al., Nucl. Phys. A 712 (2002) 247.
10. A. Dobrovolsky, et al., World Scientific, ISBN 981-238-025-6 (2002) 388.
11. F. Aksouh et al., Few Body Syst. (2003), to be publ.; F. Aksouh, thesis.
12. Conceptual Design Report: An int. accelerator facility for beams of ions and antiprotons, H.H. Gutbrod et al. (eds.), GSI Darmstadt, Germany (2001).
13. P. Egelhof et al., Proc. of the 5th Int. Conf. on Nucl. Phys. at Storage Rings STORI02, Physica Scipta (2003) to be published.
14. F.Hofmann et al., Phys. Rev. C 64, 034314 (2001); and P.G. Reinhard et al., Hyperfine Int. 127 (2000) 13.

* An article with partly similar contents will be published in the Proc. of the Int. Symp. on Physics of Unstable Nuclei, Halong Bay, Vietnam, Nucl. Phys. A (2003).

HIGH-RESOLUTION GAMMA-RAY SPECTROSCOPY WITH FAST EXOTIC BEAMS[*]

C. M. CAMPBELL, J. A. CHURCH, D.-C. DINCA, T. GLASMACHER,
P. G. HANSEN, H. OLLIVER, B. PERRY, B. M. SHERRILL AND J. R. TERRY

*Department of Physics and Astronomy and National Superconducting Cyclotron
Laboratory, Michigan State University, East Lansing, MI 48824, USA*

D. BAZIN, J. ENDERS, A. GADE, Z. HU, W. F. MUELLER

*National Superconducting Cyclotron Laboratory,
Michigan State University, East Lansing, MI 48824, USA*

P. D. COTTLE AND K. W. KEMPER

*Physics Department, Florida State University,
Tallahassee, FL 32306, USA*

L. RILEY

Ursinus College, Collegeville, PA 19426 USA

We report on in-beam gamma-ray spectroscopy of light neutron-rich nuclei produced by projectile fragmentation of ^{40}Ar and ^{48}Ca primary beams at the newly completed Coupled Cyclotron Facility (CCF) at Michigan State University. The large degree of segmentation in the segmented Germanium array (SeGA) of eighteen 32-fold-segmented high-purity germanium detectors allows for high-resolution in-beam gamma-ray spectroscopy by correcting event-by-event for the Doppler shift due to the large recoil velocities (v/c   0.3-0.4). In addition to SeGA, a position-sensitive NaI array has been used for specific Coulomb excitation experiments that require the higher photopeak efficiency traded off against inferior energy resolution.

Several successful in-beam gamma-ray spectroscopy experiments have now been performed at the CCF with an emphasis on elucidating the evolution of nuclear structure around neutron numbers $N=16$, $N=20$, and $N=28$ in the (*sd*) shell. Inelastic scattering experiments with gamma-ray detection on light and heavy targets have determined specific transition matrix elements. One- and two-particle removal reactions were used to investigate the wave functions of specific states.

The results of these experiments will be presented and prospects for high-resolution in-beam gamma-ray spectroscopy at the Coupled Cyclotron Facility and at the planned Rare Isotope Accelerator will be discussed.

---

[*] This work is supported by the National Science Foundation under grants PHY-9724299, PHY-9875122, and PHY-0110253.

1. Gamma-ray spectroscopy with fast exotic beams

Projectile fragmentation of intermediate-energy heavy-ion beams with subsequent fragment separation is an efficient method to prepare beams of exotic nuclei by physical means [1]. Experiments with standard luminosities can be carried out at intermediate beam energies with thick secondary targets (order of g/cm^2) and very low incident beam rates (order of particle/s or less) provided that inelastic scattering events can be detected. In the last several years -rays have become the probe of choice to indicate inelastic scattering events to excited bound states in thick-target experiments. Photons traverse targets with little and known attenuation. With the development of position-sensitive -ray detectors it has become possible to measure the energies and directions of photons emitted in-flight. This allows the reconstruction of the photons' energies in the frame of the moving exotic projectile with high accuracy. Experimental success in this field is strongly correlated with the development of detectors such as position-sensitive scintillation detectors or segmented Germanium detectors. In-beam gamma-ray spectroscopy of fast exotic beams has been successfully used at all projectile-fragmentation facilities and specialized techniques have been developed. Intermediate-energy Coulomb excitation [2] serves as an electromagnetic probe, inelastic protons scattering with -ray detection as a hadronic probe, in-beam fragmentation reactions populate a variety of low-lying excited states [3], and nucleon knockout reactions [4] allow for precision wavefunction spectroscopy through the measurement cross section which can be converted into spectroscopic factors.

2. The Segmented Germanium Array (SeGA) at Michigan State University

The final energy-resolution of -ray spectra reconstructed in the frame of the moving exotic projectile contains contributions from the uncertainties of the -ray scattering angle, the beam velocity, and the intrinsic energy resolution of the detector. In typical experiments the former two contribute an energy resolution of about 1%-2%, much less than the intrinsic energy resolution of scintillation detectors. Thus we have replaced the previously used position-sensitive NaI(Tl) detector array [5] with an array of eighteen 32-fold segmented high-purity germanium detectors [6-8]. Figure 1 shows the segmentation of one crystal of this new array, SeGA, which is currently the largest operational germanium detector array for gamma-ray spectroscopy with fast beams of rare isotopes. The energy resolution with this array is dominated by the uncertainty in the -ray scattering angle. This uncertainty can be reduced–for a given accuracy with which the interaction point of the photon in the detector can be determined–by moving the detector further away from the target. Therefore, in -ray spectroscopy with fast exotic beams, the experimenter has to make a choice in which detection efficiency is traded off against desired energy resolution. This argument also illustrates that doubling the accuracy of determining the first interaction point–for example through pulse shape analysis–yields a four-fold gain in detection efficiency for a given desired energy resolution.

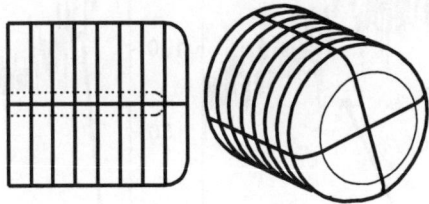

Figure 1. Schematic drawing indicating the segmentation of one SeGA germanium crystal. Each crystal is 8 cm long and 7 cm in diameter. In typical fast-beam experiments the -rays enter from the side.

3. Intermediate-energy Coulomb excitation

The in-beam performance of SeGA with fast beams was first tested with two known reactions: the intermediate-energy Coulomb excitation of the known nuclei ^{86}Kr and ^{11}Be. For these initial tests one ring of six segmented Germanium detectors was located at an angle upstream from the secondary gold target. ^{86}Kr was produced as a primary beam in the Coupled Cyclotron Facility and impinged onto a gold target. Gamma-rays were measured in coincidence with the scattered projectile and are shown in Figure 2 in the panels on the left. The laboratory spectrum shows some structure where the -rays corresponding to the transition of the first excited state to the ground state are expected, but the opening angle of the germanium detector is so large that a peak can hardly be made out. After an event-by-event Doppler-shift reconstruction of the spectrum into the projectile frame the transition becomes visible as a peak. Subsequently ^{11}Be was scattered off the same gold target and the first excited state was populated. The observation of the 320 keV -ray corresponding to the $\frac{1}{2}^{-} \rightarrow$ g.s. transition indicates that low-energy gamma-rays (at least in reactions of light projectiles with heavy targets) can be detected successfully in experiments with fast exotic beams at beam energies exceeding 100 MeV/nucleon. Correcting for the beam profile of the incoming beam and measuring and correcting for the scattering angle of the ejectile can further improve the energy resolution of the reconstructed -rays shown in Figure 2.

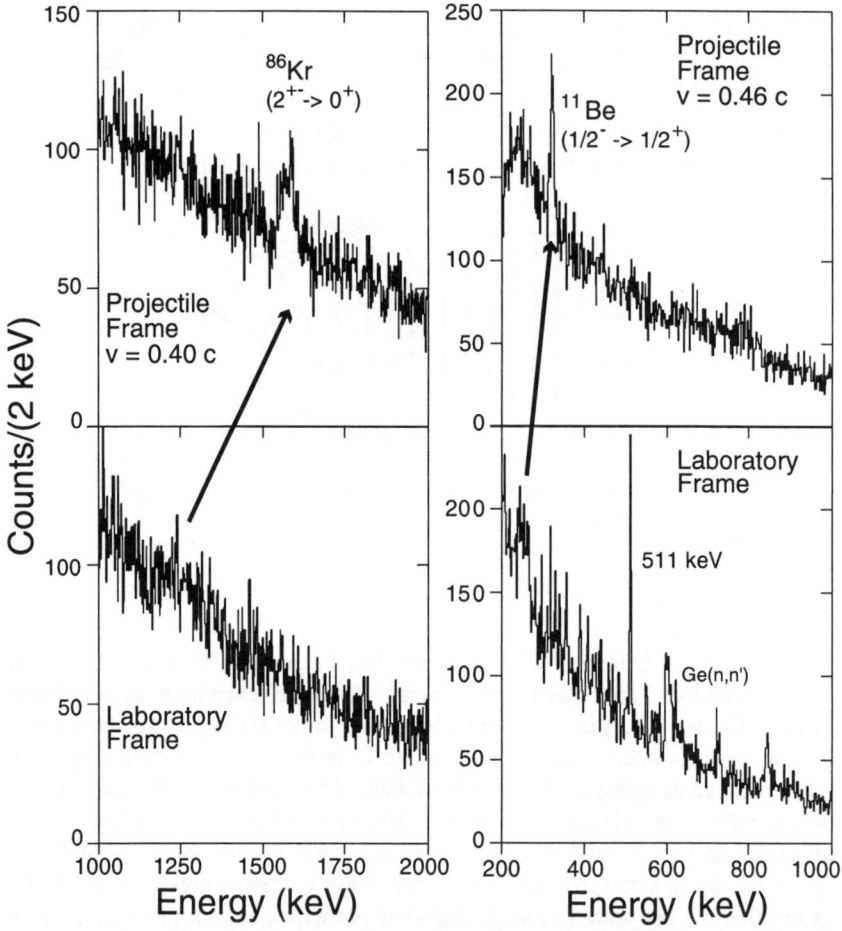

Figure 2. Online spectra from SeGA's first in-beam test experiments with a heavy nucleus (^{86}Kr) and a light nucleus (^{11}Be) impinging on a gold target at intermediate beam energies. The bottom panels show the energy spectra as they are measured in the laboratory and the top panels show the same spectra after an event-by-event Doppler-reconstruction has been applied. The -rays from the first excited states to the ground states become visible as peaks.

Following these initial tests SeGA was moved to the target position of the NSCL's S800 superconducting spectrograph for a series of intermediate-energy Coulomb excitation experiments and nucleon-knockout experiments. In these experiments the beam profile and the beam position on the target were measured in addition to the momentum of the scattered beam.

4. Nucleon knockout reactions

Several one- and two-nucleon knockout experiments were performed by impinging various exotic beams onto a beryllium target. Gamma-rays were measured in coincidence with the scattered heavy residue. SeGA detectors were arranged in two rings, one around 90° and one around 30° in the laboratory. Figure 3 shows the -ray spectra measured in the reaction ^{46}Ar(^{9}Be,^{45}Ar).

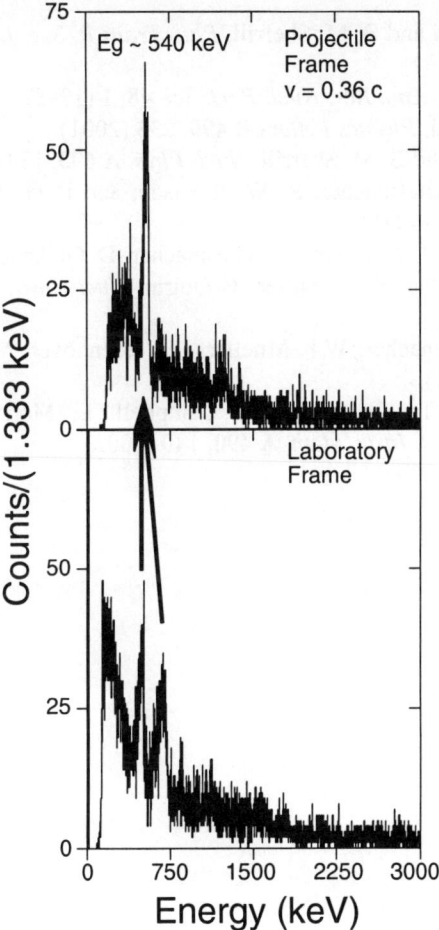

Figure 3. Gamma-ray spectra measured in the nucleon knockout reaction ^{46}Ar(^{9}Be,^{45}Ar).

5. Summary

Our first experiences with the use of highly segmented germanium detectors for in-beam spectroscopy of fast exotic beams have been very positive. The SeGA array has been successfully used for intermediate-energy Coulomb excitation experiments and nucleon-knockout experiments. The energy resolution of the reconstructed spectra is no longer dominated by the intrinsic detector resolution, but by kinematic broadening which is under the control of the experimentalist: Energy resolution can be traded off versus detection efficiency.

References

1. D. J. Morrissey and B. M. Sherrill, *Phil. Trans R. Soc. London* **A 356**, 1985 (1998).
2. T. Glasmacher, *Ann. Rev. Nucl. Part. Sci.* **48**, 1 (1998).
3. K. Yoneda et al, *Physics Letters* **B 499**, 233 (2001).
4. P. G. Hansen and B. M. Sherrill, *Nucl. Phys.* **A 693**, 133 (2001).
5. H. Scheit, T. Glasmacher, R. W. Ibbotson, and P. G. Thirolf, Nucl. Instr. Meth. **A 422**, 124 (1999).
6. W. F. Mueller, J. A. Church, T. Glasmacher, D. Gutknecht, G. Hackman, P. G. Hansen, Z. Hu, K. L. Miller, P. Quirin, *Nucl. Instr. Meth.* **A 466**, 492 (2001).
7. Z. Hu, T. Glasmacher, W.F. Mueller, I. Wiedenhöver, *Nucl. Instr. Meth.* **A 482**, 715 (2002).
8. K. L. Miller, T. Glasmacher, C. Campbell, L. Morris, W. F. Mueller, E. Strahler, *Nucl. Instr. Meth.* **A 490**, 140 (2002).

BETA DECAY STUDIES AT THE MSU/NSCL COUPLED CYCLOTRON FACILITY

P.F. MANTICA

Department of Chemistry and National Superconducting Cyclotron Laboratory
Michigan State University,
East Lansing, MI 48824, USA
E-mail: mantica@msu.edu

The experimental program utilizing fast beams of rare isotopes at the MSU/NSCL Coupled Cyclotron Facility has begun. One advantage offered by fast beams is the capability to track and identify individual isotopes on a particle-by-particle basis. A beta counting system, centered around a double-sided silicon microstrip implantation detector, has been developed with the goal of extending beta-decay half-life measurements to nuclei produced at rates of a few per day. The beta counting system has been used in conjunction with six detectors from the Segmented Germanium Array to identify delayed gamma-rays following the decay of neutron-rich $_{21}$Sc to $_{24}$Cr. The systematic variation of the first excited 2^+ states in the neutron-rich, even-even $_{24}$Cr, $_{22}$Ti, and $_{20}$Ca isotopes supports the appearance of a neutron subshell closure at N = 32.

1. Introduction

Beta decay properties of nuclei far from stability play an important role in both nuclear structure and nuclear astrophysics. The identification of new regions of deformation on the neutron-rich side of the valley of stability at N = 20 and N = 28 was aided, in part, by half-life measurements completed for ^{32}Mg$_{20}$ [1] and ^{44}S$_{28}$ [2]. Beta decay properties such as half-life, Q value, and delayed neutron probability, are also crucial input parameters for network calculations of explosive astrophysical events. The decay rates for both neutron-deficient and neutron-rich nuclei are needed to define the mass flow for nucleosynthesis. The reproduction of beta decay properties of exotic nuclei presents a difficult challenge to theoretical models. The peak in the Gamow-Teller (GT) strength is typically outside the beta-decay Q value window, and beta decay only samples the tails of the GT distribution. It is crucial, therefore, to determine these properties experimentally.

Significant progress has been made in the measurement of beta-decay half-lives of exotic nuclei. This can be directly attributed to particle-detection techniques employed with fast beams. For fast beams, energy-loss, time-of-flight, and magnetic rigidity can be used to uniquely determine the identification of each fragment on an event-by-event basis. Correlating the known fragment with its subsequent beta decay is achieved using the same methods developed to study rare proton decays [3] and superheavy elements. A double-sided silicon strip detector (DSSD) is used as both an implantation detector for the incoming fast beam and an energy-loss detector for beta particles. Beta decay half-live curves are extracted from the absolute time differences between correlated implant and decay events. This method has been applied to a variety of very neutron-rich and neutron-deficient systems. The effectiveness of this method applied to the measurement of a beta-decay half-life in a low statistics experiment is best exemplified by the case of ^{100}Sn studied at GSI. A half-life determination with 50% error was made for a total of seven implanted ^{100}Sn nuclei over an 11-day period [4].

A beta counting system [5] has been developed at the National Superconducting Cyclotron Laboratory (NSCL) to measure beta-decay properties of short-lived exotic nuclei. This system can be used as a stand alone device, or as the main trigger detector for delayed gamma-ray, delayed neutron, or beta-decay Q value measurements. Performance characteristics of the beta counting system are presented. The beta counting system has been used in conjunction with an array of germanium detectors to identify delayed gamma-rays following the decay of neutron-rich nuclides near N = 32. Results from these measurements are also presented.

2. Beta Counting System

The NSCL beta counting system (Figure 1) is centered around a 985-mm thick type BB1 DSSD from Micron Semiconductor Ltd.. The DSSD serves as both an implantation detector for high-energy (greater than 1 GeV) charged particles and an energy loss (less than 1 MeV) detector for the fast electrons emitted in beta decay. A Si PIN detector (PIN1) placed approximately one meter upstream of the DSSD is used to record energy loss and time-of-flight information for the incoming fast beam. A PPAC detector in this same location provides beam position information. The Si PIN detectors PIN3 and PIN4 are operated in high-gain mode and serve to provide secondary energy-loss information for beta particles. Detector PIN4 is a veto detector, used to reject implant events that traverse the entire stack of Si detectors.

For each valid implant, defined by an event above threshold in PIN1 and the DSSD, the energy loss and time-of-flight data from PIN1, along with the absolute time of the event, are stored in arrays indexed by the pixel location (x,y) of the implant event in the DSSD. When a subsequent beta decay is registered above threshold in pixel (x,y), with no corresponding event in PIN1, the implant information is retrieved to complete the implant-decay correlation. Beta-decay half-life curves are extracted from the absolute time differences between implant events and correlated decay events.

One of the challenges to overcome in correlating fragment implants with subsequent beta decays in the DSSD is the broad energy range of the detected particles. Dual output preampliers are used to obtain both low gain (0.1V/pC) and high gain (2 V/pC) signals from each strip of the DSSD. These preampliers, manufactured by MicroChannel Systems, contain a shaping stage that produces output signals with rise times of order 350 ns. The low gain outputs are digitized without further shaping, and provide the energy information for implant events. The high gain signals are further processed using Pico Systems Shaper/Discriminators in CAMAC. The shaper outputs from these modules are digitized and provide the energy information for decay events. The discriminator outputs are sent through a common OR gate and used to define the master trigger.

The energy resolution of the DSSD has been measured to be better than 80 keV full width at half maximum at an energy of 8.78 MeV.

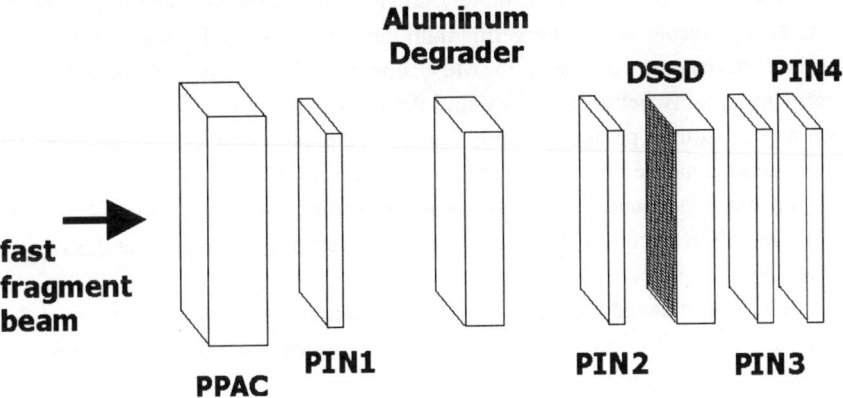

Figure 1. Schematic drawing of the NSCL beta counting system.

3. Beta-Delayed Gamma-Ray Measurements around N=32

Neutron-rich nuclides in the A~50 region were produced by fragmentation of a primary beam of ^{86}Kr accelerated to 140 MeV/nucleon using the new NSCL Coupled Cyclotron Facility. This beam impinged a 376 mg/cm^2 Be target placed at the object position of the A1900 fragment separator. Two settings of the A1900 separator were used to produce eight different parent nuclides: the first setting, $B\rho_1$ = 4.0417 Tm and $B\rho_2$ = 3.7554 Tm, resulted in a beam containing 1% [54]Sc, 14% [55]Ti, 24% [56]V, 26% [57]V, and 34% [58]Cr, while the second setting, $B\rho_1$ = 4.1261 Tm and $B\rho_2$ = 3.8417 Tm, provided a beam composed of 12% [56]Ti, 38% [57]V, 39% [58]V, and 29% [59]Cr. The secondary beam was delivered to the beta counting system, located in the S1 experimental vault. The beam was defocused to illuminate more than two-thirds of the active area of the DSSD. The average implantation rate for fragments in the DSSD was around 200 Hz for the first A1900 setting and 100 Hz for the second setting. Two time limits were established in software to reduce the chances for random fragment-beta correlations: a maximum time of ten seconds for any fragment-beta correlation and a minimum time of ten seconds for consecutive implants within a given pixel. The measured efficiency for correlated beta decays with implants was 30%.

Gamma-ray detection was accomplished using six germanium detectors from the MSU Segmented Germanium Array (SeGA) [6]. The detectors were placed in a circular geometry around the DSSD, with the long sides of the germanium cryostats parallel to the beam axis. A large volume germanium detector was also placed immediately downstream of the beta counting system. The energy resolution of the germanium detectors averaged 2.6 keV full width at half maximum for the 1.33 MeV line in [60]Co. The efficiency of the germanium array between 80 keV and 2.6 MeV was experimentally determined for a point-source geometry using standard reference sources. To account for the extended nature of the implanted secondary beam in the DSSD, a simulation of the extended source geometry was performed using the Monte-Carlo N-Particle (MCNP) code [7]. The calculated array peak efficiency at 1.33 MeV was 3.3%.

The beta-delayed gamma-ray spectra obtained for the three parent nuclides [54]Sc, [56]V, and [58]V are shown in Figure 2. Observation of the 1006-keV transition as decay of the first excited 2^+ state in [56]Cr (Figure 2b) is corroborated by results from in-beam gamma-ray spectroscopy [8]. The two intense transitions at 668 and 1006 keV are most likely the transitions observed at 700±50 keV and 1000±50 keV in a previous beta-delayed gamma-ray measurement that used high-efficiency, low-resolution BGO detectors for gamma-ray detection [9]. The 880-keV transition identified in the beta-delayed gamma-ray spectrum for [58]V (Figure 2c) was previously reported in Ref. [10,11] and tentatively assigned to the $2_1^+ \rightarrow 0_1^+$ transition in [58]Cr.

The delayed gamma-ray spectrum following the beta decay of ^{54}Sc is shown in Figure 2a. Two weak transitions are observed at 1001 keV (7 counts) and 1495 keV (6 counts). Based on intensity considerations, we have tentatively assigned the 1495 keV gamma ray to the $2_1^+ \rightarrow 0_1^+$ transition and the 1001 keV gamma ray to the $4_1^+ \rightarrow 2_1^+$ transition in ^{54}Ti$_{32}$. A recent study of higher-spin states in ^{54}Ti populated by deep-inelastic collisions [12] has confirmed these assignments.

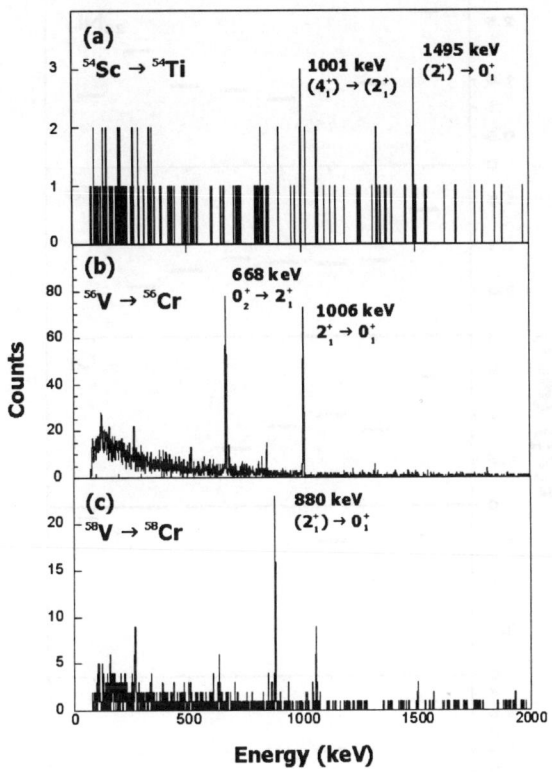

Figure 2. Beta-delayed gamma-ray spectra for decay of (a) ^{54}Sc, (b) ^{56}V, and (c) ^{58}V in the range 0 to 2000 keV.

4. Energies of the First 2^+ States in Neutron-Rich $\pi f_{7/2}$ Nuclei

The systematic variation of the first excited 2^+ states in the neutron-rich, even-even isotopes from $_{20}$Ca to $_{28}$Ni are shown in Figure 3. The first excited 2^+ state in ^{52}Ca$_{32}$ has energy 2.56 MeV [14], well above that observed in its even-even

84

neighbor, ^{50}Ca$_{30}$. This increase was attributed to an energy gap above the filled $\nu p_{3/2}$ orbital, suggesting a subshell closure at N = 32. It was proposed that a neutron spherical subshell closure at N = 32 could occur when reinforced by the Z = 20 shell closure [15]. However, a peak in the systematic variation of first excited 2$^+$ states was also observed for the $_{24}$Cr isotopes. The appearance of the N = 32 subshell closure for neutron-rich nuclides has been attributed to a

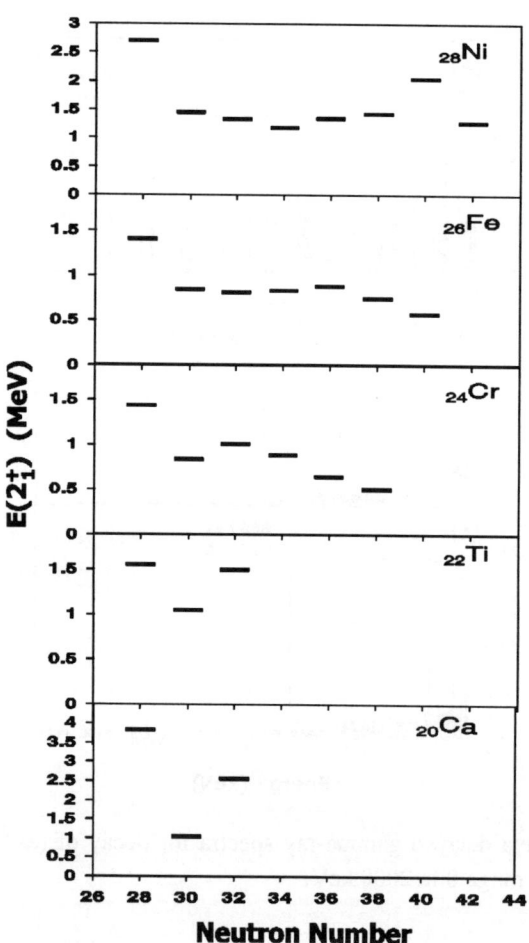

Figure 3. Excitation energies of the first 2$^+$ states for neutron-rich nuclides in the $\pi f_{7/2}$ shell. Figure is adapted from Ref. [11] and includes new data for ^{54}Ti [12] and ^{60,62}Cr [13].

diminished $\pi f_{7/2} - \nu f_{5/2}$ monopole interaction as protons are removed from the $1f_{7/2}$ orbital [11]. The monopole shift of the $\nu f_{5/2}$ orbital, in concert with the large spin-orbit splitting between the neutron $p_{3/2}$ and $p_{1/2}$ orbitals, gives rise to the subshell closure at $N = 32$.

The energies of the first 2^+ states in the even-even $_{20}$Ca to $_{28}$Ni isotopes have been calculated in the full pf-shell model space using the new interaction GXPF1 [16]. The general feature of a systematic increase in the energy of the first excited 2^+ states at $N = 32$ for the Ca, Ti, and Cr isotopes was reproduced by the calculations. These calculations for ^{54}Ti$_{32}$ have been extended to high spin, where experimental levels with angular momentum up to 10η were populated in deep-inelastic collisions and subsequent gamma-ray transitions identified using GAMMASPHERE [12]. The agreement between the calculated and experimental energy levels in ^{54}Ti was excellent. The data for ^{56}Cr, ^{54}Ti, and ^{52}Ca support a high $\nu f_{5/2}$ effective single-particle energy, and a significant $\nu p_{3/2}$- $\nu p_{1/2}$ spin-orbit splitting at $N = 32$.

One additional result of the full pf-shell model calculations with the GXPF1 interaction [16] is a significant gap in the effective single-particle energies between the $\nu f_{5/2}$ and $\nu p_{1/2}$ orbitals for Ca and Ti isotopes. This new magic number at $N = 34$ was also predicted in Ref. [17]. The low energy of the first 2^+ state in ^{58}Cr (880 keV) shows the non-existence of the $N = 34$ magic gap for the Cr isotopic chain. Experimental studies to identify excited states in ^{54}Ca and ^{56}Ti are warranted to confirm or refute the shell model predictions.

The sudden decrease in the excitation energy of the first 2^+ states observed for ^{60,62}Cr [13] is suggestive of increased quadrupole deformation in these nuclides. The shell model results using the GXPF1 interaction do not reproduce this behavior, which may indicate that the driving force for the deformation lies outside the pf-model space. As stated in Ref. [13], inclusion of intruding neutron orbital $g_{9/2}$, and possibly $d_{5/2}$, is required to reproduce the experimental data. This suggests that $N = 40$ is not a good subshell closure at $Z = 24$. Further experimental and theoretical studies are needed to fully appreciate the implications that new magic numbers would have on the beta-decay properties of neutron-rich nuclides in this region of the chart of the nuclides.

Acknowledgments

This work was supported in part by the National Science Foundation Grants PHY-01-10253, PHY-97-24299, and PHY-00-70911. A.C. Morton, J.I. Prisciandaro, C.F. Powell, and D. Seweryniak played significant roles in the development of the beta counting system. The measurement of beta-delayed gamma rays for neutron-rich nuclides around N = 32 was completed in collaboration with A.D. Davies, T. Glasmacher, D.E. Groh, S.N. Liddick, D.J. Morrissey, W.F. Mueller, H. Schatz, A. Stolz, and S.L. Tabor. The shell model calculations were completed through the efforts of B.A. Brown, M. Honma, M. Horoi, and T. Otsuka. Discussions with R.V.F. Janssens, B. Fornal, and R. Broda regarding the N = 32 shell closure and results from deep inelastic experiments are also acknowledged.

References

1. D. Guillemaud-Mueller et al., *Nucl. Phys.* **A427**, 37 (1984).
2. O. Sorlin et al., *Phys. Rev.* **C47**, 2941 (1993).
3. P.J. Sellin et al., *Nucl. Instrum. Methods Phys. Res.* **A311**, 217 (1992).
4. K. Summerer et al., *Nucl. Phys.* **A616**, 341c (1997).
5. J.I. Prisciandaro, A.C. Morton, and P.F. Mantica, "Beta Counting System for Fast Fragmentation Beams," *Nucl. Instrum. Methods Phys. Res A* in press.
6. W.F. Mueller et al., *Nucl. Instrum. Methods Phys. Res.* **A466**, 492 (2001).
7. MCNP™, A General Monte Carlo N-Particle Transport Code, Version 4C, edited by J.F. Briesmeister, LA-13709-M, 2000.
8. A.M. Nathan et al., *Phys. Rev.* **C16**, 192 (1977).
9. O. Sorlin et al., *Nucl. Phys.* **A632**, 205 (1998).
10. J.I. Prisciandaro et al., *Nucl. Phys.* **A682**, 200c (2001).
11. J.I. Prisciandaro et al., *Phys. Lett.* **B510**, 17 (2001).
12. R.V.F. Janssens et al., *Phys. Lett.* **B546**, 55 (2002).
13. O. Sorlin et al., *Eur. Phy. J.* A in press.
14. A. Huck et al., *Phys. Rev.* **C31**, 2226 (1985).
15. F. Tonduer, *CERN Report* **81-09**, 81 (1981).
16. M. Honma et al, *Phys. Rev.* **C65**, 061301(R) (2002).
17. T. Otsuka et al., *Phys. Rev. Lett.* **87**, 082502 (2001).

GROUND-STATE CONFIGURATION OF THE NEUTRON-RICH ^{11}BE AND ^{17,19,21,23}O NUCLEI STUDIED BY COULOMB BREAKUP *

R. PALIT[1], T. AUMANN[2], K. BORETZKY[3], B.V. CARLSON[4],
D. CORTINA[2], U. DATTA PRAMANIK[2†], TH.W. ELZE[1], H. EMLING[2],
H. GEISSEL[2], A. GRÜNSCHLOß[1], M. HELLSTRÖM[2], S. ILIEVSKI[1,2],
N. IWASA[2], K.L. JONES[2], L.H. KHIEM[3], J.V. KRATZ[3], R. KULESSA[5],
Y. LEIFELS[2], A. LEISTENSCHNEIDER[1], E. LUBKIEWICZ[5],
G. MÜNZENBERG[2], C. NOCIFORO[3], P. REITER[6],
C. SCHEIDENBERGER[2], H. SIMON[7], K. SÜMMERER[2], E. WAJDA[5],
W. WALUS[5]

[1] *Institut für Kernphysik, Universität Frankfurt, D-60486 Frankfurt, Germany*
[2] *Gesellschaft für Schwerionenforschung, D-64291 Darmstadt, Germany*
[3] *Institut für Kernchemie, Universität Mainz, D-55099 Mainz, Germany*
[4] *Instituto Tecnologico de Aeronautica - CTA, Sao Jose dos Campos, Brasil*
[5] *Instytut Fizyki, Uniwersytet Jagelloński, PL-30-059 Kraków, Poland*
[6] *Sektion Physik, Ludwig Maximilian Universität, D-85748 Garching, Germany*
[7] *Institut für Kernphysik, Technische Universität,D-64289 Darmstadt, Germany*

The ground-state structures of ^{11}Be and ^{17,19,21,23}O isotopes are investigated via Coulomb dissociation at beam energies around 550 MeV/u. An exclusive measurement of one-neutron removal from the projectile was carried out where the neutron, the charged fragment, and γ-rays were detected. A direct-breakup model is used to extract the spectroscopic factors for different configurations of the ground state of these nuclei.

1. Introduction

The study of the drip-line nuclei through reaction experiments has become possible in recent years due to the availability of radioactive beams[1]. Exotic structures and excitation modes were observed for these loosely bound nuclei with large neutron and proton asymmetry. One of the interesting excitation modes is the presence of low-lying dipole strength in these drip-line nuclei. The large spatial extension of the wave function of the loosely

---

*This work is supported by BMBF.
†Present address: Saha Institute of Nuclear Physics, Kolkata-64, India.

bound valence neutron gives rise to non-resonant dipole transitions to the continuum with large transition probabilities close to the neutron threshold. The fact that this dipole strength is characteristic for the structure of the projectile is used to extract the single-particle structure of the neutron in the parent nucleus[2,3]. Here, the dipole strength is measured by electromagnetic excitation of the projectile to the continuum followed by neutron decay. The coincident measurement of the charged fragment, neutron, and γ-rays allows to determine the differential cross section exclusively for the different fragment states populated. The spectroscopic factors associated with the individual configurations are obtained from the ratio of the measured cross section to the cross section calculated in a direct-breakup model based on the wave function of the concerned configuration. The results obtained are compared with transfer reactions, knockout processes, and electron magnetic scattering data as far as available from other experiments. The nuclei investigated here have a wide variation in one-neutron separation energy (0.5 MeV to 4.1 MeV) and thus will serve as a testing ground to explore the scope of this method as a spectroscopic tool. In the following sections, the experimental setup, details of the analysis, and some preliminary results for ^{11}Be and oxygen isotopes will be presented.

2. Experimental Method

Secondary beams of ^{11}Be and ^{17,19,21,23}O ions were produced in fragmentation reactions of a primary 720 MeV/u ^{40}Ar beam delivered by the synchrotron SIS at GSI. The fragments were analyzed according to their magnetic rigidity in the Fragment Separator (FRS)[4]. The secondary beams were subsequently identified event-by-event by means of energy-loss and time-of-flight measurements. Behind the secondary reaction target, fragments were deflected by a large-gap dipole magnet (ALADIN). By utilizing energy-loss, time-of-flight and position measurements before and after the dipole magnet, the charge, velocity, scattering angle, and the mass of the fragments were determined. The neutrons and γ-rays emitted from the excited projectiles were detected in the LAND and Crystal Ball detectors, respectively. By measuring the four-momenta of all breakup products of the projectile after inelastic scattering, the excitation energy of the nucleus is determined. The Coulomb dissociation cross section with a lead (1.8 g/cm^2) target was obtained after subtracting nuclear contributions determined from the data with carbon (0.573 g/cm^2) target and applying a proper scaling of the cross sections. In addition, background contributions estimated from the data taken without target were subtracted.

3. Results and Discussion

At high beam energy (~ 550 MeV/u) Coulomb breakup is dominated by electric dipole transitions and the differential cross section can be written as[5]

$$\frac{d\sigma(I_C^\pi)}{dE^*} = (\frac{16\pi^3}{9\hbar c})N_{E1}(E^*)\sum_{nlj}CS^2(I_c^\pi,nlj)\sum_m |<q|\frac{Ze}{A}rY_m^1|\psi_{nlj}(\vec{r})>|^2 \quad (1)$$

where $N_{E1}(E^*)$ is the number of equivalent dipole photons of the target Coulomb field at an excitation energy E^*. $|q>$ is the final state in the continuum and can be approximated by a plane wave or a distorted wave. $CS^2(I_c^\pi,nlj)$ represents the spectroscopic factor corresponding to one of the configurations of the projectile ground state in which the core (I_c^π) is coupled to a neutron in $\psi_{nlj}(\vec{r})$ orbital.

As the first example, we discuss the Coulomb breakup of ^{11}Be. The ground state wave function of ^{11}Be ($J^\pi = 1/2^+$) may be written as

$$|^{11}Be> = \sqrt{S_1}|2s_{1/2}\nu \otimes 0^+> + \sqrt{S_2}|1d_{5/2}\nu \otimes 2^+> +... \quad (2)$$

where S_1 and S_2 represent the spectroscopic factors for the two predominant configurations, respectively. Correspondingly, the ground state and the 2^+ of ^{10}Be are populated in a Coulomb breakup reaction where the valence neutron is lifted to the continuum. Thereby, the populated state is identified in the experiment by the coincident γ-ray measurement. As can be seen in Fig. 1, not only the ground and first excited states in ^{10}Be are populated, but also higher lying states at about 6 MeV (see also inset in Fig. 1). These states are populated after removal of a neutron from a more deeply bound p state (i.e. a core neutron) while the halo neutron acts as a spectator. About 5 % of the total cross section yields the ^{10}Be in excited states.

After subtracting the excited state contributions from the total cross section, the excitation energy spectrum of ^{11}Be corresponding solely to ground state transitions to ^{10}Be is obtained. Fig. 2 shows the differential cross section obtained after corrections for acceptance and efficiency (filled symbols). Both electromagnetic and nuclear diffraction processes contribute to the cross section. The nuclear contribution was estimated from the corresponding measurement taken with carbon target after a proper scaling (shown by open squares in Fig. 2). A scaling factor of 5.5 was obtained from the ratio of calculated diffraction cross sections[6] (^{11}Be $\rightarrow$ ^{10}Be(0^+)) for lead and carbon targets.

Finally, the ground state invariant mass spectrum was fitted with a linear sum of the simulated spectrum obtained with the direct breakup

90

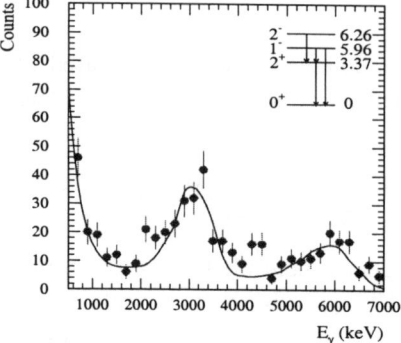

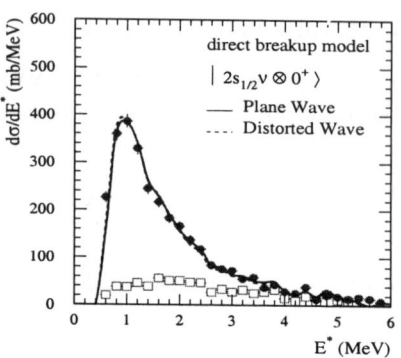

Figure 1. *Doppler-corrected γ sum-energy spectrum measured for the reaction $^{11}Be \rightarrow {}^{10}Be + n + \gamma$ (520 MeV/u). The solid curve shows a fit of simulated line shapes using GEANT transport code[10] to the experimental data which gives the partial cross sections to excited states of ^{10}Be (see inserted level scheme).*

Figure 2. *Excitation energy spectrum of ^{11}Be obtained with the lead target and for ^{10}Be in its ground state. The solid and dashed curves are fits to the experimental data (filled symbols) as explained in the text. The deduced nuclear contribution is shown by open squares.*

model calculation (Eq. 1) and the nuclear contribution derived as described above. The initial state wave function $\psi(\vec{r})$ for a 2s neutron coupled to the ground state of ^{10}Be was calculated for a Woods-Saxon potential with diffuseness parameter $a = 0.7$ fm and radius parameter $r_0 = 1.25$ fm. The fit of the calculation to the experimental spectrum yields a spectroscopic factor of 0.50(3) if the final state is approximated by an outgoing plane wave (solid curve), and 0.56(3) for the distorted wave (dashed line) calculation with parameters from Ref. 7. Both calculations reproduce the shape of the measured distribution. A change of the geometrical parameters of the Woods-Saxon potential to $a = 0.5$ fm and $r_0 = 1.15$ results in spectroscopic factors of 0.60(3) and 0.66(3) for the plane and distorted wave calculation, respectively. The spectroscopic factor of 0.56(3) to 0.66(3) (obtained with distorted wave approxiamtion) from our preliminary analysis is consistent with transfer reaction data[8], and also rather close to the shell model value[9] of 0.74.

Now we shall discuss the Coulomb breakup of the odd oxygen isotopes, where the last neutron is relatively well bound, e.g. $S_n = 4.1$ MeV for ^{17}O and $S_n = 2.3$ MeV for ^{23}O. For ^{17}O ($J^\pi = 5/2^+$) the Coulomb breakup reaction yields the ^{16}O core mainly in the ground state as expected. The relative energy spectrum of ^{16}O(0^+) and the neutron is shown in Fig. 3

after subtracting nuclear contributions and correcting for acceptance and efficiency of the set-up. A comparison of the spectrum with ^{11}Be (Fig. 2) shows that the peak cross section is much smaller (by about two orders of magnitude) and also much broader. Clearly, this reflects the fact that the valence neutron of ^{17}O is in a well bound $l = 2$ state. It clearly demonstrates the tremendous sensitivity of the Coulomb breakup cross section to a halo-like tail of the wave function. The shape of the spectrum (filled symbols in Fig. 3) is well reproduced by the plane-wave calculation for a $1d_{5/2}$ neutron. Our preliminary analysis yields spectroscopic factors of 0.6(1) and 0.8(1) for the two extremes for the geometry parameters of the Woods-Saxon potential (see above). Thus the spectroscopic factor obtained by this method is close to the value $S = 1.04(10)$ obtained from a magnetic electron scattering experiment[11].

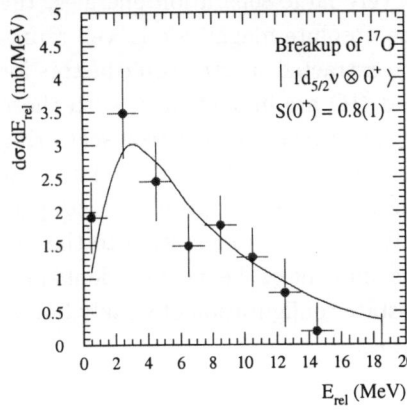

Figure 3. *Relative energy spectrum of the ^{16}O fragment in its ground state and the emitted neutron after Coulomb breakup. The excited state and nuclear contributions are subtracted. The solid curve shows a calculation in plane-wave approximation for a d neutron coupled to the ^{16}O ground state with a spectroscopic factor of 0.8(1).*

Contrary to ^{17}O, a considerable part of the breakup cross section for ^{19}O ($J^\pi = 5/2^+$) leaves the ^{18}O core in excited states. A quantitative assessment of spectroscopic factors for the individual configurations was not possible, however, since the individual states could not be resolved in the γ spectrum.

The main contributions to the ground state of ^{21}O ($J^\pi = 5/2^+$) are given by $1d_{5/2}\nu \otimes 0^+$, $1d_{5/2}\nu \otimes 2^+$, and $1d_{5/2}\nu \otimes 4^+$ configurations. From the gamma-gamma coincidence measurement the 4^+ to 2^+ and 2^+ to 0^+ cascade in ^{20}O was observed and the spectroscopic weight for the contribution $|1d_{5/2}\nu \otimes 4^+ >$ in the wave function was obtained. A preliminary analysis results in a value of 2.3(2), which is rather close to the shell model prediction of Brown *et al.* of 2.59[12].

Finally, we mention a recent experiment for ^{23}O Coulomb breakup. A first preliminary analysis[13] indicates a strong contribution of the ^{22}O ground state associated with an s-wave neutron. This observation implies a ground-state spin of $J^\pi = 1/2^+$ for ^{23}O in contrast to the interpretation of a recent $2n$-removal reaction[14]. A definite conclusion on the structure of ^{23}O, however, can only be drawn after a detailed analysis of the data.

4. Summary and Conclusion

Exclusive measurements of one-neutron Coulomb breakup reactions were performed for the first time for the halo nucleus ^{11}Be and for neutron-rich oxygen isotopes. The deduced differential cross sections $d\sigma/dE^*$ were differentiated further according to the final states populated in the fragment. A direct-breakup model was used to analyze the data and absolute spectroscopic factors were derived. Although the cross sections are much smaller for the non-halo nuclei with comparatively large separation energies, the shape of the cross section as well as the absolute magnitude is well reproduced by the direct-breakup model. The extracted spectroscopic factors for the well bound oxygen isotopes ^{17}O and ^{21}O are in agreement with those obtained in the shell model and experiments involving electron-scattering and transfer reactions. This demonstrates that Coulomb dissociation at high energies is a valid experimental method to extract ground-state spectroscopic factors. The enormous sensitivity of the cross section to the tail of the wave function makes Coulomb breakup one of the most efficient spectroscopic methods to probe the ground-state configuration of weakly bound nuclei even with very low beam intensities.

References

1. Special Issue on Research Opportunities with Accelerated Beams of Radioactive Ions, edited by I. Tanihata, Nucl. Phys. A **693**, 1-514 (2001).
2. U. Datta Pramanik *et al.* To be published in Phys. Lett. B.
3. T. Nakamura *et al.* Phys. Lett. B **331**, 296 (1994).
4. H. Geissel *et al.*, Nucl. Instr. and Meth. B **70**, 286 (1992).
5. G. Baur *et al.* Nucl.Phys. A **458**, 188 (1986).
6. K. Hencken *et al.*, Phys. Rev. C **54**, 3043 (1996).
7. M.B. Chadwick *et al.*, Nucl. Sci. Eng. **123**, 17 (1996).
8. S. Fortier *et al.* Phys. Lett. B **461**, 22 (1999).
9. T. Aumann *et al.* Phys. Rev. Lett. **84**, 35 (2000)
10. GEANT, Cern Library Long Writup W5013, (1994).
11. S. Burzynski *et al.*, Nucl. Phys. A **399**, 230 (1983).
12. B.A. Brown, Prog. in Part. and Nucl. Phys. **47**, 517 (2001).
13. K.L.Jones *et al.*, in preparation.
14. R. Kanungo *et al.*, Phys. Rev. Lett. **88**, 142502 (2002).

CHIRALITY OF ROTATING NUCLEI

V. DIMITROV AND S. FRAUENDORF

Department of Physics, University of Notre Dame,
Notre Dame, Indiana 46556, USA, and
IKH, Research Center Rossendorf, PF 510119, 01314 Dresden, Germany

The orientation of the angular momentum vector with respect to the triaxial density distribution selects a left-handed or right-handed system principal axes. This breaking of chiral symmetry manifests itself as pairs of nearly identical $\Delta I = 1$-bands. The chiral structures combine high-j particles and high-j holes with a triaxial rotor. Tilted axis cranking calculations predict the existence of such configurations in different mass regions.

1. Introduction

Chirality is a common property of molecules. Fig. 1 shows a simple example. The 2-iodubutene contains a stereo center - the C atom -, to which the four different groups are attached. If one selects the bond to the group CH_3CH_2, the three groups I, H, CH_3 form a right-handed or a left-handed screw. These two "enatiomers" are related to each other by mirror reflection. Complex bio-molecules are all chiral. Chirality is very obvious for the DNA double-helix. Although in principle two entantiomers exist, which have exactly the same binding energy, organisms synthesize only one. The reason is that the DNA blueprint provides only for one. Since the function of bio-molecules depends critical on their geometry, (e.g. the key-lock mechanism of enzymes) the other enantiomer with the mirror-geometry may not function in the organism.

Fig. 2 illustrates chirality of a mass-less particle. The spin can be parallel or anti-parallel to particle momentum. Since the spin of the particle is an axial vector, it defines the a sense of rotation and the two orientations correspond to a right-handed and a left-handed system. The neutrinos, which appear only as left-handed species, introduce chiral asymmetry into the world. The chirality of molecules is of geometrical nature, the chirality of mass-less particles is dynamical.

Nuclei have been thought as being achiral, because they consist of only

94

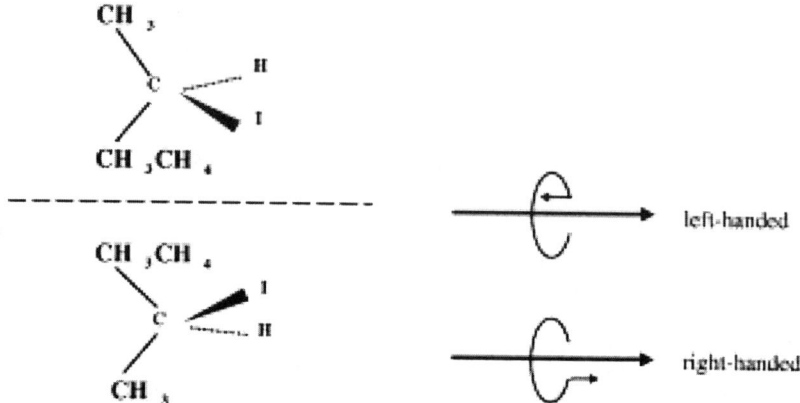

Figure 1. The two enantiomers of 2-iodubutene. The broken line indicates the mirror plane.

Figure 2. Chirality of a mass-less particle. The circle with the arrow indicates the orientation of the spin.

two species of nucleons and have relatively simple shapes as compared to molecules, However, Meng and Frauendorf [1] recently pointed out that angular momentum adds a new dimension, such that the rotation of triaxial nuclei may attain a chiral character. The lower panel of Fig. 3 illustrates this surprising possibility. We denote the three principal axes (PA) of triaxial density distribution by l, i, and s, which stand for long, intermediate and short, respectively. The angular momentum vector J introduces chirality by selecting one of the octants. In four of the octants the axes l, i, and s form a left-handed and in the other four a right-handed system. This gives rise to two degenerate rotational bands because all octants are energetically equivalent. Hence the chirality of nuclear rotation results from a combination of dynamics (the angular momentum) and geometry (the triaxial shape).

2. Symmetries of the Rotating Mean Field

If we speak about the shape of a nucleus, we mean the shape of its density distribution. The symmetry of the density distribution -spherical or deformed- decides if the spectrum will be irregular or show rotational bands. The density distribution is found by means of mean field approaches, like the various types of the Hartree-Fock calculations or the Strutinsky method. For large angular momentum one has to use the Cranking generalizations of these methods, which describe an uniformly rotating mean field. In these

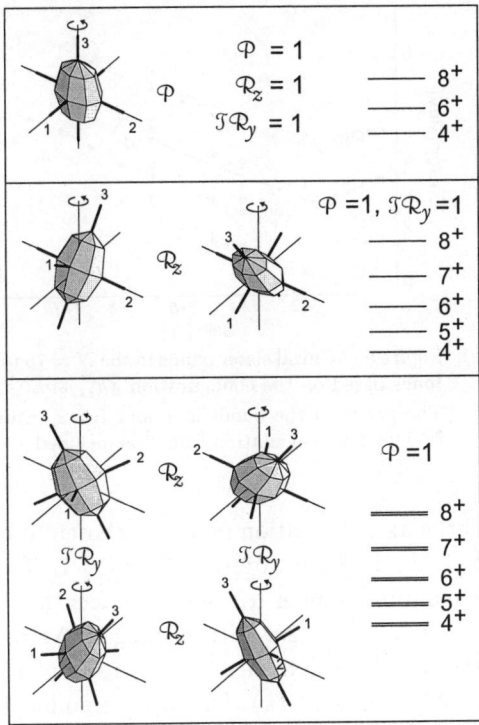

Figure 3. The discrete symmetries of the mean field of a rotating tri-axial reflection symmetric nucleus (three mirror planes). The axis of rotation (z) is marked by the circular arrow. It coincides with the angular momentum J. The structure of the rotational bands associated with each symmetry type is illustrated on the right side. The meaning of the symmetry operations is explained in the text. Note the change of chirality induced by $\mathcal{T}R_y(\pi)$ in the lowest panel.

studies, one used to assume that the axis of uniform rotation coincides with one of the principal axes of the density distribution, as it its the case for molecules. Frauendorf [2] demonstrated that nuclei are different from molecules. They may uniformly rotate about an axis that is tilted with respect to the principal axes of the density distribution. The Tilted Axis Cranking model (TAC) [2,3,4] allow us to calculate the orientation of the rotational axis. TAC consist in applying one of mean field approximations to the two-body Routhian

$$H' = H - \omega J_z. \tag{1}$$

Here, H is the two-body Hamiltonian of choice and J_z the angular momentum component along the z-axis which is the rotational axis. The new element as compared to traditional cranking calculations is to allow for all orientations of the density distribution with respect to the z-axis. Fig. 3 illustrates how changing the orientation of the rotational axis leads to different discrete symmetries, which show up in the level sequence of rota-

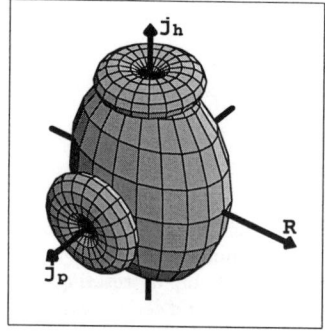

Figure 4. Orbitals of a high-j proton and a high-j neutron hole coupled to the triaxial density distribution.

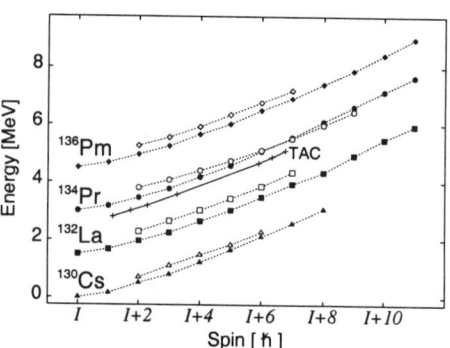

Figure 5. Chiral sister bands in the $N = 75$ isotones based on the configuration $\pi h_{11/2} \nu h_{11/2}^{-1}$. The parity of the bands is + and $I = 9$. From [8]. The TAC calculation from [6] is included.

tional bands. In the upper panel the axis of rotation (which is chosen to be z) coincides with one of the PA, i. e. the finite rotation $\mathcal{R}_z(\pi) = 1$. This symmetry implies the signature quantum number α, which restricts the total angular momentum to the values $I = \alpha + 2n$, with n integer ($\Delta I = 2$ band) [5]. In the middle panel the rotational axis lies in one of the planes spanned by two PA (planar tilt). Since now $\mathcal{R}_z(\pi) \neq 1$, there is no longer a restriction of the values I can take. The band is a sequence of states the I of which differ by 1 ($\Delta I = 1$ band). There is a second symmetry in the upper two panels: The rotation $\mathcal{R}_y(\pi)$ transforms the density into an identical position but changes the sign of the angular momentum vector J. Since the latter is odd under the time reversal operation $\mathcal{T}$, the combination $\mathcal{T}R_y(\pi) = 1$.

In the lower panel the axis of rotation is out of the three planes spanned by the PA. The operation $\mathcal{T}R_y(\pi) \neq 1$. It changes the chirality of the axes l, i and s with respect to the axis of rotation J. Since the left- and the right-handed solutions have a the same energy, they give rise to two degenerate $\Delta I = 1$ bands. They are the linear combinations of the left- and right-handed configurations , which restore the spontaneously broken $\mathcal{T}R_y(\pi)$ symmetry.

Fig. 4 illustrates how such a solution may arise. The proton aligns its angular momentum j_p with the short axis of the density distribution. This orientation maximizes the overlap of its orbital with the triaxial density, which corresponds to minimal energy, because the core-particle interaction is attractive. The neutron hole aligns its angular momentum j_h with the

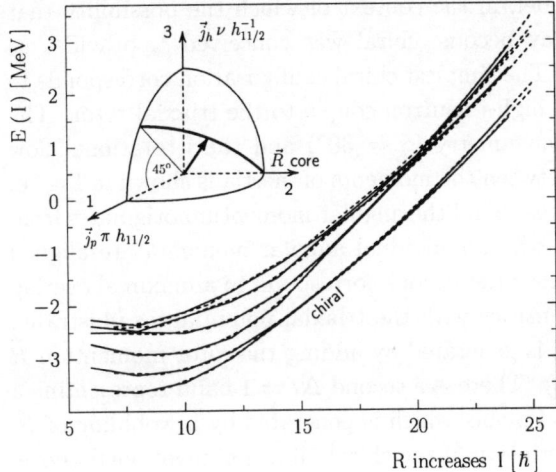

Figure 6. Rotational levels of $h_{11/2}$ particles and holes coupled to a triaxial rotor with $\gamma = 30^o$. Full lines: even I. Dashed lines: odd I. The insets show the orientation of the angular momentum with respect to the triaxial potential, where 1, 2 and 3 correspond to the short, intermediate and long principal axes, respectively. The angular momentum vector moves along the heavy arc. From 1.

long axis. This orientation minimizes the overlap of its orbital with the triaxial density, which corresponds to minimal energy, because the core-hole interaction is repulsive. The angular momentum of the core $\vec{R}$ is of collective nature. It likes to orient along the intermediate axis, which has the largest moment of inertia, because the density distribution deviates strongest from rotational symmetry with respect to this axis.

3. Appearance of Chirality in the Spectra

Dimitrov, Frauendorf and Dönau [6] found the first completely self-consistent chiral solution for $^{134}_{59}\text{Pr}_{75}$ with the maximal triaxiality of $\gamma \approx 30^o$. Fig. 5 shows that in this nucleus two $\Delta I = 1$ bands, which are based on the $[\pi h_{11/2}, \nu h_{11/2}^{-1}]$ configuration, merge forming a doublet structure [7]. Consistent with the experiment, the TAC solution attains chiral character only for $I > 15$.

Pairs of close bands with the same parity are found in the neighboring odd-odd nuclei as well [8,9,10,11,12]. Fig. 5 shows as examples the isotones of $^{134}_{59}\text{Pr}_{75}$. The bands are close but do not merge. This means that the chirality cannot be broken in a static way as in molecules. Either there is substantial tunneling between the left- and right handed configurations or chirality appears in weaker form as a slow oscillation of the angular momentum vector with respect to the l-s plane into the left- and right handed sectors. The frequency of such a chiral vibration is less than 300 keV (cf. Fig. 5).

The Particle-Rotor model, in the context of which the possibility that rotating triaxial nuclei may become chiral was conceived [1], provides us insight into the dynamics. The simplest chiral configuration corresponds to one high-j proton and one high-j neutron couple to the triaxial rotor. The generic case of maximal asymmetry ($\gamma = 30^\circ$) and the irrotational flow relation $\mathcal{J}_l = \mathcal{J}_s = \mathcal{J}_i/4$ between the moments of inertia is shown in Fig. 6. At the beginning of the lowest band the angular momentum originates from the particle and the hole, whose individual angular momenta are aligned with the s- and l-axes. These orientations correspond to a maximal overlap of the particle and hole densities with the triaxial potential, as illustrated in Fig. 4. A $\Delta I = 1$ band is generated by adding the rotor momentum $\vec{R}$ in the s-l plane (planar tilt). There is a second $\Delta I = 1$ band representing a vibration of $\vec{J}$ out of the s-l plane, which is generated by a wobbling of $\vec{R}$. This is a more precise description of the chiral vibration mentioned above. Higher in the band, $\vec{R}$ reorients toward the i-axis, which has the maximal moment of inertia. The left- and the right-handed positions of $\vec{J}$ separate. Since they couple only by some tunneling, the two bands come very close together in energy. This is the regime we called chiral rotation.

4. Predictions of chirality in various mass regions

The chiral configuration is rather special. The shape must be triaxial, and there must be high-j particles and high-j holes. The existence of such a structures and their location in the nuclear chart must by studied by means of microscopic TAC calculations. Table 1 summarizes the results of our TAC studies. The table shows only one representative nucleus for each mass region. The TAC barriers between the the left- and right handed sectors are in the order of 50- 100 keV. The Particle-Rotor calculations show that this leads to grouping into the chiral sister bands. However, for a quantitative estimate of the splitting one needs calculating the tunneling in a microscopic way, which we have not been able yet. Therefore, we only correlate the chiral TAC solutions with the appearance of pairs of bands, without trying to predict if they will be chiral vibrators or rotors. We discuss here only two regions. The other regions are discussed in [6,14].

Most TAC calculations have been carried out for the region around $^{134}_{59}\mathrm{Pr}_{75}$. Both the experiment and TAC point to an island of chirality around $Z = 59$ and $N = 75$. In the $N = 75$ chain, the shores seem to be $Z = 65$ and 51. The low-Z shoreline seems to be 72. The high-Z shore is not known yet. The center of the island is $Z \approx 60$ and $N \approx 76$. The simplest

Table 1. Representative nuclei for which TAC calculations give chiral solutions. The particle and hole orbitals, which align with the short and long axes, are indicated. The column J displays the angular momentum with the strongest chirality. The last two columns contain the shape parameters for this value of J.

Z	N	particle	hole	J	ε	γ
35	44	$\pi g_{9/2}$	$\nu g_{9/2}$	9	0.19	26
43	65	$\nu h_{11/2}$	$\pi g_{9/2}$	13	0.21	14
59	75	$\pi h_{11/2}$	$\nu h_{11/2}$	13	0.18	26
77	111	$\pi i_{13/2}$	$\nu i_{13/2}$	13	0.21	40
69	93	$\pi i_{13/2}$	$\nu h_{11/2}$	45	0.32	26

chiral configurations appear in odd-odd nuclei, where a high-j particle and high-j hole couple to the triaxial rotor. Most of the experimental chiral sister bands in the mass 134 region have this structure. For breaking the chiral symmetry, the particle and hole "legs" need not be of the same length, and they may be composed of more than one high-j orbital. As an example, we calculated the chiral configuration $[\pi h_{11/2}^2, \nu h_{11/2}^{-1}]$ in $^{135}_{60}\mathrm{Nd}_{75}$. Zhu et al. [13] found a pair of negative parity bands in this nucleus, which have this structure and come as close as 90 keV. This demonstrates that the appearance of the sister bands is not a specific property of odd-odd nuclei, but a consequence of the chiral geometry, whichever it is generated. For further discussion of this region see [14].

Another promising region contains the neutron rich nuclei near $Z = 43$ and $N = 65$. Many nuclei are predicted to be triaxial [15]. We have carried out TRS calculations [16] for the odd-odd Rh-isotopes assuming the configuration $[\nu h_{11/2}, \pi g_{9/2}]$. The most pronounced triaxiality was found for $^{110}_{45}\mathrm{Rh}_{65}$, for which we carried out TAC calculations. The energy turned out to be near-constant when the J reorients from the s-l plane into the chiral sectors. Hence, a pair of chiral sister bands is expected, which may be more like soft chiral vibrations. A pair of bands in $^{104}_{45}\mathrm{Rh}_{59}$ [17] with the configuration $[\nu h_{11/2}, \pi g_{9/2}]$ was suggested as an experimental candidate for chirality. So far, we have not carried out TAC calculations for this nucleus, because it turned out to be axial in the TRS calculations. Experimental evidence for a pair of chiral sister bands with the configuration $[\nu h_{11/2}, \nu g_{7/2}]$ was suggested for $^{106}_{42}\mathrm{Mo}_{64}$ at this conference [18]. Although the shape of this nucleus is triaxial according to the TRS calculations, our TAC calculation gave a planar geometry. The reason is that the $\nu g_{7/2}$ hole-type quasineutron does not well enough align with the long axis. It should be mentioned that the theoretical results concerning triaxiality and chirality are sensitive to the position of the single-particle levels, which are still nor very well known in this region. Hence, more spectroscopic data and comparison with the theoretical models are needed in order to reduce these uncertainties.

5. Conclusions

Chiral rotation manifest itself as a pair of nearly identical $\Delta I = 1$-bands with the same parity. Tunneling between the left- and right-handed configurations causes an energy splitting between the chiral sister bands. A weaker form chirality are the chiral vibrations, which are slow oscillations of J between the left- and right-handed configurations. They show up two $\Delta I = 1$-bands, separated by the (small) vibrational energy.

Microscopic tilted axis cranking calculations predict chirality in various mass regions. There is some systematic experimental evidence for chiral sister bands in odd-odd nuclei around mass 134. The region of neutron rich nuclei around ^{108}Tc seems to be promising for discovering more experimental examples of chirality.

References

1. S. Frauendorf, and Meng, J., 1997, Nucl. Phys. A **617**, 131 (1997)
2. S.Frauendorf, Nucl. Phys. **A557**, 259c (1993)
3. Kerman, A.K., and Onishi, N., Nucl. Phys. A **361**, 179 (1981)
4. Frisk, H., and Bengtsson, R., Phys. Lett. **196B**, 14 (1987)
5. Bengtsson, R., and Frauendorf, S., Nucl. Phys. **A327**, 139 (1979)
6. V.I.Dimitrov, S. Frauendorf, and F. Dönau, Phys. Rev. Lett. **84**, 5732 (2000)
7. C. Petrache et al. , Nucl. Phys. **A597**, 106 (1996)
8. K. Starosta, et al, Phys. Rev. Lett. **86**, 971 (2001)
9. R. A. Bark et al. Nucl. Phys. **A 691**, 577 (2001)
10. K. Starosta et al., Nucl. Phys. **A682**, 375c (2001)
11. A. A. Hecht et al., Phys. Rev. **C63**, 051302(R) (2001)
12. D. J. Hartley et al. Phys. Rev. **C64**, 31304(R) (2001)
13. S. Zhu et al. Phys. Rev. Lett., submitted
14. V. Dimitrov et al., Proc. Intern. Conf. on Frontiers in Nuclear Spectroscopy, Berkeley, 2002, APS Conference Proceedings, in press
15. J. Skalski et al. Nucl. Phys. **A617**, 282 (1997)
16. W.Nazarewicz et al. Nucl. Phys. **A435**, 397 (1985)
17. K. Koike et al., Proc. Intern. Conf. on Frontiers in Nuclear Spectroscopy, Berkeley, 2002, APS Conference Proceedings, in press
18. S. J. Zhu et al., lecture at this conference

COUPLED-CLUSTER THEORY FOR NUCLEAR SCIENCE

D. J. DEAN

Physics Division, Oak Ridge National Laboratory,
P.O. Box 2008, Oak Ridge, TN 37831-6373 USA E-mail: deandj@ornl.gov

I discuss an initial implementation of the coupled-cluster method for nuclear structure calculations and apply our method to ^{4}He. In this procedings I will discuss the effect of center-of-mass removal on our results.

1. The many-body problem and coupled cluster theory

One may follow various theoretical paths to obtain information about the properties of nuclear systems. One path originates from following a reductionist approach. One begins with some derivation of the nucleon-nucleon interaction (such as that built upon meson exchange, chiral perturbation theory, or phenomenology), and one develops computational tools for solving the many-body problem, as well as one can from this point of view. Examples of research efforts pursuing this approach include the Green's Function Monte Carlo collaboration who begin with the Argonne interaction and supplement it with effective three-body interactions[1], and the no-core shell-model collaboration who generate a G-matrix+folded-diagrams effective interaction and diagonalize in a given model space[2]. Both methods are *ab initio* from the many-body point of view: they begin with the bare nucleon-nucleon interactions. Another valid approach which has been successful requires the development of effective nuclear interactions at either the mean-field level through the use of Skyrme-like forces[3], or with the shell model by using effective interactions derived from experimental level and transition information[4].

The *ab initio* approaches, while difficult, allow one to study emergent phenomena such as deformation or vibrations of the nuclear systems from the fundamental level of the bare interactions; however, the applications of these methods are at the present time limited to light nuclear systems. The effective interactions (whether of the mean-field or shell-model variety) may be applied to various regions of the nuclear chart, but they often (especially in shell-model applications) rely on data-fitting within the region being calculated. The successful shell-model interactions, such as the $1s0d$ interaction[5] or the $0f1p$ interactions[6], all require large data sets in order to be adjusted appropriately to reproduce existant data and predict certain quantities within a given region. Herein lies the difficulty of relying on fitted interactions: the experimental data coming from the Rare Isotope Accelerator (RIA) and other facilities may not be dense enough to allow for a successful fitting of effective shell-model interactions in regions of interest. With this in mind, it becomes essential for nuclear theorists to explore methods that will allow for *ab initio* calculations of nuclei both near stability and in regions where RIA and other radio-isotopic facilities will probe.

In these Proceedings, I will discuss a many-body approach, known as coupled-cluster theory, that may prove quite useful in applications to nuclear structure[7]. Coupled-cluster theory was first introduced in nuclear physics by Coester[8] and Coester and Kummel[9]. Initial nuclear structure applications came in the mid-1970s with several papers from the Bochum group[10]. Since that time, nuclear physics applications have been rather sporadic. On the other hand, the first chemistry application was discussed by Čížek and Paldus[11], and the method became computationally feasible due to work by Pople[12] and Bartlett and Purvis[13] and has become widely used and developed in computational chemistry[14]. The interesting and desirable theoretical properties of the coupled-cluster method within computational chemistry have made it *the* method of choice in computations of many-body correlation effects in atomic, molecular, and chemical systems. While it was originally developed for the many-body ground-state, applications of the coupled-cluster method in quantum chemistry now extend to excited states and open-shell systems.

Nuclear applications of the coupled-cluster technique include approaches in coordinate-space being addressed by the Manchester group[15]. Recently, Heisenberg and Mihaila[16] have suggested a somewhat different formulation than that espoused in quantum chemistry. The reasons for the sporadic pursuit of coupled-cluster methods in nuclear structure probably arise from the lack of a good bare nucleon-nucleon interaction. In the last 10 years this problem has been effectively eliminated due to excellent nucleon-nucleon interactions that give χ^2 per degree-of-freedom of nearly one. These interactions include the phenomenological Argonne V_{18} potential[17], the meson exchange potentials such as CD-Bonn[18], and the very recent nucleon-nucleon potentials based on chiral perturbation theory [19]. Another reason that the method was not pursued was certainly the lack of computational power available in the late 1970s as compared to today.

The Coupled-cluster method is a fully microscopic theory that can be used to obtain energies and eigenstates of a given Hamiltonian. Furthermore, the theory is capable of systematic improvements through increasingly higher-order implementations of a well-defined scheme of hierarchical approximations. Coupled-cluster theory is size extensive, which means that only linked diagrams enter into a given computation[14]. The method is also size consistent. This latter property has vast implications for chemical reaction studies and is not a property of the shell model[20].

2. Approach to Coupled-Cluster Theory

The presence of a hard core in various channels of the nucleon-nucleon interaction (with repulsion on the order of 2–5 GeV) causes difficulty for theories that wish to use basis state expansion techniques. One way to overcome this difficulty is to use a renormalized effective interaction within the model space where one will actually perform computations. This model-space, dubbed the P space, is a subset of the full Hilbert space. The excluded space, dubbed the Q space, represents the remaining part of the Hilbert space and is, in principle, infinite in size. Brueckner[21] originally developed the G-matrix theory that allows for the solution of the full A-body problem in the reduced Hilbert space. The G-matrix is given by[22]

$$G(\omega) = V + V\frac{Q}{\omega - QTQ}G(\omega) \,, \tag{1}$$

where V is the bare nucleon-nucleon interaction, T is the kinetic energy operator, and ω is a starting energy. Diagrammatically, the solution of this equation amounts to generating all particle-hole ladder diagrams, with intermediate two-particle states outside the P-space, to infinite order.

We begin our calculation by first choosing the P space, as shown in Fig. 1. Within that space, we compute the G-matrix elements of the renormalized interactions. We then define a reference Slater determinant from which we perform the coupled-cluster calculation. By performing the coupled-cluster calculations only in the P-space, we insure that no double counting of many-body perturbation theory diagrams occurs. Those diagrams that we do not include in this expansion are those for which a particle below the Fermi energy in the reference Slater determinant moves to the Q-space; however, as one increases the P-space, the contribution of these diagrams to observable quantities such as the energy should become

104

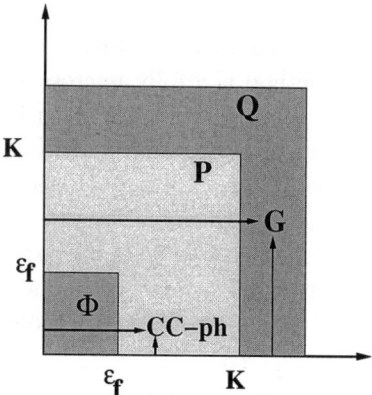

Figure 1. The choice of model space. Particle-hole excitations from the P-space (with energy cutoff K) to the Q-space are allowed during the computation of the G-matrix. Coupled-cluster computations occur only in the P-space where the Fermi energy, ε_f, is determined by the reference Slater determinant $| \Phi \rangle$.

very small.

By implementing the G-matrix formalism, we obtain as our Hamiltonian $H = \sum_{pq} K_{pq} a_p^\dagger a_q + \frac{1}{4} \sum_{pqrs} \langle pq \mid G \mid rs \rangle a_p^\dagger a_q^\dagger a_s a_r$, where K_{pq} are the one-body matrix elements of the kinetic energy operator, $K_{pq} = \langle \phi_p \mid K \mid \phi_q \rangle$, and $\langle pq \mid G(\omega) \mid rs \rangle$ are the antisymmetrized two-body matrix elements of the effective nucleon-nucleon interaction. The single-particle wave functions are the basis states of the problem, and the labels p, q, r, s represent all single-particle quantum numbers. In the following, we use the labels ijk to represent single-particle states below the Fermi surface, and labels abc to indicate single-particle states above the Fermi surface.

The basic idea of coupled-cluster theory is that the correlated many-body wave function $| \Psi \rangle$ may be obtained by application of a correlation operator, T, such that

$$| \Psi \rangle = \exp(-T) | \Phi \rangle , \qquad (2)$$

where Φ is a reference Slater determinant chosen as a convenient starting point. For example, we use the filled $0s$ state as the reference determinant for ^{4}He. This exponential ansatz has been well justified for many-body problems using a formalism in which the cluster functions are constructed by cluster operators acting on a reference determinant[23].

The correlation operator T is given by $T = T_1 + T_2 + \cdots T_A$, and represents various n-particle-n-hole (np-nh) excitation amplitudes such as

$$T_1 = \sum_{a<\varepsilon_f, i>\varepsilon_f} t_i^a a_a^\dagger a_i , \tag{3}$$

and higher-order terms for T_2 to T_A. We are currently exploring the coupled-cluster method at the T_1 and T_2 level. This is commonly referred to in the literature as Coupled-Cluster Singles and Doubles (CCSD).

We compute the expectation of the energy from

$$E = \langle \Psi_0 \mid \exp(-T) \, H \exp(T) \mid \Psi_0 \rangle . \tag{4}$$

The Baker-Hausdorf relation may be used to rewrite the similarity transformation as an expansion that terminates exactly at four nested commutators when the Hamiltonian contains, at most, two-body terms, and at six-nested commutators when three-body potentials are present. We stress that this termination is exact, thus allowing for a derivation of exact expressions for the amplitudes. The equations for amplitudes are found by left projection of excited Slater determinants so that for the $1p$-$1h$ amplitudes we must solve

$$0 = \langle \Phi_i^a \mid \exp(-T) \, H \exp(T) \mid \Phi \rangle , \tag{5}$$

and similar equations for higher-order amplitudes. The commutators also generate nonlinear terms within these expressions. To derive these equations is straightforward, but tedious, work[14].

Because of the nonlinearity of the equations, one must have a good first guess for the np-nh amplitudes. We then solve the equations by iteration. For closed-shell nuclei, we use a Moller-Plesset-like approach to generate the first guess for the iteration. Shown in Fig. 2 is the convergence of the energy of the system as a function of iteration number. For our test example, ^{4}He, we achieve convergence at the 10^{-5} level by 30 iterations in a model-space that includes seven major oscillator shells. Notice from the figure that most of the convergence is obtained within 10 iterations.

By investigating the different terms within the equations and their contributions to the energies, one is able to generate a correspondence between CCSD and many-body perturbation theory. One finds that CCSD iterates the lowest first-, second-, third-, and fourth-order many-body perturbation theory diagrams to all orders. It should be noted that the third-order diagrams are incomplete at the CCSD level of truncation, although third-order corrections may be included if they are desired[24].

106

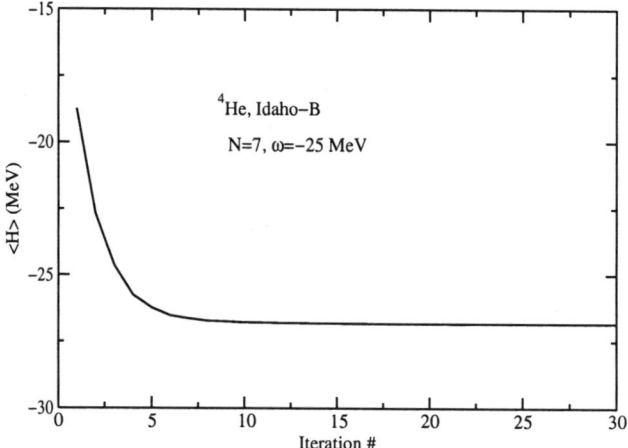

Figure 2. The convergence of the ground-state energy as a function of the CCSD iterations.

3. Initial Results

Our overall goal is to understand the structure of nuclei using coupled-cluster theory as our tool. We are at the very beginning of this effort and have a few preliminary results that we will report here. We are computing at the singles and doubles level of the coupled-cluster theory. At this level of truncation, we assume that all t_3 and higher-order amplitudes are zero. We also assume for the moment that only two-body potentials are present in the nuclear problem. We have not yet corrected these results for center-of-mass contamination, which means that they should be viewed as preliminary.

We employ the new class of chiral potentials[19] as our bare nucleon-nucleon interaction starting point. The chiral effective Lagrangians employed include one- and two-pion exchange contributions up to chiral order three and contact terms that represent the short-range force. The chiral potential reproduces the NN phase shifts below 300 MeV laboratory energy and the properties of the deuteron with high precision. We use the Idaho-B potential throughout these Proceedings.

The oscillator parameter $\hbar\omega$ is variational in our theory, and we find that the energy is minimal at $\hbar\omega = 11$ MeV for Idaho-B and the ^{4}He nucleus. As was mentioned above, the G-matrix contains a starting-energy dependence, ω. We show in Fig. 3 this dependence, along with the dependence of our results on the size of the P-space we are considering. Several interesting features emerge from this figure. The first is that as one increases the

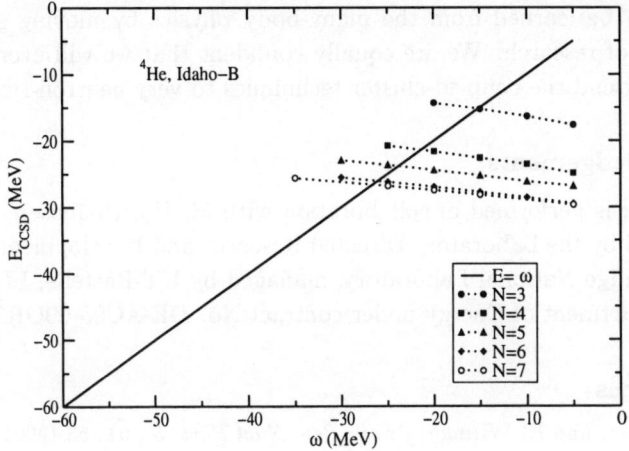

Figure 3.   The CCSD energy for ^{4}He as a function of ω for various model spaces.

P-space, the resulting energy depends less on ω. This is reasonable: if P were infinite, the solution would recover simply the bare V interaction which has no ω dependence. The second interesting feature is the rapidity of convergence of the results. Already at seven major oscillator shells one sees the onset of convergence of the total energy. In this model space we obtain the energy $E = -26.6$ MeV. We are currently investigating various possibilities for including the center-of-mass corrections.

4. Perspectives

While the results presented above indicate our first steps toward coupled-cluster theory research, they show outstanding promise. Our ^{4}He calculations show evidence of convergence using 7-8 major oscillator shells. Preliminary calculations of ^{16}O also show convergence within this model space.

We are just at the beginning of this exciting endeavor, and we first want to demonstrate the validity of the method for closed-shell systems such as ^{4}He, ^{16}O, and ^{40}Ca. For light systems, we will incorporate a center-of-mass correction. The CCSD does not include all third-order diagrams, but this deficiency can be alleviated a triples correction[24] transforming our method into CCSD[T]. We will explore methods for computing excited states and open-shell systems within CCSD. We will extend CCSD[T] to include three-body interactions. We will also explore the applicability of CCSD[T] to open shell systems and excited-state calculations. We are confident that much can be learned from the many-body physics by moving along this direction of research. We are equally confident that we will eventually be able to extend the coupled-cluster techniques to very neutron-rich nuclei.

Acknowledgements

This work is performed in collaboration with M. Hjorth-Jensen. Research sponsored by the Laboratory Directed Research and Development Program of Oak Ridge National Laboratory, managed by UT-Battelle, LLC for the U.S. Department of Energy under contract No. DE-AC05-00OR22725.

References

1. S. Pieper, and R. Wiringa, *Annu. Rev. Nucl Part. S.*, **51**, 53 (2001).
2. P. Navratil and B.R. Barrett, *Phys. Rev. C*, **57**, 3119 (1998).
3. S. Mizutori, J. Dobaczewski, G. Lalazissis, W. Nazarewicz, and P.-G. Reinhard, *Phys. Rev. C*, **61**, 044326 (2000).
4. M. Honma, T. Otsuka, B.A. Brown, and T. Mizusaki, *Phys. Rev. C*, **65**, 061301 (2002).
5. B.A. Brown and B.H. Wildenthal, *Ann. Rev. Nucl. Part. Sci.*, **38**, 29 (1988).
6. A. Poves and A.P. Zuker, *Phys. Rep.*, **70**, 235 (1985).
7. D.J. Dean and M. Hjorth-Jensen, in preparation.
8. F. Coester, *Nucl.Phys.*, **7**, 421 (1958).
9. F. Coester and H. Kummel, *Nucl. Phys.*, **17**, 421 (1960).
10. H. Kummel, K.H. Luhrmann, and J.G. Zabolitzky, *Phys. Rep.*, **36**, 1 (1978), and references therein).
11. J. Čížek and J. Paldus, *Int. J. Quantum Chem.*, **5**, 359 (1971).
12. J.A. Pople, R. Krishnan, H.B. Schlegel, and J.S. Binkley, *Int. J. Quantum Chem. Symp.*, **14**, 545 (1978).
13. R.J. Bartlett and G.D. Purvis, *Int. J. Quantum Chem.*, **14**,561 (1978).
14. T.D. Crawford and H.F. Schaefer III, *Rev. Comp. Chem.*, **14**, 33 (2000).
15. R.F. Bishop, The coupled cluster method, in *Microscopic Quantum Many-Body Theories and their Applications*, edited by J. Navarro and A. Polls, Springer-Verlag, Berlin, 1998, pp. 13–70.
16. B. Mihaila and J. Heisenberg, *Phys. Rev. C*, **61**, 054309 (2000).
17. R.B. Wiringa, V.G.J. Stoks, and R. Schiavilla, *Phys. Rev. C*, **51**, 38 (1995).
18. R. Machleidt, *Phys. Rev. C*, **63**, 035202 (2001).
19. D.R. Entem and R. Machleidt, *Phys. Lett. B*, **524**, 524 (2002).
20. R.J. Bartlett, *Ann. Rev. Phys. Chem.*, **32**, 359 (1981).
21. K.A. Bruekner, *Phys. Rev.*, **97**, 1353 (1955).
22. M. Hjorth-Jensen, T.T.S. Kuo, and E. Osnes, *Phys. Reps.*, **261**, 125 (1995).
23. F.E. Harris, H.J. Monkhorst, and D.L. Freeman, *Algebraic and Diagrammatic Methods in Many-Fermion Theory*, Oxford Press, New York, 1992.
24. K. Raghavachari, G. Trucks, J.A. Pople, and M. Head-Gordon, *Chem. Phys. Lett.*, **157**, 479 (1989).

HFB CALCULATIONS WITH HIGH ENERGY CONTINUUM COUPLING: NUCLEAR STRUCTURE AT NEUTRON DRIPLINE *

A.S. UMAR, V.E. OBERACKER, E. TERAN

Department of Physics and Astronomy,
Vanderbilt University,
Nashville, TN 37235, USA

We have developed a new Hartree-Fock-Bogoliubov (HFB) code to study ground state and pairing properties of nuclei near the neutron and proton drip lines. The unique feature of our code is that it takes into account the strong coupling to high-energy continuum states. We solve the HFB equations for deformed, axially symmetric even-even nuclei on a two-dimensional lattice using high accuracy Basis-Spline methods (Galerkin and collocation schemes). The effective N-N interaction in the p-h channel is of Skyrme-type (SLy4), and in the p-p and h-h channel it is a (modified) delta interaction. We present results for binding energies, 2-neutron separation energies, Fermi levels, pairing gaps, normal densities and pairing densities, and other observables. In particular, we will discuss neutron-rich $22O$, ^{150}Sn, and ^{102}Zr isotopes.

1. Introduction

The study of structures and reactions of nuclei far from stability has been one of the most active fields of nuclear physics in the past decade[1]. The microscopic description of such nuclei will lead to a better understanding of the interplay among the strong, Coulomb, and the weak interactions as well as the enhanced correlations present in these many-body systems. It is generally acknowledged that an accurate treatment of the pairing interaction is essential for describing exotic nuclei[2,3]. The large pairing correlations near the driplines can no longer be described by a small residual interaction. It becomes necessary to treat the mean field and the pairing field in a single self-consistent theory. Furthermore, the outermost nucleons are weakly bound, which implies a large spatial extent, and they are strongly

*This work is supported by U.S. DOE grant DE-FG02-96ER40963, and by the National Energy Research Scientific Computing Center (NERSC).

coupled to the particle continuum. Most HFB calculations to date are carried out in a truncated discrete harmonic oscillator basis[4,5,6,7,8]. This approach is quite appropriate for nuclei in the vicinity of the stability line. However, farther from stability, the continuum states become important and coordinate-based representations have numerous advantages: for example, the well-known 'French code' uses a truncated 3-D Hartree-Fock basis[9] which consists of both localized states and discretized continuum states; however, in this approach one can only include continuum states up to about 5 MeV of excitation energy. For nuclei in the vicinity of the driplines, continuum states with an equivalent single-particle energy of up to 60 MeV must be taken into account. One-dimensional calculations for spherical nuclei have been carried out in coordinate space for many years[2,3], but only recently has our Vanderbilt group succeeded in generalizing this approach to the important case of deformed axially symmetric nuclei (HFB on a 2-D lattice)[10,11]. We utilize a novel computational technique, a Basis-Spline representation of wavefunctions and operators, which allows us to accurately describe high-energy continuum states in two space dimensions (z, r).

2. Quasiparticle Wave Functions and Densities

In practice, it is to convenient to transform the standard HFB equations into a coordinate space representation and solve the resulting differential equations on a lattice. For this purpose, one defines two types of quasiparticle wavefunctions ϕ_1 and ϕ_2

$$\phi_1^*(E_\alpha, \mathbf{r}\sigma q) = \sum_i U_{i\alpha}(2\sigma)\, \phi_i(\mathbf{r} - \sigma q),$$

$$\phi_2(E_\alpha, \mathbf{r}\sigma q) = \sum_i V_{i\alpha}^*\, \phi_i(\mathbf{r}\sigma q)\,.$$

The basis wavefunctions ϕ_i depend on the position vector $\mathbf{r}$, the spin projection $\sigma = \pm\frac{1}{2}$, and the isospin projection q ($q = +\frac{1}{2}$ corresponds to protons and $q = -\frac{1}{2}$ to neutrons). From these wavefunctions we obtain the following expressions for the normal density $\rho_q(\mathbf{r})$ and the pairing density $\tilde{\rho}_q(\mathbf{r})$

$$\rho_q(\mathbf{r}) = \sum_\sigma \sum_\alpha \phi_{2,\alpha}(\mathbf{r}\sigma q)\, \phi_{2,\alpha}^*(\mathbf{r}\sigma q),$$

$$\tilde{\rho}_q(\mathbf{r}) = -\sum_\sigma \sum_\alpha \phi_{2,\alpha}(\mathbf{r}\sigma q)\, \phi_{1,\alpha}^*(\mathbf{r}\sigma q)\,.$$

The quasiparticle energy E_α is denoted by index α for simplicity. In principle, the sums go over all the energy states, but in practice a cutoff is introduced (see later). The physical interpretation of $\tilde{\rho}_q$ has been discussed in[3]: the quantity $[\tilde{\rho}_q(\mathbf{r}) \, \Delta V/2]^2$ gives the probability to find a *correlated* pair of nucleons with opposite spin projection in the volume element ΔV. The kinetic energy density $\tau_q(\mathbf{r})$ is found to be

$$\tau_q(\mathbf{r}) = \sum_\sigma \sum_\alpha |\nabla \, \phi_{2,\alpha}(\mathbf{r}\sigma q)|^2 \ .$$

3. Binding Energy and Pairing Interaction

In our calculations we utilize the Skyrme two-body effective N-N interaction. The total binding energy of the nucleus

$$E_0^{HFB} = \langle \Phi_0|H|\Phi_0 \rangle = E_{kin} + E_{Sky} + E_{Sky,LS} + E_{Coul} + E_{pair} + E_{cm}$$

consists of a kinetic energy term, various contributions from the Skyrme effective N-N interaction (including a spin-orbit term), Coulomb and pairing energy, and a center-of mass correction due to the mean-field approximation. The Coulomb energy contains the direct term as well as an exchange
 In practice, one tends to use *different* effective N-N interactions for the p-h / h-p channels and for the p-p / h-h channels. Most pairing calculations utilize a local pairing interaction of the form

$$V_p(\mathbf{r}\sigma, \mathbf{r}' - \sigma') \ = \ V_0 \, \delta(\mathbf{r} - \mathbf{r}') \, \delta_{\sigma,\sigma'} \ . \tag{3}$$

This parameterization gives rise to *volume pairing*. The pairing contribution to the nuclear binding energy is then

$$E_{pair} = \frac{V_0}{4} \int d^3 r \sum_q \tilde{\rho}_q^{\,2}(\mathbf{r}) \ . \tag{4}$$

An important related quantity is the average pairing gap for protons and neutrons which can be calculated from the general expression given in[2,3]

$$< \Delta_q > \ = \ -\frac{1}{2} \frac{V_0}{N_q} \int d^3 r \ \tilde{\rho}_q(\mathbf{r}) \ \rho_q(\mathbf{r}) \ . \tag{5}$$

where N_q denotes the number of protons or neutrons. Note that the pairing gap is a positive quantity because $V_0 < 0$.

4. HFB Equations in Coordinate Space

For certain types of effective interactions (e.g. Skyrme mean field and pairing delta-interactions) the particle Hamiltonian h and the pairing Hamiltonian $\tilde{h}$ are diagonal in isospin space and local in position space, resulting in the following HFB equations with a 4x4 structure in spin space:

$$\begin{pmatrix} (h^q - \lambda) & \tilde{h}^q \\ \tilde{h}^q & -(h^q - \lambda) \end{pmatrix} \begin{pmatrix} \phi^q_{1,\alpha} \\ \phi^q_{2,\alpha} \end{pmatrix} = E_\alpha \begin{pmatrix} \phi^q_{1,\alpha} \\ \phi^q_{2,\alpha} \end{pmatrix} \tag{6}$$

with

$$h^q = \begin{pmatrix} h^q_{\uparrow\uparrow}(\mathbf{r}) & h^q_{\uparrow\downarrow}(\mathbf{r}) \\ h^q_{\downarrow\uparrow}(\mathbf{r}) & h^q_{\downarrow\downarrow}(\mathbf{r}) \end{pmatrix}, \qquad \tilde{h}^q = \begin{pmatrix} \tilde{h}^q_{\uparrow\uparrow}(\mathbf{r}) & \tilde{h}^q_{\uparrow\downarrow}(\mathbf{r}) \\ \tilde{h}^q_{\downarrow\uparrow}(\mathbf{r}) & \tilde{h}^q_{\downarrow\downarrow}(\mathbf{r}) \end{pmatrix} .$$

The HFB equations have a mathematical structure that is similar to the Dirac equation: the spectrum of quasiparticle energies E is unbounded from above *and* below. The spectrum is discrete for $|E| < -\lambda$ and continuous for $|E| > -\lambda$. It is forbidden to choose positive and negative quasiparticle energies at the same time[4], otherwise it is impossible to satisfy the anticommutation relations for $\hat{\beta}_\alpha, \hat{\beta}^\dagger_\alpha$. For even-even nuclei it is customary to solve the HFB equations with a positive quasiparticle energy spectrum $+E_\alpha$ and consider all negative energy states as occupied in the HFB ground state. The HFB mean field Hamiltonian has the same structure as the binding energy functional. Detailed expressions for the Skyrme mean fields and the Coulomb term are given in reference[11]. The pairing interaction generates the following pairing mean field for the two isospin orientations $q = \pm\frac{1}{2}$

$$\tilde{h}_q(\mathbf{r}) = \frac{1}{2} V_0 \tilde{\rho}_q(\mathbf{r}) .$$

5. Reduction to Axial Symmetry

For numerical expediency it is advantageous to introduce axial symmetry for solving the HFB equations. It is possible to construct simultaneous eigenfunctions of the generalized Hamiltonian $\mathcal{H}$ and the z-component of the angular momentum, $\hat{j}_z$ with quantum numbers $\Omega = \pm\frac{1}{2}, \pm\frac{3}{2}, \pm\frac{5}{2}, \ldots$ corresponding to each nth energy state. The simultaneous quasiparticle eigenfunctions take the form

$$\psi_{n,\Omega,q}(\phi,r,z) = \begin{pmatrix} \phi^{(1)}_{n,\Omega,q}(\phi,r,z) \\ \phi^{(2)}_{n,\Omega,q}(\phi,r,z) \end{pmatrix} = \frac{1}{\sqrt{2\pi}} \begin{pmatrix} e^{i(\Omega-\frac{1}{2})\phi}\, \phi^{(1)}_{n,\Omega,q}(r,z,\uparrow) \\ e^{i(\Omega+\frac{1}{2})\phi}\, \phi^{(1)}_{n,\Omega,q}(r,z,\downarrow) \\ e^{i(\Omega-\frac{1}{2})\phi}\, \phi^{(2)}_{n,\Omega,q}(r,z,\uparrow) \\ e^{i(\Omega+\frac{1}{2})\phi}\, \phi^{(2)}_{n,\Omega,q}(r,z,\downarrow) \end{pmatrix} .$$

We introduce the following useful notation

$$U_{n\Omega q}^{(1,2)}(r, z) = \phi_{n,\Omega,q}^{(1,2)}(r, z, \uparrow) \ ,$$

$$L_{n\Omega q}^{(1,2)}(r, z) = \phi_{n,\Omega,q}^{(1,2)}(r, z, \downarrow) \ .$$

For axially symmetric systems it is possible to eliminate the dependence on the angle ϕ, resulting in the *reduced 2-D problem* in cylindrical coordinates[11]

$$\begin{pmatrix} (h'^q - \lambda) & \tilde{h}'^q \\ \tilde{h}'^q & -(h'^q - \lambda) \end{pmatrix} \begin{pmatrix} \phi_{n,\Omega,q}^{(1)} \\ \phi_{n,\Omega,q}^{(2)} \end{pmatrix} = E_{n,\Omega,q} \begin{pmatrix} \phi_{n,\Omega,q}^{(1)} \\ \phi_{n,\Omega,q}^{(2)} \end{pmatrix}$$

with

$$\begin{pmatrix} \phi_{n,\Omega,q}^{(1)} \\ \phi_{n,\Omega,q}^{(2)} \end{pmatrix} = \begin{pmatrix} U_{n,\Omega,q}^{(1)}(r, z) \\ L_{n,\Omega,q}^{(1)}(r, z) \\ U_{n,\Omega,q}^{(2)}(r, z) \\ L_{n,\Omega,q}^{(2)}(r, z) \end{pmatrix} .$$

Here, quantities $\tilde{h}'$, h', U and L are all functions of (r, z) only. This is the main mathematical structure that we implement in computational calculations. For a given angular momentum projection quantum number Ω, we solve the eigenvalue problem to obtain energy eigenvalues $E_{n,\Omega,q}$ and eigenvectors $\psi_{n,\Omega,q}$ for the corresponding HFB quasiparticle states. From the definitions of the normal density and pairing density we find the corresponding expressions in axial symmetry:

$$\rho_q(r, z) = \frac{1}{2\pi} \left(2 \sum_{\Omega>0}^{\Omega_{max}} \right) \times \sum_{E_n>0}^{E_{max}} \left[|U_{n\Omega q}^{(2)}(r, z)|^2 + |L_{n\Omega q}^{(2)}(r, z)|^2 \right]$$

$$\tilde{\rho}_q(r, z) = -\frac{1}{2\pi} \left(2 \sum_{\Omega>0}^{\Omega_{max}} \right) \times \sum_{E_n>0}^{E_{max}} \left[U_{n\Omega q}^{(2)}(r, z) U_{n\Omega q}^{(1)*}(r, z) \right.$$

$$\left. + L_{n\Omega q}^{(2)}(r, z) L_{n\Omega q}^{(1)*}(r, z) \right] \ .$$

6. Lattice Representation of the HFB Equations

We solve the HFB eigenvalue problem by direct diagonalization on a two-dimensional grid (r_α, z_β), where $\alpha = 1, ..., N_r$ and $\beta = 1, ..., N_z$. The four components of the spinor wavefunction are represented on the two-dimensional lattice by an expansion in basis-spline functions $B_i(x)$ evaluated at the lattice support points. Further details about the basis-spline technique are given in Ref.[12].

For the lattice representation of the Hamiltonian, we use a hybrid method[13] in which derivative operators are constructed using the Galerkin method; this amounts to a global error reduction. Local potentials are represented by the basis-spline collocation method (local error reduction). The lattice representation transforms the differential operator equation into a matrix form

$$\sum_{\nu=1}^{N} \mathcal{H}_{\mu}{}^{\nu} \psi_{\nu}^{\Omega} = E_{\mu}^{\Omega} \psi_{\mu}^{\Omega} \quad (\mu = 1, ..., N) ,$$

(11)

The calculations use as a starting point the result of a *HF+BCS* previous calculation, which makes HFB converge substantially faster. Since the problem is self-consistent we use an iterative method for the solution. At every iteration the full HFB Hamiltonian is diagonalized. Due to the axial symmetry in the intrinsic frame, the diagonalization is performed separately for each value of the angular momentum projection quantum number Ω and for the two isospin projections $q = \pm\frac{1}{2}$. Typically 20-30 iterations are sufficient for convergence at the level of one part in 10^5 for the total binding energy. Note that in this lattice approach, the number of quasiparticle states is determined by the dimensionality of the discrete HFB Hamiltonian which is $N = (4N_r N_z)^2$.

7. Results

In Table 1 we compare our 2-D HFB results for the spherical isotope ^{22}O with the 1-D radial HFB method of Ref.[2]. Corresponding results in the 2-D THO basis with 20 oscillator shells are also given. All calculations were performed with the Skyrme SLy4 force in the p-h channel and a pure delta interaction (strength $V_0 = -170 MeV fm^3$) in the p-p channel, corresponding to volume pairing. The table lists several observables: the total binding energy (for comparison, the experimental value is -162.03MeV), the Fermi level for protons and neutrons, the neutron energy gap, the *rms* radius, and the quadrupole deformation. Overall, the results of the axially symmetric code of the present work agree with the other two calculations in all the observables. We now present results for the tin isotope ^{150}Sn, a heavy nucleus far away from the valley of β-stability which is close to the two-neutron drip-line. Table 2 gives a comparison of our 2-D results (which predict a very small quadrupole deformation $\beta_2 = 0.005$) with Dobaczewski's 1-D radial HFB calculations. The box size used in the axially symmetric calculations was 20 fm in r direction and 40 fm in the z axis, whereas the 1-D code

Table 1. Calculations for ^{22}O for HFB+SLy4. The axially symmetric calculations (2D) of this work used a box size $R = 10fm$ with maximum $\Omega = \frac{9}{2}$ and an energy cutoff of 60 MeV. The spherical calculation were made with $R = 25fm$ and $j = \frac{21}{2}$. All calculations were made with a cutoff at 60 MeV.

	1-D	2-D (THO)	2-D (this work)
B. E. (MeV)	-164.60	-164.52	-164.64
λ_n (MeV)	-5.26	-5.27	-5.27
λ_p (MeV)	-18.88	-18.85	-18.16
Δ_n (MeV)	1.42	1.41	1.40
Δ_p (MeV)	0.00	0.00	0.00
R_{rms} (fm)	2.92	2.92	2.92
β_2	*	0.00002	0.0008
$E_{pair}(n)$ (MeV)	-2.85	-2.78	-2.75

had a 30 fm radial box. Also, the density of points has a different meaning in the radial code, since it uses a different grid than the one used in the B-Splines technique for our 2-D code. For these calculations the resulting mesh spacing in the 1-D code was 0.25 fm, whereas the maximum mesh spacing in the 2-D one was 1.0 fm. In the 2-D calculations an approximately 5000×5000 matrix was diagonalized for each Ω and isospin value, and for each major HFB iteration. The full calculation required about 30 HFB iterations. Like in the oxygen isotope, the agreement is very good. Table 2 also contains another interesting piece of information on ^{150}Sn:

Table 2. Comparison of calculations for spherical nucleus ^{150}Sn with HFB+SLy4. The 1-D calculations were made by using a box size $R = 30$ and a linear spacing of points of 0.25 fm, with j_{max} of $\frac{21}{2}$. Calculations by the axially symmetric HFB 2-D code were made using a box size $R = 20fm$ with $N_r = 23$, maximum $\Omega = \frac{13}{2}$. In both calculations the pairing strength V_0 was set to -170 $MeV fm^3$, and the energy cutoff to 60 MeV.

	1-D	2-D
B. E (MeV)	-1129.0	-1129.6
λ_n (MeV)	-0.96	-0.94
λ_p (MeV)	-17.54	-17.69
Δ_n (MeV)	1.02	1.00
Δ_p (MeV)	0.00	0.02
R_{rms} (fm)	5.12	5.13
β_2	*	0.01
$E_{pair}(n)$ (MeV)	-10.452	-10.057

the neutron Fermi level λ_n is located less than 1 MeV below the continuum which shows the proximity of this nucleus to the two-neutron dripline. In Table 3 we present the results of our 2-D HFB calculations in coordinate space with the results obtained by the THO method. Other observables (Fermi levels, rms-radius and deformation β_2) agree quite well. However, we find substantial differences in the energy gap values (Δ_n, Δ_p); these may be attributed to the different density of states used in the two methods.

Table 3. Comparison of calculations HFB+SLy4 for ^{102}Zr with two different methods in the axial symmetry. The configurational space calculations (THO) were made with 20 oscillator shells and pairing strength of -187.10 $MeV fm^3$. Calculations by the coordinate space HFB 2-D code were made using a box size $R = 12 fm$ with $N_r = 19$, maximum $\Omega = \frac{11}{2}$, V_0 -170 $MeV fm^3$ and the energy cutoff of 60 MeV.

	2-D (THO)	2-D (this work)
B. E. (MeV)	-859.40	-859.61
λ_n (MeV)	-5.42	-5.46
λ_p (MeV)	-12.10	-12.08
Δ_n (MeV)	0.56	0.31
Δ_p (MeV)	0.62	0.34
R_{rms} (fm)	4.58	4.58
β_2	0.429	0.431

References

1. *RIA Physics White Paper*, RIA 2000 Workshop, Raleigh-Durham, NC (July 2000), distributed by NSCL, Michigan State University.
2. J. Dobaczewski, H. Flocard and J. Treiner, *Nucl. Phys.* **A422**, 103 (1984).
3. J. Dobaczewski, W. Nazarewicz, T.R. Werner, J.F. Berger, C.R. Chinn and J. Dechargé, *Phys. Rev.* **C53**, 2809 (1996).
4. P. Ring and P. Schuck, *The Nuclear Many-Body Problem*, (Springer Verlag, New York, 1980).
5. A. Petrovici, K.W. Schmid, F. Grümmer, and A. Faessler, *Nucl. Phys.* **A517**, 108 (1990).
6. C.R. Chinn, J.-F. Berger, D. Gogny and M.S. Weiss, *Phys. Rev.* **C45**, 1700 (1992).
7. J.L. Egido and L.M. Robledo, *Phys. Rev. Lett.* **70**, 2876 (1993).
8. M.V. Stoitsov, J. Dobaczewski, P. Ring, and S. Pittel, *Phys. Rev. C* **61**, 034311 (2000).
9. J. Terasaki, P.-H. Heenen, H. Flocard, and P. Bonche, *Nucl. Phys.* **A600**, 371 (1996).
10. E. Teran, V.E. Oberacker, and A.S. Umar, (to be published in *Heavy Ion Physics*, 2002; nucl-th/0110059).
11. E. Teran, V.E. Oberacker and A.S. Umar, nucl-th/0205042, (Submitted to *Phys. Rev. C*).
12. A.S. Umar, J. Wu, M.R. Strayer, and C.Bottcher, *J. Comp. Phys.* **93**, 426 (1991).
13. D.R. Kegley, V.E. Oberacker, M.R. Strayer, A.S. Umar, and J.C. Wells, *J. Comp. Phys.* **128**, 197 (1996).

MASS TABLE MEAN–FIELD CALCULATIONS

M.V. STOITSOV[1-4], W. NAZAREWICZ[3-5], J. DOBACZEWSKI[5]

[1] *Institute of Nuclear Research and Nuclear Energy,*
Bulgarian Academy of Sciences, Sofia-1784, Bulgaria
[2] *Joint Institute for Heavy Ion Research, Oak Ridge, Tennessee 37831*
[3] *Physics Division, Oak Ridge National Laboratory, Oak Ridge, Tennessee 37831*
[4] *Department of Physics, University of Tennessee, Knoxville, Tennessee 37996*
[5] *Institute of Theoretical Physics, Warsaw University, Hoża 69, PL-00-681*
Warsaw, Poland

The mean–field methods are very successful in describing and predicting properties of nuclei across the chart of the nuclides. Results from recent large–scale Hartree-Fock-Bogoliubov calculations in configuration-space are presented for all even-even nuclei ranging from proton drip line to neutron drip line, with proton numbers $Z = 4, 6, 8, ..., 108$ using Skyrme forces and contact delta pairing interaction. Predictions of properties of exotic nuclei close to the particle drip lines are discussed.

1. Introduction

The development of experimental facilities that accelerate radioactive ion beams and the new detector technology that is accompanying them[1,2,3,4] has opened up a possibility to study the properties of nuclei very far from the valley of beta stability, all the way out to the particle drip lines.

A proper theoretical description of such weakly bound systems requires a careful treatment of the asymptotic part of the nucleonic density. An appropriate framework for is the Hartree-Fock-Bogoliubov (HFB) in the coordinate representation[5,6,7]. This method has been used extensively in the treatment of spherical systems but, is much more difficult to implement for systems with deformed equilibrium shapes[8,9].

In the absence of reliable coordinate-space solutions to the deformed HFB equations, it is useful to consider instead the configuration-space approach, whereby the HFB solution is expanded in a single-particle basis. There have been many configuration-space HFB calculations performed in a harmonic oscillator (HO) basis, either employing Skyrme forces or the Gogny effective interaction[10,11,12,13], or using a relativistic Lagrangian[14].

For nuclei at the drip lines, however, the HFB+HO expansion converges slowly as a function of the number of oscillator shells[7], producing wave functions that decrease too steeply at large distances.

An alternative approach that has recently been proposed is to expand the quasiparticle HFB wave functions in a complete set of transformed harmonic oscillator (THO) basis states[15,16,17], obtained by applying a local-scaling coordinate transformation (LST)[19] to the standard HO basis. Applications of this HFB+THO methodology have been reported both in the non-relativistic[15] and relativistic domains[17].

Recently, a new prescription for choosing the THO basis has been proposed[18]. For a given nucleus, the new prescription requires as input the results from a relatively simple HFB+HO calculation, with no variational optimization. The resulting THO basis leads to HFB+THO results that almost exactly reproduce the coordinate-space HFB results for spherical nuclei[6] and they are of comparable quality to available results for axially deformed nuclei[8].

In the present study, we report the results of HFB+THO calculations performed for all particle-bound even-even nuclei with $Z \leq 108$ and $N \leq 188$. The mass charts have been calculated with and without the Lipkin-Nogami prescription for an approximate particle number projection, followed by an exact particle number projection after the variation.

2. Hartree-Fock-Bogoliubov Theory

HFB is a variational theory that treats in a unified fashion mean–field and pairing correlations[20]. The HFB equations can be written in a matrix form as

$$\begin{pmatrix} h - \lambda & \Delta \\ -\Delta^* & -h^* + \lambda \end{pmatrix} \begin{pmatrix} U_n \\ V_n \end{pmatrix} = E_n \begin{pmatrix} U_n \\ V_n \end{pmatrix} , \tag{1}$$

where E_n are the quasiparticle energies, λ is the chemical potential, $h = t + \Gamma$, and Δ are the HF Hamiltonian and the pairing potential, respectively, and U_n and V_n are the upper and lower components of the quasiparticle wave functions.

In coordinate representation, the HFB approach consists of solving (1) as a set of integro-differential equations with respect to the amplitudes $U(E_n, \mathbf{r})$ and $V(E_n, \mathbf{r})$. The HFB continuum is discretized by putting the system in a large box with appropriate boundary conditions[7].

In the configurational approach, the HFB equations are solved by matrix diagonalization within a chosen set of single-particle basis wave functions

ψ_α with appropriate symmetry properties. The nuclear characteristics of interest are determined by the matrix elements of the density matrix and pairing tensor

$$\rho_{\alpha\beta} = \sum_{0 \le E_n \le E_{\max}} V_{\alpha n}^*(E_n) V_{\beta n}(E_n) \, ,$$

$$k_{\alpha\beta} = \sum_{0 \le E_n \le E_{\max}} V_{\alpha n}^*(E_n) U_{\beta n}(E_n) \, .$$

(2)

In configuration-space calculations, all quasiparticle states have discrete energies E_n.

3. Transformed Harmonic Oscillator Basis

Suppose $\{\varphi_\alpha(\mathbf{r})\}$ represents the complete set of HO single-particle wave functions depending on the spatial coordinate $\mathbf{r}$ and oscillator lengths $\{L_x, L_y, L_z\}$. One can introduce a LST of the three–dimensional vector space

$$\mathbf{r} \longrightarrow \mathbf{r}' \equiv \mathbf{r}'(\mathbf{r}) = \frac{\mathbf{r}}{R} f(R),$$

(3)

where R is the referent surface

$$R = \sqrt{\frac{x^2}{L_x^2} + \frac{y^2}{L_y^2} + \frac{z^2}{L_z^2}} \, .$$

(4)

The LST function $f(R)$ should have quite general mathematical properties ensuring that (3) is a valid invertible transformation of the three-dimensional space.

When one applies LST (3) to the HO set of wave functions, one obtains another set of THO single-particle wave functions

$$\psi_\alpha(\mathbf{r}) = \sqrt{\frac{f^2(R)}{R^2} \frac{\partial f(R)}{\partial R^2}} \, \varphi_\alpha \left(\frac{\mathbf{r}}{R} f(R) \right).$$

(5)

Due to the Jacobian of the LST entering Eq.(5), the THO wave functions are automatically orthonormalized. They have an asymptotic behavior

$$\psi_\alpha(r \to \infty) \sim \exp\left[-\frac{1}{2} f^2(R) \right],$$

(6)

which suggests that if the LST function satisfies the asymptotic conditions

$$f(R) = \begin{cases} R & \text{for small } R, \\ \sqrt{\kappa R} & \text{for large } R, \end{cases}$$

(7)

then the THO wave functions at small distances are identical to the HO wave functions, while at large distances they have the exponential asymptotic behavior.

In other words, the LST (3) generates, from a given complete set of HO wave functions, another orthonormal and complete set of THO wave functions (5) depending on an almost–arbitrary scalar LST function $f(R)$. The freedom in the choice of $f(R)$ provides great flexibility in the THO set $\{\psi_\alpha(\mathbf{r})\}$, and this opens up the possibility of improving on undesirable properties of the initial set. In particular, the use of the LST in THO can modify the incorrect Gaussian asymptotic properties of deformed HO wave functions.

4. Local–Scaling Transformation Function

The starting point of defining the LST function $f(R)$ is to carry out a standard HFB+HO calculation for the nucleus of interest, thereby generating an $\ell=0$ component

$$\bar{\rho}(r) = \int_0^{\pi/2} \bar{\rho}(r,\theta) \; P_{\ell=0}(\cos(\theta)) \; \sin(\theta) \; d\theta \qquad (8)$$

of the (generally deformed) HO local density $\bar{\rho}(r,\theta)$. Inspecting the density (8), one can conclude that its logarithmic derivative $\bar{\rho}'/\bar{\rho}$ exhibits a well-defined minimum near some point R_{min} in the asymptotic region. The comparison shows that the HFB+HO densities and their logarithmic derivatives are in almost perfect agreement with the coordinate-space HFB results up to R_{min} and, therefore, the HFB+HO densities are numerically reliable up to that point. The value of the density decay constant k emerging from HFB+HO calculations is also in agreement with the coordinate-space HFB results.

Beyond the point R_{min}, however, the logarithmic derivative $\bar{\rho}'/\bar{\rho}$ starts to oscillate around the coordinate-space HFB logarithmic density derivative which smoothly approaches the constant value k. As a result, the logarithmic derivative of the HFB+HO density is very close to the coordinate-space result around the midpoint $R_m = (R_{\max} - R_{\min})/2$, where $R_{\max}$ is the position of the first maximum of the logarithmic derivative for $r > R_{\min}$. Beyond the point R_m, the HFB+HO solution $\bar{\rho}(r)$ fails to capture the physics of the coordinate-space results, especially in the far asymptotic region. It is this incorrect large-r behavior that one tries to cure by introducing the THO basis.

To this end, making use of the WKB asymptotic solution of the single-particle Schrödinger equation and assuming that beyond the classical turning point only the state with the lowest decay constant k contributes to the local density, one can introduce the following approximate local density distribution

$$
\tilde{\rho}(r) = \begin{cases} \bar{\rho}(r) & \text{for } r \leq R_{\min} \\[2ex] A\, e^{-b\, r} \exp\left[-\frac{a}{r^s}\left(\frac{a\, r^3}{3-s} - \frac{2\, r^2 R_{\min}}{2-s} + \frac{r\, R_{\min}^2}{1-s}\right)\right] & \\ \qquad\qquad\qquad\qquad \text{for } R_{\min} \leq r \leq R_{\max} \\[2ex] B\, \dfrac{\exp\left[-2\int^r \sqrt{\kappa^2 + \frac{C}{r^2} + \frac{2m}{\hbar^2}\frac{Ze^2}{r}}\, dr\right]}{r^2\sqrt{\kappa^2 + \frac{C}{r^2} + \frac{2m}{\hbar^2}\frac{Ze^2}{r}}} & \text{for } r \geq R_{\max} \end{cases}
\tag{9}
$$

where $\bar{\rho}(r)$ is the HFB+HO density (8), the coefficients A and B are determined from the matching condition for the density at points $R_{\min}$ and $R_{\max}$, respectively, while the constants a and b, and the power s, are determined from the condition that the logarithmic derivative $\tilde{\rho}'/\tilde{\rho}$ and its first derivative are smooth functions at points $R_{\min}$ and $R_{\max}$. The value of C is fixed by the requirement that the logarithmic derivative of (9) coincides at the mid point R_m with the $\ell=0$ component of the HFB+HO density. The density $\tilde{\rho}(r)$ should also be normalized to the appropriate particle number.

Since Eq. (9) approximates HFB local densities fairly well for all nuclei, the next step is to define the LST function so that it transforms the HFB+HO density (8) into the density of Eq. (9). This requirement leads to the following first-order differential equation,

$$
\tilde{\rho}(r) = \frac{f^2(R)}{R^2}\frac{\partial f(R)}{\partial R}\bar{\rho}\left(\frac{r}{R}f(R)\right) ,
\tag{10}
$$

which, for the initial condition $f(0) = 0$, can always be solved for $f(R)$.

Once the LST function has been obtained, one needs simply to diagonalize the HFB matrices in the corresponding THO basis. Most importantly, no other information is required to construct the THO basis than the results of a standard HFB+HO calculation. As a consequence, one is able to systematically treat large sets of nuclei within a single calculation.

5. Numerical Example

In this section, we present the results of HFB+THO calculations performed for all the particle-bound even-even nuclei with $Z \leq 108$ and $N \leq 188$.

We have used the SLy4 Skyrme force parametrization[22] in the particle–hole channel and an intermediate (mixed) contact pairing force[22] in the pairing channel. For a given mass number A, calculations were carried out

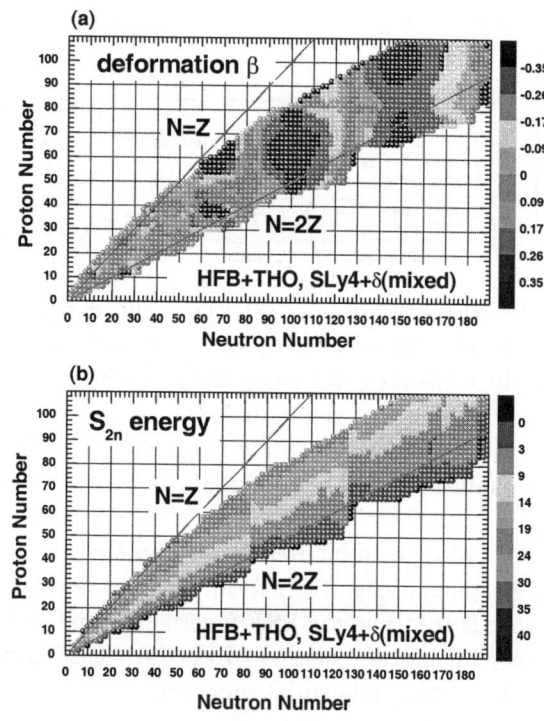

Figure 1.   Particle-bound even-even nuclei calculated within the HFB+THO method for the Skyrme SLy4 interaction and mixed contact pairing force within $N_{sh} = 20$ major shells: (a) quadrupole deformations β; (b)two-neutron separation energies S_{2n} (in MeV).

for increasing (decreasing) $N - Z$ up to the nucleus with positive neutron (proton) Fermi energy. Moreover, three independent sets of calculations were performed assuming initial wave functions to correspond to oblate, spherical, and prolate shapes. The lowest of the local minima that were found for a given nucleus was then identified with the ground-state solution. Mass charts have been calculated with and without the Lipkin-Nogami prescription for an approximate particle number projection, followed by an exact particle number projection after the variation.

The results for the ground-states of all even-even nuclei with negative Fermi energies, $\lambda_n < 0$ and $\lambda_p < 0$, are illustrated in Fig. 1. It is interesting to note from Fig. 1(a) that there are fairly large regions of nuclei far from stability with oblate shapes in their ground state. Nonetheless, it remains the case for nuclei far from stability, as for nuclei in or near the valley of stability, that there are more prolate ground states than oblate.

From Fig. 1(b), one can see that there exist numerous particle-bound even-even nuclei (i.e., nuclei with negative Fermi energies) that at the same time have negative two–neutron separation energies. Similar situation, but corresponding to negative two-proton separation energies, is predicted close to the proton–drip line. What this means is that even though these nuclei are bound against one–nucleon emission, they can nevertheless decay spontaneously by emitting two-nucleons. This is related to the fact that the HFB Fermi energies are associated with a given configuration or shape. Therefore, they tell us little about particle decays involving shape changes.

6. Concluding Remarks

In this paper, we report the application of an improved version of the configuration-space HFB method expanded in a Transformed Harmonic Oscillator basis. The method can be used reliably in systematic studies of wide ranges of nuclei, both spherical and axially deformed, extending all the way out to nucleon drip lines.

As an illustration, we carried out a systematic study of all even-even nuclei having $Z \leq 108$ and $N \leq 184$. We focused our discussion on the drip line systems, finding that in several regions of the periodic table there exist nuclei that are stable against one-particle emission but unstable against pair emission. In the description of very weakly bound systems, small changes of the effective interaction and the many-body treatment can have important consequences, determining, for example, the precise location of the drip lines. Thus, it is important to continue to improve the current HFB+THO methodology to accommodate effects not presently being included. Particularly important is the restoration of symmetries, either exact or approximate.

Acknowledgments

This work has been supported by the U.S. Department of Energy under Contract Nos. DE-FG02-96ER40963 (University of Tennessee), DE-AC05-00OR22725 with UT-Battelle, LLC (Oak Ridge National Laboratory), and DE-FG05-82ER40361 (Joint Institute for Heavy Ion Research), by the Polish Committee for Scientific Research (KBN) under Contract No. 5 P03B 014 21, and by the Interdisciplinary Centre for Mathematical and Computational Modeling (ICM) of the Warsaw University.

References

1. E. Roeckl, Rep. Prog. Phys. **55**, 1661 (1992).
2. A. Mueller and B. Sherril, Annu. Rev. Nucl. Part. Sci. **43**, 529 (1993).
3. P.-G. Hansen, Nucl. Phys. **A553**, 89c (1993).
4. J. Dobaczewski and W. Nazarewicz, Phil. Trans. R. Soc. Lond. A **356**, 2007 (1998).
5. A. Bulgac, Preprint FT-194-1980, Central Institute of Physics, Bucharest, 1980, nucl-th/9907088.
6. J. Dobaczewski, H. Flocard, and J. Treiner, Nucl. Phys. **A422**, 103 (1984).
7. J. Dobaczewski, W. Nazarewicz, T.R. Werner, J.-F. Berger, C.R. Chinn, and J. Decharge, Phys. Rev. **C53**, 2809 (1996).
8. J. Terasaki, H. Flocard, P.-H. Heenen, and P. Bonche, Nucl. Phys. **A621**, 706 (1997).
9. V.E. Oberacker and A.S. Umar, Proc. Int. Symp. on *Perspectives in Nuclear Physics*, (World Scientific, Singapore, 1999), Report nucl-th/9905010.
10. D. Gogny, Nucl. Phys. **A237**, 399 (1975).
11. M. Girod and B. Grammaticos, Phys. Rev. **C27**, 2317 (1983).
12. J.L. Egido, H.-J. Mang, and P. Ring, Nucl. Phys. **A334**, 1 (1980).
13. J.L. Egido, J. Lessing, V. Martin, and L.M. Robledo, Nucl. Phys. **A594**, 70 (1995).
14. P. Ring, Prog. Part. Nucl. Phys. **37**, 193 (1996).
15. M.V. Stoitsov, J. Dobaczewski, P. Ring, and S. Pittel, Phys. Rev. **C 61** (2000) 034311.
16. M.V. Stoitsov, P. Ring, D. Vretenar, and G.A. Lalazissis, Phys. Rev. **C58**, 2086 (1998).
17. M.V. Stoitsov, W. Nazarewicz, and S. Pittel, Phys. Rev. **C58**, 2092 (1998).
18. M.V. Stoitsov, J. Dobaczewski, W. Nazarewicz, and S. Pittel, Proceedings of the XXI International Workshop on Nuclear Theory, Heron Press, Sofia (2002), pp. 176-192.
19. I.Zh. Petkov and M.V. Stoitsov, Compt. Rend. Bulg. Acad. Sci. **34**, 1651 (1981); Theor. Math. Phys. **55**, 584 (1983); Sov. J. Nucl. Phys. **37**, 692 (1983); Ann. Phys. (NY) **184**, 121 (1988); *Nuclear Density Functional Theory*, Oxford Studies in Physics, (Clarendon Press, Oxford, 1991).
20. P. Ring and P. Schuck, *The Nuclear Many-Body Problem* (Springer-Verlag, Berlin, 1980).
21. F. Pérez-Bernal, I. Martel, J.M. Arias, and J. Gómez-Camacho, Phys. Rev. **A63**, 052111 (2001).
22. E. Chabanat, P. Bonche, P. Haensel, J. Meyer, and F. Schaeffer, Nucl. Phys. **A635**, 231 (1998).

NANO-SECOND ISOMERS IN NEUTRON-RICH NUCLEI AROUND ^{68}NI

T. ISHII, M. ASAI, M. MATSUDA, P. KLEINHEINZ AND HOU LONG*

*Advanced Science Research Center, Japan Atomic Energy Research Institute,
Tokai, Ibaraki 319-1195, Japan
E-mail: ishii@popsvr.tokai.jaeri.go.jp*

J. HORI

*Department of Fusion Engineering Research, JAERI,
Tokai, Ibaraki 319-1195, Japan*

A. MAKISHIMA

*Department of Liberal Arts and Sciences, National Defense Medical College,
Tokorozawa, Saitama 359-8513, Japan*

T. KOHNO AND M. OGAWA

*Department of Energy Sciences, Tokyo Institute of Technology,
Yokohama 226-8502, Japan*

K. OGAWA AND H. NAKADA

*Department of Physics, Faculty of Science, Chiba University,
Inage, Chiba 263-8522, Japan*

We have studied nuclear structure of neutron-rich nuclei around the doubly-magic ^{68}Ni produced in heavy-ion deep-inelastic collisions through nano-second isomer spectroscopy. New isomers in ^{68,70}Cu were identified by the relation between the mass number and the ΔE distribution of the fragments. The $B(M1)$ value between the $\pi p_{3/2}\nu p_{1/2}$ doublets in ^{68}Cu was also measured using the ^{68m}Cu source prepared by the (n,p) reaction.

---

*On leave from China Atomic Energy Institute, P. O. Box 275(67), Beijing 102413, China

1. Introduction

The ^{68}Ni nucleus, with $Z = 28$ and $N = 40$, has properties of the doubly magic nucleus, and thus, nuclear structure around ^{68}Ni provides good tests of the nuclear shell model. Furthermore, it is an important topic to investigate how the $Z = 28$ magic properties change in neutron-rich Ni region. We have been studying nuclear structure of this region produced in heavy-ion deep-inelastic collisions through γ-ray spectroscopy. Using an isomer-scope developed by ourselves,[1] we found new isomers with $T_{1/2} > 1$ ns and measured γ rays following the decay of the isomers, $e.g.$, the $(\nu g_{9/2}^2)8^+$ isomer in ^{68}Ni, the $(\pi p_{3/2} \nu g_{9/2}^2)19/2^-$ isomers in ^{69,71}Cu.[2,3,4,5,6] The isomers we have observed are summarized in Fig. 1. In this paper, we report the structure of the levels fed by new isomers in ^{68,70}Cu. The mass numbers of these isomers were identified by analyzing ΔE–E distributions measured with a high-resolution Si ΔE detector in the isomer-scope. We also report the lifetime measurement of the first excited state in ^{68}Cu using the ^{68m}Cu source, $T_{1/2} = 3.8$ min, produced by 14 MeV neutron beams.

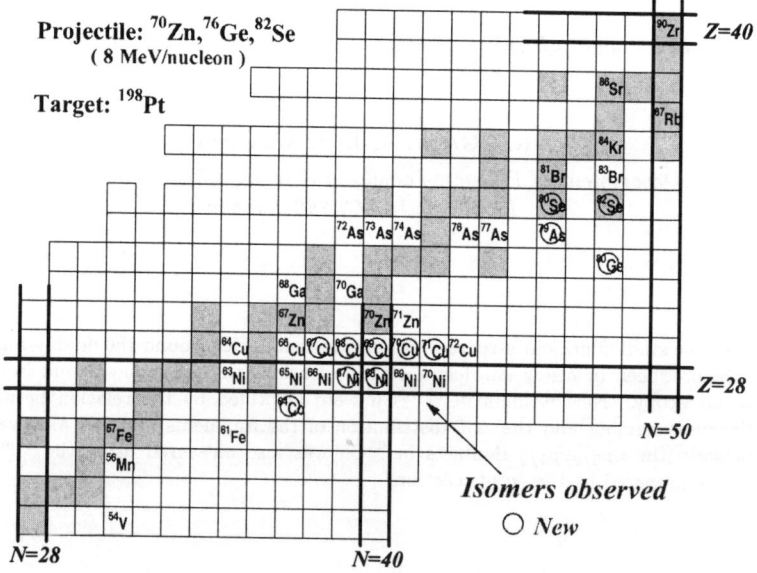

Figure 1. A nuclear chart of the neutron-rich Ni region. Stable isotopes are shown as dark squares. The nuclei depicted in the chart are those whose isomers were observed by the isomer-scope. The new isomers we have found are marked with circles.

2. Experiments

2.1. *Nano-second Isomer Spectroscopy through Deep-Inelastic Collisions*

Neutron-rich nuclei around ^{68}Ni were produced in heavy-ion deep-inelastic collisions, ^{70}Zn, ^{76}Ge or ^{82}Se beams of about 8 MeV/nucleon and a ^{198}Pt target of 4.3 mg/cm^2 thickness, at the JAERI tandem superconducting booster.[7] Gamma rays from isomers of projectile-like fragments were measured by the isomer-scope consisting of a tungsten block, Si ΔE–E detectors and Ge detectors. A schematic picture is shown in Fig. 2. The tungsten block shields the Ge detectors from prompt γ rays, and the Si detectors select projectile-like fragments by their atomic numbers. The Ge detectors measure the γ rays emitted by the stopped fragments in the Si E detector. Since the flight time between the target and the Si detector is about 1.5 ns, we can measure the isomer with a halflife longer than about 1 ns. The surface barrier ΔE detectors of 20 μm thickness were made by ourselves from selected Si wafers with good thickness uniformity. These ΔE detectors enable us to identify the mass number of a nucleus emitting an unknown γ ray.

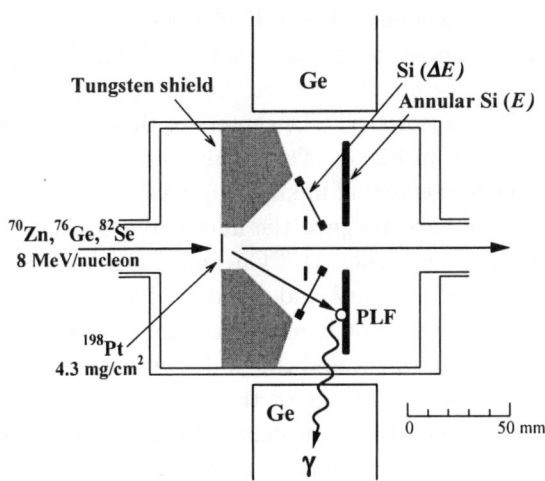

Figure 2.   Isomer-scope: a setup to measure γ rays from isomers with $T_{1/2} > 1$ ns of projectile-like fragments produced in deep-inelastic collisions.

2.2. *Decay Measurements of* ^{68m}Cu *Produced by Fast Neutrons*

Lifetimes of excited states in ^{68}Cu were measured by a standard γ-γ-t coincidence method using the ^{68m}Cu source with $T_{1/2} = 3.8$ min. The ^{68m}Cu source was prepared by the ^{68}Zn(n,p) reaction using 14 MeV neutron beams generated at FNS (Fusion Neutronics Source) in JAERI. The ^{68}Zn metal sample of 0.2 g was irradiated at a place of the 5×10^9 cm^{-2}s^{-1} neutron flux and was transported to a measurement room through a pneumatic tube periodically. Two BaF$_2$ detectors were employed for the lifetime measurement. The γ-ray intensities were also measured using a Ge detector to check the internal conversion coefficients derived from the γ-ray intensity balance.

3. Results

3.1. *Mass Assignment and New Isomers in* $^{68,70}Cu$

The energy deposited in the ΔE detector decreases almost linearly with the total energy of fragments at $E > 200$ MeV. Therefore, parallel lines corresponding to their atomic numbers appear on the ΔE–E scattered plot. Furthermore, we have observed these lines depend on the mass number. To clarify this mass dependence, the ΔE distribution was derived from the γ-ray intensities projected to the ΔE^* axis, where ΔE^* is defined as a perpendicular axis to those parallel lines on the ΔE–E plot. Figure 3 shows ΔE^* distributions for Cu isotopes. The centers of these curves are plotted against the mass number in Fig. 4. This figure permits us to identify the mass number of a nucleus emitting unknown γ rays.

Using the Fig. 4, we have assigned the nuclei emitting 693-84 keV and 231-1083-1405 keV γ-ray cascades as ^{68}Cu and ^{70}Cu, respectively. The isomer in ^{68}Cu locates at 777 keV and a halflife longer than 0.7 ns and shorter than 4 ns; the lower limit is estimated from the γ-ray yields and the flight time between the target and the Si detector, and the upper limit comes from the time resolution of the electronics. A weak transition of the 179 keV energy was also observed coincident with the 693 keV γ ray. There would be a high-lying state with a nano-second lifetime which feeds the 956 keV level. The isomer in ^{70}Cu locates at 2720 keV and the halflife of 14(2) ns. The decay schemes of the isomers in ^{68}Cu and ^{70}Cu are shown in Fig. 5 and Fig. 6, respectively.

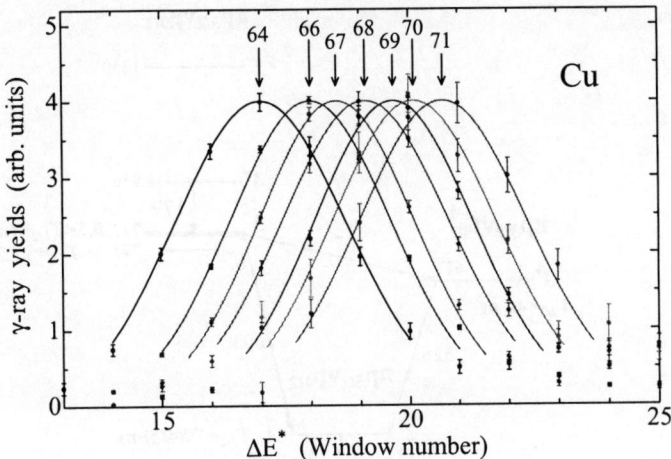

Figure 3. ΔE^* distributions for Cu isotopes. These curves were obtained from the γ-ray intensities; ^{64}Cu: 159 keV-γ, ^{66}Cu: 186 keV-γ, ^{67}Cu: 861 keV-γ, ^{68}Cu: 84 keV-γ, ^{69}Cu: 1871 keV-γ, ^{70}Cu: 231 keV-γ and ^{71}Cu: 1189 keV-γ. (See text on the definition of ΔE^*.)

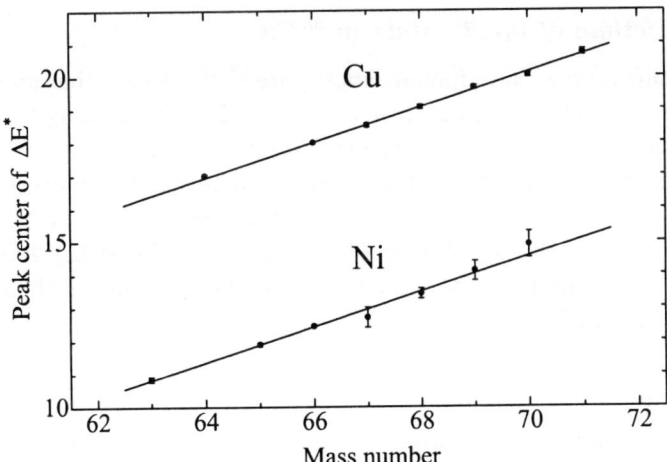

Figure 4. Relation between the mass number and the peak center of ΔE^*.

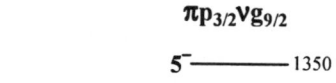

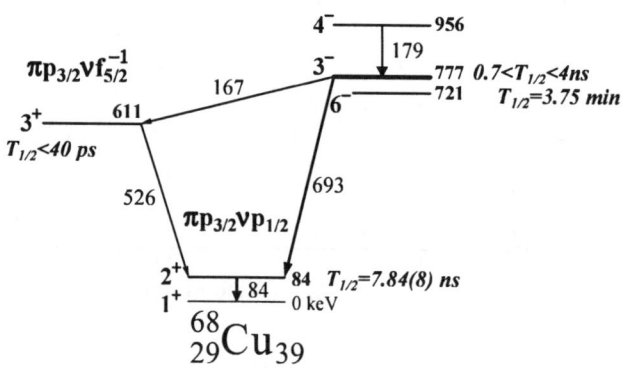

Figure 5.   Level structure of ^{68}Cu. Gamma rays decaying from the 3^- isomer are shown. The 6^- isomer produced in the (n,p) reaction decays to the 611 keV and 84 keV levels.

3.2. *Lifetime of the 2^+ state in ^{68}Cu*

The ^{68}Zn(n,p) reaction supplies almost pure ^{68m}Cu source, because another reaction channels, *e.g.*, $(n,2n)$, (n, α), only yield stable or long-lived nuclei. This fact allows us to use BaF$_2$ detectors to measure the γ rays. In the experiment, the purity of ^{68m}Cu source was monitored by measuring the γ rays by a Ge detector. Time spectra gated on some γ-ray combinations are shown in Fig. 7. All the obtained decay curves for the 84 keV level have a same slope within the errors. Therefore, we have determined the halflife of the 2^+ state as 7.84(8) ns.

4. Discussion

In $^{68}_{29}$Cu$_{39}$, low-lying levels have the $\pi p_{3/2}\nu p_{1/2}$, $\pi p_{3/2}\nu f_{5/2}^{-1}$ and $\pi p_{3/2}\nu g_{9/2}$ configurations from the viewpoint of the shell model. The ground and the first excited levels should be the $(\pi p_{3/2}\nu p_{1/2})1^+, 2^+$ states. Sherman *et al.*[8] identified the $\pi p_{3/2}\nu g_{9/2}$ multiplets to be 716, 772, 950 and 1350 keV levels

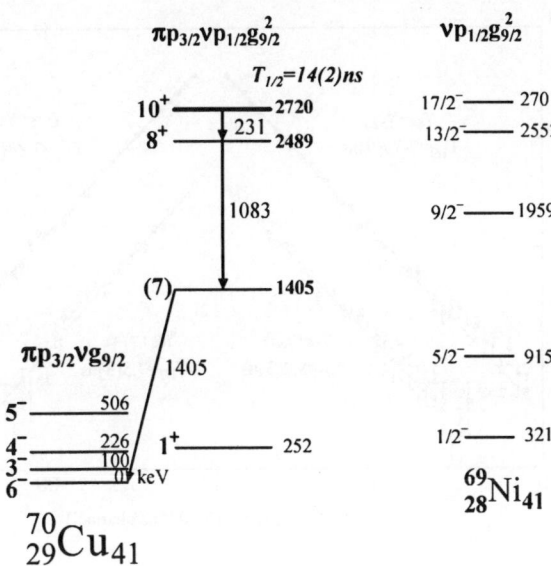

Figure 6. A decay scheme of the 10^+ isomer in ^{70}Cu. The $\pi p_{3/2}\nu p_{1/2}g_{9/2}^2$ levels in ^{70}Cu are compared to the $\nu p_{1/2}g_{9/2}^2$ levels in ^{69}Ni. The levels of the $\pi p_{3/2}\nu g_{9/2}$ multiplet and the 1^+ level at 252 keV are taken from the ref. 8 and 12, respectively.

by the $(t, ^3\text{He})$ transfer reaction. The 3.8 min 6^- isomer at 721 keV, the nano-second isomer at 777 keV and the 956 keV level are considered to be correspondent to the members of the $\pi p_{3/2}\nu g_{9/2}$ multiplet. The present result that the 777 keV level has a nano-second lifetime and decays to the 2^+ state indicates that this level has a spin of 3^- and decays through a retarded $E1$ transition. This spin assignment is consistent with our previous spin prediction of the $\pi p_{3/2}\nu g_{9/2}$ residual interactions.[4]

The 84 keV γ ray was found to be an almost pure $M1$ transition from the internal conversion coefficient. Thus, the $B(M1; 2^+ \to 1^+)$ value was obtained to be 0.00777(8) μ_N^2 or 1/230 W.u. from the measured lifetime of the 2^+ state. This value can be estimated using experimental g-factors of neighboring nuclei as,[9]

$$B(M1) = 3/(4\pi) \times (3/8) \times (g_\nu - g_\pi)^2 = 0.043\mu_N^2,$$

where g_ν and g_π are the g-factors of $\nu p_{1/2}$ in ^{67}Ni and $\pi p_{3/2}$ in ^{69}Cu,

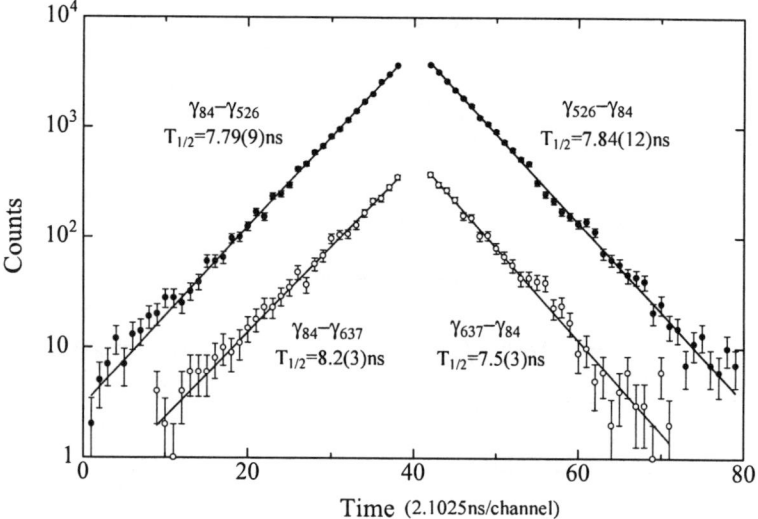

Figure 7. Decay curves for the 84 keV level in ^{68}Cu obtained from the γ-γ-t coincidence measurement with two BaF$_2$ detectors.

respectively.[10] This calculation reproduces a small $B(M1)$ value, but the experimental value is still smaller than this calculation.

To investigate the effect of the $f_{7/2}$ particle excitation on the $B(M1)$ value, we have compared the shell model calculation of the $(p_{3/2}f_{5/2}p_{1/2})^n$ model space with that of the $f_{7/2}^{-r}(p_{3/2}f_{5/2}p_{1/2})^{n+r}(r=0,1)$ model space. In the former calculation, the MSDI two-body interactions by Koops-Glaudemans were used.[11] In the latter calculation, two-body interactions derived by a G-matrix method by Kuo were used and single particle energies were adjusted to reproduce low-lying levels in ^{67}Ni, ^{68,69}Cu. The results are summarized in Table 1. The calculations show that the reduced matrix element of $\langle 2^+ \| M1 \| 1^+ \rangle$ in ^{68}Cu is sensitive to the $f_{7/2}$ particle excitation and decreases by including this excitation. This makes a contrast with the result that the g-factor of the 1^+ state is little affected by the $f_{7/2}$ excitation. Thus, the theoretical calculation including the $f_{7/2}$ orbital explains the experimental results of the small $B(M1; 2^+ \rightarrow 1^+)$ value and of the large $g(1^+)$ value.

Low-lying levels in ^{70}Cu had not been known well. Recently, Weissman et al.[12] measured the γ rays and g-factors of these states using a laser-ion source technique. They reported that the ground state is 6^- and that the

3^- and 1^+ states are β-decaying isomers. The isomer at 2720 keV observed in our experiment is considered as the $(\pi p_{3/2}\nu p_{1/2}g_{9/2}^2)10^+$ state, because this excitation energy is close to that of the $(\nu p_{1/2}g_{9/2}^2)17/2^-$ isomer in the neighboring nucleus ${}^{69}_{28}\text{Ni}_{41}$ shown in the Fig. 6.[13] Furthermore, the $B(E2;10^+ \to 8^+)$ value of 60(9) $e^2 fm^4$ is close to that of 63(3) $e^2 fm^4$ between the $(\pi p_{3/2}\nu g_{9/2}^2)19/2^-$ and $15/2^-$ states in ${}^{69}\text{Cu}$.[3] Similar level structure is also known in analogous nuclei around ${}^{90}\text{Zr}$. The ${}^{92}_{41}\text{Nb}_{51}$ nucleus has the $(\nu d_{5/2}\pi p_{1/2}\, g_{9/2}^2)11^-$ isomer at 2203 keV and ${}^{91}_{41}\text{Nb}_{50}$ has the $(\pi p_{1/2}g_{9/2}^2)17/2^-$ isomer at 2035 keV.[14]

Table 1. Shell model calculations of the $M1$ matrix elements in ${}^{67}\text{Ni}$, ${}^{69}\text{Cu}$ and ${}^{68}\text{Cu}$; Cal-1: $(p_{3/2}f_{5/2}p_{1/2})^n$ model space, Cal-2: $f_{7/2}^{-r}(p_{3/2}f_{5/2}p_{1/2})^{n+r}(r=0,1)$ model space.

Nucleus		Cal-1	Cal-2	Exp.
${}^{67}_{28}\text{Ni}_{39}$	$g(1/2^-)$	1.28	0.93	$1.20(1)^{10}$
${}^{69}_{29}\text{Cu}_{40}$	$g(3/2^-)$	2.53	2.08	$1.89(1)^{10}$
	$g(1^+)$	2.76	2.60	$2.48(2)(7)^{12}$
${}^{68}_{29}\text{Cu}_{39}$	$g(2^+)$	1.89	1.37	—
	$\langle 2^+\|M1\|1^+\rangle[\mu_N]$	1.04	0.59	0.197(2)

5. Conclusion

Nano-second isomers in neutron-rich nuclei around ${}^{68}\text{Ni}$ were produced in heavy-ion deep-inelastic collisions and were measured with the isomerscope. New isomers in ${}^{68,70}\text{Cu}$ were identified by the relation between the mass number and the ΔE distribution. We have found that the isomer in ${}^{68}\text{Cu}$ is the 3^- state of the $\pi p_{3/2}\nu g_{9/2}$ multiplet and that the isomer in ${}^{70}\text{Cu}$ is the $(\pi p_{3/2}\nu p_{1/2}g_{9/2}^2)10^+$ state. Lifetime of the $(\pi p_{3/2}\nu p_{1/2})2^+$ state in ${}^{68}\text{Cu}$ was measured by the decay of ${}^{68m}\text{Cu}$ source and obtained $B(M1;2^+ \to 1^+) = 0.00777(8)\mu_N^2$. A shell model calculation explains that the excitation of a $f_{7/2}$ particle causes the small $B(M1)$ value.

References

1. T. Ishii, M. Itoh, M. Ishii, A. Makishima, M. Ogawa, I. Hossain, T. Hayakawa and T. Kohno, *Nucl. Instrum. Methods Phys. Res.* **A395**, 210 (1997).
2. M. Asai, T. Ishii, A. Makishima, I. Hossain, M. Ogawa and S. Ichikawa, *Phys. Rev.* **C62**, 054313 (2000).

134

3. T. Ishii, M. Asai, A. Makishima, I. Hossain, M. Ogawa, J. Hasegawa, M. Matsuda and S. Ichikawa, *Phys. Rev. Lett.* **84**, 39 (2000).
4. T. Ishii, M. Asai, I. Hossain, P. Kleinheinz, M. Ogawa, A. Makishima, S. Ichikawa, M. Itoh, M. Ishii and J. Blomqvist, *Phys. Rev. Lett.* **81**, 4100 (1998).
5. A. Makisima, M. Asai, T. Ishii, I. Hossain, M. Ogawa, S. Ichikawa and M. Ishii, *Phys. Rev.* **C59**, R2331 (1999).
6. I. Hossain, T. Ishii, A. Makishima, M. Asai, S. Ichikawa, M. Itoh, M. Ishii, P. Kleinheinz and M. Ogawa, *Phys. Rev.* **C58**, 1318 (1998).
7. S. Takeuchi, T. Ishii, M. Matsuda, Y. Zhang and T. Yoshida, *Nucl. Instr. Meth.* **A382**, 153 (1996).
8. J. D. Sherman, E. R. Flynn, Ole Hansen, Nelson Stein, J. W. Sunier, *Phys. Lett.* **67B**, 275 (1977).
9. R. D. Lawson, *Theory of the Nuclear Shell Model*, (Clarendon press, Oxford, 1980) Chap. 5.
10. J. Rikovska *et al.*, *Phys. Rev. Lett.* **85**, 1392 (2000).
11. J. Koops and P. M. W. Glaudemans, *Z. Phys.* **A280**, 181 (1977).
12. L. Weissman *et al.*, *Phys. Rev.* **C65**, 024315 (2002).
13. R. Grzywacz *et al.*, *Phys. Rev. Lett.* **81**, 766 (1998).
14. R. B. Firestone and V. S. Shirley, *Table of Isotopes*, 8th. ed. (Wiley, New York, 1996).

NEUTRON-RICH NICKEL AND COBALT NUCLEI - NEW RESULTS FROM FRAGMENTATION STUDIES

R. GRZYWACZ [a,b], M. SAWICKA [b,c], M. PFÜTZNER [b], J.M. DAUGAS [d],
H. GRAWE [e], I. MATEA [c], F. BECKER [c,e], G. BELIER [d],
C.R. BINGHAM [f], R. BORCEA [g], E. BOUCHEZ [h], A. BUTA [i],
E. DRAGULESCU [i], G. GEORGIEV [c], J. GIOVINAZZO [j], F. IBRAHIM [g],
F. HAMMACHE [g], P. MAYET [e,k], V. MEOT [d], M. LEWITOWICZ [c], F. DE
OLIVIEIRA [c], F. NEGOITA [i], O. PERRU [g], O. ROIG [d],
K. RYKACZEWSKI [a,b], M.G. SAINT-LAURENT [c], J.E. SAUVESTRE [d],
O. SORLIN [g], M. STANOIU [c], I. STEFAN [i], C. STODEL [c], C. THEISEN [h],
D. VERNEY [c]

[a] ORNL, Physics Division, Oak Ridge, TN 37830, USA
[b] IFD, Warsaw University, Pl-00681 Warsaw, Hoża 69, Poland
[c] GANIL, BP 5027, 14021 Caen Cedex, France
[d] CE Bruyères-le-Châtel, DIF/DPTA/SPN, BP 12, F-91680
Bruyères-le-Châtel, France
[e] GSI, Postfach 110552, D-64220, Darmstadt, Germany
[f] University of Tennessee, Knoxville, TN 37996, USA
[g] IPN, 91406 Orsay Cedex, France
[h] CEA Saclay, DSM/DAPNIA/SPhN, F-91191 Gif-sur-Yvette Cedex, France
[i] IAP, Bucharest-Magurele P.O.Box MG6, Rumania
[j] CENBG, BP 120, F-33175 Gradignan Cedex, France
[k] IKS, University of Leuven, BG-3001 Leuven, Belgium

Beta-delayed and isomer gamma rays emitted by the very neutron-rich nuclei around ^{74}Ni have been measured using fragmentation of the ^{86}Kr beam at 58 AMeV and the new LISE 2000 spectrometer and EXOGAM germanium array. We have successfully measured beta-delayed gammas from the decay of ^{72}Co. The energies for the lowest excited states in ^{72}Ni are proposed, with a first 2^+ state at 1096 keV. These experimental findings can be related to the problem of non-observation of the 8^+ isomer in ^{72}Ni. We also measured beta decay from other neutron-rich Co isotopes including ^{70}Co. Evidence was found for a new short-lived isomer in ^{76}Ni, most likely the $I^\pi=8^+$ state.

1. Introduction

In recent years there were a variety of spectroscopic studies[1,2,3,4,5,6,7,8,9,10] seeking to develop the systematics of neutron-rich nuclei near the magic proton number Z=28. The main objective of these studies is to measure the experimental observables such as nuclear level energies, spins, and parities and compare them to current nuclear shell model calculations. There is a longstanding question of sustaining magicity for the most neutron-rich nuclei[11] and a dispute[12] of its microscopic origin. The magic very neutron-rich nickel nuclei are important from that point of view but difficult to synthesize in known nuclear reactions, and thus, the observation of their ground- and excited-state properties is an experimentally challenging problem. The isotopes of interest have been produced in multi-nucleon transfer[1], fragmentation[3], and fission[8] reactions. These studies yielded a series of surprising facts which challenge shell model calculations. In this region of nuclei the main difficulty for large-scale shell model calculations stems from the presence of the $\nu g_{9/2}$ orbital and its possible polarizing effect on the proton Z=28 shell, thus making questionable the choice of ^{56}Ni as a core. Calculations using a bigger model space with a ^{48}Ca core have been attempted recently[7]. However, the large model space may not be able to account for the changes in nuclear potential in the neutron-rich nucleonic matter, and therefore, different approaches have to be undertaken. On the other hand, the large monopole shift observed in the proton(π) - neutron (ν) multiplet[13] $\pi f_{5/2} \nu g_{9/2}$ may challenge the N=50 closure in ^{78}Ni.

1.1. *The problem of non-observed 8^+ isomers*

High spin isomers resulting from the coupling of two identical nucleons in the same orbital j to a maximum allowed spin $J = 2j - 1$ are known across the Segré chart in nuclei near closed shells. They can be related to the diagonal elements of the two-body interaction which are approximated by a delta-type interaction[14], because of the short range of nucleon-nucleon interactions.

The experiment that discovered the "isomeric island"[3] near Z=28 and N=40 resulted in the first observation of excited states in ^{70}Ni populated in the decay of the $T_{1/2}$=230(3) ns I^{π}=8^+ isomer[3,15,16]. It decays via a cascade of four stretched E2 transitions connecting the 8^+, 6^+, 4^+, 2^+ and 0^+ levels.

This isomer is interpreted by the shell model to be a $(\nu g_{9/2})^2_{J=8^+}$ excitation. The yrast 8^+ level is long lived because of the non-collective E2 transition (B(E2)=0.693(9) W.u.) and the small energy difference between the 8^+ and 6^+ levels.

Various interactions predict the isomerism of the 8^+ level in ^{70}Ni, though varying in detailed prediction of lifetimes or level energies. These 8^+ isomers are also expected to exist in the more neutron-rich nickel nuclei ^{72}Ni, ^{74}Ni and ^{76}Ni. However, it has been experimentally observed that the lifetimes of the ^{72m}Ni and ^{74m}Ni isomers places them outside the observation limits of the measurement. The number of detected ions of ^{72}Ni and ^{74}Ni allowed us to set the half-life limits for non-observation of these isomers to be $T_{1/2} < 20$ ns or $T_{1/2} > 2.5$ ms for ^{72}Ni and $T_{1/2} < 60$ ns or $T_{1/2} > 0.2$ ms for ^{74}Ni[17,18]. There was not enough experimental data to determine similar limits on on ^{76m}Ni. The shell model, which predicts isomeric lifetimes of the order of tens of microseconds[19], is in contradiction with these experimental observations. It can be concluded that there is structural feature that has not been considered which could cause the disappearance of isomerism for these mid-shell nuclei. That could happen if the strong neutron $g_{9/2}$ polarizing effect would result in breaking the Z=28 shell leading to deformation and collectivity. Such effects would possibly be accompanied by the disappearance of the N=50 closed shell, which would also result in the absence of the $\nu(g_{9/2})^{-2}_{J=8^+}$ isomer in ^{76}Ni. However, the recent observation[4] of an 8^+ isomer in ^{78}Zn suggests that its Z=28 isotone ^{76}Ni will have an isomer. It is important to confirm it experimentally. It became clear that the isomer spectroscopy in fragmentation is not a viable method to measure excited levels in ^{72}Ni and ^{74}Ni and one has to attempt to measure them with a different method. An alternative approach is beta-delayed gamma-ray spectroscopy of ^{72}Co and ^{74}Co. This task is experimentally difficult because these cobalt isotopes are even farther away from stability, and thus, more difficult to produce than their nickel isobars. The observation of the states populated in ^{72}Co decay would shed light on the prevalent type of ^{72}Ni excitations. In particular the observation of rotational type structures would suggest the onset of collectivity and provide a simple explanation for the non-existence of yrast isomers, and more importantly would be evidence for a dramatic breaking of the Z=28 shell closure. On the other hand, observation of shell-model type excitations would suggest that more subtle effects within the shell model approach should be expected.

2. Experimental technique

The experiment was performed at the GANIL laboratory using fragmentation of a 58 AMeV ^{86}Kr q=+36 beam impinging on a ^{nat}Ta target of 30 μm thickness with 125 μm carbon backing. The data were collected for about 100 hours and the average beam intensity was 50 pnA. The selection of ions was done with the new LISE2000[20] spectrometer. The standard time-of-flight and energy-loss event-by-event ion identification technique was used[21]. The A/q selection was chosen to transmit the ^{76}Ni and ^{72}Co ions with the maximum efficiency. A stack of four silicon detectors was placed at the final focus of LISE2000 with thicknesses of 300 μm, 300 μm, 1 mm, and 3.5 mm with respect to the beam direction. The ions of interest were stopped in the 1-mm detector a double-sided silicon strip detector (DSSD) with 16x16 strips of 2-mm pitch. The granularity provided a spatial correlation between the implanted ion and detected beta decay. The silicon telescope was surrounded by four clover-type EXOGAM germanium detectors. The beta detection efficiency of the DSSD, amounting to about 20%, was determined from analysis of the ^{66}Co decay, using known[22] branching ratios. The gamma detection efficiency measured with calibrated sources was 6 % at 1.3 MeV and 23 % at the maximum around 120 keV. The experiment was also designed to be sensitive to the decay of short-lived isomers using the standard technique[23].

3. Results

3.1. β^- decay of ^{70}Co

About 3300 ions of ^{70}Co were implanted into the DSSD detector. The beta-delayed gamma-ray spectrum (fig. 2) reveals gamma-ray lines observed previously in a fragmentation experiment[3] and a beta decay experiment[22]. This validates the correct ion identification and ion-decay correlation analysis technique. The half-life for the correlated beta activity was determined to be 110(9) ms, which compares with the half-life observed in Ref.[22] for the 7$^-$ state $T_{1/2}$=120(30)ms. In addition to the strong 448, 683, 970 and 1260 keV lines observed in beta decay studies by Mueller et al[22], we also identified an additional line at 916 keV. We tentatively place it at 3146 keV and feeds the 4$^+$ state at 2230 keV (fig. 1). We were also able to identify the decay of the 3$^+$ state which appears to be populated weakly in this reaction.

3.2. β^- decay of ^{72}Co

A similar data analysis was performed for the 3290 ions of ^{72}Co which were identified and implanted. The beta-delayed gamma-ray spectrum is shown in fig. 2 revealing gamma transitions of 454, 845, 1096 and 1197 keV. Based on the intensity of gamma transitions and expected similari-

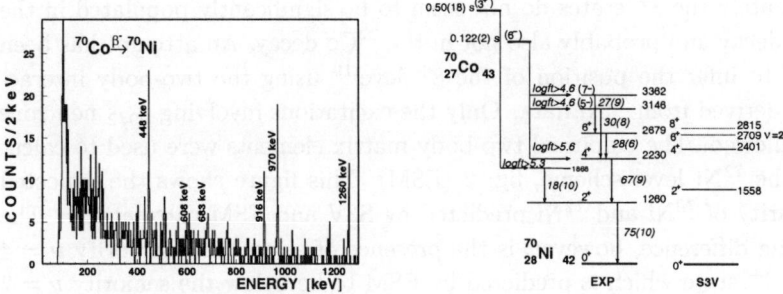

Figure 1. Beta-delayed gamma-ray spectrum of the ^{70}Co decay (left). The postulated level schemes of ^{70}Ni (right). Compare the energy levels in ^{70}Ni with shell model calculations using S3V and full fpg model space.

ties between ^{70}Co and ^{72}Co beta decays, a level scheme for ^{72}Ni has been constructed. The decay has been attributed to originate from the $(6^-,7^-)$ state. The half-life of this decay was determined to be 62(3) ms. The previously measured[9] value was 90(20) ms. We could not find conclusive evidence for the decay of the expected[22] 3^+ state.

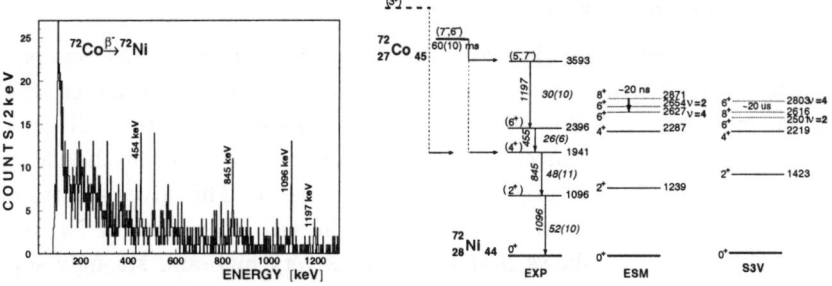

Figure 2. Beta-delayed gamma-ray spectrum of the ^{72}Co decay (left). The postulated level schemes of ^{72}Ni (right). Compare the energy levels in ^{72}Ni with shell model calculations using S3V and full fpg model space and empirical ESM residual interactions restricted to the $g_{9/2}$ shell.

We have attributed the strongest 1096 keV line to be a transition between the first excited 2^+ state in ^{72}Ni and its 0^+ ground state. The observed level scheme (fig. 2) does not reveal signatures of collective excitations, but it rather resembles the excitation pattern predicted by the shell model calculation.

The important question arises whether these experimental findings can help to resolve the problem of the disappearing 8^+ isomer in ^{72,74}Ni. Unfortunately the 8^+ states do not seem to be significantly populated in the ^{70}Co decay and probably also not in the ^{72}Co decay. An attempt has been made to infer the position of the 8^+ level[19] using the two-body interactions derived from ^{70}Ni data. Only the excitations involving $g_{9/2}$ neutrons was allowed. The empirical two-body matrix elements were used to calculate the ^{72}Ni level scheme, fig. 2 (ESM). This figure shows the expected similarity of ^{70}Ni and ^{72}Ni predicted by S3V and ESM calculations. The striking difference, however, is the presence of a low lying seniority $\nu = 4$ $J^\pi = 6^+$ state which is predicted by ESM to be below the seniority $\nu = 2$ $J^\pi = 8^+$ state, opening another deexcitation branch for the isomer with a larger B(E2)=2.7 W.u. The lifetime predicted by these calculations is about 10-20 ns, which is just at the limits of the previous experiment sensitivity. This level is placed well above the $J^\pi = 8^+$ state by the S3V calculations. Microscopically, the lowering of the $\nu = 4$ $J^\pi = 6^+$ state is related to the low $J^\pi = 2^+$ excitation energies in ^{70,72}Ni. Therefore, the present experimental results support the above mentioned ESM scenario to explain the absence of 8^+ isomerism.

3.3. Isomers in ^{76}Ni ?

In this experiment a simultaneous search for the isomer in ^{76}Ni was performed. The ^{76}Ni ions were transmitted through the spectrometer with similar efficiency as ^{72}Co. We have identified about 280 of the ^{76}Ni ions. These ions have been correlated with 44 gammas detected in the clover detectors within 1 μs after implantation, see fig. 3. This gives a ratio of F=0.16(4) gammas per ion. The comparison with other ions of Ni (e.g. F=0.021(5) for ^{75}Ni) shows that this ratio is very high and strongly suggests the presence of an isomer. The ratio of F=0.2 has been obtained for q=+27 ions of ^{70}Ni with a known 8^+ isomer. There is also evidence for a gamma transition at 930 keV which would be a good candidate for the $2^+ -> 0^+$ transition[18].

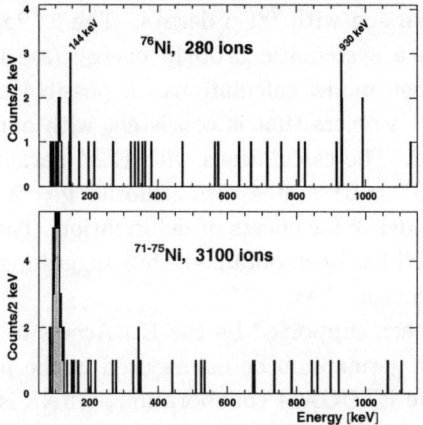

Figure 3. The gamma-ray spectrum observed within $1\mu s$ after ion implantation collected for ^{76}Ni ions (upper panel), and $^{71-75}$Ni (lower panel).

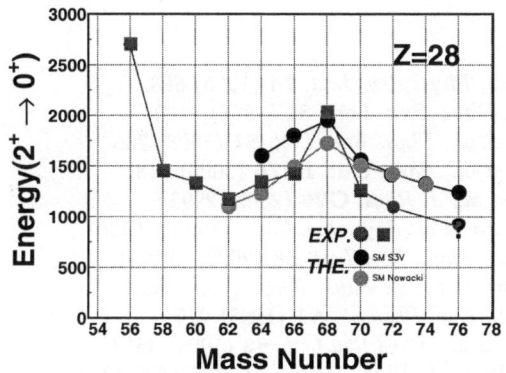

Figure 4. Systematic trend of the first excited 2^+ levels. Experimental points (dark grey squares and circles) are compared with full fpg shell model calculations using a realistic interaction with different sets of two-body matrix elements: S3V[19] (black circles) and Nowacki[7] (light grey circles). The experimental points for ^{72}Ni and ^{76}Ni originate from this study.

4. Summary

In conclusion, in an experiment performed using fragmentation of a ^{86}Kr beam, the spectroscopy of excited levels in ^{70,72}Ni populated in the decays of ^{70,72}Co isotopes was performed. For the first time the excited levels in ^{72}Ni have been identified and the spin and parities have been postulated, based on the comparison with ^{70}Co decays. The 2^+ level energy of 1096 keV in ^{72}Ni shows a systematic drop in energy (see fig. 4) that is not explained by the shell model calculations. A possible explanation for the disappearance of 8^+ isomers that is consistent with our the experimental findings is proposed. The calculations with ESM parameters predict near degeneracy of the $J^\pi = 6^+$ states with seniority $\nu = 2$ and $\nu = 4$. There is no supporting evidence for effects of deformation. The first evidence for the isomerism in ^{76}Ni has been obtained, thus suggesting the persistence of the N=50 shell closure in ^{78}Ni.

This work has been supported by the EU Access to Large Scale Facilities Program. The germanium detectors used in the present experiment were founded by the EXOGAM collaboration. ORNL is managed by UT-Battelle, LLC, for the U.S. Department of Energy under Contract DE-AC05-00OR22725. Nuclear Physics research at the University of Tennessee is supported by the US Department if Energy under contract DE-F602-96ER40983.

References

1. R. Broda *et al.*, *Phys. Rev. Lett.* **74** (1995) 868.
2. T. Ishii *et al.*, Phys. Rev. Lett. **81** (1998) 4100.
3. R. Grzywacz *et al.*, *Phys. Rev. Lett.* **81** (1998) 766.
4. J.M. Daugas *et al.*, *Phys. Lett.* **B476** (2000) 213.
5. G. Georgiev *et al.*, *J. Phys.* **G28** (2002) 2993
6. J.I. Prisciandaro *et al.*,*Phys.Rev.* **C60**, (1999) 054307 .
7. O. Sorlin *et al.*, *Phys. Rev. Lett.* **88** (2002) 092501.
8. Ch. Engelmann *et al.*, *Z. Phys.* **A352** (1995) 351.
9. F. Ameil *et al.*, *Eur. Phys. J.* **A1** (1998) 275.
10. W.F. Mueller *et al.*, *Phys.Rev.Lett.* **83** (1999) 3613.
11. J. Dobaczewski *et al.*, Phys. Rev. Lett. **72**(1994) 72.
12. B.A. Brown *Nucl. Phys.* **A 704** (2002) 11c.
13. S. Franchoo *et al.*, *Phys.Rev.* **C64** (2001) 054308
14. N. Anantaraman and J.P. Schiffer, *Phys. Lett.* **37B** (1971) 229.
15. J.-M. Daugas *et al.*, PhD Thesis, Caen University (1999).
16. M. Lewitowicz *et al.*, *Nucl. Phys.* **A 682** (2001) 175c.
17. R. Grzywacz *et al.*, Second International Conference on Fission and Neutron-rich Nuclei, St. Andrews, Scotland 28th June - 2nd July, 1999, p. 38.
18. M. Sawicka *et al.*, to be published
19. H. Grawe *et al.*, *Nucl.Phys.* **A704** (2002) 211c
20. R. Anne, Preprint GANIL, P 02 01, 2001
21. D. Bazin *et al.*, *Nucl. Phys.* **A 515** (1990) 349
22. W.F. Mueller *et al.*, *Phys.Rev.* **C 61** (2000) 054308.
23. R. Grzywacz *et al.*, *Phys. Lett.* **B355**(1995) 439.

FISSION FRAGMENT ORIENTATION AND γ-RAY EMISSION ANISOTROPY

P. ADRICH[a], YU.N. KOPATCH[b], E. LUBKIEWICZ[a], H.J. WOLLERSHEIM

GSI-Gesellschaft für Schwerionenforschung mbH.
Planckstrasse 1,
D-64291 Darmstadt, Germany
E-mail:h.j.wollersheim@gsi.de

M. MUTTERER

Institut für Kernphysik, Technische Universität.
D-64289 Darmstadt, Germany

Angular correlations of prompt γ-rays in binary spontaneous fission of ^{252}Cf were measured with a GSI super clover detector. The position-sensitive γ-ray spectrometer was combined with an efficient detection system for fission fragments. For the study of γ-ray angular correlations the intensities of individual γ-transitions were measured relative to the fission axis. Surprisingly, the measured γ-ray angular correlations for stretched E2 transitions show a large anisotropy, which can be described by a complete alignment of the initial fragment spins.

1. Introduction

It is well known that fragment spins are rather high ($<J>\sim7$-$8\hbar$) even for the spontaneous fission of a zero spin nucleus like ^{252}Cf. The fragment spins are aligned in the plane perpendicular to the fission axis causing an anisotropy of γ-rays emitted from the fragments. The most probable reasons for the fragment spins and their alignment are the excitation of collective vibrational modes like bending or wriggling at the saddle - to - scission stage and subsequent Coulomb excitation. Both effects lead to a K-distribution in the fragment spin after scission. Although vibrational modes are being discussed for a long time [1,2] their possible influence on the formation of mass and energy distributions of the fragments is not really taken into account (see, e.g., [3,4]). Similarly, vibrational modes are ignored in the A. Bohr model [5] for fission fragment angular distribution since they violate the axial symmetry of fissioning nucleus at the descent stage and thus can perturb the distribution of the K quantum number formed at the saddle point.

143

Recently, experimental data on angular distributions of prompt γ-rays emitted from spontaneous fission of ^{252}Cf have been published by Pilz and Neubert [6] and Kopach et al. [7]. In both experiments NaI crystals were used as γ-detectors. Since the γ-ray energy resolution was limited to ~90keV, the γ-ray angular correlation was analyzed for different energy ranges and not for individual γ-ray transitions. For binary and ternary fission, the largest anisotropy ratio A=0.25, defined as $A = W(0^0)/W(90^0) - 1$, was observed for γ-energies between 0.5 and 0.7MeV. Since in both experiments different multipole transitions were not discriminated from each other, no conclusion can be drawn for the K-distribution of the fragment spins. Therefore, the interpretation of γ-ray angular correlations are based on high-resolution spectroscopy in the present paper, in which Yrast transitions from specific even-even fragments were measured.

2. Fission Fragment γ-Ray Angular Correlation Measurement

In the experiment, a VEGA Ge-detector was combined with an efficient 4π detector system for the fission fragments. The clover detector consists of 4 segmented Ge-crystals, which are 14cm long and especially suited for the detection of high-energetic γ-rays. The high segmentation allowed us to place the Ge detector as close as 9cm to the fission source, still obtaining an energy resolution of 1% after Doppler shift correction. The efficiency of the VEGA detector is 2.5%. For the detection of the fission fragment pairs, which are emitted from a ^{252}Cf source mounted on a thin backing, a double ionization chamber [8,9] was used. The kinetic energies of both fragments, which are stopped in the methane gas, are measured in both halves of the chamber from which fragment masses can be deduced. The drift time of the ionization electrons towards the anodes is used to determine the polar angle of the fragment directions, when the track lengths of the fragments is known as a function of mass and energy. For determining the azimuthal angles of the fission axes, signals from the four sectors were used, in which each side of the cathode plate was subdivided. Figure 1 displays two typical γ-ray spectra after Doppler shift correction, obtained for selected heavy-fragment masses. The γ-ray lines from the $4^+ \rightarrow 2^+$ transitions in ^{138}Xe and ^{144,146}Ba are marked in the figure. The $4^+ \rightarrow 2^+$ transitions in ^{144}Ba (331keV) and in ^{146}Ba (332keV) could not be resolved, but were analyzed as a common transition. Due to the large solid angles of the fragment detector $\left(0^0 \leq \vartheta_p \leq 80^0\right)$ and Ge-detector

$\left(\mathcal{9}_{\gamma} \leq 34^{0}\right)$ the fission fragment γ-ray angular correlation could be investigated between 0^{0} and 110^{0}.

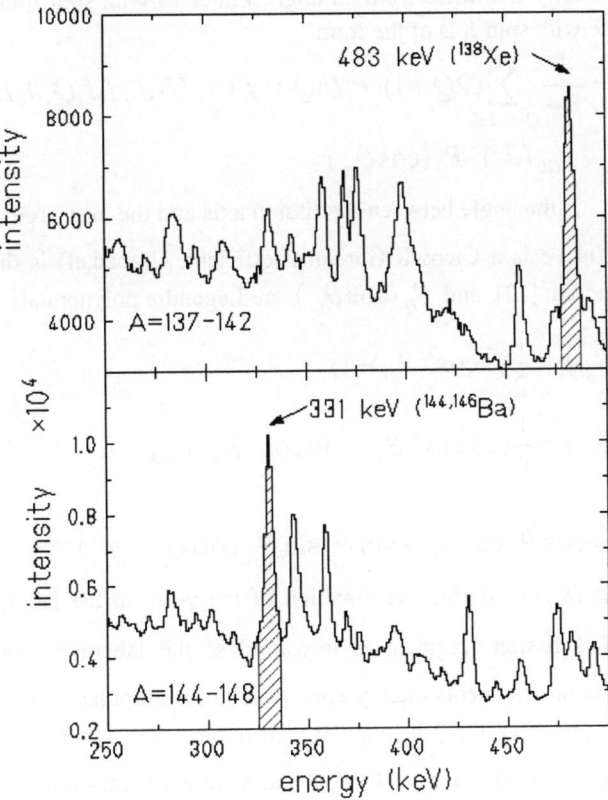

Figure 1. Doppler-shift corrected γ-ray spectra for the two selected fragment mass windows indicated. The γ-ray lines associated with the $4^{+} \rightarrow 2^{+}$ transitions in ^{138}Xe and ^{144,146}Ba fission fragments are marked.

3. γ-Ray Emission from Aligned Nuclei

According to the standard formalism [10,11] the angular distribution of γ-rays of multipolarity L emitted from an aligned nucleus with spin J_i in its transition to the state with spin J_f is of the form

$$\frac{dW}{d\Omega_{rest}} = \frac{1}{4\pi} \sum_{Q=0,2,K} (2Q+1) \cdot < L1Q0 \mid L1 > \cdot U(J_f LJ_i Q, J_i L)$$

$$\cdot \tau_{Q0}(J_i) \cdot P_Q(\cos\vartheta_{\gamma p}) \tag{1}$$

where $\vartheta_{\gamma p}$ is the angle between the fission axis and the measured γ-ray (see Eq. 4), $<L1Q0|L1>$ is a Clebsch Gordan coefficient, U(abcd,ef) is the normalized Racah function [12], and $P_Q(\cos\vartheta_{\gamma p})$ are Legendre polynomials

$$P_2(\cos\vartheta_{\gamma p}) = \frac{1}{2}(3\cos^2\vartheta_{\gamma p} - 1) \tag{2}$$

$$P_4(\cos\vartheta_{\gamma p}) = \frac{1}{8}(35\cos^4\vartheta_{\gamma p} - 30\cos^2\vartheta_{\gamma p} + 3) \tag{3}$$

with

$$\cos\vartheta_{\gamma p} = \cos\vartheta_\gamma \cos\vartheta_p + \sin\vartheta_\gamma \sin\vartheta_p \cos(\varphi_\gamma - \varphi_p) \tag{4}$$

The angles $\vartheta_\gamma, \varphi_\gamma$ define the direction of the γ-ray in the laboratory system. The excited fission fragment is measured at the laboratory angles $\vartheta_p \varphi_p$. Excited fission fragments mainly emit γ-rays of multipolarity E1, M1, and E2, so $L \le 2$ and, therefore, the angular distribution $dW/d\Omega_{rest}$ consists only of the terms corresponding to Q=0, 2 and 4. In each transition $J_i \to J_f$, the spin-tensors of nuclear orientation for second and fourth ranks have the explicit expressions

$$\tau_{20}(J) = \left[\frac{J(J+1)}{(2J-1)(2J+3)}\right]^{1/2} \left[3\frac{<K^2>}{J(J+1)} - 1\right] \tag{5}$$

$$\tau_{40}(J) = \left[\frac{J^3(J+1)^3}{(2J-3)(2J-2)(2J-1)(2J+3)(2J+4)(2J+5)}\right]^{1/2}$$

$$\cdot\left[35\frac{<K^4>}{J^2(J+1)^2} - 30\frac{<K^2>}{J(J+1)}\left(1 - \frac{5}{6J(J+1)}\right) + 3\left(1 - \frac{2}{J(J+1)}\right)\right] \tag{6}$$

where

$$< K^n > = \sum_K K^n \gamma_K \qquad (7)$$

To study the sensitivity of fission fragment γ-ray angular correlation to the spin alignment we take the K-distribution in the Gaussian form

$$\gamma_K = \exp\left(-\frac{K^2}{2\sigma^2}\right) \Big/ \sum_{K'} \exp\left(-\frac{K'^2}{2\sigma^2}\right) \qquad (8)$$

Figure 2 shows the sensitivity of a calculated $4^+ \to 2^+$ γ-ray angular correlation for different variances of the Gaussian distribution. A variance of σ=0. corresponds to complete alignment of the fragment spin.

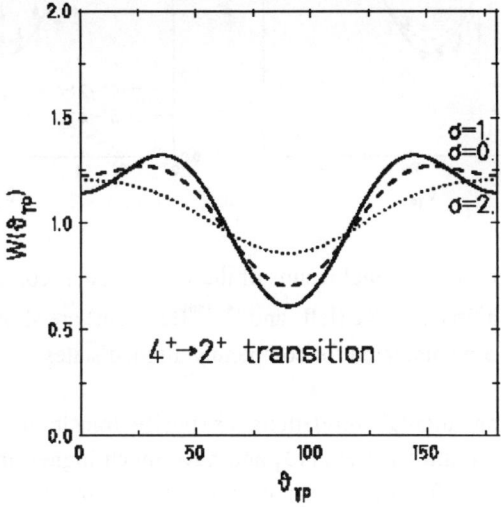

Figure 2. The calculated γ-ray angular correlation in the transition $4^+ \to 2^+$ corresponds to a variance of the Gaussian distribution σ=0.0 (solid line), 1.0 (dashed line) and 2.0 (dotted line). The calculations are performed in the rest system of the excited nucleus.

4. Fission Fragment γ-Ray Angular Correlations

The experimental results of the γ-ray angular correlations of the $4^+ \to 2^+$ transitions in ^{138}Xe and ^{144,146}Ba are plotted in Figure 3. The data are transferred in the rest frame of the excited fragment by correcting for the solid angle aberrations and the Doppler shifts in γ-ray energies. The lines are calculated for completely aligned states (σ=0). Since the population pattern of the Yrast band is not known experimentally, it was assumed that the 8^+ state collects all the

148

feeding intensity. From the almost identical $4^+ \rightarrow 2^+$ and $6^+ \rightarrow 4^+$ angular correlations one can conclude, that the determination of the anisotropy is only slightly influenced by the feeding effect.

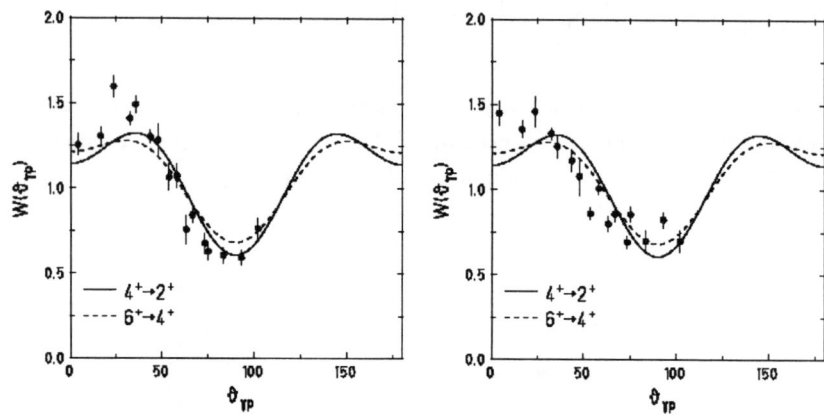

Figure 3. The experimental results of the γ-ray angular correlations of the $4^+ \rightarrow 2^+$ transitions in ^{138}Xe (left) and ^{144,146}Ba (right) are shown. The solid and dashed lines are calculated for completely aligned states.

The measured γ-ray angular correlations for specific transitions are in agreement with those of Wolf and Cheifetz [13] and have much higher anisotropy (A~1.) than those observed for the gross unresolved γ-ray spectrum [6,7]. From the comparison with calculated γ-ray angular correlations one observes direct evidence that the angular momenta of the primary fragments seem to be completely aligned perpendicular to the fission axis, since the evaporation of one or two neutrons can only cause de-alignment.

Since light charged particles were also measured in a recent experiment [14], γ-ray angular correlations of fragments from binary and ternary fission will be compared in the future. In most cases α particles are emitted in ternary fission, which may correlate with high alignment of the fission fragments. Therefore, results for discrete γ-ray transitions may shine new light on the excitation of collective modes during the saddle – to - scission stage.

Appendix

In this chapter the most important formulas are summarized for the γ-ray angular correlations of stretched E2 transitions in even-even nuclei. The Clebsch Gordan coefficients entering in Eq.1 for L=2 and $J_f = J_i - 2$ are

$$< 2120 \,|\, 21 > = -\sqrt{\frac{1}{14}} \qquad < 2140 \,|\, 21 > = -\sqrt{\frac{8}{63}} \qquad \text{(A.1)}$$

and the Racah functions are of the form (Q=2,4)

$$U(J_f 2J_i 2, J_i 2) = \left(\frac{2(J_i + 1)(2J_i + 3)}{7J_i(2J_i - 1)} \right)^{1/2} \qquad \text{(A.2)}$$

$$U(J_f 2J_i 4, J_i 2) = \frac{1}{6} \left(\frac{2(J_i + 1)(J_i + 2)(2J_i + 3)(2J_i + 5)}{7J_i(J_i - 1)(2J_i - 1)(2J_i - 3)} \right) \qquad \text{(A.3)}$$

The γ-ray angular correlation of multipolarity L=2 emitted from an aligned nucleus with spin J_i in its transition to the state with spin $J_f = J_i - 2$ is given by

$$\frac{dW}{d\Omega}\bigg|_{rest} = a_0 \left[1 + \frac{a_2}{a_0} P_2(\cos \vartheta_{\gamma p}) + \frac{a_4}{a_0} P_4(\cos \vartheta_{\gamma p}) \right] \qquad \text{(A.4)}$$

with

$$a_0 = \frac{1}{4\pi} \qquad \text{(A.5)}$$

$$\frac{a_2}{a_0} = \frac{5}{7} \frac{J_i + 1}{2J_i - 1} \cdot (-1) \left[3 \frac{< K^2 >}{J_i(J_i + 1)} - 1 \right] \qquad \text{(A.6)}$$

$$\frac{a_4}{a_0} = -\frac{3}{7} \frac{(J_i + 1)(J_i + 2)}{(2J_i - 3)(2J_i - 1)}$$

$$\cdot \left[35 \frac{< K^4 >}{J_i^2(J_i + 1)^2} - 30 \frac{< K^2 >}{J_i(J_i + 1)} \left(1 - \frac{5}{6J_i(J_i + 1)} \right) + 3 \left(1 - \frac{2}{J_i(J_i + 1)} \right) \right] \qquad \text{(A.7)}$$

$$: \left(3 - \frac{6}{J_i(J_i + 1)} \right)$$

References

1. J.O. Rasmussen et al., *Nucl. Phys.* **A136**, 465 (1969).
2. L.G. Moretto et al., *Nucl. Phys.* **A502**, 453c (1989).
3. G.D. Adeev and V.V. Pashkevich, *Nucl. Phys.* **A502**, 405c (1989)
4. U. Brosa et al., *Nucl. Phys.* **A502**, 423c (1989)
5. A. Bohr, *in Proc. Int. Conf. Peaceful Uses Atom. Energy* (Geneva, 1955) Vol.2, UN, New York, 151 (1956)
6. W. Pilz and W. Neubert, *Z. Phys.* **A338**, 75 (1991)
7. Yu.N. Kopach et al., *Phys. Rev. Lett.* **82**, 303 (1999)
8. M. Mutterer et al., *in Proc. Int. Conf. DANF96, Častá Papiernička, Slovakia,* 1996, JINR Dubna, Russia, p.250
9. P. Singer et al., *Z. Phys.* **A359**, 41 (1997)
10. S.R. de Groot and H.A. Tolhoek, *Beta and gamma-ray spectroscopy* ed. K. Siegbahn, North-Holland, Amsterdam, 613 (1955)
11. A.L. Barabanov, *12th Meeting on Physics of Nuclear Fission, Obninsk, Russia, 1993. Preprint IAE-5670/2,* (1994) http://arxiv.org/pdf/nucl-th/9402024
12. H.A. Jahn, *Proc. Roy. Soc* **A205**, 192 (1951)
13. A. Wolf and E. Cheifetz, *Phys. Rev.* **C13**, 1952 (1976)
14. Yu.N. Kopach et al., *in Symposium on Nuclear Clusters, Rauischholzhausen, Germany, 2002, in press*

IS THERE A SHELL CLOSURE AT N=32? SPECTROSCOPY OF NEUTRON-RICH NUCLEI NEAR ^{56}CR

D.E. APPELBE[1], R.A.E. AUSTIN[2], C.W. BEAUSANG[3], C.J. BARTON[1], J.A. CAMERON[2], M.A. CAPRIO, J.R. COOPER[3], C. MALSOLMSON[2], M.H. MUIKKU[1], J.R. NOVAK[3], J. SIMPSON[1], J.C. WADDINGTON[2], D.D. WARNER[1], F.R. XU[4], N.V. ZAMFIR

1 CLRC Daresbury Laboratory, Keckwick Lane, Daresbury, Warrington, WA4 4AD, U.K.
2 Dept. Physics and Astronomy, McMaster University, Hamilton, Ontario L8S 4M1, Canada.
3 Wright Nuclear Structure Laboratory, Yale University, 273 Whitney Avenue, New Haven, CT 06511.
4 Department of Technical Physics, Peking University, Beijing 100871, China.

In recent years a large degree of experimental and theoretical effort has been directed towards characterizing and understanding the chanes in nuclear structure which take place at ever-increasing distances from the line of stability. The development of a significant excess of one type of nucleon has been shown to produce surprising alterations in the anticipated shell structure, evidenced, for example, by the appearance of large deformations about ^{80}Zr [1] and ^{32}Mg [2]. Such changes are, perhaps, to be expected given the fact that the only direct empirical determination of single particle structure has been performed with stable nuclei. Nevertheless, the observation of unexpected phenomena such as those cited above signals the first indication of major changes in the underlying shell structure and hence provides crucial tests of the predictive power of self consistent theories of the nuclear mean field.

Recently Otsuka *et al.*, [3] have investigated the role of the nucleon-nucleon interaction in generating shell closures in light mass (A≈20) nuclei. They found that the n=16 shell closure was well described by the use of a strong monopole interaction between protons and neutrons in orbitals with $l + ½$ and $l - ½$. Using this explanation, Otsuka *et al.*, have predicted [3] the presence of a shell closure at N=34, brought about by the Π $f_{7/2}$-$vf_{5/2}$ interaction.

Following a recent experiment using the YRASTBall [5] γ-ray spectrometer, the level scheme of ^{56}Cr has been elucidated using the reaction ^{48}Ca(^{11}B,pxn)$^{58-x}$Cr at a beam energy of 35 MeV. This has allowed the spins and paritites for many of the observed states in ^{55}Cr and ^{56}Cr to be unambiguously ascertained. Plotting the E(4^+)/E(2^+) ratios for the chromium isotopes provides additional confirmation of this anomaly at N=32. This experiment has also confirmed the position of $\nu 1g_{9/2}$ intruder band in ^{55}Cr.

The evidence of a possible shell closure at N=32 and the presence of the $\nu 1g_{9/2}$ intruder orbital will be discussed.

References

1. S.M. Fischer *et al.*, Phys. Rev. Lett. **87**, 132501 (2001).
2. V. Chiste *et al.*, Phys. Lett. **B 514**, 223 (2001).
3. T. Otsuka, R. Fujimoto, Y. Utsuno, B.A. Brown, M. Honma and T. Mizusaki, Phys. Rev. Lett. **87**, 082502 (2001).
4. J.I. Prisciandaro *et al.*, Phys. Lett. **B 510**, 17 (2001).
5. C.W. Beausang *et al.*, Nucl. Instrum. Meth. Phys. Res. **A452**, 431, (2000).

STUDY ON NEUTRON-RICH NUCLEI NEAR ^{132}SN FROM THE SPONTANEOUS FISSION OF ^{252}CF

J.K. HWANG[1], A.V. RAMAYYA[1], J.H. HAMILTON[1], Y.X. LUO[1,2,3], P. M. GORE[1], C.J. BEYER[1], X.Q. ZHANG[1], J. KORMICKI[1], J.O. RASMUSSEN[3], S.C. WU[3], C.M. FOLDEN III[3], P. FALLON[3], P. ZIELINSKI[3], K.E. GREGORICH[3], A.O. MACCHIAVELLI[3], I-YANG LEE[3], T.N. GINTER[4], M. STOYER[5], J.D. COLE[6], R. DONANGELO[7], S.J. ASZTALOS[8], G.M. TER-AKOPIAN[9], YU.TS. OGANESSIAN[9], A.V. DANIEL[9], A. COVELLO[10], A. GARGANO[10] AND S.C. PANCHOLI[11]

[1]Department of Physics and Astronomy, Vanderbilt University, Nashville, TN 37235
[2]Joint Institute for Heavy Ion Research, Oak Ridge, TN 37830
[3]Lawrence Berkeley National Laboratory, Berkeley, CA 94720
[4]National Superconducting Cyclotron Laboratory, Michigan State University, East Lansing, MI 48824
[5]Lawrence Livermore National Laboratory, Livermore, CA 94550
[6]Idaho National Environmental and Engineering Laboratory, Idaho Falls, ID 83415
[7]Universidade Federal do Rio de Janeiro, CP 68528, RG Brazil
[8]assachusetts Inst. of Technology, Cambridge, MA 11830
[9]Flerov Laboratory for Nuclear Reactions, JINR, Dubna, Russia
[10]ipartimento di Scienze Fisiche, Universita di Napoli Federico II and Istituto Nazionale di Fisica Nucleare, I-80126 Napoli, Italy
[11]Department of Physics and Astrophysics, University of Delhi, Delhi 110 007, India

Several new states in ^{121,123}Cd and ^{115,117}Ag near ^{132}Sn produced in the spontaneous fission of ^{252}Cf are identified by using Gammasphere. Excited states built on the 11/2$^-$ isomeric states in neutron-rich ^{121,123}Cd are identified. The new data indicate that the 11/2$^-$ isomers in odd-A Cd isotopes with N>74 have spherical shapes similar to those in neighboring Te isotopes while the even-A Cd isotopes are likely deformed. π7/1[413] rotational bands are identified in ^{115,117}Ag, along with higher bands of uncertain structures. A doublet structure in some bands is suggestive of softness toward triaxiality.

1. High spin states in ^{121,123}Cd

The Z=48 Cd nuclei are close to the spherical Sn nuclei. The $h_{11/2}$ decoupled neutron bands have been observed in $^{111-119}$Cd [1] built on the isomeric states. From our SF work [2], the $h_{11/2}$ decoupled bands of ^{121,123}Cd are established (see Fig. 1). The near constant values of R = E(4$^+$)/E(2$^+$) $\cong$ 2.2 with β_2 ~0.12. [3] for N=78,80, ^{126,128}Cd isotopes indicate the possible quenching of the N=82 spherical shell gap. It is interesting to investigate the heavier odd -A Cd isotopes.

154

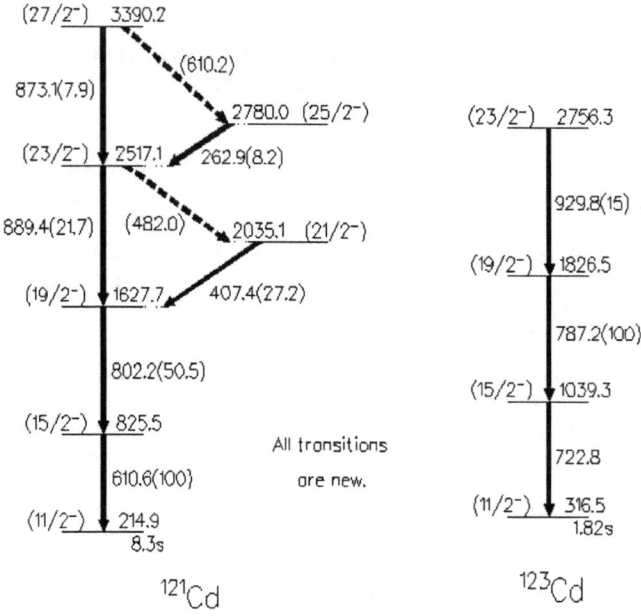

FIGURE 1. Level scheme of ^{121,123}Cd[2]

The ratios of R versus N for odd N Pd, Cd, Te and Xe are shown in Fig. 2. The odd –A and even-A Te isotopes have nearly spherical shapes, where the value of R is $\cong 2.0$-2.1. The even-even Cd isotopes show only a small decreasing trend for R with increasing N. The $h_{11/2}$ decoupled bands built on the isomeric states in odd -A Cd isotopes have a much larger value of $R \cong 2.5$ for $N \le 71$, characteristics of deformation. but strongly decreasing value of R to $R \cong 2.1$ for ^{123}Cd. This indicates a shape change with the 11/2⁻ isomer becoming spherical. The ground states of even and odd -A Cd isotopes may have non-zero deformations near A=130 as indicated by the near constant R for the even A isotopes out to ^{128}Cd.

Our data suggest that the isomers in ^{125,127,129}Cd could be nearly spherical with β_2 much less than 0.1 compared to ^{126,128}Cd with β_2 =0.12 [3]. If this suggestion is correct, this shape coexistence effect can help remove the discrepancies between experiments and calculations in the solar r-process abundances in this region. It is suggested that the backbending observed at $\eta\omega \approx 0.45$ MeV in ^{121}Cd is related to the alignment of the $h_{11/2}$ neutron pair as seen in the lighter A Cd nuclei [3].

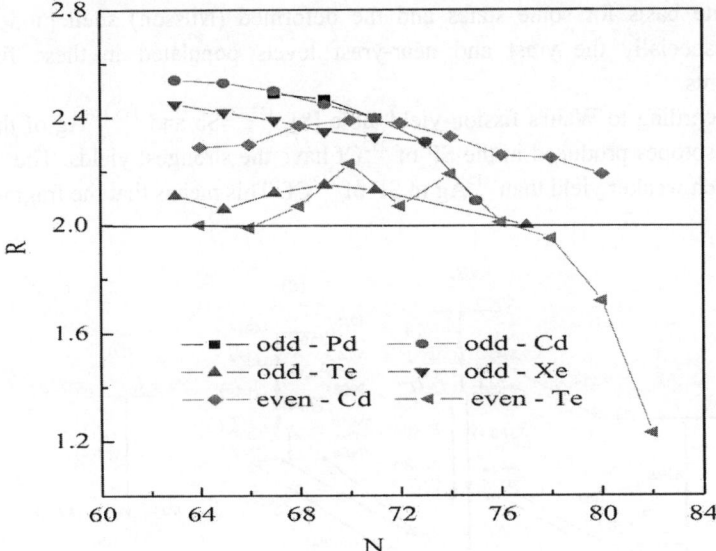

FIGURE 2. Ratios, R, of $(E_{19/2} - E_{11/2})/E_{15/2} - E_{11/2})$
for the odd N isotopes and $E(4^+)/E(2^+)$
for the even N isotopes[2].

The A=120 region with Z < 50 and N < 82 is very important in astrophysics because of inconsistencies between the observed solar isotopic r-process abundances and the model calculations [4,5]. The center of the r-process path for Z=48 is ^{130}Cd [4,5]. However, nuclei to either side influence the r-process. Kratz et al. [4] suggested that presently inaccessible ^{120}Mo, $^{119-123}$Tc and ^{122,124}Ru may exhibit shape coexistence. Such shape coexistence will influence the r-process abundances in this region.

2.   High spin states in ^{115,117}Ag

Earlier studies on the odd-Ag isotopes were carried out from the beta-decay work of the Pd isotopes [6]. Isomeric states with spin and parity of $7/2^+$ were observed in ^{115,117}Ag [7]. The Ag isotopes are the partners of Sb in the spontaneous fission (SF) of ^{252}Cf.We have used this relation to identify the levels built on the $7/2^+$ isomeric states in ^{115,117}Ag. This 7/2[413] rotational bands are interpreted by using the Nilsson deformed shell model, since the rotational band spacings are sufficiently small. We recognize that these Ag nuclei may be in a region of shape coexistence, where the spherical shell model with phonons may be the more

156

appropriate basis for some states and the deformed (Nilsson) shell model for others, especially the yrast and near-yrast levels populated in these fission experiments.

According to Wahl's fission-yield table [8], ^{133,132}Sb and ^{115,116}Ag of the Sb and Ag isotopes produced in the SF of ^{252}Cf have the strongest yields. The ^{114}Ag has a much weaker yield than ^{115}Ag in SF of ^{252}Cf. This means that the fragment

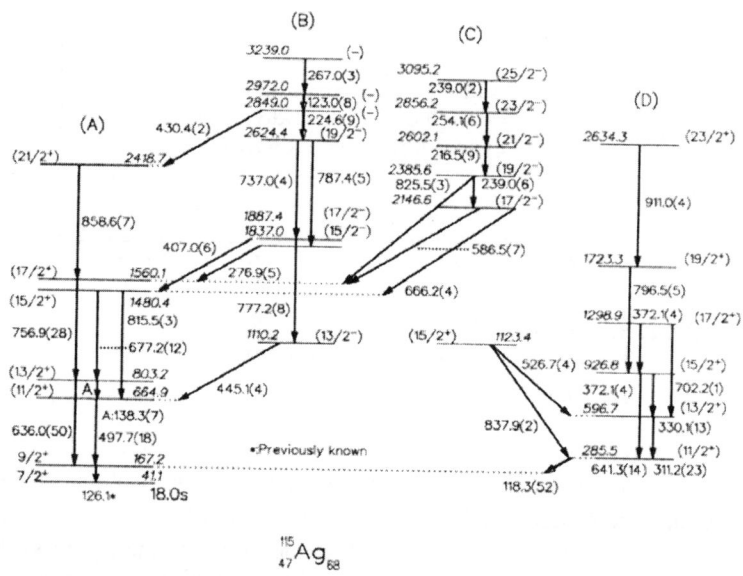

FIGURE 3. Level scheme of ^{115}Ag[12]

pair of ^{133}Sb-^{115}Ag with 4 neutrons emitted is more strongly populated than the fragment pair of ^{133}Sb-^{114}Ag with 5 neutrons emitted. The average neutron multiplicity in the spontaneous fission of ^{252}Cf is about 3.5 [9,10]. By using these informations and the coincidence relationships of the gamma trantions, new level schemes in ^{115,117}Ag are shown in Figs. 3 and 4 [12]. The spins and parities of band-A built on the $7/2^+$ isomeric state with a half-life of 18.0 seconds in ^{115}Ag are assigned based on the first two states having known spins and parities of $9/2^+$ and $7/2^+$ [11]. The spins and parities of band-A in ^{117}Ag built on the $7/2^+$ isomeric state with a half-life of 5.34 second [11] are, also, assigned by its similarity to band-A in ^{115}Ag.

The remarkable bunching of levels into doublets may be a consequence of a softness for triaxiality, or breaking of cylindrical symmetry of the well. The level schemes of the even-even palladium neighbors show the second 2^+ excited states to be fairly low, an indicator of triaxiality, which could well be accentuated in the odd neighbors by mixing of nearby Nilsson states differing in projection K by two units; such mixing does not require breaking of pairs in odd-A nuclei. The triaxial field, in turn, may be thought of as partially quenching the orbital motion of the odd nucleon, leaving only its intrinsic spin of 1/2 to couple to excited rotational states of an even-even palladium rotor. The doublet bunching may also be derived by second-order band-mixing extending to the anomalously spaced K=1/2 Nilsson state of the $g_{9/2}$ orbital. The paper of Skalski, Mizutori and Nazarewicz [11]

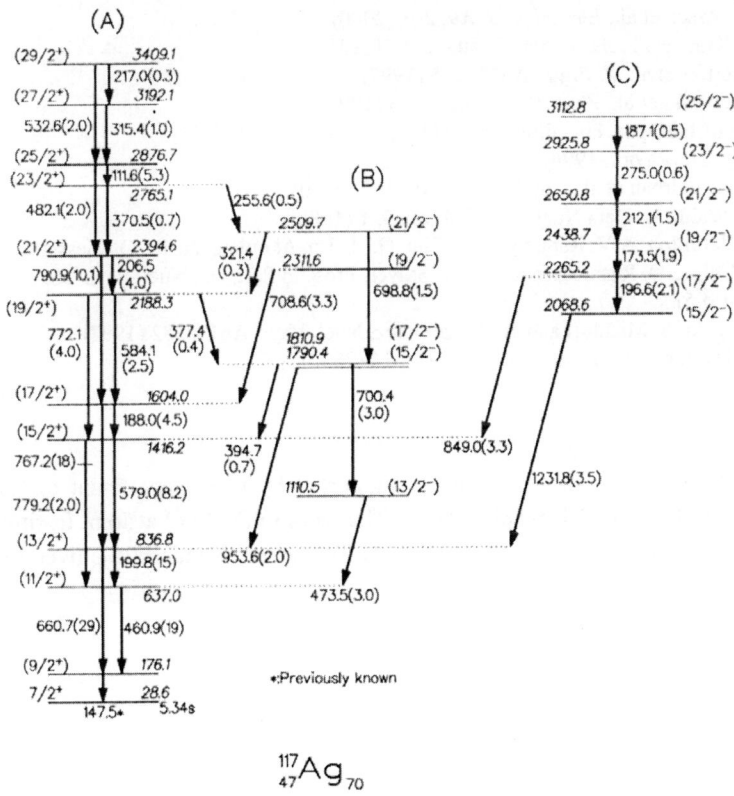

FIGURE 4. Level scheme of ^{117}Ag[12]

explores triaxial tendency in this region and gives a good Nilsson level diagram for the A=100 region.

In summary, The 34 new transitions with 24 new derived levels in ^{115}Ag are identified based on the 18.0 sec E3 isomeric state. The 32 new transitions with 22 new derived levels in ^{117}Ag are identified based on the 5.34 sec E3 isomeric state. The band built on the $7/2^+$ state is proposed to be a proton 7/2[413] rotational band. There is evidence from doublet structure in bands for softness toward tri-axiality.

References

1. N. Fotiades et al., *Phys. Rev.* **C61**, 064326 (2000).
2. J.K. Hwang etal., *J. of Phys.* **G28**, L9 (2002).
3. T. Kautzsch et al., *Eur. Phys. J.* **A9**, 201 (2000)
4. K.L. Kratz et al., *Astrophys. J.* **403**, 216 (1993).
5. B. Pfeiffer et al., *Z. Phys.* **A357**, 235 (1997).
6. J. Rogowski et al., Phys. Rev. **C42**, 2733 (1990).
7. **Table of Isotopes**, 8th edition, edited by R.B. Firestone et al. (Wiley, New York, 1996).
8. T. Rzaca-Urban et al., Eur. Phys. J. **A9**, 165 (2000).
9. A.C. Wahl, At. Data Nucl. Data Tables **39**, 1 (1988).
10. J.H. Hamilton, A.V. Ramayya, S.J. Zhu, G.M. Ter-Akopian, Yu.Ts. Oganessian, J.D. Cole, J.O. Rasmussen and M.A. Stoyer, Prog. in Part. and Nucl. Phys. **35**, 635 (1995).
11. J. Skalski, S. Mizutori and W. Nazarewicz, Nucl. Phys. **A617**, 282 (1997).
12. J.K. Hwang et al., Phys. Rev. **C65**, 054314 (2002).

Acknowledgments

The work at Vanderbilt University is supported by U.S. Department of Energy under Grant No. DE-FG05-88ER40407. The work (Y.X. Luo) at Joint Institute for Heavy Ion Research is supported by U. of Tennessee, Vanderbilt University and U.S. DOE.

EXOTIC CLUSTER STRUCTURE IN LIGHT
NEUTRON-RICH NUCLEI

N. ITAGAKI,* S. HIROSE AND T. OTSUKA

Department of Physics, University of Tokyo
7-3-1 Hongo,
Tokyo 113-0033, Japan

S. OKABE

Center for Information and Multimedia Studies, Hokkaido University,
Sapporo 060-0810, Japan

K. IKEDA

The Institute of Physical and Chemical Research (RIKEN),
2-1 Hirosawa, Wako, Saitama 351-0198, Japan

The triaxial deformation in ^{10}Be is investigated using a microscopic $\alpha+\alpha+n+n$ model. The states of two valence neutrons are classified based on the molecular-orbit (MO) model, and the π-orbit is introduced about the axis connecting the two α-clusters for the description of the rotational bands. There appear two rotational bands comprised mainly of $K^\pi = 0^+$ and $K^\pi = 2^+$, respectively, at low excitation energy, where the two valence neutrons occupy $K^\pi = 3/2^-$ or $K^\pi = 1/2^-$ orbits. The triaxiality and the K-mixing are discussed in connection to the molecular structure, particularly, to the spin-orbit splitting. The extent of the triaxial deformation is evaluated in terms of the electro-magnetic transition matrix elements (Davydov-Filippov model). The obtained values turned out to be $\gamma = 15^o \sim 20^o$. Furthermore, the mirror symmetry breaking of the triaxial deformation of ^{10}C is investigated. The triaxiality of protons in ^{10}C is much larger than that of neutrons in ^{10}Be due to the weak binding nature of the valence particles. This agrees with the recent experiment which suggests the mirror symmetry breaking of these nuclei through the β-decay transition strength.

1. Triaxial deformation in ^{10}Be

Recently, exotic structures of the Be isotopes have been theoretically and experimentally studied, and many new phenomena have been discussed.

---

*e-mail: itagaki@phys.s.u-tokyo.ac.jp

One of them is the appearance of the cluster rotational band structure in the excited states of ^{10}Be ($\alpha+^6$He) and ^{12}Be ($\alpha+^8$He, ^{6}He$+^6$He). We have suggested based on the molecular-orbit (MO) model, that the development of α-α cluster structure depends on the neutron orbits located around the core comprised of α-clusters. In the second 0^+ state of ^{10}Be, an anomalously prolonged α-α clustering structure emerges due to the valence neutrons located along the α-α axis[1].

The nucleus ^{10}Be has been known to have a strong β deformation due to the α-α core. Here, we discuss another aspect, a triaxial deformation[2]. The triaxial deformation is possible as a result of the dynamics of the two valence neutrons. The triaxiality of ^{10}Be has been theoretically discussed based on the deformed oscillator model, in which a γ distortion of 34.8^o has been predicted. According to Davydov-Filippov model[3], this γ value suggests that the excitation energy of the second 2^+ state is 2.3 times higher than that of the first 2^+ state. In ^{10}Be, the first 2^+ state is observed at 3.358 MeV, and several 2^+ states have been observed around 8 MeV region.

We introduce a microscopic $\alpha+\alpha+N+N$ model for ^{10}Be and ^{10}C. The configurations of valence nucleons are introduced based on the molecular orbit (MO) picture[4]. The total wave function is fully antisymmetrized and expressed as a superposition of Slater determinants with various configurations of the valence neutrons. The Slater determinants are also superposed with respect to different relative distances between the two α clusters. The projection to the eigen states of angular momentum J is numerically performed. All nucleons are described by Gaussians with a common oscillator parameter. Two α-clusters are introduced on the z-axis, and the wave function of the $i-$th valence neutron ($\phi_i\chi_i$) is expressed by a linear combination of local Gaussians:

$$\phi_i\chi_i = \sum_j g_j G_{R_j}\chi_i, \tag{1}$$

where $\vec{R}_j$ is a parameter corresponding to the Gaussian center.

In the MO model, the states of the valence neutrons are expressed by a linear combination of orbits around two α clusters. The lowest orbit has one node and negative parity, that is the p-orbit. We take the harmonic-oscillator-type wave function for the p-orbit. Such p-orbits in the x-, y-, and z-direction are denoted as ψ_x, ψ_y, and ψ_z, respectively. We now classify single particle orbits for the valence neutrons in terms of K quantum numbers:

$$\psi_x + i\psi_y = (x + iy)\exp[-\nu r^2] \propto r Y_{11}\exp[-\nu r^2], \tag{2}$$

$$\psi_z = z \exp[-\nu r^2] \propto r Y_{10} \exp[-\nu r^2], \tag{3}$$

$$\psi_x - i\psi_y = (x - iy) \exp[-\nu r^2] \propto r Y_{1-1} \exp[-\nu r^2]. \tag{4}$$

These p-orbits have $K = 1$, 0, and -1, respectively.

Now we construct MO from these p-orbits. Since each valence neutron can move around one of the two α-clusters, there should be two sets of the p-orbits defined with respect to those two α-clusters. Thus, the $(\psi_x \pm i\psi_y)$ orbit whose center is shifted at $+a$ $(-a)$ on the z-axis is denoted as $(\psi_x \pm i\psi_y)_{+a}$ $((\psi_x \pm i\psi_y)_{-a})$. As linear combinations of these orbits, the lowest MOs are expressed as:

$$\psi_1^{MO} = (\psi_x + i\psi_y)_{+a} + (\psi_x + i\psi_y)_{-a}, \tag{5}$$

$$\psi_{-1}^{MO} = (\psi_x - i\psi_y)_{+a} + (\psi_x - i\psi_y)_{-a}. \tag{6}$$

These orbits clearly have $K = 1$ and $K = -1$, respectively. In these cases, the classical picture of p-orbit is a circular motion about the α-α (z) axis. This is the so-called π-orbit.

If we take a linear combination of $(\psi_z)_{+a}$ and $(\psi_z)_{-a}$, the orbit becomes the so-called σ-orbit, just along the α-α (z) axis. This orbit is a higher-nordal orbit and only relevant to the second 0^+ state, discussed also in antisymmetrized molecular dynamics (AMD) calculations. and in stochastic variational method (SVM) calculations. and now we disregard this orbit.

These $(\psi_x)_{+a}$, $(\psi_x)_{-a}$, $(\psi_y)_{+a}$, and $(\psi_y)_{-a}$ orbits can be approximated by a combination of two local Gaussians, whose centers are shifted by a variational parameter b perpendicular to the z-axis. We thus use the following wave functions,

$$(\psi_x)_{+a} \propto G_{a\vec{e}_z + b\vec{e}_x} - G_{a\vec{e}_z - b\vec{e}_x}, \quad (\psi_y)_{+a} \propto G_{a\vec{e}_z + b\vec{e}_y} - G_{a\vec{e}_z - b\vec{e}_y}, \quad \cdots \cdots (7)$$

The values of these parameters a and b are variationally determined by using the cooling method in AMD. independently for each α-α distance. Since the rotational symmetry about the z-axis is broken with Eq. (10), the wave functions do not have exactly conserved quantum number K. On the other hand, the parameter b is small enough usually, and the K^π number is preserved to a good extent, and is expressed hereafter as $\bar{K}^\pi$.

When these orbital part of the wave functions are coupled with the spin part, the orbits with $\bar{K}^\pi = 3/2$, $1/2$, $-1/2$, and $-3/2$ are introduced, as follows:

$$|3/2^-\rangle = \psi_1^{MO}|n \uparrow\rangle, \quad |1/2^-\rangle = \psi_1^{MO}|n \downarrow\rangle, \tag{8}$$

and their time reversal,

$$| -3/2^- \rangle = \psi^{MO}_{-1} |n \downarrow \rangle, \quad | -1/2^- \rangle = \psi^{MO}_{-1} |n \uparrow \rangle. \tag{9}$$

The effective nucleon-nucleon interaction is Volkov No.2 for the central part and the G3RS spin-orbit term for the spin-orbit part, as follows:

$$V_{ij} = \{V_1 e^{-a_1 r_{ij}^2} + V_2 e^{-a_2 r_{ij}^2}\}\{W - MP^\sigma P^\tau + BP^\sigma - HP^\tau\}, \tag{10}$$

$$V_{ij}^{ls} = V_0^{ls}\{e^{-a_1 r_{ij}^2} - e^{-a_2 r_{ij}^2}\}\vec{L} \cdot \vec{S} P_{31}, \tag{11}$$

where P_{31} is a projection operator onto the triplet odd state. The parameters are $V_1 = -60.650$ MeV, $V_2 = 61.140$ MeV, $a_1 = 0.980$ fm^{-2} and $a_2 = 0.309$ fm^{-2} for the central interaction, and $V_0^{ls} = 2000$ MeV, $a_1 = 5.00$ fm^{-2}, and $a_2 = 2.778$ fm^{-2} for the spin-orbit interaction. We employ the Majorana exchange parameter, $M = 0.6$ ($W = 0.4$), the Bartlett exchange parameter, $B = 0.125$, and the Heisenberg exchange parameter, $H = 0.125$, for the Volkov interaction. All of these parameters are determined from the $\alpha + n$ and $\alpha + \alpha$ scattering phase shifts and the binding energy of the deuteron.

Using this framework, the binding energy of one α-cluster is calculated to be 27.5 MeV, and the threshold energy of free $\alpha + \alpha + n + n$ system is $2 \times (-27.5) = -55.0$ MeV. Experimentally, the ground state of ^{10}Be is lower than this energy by 8.4 MeV. As $K = 0$ levels, the ground 0^+ state is calculated at -60.5 MeV, the 2^+ state appears at $E_x = 3.6$ MeV, and the 4^+ state appears at $E_x = 12.9$ MeV, and the dominant component is $(3/2^-)^2$ for the two valence neutrons. As $K = 2$ states the 2^+ state at $E_x = 5.6$ MeV, the 3^+ state at $E_x = 9.4$ MeV, and the 4^+ state at $E_x = 14.4$ MeV form a rotational band structure. The $K = 2$ band dominantly has a component of $(3/2^-)(1/2^-)$ for the two valence neutrons, where one of them feels the spin-orbit interaction attractively and the other feels repulsively. After the mixing of these two bands, the first 2^+ state becomes $E_x = 3.1$ MeV (experimentally 3.358 MeV), is slightly pushed up ($E_x = 5.6$ MeV, experimentally 5.958 MeV). The first 2^+ state has the squared overlap with the $K = 0$ state by 0.96, and the second 2^+ state has the squared overlap with the $K = 2$ state by 0.92. Therefore, the second 2^+ state has the component of $K = 0$ by 8%, and it can be considered as an indication of a triaxial deformation.

This triaxiality of the 2^+ states is reflected in the electro-magnetic transition rate. The E2 transition between the first 2^+ state and the ground state is calculated as B(E2: $2_1^+ \rightarrow 0_1^+$) = 11.8 e^2fm^4, which agrees with the experimental value of 10.04 ± 1.2 e^2fm^4. Furthermore, the inter-band transition is calculated: B(E2: $2_1^+ \rightarrow 2_2^+) = 3.99$ e^2fm^4. If the system is axially symmetric, this transition between different K-values is more suppressed. Therefore, the value indicates the 2^+ states have component of a triaxial deformation.

Table 1. The electro-magnetic transition probability (B(E2) (e^2fm^4)) from first 2^+ state to second 2^+ calculated by changing the strength of the spin-orbit interaction (V_0^{ls}). The Majorana parameter (M) for the central interaction is also changed to keep the calculated ground 0^+ state energy constant.

V_0^{ls} (MeV)	1000	1500	2000	2500
M	0.58	0.59	0.60	0.61
0_1^+ (MeV)	-60.48	-60.38	-60.51	-60.75
2_1^+ (MeV)	-57.69	-57.30	-57.26	-57.33
2_2^+ (MeV)	-56.71	-55.78	-54.88	-53.93
B(E2: $2_1^+ \rightarrow 2_2^+$)	17.53	9.53	3.99	2.20

Table 2. The calculated electro-magnetic transition probability (B(E2)) of ^{10}C and ^{10}Be. ^{10}C(p) and ^{10}Be(p) are proton B(E2) values calculated with the charge value of $(p, n) = (e, 0)$, whereas ^{10}C(n) and ^{10}Be(n) are neutron B(E2) values calculated with the charge value of $(p, n) = (0, e)$. The values in parenthesis are experimental values. All units are e^2fm^4.

B(E2)	^{10}C(p)	^{10}C(n)	^{10}Be(p)	^{10}Be(n)
$2_1^+ \rightarrow 0_1^+$	12.8 (12.3 $\pm$2.0)	12.1	11.8 (10.04$\pm$1.2)	11.4
$2_2^+ \rightarrow 0_1^+$	1.3	1.1	0.7	1.7
$2_1^+ \rightarrow 2_2^+$	17.0	6.2	4.0	13.7

Using Davydov-Filippov model[3], we can estimate the degree of the triaxiality as a function of the γ angle. The ratios $\frac{\text{B(E2: } 2_2^+ \rightarrow 0_1^+)}{\text{B(E2: } 2_1^+ \rightarrow 0_1^+)}$ and $\frac{\text{B(E2: } 2_2^+ \rightarrow 2_1^+)}{\text{B(E2: } 2_1^+ \rightarrow 0_1^+)}$ are given in Davydov-Filippov model as follows:

$$\frac{\text{B}(E2 : 2_2^+ \rightarrow 0_1^+)}{\text{B}(E2 : 2_1^+ \rightarrow 0_1^+)} = \frac{1 - \frac{3-2\sin^2(3\gamma)}{\sqrt{9-8\sin^2(3\gamma)}}}{1 + \frac{3-2\sin^2(3\gamma)}{\sqrt{9-8\sin^2(3\gamma)}}}, \tag{12}$$

$$\frac{\text{B}(E2 : 2_2^+ \rightarrow 2_1^+)}{\text{B}(E2 : 2_1^+ \rightarrow 0_1^+)} = \frac{\frac{20}{7}\frac{\sin^2(3\gamma)}{9-8\sin^2(3\gamma)}}{1 + \frac{3-2\sin^2(3\gamma)}{\sqrt{9-8\sin^2(3\gamma)}}}. \tag{13}$$

The ratio $\frac{\text{B(E2: } 2_2^+ \rightarrow 2_1^+)}{\text{B(E2: } 2_1^+ \rightarrow 0_1^+)}$ calculated with the Davidov-Filippov model. The ratio becomes 0.34 in our calculation, and it has a crossing point with

Davydov-Filippov model around $\gamma = 19°$. The ratio $\frac{B(E2: 2_2^+ \to 0^+)}{B(E2: 2_1^+ \to 0_1^+)}$ calculated with the Davidov-Filippov model, and it crosses with our result of 0.059 around $\gamma = 17°$ and $22°$. These results strongly suggest that ^{10}Be has a triaxial deformation of $\gamma = 15° \sim 20°$. Although the α-α core is of axial symmetry and electric charge are only in the α's, the recoil effect gives rise to a change from the axial symmetry to the triaxial shape.

We discuss the electro-magnetic transition between the two 2^+ states (K-mixing effect between the two 2^+ states) by artificially changing the strength parameter V_0^{ls} for the spin-orbit term. The Majorana parameter M for the central term is simultaneously changed to keep the calculated ground 0^+ state energy constant. As shown in Table 1, using the original interaction ($V_0^{ls} = 2000$ MeV and $M = 0.6$), the B(E2: $2_1^+ \to 2_2^+$) value is predicted as 3.99 e^2fm^4. When we increase the V_0^{ls} value, the excitation energy of the second 2^+ state becomes higher, since one of the valence neutrons repulsively feels the spin-orbit interaction. The original interaction gives $E_x = 5.70$ MeV for the 2_2^+ state, but the interaction with $V_0^{ls} = 2500$ MeV and $M = 0.61$ gives $E_x = 6.82$ MeV. The B(E2: $2_1^+ \to 2_2^+$) value then decreases from 3.99 e^2fm^4 to 2.20 e^2fm^4. Therefore, it is considered that with increasing LS strength, the K quantum number of each 2^+ state approaches to a good number, where the E2 transition between the two 2^+ states is suppressed. On the other hand, when the spin-orbit interaction becomes weaker, the transition rapidly increases. The V_0^{ls} value of 1500 MeV gives the B(E2: $2_1^+ \to 2_2^+$) values of 9.53 e^2fm^4, and when we adopt $V_0^{ls} = 1000$ MeV, the value becomes 17.53 e^2fm^4. Here, the orbits of the valence neutrons deviate from ones with good K quantum numbers, and a triaxial $\alpha+\alpha+$di-neutron clustering configuration where the two valence neutrons are strongly correlated becomes important.

We can intuitively interpret this behavior as follows: when the spin-orbit is weak enough, the two valence neutrons form di-neutron, in which the attractive interaction between them strongly contributes. However, when the spin-orbit interaction significantly acts and becomes more important, the di-neutron is broken and each valence neutron rotates around the core in opposite direction with definite K-values. Note that the spin-orbit interaction does not act to di-neutron with $S = 0$. This is close to jj-coupling picture and axial symmetry of the system is restored.

2. Mirror symmetry breaking between ^{10}Be and ^{10}C

Using this model, the ground 0^+ state of ^{10}C is calculated at -56.3 MeV, which is higher by 4.2 MeV than ^{10}Be (experimentally 4.66 MeV), and as a results of K-mixing effect, two 2^+ states appear at $E_x = 3.5$ MeV (Exp. $E_x = 3.35$ MeV) and $E_x = 6.0$ MeV. The electro-magnetic transition rate, B(E2) values among these states are summarized in Table 2, together with those of ^{10}Be. Here, not only the standard B(E2), namely the B(E2) value for the protons, neutron B(E2) values are also listed. The proton and the neutron B(E2) values are calculated with the use of the charge values of $(p, n) = (e, 0)$ and $(0, e)$, respectively.

The E2 transition between the first 2^+ state and the ground state of ^{10}C is calculated as B(E2: $2_1^+ \rightarrow 0_1^+$) = 12.8 e^2fm^4, which well agrees with the experimental value of 12.3$\pm$1.2 e^2fm^4. This value is experimentally slightly larger than that of ^{10}Be (10.04 $\pm$ 1.2), and our calculation reproduces this tendency, which has been known to be not trivial in previous analyses.

Experimentally, although two valence neutrons are bound from the ^{8}Be$+n+n$ threshold by 8.4 MeV in ^{10}Be, the two valence protons are bound by only 3.8 MeV in ^{10}C, due to the Coulomb energy shift. Therefore, triaxiality in ^{10}C is expected to be more important than in ^{10}Be, and it would be reflected in the mirror symmetry breaking of the B(E2) values. We compare the calculated proton B(E2) value of ^{10}C with neutron B(E2) of ^{10}Be. The proton B(E2) value of ^{10}C (12.8 e^2fm^4) is almost the same as the neutron B(E2) of ^{10}Be (11.9 e^2fm^4), and mirror symmetry seems to hold judging from the transition probability between the yrast states. However, the inter-band transition, B(E2: $2_1^+ \rightarrow 2_2^+$) values show significant difference. The neutron B(E2: $2_1^+ \rightarrow 2_2^+$) value of ^{10}C is calculated to be 6.2 e^2fm^4, which is about twice the proton B(E2) of ^{10}Be (4.0 e^2fm^4). Also, the proton B(E2: $2_1^+ \rightarrow 2_2^+$) value of ^{10}C (17.0 e^2fm^4) is larger than neutron B(E2) of ^{10}Be (13.7 e^2fm^4). If the system is axially symmetric, the transition between different K-values is more suppressed. Since the present value is much larger than that of ^{10}Be, the triaxiality of ^{10}C is considered to be much larger than that in ^{10}Be. Using the Davydov-Filippov model, our proton B(E2) values are found to correspond to the γ deformation of $\gamma = 28^o$. If the B(E2) value of protons would be measured, it provides us the direct information for the γ deformation of ^{10}C. On the contrary in ^{10}Be, charge distribution of α-α is originally axial symmetric, and inter-band transition probability which would be experimentally observed only tells us the recoil effect, however, in the case of ^{10}C, valence particles have charge.

This result indicates that the inter-band transition, which appears as a result of mixing of different K-values is good tool and sensitive physical quantity to see the mirror symmetry breaking of the triaxiality. In the present case, due to the Coulomb energy shift of about 4 MeV, the valence protons of ^{10}C are more weakly bound and contribution of the spin-orbit interaction from the core nucleus becomes weaker. We can prove this by artificially increasing the binding energy. By employing the Majorana parameter of $M = 0.58$ of the Volkov No.2 interaction, ^{10}C has almost the same binding energy as ^{10}Be calculated with the original interaction ($M = 0.6$), and the B(E2: $2_1^+ \rightarrow 2_2^+$) value decreases to 12.1 e^2fm^4. Actually, this value is smaller than the neutron B(E2) of ^{10}Be. Therefore, it would be said that two valence neutrons without the Coulomb interaction between them have more di-neutron-like configuration than two valence protons, if the binding energy from $\alpha+\alpha+N+N$ are adjusted to be the same. However, because of the difference of the binding energies, ^{10}C is more triaxial than ^{10}Be, and mirror symmetry is broken. Therefore, the $\alpha+\alpha+p+p$ system is more SU(3) like structure than the $\alpha+\alpha+n+n$ system, where the neutron motion is closer to the single-particle motion. The mirror symmetry between these nuclei is broken. Recently, mirror symmetry breaking between ^{10}C and ^{10}Be is experimentally discussed through the β-decay probability to ^{10}B[5]. The transition strength from the 2^+ states of ^{10}Be and those from ^{10}C show significant difference, which is consistent with our analysis.

References

1. N. Itagaki and S. Okabe, Phys. Rev. C **61** 044306, (2000).
2. N. Itagaki, S. Hirose, T. Otsuka, S. Okabe, and K. Ikeda, Phys. Rev. C**65**, 044302 (2002).
3. A.S. Davydov and G.F. Filippov, Nucl. Phys. 8, 237 (1958).
4. Y. Abe, J. Hiura, and H. Tanaka, Prog. Theor. Phys. **49**, 800 (1973).
5. I. Daito *et al.*, Phys. Lett **B418**, 27 (1998).

NONLINEAR ASPECTS OF QUADRUPOLE EXCITATIONS IN NUCLEI WITH A=100-170[*]

K.A. GRIDNEV [1,2], V.G. KARTAVENKO [2,3] AND W. GREINER [2]

1 *Vanderbilt University, Nashville, Tennessee, USA*

2 *J. W. Goethe Universität Frankfurt am Main, Germany*

3 *Joint Institute for Nuclear Research, Dubna, Moscow District, Russia*

The aim of the present report is to analyze the dependence of the reduced electric quadrupole probabilities from N and Z, the difference in the deformation lengths for different projectiles and also for a fixed projectile incident on different isotopes of an element. As the framework, we use a non traditional in nuclear physics approach — the theory of nonlinear differential equations. We exploited the notion, that the perturbation, which is responsible for the inelastic transition in inelastic α-particle and proton scattering has a soliton shape. Then we took from soliton theory another idea: The dynamic of the potential of the Schrödinger equation is governed by the Korteweg-de Vries equation. Using the basic properties of these solitons, it was possible to understand specific features of the inelastic scattering of protons and α-particles on different nuclei and their isotopes. The corresponding deformation lengths for different projectile-target systems are compared with experimental data. We explain the difference in the data obtained with the inelastic scattering of alpha-particles and protons, also the behavior of the deformation length parameters for the even-even isotopes of the elements with A=100–170.

1. Introduction

In many cases quadrupole excitations in nuclei were investigated by means of Coulomb excitations and inelastic scattering. Knowing a mechanism of reaction and comparing a theory with experimental data it is possible to obtain some knowledge about reduced probabilities of electromagnetic transitions. As a description of reactions with excitations of the states with quadrupole momenta usually the couple channels method (CCM) or distorted wave Born approximation (DWBA) is used [1]. Apart from a big

[*]Work supported in part by Deutsche Forschungsgemeinschaft, Russian Foundation for Basic Research and the Heisenberg-Landau program.

array of accumulated data there is a set of unexplained data in the region of A=100–170 nuclei. As an example, it is on the first sight regular dependence of the deformation parameters upon filling of nuclear shells [2]. The same behavior of the reduced electric quadrupole probabilities with N and Z was found by Prof. J. Hamilton [3].

To analyze of the dependence of reduced electric quadrupole probabilities from N and Z we decided to use a non-traditional approach — the theory of nonlinear differential equations.

2. Application of Soliton theory to the nuclear inelastic scattering

It is worth to recall some points concerning the interaction employed in nuclear physics to describe the inelastic scattering process. Up to first order of the perturbation theory the inelastic transition potential has the shape of the first derivative of the Woods-Saxon potential [4]. Expansion of the interaction in a Taylor series allows to write the transition potential in terms of the deviations from the spherical shape δR as

$$U(r, \delta R) = U(r) + \frac{\partial U(r)}{\partial r} \delta R + \cdots \tag{1}$$

The perturbation, which gives rise to the inelastic transitions, is taken to have the form of the derivative of the Woods-Saxon distribution, and the relation

$$\frac{1}{\cosh^2\left(\frac{r-R}{2a}\right)} = -\frac{a}{2} \cdot \frac{d}{dr} f, \qquad f \equiv \left(1 + \exp\left(\frac{2(r-R)}{a}\right)\right)^{-1} \tag{2}$$

can easily be verified. The time evolution of such a radial perturbation may be described by the Korteweg de Vries (KdV) equation [5],

$$u_t - 6u\, u_x + p \cdot u_{xxx} = 0 \tag{3}$$

The soliton solution at t=0 of this equation is exactly an interaction, which is responsible for the excitation of quadrupole states in inelastic scattering:

$$U_0 = \frac{A_0}{\cosh^2((r-R)/a)}, \qquad A_0 \equiv \frac{V_0\, R_0\, \beta_L}{4\, a}. \tag{4}$$

Here the solutions amplitude A_0 is connected through the deformation length $\beta_2 R_0$ with the reduced electric probabilities as

$$B(EL) = 4\pi^2 e^2 |3Z\beta_L R_0| \tag{5}$$

The following features of the KdV equation are of special interest to us and will be subsequently discussed in more detail:

(1) The similarity principle, as introduced by BEREZIN and KARP-MAN [6].

(2) The connection of the KdV equation with the Schrödinger equation, which is established via the inverse scattering problem [7]. (See also the review in Ref. [8]).

Before applying the KdV equation to nuclear physics problems we shall remember Miura's theorem [7]. Namely, *If v is a solution of the following, so called modified KdV equation:*

$$Q\,v = v_t - 6\,v^2\,v_x + p \cdot v_{xxx} = 0 \qquad (6)$$

then $u = v^2 + \sqrt{p} \cdot v_x$ is a solution of the KdV equation (3).

$$P\,u = u_t - 6\,u\,u_x + p \cdot u_{xxx} = 0 \qquad (7)$$

This is easily confirmed by inserting u into (7) to yield:

$$P\,u = \left[2\,v + \sqrt{p}\,\frac{\partial}{\partial x}\right] Q\,v \qquad (8)$$

With the help of this transformation we can go over to the inverse scattering problem. To this end the standard substitution $v = \sqrt{p} \cdot \psi_x/\psi$ for the solution of a Riccati equation $u = v^2 + \sqrt{p} \cdot v_x$ is performed to yield the Schrödinger equation

$$p \cdot \psi_{xx} - u\,\psi = 0 \qquad (9)$$

where the time is a parameter. The constant p is identified as $p = \hbar^2/2\mu$. Here we additionally assume the Galilean invariance for u: $\tilde{u} = u + E$.

If we look at Eqs. (7) and (9) we can see that the time evolution of the potential u in the Schrödinger equation is given by the KdV equation. Equation (7) establishes the relation between nonlinearity and dispersion $u^2/2 \sim -pu_{xx}$. From this expression one can obtain the similarity constant. $u\,\Delta^2/(\hbar^2/2\mu) = const$. The last expression means that the higher the soliton the narrower it is. For further applications the best way to present the KdV equation and some of its properties is to use the dimensionless form. Substituting

$$\epsilon = E\Delta^2/p, \qquad v = a_0\Delta^2/p, \qquad z = x/\Delta \qquad (10)$$

into Eq. (9) we obtain the Schrödinger eqiation

$$\psi_{zz} + (\epsilon + v\cosh^{-2} z)\psi = 0, \qquad (11)$$

Table 1. Deformabilities β and deformation lengths and soliton amplitude A_0 of the first three 2^+-states in even Pd-isotopes, excited by 30 MeV α-particle scattering [15]

	^{104}Pd	^{106}Pd	^{108}Pd	^{110}Pd
$E_{exc}(2_1^+)$ [MeV]	0.556	0.512	0.434	0.374
$\beta(2_1^+)$	0.116	0.146	0.158	0.163
$\beta \cdot R$ [fm]	0.856	1.050	1.129	1.148
A_0 [MeV]	36.29	41.05	53.46	53.65
a_n [MeV]	4.03	4.59	5.30	4.72
$E_{exc}(2_2^+)$ [MeV]	1.342	1.129	0.930	0.814
$\beta(2_2^+)$	0.199*	0.170*	0.139	0.179
$\beta \cdot R$ [fm]	1.469	1.223	0.993	1.261
$E_{exc}(2_3^+)$ [MeV]	1.796	1.557	1.444	1.209
$\beta(2_3^+)$	0.089	0.227	0.079	0.162
$\beta \cdot R$ [fm]	0.657	1.633	0.564	1.141

where we used the solitonic potential. Let the initial conditions of the system be given by

$$u(x,0) = u_0 \cdot \phi(x/\Delta) \qquad (12)$$

where u_0 and Δ are the characteristic amplitude and width of the initial perturbation. $\phi(x/\Delta)$ is a dimensionless function describing its profile. Varying u_0 and Δ a family of similar initial conditions is obtained which are characterized by $\phi(x/\Delta)$.

Let us introduce the dimensionless variables

$$\xi = \frac{x}{\Delta}, \qquad \tau = \frac{u_0}{\Delta}, \qquad \eta(\xi,\tau) = \frac{u}{u_0}. \qquad (13)$$

In terms of these reduced variables the KdV equation (3) becomes

$$\eta_\tau - 6\eta \cdot \eta_\xi + \frac{1}{\sigma^2} \cdot \eta_{\xi\xi\xi} = 0, \qquad (14)$$

where $\eta(\xi,0) = \phi(\xi)$ and the nonlinearity index or the similarity parameter σ of the medium

$$\sigma = \Delta \sqrt{\frac{u_0}{p}}. \qquad (15)$$

Its value is characteristic for the type of solution to be obtained; i.e. identical initial conditions but different σ yield qualitatively different results. The number of solitons which emerge from the evolution of the initial perturbation is determined by the similarity parameter and the form of the initial profile $\phi(\xi)$.

Table 2. Deformabilities β and deformation lengths and soliton amplitude A_0 in even Pd-isotopes, excited by 26 MeV proton scattering [11]

	^{104}Pd	^{106}Pd	^{108}Pd	^{110}Pd
$E_{exc}(2_1^+)$ [MeV]	0.556	0.512	0.434	0.374
$\beta(2_1^+)$		0.220		0.223
$\beta \cdot R$ [fm]		1.215		1.276
A_0 [MeV]	17.08	17.85	20.63	24.47
a_n [MeV]	1.09	1.03	1.03	0.85
$E_{exc}(2_2^+)$ [MeV]	1.342	1.129	0.930	0.814
$\beta(2_2^+)$		0.319*		0.305
$\beta \cdot R$ [fm]		1.750		1.745
$E_{exc}(2_3^+)$ [MeV]	1.796	1.557	1.444	1.209
$\beta(2_3^+)$		0.551		0.526
$\beta \cdot R$ [fm]		3.040		3.004

The amplitude of the solitary waves arising from the initial perturbation is determined e.g. by the discrete spectrum of the Schrödinger equation [7,9]

$$\psi_{\xi\xi}(\xi,0) + \sigma^2 \cdot [\epsilon + \phi(\xi)] \, \psi(\xi,0) = 0 \tag{16}$$

e.g. the reduced amplitude of the soliton $a_n = -2A_0\epsilon$ and this initial reduced amplitude is steady for every medium. The cross section of the inelastic scattering in terms of the DWBA can be presented in the following factorized form

$$\frac{d\sigma}{d\Omega} = \text{const} A_0^2 \sigma_{\text{calc}} \tag{17}$$

Thus we have the starting perturbation in the soliton shape. We already noted that the changing of the nuclear density takes place in the radial direction, so that the further dynamics of the perturbation is governed by the KdV equation.

Expression (22) provides the key to the understanding of the behaviour of the deformation lengths for different targets and for different projectiles. In the paper by KOSTER et al. [10] the empirical formula

$$(\beta_2 r_0)_{\text{nucleons}} = \frac{Z}{A_T} (\beta_2 r_0)_{\text{protons}} + \frac{N}{A_T} (\beta_2 r_0)_{\text{neutrons}} \tag{18}$$

is given to explain the different values of the deformation lengths for even isotopes of Pd (see Ref. [11]). In Ref. [2] the increase of the deformation parameter with the filling of the n-shells was shown. It is necessary to say that the experimental features of even-even isotopes of Pd, Xe, Ba and other nuclei have been well described by the IBM using a Hamiltonian close to the O(6) symmetry [12]. The O(6) symmetry provides only an approximate

172

Table 3. Similarity parameters of the soliton-like excitation of the first 2^+-states in some even Palladium isotopes.

	^{104}Pd	^{106}Pd	^{108}Pd	^{110}Pd
σ (α–scattering)	1.33	1.54	1.78	1.85
σ (proton–scattering)	1.54	1.57	1.71	1.64

description of the nuclei investigated.

The compatibility condition for solitons in different media reads [13]

$$A\,p = A'\,p' = \text{const}, \qquad (19)$$

where A and A' correspond to the formula (4) and

$$p = \frac{\hbar^2}{2\,M} \qquad p' = \frac{\hbar^2}{2\,M'}$$

where M and M' denote the reduced masses. The product $A\,p$ should be the same for different isotopes of the target nucleus. The resulting prediction is that the deformation lengths of the first 2^+ states in the case of inelastic scattering of α-particles on different Pd isotopes should be largest for the heaviest one. The order of magnitude of the deformation lengths in the case of inelastic scattering of protons is close to the set of the deformation lengths for Pd isotops in the case of α–particles.

In table 1 and table 2 we present the properties of the first three 2^+-states, excited in the inelastic scattering of α-particles with an energy of 30 MeV and protons with energy 25 Mev [11,15] on even isotopes of Pd. Results of first order coupled channels parameter search calculations on the level system $<$gs $\ 2_1^+ \ \ 2_2^+ \ \ 2_3^+ >$ are given. Initially we used the same matrix elements as given by TAMURA in Ref.[14]. The potential parameters, taken from Ref.[11] for α-particles and for protons, taken from Refs. [15], have not been changed.

The ratio of amplitudes is supported by the vibrational model. The level structures between the extreme limits of harmonic vibrator and symmetric rotor $(E_{4_1}^+/E_{2_1}^+ = 2.0 - 3.33)$ demand some anharmonicity [16,17]. From these results it follows that the solitons can be considered to represent phonons. Since the initial perturbation decays into different solitons, it has the asymptotic form

$$U = U_0 + U_1 + U_2 + U_3 + U_4 + \cdots \qquad (20)$$

Following the work of KARPMAN [18] the maximum number of solitons into which the initial perturbance decays can be determined according to the

Table 4.   The 2^+ level energy (keV) for certain isotopes.

N	Ba	Ce	Nd	Sm	Gd	Dy	Er	Yb	Hf	W
82	1435.8	1596.2	1575.8	1660.2	1972.0	1677.3	1578.9			
84	602.4	641.3	696.5	747.1	784.4	803.0	807.3	821.3		
86	359.5	397.2	453.8	550.3	638.0	613.8	560.0	536.4		
88	199.0	285.5	301.7	333.9	344.3	334.6	344.5	358.2	389.6	
90	181.1	158.5	130.3	121.8	123.1	137.8	192.2	243.1	285.0	332.7
92	141.8	97.2	73.1	82.0	89.0	98.9	125.8	166.9	211.1	251.7
94		81.7	71.1	75.9	79.5	86.8	102.0	123.4	158.5	199.3
96			67.2	72.8	75.3	80.7	91.4	102.4	124.0	156.7
98				70.6	72.1	73.4	80.6	87.7	100.8	123.2
100					73.7	76.6	79.8	84.3	95.2	113.0
102						78.7	78.7	91.0	107.8
104							76.5	88.4	106.1
106							82.1	93.2	103.6
108							84.0	93.3	100.1
110								97.8	111.2
112								107.4	122.6

relation

$$N = \Delta \cdot \sqrt{\frac{U_0}{p}}, \qquad \Delta = 2a \tag{21}$$

As shown above the constant p is in the present case given by (12). For the scattering of α-particles with a kinetic energy of e.g. 30 MeV on Pd isotopes we obtain the value $N \approx 1$ to 3. This corresponds to the number of excited states with a given parity and given spin, e.g. 2^+, as formed in the potential well of (4) in the region of the nuclear surface.

The ratio of reduced amplitudes for the alpha-particles and protons (see Tables 1,2) shows that the number of solitons (28) excited in the inelastic scattering of alpha-particles is two times more then for the protons.

According to this qualitative nonlinear approach, where in an inelastic event the initial perturbation in the shape of a soliton in the initial channel (i.e. medium 1) decays into smaller solitons in the exit channel (medium 2), it is necessary to note, that for the states with the same parity and with the same spin a sum rule holds.

From the Tables 1,2 one can see the approximately equal values of the soliton amplitudes for the light and heavy isotopes of Pd. This is a consequence of the increasing deformation lengths. Soliton amplitudes were calculated according to the formula (4), were we used dimensionless values ϵ (15).

The decay of a soliton (in medium 1) into a few smaller solitons (in medium 2) the overall "volume" is conserved and the individual (soli-

174

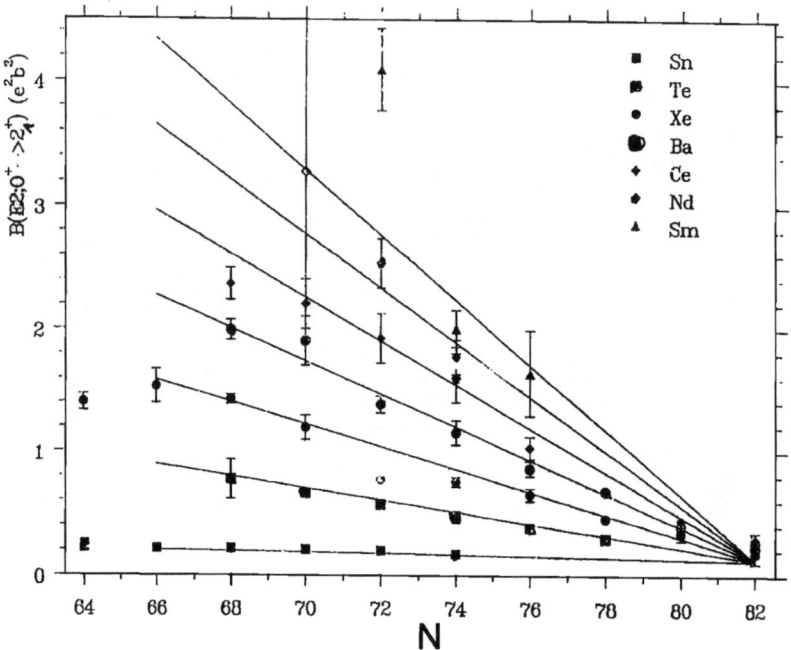

Figure 1. Dependence of the reduced electric probabilities for the isotopes Ba, Ce, Nd, Sm, Gd, Dy, Er, Yb, Hf, W from N and Z.

ton) decay products themselves assume (for large times) as well sech² - shapes. Moreover, it is possible to observe, that $E_{exc}(2_2^+) \approx 2 \cdot E_{exc}(2_1^+)$ and $E_{exc}(2_3^+) \approx 3 \cdot E_{exc}(2_1^+)$. This is in accordance with the similarity principle for solitons as well as with the assumptions of the harmonic vibrational model.

For the special case of inelastic scattering of protons and α-particles it is now very simple to evaluate the similarity parameter σ (See Table 3):

$$\sigma = \Delta \cdot \sqrt{\frac{U_0}{p}} = 2\,a\,\sqrt{\frac{A}{p}} \qquad (22)$$

The nuclear potential parameters as well as the 2^+ - deformabilities have been taken from experiments on α-particle and proton scattering [11,19]. A distance-of-interaction parameter was used according to HENDRIE [20]:

$$R_0 = r_0 \cdot \{A_{\text{target}}^{1/3} + A_{\text{projectile}}^{1/3}\}$$

Considering the tidal nature of nuclear solitons — the reduced mass μ in $p = \hbar^2/2\mu$ is the reduced nucleon mass — we should obtain the same value of the similarity parameter for protons and α-particles for every isotope of palladium. In fact, this is shown to be true within several percent.

By fixing the similarity parameters we confirmed the hypothesis of HEN-DRIE [20], that the heavier projectile yields the smaller parameter of deformation. The behaviour of the deformation lengths and the decrease of the energy of the first 2^+ states in the isotopes of Pd (from light isotopes to the heavy ones) is understandable in terms of soliton theory. There is increasing in the widths the soliton, also the decreasing of the amplitudes of the soliton, in other words, the increasing of effective soliton masses.

In terms of solitons the behavior of reduced electric probabilities for the isotopes of Ba, Ce, Nd, Sm, Gd, Dy, Er, Yb, Hf, W from N and Z is presented on Fig. 1 and Table 4, where one can see the first 2^+ states of the isotopes of considered elements together with Fig. 1 confirms the soliton-like character of quadrupole excitations in nuclei with A=100-170.

3. Conclusion

The purpose of the analysis of the present paper has been to clarify the difference in the deformation lengths for different projectiles and also for a fixed projectile incident on different isotopes of an element. The present paper is phenomenological in the sense that we used the phonon description for high excitations. Thus this analysis is a preliminary one, having so far a rather qualitative character.

We exploited the notion, that the perturbation, which is responsible for the inelastic transition, has a soliton shape. Then we took from soliton theory another idea: The dynamic of the potential of the Schrödinger equation is governed by the KdV equation. The properties of solitons help to understand some specific features of the inelastic scattering of protons and α-particles on the even-even isotopes of the elements with A=100–170.

References

1. G. R. Satchler, Direct Nuclear Reactions, Oxford University Press, 1983.
2. I. Yan, O. Vogel, P. von Brentano, and A. Gelberg, Phys. Rev., 48C (1993) 1046
3. J.H. Hamilton, A.V. Ramayya et al., ENAM-2001 Conf., July 2–7, 2001, Hameenlinna, Finland, Abstracts, H2–5.
4. N. Austern, Direct Nuclear Reaction Theories, Wiley, NY, 1970.
5. D.J. Korteweg, G. de Vries, Phil.Mag. 39 (1895) 422

6. Yu.A. Berezin, V.I. Karpman, Sov. Phys. JETP, 24 (1967) 1049
7. R.M. Miura, SIAM Review, 18 (1976) 412
8. E.F. Hefter, I.A. Mitropolsky, J. Mosc. Phys. Soc., 1 (1991) 99
9. C.S. Gardner, J.M. Greene, M.D. Kruskal, R.M. Miura, Phys. Rev. Lett., 19 (1967) 1095
10. J.E. Koster et al., Phys. Rev., C49 (1994) 710
11. V. Riech, R. Scherwinski, G. Lindström, E. Fretwurst, K. Gridnev, P.P. Zarubin, R. Kolalis, Nucl. Phys., A542 (1992) 61
12. Ka–Hae Kim, A. Gelberg, T. Mizusaki, T. Otsuka, P. Von Brentano, Nucl. Phys., A604 (1996) 163
13. K.A. Gridnev, and E.F. Hefter, Phys. Lett. 177A (1980) 490
14. T.Tamura, Rev. Mod. Phys. 37 (1965) 679
15. J. Bang, K.Y. Chan, A.B. Kurepin, Ö. Saethre, Nucl. Phys., A122 (1968) 34
16. A. Wolf, R.F.Casten, N.V. Zamfir, D.S. Brenner, Phys. Rev., C49 (1994) 802
17. R.F. Casten, N.V. Zamfir, D.S. Brenner, Phys. Rev. Lett. 71 (1994) 227
18. V.I. Karpman, Phys.Lett., 25A (1967) 708
19. F.D. Becchetti and G.W. Greenless, Phys. Rev., 1969 182 1190
20. D.L. Hendrie, Phys. Rev. Lett., 31 (1973) 478

DEFORMED TWO-CENTER SHELL MODEL

R. A. GHERGHESCU[1,2] D. N. POENARU[1,2,3], W. GREINER[2]

[1] *National Institute of Physics and Nuclear Engineering,*
PO Box MG-6, RO-76900 Bucharest, Romania
E-mail: rgherg@ifin.nipne.ro
[2] *Institut für Theoretische Physik der Universität,*
Pf 111932, D-60054 Frankfurt am Main, Germany
[3] *Department of Physics, Vanderbilt University, Nashville, TN 37235, USA*

The model yields the energy levels of a binary nuclear system. Two deformed oscillator wells are in direct correspondence with the spheroidally deformed fragments. A third potential is introduced to account for the influence of the neck between the fragments. Spin-orbit and l^2 energy contributions are calculated with shape dependent operators. Results are presented for two mass asymmetries in the splitting of $^{306}122$.

1. Introduction

In the 1970's, W. Greiner and the Frankfurt school developed the asymmetric two-center shell model [1]. It was an important step for the study of mass asymmetry in binary nuclear fission. The model considered two spherical asymmetric fragments subjected to two-oscillator type potential. Applications for nuclear molecular states experimentally observed have been developed in light excited systems [2], $^{17}O + ^{12}C$ reactions with emphasis on Landau-Zener effect [3], and many other asymmetric systems [4]. The importance of an adequate description of cold fission, cluster radioactivities and alpha decay in terms of an asymmetric and *deformed* single particle shell model with more realistic shapes during fission and fusion processes was repeatedly stressed [5]. Recently, attempts have been made to use the two-center shell model in synthesis and decay of superheavy nuclei. The only way a superheavy element can survive is due to the shell effects [6]. Up to now, all the variants of two-center shell models used spherical nuclei. A deformed two-center shell model (DTCSM) is proposed, where the main part of the potential consists of two ellipsoidaly deformed Nilsson type oscillators for axially symmetric shapes. Calculations are presented for exit channels of superheavy nucleus $^{306}122$.

2. Shapes

Fig. 1 shows the main geometrical parameters defining the axially symmetric shape family DTCSM is dealing with. Two ellipsoids (the deformed fragments) with semiaxes a_1, b_1 and a_2, b_2 are, at a certain moment, separated at a distance R between the two centers O_1 and O_2. A sphere centered in O_3 with radius R_3 is rolling around the symmetry axis, being tangent all the time to the two ellipsoids. The necking region, is generated in this way. Thus we have five independent parameters to design the deformation space: two fragment shape asymmetries $\chi_1 = b_1/a_1$, $\chi_2 = b_2/a_2$ (if a_1 and a_2 are given as the correspondent of β_2 for every A_1 and A_2, the other semiaxes are calculated from the total volume conservation condition), mass asymmetry $(A_1 - A_2)/(A_1 + A_2)$, the neck radius R_3 and the distance between centers R.

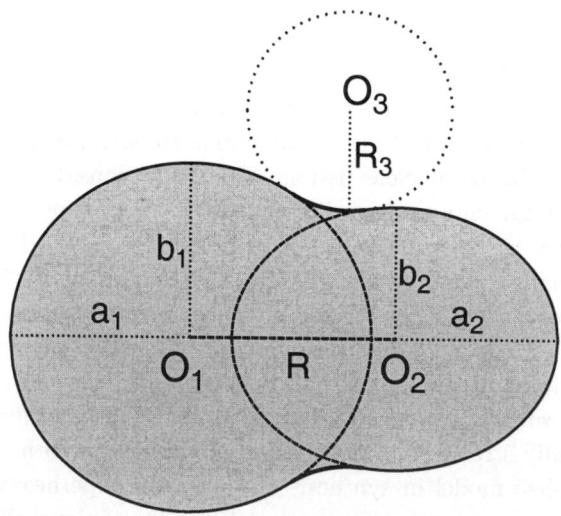

Figure 1. Fission like nuclear shape for two spheroidally deformed fragments with neck region. Besides the mass and charge asymmetry, the geometrical parameters of the model are: the two ratios of the spheroids semiaxes: b_1/a_1 and b_2/a_2, the neck sphere radius R_3 and the distance between centers R.

3. The potential

The deformed two center oscillator potential which is proposed here reads:

$$V_{DTCSM} = \begin{cases} V_1 = & \frac{1}{2}m_0\omega_{\rho_1}^2\rho^2 + \frac{1}{2}m_0\omega_{z1}^2(z+z_1)^2 & ,v_1 \\ V_{g1} = 2V_0 - [\frac{1}{2}m_0\omega_g^2(\rho-\rho_3)^2 + \frac{1}{2}m_0\omega_g^2(z-z_3)^2] & ,v_{g1} \\ V_{g2} = & V_0 & ,v_{g2} \\ V_2 = & \frac{1}{2}m_0\omega_{\rho_2}^2\rho^2 + \frac{1}{2}m_0\omega_{z2}^2(z-z_2)^2 & ,v_2 \end{cases} \quad (1)$$

where v_1, v_{g1}, v_{g2} and v_2 are the regions where the corresponding potentials are active and are calculated from the matching conditions starting with:

$$V_1(\rho, z) = V_2(\rho, z) \quad (2)$$

and leading to:

$$\frac{\rho^2}{\dfrac{\omega_{z1}^2\omega_{z2}^2 R^2}{(\omega_{\rho_2}^2 - \omega_{\rho_1}^2)(\omega_{z2}^2 - \omega_{z1}^2)}} + \frac{(z+k)^2}{\dfrac{\omega_{z1}^2\omega_{z2}^2 R^2}{(\omega_{z2}^2 - \omega_{z1}^2)^2}} = 1 \quad (3)$$

which defines a matching ellipsoid. The heavy deformed fragment O_1 is acting through $V_1(\rho, z)$ outside the matching potential ellipsoid (MPE). The light fragment part is contained inside MPE. Hence its action on the Hilbert space, $V_2(\rho, z)$, goes all over the matching ellipsoid volume. The same is valid for $V_1(\rho, z) = V_{g1}(\rho, z)$, and $V_2(\rho, z) = V_{g2}(\rho, z)$, as far as the neck potential shares its action space with the fragments.

4. Total Hamiltonian

The total Hamiltonian:

$$H_{DTCSM} = -\frac{\hbar^2}{2m_0}\Delta + V_{DTCSM}(\rho, z) + V_{\Omega s} + V_{\Omega^2} \quad (4)$$

is not separable. A basis is needed and diagonalization of oscillator potential differences and of angular momentum dependent operators has to be performed.

A separable oscillator Hamiltonian is obtained if one takes $\omega_{\rho_1} = \omega_{\rho_2} = \omega_1$ hence:

$$V^{(d)}(\rho, z) = \begin{cases} V_1^{(d)}(\rho, z) = \frac{1}{2}m_0\omega_1^2\rho^2 + \frac{1}{2}m_0\omega_1^2(z+z_1)^2 & ,z \leq 0 \\ V_2^{(d)}(\rho, z) = \frac{1}{2}m_0\omega_1^2\rho^2 + \frac{1}{2}m_0\omega_2^2(z-z_2)^2 & ,z \geq 0 \end{cases} \quad (5)$$

This is a two-center potential for a sphere ($z \leq 0$) intersected with a vertical spheroid. As a result of variable separation, an eigenfunction system and three differential equations are obtained for harmonic functions, Laguerre polynomials and Hermite functions dependent solutions [7].

The eigenvalues for the diagonalized Hamiltonian are:

$$E_{osc}^{(d)} = \hbar\omega_1(2n_\rho + \mid m \mid +1) + \hbar\omega_{z_1}(\nu_1 + 0.5) \tag{6}$$

To obtain the DTCSM energy levels, the matrix of the non-diagonal elements should be constructed. Then, adding $E_{osc}^{(d)}$, after diagonalization we obtain the eigenvalues. The difference $V_1(\rho, z) - V^{(d)}(\rho, z)$ defined in the region v_1 mentioned above and added to $E^{(d)}(n_\rho, \nu, m)$ produces the suitable matrix elements for V_1 and similarly for V_2 on v_2. The neck potential $V_g(\rho, z)$ is filling its volume v_g without any subtraction $\Delta V_g = V_{g1} + V_{g2}$.

The anticommutators are used to assure hermicity for the following operators:

$$V_{\mathbf{\Omega s}} = V_{\mathbf{\Omega s}}(v_1) + V_{\mathbf{\Omega s}}(v_2) + V_{\mathbf{\Omega s}}(v_g) \tag{7}$$

$$V_{\mathbf{\Omega s}}(v_1) = -\frac{\hbar}{m_0\omega_{01}}\kappa_1\{\mathbf{\Omega s}, (v_1)\} \tag{8}$$

$$V_{\mathbf{\Omega s}}(v_2) = -\frac{\hbar}{m_0\omega_{02}}\kappa_2\{\mathbf{\Omega s}, (v_2)\} \tag{9}$$

$$V_{\mathbf{\Omega s}}(v_g) = -\frac{\hbar}{m_0\omega_{01}}\kappa_1\{\mathbf{\Omega s}, (v_g^{(1)})\} - \frac{\hbar}{m_0\omega_{02}}\kappa_2\{\mathbf{\Omega s}, (v_g)^{(2)}\} \tag{10}$$

For the operators we have:

$$\mathbf{\Omega}^+(v_1) = -e^{i\varphi}\left[m_0\omega_{\rho_1}^2\rho\frac{\partial}{\partial z} - m_0\omega_{z_1}^2(z+z_1)\frac{\partial}{\partial\rho} - \frac{i}{\rho}m_0\omega_{z_1}^2(z+z_1)\frac{\partial}{\partial\varphi}\right]$$

$$\mathbf{\Omega}^-(v_1) = e^{-i\varphi}\left[m_0\omega_{\rho_1}^2\rho\frac{\partial}{\partial z} - m_0\omega_{z_1}^2(z+z_1)\frac{\partial}{\partial\rho} + \frac{i}{\rho}m_0\omega_{z_1}^2(z+z_1)\frac{\partial}{\partial\varphi}\right]$$

$$\mathbf{\Omega}_z(v_1) = -im_0\omega_{\rho_1}^2\frac{\partial}{\partial\varphi} \tag{11}$$

with

$$\mathbf{\Omega s}(v_1) = \frac{1}{2}(\mathbf{\Omega}^+(v_1)\mathbf{s}^- + \mathbf{\Omega}^-\mathbf{s}^+) + \mathbf{\Omega}_z(v_1)\mathbf{s}_z \tag{12}$$

The same aplies for (v_2) and (v_g) spin-orbit operators terms. Now the total matrix elements for DTCSM can be calculated as:

$$\langle i|DTCSM|j\rangle = E_{osc}^{(d)}(n_\rho, |m|, \mu) + \langle i|\Delta V_1|j\rangle + \langle i|\Delta V_2|j\rangle +$$
$$\langle i|V_g|j\rangle + \langle i|V_{\mathbf{\Omega s}}|j\rangle + \langle i|V_{\mathbf{\Omega}^2}|j\rangle \tag{13}$$

5. Level schemes and shell effects

DTCSM spectra are computed for the superheavy fission reaction channel $^{306}122 \to^{198} W +^{108} Cd$, with the nuclei deformations $\chi_{122} = 0.9$, $\chi_W = 1$ and $\chi_{Cd} = 0.83$ as a function of elongation $(R - R_i)/(R_f - R_i)$. Fission of $^{306}122$ through two channels is studied: $^{306}122 \to^{198} W +^{108} Cd$ and $^{306}122 \to^{154} Gd +^{152} Ce$. The two corresponding level schemes are plotted on the upper side of Fig. 2.

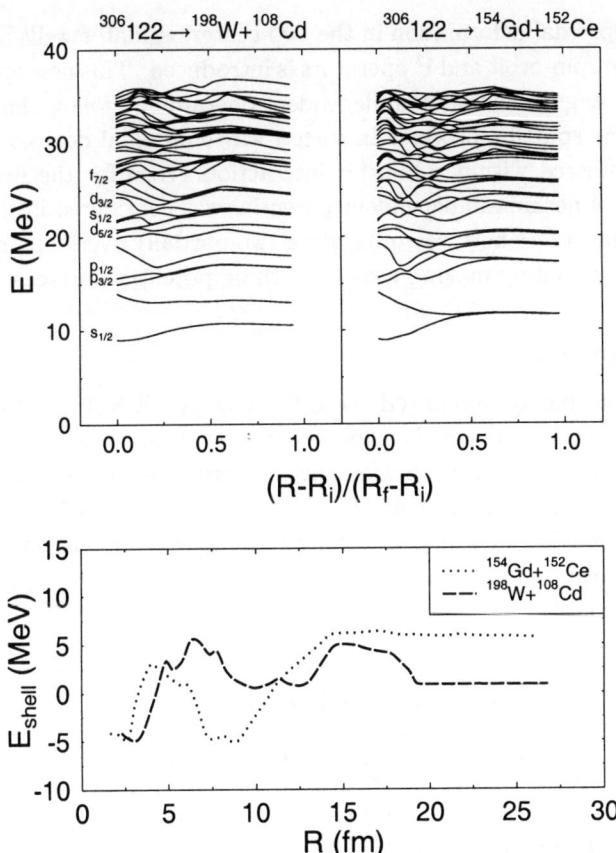

Figure 2. Two level-spectra with different mass asymmetry for the same superheavy parent $^{306}122$ are presented. The lower part of the figure represents the corresponding shell corrections as a function of R.

Calculations in this case preserved $R_3 = 4$ fm. Shells are more clearly visible at $(R-R_i)/(R_f-R_i) = 1$ in the symmetric case, since asymptotically the levels are practically coincident for the two quasi-equal fragments. The lower part of the figure represents the shell corrections calculated using Strutinsky's method with the above level schemes as an input. Asymptotically there is a $\Delta E \approx 4$ MeV difference in between the two channels, suggesting a possible fission path along the mass asymmetry degree of freedom.

6. Conclusions

Fragment ellipsoidal deformation in the two-center oscillator wells and further on, in the spin-orbit and $\mathbf{l}^2$ operators is introduced. This new approach allows for the angular momentum dependent operators to follow the shapes throughout the splitting process. In such a way ellipsoidal degrees of freedom are considered within spin-orbit interaction. Also for the first time the influence of necking-in dependent potential was considered in the level scheme calculation. A new way of treating two partially overlapping nuclei has been introduced by making use of matching potential surfaces.

Acknowledgments

This work was partly supported by UNESCO (UVE-ROSTE Contract 875.737.2), by the Centre of Excellence IDRANAP under contract ICA1-CT-2000-70023 with European Commission, Brussels, by Bundesministerium für Bildung und Forschung (BMBF), Bonn, Gesellschaft für Schwerionenforschung (GSI), Darmstadt, and by Ministry of Education and Research, Bucharest.

References

1. J. Maruhn and W. Greiner, Z. Phys. **251**, 431 (1972).
2. W. von Oertzen, Z. Phys. **354**, 37 (1996).
3. A. Thiel, J. Phys. G: Nucl. Part. Phys. **16**, 867 (1990).
4. J. W. Park, W. Scheid and W. Greiner, Phys. Rev. **C21**, 958 (1980).
5. D. N. Poenaru and W. Greiner, *Nuclear Decay Modes* (Institute of Physics Publishing, Bristol, England, ch.6, 1996)
6. R. A. Gherghescu, J. Skalski, Z. Patyk and A. Sobiczewski, Nucl. Phys. **A651**, 237 (1999).
7. E. Badralexe, M. Rizea and A. Sandulescu, Rev. Roum. Phys. **19**, 63 (1974).
8. A. C. Wahl, Atomic Data Nucl. Data Tables, **39**, 1 (1998).

Gamma rays from different Fission Sources

W.R.Phillips
Department of Physics and Astronomy, University of Manchester,
Manchester , M13 9PL, U.K.

Abstract

This paper briefly discusses yields of secondary fragments from fission processes which produce the most neutron-rich nuclei whose near-yrast structures can be studied with present γ-ray arrays. The observed yields of the strongest even-even secondary fragments may be closely fitted using semi-empirical models. Fission of ^{235}U by thermal neutrons produces observable nuclei with very similar neutron to proton ratios to those given by spontaneous fission sources. Fission of^{238}U by neutrons with energies just above threshold gives observable nuclei with slightly higher N/Z ratios. The angular momentum in the fragments from the different sources is briefly discussed. The average angular momentum in ^{235}U and ^{238}U secondary fragments is similar to that found in the SF sources.

1. Introduction

Independent yields from fission processes which can most usefully be used to study the structure of neutron-rich nuclei are briefly reviewed, and the angular momentum in secondary fragments discussed. Attention is restricted to situations where large γ-ray arrays can profitably be used to establish yrast or near-yrast structures in secondary fragments nearest the neutron drip line. With present techniques, this effectively limits the discussion to fragments formed in the fission of ^{235}U and ^{238}U induced by low-energy neutrons, and to the spontaneous fission (SF) sources ^{248}Cm and ^{252}Cf. The structures up to medium spins of nuclei nearest the neutron drip-line have been studied [1,2] using the SF sources and other structure aspects investigated [3,4] using low-energy fission of the uranium isotopes. Fission of nuclei at high excitation, such as formed by charged particle bombardment, although giving slightly higher average angular momentum in fragments [1], results in more neutron emission than fission of nuclei at low excitation, and produces less neutron-rich secondary fragments.

Spallation produced by high-energy protons has been used [5] to produce a wide variety of nuclei. The β decays of separated products have been examined to reveal details of nuclei as neutron-rich as those seen in fission. However, β decays usually populate low-spin levels in daughter nuclei and these experiments will not be discussed here.

2. Independent yields

Primary fragments are formed at scission and these emit zero, one, two

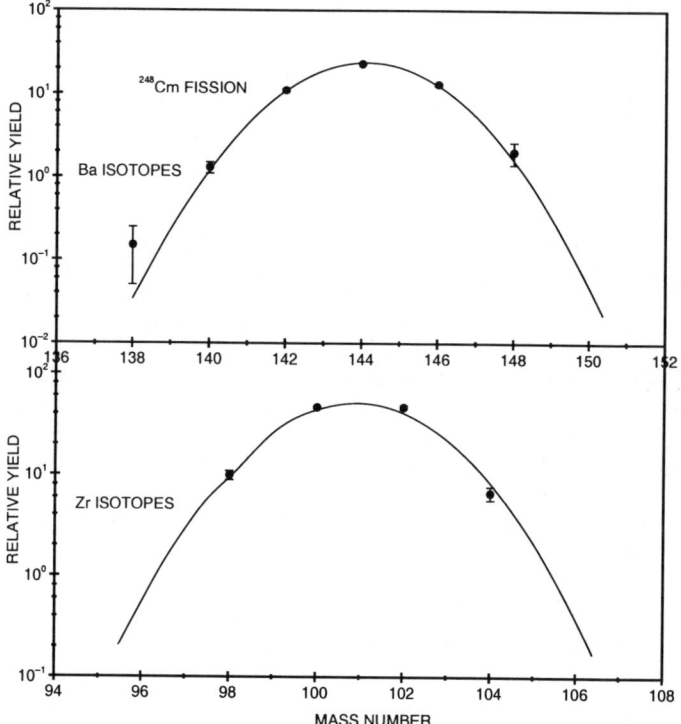

Figure 1: Experimental values of the relative yields of secondary Ba and Zr fragments in the SF of ^{248}Cm. For Ba fragments the full line corresponds to $A_P(56)=144.1$ and $\sigma_{A'}=1.65$, in the notation of Ref [6], and for Zr, $A_P(40)=101.1$ and $\sigma_{A'}=1.6$.

or more neutrons before γ-ray cascades lead to the ground states of *secondary fragments*. The yield of a secondary fragment is called its *independent yield* and is the yield which is important when discussing the extent of neutron-rich nuclei which can be studied as fission fragments using the techniques of γ-ray spectroscopy.

For a fissioning nucleus with mass number A_F and atomic number Z_F, the distribution of masses for particular secondary fragments with atomic number Z has been parametrized [6,7] using the most probable mass $A'_P(Z)$ of the primary fragments. The mass $A'_P(Z)$ within the model is taken to be that given by the unchanged charge model prediction with a small correction $\Delta A'_P$ for the variation of the ratio N/Z with mass.

$$A'_P(Z) = Z \left(\frac{A_F}{Z_F}\right) + \Delta A'_P.$$

The corresponding distribution of secondary fragment masses $A(Z)$ can be calculated in the model in terms of the difference between two error functions, giving an approximately Gaussian distribution around a most probable mass given by

$$A_P(Z) = A'_P(Z) - \overline{\nu_A}$$

for masses within three or four mass units either side of $A_P(Z)$. Here, $\overline{\nu_A}$ is the average number of neutrons emitted in the formation of secondary fragments of mass $A_P(Z)$. The full width at half-height of the distribution for different systems for which data are available is about 4 mass units.

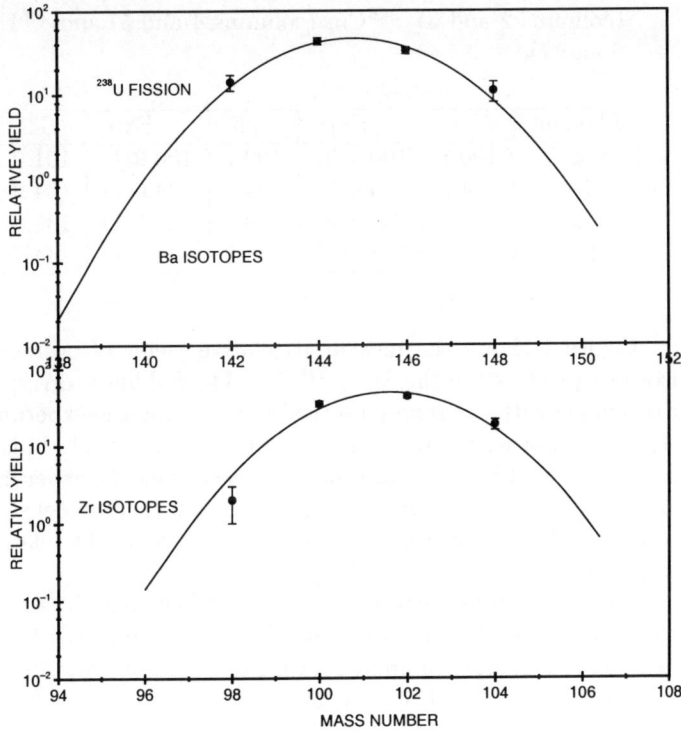

Figure 2: Experimental values of the relative yields of secondary Ba and Zr fragments in the fission of ^{238}U just above threshold. For Ba fragments the full line corresponds to $A_P(56)=144.8$ and $\sigma_{A'}=1.7$ and for Zr, $A_P(40)=101.6$ and $\sigma_{A'}=1.6$.

The model has been used to fit early information from the fission of ^{252}Cf and ^{235}U, and predictions agree closely with measured independent yields of the strongest even-even fragments from recent γ-ray array data. The model also predicts yields of odd-A and odd-odd nuclei, and where such yields can be obtained with reasonable accuracy from arrays there is again satisfactory agreement.

Table 1 shows $A_P(Z)$ values used with the model [6,7] and experimentally observed values for Zr, Te, Xe and Ba isotopes formed in the fission of ^{235}U, ^{248}Cm and ^{238}U, for which data have recently become available [4]. The mass A_P is a measure of the degree of neutron-richness of the secondary fragments and is higher for the fission of ^{238}U by neutrons of energy just above the threshold than for the other fission processes.

Table 1: Observed and model predicted [6,7] mean mass numbers of secondary fragments of nuclei formed in the fission of ^{235}U (columns 2 and 3), ^{248}Cm (columns 4 and 5) and ^{238}U (columns 6 and 7).

Element	Th.	Exp.	Th.	Exp.	Th.	Exp.
Zr	100.1	100.1(3)	100.7	100.9(1)	101.2	101.6(2)
Te	134.0	133.9(3)	134.5	134.6(1)	134.9	135.4(2)
Xe	138.8	138.4(2)	139.2	139.5(1)	139.8	140.2(2)
Ba	143.8	144.1(2)	143.8	144.1(1)	144.7	144.8(2)

Figure 1 shows the measured relative yields of even-even Ba and Zr isotopes produced in the SF of ^{248}Cm. The full lines corresponds to predictions made with the semi-empirical model using the experimental values of $A_P(Z)$ and other parameters chosen to give a best fit. Independent yields in the fission of ^{238}U can also be predicted using the experimental values of $A_P(Z)$ with the remaining parameters of the yield model sensibly adjusted to fit the observed isotopic yields. Figure 2 shows the relative yields of Ba and Zr isotopes and the model predictions.

The model gives satisfactory fits to all the yield data and may be used to estimate the degree to which ^{238}U fission enhances the production of neutron-rich species compared with SF. Figure 3 shows the isotopic yields as a fraction of the yields at the most probable secondary fragment mass A_P (on the neutron-rich side of A_P) for Ba and Zr nuclei in the two processes. The ratio of the yield of so far unobserved ^{150}Ba to the yield at A_P is about a factor of four greater in ^{238}U fission than in the fission of ^{248}Cm; the ratio of the yield of so far unobserved ^{106}Zr to the yield at A_P is about a factor of five greater.

For very weakly populated isotopes (with yield less than 0.01 of the yield of the strongest isotope) the yield falls off more rapidly than predicted by a Gaussian function. The logarithm of the yields of very weakly populated nuclei in the thermal neutron-induced fission of ^{235}U have been shown [8] to depend linearly on the Q-value for the fission split. So far it has not been

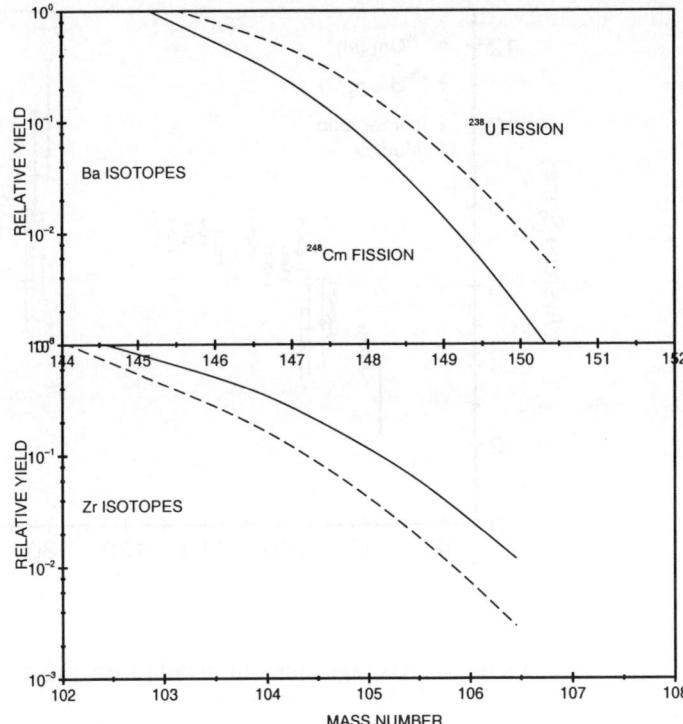

Figure 3: Predicted yields of neutron-rich Ba and Zr isotopes, as a fraction of the yields at the most probable mass A_P, for fission of ^{248}Cm and ^{238}U.

possible to extend spectroscopic studies to such weakly populated species.

3. Angular momentum in fragments

The spin distributions in fragments, together with the characteristics of the γ-ray array used, determine the range of spins over which level structures can be observed in any particular fragment.

The average angular momentum in secondary fragments is closely similar for ^{235}U and the SF sources. For ^{235}U and ^{248}Cm sources the populations of levels which emit discrete γ rays have been used [3] to determine the spin distributions after using statistical models to make small corrections for the emission of unobserved statistical γ rays. A further model-dependent correction [9] for the emission of neutrons directly after scission gives the spin distributions in primary fragments.

Figure 4 shows average spins in primary fragments determined in this way. The data for the two sources ^{235}U and ^{248}Cm closely overlap. The global average is close to $\sim 6\hbar$ and the trend is for the average spin to

188

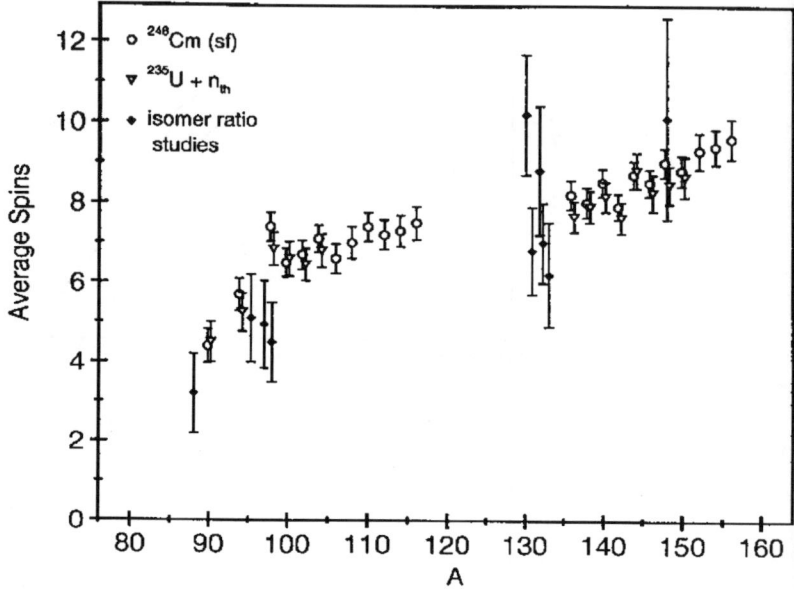

Figure 4: Average spins in primary fragments.

increase with mass number. It is noteworthy that early pioneering work [10] on fragment spins gave results for SF in agreement with these conclusions. It may also be noted that memory of the channel spins of 3 or 4 in the case of ^{235}U fission is lost and the results for the above three sources are consistent with suggested mechanisms for spin generation at scission, *viz* thermal excitation [11,12], macroscopic bending and wriggling modes [13,14] or orientation pumping [15]. All these mechanisms may play a part and all are accompanied by Coulomb excitation.

Present data [4] on fragments from the fission of ^{238}U do not allow accurate determinations of average spins. However, there is evidence that the spin distributions for fission of ^{238}U just above threshold are similar to those for the other processes. Figure 5 shows the relative intensities of $8 \to 6$, $6 \to 4$ and $4 \to 2$ γ rays in the decays of ^{100}Zr, ^{140}Xe, ^{104}Mo and ^{144}Ba in both the fission of ^{248}Cm and that of ^{238}U. The observation that the ratios are the same within the errors suggests that the spin distributions in both cases are much the same.

4. Uranium fission

The considerations of Section 2 suggest that study of ^{238}U fission may extend knowledge of neutron-rich nuclei. This section outlines some con-

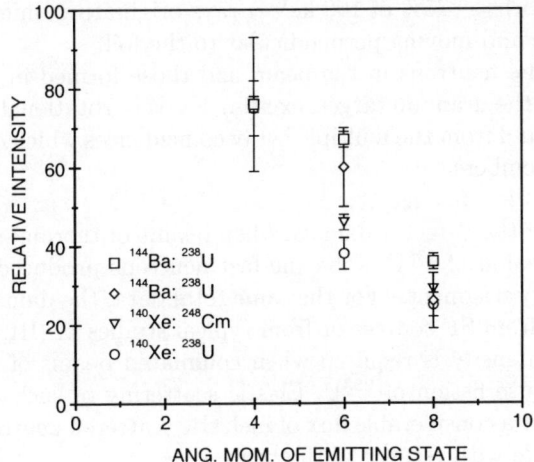

Figure 5: The relative intensities of γ rays de-exciting levels in the ground-state bands of secondary fragments following the fission of ^{238}U and ^{248}Cm.

siderations relevant to γ-ray experiments involving neutron-induced fission of uranium.

a Slowing down times

The kinetic energy of fragments is around 1 MeV per nucleon, an energy below the Bragg peak in the specific energy loss curve. The slowing down time in this regime is roughly inversely proportional to the density of the stopping medium, and uranium metal targets used in the fission of uranium nuclei give faster slowing down times of fragments than the low Z media in which SF sources have usually been embedded. Hence, lines corresponding to transitions from high-spin states of short lifetimes are sharper in uranium fission than in SF fission. Uranium fission experiments thus facilitate yield measurements and can pursue to higher spin states studies of lifetimes [16] based on line-shape measurements and of electromagnetic moments [17] based on transient and permanent field effects.

b. Target-beam requirements

A collimated beam of 10^5 thermal neutrons per second incident on 100 mg cm^{-2} of uranium highly enriched in ^{235}U gives more than 10^4 fissions per second. Energetic, collimated beams of neutrons are required to cause fission in ^{238}U, for which the fission cross section with neutrons of energies between 2 and 5 MeV is about 0.6 barns. To provide more than 10^4 fissions per second using a target of thickness 100 mg cm^{-2} requires a beam intensity of $\sim 10^8$ neutrons per second. Increasing the target thickness by a factor of ten reduces the beam intensity requirement, however, uranium targets strongly absorb low-energy γ rays. A 1 g cm^{-2} target (0.5 mm

190

thick) absorbs $\sim 75\%$ of 100 keV γ rays originating uniformly throughout its volume and moving perpendicular to the foil.

Both the neutrons in the beam and those formed in the fission events scatter off the uranium target, exciting levels in rotational bands and giving a background from the multiple γ-ray coincidences which occur in the decay of band members.

c Detector damage

Most of the detector damage when beams of thermal neutrons are used to cause fission of ^{235}U is via the fast neutrons produced in the decays of the primary fragments. For the same total data, the damage is comparable with that from SF sources or from typical studies of (HI, xn) reactions.

High intensity is required when collimated beams of fast neutrons are used to cause fission of ^{238}U. Elastic scattering of such beams is forward peaked, but a considerable flux of inelastic scattered neutrons hits detectors at any angle with respect to the beam.

The data from which these observations are culled were taken in collaboration with many people from various institutions. I am grateful to all colleagues, among whom Irshad Ahmad must have special mention, for enlightenment and enjoyment during my participation in this research.

References

[1]. I.Ahmad and W.R.Phillips, Rep.Prog. Phys. **58**, 1415 (1995)

[2.] J.H.Hamilton *et al*, Prog. Part. Nucl. Phys. **35**, 635 (1995)

[3.] J.A.Shannon, Ph. D. thesis, Manchester (1997)

[4.] W.R.Phillips, A.P.Byrne, G.D.Dracoulis, G.J.Lane, T.R.McGoram and R.Newman, Eur. Phys. J. **A3**, 205 (1998)

[5.] J.Shergur *et al*, Phys. Rev. **C65**, 03413 (2002)

[6.]A.Wahl, At. Data and Nucl. Data Tables **39**, 1 (1988)

[7] T.R.England and B.F.Rider, LA-UR-94-3106: ENDF-349 (1993)

[8.] H.Faust and G.Fioni, in *Workshop on Nuclear Fission and Fission Product Spectroscopy*, eds. H.Faust and G.Fioni, (ILL report 94FA05T,1994,p.91)

[9.] Y. Abdelrahman *et al*, Phys Lett. **199B**, 504 (1987)

[10.] J.B.Wilhelmy, E.Cheifetz, R.C.Jared, S.G.Thompson, H.R.Bowman and J.O.Rasmussen, Phy.s Rev. **C5** 2041 (1972)

[11.] L.J.B.Goldfarb, Nucl. Phys. **A465**, 529 (1987)

[12.] H.Rossner, J.R.Huizenga and W.U.Schroder, Phys. Rev. **C33**, 560 (1986)

[13.] J.R.Nix and W.J.Swiatecki, Nucl. Phys. **71**, 1 (1965)

[14.] J.O.Rasmussen, W.Norenberg and H.J.Mang, Nucl. Phys. **A136**, 465 (1969)

[15.] I.N.Mikhailov and P.Quentin, in *Proc. 2nd Int. Conf. on Fission and Properties of Neutron-rich Nuclei* eds. J.H.Hamilton, W.R.Phillips and H.K.Carter(World Scientific, 2000, p 384)

[16.] A.G.Smith *et al*, Phys. Rev. Lett. **73**, 2540 (1994)

[17.] D.Patel *et al*, J.Phys. G:Nucl. Part. Phys. **28**, 649 (2002)

COLLECTIVE BANDS IN Mo, Ru, AND Pd NUCLEI AND CHIRAL BANDS IN ^{106}Mo

S. J. ZHU[1,2,3], J.H. HAMILTON[2], A.V. RAMAYYA[2], P.M. GORE[2], J.O. RASMUSSEN[4], E.F. JONES[2], J.K. HWANG[2], R.Q. XU[1], L.Y. YANG[1], K. LI[1,2], Z. JIANG[1], Z. ZHANG[1], S.D. XIAO[1], X.Q. ZHANG[2], J. KORMICKI[2], Y.X. LUO[2,4], L. CHATURVEDI[1,5], W.C. MA[6], J.D. COLE[7], M.W. DRIGERT[7], I.Y. LEE[4], C. FOLDEN[4], P. FALLON[4], P. ZIELINSKI[4], K.E. GREGORICH[4], A.O. MACCHIAVELLI[4], M.A. STOYER[8], T.N. GINTER[9], G.M. TER-AKOPIAN[10], A.V. DANIEL[10], YU. TS. OGANESSIAN[10], AND R. DONANGELO[11]

[1] *Tsinghua University, Beijing, China*
[2] *Vanderbilt University, Nashville, TN 37235*
[3] *Joint Institute for Heavy Ion Research, Oak Ridge, TN 37835*
[4] *Lawence Berkeley National Laboratory, Berkeley, CA 94720*
[5] *Banaras Hindu University, Varanasi 221 005, India*
[6] *Mississippi State University, MS 39762*
[7] *Idaho National Environmental and Engineering Laboratory, Idaho Falls, ID 83415*
[8] *Lawrence Livermore National Laboratory, Livermore, CA 94551*
[9] *Michigan State Univ, East Lansing, MI 48824*
[10] *Joint Institute for Nuclear Research, RU-141980 Dubna, Russia*
[11] *Universidade Rio de Janeiro, Brazil*

Gamma-vibrational bands are extended to 14$^+$ or 17$^-$ in ^{104,106}Mo, ^{108,110,112}Ru, and ^{112,114,116}Pd. Quite different staggering patterns are observed from the odd spin member pushed up relative to the even member and visa-versa. The one- and two-phonon γ-vibrational bands in ^{104,106}Mo have nearly identical γ energies. A β-vibrational band is identified in ^{106}Mo. Several two quasi-particle bands are found in ^{104,106}Mo with moments of inertia essentially constant near the rigid body value. The first evidence for chiral doublet bands is found in an even-even nucleus based on a two quasi-neutron configuration with the υ $g_{7/2}$ orbital nearly filled and the υ $h_{11/2}$ one just filling.

1. Introduction

A variety of collective band structures is now observed to high spin in neutron-rich Mo, Ru, and Pd nuclei. These results were made possible by new high statistics data, 5.7 x 10^{11} triple and higher fold coincidences, obtained with 102 Ge detectors in Gammasphere and a 62 μCi spontaneous fission (SF) source of ^{252}Cf. Totally different staggering patterns are observed for the γ-vibrational bands in ^{104,106}Mo, ^{108,110,112}Ru, and ^{112,114,116}Pd. These differences present new challenges for microscopic descriptions that include significant quadrupole deformations and triaxiality in these nuclei. In extending them to higher spins, the two-phonon γ-vibrational bands in ^{104,106}Mo are found to have transition energies essentially identical to their one-phonon γ-bands. A new β-vibrational band based on the low lying known 0 state is identified. Numerous quasiparticle bands are observed in ^{104,106}Mo. Two bands with $\Delta I=1$ sequences, assigned a two quasineutron configuration, have all the properties expected for chiral doublet bands, including significant deformation, triaxial shapes, and a neutron $g_{7/2}$ orbital nearly filled and neutron $h_{11/2}$ one just beginning to fill. This is the first example of chiral doublets in an even-even nucleus.

2. γ-vibrational bands in ^{104,106,108}Mo, ^{108,110,112}Ru, and ^{112,114,116}Pd

The quality of the high statistical data in our experiment is shown in Fig. 1. This shows clearly the transitions in the γ-band in ^{112}Ru to 17^+ and the transitions feeding out to the ground band when gating on the new $13^+ \rightarrow 11^+$ and $11^+ \rightarrow 9^+$ transitions in the band. The one-phonon γ-vibrational bands have been extended to 14^+ in ^{104,106}Mo, to 13^+, 17^+ in ^{110,112}Ru and to 11^+, 15^+, 15^+ in ^{112,114,116}Pd.

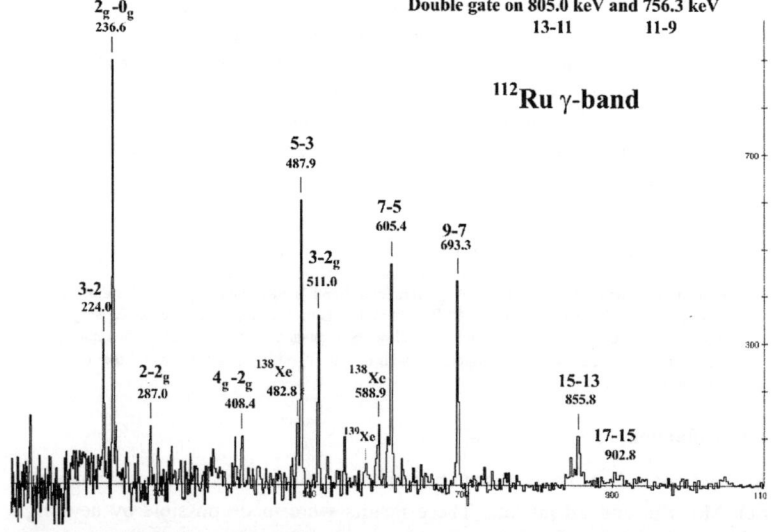

Figure 1. Spectrum from Gammasphere gated upon the 13^+ to 11^+ and 11^+ to 9^+ transitions of the γ-band of ^{112}Ru.

The 2^+ ground energies of 192.4, 171.8, 192.7, 242.1, 240.7, 236.6, 348.8, 332.6, and 340.5 keV in $^{104-108}$Mo, $^{108-112}$Ru, and $^{112-116}$Pd, respectively, show significant quadrupole deformation in the Mo nuclei with systematic decrease with increasing Z. The $2^+(\gamma)$ bandhead energies of 812.1, 710.5, 586.0, 707.9, 612.4,

523.6, 736.8, 694.8, and 738.3 keV, respectively, for ^{104}Mo to ^{116}Pd show a systematic decrease to a minimum in ^{112}Ru and then a sharp increase. These energies and the staggering patterns of the odd and even spin state energies indicate the presence of triaxial shapes in these nuclei. The marked variations in the energy staggering patterns for the odd and even spin state energies are illustrated in plots of the energy difference, e.g., $E(3_\gamma^+) - E(2_\gamma^+)$, $E(4_\gamma^+) - E(3_\gamma^+)$, etc., up the bands as shown in Fig. 2a., b., c. Several striking similarities and differences are seen in these figures. First, one notes that the lowest member of each triplet, ^{104}Mo, ^{108}Ru, and ^{112}Pd with N=62, 64, 66, shows the same staggering where the odd spin 3^+, 5^+, ... states are pushed up to be nearer the 4^+, 6^+, ... states, respectively.

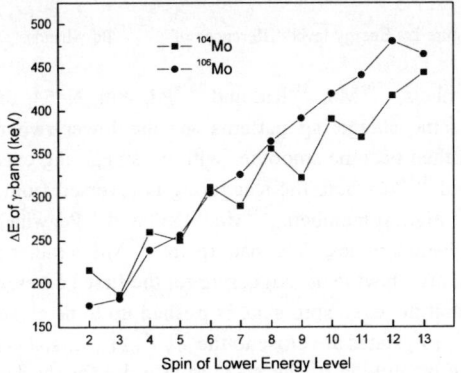

Figure 2a. Energy level differences of ^{104}Mo and ^{106}Mo γ-bands.

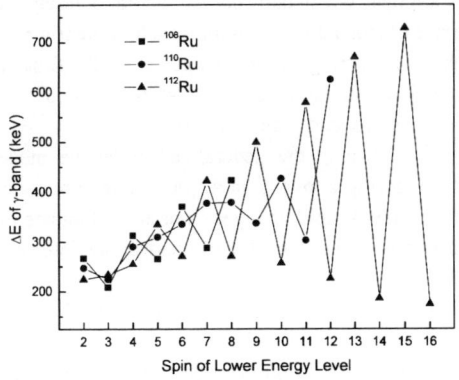

Figure 2b. Energy level differences of $^{108, 110, 112}$Ru γ-bands.

194

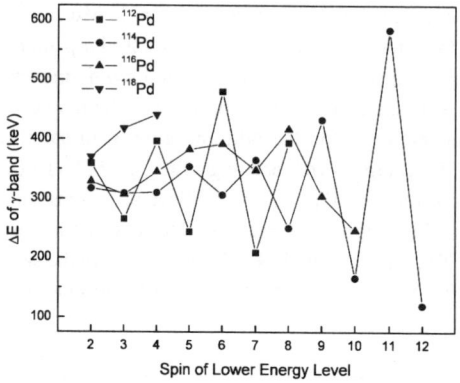

Figure 2c. Energy level differences of $^{112-118}$Pd γ-bands.

The middle members, ^{106}Mo, ^{110}Ru, and ^{114}Pd, with N=64, 66, and 68, show somewhat the same staggering patterns for the lower two or three energy differences and then become smoother with no staggering except at the higher spin in ^{110}Ru and ^{114}Pd where the staggering is reversed from that of ^{104}Mo – ^{112}Pd. The three highest members, ^{108}Mo, ^{112}Ru, and ^{116}Pd, with N=66, 68, and 70, have different characteristics. The pattern for ^{108}Mo is rather similar to that of ^{106}Mo but ^{112}Ru, after having no staggering for the first 3 cases, exhibits opposite staggering, so that the even spin state is pushed up to be close to the odd spin state, e.g., $6^+, 8^+, 10^+$, states are close to the $7^+, 9^+, 11^+, \ldots$ states in sharp contrast to ^{108}Ru and the magnitude of the staggering is by far the largest of all nine cases. Note ^{112}Ru also has the lowest $2^+(\gamma)$ energy of any of these nuclei while the $2^+(\gamma)$ energies are nearly constant for ^{112,114,116}Pd. Clearly these data show a rapidly changing role of triaxial shapes within each of the triplets. It is not simply the neutron member which triggers the changes since the first three members of each triplet which look so similar have N=62, 64, 66 and middle members of each triplet with different patterns start with N=64, 66.

A study of ^{108,110,112}Ru in the generalized collective model approach [1] indicated similar staggering above 6^+ in all three nuclei but such are not found now. These significant and rapid changes in patterns indicate the need for more microscopic descriptions of the interplay of quadrupole deformation and triaxial shapes in these nuclei.

Several examples of the J moment of inertia, which are very sensitive to structure changes, are shown in Figs. 3 a.,b.,c. The J_2s for the ground band and the one-phonon γ-bands are strikingly similar. When sharp changes occur in the ground band from backbending, similar sharp changes occur in J_2 for the one-phonon γ-bands. In ^{106}Mo, ^{108,110,112}Ru, and ^{116}Pd, the J_1s for the ground and one-phonon γ-bands are essentially identical below the backbends.

A β-vibrational band is identified from 0^+ to 10^+ in ^{106}Mo. The 0^+ bandhead is known from beta decay work [4]. This is the first β-type vibrational band to be reported in neutron-rich nuclei and provides an important additional structure for testing nuclear models. Its J_1 moment of inertia increases rather rapidly with spin in contrast to the ground and γ-vibrational bands.

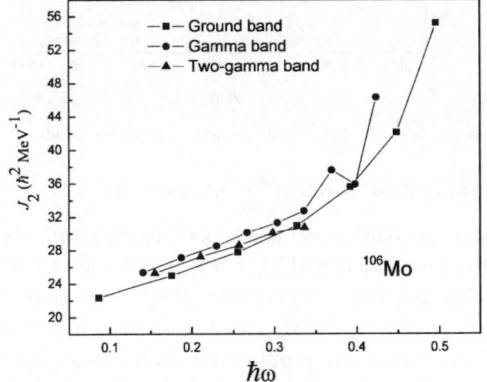

Figure 3a. J_2 vs. $\hbar\omega$ for ^{106}Mo ground, 1-phonon, and 2-phonon bands.

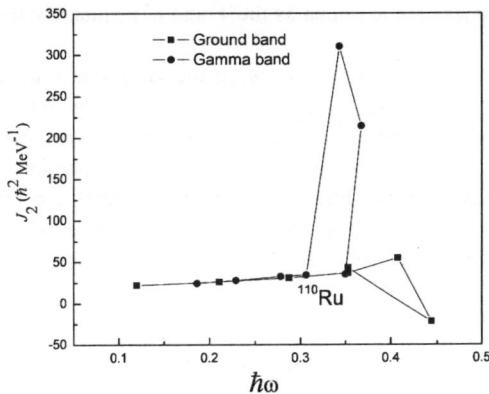

Figure 3b. J_2 vs. $\hbar\omega$ for ^{110}Ru ground and 1-phonon bands.

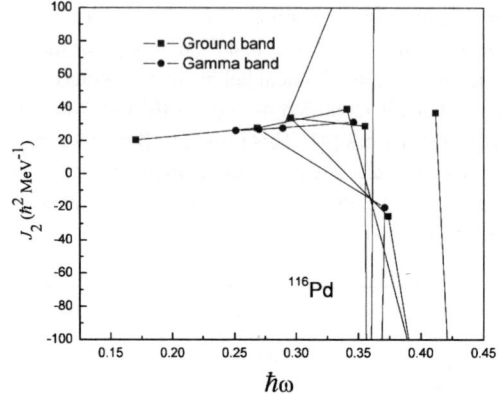

Figure 3c. J_2 vs. $\hbar\omega$ for [1116]Pd ground and 1-phonon bands.

3. Two-phonon γ-vibrational bands in [104,106]Mo and [112]Ru

One- and two-phonon γ-vibrational bands were reported earlier in [104,106]Mo [2,3] with $2^+(\gamma)$ and $4^+(\gamma\gamma)$ bandheads at 812.1, 710.5, 1683.4, and 1434.9 keV, extending to $9^+(\gamma)$, $7^+(\gamma\gamma)$, $8^+(\gamma)$, and $7^+(\gamma\gamma)$, respectively. We have extended these bands to $14^+(\gamma)$, and $9^+(\gamma\gamma)$, $12^+(\gamma\gamma)$ in [104,106]Mo (see Fig. 4) as well as identifying 5 new cascades and 20 new transitions feeding out of these bands. The two-phonon γ-bandhead energies are very close to twice the one-phonon γ-bandhead. In our recent [112]Ru work we have identified two states at 1183 and 1759 keV which primarily feed into the one-phonon γ-band. On the basis of the energies and decay patterns, these are assigned as the 4^+ and 6^+ members of the two-phonon γ-band in [112]Ru.

The energies of the transitions between the same spin state, e.g., $6^+ \to 4^+$, $7^+ \to 5^+$, ... in the one- and two-phonon γ-bands differ by only 0 up to 4% as shown in Fig. 4. The new $6^+ \to 4^+$ two-phonon γ-band transition assigned in [112]Ru is only 2% lower than the same transition in the one-phonon γ-band. The J_2 moments of inertia of the two-phonon γ-band likewise track very closely those of the ground and one-phonon γ-band as shown in Fig. 3 a.,b.,c. where $\hbar\omega$ includes K. Indeed the one- and two-phonon γ-bands are strikingly similar.

4. Chiral bands in [106]Mo

A number of two quasi-particle bands are observed to reasonably high spins in [104,106]Mo. In every case, their J MOI are essentially constant and close to the rigid body value. These large, constant MOI could indicate blocking of the pairing interaction.

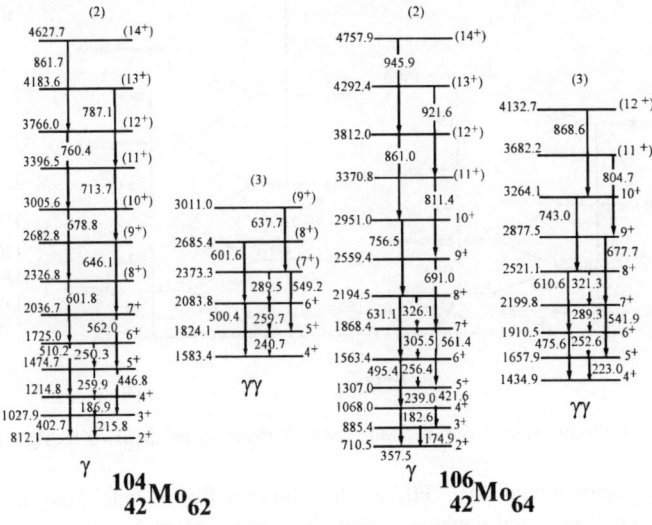

Figure 4. One- and two-phonon γ-bands of ^{104}Mo and ^{106}Mo.

Two pairs of bands in ^{106}Mo are remarkably similar. The quadrupole deformation is clearly significant in ^{106}Mo, as indicated by its low 2^+ energy, and triaxiality is indicated by the γ-vibrational bands. From cranked shell model calculations, one notes that for N=64 one has the neutron $g_{7/2}$ orbital nearly filled, the nearby neutron $g_{9/2}$ orbital filled, and the neutron $h_{11/2}$ orbital just beginning to fill. Thus a two quasi-neutron band built on two of these orbitals (more likely $g_{9/2}$-$h_{11/2}$) could have all of the characteristics to give rise to chiral doublet bands [5]. The two bands are shown separately in Fig. 5 where one notes their striking similarity. We have simply called these bandhead spins I and separated the I, I+2, ..., and I+1, I+3, ... bands. Based on the decays out of these bands, their bandheads are likely 4. The energy spacings and cascade and crossing transition patterns are as expected for chiral doublets. Moreover, it is expected that chiral doublets would have crossing transitions at the low spins, and that as the spins increase up the band, the energy difference ΔE between the level energies for each even spin (e.g., E(I)$_a$ − E(I)$_b$) and odd spin [e.g., E(I+1)$_a$ − E(I+1)$_b$] should decrease as you go up in spin.

198

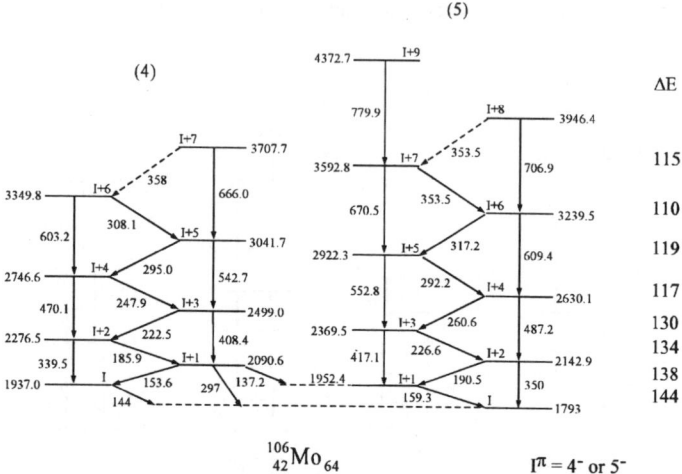

Figure 5. Collective chiral bands in ^{106}Mo. Energy differences are shown on the right in keV.

The ΔEs starting with $E(I)_a - E(I)_b$ are also shown in Fig. 5. Indeed they decrease as expected for chiral doublets. Thus these two sets of doublets meet every criterion for chiral doublets. These are the first chiral doublets reported in an even-even nucleus. This is a further example of how neutron-rich nuclei are providing new insights into nuclear structure.

Work at Tsinghua is supported by the Major State Basic Research Dev. Program under Contract No. G2000077400 and the Chinese National Natural Science Foundation under Grant No. 19775028. Work at Vanderbilt and Mississippi State was supported in part by the U.S. Department of Energy under Grant Nos. DE-FG-05-88ER40407, DE-FG05-95ER40939; Idaho National Engineering and Environmental Lab, Lawrence Berkeley National Lab, and Lawrence Livermore National Lab are supported in part by the U.S. DOE under Nos. DE-AC07-761DO1570, DE-AC03-76SF00098, and W-7405-ENG48. JIHIR was supported by its member institutions and DOE.

1. D. Troltenier, et al, *Nucl. Phys.* **A601**, 56 (1996).
2. A. Guessous, et al., *Phys. Rev.* **C53**, 1191 (1996).
3. A. Guessous, et al., *Phys. Rev. Lett*. **75**, 2280 (1995).
4. *Table of Isotopes*, 8th ed., R.B. Firestone and V.S. Shirley eds. (NY: Wiley & Sons) 1996.
5. S. Frauendorf, *Rev. of Mod. Phys.* **73**, No.2, April 2001 and this volume.

SHAPE COEXISTENCE, SHAPE TRANSITIONS, AND TRIAXIAL DEFORMATION IN NEUTRON-RICH A∼100 NUCLEI POPULATED BY FUSION-FISSION REACTIONS

C.Y. WU, H. HUA, D. CLINE, A.B. HAYES, AND R. TENG

NSRL, Department of Physics,
University of Rochester, Rochester, NY 14627, USA
E-mail: wu@nsrl.rochester.edu

R.M. CLARK, P. FALLON, A. GOERGEN, A.O. MACCHIAVELLI, AND K. VETTER

Nuclear Science Division,
Lawrence Berkeley National Laboratory, Berkeley, CA 94720, USA

Neutron-rich nuclei with mass∼100 were populated by the ^{238}U(α,f) fusion-fission reaction. The deexcitation γ rays were detected by Gammasphere in coincidence with the detection of both fission fragments by the Rochester 4π heavy-ion detector array, CHICO. This technique allows Doppler-shift corrections to be applied for the observed γ rays thus establishing the origin of γ rays from either fission fragment. In addition, it allows observation of γ-ray transitions from states with short lifetimes and offers the opportunity to study nuclear species beyond the reach of the spontaneous fission process. With these advantages, one can extend the spectroscopic study to higher spins than those derived using the thick-target technique, and to more neutron-rich nuclei than those derived from spontaneous fission. Among the new and interesting phenomena identified in this rapid shape-changing region, the most distinct result is evidence for a prolate-to-oblate shape transition occurring at both Ru and Pd isotopes. This represents the first observation of a transition from prolate to oblate shape in a given isotope chain.

1. Introduction

There is a long history of studies of nuclear structure for neutron-rich nuclei by identifying the prompt γ rays from fission fragments. Most recent studies [1,2] were performed using modern γ-ray detector arrays and spontaneous fission sources, either ^{248}Cm or ^{252}Cf, on thick backings to stop the recoiling fission fragments. This enhances the resolving power by maximizing the energy resolution available using the stopped component of emitted deexcitation γ rays. The disadvantage of this approach is that the origin of the γ rays from either fission fragment cannot easily be established without having mass or energy measurement of the coincident fission fragments emitting the γ rays. In addition, only states with lifetimes longer than the stopping time for the recoiling fission fragments can be studied because of large Doppler broadening effects.

The shortcomings of the thick-target technique can be eliminated by the fission-fragment γ-ray coincident technique [3,4,5], which adds mass selectivity as well as identifying from which fission fragment the γ rays originate. This significantly improves the sensitivity for the study of neutron-rich nuclei. For example, in our early work with a ^{252}Cf fission source [3,4], the ground-state band of ^{104}Mo was extended from 14^+ at 4.114 MeV to 20^+ at 7.282 MeV and the γ band from spin 10^+ at 3.004 MeV to spin 16^+ at 5.591 MeV. Both bands exhibit the upbend phenomenon at a rotational frequency $\approx$400 keV, which is beyond the reach of the thick-target technique. Extending the spectroscopic study to high-spin states, where the band crossing occurs, presents an opportunity to study the interplay between the single-particle and the collective degrees of freedom.

In this presentation, we report the results of our study of neutron-rich A$\sim$100 nuclei using the ^{238}U(α,f) fusion-fission reaction with the fission-fragment γ-ray coincident technique. Nuclei in this mass region exhibit unusually rich nuclear structure phenomena, such as the sudden onset of quadrupole deformation in the Zr isotopes, the emerging importance of the γ degree of freedom in the Mo-Ru region, and the predicted prolate-to-oblate shape transition in the Pd isotopes. Even though these nuclei have been studied extensively, a comprehensive understanding of their nuclear structure has not been made, especially, in the high-spin region. A description of the details of this experiment is given followed by the analyses and results for the neutron-rich Zr-Mo-Pd nuclei.

2. Experiment and Analysis

The present experiment was performed at the 88-inch cyclotron facility of Lawrence Berkeley National Laboratory by bombarding a ^{238}U target with an α beam at E_{lab}=30 MeV. Neutron-rich nuclei were populated by the ^{238}U(α,f) fusion-fission reaction. A $\approx$300 μg/cm^2 ^{238}U target on a $\approx$30 μg/cm^2 thickness carbon backing was used. The Rochester 4π, highly-segmented heavy-ion detector array, CHICO [5], was used to detect the fission fragments in coincidence with the detection of the deexcitation γ rays using Gammasphere. A total of $\approx$600 M events with at least five-fold coincidence,

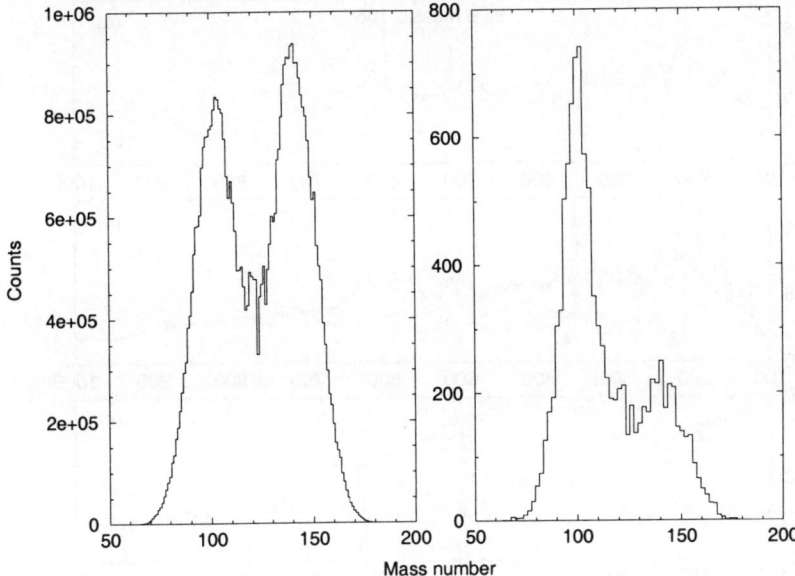

Figure 1.   The fission-fragment mass distribution deduced from ^{238}U$(\alpha,$f$)$ fusion-fission reaction (left) and with the gates on the $2^+ \rightarrow 0^+$ and $4^+ \rightarrow 2^+$ transitions of ^{100}Zr (right).

fragment-fragment and at least three γ rays, were collected.

The masses, plus velocity vectors, of the fission fragments were deduced from the measured angles of both fission fragments and their time-of-flight difference, assuming the fragments have the same kinetic energies as those for ^{240}Pu spontaneous fission [6]. The latter assumption, that the prompt fission originates from the Pu-like compound nucleus, was supported by the observed γ-ray cross correlation between the partner fragment pairs. The deduced mass distribution, shown in Fig. 1, has a mass resolution of 12 mass units, which reflects the time resolution $\approx$500 ps. The achieved position resolution, $\approx$1° in θ and 4.6° in ϕ, are consistent with prior CHICO performance [7,8,9,10].

To increase the sensitivity in the channel selection, mass-gated events with three or higher γ rays were used to develop the level schemes of nuclei. An example is shown in Fig. 2.(c), where the Doppler-shift corrected γ-ray spectrum gated by the mass and the known γ-ray transitions in ^{100}Zr is given. The resulting energy resolution is better than 1%, limited primarily by the finite size of Ge detectors. Since the origin of γ rays from either fission fragment was established, after making the proper Doppler-

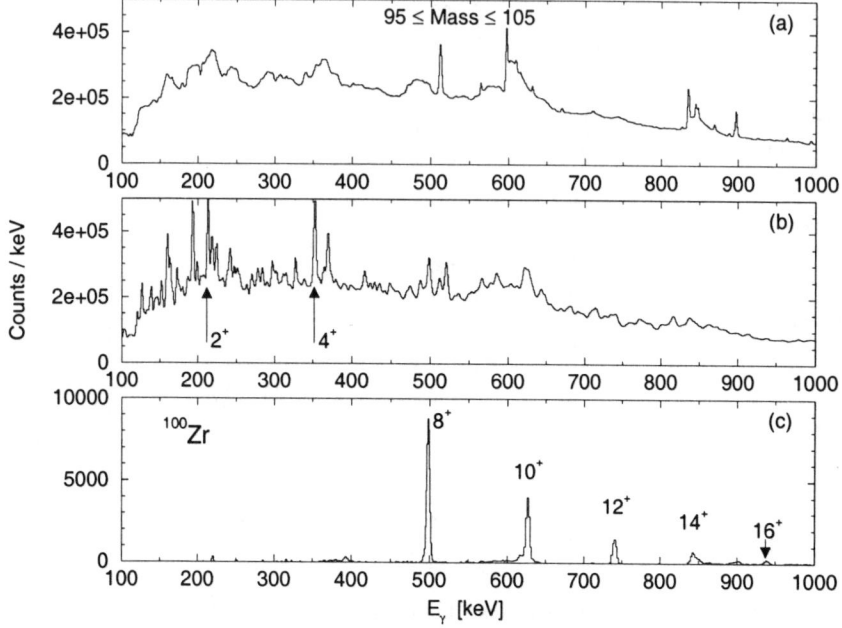

Figure 2. The mass gated γ-ray spectrum with no correction (a), with the Doppler-shift correction (b), and with additional gates on the γ-ray transitions of ^{100}Zr (c). The labeled transitions are ΔI=2 transitions. The gating γ-ray transitions are indicated by arrows in (b).

shift correction, no γ-ray transitions from the partner Xe isotopes are visible in the spectrum. The drastic simplification from the raw and the complex Doppler-shift corrected spectra can be observed in Fig. 2.(a) and (b). The resultant spectrum is clean and straightforward to interpret. The yrast states up to spin 24$^+$ at 8.872 MeV were identified in ^{112}Ru.

3. Results and Discussion

For neutron-rich A∼100 nuclei, the valence nucleons begin to fill the $h_{11/2}$ neutron and $g_{9/2}$ proton orbitals. The nuclear structure in this mass region is sensitive to the occupancy level of the single-particle configurations, which can be illustrated by the rich variety of observed spectroscopic properties, such as shape coexistence, shape transitions, and the emerging importance of the γ degree of freedom. The diversity and rapid change in the nuclear structure makes neutron-rich A∼100 nuclei an ideal testing ground for various theoretical models [11]. In the following, we present our recent results on the band crossing phenomena in neutron-rich Zr-Mo-Pd nuclei to address the interplay between the shape degrees of freedom and those single-particle configurations.

Zirconium. The sudden onset of the quadrupole deformation occurring at neutron-rich Zr isotopes is a well-known phenomenon and can be understood in the framework of shape coexistence [12]. Two shapes with very different deformation coexist in the Zr isotopes and reverse their roles between the ground and excited states at ^{100}Zr. The mixing between the strongly deformed ground state and the weakly deformed excited 0^+ state in ^{100}Zr had been studied by a two-state-mixing model using the measured B(E0) and B(E2) strengths [13,14], implying a mixing strength of $\approx 14\%$ and a downward energy shift of 46.9 keV for the ground state. This mixing model has been expanded by introducing an intrinsic E2 matrix element between those two 0^+ states. New calculations give a mixing strength of $\approx 7.7\%$ and a downward energy shift of ≈ 26 keV [15], which is consistent with the observed energy shift of ≈ 21 keV [2].

Molybdenum. In contrast to the Zr isotopes, low-lying $K^\pi = 2^+$ bands were identified in Mo isotopes, which indicates the emerging importance of the γ degree of freedom. The latter also is manifested by the observation of a spin-dependent triaxial deformation [16] and a harmonic two-phonon γ-vibrational motion [17,18]. Level schemes of Mo isotopes were established up to the band crossing region by the present work, which is evident from the moment of inertia plot for ^{102,104,106,108}Mo in Fig. 3. The observed band crossing is ascribed to the alignment of a pair of $h_{11/2}$ neutrons, which is similar to that of the Zr isotopes and is described in a recent publication [19]. One striking feature in this figure is the disparity for the rate of the spin alignment and the crossing frequency between ^{104}Mo and ^{106}Mo despite the fact that they have very similar quadrupole and γ deformations. The most likely explanation for the delayed crossing in ^{106}Mo is the subshell closure effect at neutron number 64. According to the macroscopic-microscopic calculations based on the deformed Woods-Saxon potential [11], the transitional nucleus ^{102}Mo is very soft and the isotopes ^{104,106,108}Mo have triaxial ground-state minima. The similarity of the observed J_2 moment between ^{102}Mo and ^{104}Mo indicates that the spin-alignment rate and crossing frequency are insensitive to the softness degrees of freedom.

Palladium. The assignment of the $g_{9/2}$ proton-pair alignment responsible for the observed band crossing in heavy neutron-rich nuclei [2] was based on the assumption of an oblate shape for $^{112-116}$Pd, which is predicted by the macroscopic-microscopic calculations [20]. In the later studies of the

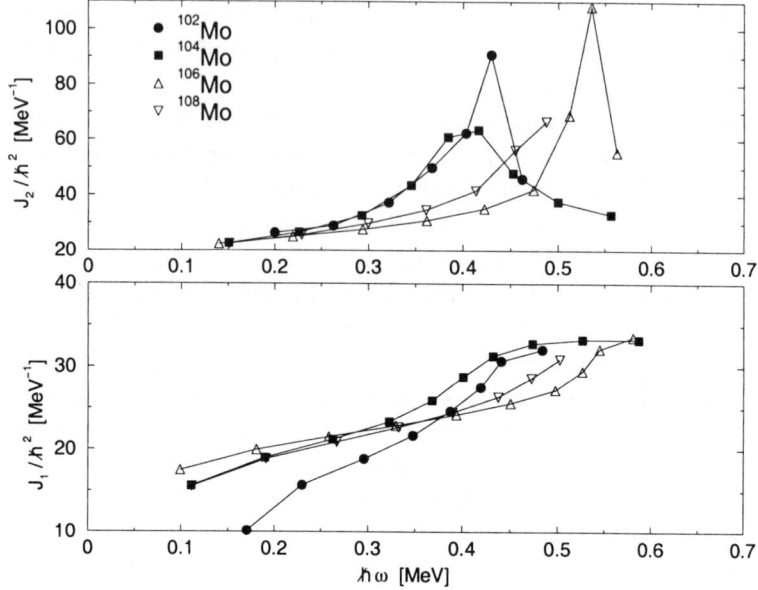

Figure 3. The kinematical and dynamical moment of inertia as a function of the rotational frequency for the yrast states of ^{104,106}Mo.

odd-A Pd isotopes [4,21,22,23], the blocking effect was observed for the $\nu h_{11/2}$ band in ^{111,113,115}Pd, which supports the fact that the alignment of a pair of $h_{11/2}$ neutrons for a prolate shape is responsible for the band crossing in the even Pd isotopes. The predicted prolate-to-oblate shape transition for the heavy neutron-rich Pd isotopes had not been confirmed experimentally.

The current study has extended the yrast states up to spin 22^+ at 8362.8 and 7818.8 keV for ^{114}Pd and ^{116}Pd, respectively. The moment of inertia plotted as a function of the rotational frequency for $^{113-118}$Pd is shown in Fig. 4. The second band crossing in ^{116}Pd occurs unexpectedly at a rotational frequency only ≈ 50 keV above that of the first band crossing. The rotational alignment of a pair of $g_{9/2}$ protons is believed to be responsible for the second band crossing judging by the amount of angular momentum gained in the alignment and a similar crossing frequency observed for the rotational band based on the $h_{11/2}$ neutron orbital in ^{117}Pd. The emergence of oblate shape is the most plausible explanation for the closeness of the crossing frequency between the first and second band crossings, based on the Cranking Shell Model calculations. The latter suggests that in a transition from prolate to oblate shape, the band crossing shifts to a lower

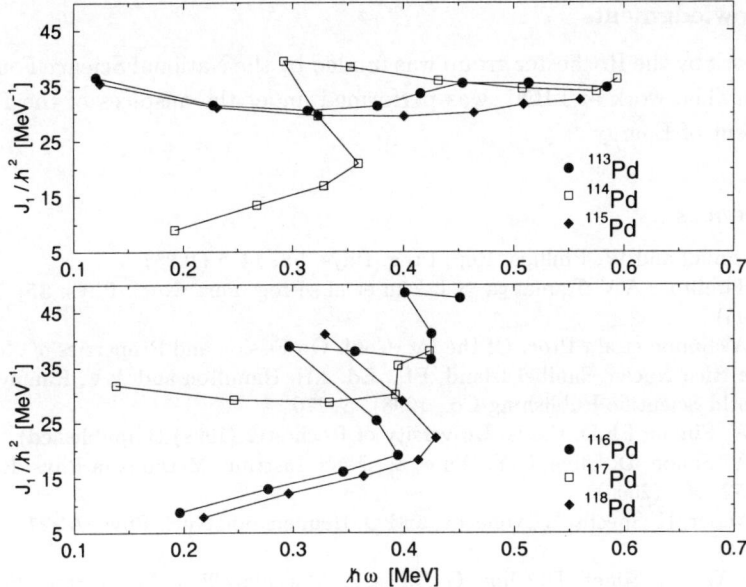

Figure 4. The kinematical moment of inertia as a function of the rotational frequency for the yrast states of ^{113,114,115}Pd (upper) and ^{116,117,118}Pd (lower).

rotational frequency for the aligned $g_{9/2}$ proton pair and is close to that for the aligned $h_{11/2}$ neutron pair, which remains relatively unchanged. Similar band crossing behavior was observed in ^{111}Ru and ^{112}Ru. This represents for the first time the observation of a transition from prolate to oblate shape in a given isotope chain. The shape transition occurring at ^{116}Pd, which is consistent with the prediction of Ref. 11 but not with that of Ref. 20, may have important implications to our understanding of the shell structure in neutron-rich nuclei.

In summary, the neutron-rich A~100 nuclei have been studied using the ^{238}U(α,f) fusion-fission reaction with the fission-fragment γ-ray coincident technique. The achieved sensitivity in the channel selection has enriched our understanding of nuclear structure in this mass region, evident by such findings as the insensitivity of the spin-alignment rate and crossing frequency to the softness degrees of freedom in the Mo isotopes and the prolate-to-oblate shape transition occurring at ^{116}Pd. This technique provides an important complementary method to others for the study of neutron-rich nuclei produced by fission.

206

Acknowledgments

The work by the Rochester group was funded by the National Science Foundation. The work at LBNL was performed under the auspices of the Department of Energy.

References

1. I. Ahmad and W. Phillips, Rep. Prog. Phys. **58**, 1415 (1995).
2. J. Hamilton, A.V. Ramayya, S.J. Zhu et al., Prog. Part. Nucl. Phys. **35**, 635 (1995).
3. M.W. Simon et al., Proc. Of the Int. Conf. On Fission and Properties of Neutron Rich Nuclei, Sanibel Island, FL., Ed. J.H. Hamilton and A.V. Ramayya (World Scientific Publishing Co., 1998), p. 270.
4. M.W. Simon, Ph.D. thesis, University of Rochester (1999) (unpublished).
5. M.W. Simon, D. Cline, C.Y. Wu et al., Nucl. Instrum. Methods in Phys. Res. **A452**, 205 (2000).
6. J. Weber, H. Specht, E. Konecny, and D. Heunemann, Nucl. Phys. **A221**, 414 (1974).
7. C.Y. Wu, M. Simon, D. Cline, G. Davis, A. Macchiavelli, and K. Vetter, Phys. Rev. C **57**, 3466 (1998).
8. K. Vetter et al., Phys. Rev. C **58**, 2631 (1998).
9. C.Y. Wu, D. Cline, M.W. Simon et al., Phys. Rev. C **61**, 021305(R) (2000).
10. C.Y. Wu, D. Cline, M.W. Simon et al., Phys. Rev. C **64**, 064317 (2001).
11. J. Skalski, S. Mizutori, and W. Nazarewicz, Nucl. Phys. **A 617**, 282 (1997).
12. J.L. Wood, K. Heyde, W. Nazarewicz, M. Huyse, and P. Van Duppen, Phys. Rep. **215** (1992) 101 and references therein.
13. H. Mach, M. Moszynski, R.L. Gill, F.K. Wohn, J.A. Winger, J.C. Hill, G. Molnar, and K. Sistemich, Phys. Lett. B **230**, 21 (1989).
14. H. Mach, M. Moszynski, R.L. Gill, G. Molnar, F.K. Wohn, J.A. Winger, and J.C. Hill, Phys. Rev. C **41**, 350 (1990).
15. C.Y. Wu, H. Hua, and D. Cline, Phys. Lett. B **541**, 59 (2002).
16. A.G. Smith, J.L. Durell, W.R. Phillips et al., Phys. Rev. Lett. **77**, 1711 (1996).
17. A. Guessous, N. Schulz, W.R. Phillips et al., Phys. Rev. Lett. **75**, 2280 (1995).
18. A. Guessous, N. Schulz, M. Bentaleb et al., Phys. Rev. C **53**, 1191 (1996).
19. H. Hua, C.Y. Wu, D. Cline et al., Phys. Rev. C **65**, 064325 (2002).
20. P. Möller, J.R. Nix, W.D. Myers, and W.J. Swiatecki, At. Data Nucl. Data Tables **59**, 185 (1995).
21. R. Krücken et al., Phys. Rev. C **60**, 031302 (1999).
22. M. Houry et al., Eur. Phys. J. A **6**, 43 (1999).
23. X.Q. Zhang et al., Phys. Rev. C **61**, 014305 (1999).

GAMMA-RAY EMISSION FROM FISSION OF HEAVY NUCLEI

L. KRUPA, G.N. KNIAJEVA, A.A. BOGATCHEV, G.M. CHUBARIAN[5],
O. DORVAUX[2], I.M. ITKIS, M.G. ITKIS, S. KHLEBNIKOV, J. KLIMAN,
N.A. KONDRATIEV, E.M. KOZULIN, V. LYAPIN[3,4], T. MATERNA[1],
W. RUBCHENIA[3,4], I.V. POKROVSKY, W. TRZASKA[3], D. VAKHTIN[4],
V.M. VOSKRESENSKI

Flerov Laboratory of Nuclear Reactions, JINR, 141980 Dubna, Russia
[1]Universite Libre de Bruxelles, Bruxelles, Belgique
[2]Institut de Recherches Subatomiques, CNRS-IN2P3, Strasbourg, France
[3]Deparment of Physics, University of Jyväskylä, FIN-40351, Jyväskylä, Finland
[4]V.G. Khlopin Radium Institute, St.-Petersburg 194021, Russia
[5]Texas A&M University, Cyclotron Institute, College Station, Texas, 77843-3366, USA

Average γ-ray multiplicity $<M_\gamma>$, average energy $<E_\gamma>$ emitted by γ-rays and average energy per one gamma quantum $<\varepsilon_\gamma>$ as a function of mass and total kinetic energy (TKE) of fission fragments were measured in proton induced reactions $p+^{242}Pu->^{243}Am$, $p+^{238}U->^{239}Np$ (at proton energy $E_p=13$, 20 and 55 MeV) and spontaneous fission of ^{252}Cf. Complex structures in $<M_\gamma>$, $<E_\gamma>$ and $<\varepsilon_\gamma>$ (as function of mass and TKE) are observed in all studied nuclei.

1. Introduction

At fission of nuclei the main part of the energy set free in the fission process is released in the form of kinetic energy of the fragments. Relatively large amount of energy remains just after scission as excitation of the fragments. The excitation being stored in the forms of fragment deformation, rotation and vibration and also as internal heat due to dissipation process from saddle to scission is released by evaporation of neutrons and emission of gamma rays. The study of integral characteristics of γ-ray emission in fission is necessary in order to obtain information on final steps of the de-excitation process of fission fragments.

A spontaneous or excited compound nucleus, after neutron evaporation, decays towards the yrast line mostly by statistical dipole (E_1) and collective quadrupole (E_2) γ-emission. The first one is responsible for the major part of the cooling of the nucleus, while the second one slows down the rotation. The behaviour of the γ-flow is governed to a large extent by the interplay of these two decay modes. By analyzing the multiplicity and energy of gamma rays one can derive information about this interplay and knowing the neutron multiplicity evaluate the absolute values of fragment spins. The dependence of gamma rays on the mechanism of formation and decay of compound heavy systems has been studied in this work in proton induced reactions p+^{242}Pu->^{243}Am, (with proton energy E_p=13, 20 and 55 MeV), p+^{238}U->^{239}Np (E_p=20 and 55 MeV) and spontaneous fission of ^{252}Cf.

2. Experimental techniques

The measurements were carried out at the Accelerator Laboratory, University of Jyväskylä, (Finland) using a setup that included:
- the two-armed time-of-flight reaction products spectrometer CORSET built with the use of microchannel plates (MCP) [1];
- a 8 detector time-of-flight neutron spectrometer DEMON using scintillation modules [2];
- a High Efficiency Neutron DEtection System (HENDES) facility [3];
- a 6 scintillation detector γ-ray multiplicity spectrometer.

The spectrometer for γ-ray multiplicity consisted of six 76.2×76.2 mm NaI(Tl) detectors placed in the lower hemisphere at 24 cm distance from the target (see Fig. 1). The threshold for γ-ray registration was set to 120 keV.

In order to obtain reasonable statistics the data from all six detectors have been summed. The data processing of γ-rays was carried out with the use of response matrix technique [4]. This method makes it possible to get except average γ-ray multiplicity $<M_\gamma>$ an average total energy $<E_\gamma>$ emitted by γ-rays as well. The total efficiency as a function of energy and response matrix of the NaI(Tl) detectors was taken from the simulation using EGSnrc code [5].

A contribution from neutrons to γ-ray multiplicities was estimated to be less than 5 percent and the accidental coincidences were about 1 percent. The error bars shown on subsequent figures of the data represent only the statistical counting uncertainty, which is the relevant quantity, when assessing the systematic trends seen in the data.

3. Experimental results

In Figure 1 a total γ–ray spectra (normalized to one) for spontaneous fission of ^{252}Cf and proton induced reactions are shown. We calculated average γ-ray multiplicity $<M_\gamma>$, average energy $<E_\gamma>$ emitted by γ–rays, average energy per one gamma quantum $<\varepsilon_\gamma>$ and temperature of the nucleus after neutron emission T_γ (see Table 1) using the formula $E_\gamma \sim E_\gamma^3 * \exp(-E_\gamma/T_\gamma)$. One can see that with increasing proton energy E_p (excitation energy of compound nucleus) the $<M_\gamma>$ and $<E_\gamma>$ increases. But after E_p=20 MeV the enhancement is small and

T_γ is almost constant, what you can see from the γ–ray spectra too. This means that the most of the excitation energy is released by neutron evaporation and the remaining part of the excitation energy is converted mainly to the deformation of fission fragments.

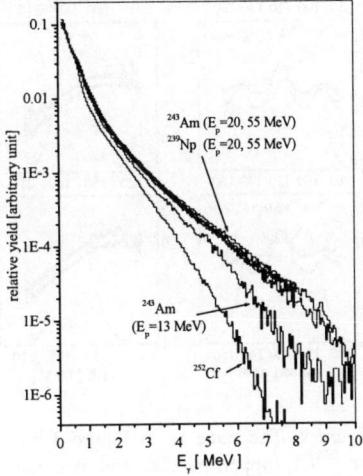

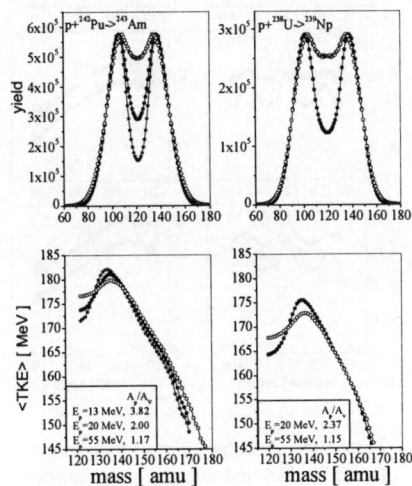

Figure 1. Experimental γ–ray spectra (normalized to one) measured in coincidence with fission fragments for indicated nuclei.

Figure 2. Mass distribution of fission fragments (upper panels) and average <TKE> as a function of fission fragment mass (lower panels) for reaction p+^{242}Pu->^{243}Am (left panels) and p+^{238}U->^{239}Np (right panels) at proton energy E_p=13 MeV (stars), 20 MeV (solid circles) and 55 MeV (open circles). The peak to valley ratio A_p/A_v is shown too.

Table 1. Average γ-ray multiplicity <M_γ>, average energy <E_γ>, emitted by γ–rays, average energy per one gamma quantum <ε_γ> and temperature of the nucleus after neutron emission T_γ.

	<M_γ>	<E_γ> [MeV]	<ε_γ> [MeV]	T_γ [MeV]
^{252}Cf	8.1	6.8	0.84	0.44
^{243}Am (E_p=13 MeV)	8.1	8.1	1.00	0.48
^{243}Am (E_p=20 MeV)	9.0	10.4	1.16	0.52
^{243}Am (E_p=55 MeV)	9.5	11.0	1.16	0.52
^{239}Np (E_p=20 MeV)	8.4	9.2	1.10	0.52
^{239}Np (E_p=55 MeV)	9.0	9.8	1.09	0.52

210

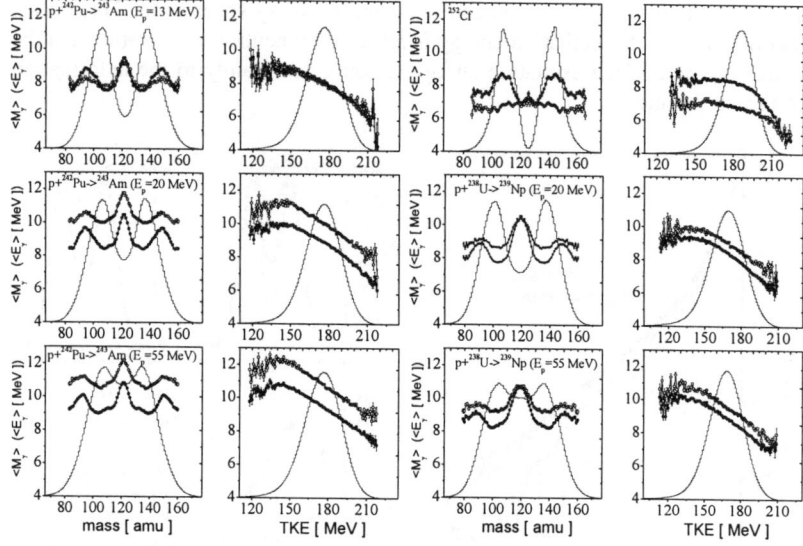

Figure 3. Average γ–ray multiplicity (solid circles) and average γ–ray energy (open circles) as a function of mass (left panels) and TKE (right panels) for reaction p+^{242}Pu->^{243}Am at proton energy E_p = 13, 20 and 55 MeV. The mass and TKE distribution of fission fragments are also shown here.

Figure 4. The same as in Figure 3 but for ^{252}Cf (upper panels) and reaction p+^{238}U->^{239}Np (middle and lower panels) at proton energy E_p = 20 and 55 MeV.

In Figure 2 a mass distribution of fission fragments (upper panels) and average TKE as a function of mass (lower panels) for proton induced reactions are shown. The peak to valley ratio A_p/A_v is shown too.

In Figure 3 an average γ-ray multiplicity and average energy as a function of mass (left panels) and TKE (right panels) for reaction p+^{242}Pu->^{243}Am is displayed. One can see the distinct structures in $<M_\gamma>$ and $<E_\gamma>$ which remain almost the same with increasing E_p. The average $<M_\gamma>$ and $<E_\gamma>$ increase only (in the symmetric part we have maximum and in the asymmetric one minimum). In the case of TKE they decrease with increasing TKE which reflects the fact that for fission fragments with higher TKE the excitation energy is lower and they become more compact in shape. For lower TKE the $<M_\gamma>$ and $<E_\gamma>$ reach maximum and then decrease slowly.

The same one can see in the case of ^{238}U target (Figure 4, lower and middle panels), but the maximum in symmetric part is higher and minimum lower.

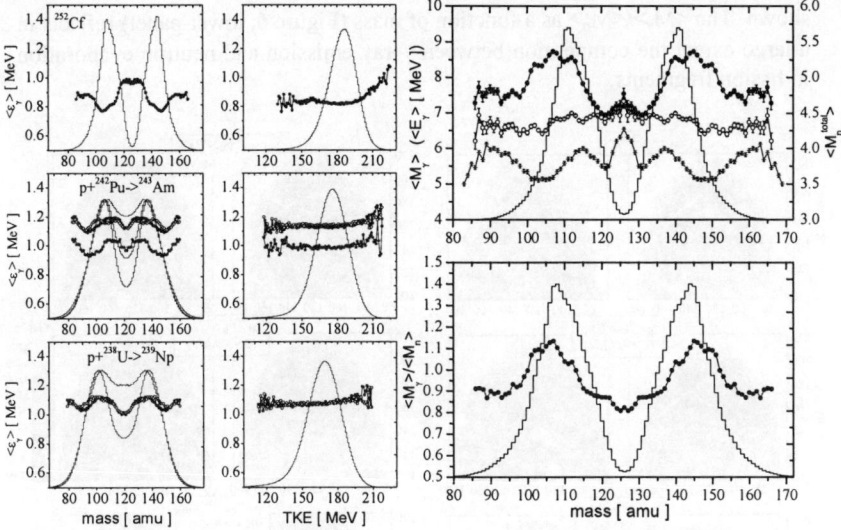

Figure 5. Average energy per one gamma quantum $<\varepsilon_\gamma>$ as a function of mass (left panels) and TKE (right panels) for ^{252}Cf (upper panels) and proton induced reactions (middle and lower panels). E_p=13 MeV (stars), E_p=20 MeV (solid circles), E_p=55 MeV (open circles). The mass and TKE distribution of fission fragments (normalized) are also shown here.

Figure 6. Upper panel: Average γ–ray multiplicity (solid circles), average γ–ray energy (open circles) and average total neutron multiplicity (crosses) taken from the work by Budtz- Jørgensen and H.-H. Knitter [6] as a function of mass are shown. Lower panel: $<M_\gamma>/<M_n>$ as a function of mass normalized to one is shown.

For spontaneous fission of ^{252}Cf (Figure 4, upper panels) one can see additional structures in $<M_\gamma>$ as a function of mass as compared with proton induced reactions. Further information we can obtain from average energy per one gamma quantum $<\varepsilon_\gamma>$ (Figure 5) which reflects the competition between statistical dipole and collective quadrupole γ-rays. In the case of ^{252}Cf the $<\varepsilon_\gamma>$ has maximum in symmetric part and minimum in asymmetric mass division which is opposite to proton induced reactions (Figure 5, middle and lower panels). With increasing E_p the minimum and maximum in $<\varepsilon_\gamma>$ as a function of mass move to the asymmetric mass division. The $<\varepsilon_\gamma>$ as a function of TKE is almost constant and increases for high TKE tail only (most of all for ^{252}Cf), since the fission fragments become more compact and cooler.

In Figure 6 (upper panel) average γ–ray multiplicity, average γ–ray energy and average total neutron multiplicity [6] for ^{252}Cf as a function of mass are

212

shown. The $\langle M_\gamma \rangle / \langle M_n \rangle$ as a function of mass (Figure 6, lower panel) reflects in a large extent the competition between γ–ray emission and neutron evaporation in fission fragments.

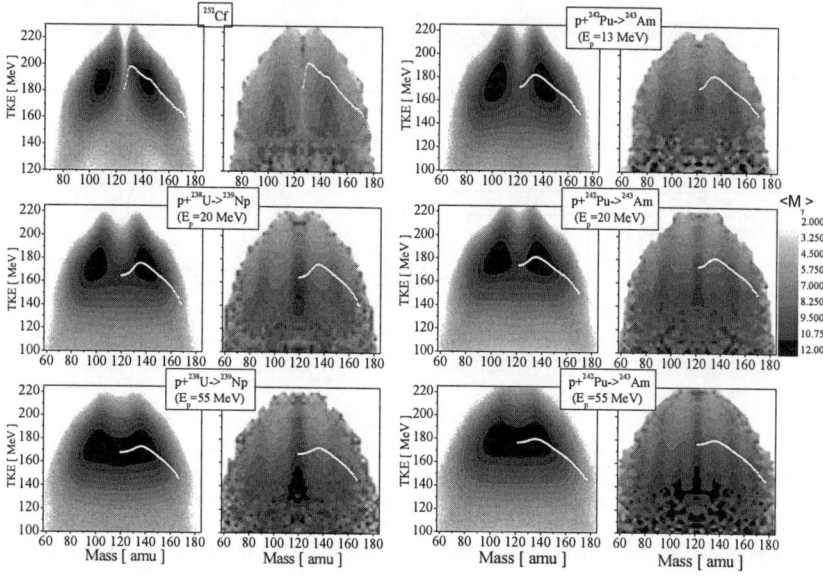

Figure 7. Two-dimensional matrices of fission fragment mass-yield (left panels) and corresponding average γ-ray multiplicities (right panels) for ^{252}Cf (upper panels) and reaction p+^{238}U->^{239}Np (lower panels) at proton energy E_p=20 and 55 MeV. Average TKE as a function of mass is shown too (open circles).

Figure 8. The same as in Figure 7 but for reaction p+^{242}Pu->^{243}Am at proton energy E_p = 13, 20 and 55 MeV. The scale for $\langle M_\gamma \rangle$ is shown too.

More information we can acquire from matrices. In Figure 7 and 8 a mass-energy distribution of fission fragments (left panels) and corresponding $\langle M_\gamma \rangle$ (right panels) for studied nuclei are shown. The white curves represent the average TKE as a function of mass. Bins were summed up over 3 amu and 3 MeV. The difference in energy between the white curves and the maximum value of $\langle M_\gamma \rangle$ for given mass corresponds approximately to the energy released by neutrons. One can see the drastic changes in $\langle M_\gamma \rangle$ for given mass as a function of TKE which probably reflects the shape changes in fission fragments. In Figure 9 some examples of average γ–ray multiplicities as a function of mass for given TKE are displayed. Again one can see complex structures in $\langle M_\gamma \rangle$.

Some local minima and maxima appear and some disappear with decreasing TKE which reflects the great influence of shell effects in fission fragments. Hence the fission fragments can obtain various deformations.

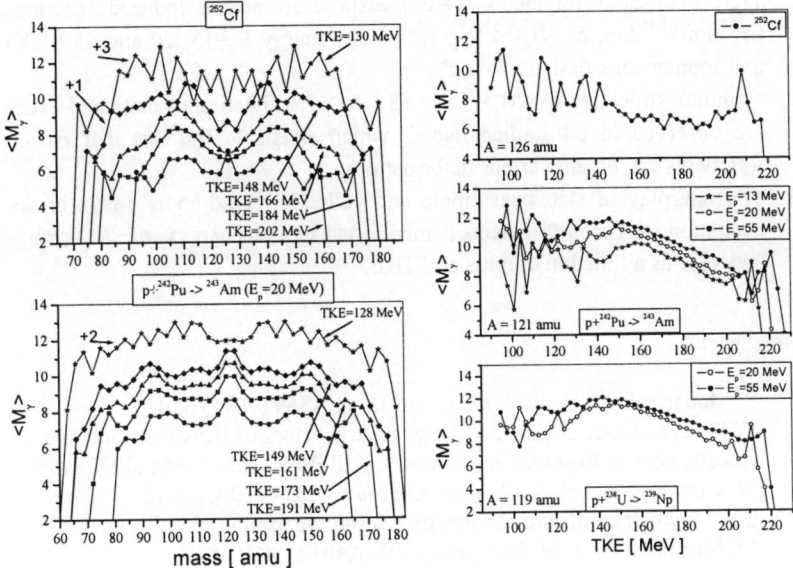

Figure 9. Average γ–ray multiplicities as a function of mass for some TKE. Upper panel - ^{252}Cf; lower panel - ^{243}Am. For some multiplicity curves the <M$_\gamma$> is increased by +1, +2 and +3 units due to better visualization.

Figure 10. Average γ–ray multiplicities as a function of TKE for symmetric mass division of studied nuclei: ^{252}Cf (upper panel), ^{243}Am (middle panel), ^{239}Np (lower panel).

In the end we would like to refer to large enhancement in <M$_\gamma$> for symmetric part of mass division at high TKE (see Figure 10 and matrices in Figures 7 and 8). This phenomenon is observed in all studied nuclei but the most distinct is in the case of ^{252}Cf (Figure 10, upper panel). Probably the neutron emission is forbidden due to small excitation energy, thus all excitation energy is converted to γ-ray emission. Due to small statistics is difficult to say if this energy is transferred mainly to deformation or heating of fission fragments.

4. Conclusions

1. Average γ-ray multiplicity $<M_\gamma>$, average energy $<E_\gamma>$ emitted by γ–rays and average energy per one gamma quantum $<\varepsilon_\gamma>$ as a function mass and TKE of fission fragments were measured in proton induced reactions $p+^{242}Pu->^{243}Am$, $p+^{238}U->^{239}Np$ (at proton energy $E_p=13$, 20 and 55 MeV) and spontaneous fission of ^{252}Cf.

2. Complex structures in $<M_\gamma>$, $<E_\gamma>$ and $<\varepsilon_\gamma>$ (as function of mass and TKE) are observed in all studied nuclei, which attests to the fact that fission fragments can obtain various deformations.

3. The interplay of statistical dipole and collective quadrupole γ-rays in de-excitation of fission fragments is manifested in many aspects of $<M_\gamma>$, $<E_\gamma>$ and $<\varepsilon_\gamma>$ as a function of mass and TKE.

References

1. S. Mouatassim et al, *Nucl. Instr. and Meth.* **A359** (1995) 330.
2. W.H. Trzaska et al., in: J.L. Duggan, I.L. Morgan (Eds.), Application of Accelerators in Research and Industry, AIP Press, New York, (1977) 1059.
3. M. Guttormsen et al, *Nucl. Instr. and Meth.* **A374** (1996) 371.
4. http://www.irs.inms.nrc.ca/inms/irs/EGSnrc/distribution.html.
5. M. Morhac et al, *Nucl. Instr. and Meth.* **A401** (1997) 385.
6. Budtz-Jørgensen and H.-H. Knitter, *Nucl. Phys.* **A** (1988) .

STRUCTURE STUDIES OF FISSION PRODUCTS AT IGISOL-FACILITY*

A. JOKINEN†, A. NIEMINEN, S. RINTA-ANTILA, Y. WANG AND J. ÄYSTÖ

Department of Physics, University of Jyväskylä
POB 35 (YFL),
FIN-40351 Jyväskylä, FINLAND
E-mail: ari.s.jokinen@phys.jyu.fi

P. CAMPBELL

Schuster Laboratory, University of Manchester
Manchester M13 PL, UNITED KINGDOM

Ion guide technique and charge particle induced fission of uranium have been applied to produce neutron-rich transitional nuclei, which are studied by means of beta gamma spectroscopy and optical spectroscopy. New data on the level structures of ^{116,118,120}Cd are presented. The results are discussed in the framework of anharmonic vibrator picture. The optical studies on neutron-rich Zr isotopes are reviewed.

1. Introduction

Neutron-rich nuclei with A=100-120 provides an interesting landscape to study a variety of structural effects. These nuclei are vastly produced in fission reaction, especially in charged particle induced fission, which results in a more symmetric mass distribution compared to low energy neutron-induced fission or spontaneous fission. Most of the isotopes in question are refractory elements, which can be efficiently extracted as low-energy ion beams by applying ion guide technique. Here we will report on the recent studies on fission products performed in the IGISOL-facility in the University of Jyväskylä [1], while a broad review of the applications of ion guide technique can be found in [2]

2. Gamma-spectroscopic studies

Neutron-rich Ru, Pd and Cd fill the transitional region between strongly deformed Mo, Zr and Sr and spherical closed shell Sn nuclei. Beta-decays of the odd-odd Tc and Rh nuclei have earlier been studied providing a wealth of

---

* This work is supported by the Academy of Finland
† Present address: CERN/PS-division, CH-1211 Geneva, Switzerland

216

information on low-lying collective structures in ^{110,112}Ru [3] and $^{110-118}$Pd isotopes [4,5,6]. These data also provide important experimental information for the prompt fission fragment spectroscopy, which extends the low-lying structures to higher spins. As a continuation to our work on Ru and Pd isotopes, we have explored the structure of Cd isotopes as reported here.

Cadmium isotopes are good examples of nearly spherical, vibrational nuclei, which can be described as quadrupole vibrators in collective model. The observation of additional 0^+ and 2^+ states around neutron midshell cannot be explained in such a model. Thus the role of the proton intrudes configurations have been investigated. Here we would like to report on the recent investigation on structures of ^{116}Cd and ^{118,120}Cd isotopes obtained from the beta decay of ^{116,118,120}Ag.

Ag-parent nuclei were produced in charged particle induced fission of ^{238}U. Energy and intensity of H$_2$ ions were 50 MeV and 6-10 μA, respectively. The H$_2$ beam was impinging 15 mg/cm^2 thick ^{nat}U target tilted to 7° angle in respect to primary beam, thus increasing the effective thickness of the target to 120 mg/cm^2. Reaction products were extracted by applying the ion guide technique, mass-separated and transported to the experimental station, where they were implanted to the collection tape surrounded by a 4π beta detector and viewed by four EUROGAM type Ge-detectors.

2.1. Three-phonon multiplets in ^{116,118,120}Cd

Prior to this work three-phonon multiplets have been identified for neutron rich Cd isotopes $^{108-114,118}$Cd, but such information is missing in the case of ^{116}Cd. In order to fill this gap we have studied the beta decay of ^{116}Ag populating the states in ^{116}Cd [7].

In general, ^{116}Cd fulfills the criteria for vibrational character. The obtained ratio of the first 4^+ state and the first 2^+ state, $R_{4/2} = E(4^+)/E(2^+)$, is 2.37, which is slightly higher than the expected ratio of 2 for an ideal vibrator. However, as pointed out in [8], the typical values for a good vibrator are in the range of ~ 2.3. The energy spread of the two-phonon triplet members is only 69.5 keV, which is even smaller than 95.5 keV observed in ^{114}Cd. Finally, the B(E2) ratio for the second 2^+ state at 1213.1 keV is 20, which is close to the value of 30 indicating good vibrational character.

Switching to three-phonon multiplet, we propose full quintuplet as listed in the Table 1. The selection of states results from overall preference for one-phonon decay to N=2 states. The obtained set of states can be compared to the prediction for anharmonic vibrator [9]. The predicted energies are given also in the Table 1. The overall agreement is only qualitative suggesting that Hamiltonian used in the model is partially unrealistic. A more realistic interpretation of the level structure of neutron-rich Cd isotopes may require a configuration mixing between normal states and intruder states.

The existing information on the heavy Cd-isotopes, like ^{118}Cd and ^{120}Cd is solely based on different fission experiments. The prompt fission fragment spectroscopy has resulted in a tentative yrast band up to the spin 16^+ and several groups have studied beta decay of ^{118,120}Ag isotopes produced in neutron-induced fission. A comprehensive study of Abrahamian et al proposed ^{118}Cd as a first candidate for a perfect vibrator [10], which was contradicted by level lifetime measurements [11]. We have provided new experimental information to identify multi-phonon vibrational and intruder states in ^{118}Cd and ^{120}Cd. We have largely confirmed the level scheme of [10], but instead of the proposed 2^+ state at 1915.8 keV, we have introduced a new level at 2023.6 keV, which replaces earlier assignment as 2^+ member of three-phonon quintuplet. For ^{120}Cd, we propose 1899.0, 1920.5, 1997.9 and 2082.8 keV states as candidates for three phonon multiplet with spins and parities 3^+, 2^+, 4^+ and 6^+, respectively. Again the main argument of new results in ^{118}Cd and in case of ^{120}Cd are based on the preferred one-phonon decay to N=2 states.

Table 1. Members of the three-phonon quintuplet in ^{116}Cd. Second column list predicted energies in anharmonic vibrator model [9] and the last column compiles experimental data [7].

L^{π}	$E_{U(5)}$ [keV]	E_{expt} [keV]
0^+_3	2098.8	1928.4
3^+_1	2104.3	1916.0
4^+_2	2107.9	1896.7
6^+_1	2118.0	2026.8
2^+_4	2202.7	1951.4

2.2. *Beta decay to two quasi neutron states*

Earlier studies on the beta decay of neutron rich Rh isotopes have revealed a distinct set of levels situated above the pairing gap in the Pd daughter nuclei, which are strongly populated by the allowed beta decay. These decays are mediated by the $vg_{7/2} \rightarrow \pi g_{9/2}$ allowed transition, inferring that two quasi neutron states are mainly composed of $g_{7/2}$ neutron hole coupled to $d_{5/2}$, $d_{3/2}$, $s_{1/2}$ of $h_{11/2}$ orbitals. Situation is rather similar in case of Cd isotopes and indeed in all studied cases in this work, a set of strongly populated states above 2 MeV in excitation energy are identified. In the case of ^{116}Cd and ^{118}Cd the states at 2958.5 keV and 3031.9 keV with associated logft values 4.9 and 4.8, respectively. The similarity of the decays implies very similar configurations in these two cases. For ^{120}Ag decay, the global strength is more spread and the most strongly populated state with logft 4.8 is situated higher in energy, at 3773.2 keV.

3. Collinear laser spectroscopy of cooled and bunched Zr-ions

Optical spectroscopy provides a sensitive method to probe the nuclear shape. The combination of the ion guide technique, recently developed ion cooling and bunching method and collinear laser spectroscopy has made for the first time possible to apply optical studies for neutron-rich Zr isotopes with A=96-102 [12]. Here we will report on the mean-square charge radii, magnetic moments and quadrupole moments across the N=60 shape change.

An inherent problem applying collinear laser spectroscopy in the ion guide technique is the wide energy spread of the extracted ion beam when an ion guide is optimized to yield the highest possible intensities for mass separated reaction products. To overcome this problem, an ion cooler and buncher was introduced [13]. It is a buffer gas filled radio frequency quadrupole (RFQ). In an RFQ, the ions are transversally confined by the RF-field and in longitudinal direction gentle axial potential is generated by applying slightly varying DC-component on top of the RF-voltage. Ions loose energy in ion-atom interactions. The mutual action of energy loss and confining potential results in a cooled ion ensemble centered to the optical axis of the device. The ion optical properties of such an ion cloud, when extracted as an ion beam or an ion bunch are decoupled from the ion optical characters of the incoming beam. Thus, it is possible to optimize the ion guide for the highest intensity while cooling taking place in the RFQ provides small emittance and small energy spread, which are necessary for high sensitivity of the optical spectroscopy.

Neutron-rich Zr isotopes were produced in fission of uranium, induced by 3.5 μA of 25 MeV H$^+$ beam. The maximum production, 3000 ions/s, was obtained for ^{100}Zr isotope and the minimum ion rates used were around 500 ions/s. The DC beam from an ion guide was cooled and bunched in 200 ms periods and released as 15-20 microsecond wide bunches to an optical spectroscopy station, where the ions are Doppler tuned to the resonance with an additional acceleration voltage, while the laser frequency is locked to the fixed value. Table 2 summarizes the nuclear magnetic moments, quadrupole moments and changes in mean-square charge radii, as extracted in the data.

The mean-charge radii of Zr-isotopes are steadily increasing with increasing neutron number until the known shape change at N=60. If the change in mean-square charge radius is solely due to the change in mean-square quadrupole deformation, $\langle \beta_2 \rangle$, one can make a comparison to trends in $\langle \beta_2 \rangle$ derived from nuclear lifetime measurements. In a region of strong deformation, beyond N=60, a consistent picture is obtained, but for N<56 zirconium isotopes the approximations, in both methods, are poor [12].

Table 2. Compilation of the extracted data from optical spectroscopy for neutron-rich Zr isotopes. The last column showes the changes in mean-square charge radii. The statistical and full systematical errors are separated.

A	I	μ [n.m.]	Q_s [b]	$\delta<r^2>^{90,A}$ [fm^2]	
91	5/2	11.30362(3)[a]	-0.176(3)[b]	+0.128	[6][c]
92				+0.310	[16][c]
94				+0.537	[27][c]
96				+0.702	[35][c]
97	1/2	-0.937(2)		+0.813(5)	[41]
98				+0.980(5)	[49]
99	1/2	-0.930(1)		+1.089(5)	[54]
100				+1.665(4)	[83]
101	3/2	-0.272(1)	+0.812(56)	+1.843(5)	[92]
102				+1.979(5)	[99]

Calibration data from refs. [a][14], [b][15], [c][16]

4. Outlook

Ion guide technique has shown its power for a wide variety of spectroscopic studies on fission products. However, the present system is reaching the limit of the vanishing yields and other experimental difficulties. Thus an important project to upgrade the faculty was, as described elsewhere in this proceedings [17]. It is hoped that with an upgrade of the IGISOL-facility, another step further from the line stability can be taken in the coming years to extend on going program of structure studies of neutron-rich transitional nuclei.

Although one of the strongest points of the ion guide technique is its ability to provide ion beam of all elements, it has also disadvantage of creating isobaric contaminant beams. The production rates tend to decrease rapidly when going further from the stability. Increase of the efficiency of the device either by consolidating the ion guide itself or increasing the intensity of primary beam results in a higher production of exotic species. Unfortunately, the amount of the isobaric background increases with the same rate. Thus the access to isobarically purified beams, as described in a separate contribution to this conference [18], is more than welcome to increase the sensitivity of the spectroscopic studies. In an extreme case, one may hope to apply the same technique to separate isomers and ground states. In addition, the availability of a high-precision Penning trap provides natural means to explore the mass surface of transitional region in detail and allows making good use of the strong magnetic field in dedicated spectroscopic setups inside the trap, especially for charged particles. For example, betas and conversion electrons can be guided further from the source with high efficiency, while gammas and X-rays are observed near the source [19].

Introducing the RFQ ion cooler and buncher has increased the sensitivity of the optical spectroscopy remarkably. The present system should allow extending optical studies to $^{103-105}$Zr and the unexplored region of transitional nuclei can easily be reached.

Acknowledgments

This work has been supported by the European Union within the EXOTRAPS RTD project under Contract No. ERBFMGECT980099 and by the Academy of Finland under the Finnish Centre of Excellence Programme 2000-2005 (Project No. 44875).

References

1. H. Penttilä, et al., *Nucl. Instr. Meth. Phys. Res. B* **126** (1997) 213
2. J. Äystö, *Nucl. Phys. A* **639** (2001) 477
3. J. Äystö et al., *Nucl. Phys. A* **515** (1990) 365
4. J. Äystö et al., *Nucl. Phys. A* **480** (1988) 104
5. A. Jokinen et al., *Eur. Phys. J. A* **9** (2000) 9
6. Y. Wang et al., *Phys. Rev. C* **63** (2001) 024309
7. Y. Wang et al., *Phys. Rev. C* **64** (2001) 054315
8. J. Kern et al., *Nucl. Phys. A* **593** (1995) 21
9. D.M. Brink et al., *Phys. Lett.* **19** (1965) 413
10. A. Abrahamian et al., *Phys. Rev. Lett.* **59** (1987) 535
11. H. Mach et al., *Phys. Rev. Lett.* **63** (1989) 143
12. P. Campbell et al., *Phys. Rev. Lett.* **89** (2002) 082501
13. A. Nieminen et al., *Nucl. Instr. Meth. A* **469** (2001) 244
14. P. Pyykkö, *Mol. Phys.* **99** (2001) 1617
15. G. Fricke et al., *At. Data Nucl. Data Tables* **60** (1995) 177
16. P. Ragahavan, *At. Data Nucl. Data Tables* **42** (1989) 189
17. H. Penttilä et al., *This proceedings* (200x)
18. J. Szerypo et al., *This proceedings* (200x)
19. J. Äystö and A. Jokinen, *J. Phys B.* (2003), in print.

DECAY OF THE NEUTRON RICH NUCLEUS ^{116}Ag

J. C. BATCHELDER[1], J. -C. BILHEUX[2], C. R. BINGHAM[2,3], H. K. CARTER[1], J. D. COLE[4], D. FONG[5], P. E. GARRETT[6], R. GRZYWACZ[2], J. H. HAMILTON[5], D. J. HARTLEY[3], J. K. HWANG[5], W. KROLAS[5,7,8], D. KULP[9], Y. LAROCHELLE[3], A. PIECHACZEK[10], A. V. RAMAYYA[5], K. P. RYKACZEWSKI[2], E. H. SPEJEWSKI[1], D. W. STRACENER[2], M. N. TANTAWY[3], J. A. WINGER[11], J. WOOD[9], E. F. ZGANJAR[10], A. CASARES[2,3], D. RADFORD[2], AND H. WOLLNIK[2,12]

1. UNIRIB/Oak Ridge Associated Universities, Oak Ridge TN 37831
2. Physics Division, Oak Ridge National Laboratory, Oak Ridge TN 37831
3. University of Tennessee, Knoxville TN 37996
4. Idaho National Engineering Laboratory, Idaho Falls, ID 83415-2114
5. Vanderbilt University, Nashville TN 37235
6. Lawrence Livermore National Laboratory, Livermore CA 94550
7. Joint Institute for Heavy Ion Physics, Oak Ridge TN 37831
8. H. Niewodniczanski Institute of Nuclear Physics, PL-31342, Krak w, Poland
9. Georgia Institute of Technology, Atlanta GA 30332
10. Louisiana State University, Baton Rouge, LA 70803
11. Mississippi State University, Mississippi State, MS 39762
12. U. Giessen, Heinrich-Buff-Ring 16, Giessen, D-6300 Germany

The decay of the neutron-rich nucleus ^{116}Ag was studied. Through the use of conversion electron, beta and gamma spectroscopy, we have discovered a new short-lived isomer in ^{116}Ag. The conversion electron data shows previously unknown electron peaks at 22.5, 44.3, and 47.2 keV with half-lives of ~16 seconds each. These are interpreted as the K, L, and M conversion electron lines resulting from a 48 keV transition associated with the decay of a third isomer of ^{116}Ag.

1. Introduction

The cadmium isotopes near their mid-neutron shell, *i.e.*, N=66, exhibit one of the best-known examples of shape coexistence [1]. These nuclei span the "bottom of the parabola" that is characteristic of intruding particle-hole configurations that underlie shape coexistence. The low-lying levels of ^{112,114}Cd were explained by Heyde *et al.* [2] as mixtures of vibrational and intruder configurations.

Information on the structure of levels in neutron-rich Ag isotopes comes mostly from beta-decay studies of the Pd isotopes. In the case of even mass Pd isotopes, since the ground state is 0+, only low-spin states will be populated in beta decay to any significant extent. In the case of beta-decay of ^{116}Pd to states in ^{116}Ag, five low-spin states (1+ or 2+) have been reported [3,4]. The only other excited state known arises from the decay of a short-lived ($t_{1/2}$ = 10s) isomer of ^{116}Ag [4,5,6]. The ground state of ^{116}Ag is reported to be (2-) [5]. There are other high-spin isomers known in the region of ^{116}Ag. Both ^{116}In and 118In are reported to have 1+ ground states and two isomers with spins of 5+ and 8-. With this as our motivation, we have measured the decay of ^{116}Ag.

2. Experimental Method

Silver-116 was produced at the Holifield Radioactive Ion Beam Facility (HRIBF) via the proton-induced fission of Uranium. Forty MeV protons with an intensity of 25nA were bombarded on a UC_x target [7] installed at the UNISOR separator. The proton induced fission products were then separated by mass (A) by the UNISOR separator and deposited on a moving tape collector (MTC) and subsequently moved to the counting position. The counting position was located at the center of an array consisting of three segmented clover Ge detectors, plastic scintillators, and a high resolution (~ 1 keV) Si conversion electron spectrometer. The detectors were mounted in a close geometry surrounding the MTC giving an efficiency for conversion electrons of ~1% and 4% for gamma rays at 250 keV. The array has been named CARDS (Clover Array for Radioactive Decay Spectroscopy) and was designed for decay studies of neutron-rich nuclei at the UNISOR separator (see ref [8] for more details. The conversion electron spectrometer named BESCA (Bellows Electron Spectrometer for the CARDS Array) consists of a 5 mm thick 200 mm^2 liquid nitrogen cooled SiLi detector. The detector is mounted on a moveable cold finger, and the entire Si detector assembly can be pulled back and isolated as needed, providing BESCA with an independent vacuum system. The data acquisition system uses a digital spectroscopy system based on DGF-4C modules (produced by X-ray Instrumentation Associates) [9]. These modules incorporate 40 MHz flash ADC's, and serve as a replacement for amplifiers, discriminators, conventional ADC's and TDC's. Signals from the preamps are input directly into the DGF modules. The module then analyzes the preamp signals via the on-board processors. Each preamp signal is analyzed to determine its amplitude by fitting the waveform and is time-stamped with a continuously running clock. The information is then stored in the onboard buffers until readout. The dead time of this system is determined by the readout time of the modules. The processors can also be programmed to reject pileup events, thus allowing a higher count rate to be used.

There were three sets of data taken at tape speeds relevant for the half-lives (8 s and 2.7m) of the two previously reported isomers. These data sets were as follows: 1. ~20 hours of data taken in singles with a tape movement every 15 seconds, 2. ~8 hours of singles data with a tape cycle of 5 minutes, and 3. ~35 hours of data taken in coincidence with the BESCA electron spectrometer with a tape cycle of 15 seconds. As data analysis is ongoing at the time of this writing, and space is limited, only a few highlights will be presented in this paper.

3. Experimental Results

A total of ~55 hours of data were taken at the shorter tape cycle of 15 seconds. In this scheme, data were collected on an aluminized 35 mm tape and then moved ~30 cm to the center of the CARDS array. The tape transport time was 700 ms and the data acquisition was turned off during this time to avoid any noise in the detectors associated with the tape movement. Gamma-gamma coincidences were used to construct a level scheme in [116]Cd after the beta decay of [116]Ag, while conversion electron data was used to determine the multipolarity of a given gamma-ray transition. In addition, the conversion electron spectra contained peaks arising from de-excitation of levels in [116]Ag (*i.e.* isomers of [116]Ag).

Figure 1a shows the low energy portion of the conversion electron spectrum. It clearly shows peaks at 55.2, 77.2 and 80.1 keV arising from the K, L, and M conversion electrons from the previously reported 81 keV transition of a 10.5(5) s isomer of [116]Ag to the ground state. In addition to this are X-rays (from Cd and Ag) and large peaks at 44.3 and 47.2 keV. A comparison of the ratio of counts and difference in the peak energies between these two peaks and the L and M 81-keV peaks indicates that the two new peaks arise from L and M conversion electrons of a 48 keV transition. The corresponding K electron to these two peaks would be at 22.5 keV. A comparison of the background subtracted half-lives of these peaks reveals two groups, with the K, L, and M electrons from the 81-keV transition having half-lives of ~9 s and the two new peaks with half-lives of ~16s, supporting the assignment of these peaks to a new transition of 48 keV. The half-life curves are displayed in Figure 2, and a summary of the energies and half-lives is given in Table 1. None of these peaks are in coincidence with any gamma rays following the beta decay of this nucleus.

The resolution of the BESCA detector was not sufficient in these runs to unambiguously separate a 22.5 keV electron from the Ag K_α x-rays. To resolve this, a second experiment was performed when increased resolution and a lower energy threshold in the BESCA detector was achieved. The spectrum achieved is displayed in Fig 1b. An expansion of the spectrum is shown in the inset. It shows a peak on the shoulder of the Ag K_α x-ray peak with an energy of 22.5 keV. Based on this and the half-life information, we assign the peaks at 22.5, 44.3, and 47.2 keV as the K, L, and M conversion electrons of a 48 keV transition arising from a new 16s isomer in [116]Ag.

A comparison of the rates of K to L conversion electrons from a given transition allows one to determine the multipolarity of the transition. For the 81 keV transition, the K/L ratio is 0.76(2). This value compares well with the calculated conversion electron value [10] of 0.80 for a 81-keV E3 transition. The experimental value for the 48-keV K/L ratio is 0.19(4). The calculated values for this transition are 0.38 for a M4 transition, 0.04 for E4 and 0.19 for

Peak Energy	Energy (+B.E.)	T1/2 (s)	# counts		48 keV	81 keV
22.5 keV	+25.5(K) = 48.0 keV......		2300(450)			
44.3 keV	+3.8(L) = 48.1 keV	15.9 s	12100(300)	K/L	0.19(4)	0.76(2)
47.2 keV	+0.8(M) = 48.0 keV	14.5 s	2700(250)			
55.2 keV	+25.5(K) = 80.7 keV	9.2 s	13200(300)	M/L	0.22(2)	0.24(1)
77.2 keV	+3.8(L) = 80.8 keV	8.8 s	17300(300)			
80.1 keV	+0.8(M) = 80.9 keV	8.5 s	4100(80)			

Table 1. Summary of energies and half-life values of the conversion electrons arising form the two short-lived isotopes of [116]Ag (taken from Figure 1b).

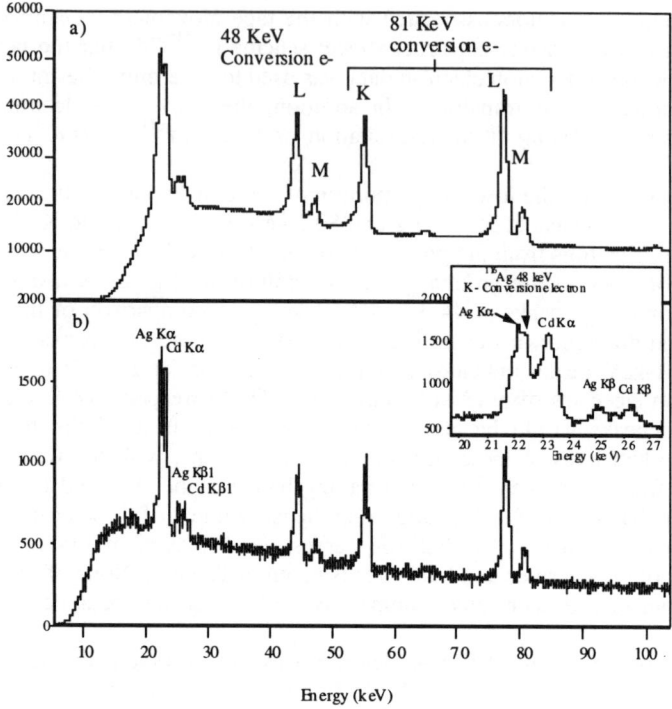

Figure 1. Conversion electron spectra from the decay of ^{116}Ag (see text).

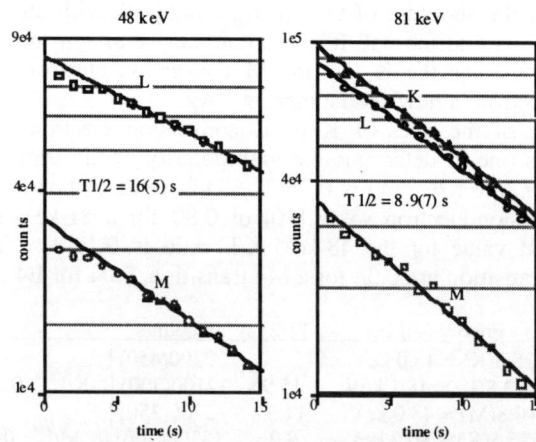

Figure 2. Half-life curves of conversion electron peaks displayed in Figure 1.

E3. All other multipolarities give values that are either much larger or smaller than the experimental value. We therefore assign the 48-keV transition a multipolarity of E3.

If both of these isomers decayed directly to the ground state of [116]Cd, a 33-keV E1 transition between the two would be much faster than 81-keV E3 transition. Such a transition would be expected to be highly converted. While the K electron would be too low in energy to observe, the L electron would be at 29.2 keV. No evidence of this transition has been observed in the data. A close inspection of the half-life curves in Figure 2 reveals that the 81-keV lines can be fit well with a straight line on a log plot, and the 48-keV lines exhibit a small "grow-in" curve. This indicates that the 81-keV conversion electron decays into the 48-keV state. If one isomer was decaying into another, it would be expected to have a small effect on the shape of the decay curve as the 81-keV emitting state has been previously reported [4] to decay by §- decay (94%) and IT decay (6%). In addition, both isomers and the longer-lived ground state will be made directly in the induced fission of UC. A proposed partial decay scheme of [116]Ag is shown in Figure 3.

For [116]Ag the likely orbitals available are $1g_{7/2+}$, $1h_{11/2-}$, $2d_{3/2+}$ and $3s_{1/2+}$ for the neutrons, and $1g_{9/2+}$, $2p_{1/2-}$, and $1f_{5/2-}$ for the protons. Based on these, we propose as possible configurations $\nu 1/2+[411] \otimes \pi 9/2+[404]$ for the 5+ isomer and $\nu 7/2-[523] \otimes \pi 9/2+[404]$ for the 8- isomer.

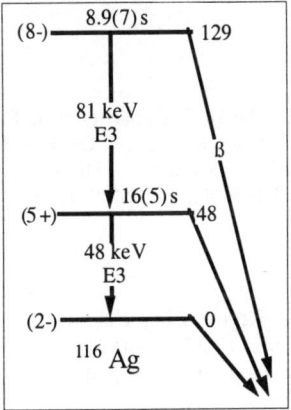

Figure 3. Proposed decay scheme of §- and IT decaying isomers of [116]Ag.

4. Future Work

The induced fission of a UC target by 40 MeV protons produces high-yield neutron rich nuclei ranging from Cu to Dy with 10-15 isotopes of each element. A systematic detailed spectroscopic study using our state of the art detectors and electronics of these nuclei at and near the closed shells will allow a more complete understanding of both the evolution of shell structure in neutron-rich nuclei and the r-process. The experimental difficulty with the proposed studies is the fact that nuclei closer to stability (background) are produced in quantities several orders of magnitude higher. A moving tape system can be used to greatly favor the shorter half-lives of the exotic nuclei we want to study. However, even with the aid of the tape system, the copious amounts of background will overwhelm the signals from the more interesting nuclei. For example, ^{126}Ag can be produced and delivered to an experimental station at rate of $\sim 10^5$ Hz. Even with the most favorable tape speed however, the "background" from other nuclei would be 2×10^9 disintegrations per sec., *i.e* ^{126}Ag would only be 0.01% of all decays recorded. The situation gets even worse with ^{128}Ag (10^3 Hz) where the interesting events would be only 0.0003% of all decays. Clearly these studies would be impossible without some kind of isobaric separation.

We propose to address this issue through the use of a multi-pass time-of-flight separator (MTOF). The MTOF works by reflecting ions between two electrostatic mirrors repeatedly to achieve long flight times with resulting high mass resolving powers. A prototype system with ~ 0.4m between the two reflectors, has achieved a $M/\Delta M$ ratio of $\sim 17,000$. A larger system with ~ 1m between the reflectors will be able to achieve separations of 50,000 to 100,000. Through the use of electrostatic deflectors to remove the background Z nuclei, it will become possible to study the most exotic nuclei produced at HRIBF, and make possible our proposed studies along the r-process path.

5. Acknowledgments

This work has been supported by the U. S. Department of Energy under contracts DE-AC05-76OR00033, DOE-AC05-00OR22725, DE-FG05-88ER40330, DE-AC05-96OR22464, No. DE-FG02-96ER41006, DE-FG05-88ER40407, DE-FG02-96ER40983, W-7405-ENG-48, DE-FG02-96ER40978, and DE-AC07-99ID13727.

6. References

1. J. L. Wood, *et. al.*, Phys. Repts. **215**, 101 (1992).
2. K. Heyde, *et al.*, Phys. Rev. C **25**, 3160 (1982).
3. V. Koponen, *et al.*, Z. Phys A**333**, 339 (1989).
4. B. Fogelberg, *et al.*, Z. Phys A**337**, 251 (1990).
5. T. Bjornstad and J. Alstad, J. Inorg. Nucl. Chem., **36**, 2159 (1974).
6. Y. Wang, *et al.*, Phys.Rev. C **64**, 054315 (2001).
7. D. W. Stracener, in *Proceedings of the Sixteenth International Conference on the Application of Accelerators in Research and Industry,* edited by J.L. Duggan and I.L. Morgan, CP576, AIP Press, New York (2000) pp. 257-260.
8. J. C. Batchelder, *et al.*, in press, Nucl. Instr. Meth. Phys. Res. A.
9. B. Hubbard-Nelson, *et al.*, Nucl. Instr. Meth. Phys. Res. B **422**, 411, (1999).
10. 10. HISSC program at http://www.nndc.bnl.gov/nndc/physco

GAMMA-RAY SPECTROSCOPY FOR NEUTRON-RICH $A\approx160$–170 NUCLEI: THE β^- DECAY OF ^{159}PM, $^{160-162}$SM, ^{162}EU, $^{164-166}$GD, AND $^{166-168}$TB

M. ASAI, S. ICHIKAWA, K. TSUKADA, A. OSA, I. NISHINAKA,
AND Y. NAGAME

Advanced Science Research Center,
Japan Atomic Energy Research Institute,
Tokai, Ibaraki 319-1195, Japan
E-mail: asai@tandem.tokai.jaeri.go.jp

Y. KOJIMA

Graduate School of Engineering,
Hiroshima University,
Higashi-Hiroshima 739-8527, Japan

M. SHIBATA

Facility for Nuclear Materials,
Nagoya University,
Nagoya 464-8603, Japan

β^- decays of neutron-rich $A\approx160$–170 nuclei ^{159}Pm, $^{160-162}$Sm, ^{162}Eu, $^{164-166}$Gd, and $^{166-168}$Tb have been studied through γ-ray spectroscopy using an on-line isotope separator. These nuclei were produced by the 15–20 MeV proton induced fission of ^{238}U. Excited states of their daughter nuclei have been established.

1. Introduction

Experimental studies for neutron-rich $A\approx160$–170 nuclei are limited to the vicinity of stable nuclei because of difficulties in their productions and observations. In previous studies, neutron-rich $A\approx160$–170 nuclei were produced by the ^{235}U(n_{th}, f) reaction[1] or the spontaneous fission of ^{252}Cf.[2,3] The short-lived fission products were mass-separated by on-line isotope separators, and their β^- decays were measured. The heaviest nuclei observed in those studies were ^{162}Eu[1] and ^{164}Gd,[2,3] but no spectroscopic data were

228

obtained for these nuclei except for β-decay half-lives.

Recently, prompt γ-ray spectroscopy using a spontaneous fission source and a large Ge array has made great progress in studying neutron-rich nuclei.[4,5] This technique can populate high-energy, high-spin, and yrast excitations, which is complementary to β-decay studies. In the $A\approx160$–170 region, ^{156,158,160}Sm and ^{160,162,164}Gd were studied with this technique.[6] In particular, excited states of ^{160}Sm and ^{164}Gd were established for the first time. This fact demonstrates a potential power of this technique to study neutron-rich $A\approx160$–170 nuclei.

In this work, we have studied β^- decays of neutron-rich $A\approx160$–170 nuclei produced in the proton-induced fission of ^{238}U. Fission yields of the proton-induced fission of ^{238}U in the $A\approx160$–170 region are comparable to those of the spontaneous fission of ^{252}Cf. The fission products were mass-separated by a gas-jet coupled on-line isotope separator (ISOL), and γ rays following the β^- decay of ^{159}Pm, $^{160-162}$Sm, ^{162}Eu, $^{164-166}$Gd, and $^{166-168}$Tb were measured.

2. Experiments

A stack of four or eight ^{nat}U targets with a thickness of 4–5 mg/cm^2 each was bombarded with a 15–20 MeV proton beam from the JAERI tandem

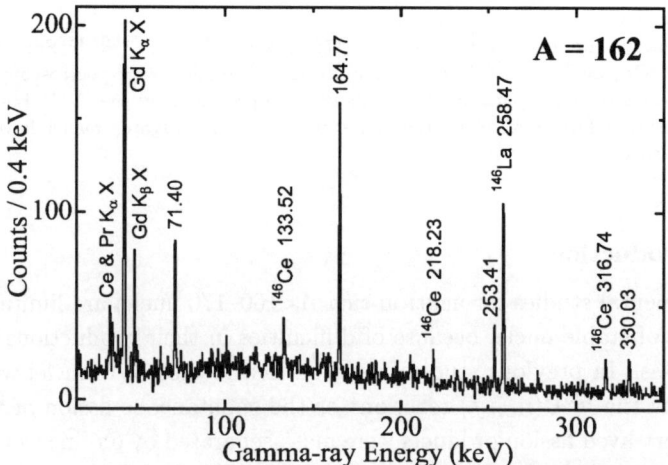

Figure 1. Gamma-ray spectrum observed at the mass 162 fraction. Gd K X rays and γ rays indicated by their energies in keV are associated with the β^- decay of ^{162}Eu. Gamma rays from the monoxide ions of ^{146}LaO$^+$ and ^{146}CeO$^+$ are also seen.

accelerator. Fission products emitted from the targets were thermalized in He gas loaded with PbI_2 aerosol clusters, and transported into an ion source of the ISOL with a gas-jet transport system. The nuclei of interest were mass-separated as monoxide ions[7] except for ^{162}Eu. Mass-separated ions were implanted into an aluminum-coated Mylar tape, and periodically transported to a measuring position equipped with a sandwich-type plastic scintillator and two Ge detectors. Gamma-ray singles, β-γ(-γ), and γ-γ coincidence events were accumulated together with time information.

3. Results and Discussions

The nuclei ^{159}Pm, ^{161,162}Sm, ^{165,166}Gd, and $^{166-168}$Tb are new isotopes whose half-lives and γ transitions have been measured for the first time in this work. Excited states of their daughter nuclei have also been established for the first time except for ^{166}Dy. These results have already been reported elsewhere.[7-10] In this paper, we present other new results on the β^- decay of ^{160}Sm, ^{162}Eu, and ^{164}Gd.

3.1. ^{162}Eu

The β^- decay of ^{162}Eu was previously measured by Mach et al.[1] and Greenwood et al.[2] They observed only Gd K X rays and no γ transitions. The half-life value of 10.6(10) s was determined from the decay of Gd K X rays.[2] Excited states of the daughter nucleus ^{162}Gd were first established by the (t,p) reaction,[11] and recently through the prompt γ-ray spectroscopy using a ^{252}Cf source.[6]

We measured γ rays following the β^- decay of ^{162}Eu in the mass 162 and 162+16 fractions. Usually, ^{162}Eu cannot be observed in the mass 162 fraction because of severe contamination of this fraction by ^{146}LaO$^+$ monoxide ions; it is difficult to control the ion source condition to reduce monoxide ions. The γ-ray spectrum shown in Fig. 1 was obtained under the best condition ever attained. Although γ rays associated with the ^{146}LaO$^+$ and ^{146}CeO$^+$ ions were still visible, Gd K X rays and seven γ transitions attributable to the β^- decay of ^{162}Eu were clearly observed. Figure 2 shows a partial decay scheme of ^{162}Eu compared with the levels established in the previous works.[6,11] The 4$^+$ and 6$^+$ states are populated strongly, indicating that the spin of the ground state of ^{162}Eu would be 5 or 6. Taking account of the proton and neutron configurations of the ground states of neighboring odd-mass nuclei, the $\pi 5/2[413]\nu 7/2[633]$ configuration is plausible for the ground state of ^{162}Eu. This makes the lowest-energy 1$^+$ or

6^+ state. However, the 6^+ assignment contradicts the β feeding to the 4^+ state. Further experimental studies, especially with good statistics, are needed to solve this problem.

β^- decays of ^{162}Eu were also observed in the mass 162+16 fraction in the experiment to identify ^{162}Sm.[10] Since the Eu element is hardly ionized as a monoxide ion, ^{162}Eu is not directly mass-separated at the mass 162+16 fraction. Thus, the ^{162}Eu observed in the mass 162+16 fraction must be those generated via the β^- decay of ^{162}Sm whose spin-parity is 0^+. Gamma-ray intensities relative to the 164.8 keV one in both the fractions are compared in Fig. 2. Although their uncertainties are large, the intensities of the γ rays depopulating 2^+ and 3^+ states are about two times larger than those obtained at the mass 162 fraction. This fact suggests the

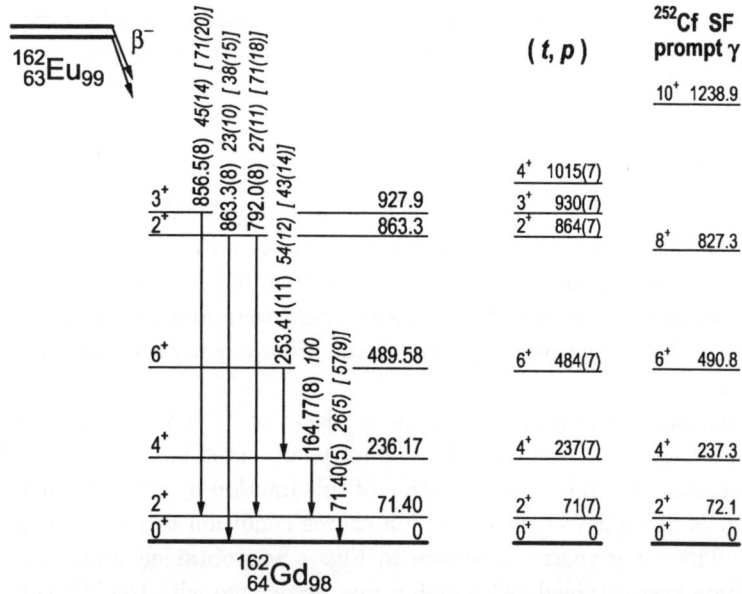

Figure 2. Partial decay scheme of ^{162}Eu. Experimental uncertainties of energies and intensities are given in parentheses. Relative γ-ray intensities are depicted in italic. The intensities deduced from the data obtained at the mass 162+16 fraction, in which the ^{162}Eu nuclei are generated only through the β^- decay of ^{162}Sm, are given in square brackets, compared with those obtained at the mass 162 fraction. The 330.03(11) keV transition with I_γ=62(13) for the mass 162 fraction is also attributed to the β^- decay of ^{162}Eu, but not placed in the decay scheme. This transition is not seen in the mass 162+16 fraction because of less statistics. Levels previously established in ^{162}Gd are also depicted in the figure.

existence of a β-decaying isomer in ^{162}Eu.

3.2. ^{160}Sm

The decay of ^{160}Sm was first reported by Mach et al.[1] Only one γ transition at 109.7 keV with $T_{1/2}$=8.7(14) s was found. Greenwood et al.[2] also determined the half-life value of 9.6(3) s, and found five γ transitions. However, they reported no decay scheme information.

In this work, the half-life of ^{160}Sm was determined to be 9.2(7) s from the decay curve of the 109.6 keV γ rays. Five γ transitions were attributed to the β^- decay of ^{160}Sm, four of which were incorporated in the decay scheme given in Fig. 3(a). The rest is the 158.2(2) keV transition with I_γ=26(4) which is not coincident with any γ and X rays. Multipolarities of the 109.6 and 76.2 keV transitions were determined to be E1 and M1/E2, respectively, from the internal conversion coefficients of $\alpha_K(110)$=0.25(11) and $\alpha_K(76)$=3.0(8) deduced from the K X to γ intensity ratios in the γ-γ coincidence spectra.

3.3. ^{164}Gd

The nucleus ^{164}Gd was first identified by Greenwood et al.[3] They determined the half-life value of 45(3) s from the decay curves of Tb K X rays, but found no γ transitions.

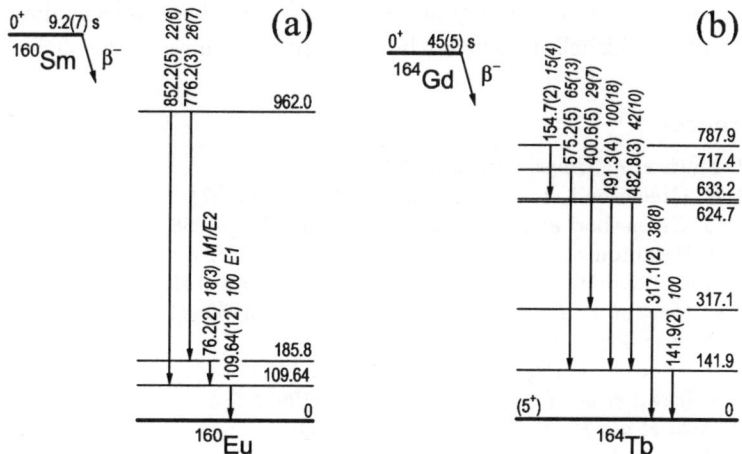

Figure 3. Partial decay schemes of (a) ^{160}Sm and (b) ^{164}Gd.

In the present experiment, many γ transitions were identified through half-life and γ-γ coincidence analyses. The half-life value of 45(5) s was determined from the decay curve of the 141.9 keV γ rays. Figure 3(b) shows the proposed decay scheme of ^{164}Gd. Three γ transitions with energies of 38.2(3), 366.2(3), and 645.5(3) keV and intensities of 16(4), 29(7), and 73(14), respectively, are not incorporated in this decay scheme; the 38.2 and 645.5 keV transitions are in coincidence with each other and with Tb K X rays, but with no other γ rays. The 366.2 keV transition is not coincident with any γ and X rays.

The β feeding to the 141.9 keV level is negligibly small, and that of 317.1 keV is also small. This decay pattern is consistent with the previous (5^+) assignment[12] to the ground state of ^{164}Tb because of a large spin difference between the 0^+ ground state of ^{164}Gd and the (5^+) one of ^{164}Tb.

4. Summary and Outlook

β^- decays of neutron-rich $A\approx160$–170 nuclei ^{159}Pm, $^{160-162}$Sm, ^{162}Eu, $^{164-166}$Gd, and $^{166-168}$Tb produced in the 15–20 MeV proton induced fission of ^{238}U have been studied through γ-ray spectroscopy using an on-line isotope separator. Partial decay schemes for the β^- decay of ^{160}Sm, ^{162}Eu, and ^{164}Gd were proposed. These results are, however, still insufficient to discuss their nuclear structure. In these studies, the proton beam intensity was only 1–3 μA. Now we can use 10 μA proton beam in our tandem accelerator facility. This enables us to study these nuclei in more detail and study more neutron-rich heavy nuclei, e.g., ^{170}Tb, whose β-decay daughter is ^{170}Dy, the midshell nucleus for both the proton and neutron shells.

References

1. H. Mach et al., Phys. Rev. Lett. **56**, 1547 (1986).
2. R. C. Greenwood et al., Phys. Rev. **C35**, 1965 (1987).
3. R. C. Greenwood et al., Radiochim. Acta **43**, 129 (1988).
4. J. H. Hamilton et al., Prog. Part. Nucl. Phys. **35**, 635 (1995).
5. I. Ahmad and W. R. Phillips, Rep. Prog. Phys. **58**, 1415 (1995).
6. E. F. Jones et al., ENAM98, AIP Conf. Proc. **455**, 523 (1998); J. H. Hamilton et al., Eur. Phys. J. **A15**, 175 (2002).
7. M. Asai et al., J. Phys. Soc. Jpn. **65**, 1135 (1996).
8. S. Ichikawa et al., Phys. Rev. **C58**, 1329 (1998).
9. M. Asai et al., Phys. Rev. **C59**, 3060 (1999).
10. M. Asai et al., JAERI-Review **2001-030**, 13 (2001).
11. G. Løvhøiden et al., Phys. Scripta **34**, 691 (1986).
12. N. Kaffrell and G. Herrmann, Phys. Lett. **34B**, 46 (1971).

Q_β MEASUREMENTS OF NEUTRON-RICH RARE EARTH NUCLEI PRODUCED WITH PROTON INDUCED FISSION OF ^{238}U

M. SHIBATA, O. SUEMATSU AND K. KAWADE

*Department of Energy Engineering and Science, Nagoya University, Nagoya, 464-8603
JAPAN*
E-mail: i45329a@nucc.cc.nagoya-u.ac.jp

M. ASAI, S. ICHIKAWA, Y. NAGAME, A. OSA AND K. TSUKADA

Japan Atomic Energy Research Institute, Tokai, 319-1195 JAPAN

Y. KOJIMA

*Graduate School of Engineering, Hiroshima University, Higashi-Hiroshima, 739-8527
JAPAN*

A. TANIGUCHI

Research Reactor Institute, Kyoto University, Kumatori, 590-0494 JAPAN

The β-decay energy (Q_β) of neutron-rich rare earth nuclei ^{158,159}Pm, ^{159,161}Sm and ^{166}Tb, produced with the 16 MeV proton induced fission of ^{238}U, were measured using a total absorption detector installed in an on-line mass separator. By means of the measurement having time information, their half-lives were also measured, simultaneously, from the decay curves. We propose a measuring method of atomic masses of rare nuclei, and indicate that the predicted half-lives by some models are longer than the experimental ones if the models adopt the present Q_βs in this region.

1. Introduction

Both atomic masses and half-lives are essential constants in nuclear physics. Some nuclear quantities related to the stability of nuclei and nuclear structure, e.g. two neutron or two proton separation energies are directly derived from the atomic masses. The Q_β measurement is one of the precise methods for atomic mass determinations. The theoretical prediction of half-lives strongly depends on the Q_β, therefore, the experimental determination of the Q_β, also atomic mass, is more essential.

For the nuclei far from the stability, however, in usual case, only the half-lives can be determined, the Q_βs cannot be reported owing to their low yields and low detection efficiency. If total absorption is realized, the highest

233

point of the energy spectrum indicates the Q_β. Then, we developed a total absorption detector and measured Q_βs of ^{151}Ce [1] and ^{144}La [2] of fission products of ^{235}U, within the uncertainties of 0.1~0.2 MeV, independently of the decay scheme information using the on-line mass-separator in Kyoto University Reactor (KUR-ISOL). Since the reliabilities of the mass formulae are considered to be about 0.5 MeV, it is useful to deduce the Q_β below 0.5 MeV for the nuclei, having no experimental values, far off the stability.

We had already identified some neutron-rich rare earth isotopes, ^{159}Pm, ^{161}Sm, $^{166-168}$Tb, using the on-line mass separator JAERI-ISOL [3-6] in the Japan Atomic Energy Research Institute. Then, we aimed at determination of the Q_β of them with uncertainties below 0.3 MeV at most.

2. Total absorption detector

The total absorption detector is composed of twin BGO scintillation detectors ($12\text{cm}^\phi \times 10\text{cm}^l$) that are arranged at 180° with the distance of 3 mm, and energy sum signals of the two scintillators are measured. The ^{207}Bi free BGO crystals were adopted for background reduction. The simulated peak to total ratios for the mono-energetic electrons and γ-rays at 8 MeV are 90% and 75%, respectively, by the Monte Carlo code EGS4. This detector can absorb almost all γ-rays and β-rays including their backscattering and bremsstrahlung from the sandwiched radioactive sources, therefore, the end-points of the measured spectra are expected to give the Q_β, so that the end-point of the energy spectra gives the Q_β directly without detailed information of the decay scheme.

3. Experiments

The nuclei of interest were produced with the ^{238}U(p,f) reaction. A stuck of eight ^{238}U targets (4 mg/cm^2 thick each) was bombarded with 16 MeV proton beams with the intensity of about 1-3 µA at the JAERI tandem accelerator facility. The fission products were transported into a thermal ion source by the He gas-jet system containing PbI$_2$ aerosols and they were mass-separated by the JAERI-ISOL [4]. The mass-separated beam was implanted in a computer controlled aluminized Mylar tape, which moved to a counting position periodically. Ten fission products of $^{91-93}$Rb, $^{139-141}$Cs, $^{142-144}$La, ^{148}Pr, having well-determined Q_βs energy range between 3.4 and 8.1 MeV, were measured for the on-line energy calibration. The counting rates were always kept below 1.5 kcps in order to avoid pulse pile-up. The detector was shielded with 10 cm thick lead blocks, 10 cm thick paraffin ones and borated rubber sheets to stop the streaming neutrons. The counting rate of the background was about 30 cps, which originated from the ^{207}Bi radioactivity in the BGO crystals themselves mainly, and the prompt γ-rays from the surroundings by the fission neutrons.

The collecting, cooling and counting periods were predetermined properly for each isotope to reduce daughter activities. The Q_βs of ^{158,159}Pm, ^{159}Sm, and new isotopes of ^{161}Sm and ^{166}Tb were measured with event-by-event mode including timing information to determine the half-lives. The overall efficiency of the ISOL is smaller than 1% for rare earth elements at A~160, then, it resulted in the source intensity in the order of 10 decays per second, approximately.

4. Results and Discussion

Since the detector has high detection efficiency, a conventional square root plot method, which is often adapted for the end-point analysis for a plastic scintillation detector, gave adequate precision in this experiment. The analyzing procedure was described in elsewhere [1,2]. The on-line energy calibration line, which was obtained with the least square fitting between the obtained end-points and the evaluated Q_βs by Audi et al. [7] for the ten nuclei, is shown in Figure 1. It shows good linearity between 3 and 8 MeV. It means this detector can be measured Q_βs without detailed decay scheme information.

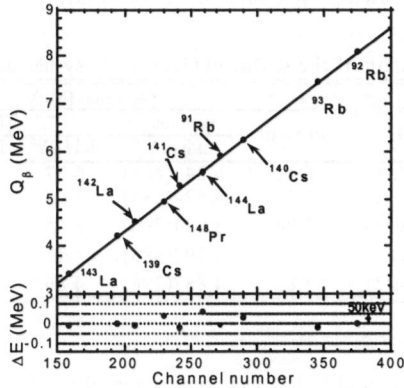

Figure 1. On-line energy calibration obtained from ten nuclei

Typical total absorption spectrum of ^{166}Tb measured for 22 hours and deduced half-lives by following the counts of near the end-points are shown in Figure 2. The measured Q_βs of ^{158,159}Pm, ^{159,161}Sm and ^{166}Tb are listed in Table 1 together with the systematics by Audi et al. [7], and three theoretical values [8,9]. So far the uncertainties were evaluated to be 0.1 MeV, approximately, on the basis of the accuracy of the energy calibration line (~50 keV) and from the statistics of each spectrum. The precise analysis is in progress. The systematics by Audi et al. is in agreement with the present values within the uncertainties. The possibilities of high-energy long-lived isomers are probably low in this region, therefore, the present results are considered to be probable. Since the analysis of the half-lives depend on the subtraction of the background spectrum, the values are evaluated to have larger uncertainties than those by the statistics themselves. These results are almost in agreement with the result (25.6(22) s) by following the decay of the Tb Kx-ray [10].

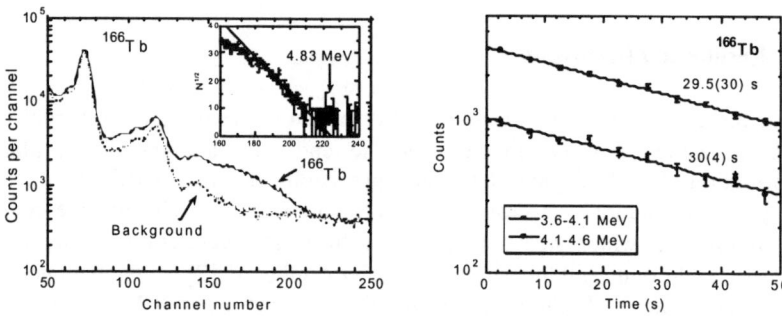

Figure 2. Total absorption spectrum of ^{166}Tb and its square root plot (inset), and the deduce half-life following the counts of corresponding energy regions (inset).

Table 1. Comparison of the deduced Q_βs with the systematics and theoretical ones.

Nuclei	Half-life* (s)	Q_β-value (MeV)				
		Present	Systematics	Theoretical		
			Audi et al.[1]	KUTY[8]	FRDM[9]	FRLDM[9]
^{158}Pm	4.8(5)	6.12	6.243(408)	6.27	6.58	6.63
^{159}Pm	1.47(15)	5.66	5.524(585)	5.08	5.60	5.66
^{159}Sm	11.37(15)	3.84	3.834(298)	3.45	3.53	3.62
^{161}Sm	4.8(8)	4.92	4.797(585)	4.43	4.57	4.65
^{166}Tb	25.6(22)	4.83	4.887(298)	4.86	4.96	4.97

* ref. 10.

We determined Q_βs and half-lives experimentally. Then, it is interesting to compare the experimental half-lives to the predicted ones for different Q_βs in some theoretical models. In Figure 3 [11], the comparisons of them in the Gross Theory (GT) and the pn-QRPA model are shown. They show that the GT2-1996 [12] predicts the half-lives and Q_βs much better in two nuclei, while, the other models predict half-lives longer if the present Q_βs are adopted. The similar properties were also observed in ^{159}Pm. The experimental Q_βs restrict in the prediction of the half-lives by theoretical models.

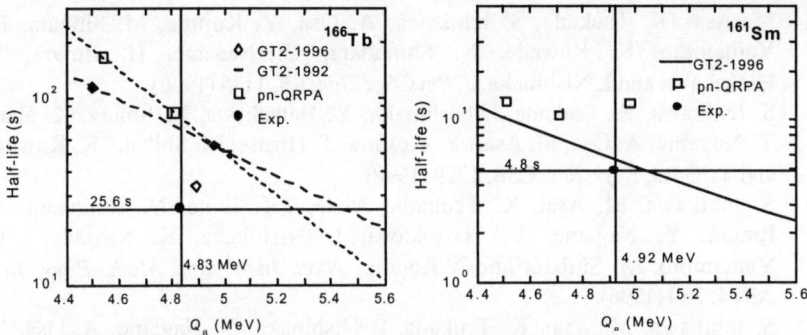

Figure 3. Comparisons of experimental half-lives and Q_βs and the theoretical ones for ^{166}Tb and ^{161}Sm.

5. Conclusion

The Q_β values of the short-lived neutron-rich rare earth nuclei of ^{158,159}Pm, ^{159,161}Sm and ^{166}Tb produced with the ^{238}U(p,f) reaction with 16 MeV protons were measured using the total absorption detector with the JAERI-ISOL. In this region, the systematics by Audi *et al.* still predicts the Q_βs well and the present results propose a direction of the improvements of the nuclear models concerning masses and half-lives. The measurements for other nuclei to A~170 will be continued systematically.

Acknowledgements

We would like to thank the crew of the JAERI tandem accelerator for generating an intense proton beam and providing stable operation. This work was partially supported by the University-JAERI collaboration. One of the authors (M. S.) would like to express his appreciation to Research Foundation for the Electrotechnology of Cyubu.

References

1. M. Shibata, Y. Kojima, H. Uno, K. Kawade, A. Taniguchi, Y. Kawase, S. Ichikawa, F. Maekawa and Y. Ikeda, *Nucl. Instr. and Meth. Phys. Res.* **A459**, 581 (2001).
2. M. Shibata, T. Shindou, A. Taniguchi, Y. Kojima, K. Kawade, S. Ichikawa and Y. Kawase, *J. Phys. Soc. Jpn.* **71**, 1401 (2002).

238

3. M. Asai, K. Tsukada, S. Ichikawa, A. Osa, Y. Kojima, M. Shibata, H. Yamamoto, K. Kawade, N. Shinohara, Y. Nagame, H. Iimura, Y. Hatsukawa and I. Nishinaka, *J. Phys. Soc. Jpn.* **65**, 1135 (1996).

4. S. Ichikawa, K. Tsukada, I. Nishinaka, Y. Hatsukawa, H. Iimura, K. Hata, Y. Nagame, A. Osa, M. Asai, Y. Kojima, T. Hirose, M. Shibata, K. Kawade and Y. Oura, *Phys. Rev.* **C58**, 1329 (1998).

5. S. Ichikawa, M. Asai, K. Tsukada, A. Osa, T. Ikuta, N. Shinohara, H. Iimura, Y. Nagame, Y. Hatsukawa, I. Nishinaka, K. Kawade, H. Yamamoto, M. Shibata and Y.Kojima, *Nucl. Instr. and Meth. Phys. Res.* **A374**, 330 (1996).

6. S. Ichikawa, M. Asai, K. Tsukada, I. Nishinaka, Y. Nagame, A. Osa, Y. Kojima, M. Shibata, K. Kawade, M. Sakama and Y. Oura, in *Proc of the 2nd Int. Conf. on Fission and Properties of Neutron-rich Nuclei*, St. Andrews, Scotland, June 28-july 3, 1999. (World Scientific, Singapore 2000) p. 203.

7. G. Audi and A. H. Wapstra, *Nucl. Phys.* **A595**, 409 (1995).

8. H. Koura, M. Uno, T. Tachibana, and M. Yamada, *RIKEN-AF-NP*-394.

9. P. Möller, J.R. Nix and W.J. Swiatecki, *At. Data Nucl. Data Tables* **59**, 185 (1995).

10. ibid. M. Asai, S. Ichikawa, K.Tsukada, A. Osa, I. Nishiknaka, Y. Nagame, Y. Kojima and M. Shibata.

11. This figure for ^{166}Tb is taken from Ref. 3. (References are therein, e.g. A. Staudt, E. Bender, K. Muto and H. V. Klapdor-Kleingrothaus, *At. Data and Nucl. Data Tables* **44**, 79 (1990).)

12. T. Tachibana, private communication.

POPULATION OF NEUTRON-RICH FISSION FRAGMENTS FOLLOWING HEAVY-ION FUSION REACTIONS

JOLIE A. CIZEWSKI

Department of Physics and Astronomy, Rutgers University, New Brunswick, New Jersey 08903

NIKOLAOS FOTIADES

Los Alamos National Laboratory, Los Alamos, New Mexico 87545

Fusion-evaporation reactions forming compound nuclei with $A \sim 200$ have been used to populate high-spin excitations in nuclei near the line of stability as fragments following fission of the compound system. The high-spin excitations in $^{92-95}$Zr are used to illustrate the power of prompt γ-ray spectroscopy of fission fragments from compound nuclei. The level spectra are compared with shell model calculations.

1. Introduction

Relatively little is known about high-spin excitations in nuclei near stability, or only slightly neutron rich. Most information on high-spin excitations comes from fusion-evaporation reactions which populate neutron-deficient species. High-spin excitations in neutron-rich nuclei have been studied by prompt γ-ray spectroscopy of fragments from spontaneous fission sources or light-ion or neutron-induced fission of actinide targets. In contrast, most knowledge of high-spin excitations in nuclei near stability comes from heavy-ion induced Coulomb excitation, which is limited to stable species and to excitations connected to the ground state by strong matrix elements. The present study of fragments following fission of compound nuclei populated in heavy-ion fusion reactions enables the study of high-spin excitations in nuclei near stability, including unstable odd-mass isotopes, as well as slightly neutron-rich species.

2. Experimental Techniques

The nuclei were populated as fragments following fission of compound nuclei near $A \sim 200$ formed in heavy-ion fusion reactions. Most of the results come from the population of ^{197}Pb compound nuclei with the ^{24}Mg + ^{173}Yb reaction with 134.5-MeV beams from the 88-Inch Cyclotron accelerator at Lawrence Berkeley National Laboratory; other studies involved ^{199}Tl and ^{226}Th compound (CN) systems. Gold-backed targets were used in the Pb and Tl CN experiments, resulting in fusion reactions with ^{197}Au, and ^{221}Pa and ^{220}Th compound systems. Gamma-ray spectroscopy was enabled by the Gammasphere array of Compton-suppressed Ge detectors; typically 90-100 detectors were in the array.

The nuclei for which new level information have been deduced are summarized in Fig. 1. In particular, high-spin excitations in nuclei near $Z = 40$ and $N = 50$, nuclei which are stable or only slightly neutron rich, have been identified. A more detailed summary of the distribution of fragments was presented in Ref. [1] and the contribution of Fotiades [2] and coworkers in these proceedings.

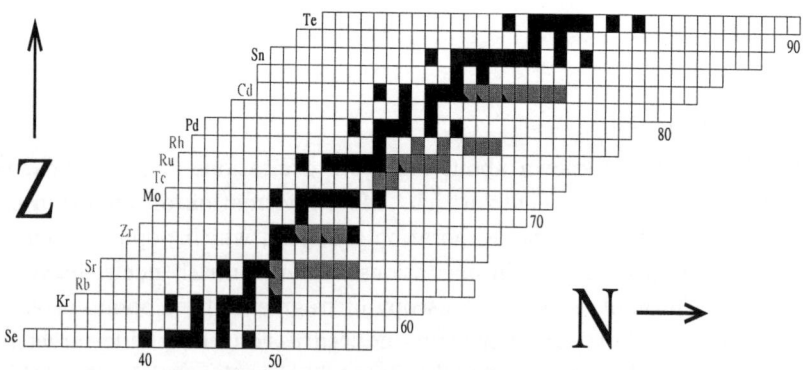

Figure 1. Section of chart of the nuclides indicating the nuclei (in gray) for which new levels were identified in our measurements of fragments following compound nuclear fission. The stable isotopes are indicated by black squares or triangles.

In the present measurements new transitions are assigned to a specific nucleus by gating on the complementary fragment following the *symmetric* fission of the compound system. Typically 5-8 neutrons are emitted as the compound nucleus undergoes fission, which competes with neutron evaporation.

3. High-spin excitations in $^{92-95}$Zr

Near the subshell closure at $Z = 40$ and shell closure at $N = 50$ the valence protons are in the $\pi(p_{1/2}, g_{9/2})$ orbitals and neutrons are filling the $\nu d_{5/2}$ orbital with higher spin excitations involving the $\nu h_{11/2}$ orbital. High-spin excitations in this region have been calculated by Gloeckner [3] using the limited space of $\pi(p_{1/2}, g_{9/2})$ and $\nu(d_{5/2}, s_{1/2})$ and by Holt and coworkers [4] using the full 50-82 neutron shell $\nu(d_{5/2}, g_{7/2}, s_{1/2}, h_{11/2}, d_{3/2})$ and a realistic nucleon-nucleon interaction.

The level schemes of ^{92,94}Zr have been identified up to 18$\hbar$ in spin and over 9 MeV in excitation energy [5]. Using fission from a compound system populated in a fusion-evaporation reaction is an effective way to populate high-spin excitations. For example, the relative intensities of the transitions in ^{92}Zr at the highest spins are about an order of magnitude more intense than relative intensities in ^{146,148}Ba at 16$\hbar$ [6].

The level schemes for ^{92,94}Zr up to 12$\hbar$ with dominant shell model configurations are displayed in Fig. 2. The low-lying states are predominantly $(\nu d_{5/2})^2_{0+,2+,4+}$ with the 6^+ and 8^+ states predominantly $(\pi g_{9/2})^2$. The 10^+ and 12^+ states are predominantly $(\nu d_{5/2})^2_{2+,4+}$ excitations coupled to the $(\pi g_{9/2})^2$. The negative-parity 5^- state is predominantly $(\pi g_{9/2} p_{1/2})$ with the 7^- state $(\nu d_{5/2})^2$ coupled to the two-proton state.

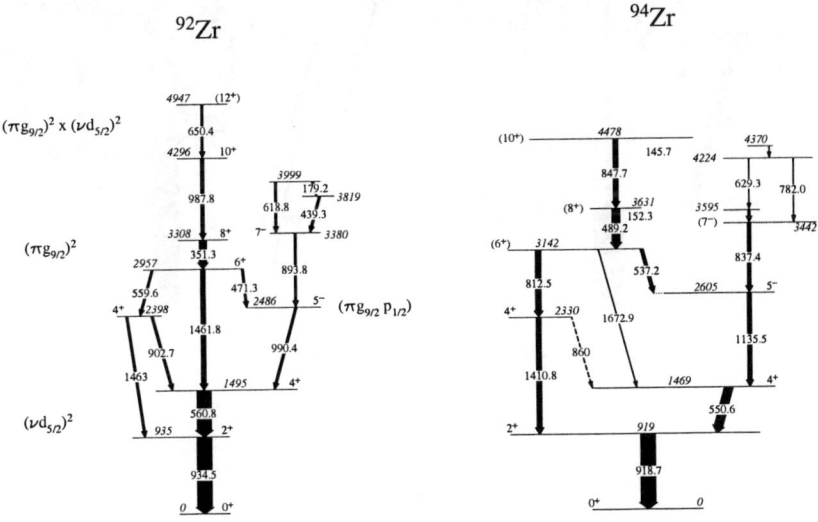

Figure 2.   Level spectra for ^{92,94}Zr with dominant shell model configurations.

The comparison between the experimental levels of ^{94}Zr and the shell model calculations of Gloeckner [3] and Holt and coworkers [4] was displayed in Fig. 5 of Ref [5]. While the Gloeckner calculations reproduce much of the observed high-spin structures, the work of Holt et al., with the complete 50-82 neutron shell, predicts a spectrum considerably compressed compared to experiment.

The high-spin excitations in ^{93,95}Zr are displayed in Fig. 3. These are the first high-spin excitations observed in these nuclei, except for (d,p) measurements [7] which identified $\ell=5$ transfer to excitations near 2.37 and 2.02 MeV in ^{93,95}Zr, respectively. The high-spin excitations in ^{93}Zr have similar spacings to excitations in the ^{92,94}Zr cores, suggesting that much of the structure comes from coupling the $d_{5/2}$ neutron to core excitations. The one difference is the relatively low-lying $13/2^+$ state in ^{93}Zr. Since a $13/2^+$ state cannot be formed from $(\nu d_{5/2})^3$, that the $13/2^+$ state is at 1655 keV, reflects the importance of configurations outside of $(\nu d_{5/2})^2$ in the 4^+ states of the cores.

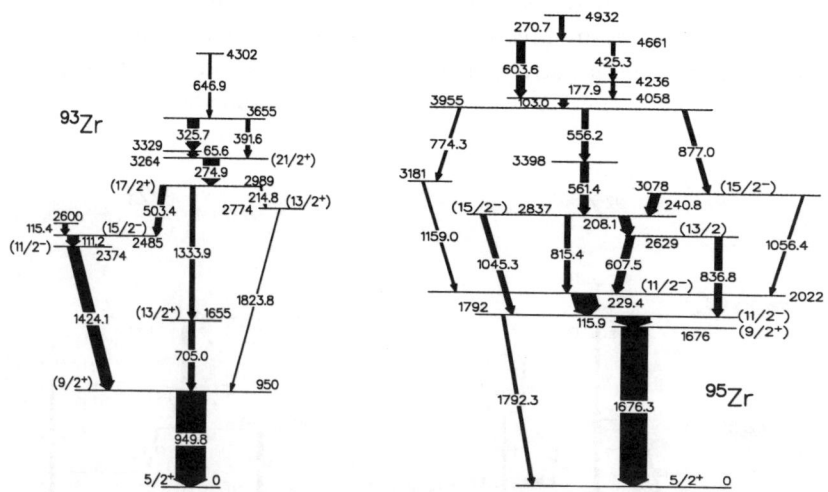

Figure 3.   Level spectra of ^{93,95}Zr.

4. Conclusions

The present report is only a fraction of the results which have been obtained from the prompt γ-ray spectroscopy of fragments following fission of compound nuclei populated in heavy-ion fusion reactions. These studies have enabled the identification of high-spin excitations in nuclei near stability, measurements which are difficult to obtain with other reactions or asymmetric fission of actinides. The experimental challenge is to obtain more definitive spin-parity assignments from such measurements. These are especially important since it is difficult to assign even tentative J^{π} values to spherical nuclei based on systematics. Unfortunately, the presently available data sets do not have sufficient statistics to enable angular correlation measurements to deduce multipolarities.

The final challenge is to theory. The present measurements have identified high-spin excitations in nuclei near the $Z = 40$ and $N = 50$ subshell closures – excitations which should be amenable to shell model calculations. The hope is that the present work will motivate new calculations.

Acknowledgements

We are indebted to our colleagues from Lawrence Livermore and Lawrence Berkeley National Laboratories for their invaluable help with the data acquisition and data analysis, and in particular to Dr. Dennis McNabb who played a leadership role in the analysis of the data set of the ^{197}Pb compound nucleus. This work was supported in part by the National Science Foundation (Rutgers) and the Department of Energy (LANL) under Contract No. W-7405-ENG-36 (LANL).

References

1. N Fotiades et al., Physica Scripta Vol. **T88**, 127 (2000).
2. N. Fotiades et al., these proceedings.
3. D. H. Gloeckner, Nucl. Phys. **A253**, 301 (1975).
4. A. Holt, T. Engeland, M. Hjorth-Jensen, and E. Osnes, Phys. Rev. C **61**, 064318 (2000) and private communication.
5. N. Fotiades, et al., Phys. Rev. C **65**, 044303 (2000).
6. W. Urban et al., Nucl. Phys. **A613**, 107 (1997).
7. C. M. Baglin, Nucl. Data Sheets **80**, 1 (1997) and T. W. Burrows, Nucl. Data Sheets **68**, 1 (1993).

BAND STRUCTURES IN FISSION FRAGMENTS OF RHODIUM AND NEIGHBORS

J.O. RASMUSSEN, J. GILAT, S.C. WU

Lawrence Berkeley National Laboratory
Berkeley, CA 94720 USA
E-mail: JORasmussen@lbl.gov

Y.X. LUO, J.K. HWANG, A.V. RAMAYYA, J.H. HAMILTON

Vanderbilt University
Nashville, TN 37235 USA
E-mail: YXLuo@lbl.gov

S.J. ZHU, R.Q.XU

Tsinghua University
Beijing 100084, People's Republic of China
E-mail: zhushj@mail.tsinghua.edu.cn

R. DONANGELO

Universidade Federal do Rio de Janeiro (UFRJ)
21945-970 Rio de Janeiro, Brazil
E-mail: donangel@if.ufrj.br

We present level schemes from fission-gamma spectroscopy in the region of triaxiality around [110-113]Rh, studied by our Gammasphere collaboration. We will here emphasize features involving one or more unpaired nucleons in high-j orbitals, $g_{9/2}$ protons and $h_{11/2}$ neutrons. Backbending and bifurcation is observed between rotational frequencies of 0.32 and 0.39 MeV for the odd-even nuclei but not for the odd-odd. Possible chiral doublets in 3-quasiparticle states above the backbend in [111]Rh and in [113]Rh are noted. We note that signature-splitting patterns in odd-even, $g_{9/2}$ bands of the region indicate a shape somewhat on the prolate side of maximum triaxiality for the $g_{9/2}$ ground bands of [109,111,113]Rh. The more deformed K=1/2 intruder bands from proton orbitals above the Z=50 shell are probably prolate spheroidal, a case of shape coexistence.

1. Introduction

The nuclei in the approximate range of 41<Z<48 and 58<N<69 have been of interest for many years as being "gamma-soft" or having stable triaxial shapes. The ground bands of even-even nuclei in the region are believed to be prolate spheroidal in shape. The "gamma-softness" is indicated by the generally low-lying $K=2^+$ bands (gamma-vibrational bands.) A good example of this is [106]Mo, where early beta-decay studies [1] found the beginnings (spins 2 and 3) of a presumed gamma vibrational band with band-head only slightly above the 4^+ member of the ground band. A collaboration [2] at the Eurogam detector array extended this level scheme using prompt-gamma spectroscopy from fission. In particular the 2-phonon gamma bandhead with $K=4^+$ is identified.

Theoretical analyses of triaxiality in this region have been given by Skalski et al. [3] and by Troltenier et al. [4]. We shall later cite the presence of wobbler bands and signature splitting in high-j orbitals for evidence on the triaxiality. We will show below level schemes of rhodium isotopes from our year 2000 Gammasphere data, the rhodium data, which were mainly analyzed at Lawrence Berkeley National Laboratory.

2. Odd-A Rhodiums (Z=45)

We present here two unpublished rhodium level schemes, deduced using Radford's software [5] with "Radware cubes" (3-dimensional histograms of triple-coincidences from 3-fold events and all possible triple coincidences from higher-fold events.) Details of our Gammasphere experiment and analyses have previously been given in several publications [6]. We note that fission-gamma spectroscopy induced by heavy-ion reactions is beginning to complement and compete with spontaneous fission in important respects. The work of Fotiades et al. [7] on rhodium isotopes is an example. The yield patterns are somewhat different. Both spontaneous and their induced fission work overlap most at [111]Rh and neighboring odd-odd nuclei, but their induced fission work in Gammasphere has higher relative yield at [109]Rh, and our spontaneous fission work has higher yield at the heavier [113]Rh.

We have made much use of analogies with the heavy-ion reaction work of Venkova et al. [8] on [107]Rh and [109]Rh. We have chosen to keep their numbering system for the various bands. The 1998 beta- decay study in the paper of Lhersonneau et al. [9] on the lower-spin levels of [111]Rh, and the 1998 and 2002 beta-decay papers of Kurpeta et al. [10] on [113]Rh have also been most valuable.

In Figures 1 and 2 we show our proposed level schemes for [111]Rh and[113]Rh, respectively.

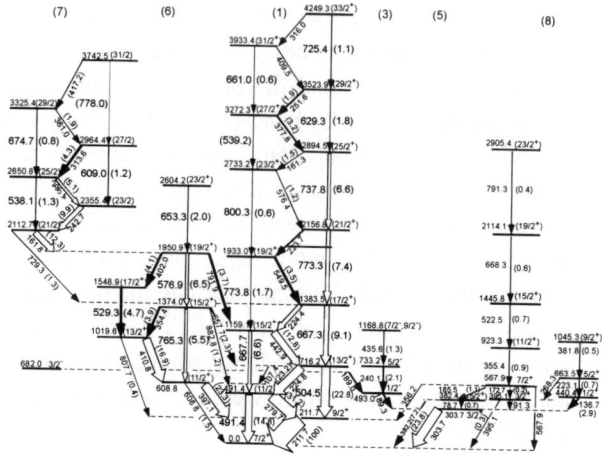

Figure 1. Proposed level scheme for ^{111}Rh, showing relative intensities (in parentheses) for prompt (<1 μ sec) gamma transitions from spontaneous fission of ^{252}Cf.

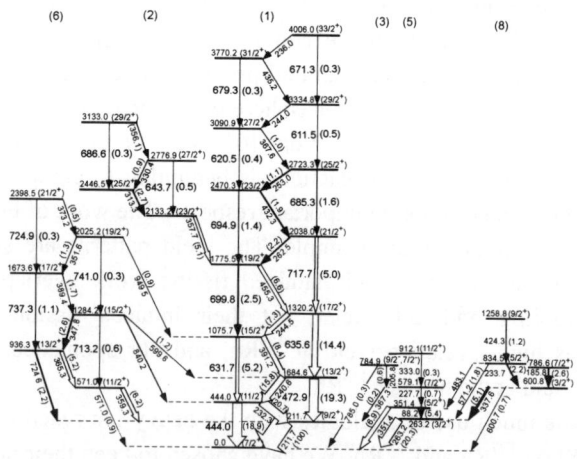

Figure 2. Same as Fig. 1 caption except this is for ^{113}Rh

There is a band-crossing (backbending) in band (1) that begins above the spin 19/2$^+$ level. Our assignment of spin 23/2 to the lowest level in band (2) of ^{113}Rh, making it yrast, mainly rests on the lack of a transition from band (2) to the 17/2$^+$ level of band (1). We believe the backbending indicates the alignment of a pair of $h_{11/2}$ neutrons, since backbending is not observed in ^{112}Rh up to our limits of observation of $\hbar\omega$ of 0.39 MeV, while it occurs in ^{113}Rh and ^{111}Rh at the lower frequencies of 0.35 and 0.37 MeV, respectively. The aligned angular momentum, as determined graphically from a plot of spin vs. rotational frequency is around 8 units. With our high-statistics data we are able to add one more transition at the top of band (1) in [109]Rh as given in the work of Venkova et al.[8] That transition constitutes a backbend starting around 0.39 MeV for [109]Rh, but the band is not determined high enough to fix the amount of aligned spin. For the adjacent even element, Pd (Z=46), Zhang et al. [11] have extended the ground band of [118]Pd to 14[+] and observed backbending at a frequency hω of ~0.36 MeV, similar to backbending earlier found for even-even [112,114,116]Pd. Their plots "indicate a backbending at a rotation frequency of hω ~ 0.35-0.38 MeV and with a gain in alignment of ~10h." The plots of Zhang et al. [11] show a gentler upbending alignment at higher frequencies in some of the odd-A Pd isotopes. Similarly for isotopes of Cd (Z=48) Fotiades et al.[12] show the systematics of backbending, which they also attribute to $h_{11/2}$ neutron pair alignment, showing a gain in alignment of ~8h.

3. Odd-odd [110,112]Rh

We show in Fig. 3 the levels we identify in [110]Rh and [112]Rh.

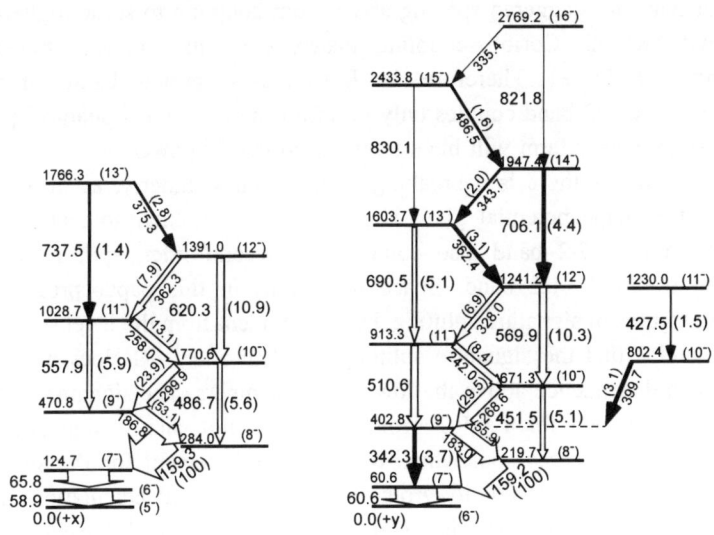

Figure 3. Proposed level schemes in [110]Rh (left) and [112]Rh (right).

A remarkably long series of odd-odd rhodium (Z=45) isotopes and some silver (Z=47) isotopes exhibit a strikingly similar band above the 8[-] level. Duffait et al. [13] in their Fig. 10 show the series of odd-odd rhodium and silver isotopes for neutron numbers 55, 57, and 59 going up to spins 13 or 14. Fotiades et al. [7] extend the rhodium series through neutron numbers 61, 63, 65, and 67, going up to spin 13. From our ^{252}Cf Gammasphere data we are able to extend the N=67 band (^{112}Rh) up another three levels to spin 16, and we observe a side-band of two levels, decaying into the 9⁻ level. Backbending is not observed. In the preceding section we used the argument of lack of backbending in ^{112}Rh to make the case for the backbending in neighboring odd-A rhodiums as due to alignment of an $h_{11/2}$ neutron pair. Also we add two levels below the 8⁻ to a bandhead of 6⁻. It should be pointed out that the spin-8 level is not the real bandhead, since in the odd-odd rhodiums there are often a few low-energy cascading transitions down to spin 5 or 6. In the case of the larger neutron numbers, i.e. the deformed region, these lower-spin, closely-spaced levels are the unaligned beginning of the band, and by spin 8 both the odd proton and odd neutron are aligned with the rotational angular momentum.

4. Signature splitting of odd-A nuclei in the range of Z 44-47

A basic way of understanding signature splitting in spheroidal odd-A nuclei is to consider that this staggered spacing arises from coupling to some higher-lying band with K=1/2. Coriolis-coupling matrix elements connect only bands differing in K by ±1. Therefore, for K=7/2, as in ground bands of odd-A rhodiums, a K=1/2 band couples only in third order, and the signature-splitting perturbation energy term will have terms up to the 7th power of the spin I. On the other hand, if there is a breaking of the axial symmetry, as in a triaxial nucleus, the shape potential gives rise to K= ±2 coupling, so that signature splitting in a K=7/2 band also comes from second-order, spin-independent coupling to the K=1/2 band. Space limitations in this paper preclude our showing figures of signature splitting, but it is evident from the level schemes of Figs. 1 and 2 that the signature splitting in ^{111}Rh and ^{113}Rh does not have a strong spin dependence above the lowest two members. The signature splitting has the proper sign for a j=9/2 orbital and does not change sign above the backbend but decreases somewhat. We find reinforcement for the signature-splitting evidence of triaxiality from the nice theoretical calculations in the paper of Gelberg et al. [14]. They carry out calculations for ^{125}Xe with an odd $h_{11/2}$ neutron orbital in various shapes ranging from prolate through triaxial to oblate. They find two triaxiality values, $\gamma = 24°$ and $\gamma = 50°$ which match ^{125}Xe, a signature splitting similar but slightly greater than that in the rhodiums. (Of course, the sign is opposite for the j=11/2 odd nucleon.) They rule out the more oblate solution $\gamma = 50°$ from the signature splitting of the yrare ("wobbler") band, designated band (6) in the work of ours and of Venkova et al.[8]. Experimentally there is only a small signature splitting in the yrare bands of ^{125}Xe and the odd-A rhodiums, with the sign opposite that of the main ground band. Since the rhodium signature splitting is somewhat less than the xenon, the rhodium triaxiality parameter γ is probably somewhat less than 24°.

The theoretical work of Gelberg et al. [14] shows the signature splitting at maximum triaxiality of $\gamma = 30°$ going to the limit of degenerate doublet structure. That degenerate doublet structure is closely approached in the $g_{9/2}$ band of neighboring ^{115,117}Ag (Z=47), as reported by Hwang et al. [15]. There is backbending above spin 21/2 in ^{117}Ag, and observations do not extend above 21/2 in ^{115}Ag. In the case of ^{115}Ag they also observe an yrare band with small signature splitting, as in the rhodiums. In Fig. 14 of Hwang et al. [16] we charted the $g_{9/2}$-band signature splitting in various odd-even nuclei with Z=39, 41, 43. As we discussed in ref. [16] the relative constancy of splitting with increasing spin calls for the spin-independent mixing from a triaxial potential well. The lower-Z, lower-N examples show the least signature splitting, probably because they have the nearly cylindrical symmetry of spheroids.

5. Chiral Possibilities

Frauendorf and Meng [17] have brought out a whole new theoretical perspective on nuclear structure possibilities of triaxial nuclei. They have pointed out that near the beginning of the filling of a higher-j shell an unpaired nucleon will preferentially align its angular momentum along the shortest axis. Near the end of the filling of a higher-j shell the unpaired nucleon hole will prefer to align its angular momentum along the longest axis. The largest of the moments-of-inertia along the three principal axes will be that along the axis of intermediate length. For example, consider that odd-odd nuclei like ^{104}Rh may have an $h_{11/2}$ neutron align along the longest axis and the $g_{9/2}$ proton along the shortest axis, with the rotational angular momentum vector mainly along the intermediate axis. Such a structure has chirality, that is, it has right-handed left-handed forms, which cannot be rotated into one another. Frauendorf and Meng [17] show that in this situation one may get a doubling of bands, which will have the same spins and parities. In actual nuclei we expect to see two linear combinations of right- and left-handed forms, one connected with a plus sign and the other with a minus sign. Starosta et al. [18] claim evidence shows chiral pairs in several odd-odd N=75 isotones.

It is appropriate to look for evidence of chiral doublets (two bands with the same spins and parities) in this region. It is a remarkable feature of nuclei of this region that levels of the same spin and parity often occur quite close in energy to one another. That alone is insufficient evidence to invoke chirality features, for so-called "wobbler bands" can be quite low-lying in regions of triaxial shapes, especially where there is an odd number of protons or neutrons. For example, note the bands (6) in Figs. 1 and 2, with bandhead $11/2^+$. Many states of bands (6) lie close in energy to members of the ground bands with the same spins and parities. This occurs because the two units of core rotational angular momentum along the longest axis, the 1-axis, forming the $11/2^+$ bandhead involve about the same amount of rotational energy as the larger amount of rotational angular momentum perpendicular to the long axis required to form the $11/2^+$ member of ground band (1). The added energy is nearly equal because the rotational moment-of-inertia along the longest axis is smaller than that along the perpendicular axes.

Our search for chiral candidates is drawn to the apparent bifurcation of the ground band (1) around 21/2 or 23/2 into band (7) for ^{111}Rh or band (2) for ^{113}Rh. This is also the spin value where backbending is occurring in ground bands (1), though both side-bands could have spins one unit lower, making the yrast bands those designated (1)..

For ^{111,113}Rh in the 3-quasiparticle bifurcations of the ground band above the bandcrossing we have Frauendorf's four conditions favoring chiral doubling. That is, (1) the ground bands of the odd-A rhodiums probably have a triaxial-shaped core; (2) the angular momentum of the proton hole in the nearly-filled $g_{9/2}$ subshell should prefer alignment along the longest axis, (3) the two aligned $h_{11/2}$ neutrons in the less-than-half-filled $h_{11/2}$ subshell should prefer alignment along the shortest axis, and (4) the rotational angular momentum will preferentially be along the intermediate axis where there is the largest rotational moment-of-inertia of the three principal axes. In estimating the base spin of such a chiral enantiomer we must bear in mind that there are probably three holes in the proton $g_{9/2}$ subshell, and there may be three or five particles in the $h_{11/2}$ neutron subshell. A closely analogous situation is seen in ^{101}Tc$_{58}$, as shown in Fig. 8 of Fotiades et al. [19]

It remains to be proved what the nature of the bifurcation above the backbend really is. There are four (or five) angular momentum vectors being coupled above the backbend, a $g_{9/2}$ proton hole, two $h_{11/2}$ neutrons and the collective core rotational angular momentum (or two core rotations, along the long axis or perpendicular to it.) The non-chiral possibilities include (1) stretched, or nearly-stretched, coupling, and (2) tilted rotor coupling with particle j vectors and rotational angular momentum in the same plane. Further theoretical and experimental work is clearly needed to resolve these fascinating questions.

6. Acknowledgments

This research was partially supported by U.S. Department of Energy at LBNL through Contract W-7405-ENG48, at Vanderbilt University through Grant No. DE-FG05-88ER40407, at Tsinghua University by the National Natural Science Foundation and the Science Foundation for Nuclear Industry, China, and at UFRJ through a grant from the Brazilian CNpQ Science Foundation.

References

1. R.B. Firestone et al, eds. *Table of Isotopes*, 8th ed., (Wiley, New York, 1996.)
2. A Guessous et al., *Phys. Rev. Lett.*, **75**, 2280 (1995) See also A.G. Smith et al., *Phys. Rev. Lett.*, **77** 1711 (1996).
3. J. Skalski, S. Mizutori, and W. Nazarewicz, *Nucl. Phys.* **A617**, 282 (1997).
4. D. Troltenier, et al., *Nucl. Phys.*, **A601**, 56 (1996).
5. D. C. Radford, *Nucl. Instrum. Methods in Phys. Research*, **A361**, 297 (1995).
6. cf. S. J. Zhu et al., *Phys. Rev.* **C65**, 014307 (2001).
7. N. Fotiades et al., *Abstract Book, Nuclear Triangle Conference, Jackson Hole, Wyoming, May 2002*, "High-Spin states in neutron-rich Rh isotopes" (2002).
8. Ts. Venkova et al., *Euro. Phys. J.* **A6**, 405 (1999).
9. G. Lhersonneau et al., *Euro. Phys. J.*, **A1**, 285 (1998).
10. J. Kurpeta et al., *Euro. Phys. J.*, **A13**, 449 (2002).
11. X.Q. Zhang, et al., *Phys. Rev.* **C63**, 027302 (2001).
12. N. Fotiades et al, *Phys. Rev.*, **C61**, 064326 (2000).
13. R. Duffait et al., *Nucl. Phys.*, **A454**, 143 (1986).
14. A. Gelberg et al., *Nucl. Phys.* **A557**, 439c (1993).
15. J.K. Hwang et al, *Phys. Rev.* **C65**, 054314 (2002).
16. J.K. Hwang et al., *Phys. Rev.*, **C58**, 3252 (1998).
17. S. Frauendorf and J. Meng, *Nucl. Phys.* **A617**, 131 (1997).
18. K. Starosta et al., *Phys. Rev. Let.* **86**, 971 (2001).
19. N. Fotiades et al., *Phys. Scr.*, **T188**, 127 (2000).

NUCLEAR STRUCTURE OF NEUTRON-RICH N=83 ISOTONES BY ^{252}CF FISSION AT GAMMASPHERE

Y.X. Luo[1,2], J.O. Rasmussen[2], J.H. Hamilton[1], A.V.Ramayya[1], J.K. Hwang[1], C.J. Beyer[1], S.J. Zhu[3], J. Kormicki[1], X.Q. Zhang[1], E.F. Jones[1], P.M. Gore[1], T.N. Ginter[2], I.Y. Lee[2], K.E. Gregorich[2], A.O. Macchiavelli[2], G.M. Ter-Akopian[4], Yu.Ts. Oganessian[4], A.V. Daniel[4], M.A. Stoyer[5], S.J. Asztalos[5], J.D. Cole[6], R. Donangelo[7], and S.C. Wu[2]

[1]*Vanderbilt University, Nashville, TN 37235*
[2]*Lawence Berkeley National Laboratory , Berkeley, CA 94720*
[3]*Tsinghua University, Beijing, China*
[4]*Joint Institute for Nuclear Research, RU-141980 Dubna, Russia*
[5]*Lawrence Livermore National Laboratory, Livermore, CA 94551*
[6]*Idaho National Environmental and Engineering Laboratory*
[7]*Federal University of Rio de Janeiro (UFRJ), 21945-970 Rio de Janeiro, Brasil*

Structure of neutron-rich N=83 isotones was studied with ^{252}Cf source and Gammasphere. High spin level schemes of ^{139}Ba (Z=56) and ^{135}Te (Z=52) were proposed, and those of ^{137}Xe (Z=54) and ^{136}I (Z=53) were extended. Spins, parities and configurations are tentatively assigned to the levels. Rich spectroscopic information characteristic of spherical shell model excitations and enhanced E3 cascade were observed in ^{135}Te, ^{137}Xe and ^{136}I, and resemblance between ^{132}Sn and ^{208}Pb region was observed in the isotones. But ^{139}Ba exhibits different structure which is interpreted by the difference between particle-particle and particle-hole excitations. Based on a detailed measurement of a regular band consisting of intense M1 and very weak cross-over E2 high-lying in ^{135}Te, the band is interpreted as a magnetic rotation band.

1. Introduction

Structure of nuclei near N=82 closed shell are interesting because it spans a broad range of Z. Straightforward and important question is "To what extent is the spherical shell model (SSM) valid for nuclei near N=82?" The answers to these questions provide stringent tests of the basic ingredients of the spherical shell model calculations. With one valence neutron outside the N=82 closed shell and a few valence protons beyond the Z=50 closed shell, the N=83 isotones under consideration are near ^{132}Sn, and their yrast spectroscopy is of significance for the information about empirical proton-neutron and proton-proton interactions. The resemblance of structure between the ^{132}Sn and ^{208}Pb [1] regions is one of the interesting topics, and also provides important information like nucleon-nucleon interactions for the calculations in ^{132}Sn region.

Beta-decay of fission products [2] and (α, xn) and heavy ion reactions were used to populate low spin states of the N=83 isotones [3]. However, broad and detailed exploration of the high spin yrast and near-yrast spectroscopy in the region had not been realized until the advent of large γ–detector arrays [4]. Before the present work, some progress was made at Gammasphere and Eurogam II on N= 83 isotones such as [134]Sb, [135]Te, [136]I [5], and [137]Xe [6].

A total of 5.7*10[11] 3- and higher-fold coincidence events were accumulated in our experiment using a 62 μCi [252]Cf at Gammasphere. The high statistics data make it possible for us to extend our understanding to very neutron-rich nuclei with low-production rates, and weakly populated bands.

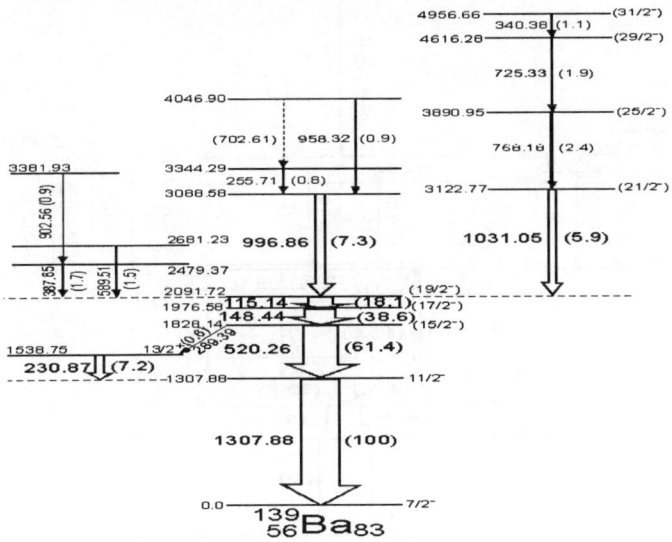

Figure 1 Level scheme of [139]Ba proposed by our collaboration [7].

2. Level schemes of the N=83 isotones: [139]Ba, [137]Xe, [136]I, and [135]Te

High spin states of [139]Ba from fission were proposed for the first time [7] (see Figure 1); the previously reported level scheme of [137]Xe [6] was extended from excitation energy of ~4.5 MeV to 6.690 MeV with 11 new levels (see Figure 2); that of [136]I [5] was extended from ~4.0 MeV to 5.433 MeV with 21 new levels (see Figure 3); and a well established level scheme of [135]Te was established by us independently [7] (see Figure. 4)], but a similar one was first published by Fornal et al.[8]. Our special attention was paid to the measurements of the intense M1 and very weak cross-over E2 transitions in the upper part of the level scheme of [135]Te, and this led to the identification of an

254

interesting regular band high-lying on several high energy transitions (see Figure 4).

Based on the previous assignments to the lower-lying states for ^{139}Ba [3], spins and parities are tentatively assigned to the new levels of ^{139}Ba [7]. For ^{135}Te, $25/2^+$, $27/2^+$ and $31/2^-$ were assigned, respectively, to the 3233.7, 4591.1 and 5641.7 keV levels and confirmed by shell model calculations [9], identifying two E3 transitions of 1678.4 and 2408.0 keV in cascade; and the spin and parities of $(19/2^-)$, $(21/2^-)$...$(33/2^-)$ were assigned to the newly observed band.

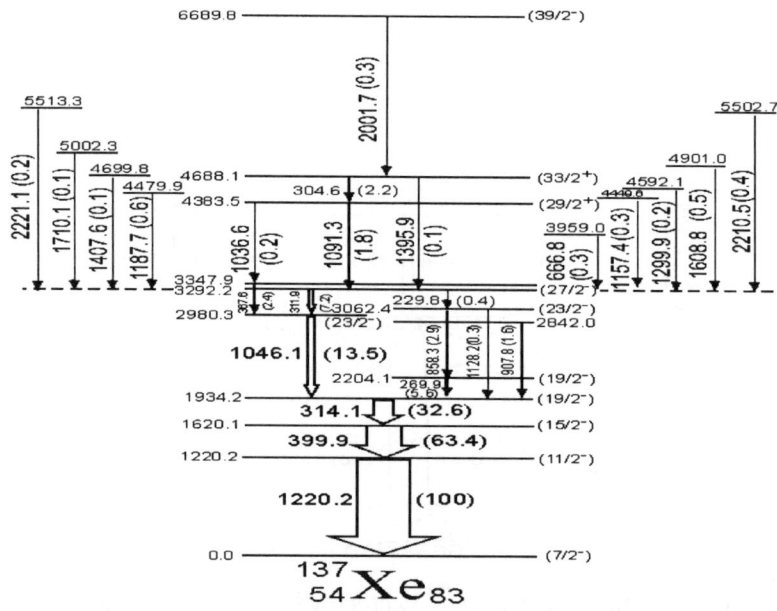

Figure 2 Level scheme of ^{137}Xe proposed by us extending that of Daly $et\ al.$ [6]

Spin-parity assignments were tentatively made for ^{137}Xe following [6] in which the assignments were made by means of angular correlation and linear polarization measurements; and $39/2^-$ is tentatively assigned to the new 6689.8 keV level considering the similar level pattern and spacing to those of ^{135}Te. The assignments for ^{136}I are made based on those in [5], and 17^- is tentatively assigned to the new 5114.9 keV level based on the analogy of level patterns.

3. Interpretations and discussions for level structure of the N=83 isotones

In view of the six protons in the $g_{7/2}$ subshell in ^{139}Ba, configurations of a pair of $g_{7/2}$ proton hole coupled to a single-particle $f_{7/2}$ neutron, $(\pi g^{-2}_{7/2}\upsilon f_{7/2})_{(7/2, 11/2, 15/2, 19/2)^-}$, were assiged to the lower states, and $(\pi g^{-3}_{7/2}d_{5/2}\upsilon f_{7/2})_{(21/2, 25/2, 29/2, 31/)^-}$ and $(\pi g^{-2}_{7/2}\upsilon i_{13/2}$ or $\pi g^{-2}_{7/2}\upsilon f_{7/2}*3)_{13/2^+}$ to the other levels in ^{139}Ba [7].

The level structure of ^{137}Xe is attributed to the coupling between the

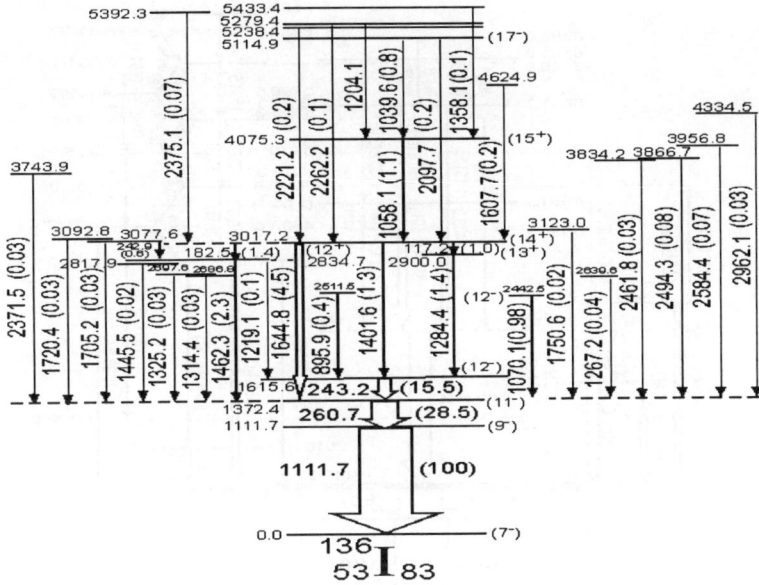

Figure 3 Level scheme of ^{136}I extending that of Bhattacharyya *et al.* [5].

configurations of ^{136}Xe with an additional $f_{7/2}$ valence neutron, the latter being assigned by means of SM calculations as four valence protons with related single particle excitations $(\pi g^4_{7/2}, \pi g^3_{7/2}d_{5/2}$ and $\pi g^3_{7/2}h_{11/2})$ [6]. The newly identified 6689.8 keV level in ^{137}Xe is assumed to come from $(\pi g^3_{7/2}h_{11/2} \upsilon i_{13/2})_{39/2^-}$; as in ^{135}Te an enhanced E3 cascade 2001.7 keV – 1395.9 keV is observed in ^{137}Xe.

The calculations in [5] suggested the interpretation of the ^{136}I levels up to 1615.6 keV as members of the $\pi g^3_{7/2}\upsilon f_{7/2}$ and $\pi g^2_{7/2}d_{5/2}\upsilon f_{7/2}$ multiplets, and levels around 2.9 MeV as $\pi g^2_{7/2}h_{11/2}\upsilon f_{7/2}$. We feel reasonable to assign $(\pi g^2_{7/2}h_{11/2}\upsilon i_{13/2})_{17^-}$ to the level of 5114.9 keV of ^{136}I observed in the present work, so an E3 cascade, 2097.7 keV—1644.8 keV, is also observed in ^{136}I.

Based on shell model calculations and comparison of level patterns with that of ^{134}Te, single proton-neutron excitations, $\pi g^2_{7/2}\upsilon f_{7/2}$, $\pi g_{7/2}d_{5/2}\upsilon f_{7/2}$,

256

$\pi g_{7/2}h_{11/2}\upsilon f_{7/2}$, and $\pi g_{7/2}h_{11/2}\upsilon i_{13/2}$ etc., are assigned to levels in ^{135}Te. The particularly interesting high-lying regular sequence was interpreted by Fornal *et al.* as core excitations $(\pi g^2_{7/2}\upsilon f^2_{7/2}h^{-1}_{11/2})$ [8]. We concur with the interpretation, and based on our identifications and intensity measurements of the four weak-cross overs, we would further interpret this regular band as tilted rotation [10].

It is interesting to make a comparison for the structure of the N=83 isotones studied. The richness in structure, especially that observed in ^{135}Te, is impressive

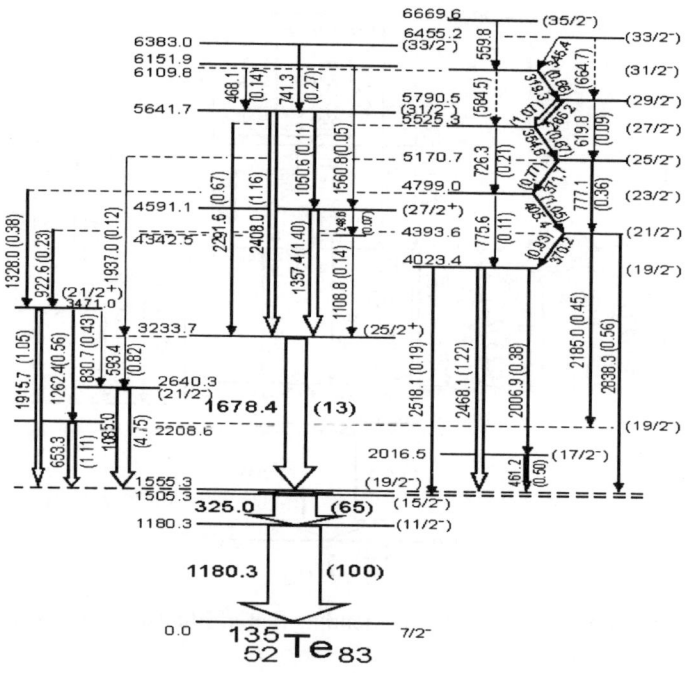

Figure 4 Level scheme of ^{135}Te proposed and close to that of Fornal *et al.* [8]

and the pronounced irregularities of the level patterns observed in the isotones with fewer valence protons show typical single-particle character. The enhanced high energy E3 transition cascade is observed in ^{135}Te, ^{136}I and ^{137}Xe, respectively. The resemblance of spectroscopy and the counterparts are observed between ^{135}Te and ^{211}Po, ^{137}Xe and ^{213}Rn (see Figure 5, and [8,11]), and ^{136}I and ^{212}At. In Figure 5 similar level patterns, opposite parities and regular differences in spins are seen for corresponding states in each pair of nuclei. However, the level structure of ^{139}Ba exhibits remarkable differences. A 17/2⁻ level interposes itself between the 19/2⁻ and 15/2⁻ states. Even more notably, at the higher spin part the level scheme of ^{139}Ba is very sparse, in a striking contrast with the richness in ^{135}Te and other isotones. The enhanced high energy E3 transitions do not show up in ^{139}Ba. This interesting diversity may be partly attributed to the difference of particle-particle coupling in ^{135}Te (two valence protons), ^{136}I (three protons), and ^{137}Xe (four protons), and particle-hole coupling in ^{139}Ba (a pair of proton-holes), respectively. The particle-hole coupling in ^{139}Ba case would lower the level of spin one less than the stretched member of the multiplet, which just explains why the 17/2 level is lower in energy than the 19/2 level in ^{139}Ba. This lowering may be also responsible for the 'short circuit' and non-observance of the enhanced high energy E3 transitions in ^{139}Ba.

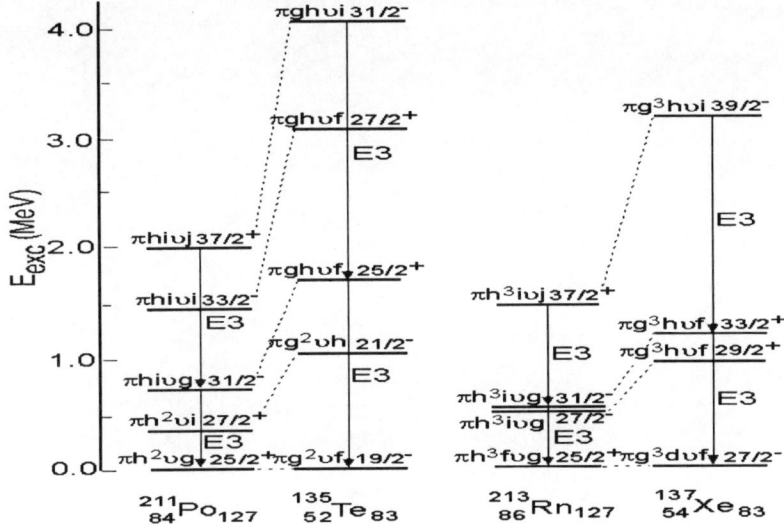

Figur 5   Comparison of yrast three-particle states in ^{135}Te and ^{211}Po, and of possible yrast five-particle states in ^{137}Xe and ^{213}Rn [11]. See text.

Works at Vanderbilt University, LBNL, and LLNL were supported by the U.S. DOE under Grants No. DE - FG05 - 88ER40407, DE- AC07-761DO1570 and DE - AC03 - 76SF00098, and W-7405-ENG-48, respectively.

References

1.   J. Blomqvist, in *Proceedings of the 4th Inter. Conf. on Nucl. Far From Stability*, Helsingoer, Denmark, 1981, p.536.

2.   *Table of Isotopes*, 8th ed., edited by R.B. Firestone and V.S. Shirley, (Wiley, New York, 1996).

3   H. Prade *et al., Nucl. Phys.* **A472**, 381(1987).

4.   J.H. Hamilton *et al.*, *Prog. Part. Nucl. Phys.* **35**, 635(1995).
5.   P. Bhattacharyya et al., *Phys. Rev.* **C56**, R2363(1997).
6.   P.J. Daly *et al.*, *Phys. Rev.* **C59**, 3066(1999).
7.   Y.X. Luo *et al.*, *Phys. Rev.* **C64**, 054306-1(2001).
8.   B. Fornal *et al.*, *Phys. Rev.* **C63**, 024322(2001).
9.   B. Fornal et al., in *Proceedings of the Inter. Conf. "Nucl. Phys. Close to the Barrier"*, Warszawa, Poland, June 30, 1998, p.1219.
10. S. Frauendorf, *Nucl. Phys.* **A557**, 259c(1993); ibid. **A677**, 115(2000).
11. A.E. Stuchbery *et al.*, *Nucl. Phys.* **A482**, 692(1988).

A STUDY OF NEUTRON-RICH CL ISOTOPES BY USING DEEP-INELASTIC REACTIONS

X.LIANG¹, R.CHAPMAN¹, F.HAAS², K.-M.SPOHR¹, P.BEDNARCZYK³
S.M.CAMPBELL¹, P.J.DAGNALL⁶, M.DAVISON¹, G.DE ANGELIS⁴,
G.DUCHÊNE², TH.KRÖLL⁵, S.LUNARDI⁵, S.NAGULESWARAN² AND
M.B.SMITH¹

1 Electronic Engineering and Physics, University of Paisley, Paisley PA1 2BE, UK
2 Institut de Recherches Subatomiques, UMR 7500, CNRS-IN2P3 and Université Louis Pasteur, F--67037 Strasbourg Cedex 2, France
3 Institut de Recherches Subatomiques, F--67037 Strasbourg Cedex 2, France and The Niewodniczanski Institute of Nuclear Physics PL-31342 Krakow, Poland
4 Istituto Nazionale di Fisica, Laboratori Nazionali di Legnaro, I--35020 Legnaro, Italy
5 Dipartimento di Fisica and Istituto Nazionale di Fisica, Sezione di Padova, I-35131 Padova, Italy
6 Department of Physics and Astronomy, University of Manchester, Manchester M13 9PL, UK

The yrast decay sequences of neutron-rich Cl isotopes, populated in deep-inelastic processes produced by the interaction of 234 MeV ^{37}Cl ions with a ^{160}Gd target, were studied using the EUROBALL IV gamma-ray detector array. The observed energy separation of the $3/2^+$ and $1/2^+$ levels in odd-A Cl isotopes decreases (with an eventual crossing of levels) as the occupation of the $1f_{7/2}$ neutron shell increases, in agreement with the predictions of shell model calculations.

1. Introduction

Studies of the properties of nuclei far from stability provide a unique opportunity to increase our understanding of nuclear interactions in extreme conditions and often challenge our theoretical models. Neutron-rich sdf-shell nuclei reveal interesting phenomena which are of particular interest to the practitioners of large-scale shell model calculations. Theoretical calculations[1] by Werner et al. and experimental measurements[2] of B(E2) have indicated the possibility of deformation in the region near (N =26) ^{42}S. However, a challenge lies in their production, since conventional reaction mechanism do not populate neutron-rich isotopes. Deep-inelastic reactions can access nuclei with appreciably high N/Z ratios[3], and this has resulted in an increasing interest in the use of such reactions to populate neutron-rich sdf-shell nuclei[4,5,6]. The experimental quadrupole deformation β_2 values obtained from B(E2;$0^+ \rightarrow 2^+$)

measurements for ^{40}S(0.284)[7] and ^{42}Ar (0.273)[8] provide evidence for the deformation of these isotones of ^{41}Cl. However, there is little spectroscopic information available currently for the neutron-rich isotopes of Cl with N > 23. Spectroscopic measurements can reveal details of the underlying microscopic structure and have proved essential for understanding properties of nuclei far from stability. Furthermore, they potentially provide a stringent test of modern large scale shell-model calculations.

2. Experiment

In the present experiment, many target- and projectile-like nuclei, including neutron-rich Cl isotopes were populated in the interaction of a 234 MeV beam of ^{37}Cl ions, delivered by the VIVITRON at IReS, Strasbourg, with a ^{160}Gd target. The target, isotopically enriched to 98.2% in ^{160}Gd, was of thickness 12mg·cm^{-2} and was backed with 40mg·cm^{-2} of isotopically enriched ^{208}Pb (99.47%). The backed target was sufficiently thick to stop all forward-recoiling reaction fragments. Gamma rays were detected using the highly-sensitive EUROBALL IV array[9]. Experimental data were taken during a three day run. The electronic trigger condition was such that when seven Ge signals before Compton suppression or five Ge signals before Compton suppression and nine BGO inner ball element signals were in time coincidence, the "event" was accepted and subsequently written to tape. The data were sorted off-line into a γ-γ-γ cube and analysed using the RADWARE code[10].

3.   Results and Discussion

3.1 Population Yields

In the deep-inelastic reaction discussed here, many product nuclei were populated to moderately high spins and energies in both the target and projectile region, and this raises questions concerning the relative cross sections for the processes involved. The relative isotopic yields for target-like and odd-Z projectile-like nuclei are presented graphically in figure1. In this figure the area of each circle is proportional to the relative isotopic yields for that particular isotopes. The relative yields of odd-A target-like nuclei are small because, in general, several yrast and near-yrast rotational bands appear to be populated with comparable strength and this makes yield estimates difficult to determine. However, relative isotopic yields for even-even target-like and of odd-Z projectile-like nuclei give us reasonable information on relative cross sections

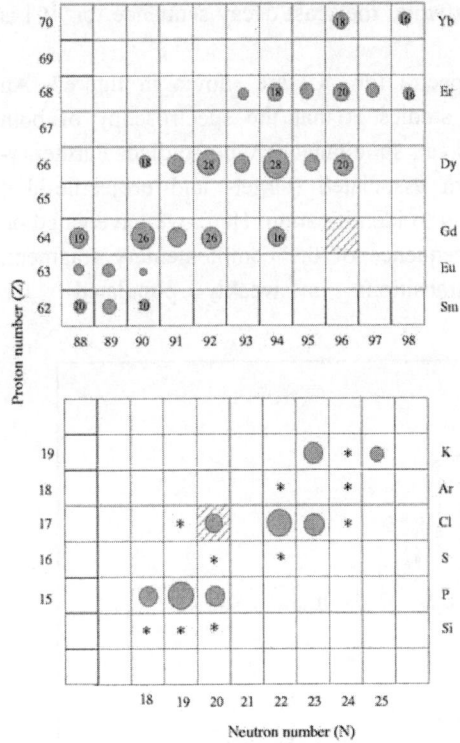

for deep-inelastic processes. The numbers inside the circles in figure 1 correspond to the highest angular momenta (in unit of h) observed in the present work for even-even target-like nuclei. We can see that, in general, the stronger the population of the nucleus, the higher we are able to see in spin. It is apparent that the transfer of protons from projectile to target and of neutrons from target to projectile are generally favoured. In this reaction the ratio N/Z is substantially larger for the target (1.50) than for the projectile (1.18), and so these results are in keeping with the expected tendency[3] toward N/Z equilibration of the colliding ions.

Figure 1. Relative yields for target-like and odd-Z projectile-like nuclei populated in this work (after neutron evaporation) are represented by circles having areas directly proportional to the corrected γ-γ coincidence intensities. Target and projectile nuclei are indicated by crosshatching, while stars label projectile-like nuclei that were identified in the present work but for which reaction yields could not be measured.

3.2 Neutron-rich Cl isotopes

In the present work, yrast levels schemes of ^{36,37,39,40,41}Cl have been established. Obviously, ^{38}Cl should also be populated. The non-observation of this nucleus is due to the severe Doppler broadening of the γ-ray photopeaks corresponding to the decay from the previously observed[11]3$^-$ (half-life 220 fs) and 4$^-$ (half-life

370 fs) excited states. In the present work, the yrast decay sequence for ^{41}Cl is observed for the first time.

Examples of double-gated γ-ray spectra for ^{41}Cl are shown in figure2. An important aspect of deep-inelastic studies is that the spectroscopy of both projectile- and target-like fragments can simultaneously be studied. Further, γ-rays observed in coincidence from associated (target- and projectile-like) fragments can often be used as an aid to identification. Here, we have gated on γ-ray transitions within the yrast sequence of the complementary fragment, ^{152}Gd, in order to identify transitions in weakly populated ^{41}Cl.

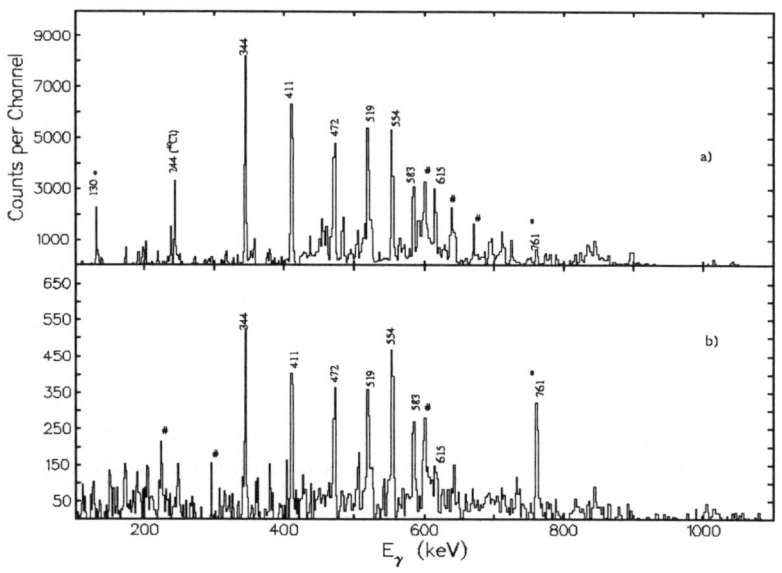

Figure2. Double-gated γ-ray spectra for ^{41}Cl and the associated target-like nucleus ^{152}Gd.

In figure2, spectrum (a) corresponds to a sum of double gates involving all combinations of γ-ray transitions at 344, 411, 472, 519, 554 and 583 keV in the yrast decay sequence of ^{152}Gd; a photopeak at an energy of 130keV is clearly seen along with the lowest transition in ^{40}Cl (244 keV) and the yrast decay sequence of ^{152}Gd. Spectrum (b) corresponds to a sum of double gates with the first gate set at 130 keV and second gates at 344, 411, 472, 519, 554 and 583 keV in the yrast decay sequence of ^{152}Gd; a peak at an energy of 761 keV together with the yrast decay sequence of ^{152}Gd can be seen The peaks marked by "*" correspond to the observed gamma-ray transitions in ^{41}Cl. The peaks marked by "#" are assigned as impurities by examining other gated spectra. The observation of 130 keV and 761 keV gamma-rays in the β-decay of $^{41}Cl^{[12]}$ confirms the assignment of this decay sequence to ^{41}Cl. The ordering of these two transitions is based on relative intensity measurements after correction for detection efficiency.

The systematics of the level structures of the odd-A Cl isotopes with N= 20, 22 and 24 are shown in figure3. This figure presents the yrast level schemes for ^{37}Cl, ^{39}Cl and the level scheme of ^{41}Cl from the present work. The known $1/2^+$ levels[12], which are not yrast states, are also presented for $^{37,39}Cl$.

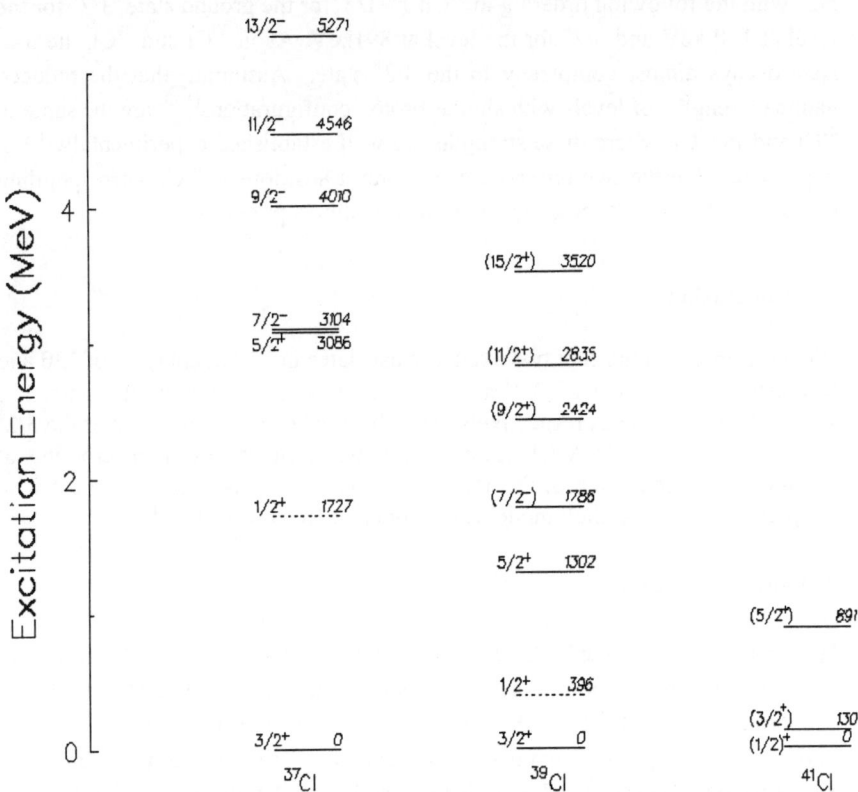

Figure 3 Systematic yrast level structures for N=20, 22 and 24 Cl isotopes

The results of shell-model calculations[13] predict that the $3/2^+$ and $1/2^+$ levels of the odd-A Cl isotopes come closer in energy and eventually cross as the occupation of the $1f_{7/2}$ neutron shell increases. Indeed, the near-degeneracy of the $1d_{3/2}$ and $2s_{1/2}$ orbitals is the determining factor for the onset of deformation in neutron rich sulfur isotopes with N=24 and 26[13]. Shell model calculations[13] indicate that for the Cl chain, the $1/2^+$ state becomes the ground state for the N = 26, ^{43}Cl isotope; for the N = 24, ^{41}Cl isotope the predicted separation energy between the $1/2^+$ and $3/2^+$ states is 60keV (12 keV in the calculation of Woods[14]). However, since the deep-inelastic reaction is known to populate selectively yrast states such as those of ^{37,39}Cl shown in figure 3 and taking into account the relative intensities measurement, the present work would suggest that the first three states in this odd-A Cl chain have $J^\pi=1/2^+$, $3/2^+$ and $5/2^+$ with the following ordering in ^{41}Cl: $J^\pi=1/2^+$ for the ground state, $3/2^+$ for the level at 130 keV and $5/2^+$ for the level at 891keV. As in ^{37}Cl and ^{39}Cl, the $5/2^+$ state decays almost completely to the $3/2^+$ state. Assuming that the reduced gamma strengths of levels with similar proton configurations[1,13] are the same in ^{41}Cl and in ^{37}Cl, where these strengths are well established experimentally[11], it is proposed that the two observed gamma-ray transitions in ^{41}Cl, corresponding to a $5/2^+ \rightarrow 3/2^+ \rightarrow 1/2^+$ cascade, are probably almost pure M1.

3. Conclusion

We have observed the first two excited yrast states in ^{41}Cl at energies of 130 and 891 keV. J^π values of $1/2^+$, $3/2^+$ and $5/2^+$ are proposed for the ground state, 130 keV and 891 keV states respectively. The observed energy separation of the $3/2^+$ and $1/2^+$ levels in odd-A Cl isotopes decreases (with an eventual crossing of levels) as the occupation of the $1f_{7/2}$ neutron shell increases, in agreement with the predictions of the shell model calculations of Retamosa et. al.[13].

Acknowledgements

We would like to thank the technical staff of the VIVITRON accelerator laboratory for their support during the course of the experiment. This work was supported by the EPSRC (UK). We also would like to acknowledge support under the European Commission Programme Training and Mobility of Researchers: Access to Large-Scale Facilities: Vivitron Accelerator and Associated Detectors" contract number ERBFMGECT980145.

References

1. T.R.Werner et. al., Nucl. Phys. **A597**, 327 (1996).
2. T.Glasmacher et. al., Phys .Lett. **B 395**, 163 (1997).
3. H.Freiesleben and J.V.Kratz, Phys.Rep. **10**, 1 (1984).
4. B.Fornal et. al., Phys.Rev. **C49**, 2413 (1994).
5. X.Liang et. al., Phys. Rev. **C66,** 014302 (2002).
6. X.Liang et. al., Phys. Rev. **C66,** 037301 (2002).
7. H.Scheit et. al., Phys. Rev. Lett. **77,** 3967 (1996).
8. S.Raman et. al., At. Data Nucl. Data Tables **36,** 1 (1987).
9. J.Simpson, Z. Phys. **A358,** 139 (1997).
10. D.C.Radford, Nucl. Instrum. Methods Phys. Res. **A361,** 297 (1995).
11. P.M.Endt, Nucl. Phys. **A521** 1 (1990); Nucl. Phys. **A633,** 1 (1998).
12. J.A.Winger et. al., Proc.Conf on Exotic Nuclei and Atomic Masses, Bellaire, Michigan, June 23-27, 1998, 606 (1998).
13. J.Retamosa et. al., *Phys. Rev.* **C55**, 1266 (1997).
14. C.L. Woods, Nuclear Physics **A451** 413 (1986).

THE YRAST SPECTROSCOPY OF NEUTRON-RICH NUCLEI PRODUCED IN DEEP-INELASTIC PROCESSES*

J. OLLIER†, R. CHAPMAN, X. LIANG, M. LABICHE, K.-M. SPOHR AND M. DAVISON

Electronic Engineering and Physics, University of Paisley
Paisley, PA1 2BE, UK

G. De ANGELIS, M.AXIOTIS, T. KRÖLL, D. R. NAPOLI AND T. MARTINEZ

INFN – Laboratori Nazionali di Legnaro,
Via Romea 4,35020 Legnaro (Padova), Italy

D. BAZZACCO, E. FARNEA, AND S. LUNARDI

INFN Sezione di Padova and Dipartimento di Fisica dell' Università,
Via F. Marzolo 8, 35131 Padova, Italy

A. G. SMITH

Schuster Laboratory, Department of Physics and Astronomy,
University of Manchester, Manchester, M13 9PL, UK

The highly sensitive GASP array at the INFN Legnaro Laboratory was used to study the γ–ray de-excitation of neutron-rich nuclei produced in the deep-inelastic processes which occur when 230MeV ^{36}S ions interact with a target of ^{176}Yb. Yrast decay schemes were identified in over forty target-like fragments and in over twenty projectile-like fragments. Extensions have been made to the yrast decay sequences of ^{176}Hf, ^{166}Er, ^{172}Yb, and ^{152}Sm in the target-like species. New transitions have also been observed in a number of projectile-like species including ^{34}P and ^{41}Cl. The level structures of ^{34}P and ^{41}Cl are compared with the results of shell-model calculations.

1. Introduction

The study of neutron-rich nuclei is currently a major topic of interest in nuclear structure physics. Such nuclei are predicted to reveal new aspects of nuclear structure that challenge established theoretical models. In particular, neutron-rich nuclei around the shell-model magic numbers of N = 20 and 28 exhibit properties inconsistent with shell closure. The first experimental evidence for this around N = 20 came from mass measurements of neutron-rich Mg isotopes [1, 2]. Further evidence was based on the subsequent

---

* This work is supported by the EPSRC (UK).
† E-mail: olli-ph0@wpmail.paisley.ac.uk.

measurements of the excitation energy of the first 2^+ state [3] of ^{32}Mg, which was much lower than expected, and measurements of the B(E2; $0^+ \rightarrow 2^+$) value [4].

In the N=28 region, measurements by intermediate-energy Coulomb excitation of the B(E2; $0^+ \rightarrow 2^+$) values involving the first 2^+ excited states of ^{44}S [5] and ^{46}Ar [6] also show evidence of collectivity for these magic number nuclei. Recent work by Azaiez et al [7][8] has provided evidence for shape coexistence in these nuclei; the second excited 2^+ and 0^+ states would appear to have spherical shapes.

Shell-model studies in the N=20 region [9] have shown that the observed anomalies can be understood within calculations which consider the promotion of neutrons from the sd to the fp shell. The residual neutron-proton interaction lowers the energy of configurations with neutrons in the $f_{7/2}$ subshell and protons in the $d_{5/2}$ subshell. This leads to low-lying intruder states.

The production of nuclei in this region is difficult, however, due to the limitations of traditional reaction mechanisms. New techniques have been developed recently for the population of neutron-rich nuclei and the deep-inelastic process, for example, has become a reliable way to populate such nuclei.

2. Experimental Technique.

The combination of Tandem-XTU and ALPI accelerators at the INFN Legnaro Laboratory, Italy, was used to deliver a beam of ^{36}S ions at an energy of 230 MeV (6.39 MeV/u) onto a target of ^{176}Yb. The target was isotopically enriched to 97.75% and was of thickness 14mg·cm^{-2} with an isotopically enriched ^{208}Pb (98.70%) backing of thickness 35mg·cm^{-2}. ^{36}S ions entered the ^{208}Pb backing with an energy of 160 MeV (4.4 MeV/u). The backing was sufficiently thick to stop all forwards-moving reaction fragments. The present experiment was designed to study the γ–decay of excited nuclear states with lifetimes longer than the slowing-down time of the recoiling nuclei in the composite target (~1ps) and thus no Doppler correction was necessary when sorting the data. The GASP array was used to measure γ–rays emitted from the de-exciting target-like and projectile-like fragments populated in the deep-inelastic processes. The array is composed of 40 HPGe Compton-suppressed detectors and an inner ball of 80 BGO crystals that can act as a calorimeter. The total photopeak efficiency is about 3% at a γ-ray energy of 1332keV (^{60}Co) and the mean peak to total ratio about 60-65% at 1332 keV for the Compton suppressed γ–ray spectra [10]. The electronic trigger conditions were set so that if three or more Ge signals (unsuppressed) and two or more BGO signals were in time coincidence then the event was accepted by the acquisition system and written to tape. Data were collected during six days of beam time. Gain-matching of the detectors and data sorting were performed off-line. A γγγ–cube with no conditions was constructed. Analysis of this cube was undertaken using the RADWARE code [11]. Yields from deep-inelastic processes are small, ranging typically from about 10mb to sub-mb, and this can increase the experimental difficulties in studying γ–de-excitations using this type of reaction.

3. Results and Conclusions.

In this work we have identified over forty target-like nuclei and over twenty projectile-like nuclei populated through deep-inelastic processes. For target-like fragments, extensions to yrast sequences in a number of nuclei were established. An example can be seen in the γ-ray spectrum of figure 1 which shows the yrast decay sequence of ^{176}Hf. This figure shows three possible extensions to the rotational band at energies of 736-, 771- and 802 keV.

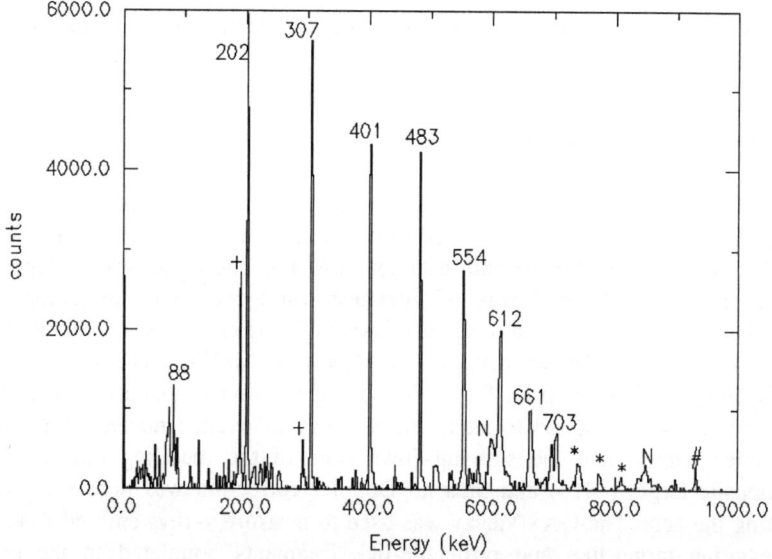

Figure 1. γ-ray spectrum corresponding to a sum of double gates for the yrast decay sequence of ^{176}Hf. New transitions are marked *, neutron peaks, N, ^{176}Yb transitions, +, and ^{34}Si, # (the complementary projectile-like fragment).

Extensions to yrast decay schemes were also observed in the projectile-like fragments. Recent experimental work by Liang *et al* [12], involving studies of the γ-de-excitation of fragments from deep-inelastic processes, has established excited states of ^{41}Cl at 130 keV and 891 keV. By double gating on these transitions we have been able to extend the yrast decay sequence by a further two transitions with energies of 554- and 1006 keV (see figure 2). Measurements of

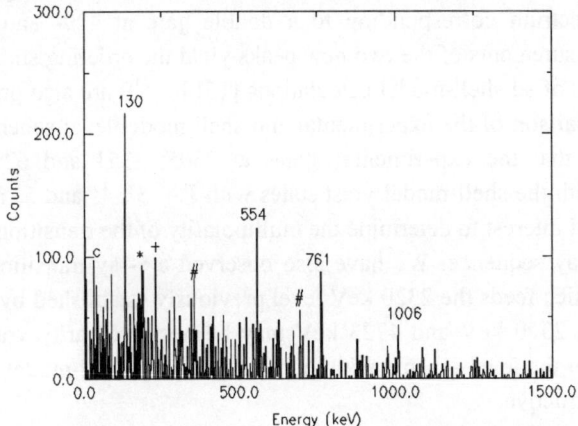

Figure 2. Sum of double gates set on the 130-, 761- and 554 keV transitions in ^{41}Cl. Also observed are transitions from ^{169}Tm (+), ^{167}Tm (#) and ^{176}Yb (*) and contaminants (c).

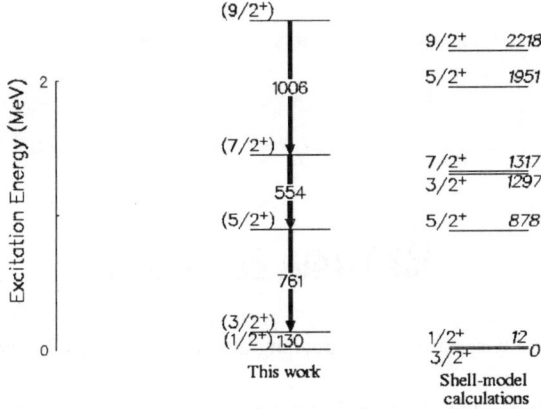

Figure 3. The Yrast level scheme of ^{41}Cl with the results of shell-model calculations using the Wildenthal universal interaction [13].

the relative intensities of these two photopeaks were used to order the transitions (see figure 3). Also shown in figure 3 are the results of shell-model calculations for ^{41}Cl based on the Wildenthal universal interaction [13]. Comparison of the experimental and shell-model level schemes would suggest J^{π} assignments of $7/2^{+}$ and $9/2^{+}$ for the new levels at excitation energies of 1445 keV and 2451 keV respectively.

Extensions to the yrast decay sequence of ^{34}P were also established by gating on the known yrast transitions at 429- and 1876 keV [14]. Two new photopeaks were observed at energies of 1046- and 2885 keV. Figure 4 shows the γ-ray spectrum corresponding to a double gate at 429- and 1046 keV. Intensity measurements of the two new peaks yield the ordering shown in figure 5. The results of sd shell-model calculations [15] for ^{34}P are also present in this figure. Comparison of the experimental and shell-model level schemes leads us to speculate that the experimental states at 2305, 3351 and 6236 keV are associated with the shell-model yrast states with $J^{\pi} = 3^{+}$, 4^{+} and 5^{+} respectively. It would be of interest to determine the multipolarity of the transitions within the ^{34}P yrast decay sequence. We have also observed a γ-ray transition of energy 2403 keV which feeds the 2320 keV level previously established by Liang [16]. The levels at 2320 keV and 4723 keV may be negative-parity yrast states. A comparison of the level scheme with the results of sdfp shell-model calculations would be instructive.

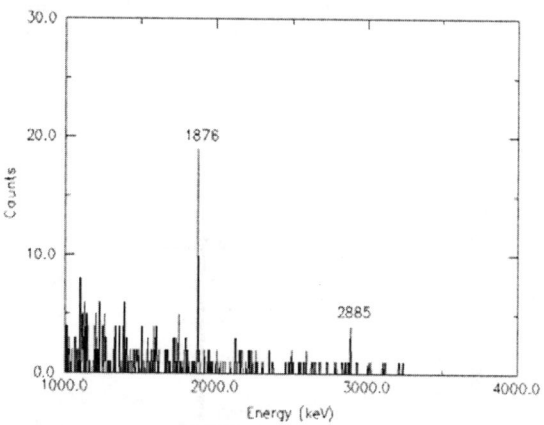

Figure 4: The γ-ray spectrum corresponding to a double gate at 429- and 1046 keV in ^{34}P.

In summary, the deep-inelastic process is a very useful way to populate yrast decay sequences in neutron-rich target-like and projectile-like nuclei.

This work is supported by the EPSRC (UK). We also acknowledge the support under the European Commission Programme ``Transnational Access to major Research Infrastructures - Improving the Human Research Potential and Socio-Economic Knowledge Base" contract number HPRI-1999-CT-00083. One of us (J. O.) acknowledges the receipt of an EPSRC postgraduate studentship during the course of this work.

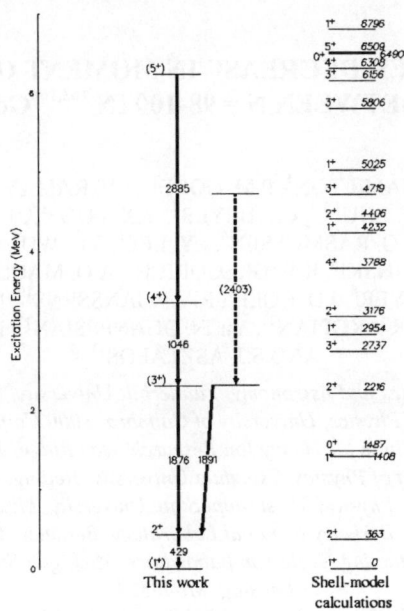

Figure 5. The experimental yrast decay level scheme of ^{34}P and the results of sd shell-model calculations [15].

References

1. C. Thibault *et al*, Phys. Rev. C **12** 644 (1975).
2. C. Détraz *et al*, Phys. Rev. C **19** 164 (1979).
3. D. Guillemaud-Mueller *et al*, Nucl. Phys. A **426**, 37 (1984).
4. T. Motobayashi *et al*, Phys. Lett. B **346**, 9 (1995).
5. T. Glasmacher *et al*, Phys. Lett. B **395** 163 (1997).
6. H. Scheit *et al*, Phys. Rev. Lett. **77** 3967 (1996).
7. F. Azaiez, Nuc. Phys. A **704**, 37c (2002).
8. D. Guillemaud-Mueller, Eur. Phys. J. A **13**, 63 (2002).
9. A.Watt *et al*, J. Phys. G **7**, L145 (1981).
10. D. Bazzacco, Proc. International Conference on Nuclear Structure at High Angular Momentum, Ottawa, Canada. p.376 AECL-10613, May 18-20 (1992).
11. D.C.Radford, Nucl. Instrum. Methods Phys. Res. A **361**, 297 (1995).
12. X. Liang *et al*, Phys. Rev. C **66** 037301 (2002).
13. C. L. Woods, Nucl. Phys. A **451**, 413 (1986).
14. B. Fornal *et al*, Phys. Rev. C **49** 2413 (1994).
15. B. A. Brown, http://www.nscl.msu.edu/brown/sde.htm.
16. X. Liang, PhD Thesis, University of Paisley (2002), unpublished.

UNEXPECTED DECREASE IN MOMENT OF INERTIA
BETWEEN N = 98-100 IN ^{162,164}Gd

E.F. JONES[1,2], J.H. HAMILTON[1], P.M. GORE[1], A.V. RAMAYYA[1], J.K. HWANG[1], A.P. DELIMA[1,2], S.J. ZHU[1,3,4], C.J. BEYER[1], Y.X. LUO[1,3,6], J. KORMICKI[1], X.Q. ZHANG[1], W.C. MA[5], J.O. RASMUSSEN[6], I.Y. LEE[6], S.C. WU[6], C.M. FOLDEN III[6], P. FALLON[6], P. ZIELINSKI[6], K.E. GREGORICH[6], A.O. MACCHIAVELLI[6], T.N. GINTER[7], M. STOYER[8], J.D. COLE[9], R.V.F. JANSSENS[10], I. AHMAD[10], A.V. DANIEL[11], G.M. TER-AKOPIAN[11], Yu.Ts. OGANESSIAN[11], R. DONANGELO[12], AND S.J. ASZTALOS[13]

[1]*Department of Physics and Astronomy, Vanderbilt University, Nashville, TN 37235*
[2]*Department of Physics, University of Coimbra, 3000 Coimbra, Portugal*
[3]*Joint Institute for Heavy Ion Research, Oak Ridge, TN 37830*
[4]*Department of Physics, Tsinghua University, Beijing, P.R. China*
[5]*Department of Physics, Mississippi State University, Mississippi 39762*
[6]*Lawrence Berkeley National Laboratory, Berkeley, CA 94720*
[7]*National Superconducting Cyclotron Laboratory, Michigan State University, East Lansing, MI 48824*
[8]*Lawrence Livermore National Laboratory, Livermore, CA 94550*
[9]*Idaho National Environmental and Engineering Laboratory, Idaho Falls, ID 83415*
[10]*Argonne National Laboratory, Argonne, IL 60439*
[11]*Flerov Laboratory for Nuclear Reactions, JINR, Dubna, Russia*
[12]*Universidade Federal do Rio de Janeiro, CP 68528, RG Brazil*
[13]*Massachusetts Institute of Technology, Cambridge, MA 11830*

From γ-γ-γ coincidence studies of the prompt γ-rays emitted in the spontaneous fission (SF) of ^{252}Cf with Gammasphere, the yrast energy levels in neutron-rich ^{162,164}Gd were identified for the first time from 2^+ to 16^+ and from 2^+ to 14^+, respectively. The 2^+ level energies are 71.6 and 73.3 keV in ^{162,164}Gd, respectively. The transition energies from every level in ^{164}Gd are higher than those from the same levels in ^{162}Gd. There is a systematic decrease at every level of the moment of inertia (MOI) and similarly β_2 deformation in N = 100 ^{164}Gd compared to N = 98 ^{162}Gd. The minimum in E(2^+) and maxima in MOI and β_2 occur at N = 98. This behavior is unexpected compared to the E(2^+) and β_2 trends in Er, Yb, and Hf nuclei where the minima and maxima occur at neutron midshell, N = 104.

1. Introduction

Theoretically, it is expected that the minimum in E(2^+) and maximum in β_2 in nuclei should occur when the neutrons and protons are at midshell between two spherical magic numbers. It was noted at the International Conference on Mapping the Triangle in May 2002 in discussing rotational behavior in rare earth nuclei that ^{170}Dy with Z = 66 and N = 104 should be the most rotational nucleus possible since it has both protons and neutrons exactly at midshell between the spherical Z = 50, 82 and N = 82, 126 closed shells. Indeed, a minimum in $E(2^+)$ and maximum in β_2 at N = 104 are found in Er, Yb, and Hf nuclei.

In a 1995 γ-γ-γ coincidence study of SF of ^{252}Cf, we identified levels in ^{162}Gd [1] to 14^+. In our 2000 γ-γ-γ study with much higher statistics, we searched for transitions in ^{164}Gd, whose yield is $(2.9\pm0.3)\times10^{-2}$ %, compared to $(5.2\pm0.6)\times10^{-2}$ % for ^{162}Gd [2], from the SF of ^{252}Cf. Based on the expectation of continued decrease in $E(2^+)$ out to N = 104, careful searches were made for γ-rays with energies somewhat lower than those in ^{162}Gd in coincidence with the Se partners of Gd in ^{252}Cf SF, but none were found. A series of γ-rays with energies just above those in ^{162}Gd in coincidence with the Se partners was found and through many cross checks was established as the yrast cascade in ^{164}Gd. Unexpectedly, all the ^{164}Gd γ-transitions from $2^+ \rightarrow 0^+$ to $14^+ \rightarrow 12^+$ are higher in energy than those in ^{162}Gd. These data clearly show that the MOI and β_2 (based on the relationship between $E(2^+)$ and β_2) [3] unexpectedly decrease in N = 100 ^{164}Gd compared to N = 98 ^{162}Gd.

2. Experimental Methods

Our 2000 Gammasphere γ-γ-γ coincidence study of the SF of ^{252}Cf with 102 detectors and a 62 μCi ^{252}Cf source yielded 5.7×10^{11} triples and higher-fold coincidences with much higher statistics data. By using uncompressed spectra and both internal and external energy standards, improved energies were obtained particularly at low energies. Further experimental details are found in Luo, et al. [4]. The powerful advantage of our high statistics data is evidenced by the fact that clean spectra are obtained for one channel gates even when both gate transitions differ by only about 1 keV as for transitions in ^{156}Sm and ^{160}Gd.

From double gating on known 1455.1 keV $2^+ \rightarrow 0^+$ and 667.1 keV $4^+ \rightarrow 2^+$ transitions in ^{84}Se, the 6n and 4n SF partners of ^{162,164}Gd, respectively, two sequences of transitions separated by a few percent were assigned to ^{162,164}Gd, shown in Figure 1(a.). In addition, we find a 165 keV $6^+ \rightarrow (5^+)$ transition in ^{84}Se which enhances the 164.8 keV peak in Figure 1(a.). With one gate on a background energy, as in Figure 1(b.), these two new close-lying sequences of transitions in ^{162,164}Gd are not seen.

Double gates on the $6^+ \rightarrow 4^+$ transitions in ^{164}Gd and ^{162}Gd, and the $2^+ \rightarrow 0^+$ transition of ^{84}Se, show transitions up to $14^+ \rightarrow 12^+$ in each isotope, and the $4^+ \rightarrow 2^+$ transition in ^{84}Se, in Figure 2(a.,b.). These two very clean spectra, each of which forms a background for the other (e.g. the $6^+ \rightarrow 4^+$ transition energies are very close so the gate on 253.6 keV is a Compton background for the 261.2 keV transition and vice versa, so peaks in Figure 2(a.) are not seen in Figure 2(b.) and

274

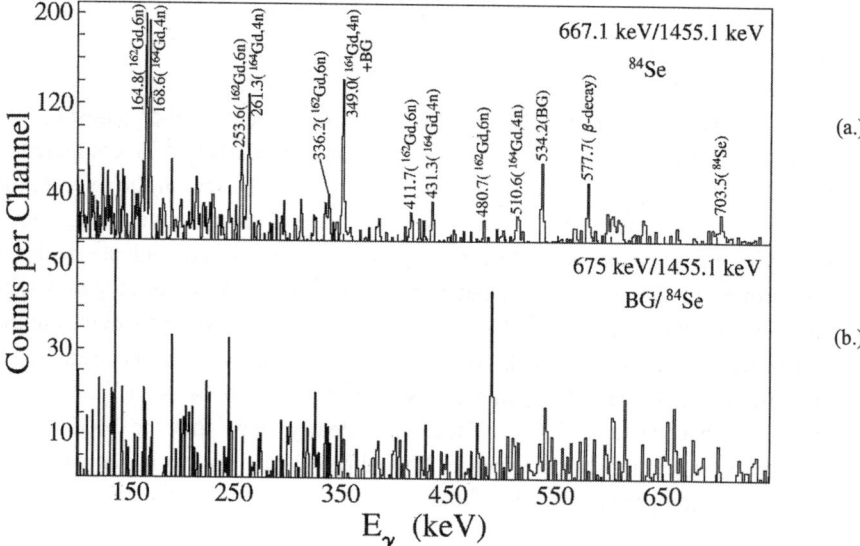

Figure 1. (a.) Double gate on 667.1 and 1455.1 keV in ^{84}Se. (b.) Double gate on background and ^{84}Se.

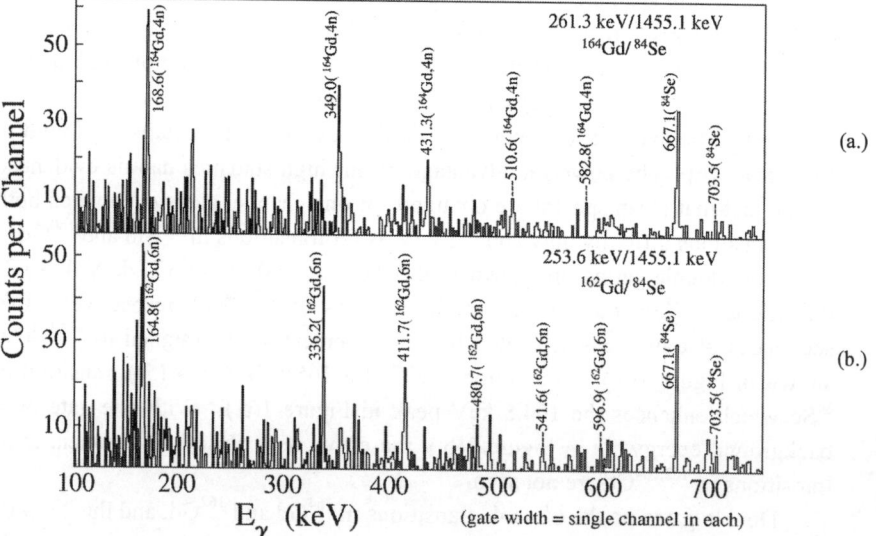

Figure 2. (a.) Double gate on ^{164}Gd 261.3 keV and ^{84}Se 1455.1 keV. (b.) Double gate on ^{162}Gd 253.6 keV and ^{84}Se 1455.1 keV.

vice versa), help establish assignments of the respective transitions in ^{162}Gd and ^{164}Gd.

3. Conclusions

Level schemes of ^{162,164}Gd based on our coincidence analysis studies are shown in Figure 3. Transitions in ^{162}Gd from the 2^+ to 14^+ levels were first observed in our 1995 Gammasphere data [1]. The ^{164}Gd transitions and the 14^+ to 16^+ transition in ^{162}Gd are from our 2000 Gammasphere data, with improved energies from the uncompressed data. In Figure 4 are plots of J_1 for the most neutron rich known Sm, Gd, and Dy isotopes. The energies from $2^+ \rightarrow 0^+$ to $14^+ \rightarrow 12^+$ all decrease from N = 94 to 98 and the J_1 and J_2 values increase in a systematic pattern from ^{156}Sm to ^{160}Sm. The J_1 and J_2 for ^{164}Gd fall between those of ^{160,162}Gd from the $2^+ \rightarrow 0^+$ to $10^+ \rightarrow 8^+$ transitions, then become less than those of even ^{160}Gd above 10^+. So ^{164}Gd is more rigid and has less stretching with rotation. The N = 98, 100 ^{164,166}Dy [5] transition energies likewise increase from N = 98 to 100, and the J_1 and J_2 values of ^{166}Dy similarly fall between those of ^{162,164}Dy from $2^+ \rightarrow 0^+$ up to $12^+ \rightarrow 10^+$, then become less than those of ^{162}Dy at 12^+. However, Asai et al. [6] found that the 2^+ and 4^+ energies in ^{168}Dy are lower than those of ^{166}Dy, so the J_1 values of ^{168}Dy for N = 102 fall above the N = 100 values but still below the N = 98 values.

In Figure 5, Gd and Dy isotopes have an $E(2^+)$ minimum and presumably a β_2 maximum at N = 98, while Yb and Hf nuclei have an $E(2^+)$ minimum at N = 104 with the $E(2^+)$ values for Er isotopes following this trend. The lowest known $E(2^+)$s in this region for N = 92 – 110 are for Z = 60 Nd, followed by Z = 62 Sm and then Z = 64 Gd, with Z = 58 Ce $E(2^+)$s curiously falling between the Gd and Dy values at N = 92 and 94.

Theoretical β_2 values [7] for every Z steadily increase toward midshell and where calculated peak around midshell at N = 104. Experimental β_2 values [3] where known are generally larger than the theoretical values above N = 88. They show the same increasing trend as N increases and then saturate around midshell for Er, Yb, and Hf.

The β_2 values do not absolutely track the $E(2^+)$ values in two ways. First, for Er, Yb, and Hf, the measured β_2 values reach an essentially constant maximum value for N = 96 – 102 (Er), 98 – 104 (Yb), and 102 – 106 (Hf), while the $E(2^+)$ values continue to decrease to a minimum at 104. The largest known experimental β_2 values occur for $E(2^+)$ at 79.5, 75.3 and 73.4 keV in ^{158,160}Gd and ^{162}Dy, respectively. Now that we know that both the Gd and Dy $E(2^+)$s start to increase from N = 98 to 100, the most neutron rich ^{156}Nd and ^{160}Sm surprisingly have the lowest $E(2^+)$s known in this region but their theoretical β_2

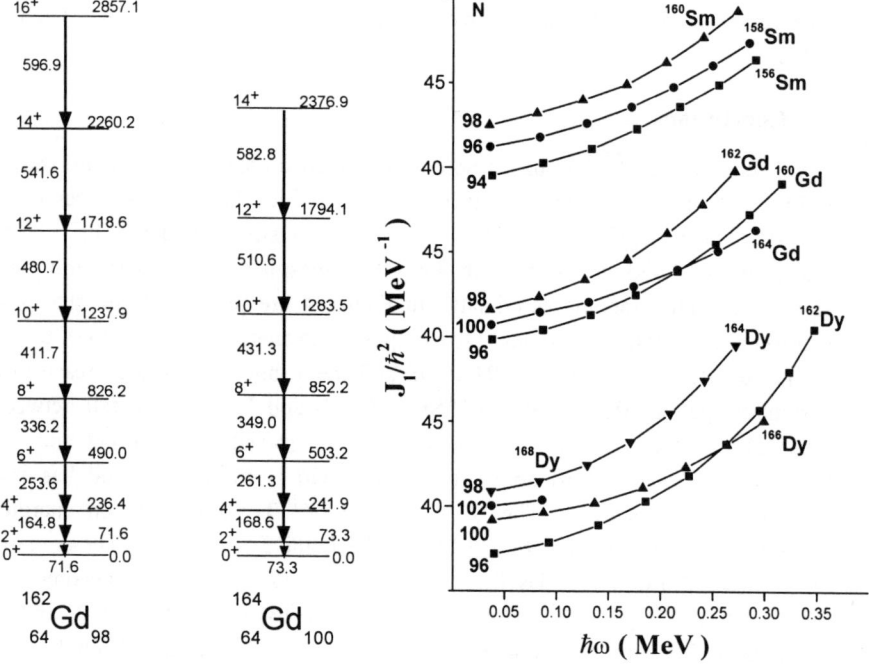

Figure 3. ^{162,164}Gd level schemes.

Figure 4. J_1/η^2(MeV^{-1}) vs. $\eta\omega$(MeV) for
^{156,158,160}Sm, ^{160,162,164}Gd, and

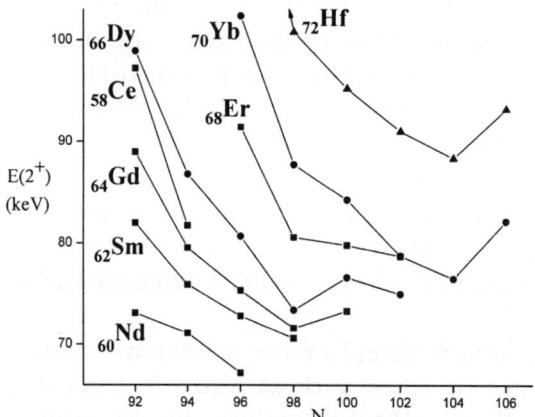

Figure 5. 2^+ level energies vs. N.

values of 0.28, 0.29 are slightly lower than the largest theoretical values of 0.29, 0.30. If we scale their $E(2^+)$s by the MOI to A $\approx$ 172, then the ^{156}Nd, ^{160}Sm energies would be even smaller compared to the Er or Yb $E(2^+)$s. The Nd isotopes, with Z = 60, are well removed from the proton midshell at Z = 66, and the most neutron rich N = 96 is 8 neutrons away from midshell. Our ^{162,164}Gd data, along with the ^{164,166,168}Dy data, raise a new question of why the most neutron rich known Z = 60, 62 Nd, Sm isotopes have the lowest $E(2^+)$s in the deformed region bounded by Z = 50 – 82 and N = 82 – 126.

In summary, the Gd and Dy $E(2^+)$s unexpectedly increase from N = 98 to 100, which suggests that their β_2 values decrease from N = 98 to 100. Their MOIs and $E(2^+)$s have a different pattern from those of Er – Hf, which decrease to a minimum at N = 104. It would be interesting to know whether or not Sm and Nd follow the trends of Gd and Dy with a minimum $E(2^+)$ at N = 98. These data present a new challenge for microscopic descriptions since they oppose longstanding expectations that the $E(2^+)$ minimum and β_2 maximum should occur around N = 104 as found in Er, Yb, and Hf nuclei.

Work at VU, INEEL, LBNL, LLNL, MSU, and ANL was supported by U.S. DOE grants and contracts DE-FG05-88ER40407, DE-AC07-76ID01570, DE-AC03-76SF00098, W-7405-ENG-48, DE-FG05-95ER40939, and W-31-109-ENG-38; Tsinghua by Nat'l. Nat. Sci. Found. of China and Sci. Found. for Nucl. Ind. The JIHIR is supported by U.TN, VU, ORNL, and U.S.DOE. The authors are indebted for the use of ^{252}Cf to the office of Basic Energy Sciences, U.S.DOE, through the transplutonium element production facilities at ORNL.

References

1. E.F. Jones, et al, 2nd Int. Conf. on Exotic Nuclei and Atomic Masses (ENAM98), B.M. Sherrill, D.J. Morrissey, and C.N. Davids, eds., New York: Am. Inst. Phy. (1998) pp. 523-526.
2. A.C. Wahl, *Atomic Data and Nuclear Data Tables*, **39**, 1 (1988).
3. S. Raman, et al., *Atomic Data and Nuclear Data Tables*, **36**, 1 (1987).
4. Y.X. Luo, et al., *Phys. Rev.* **C64**, 054306 (2001).
5. C.Y. Wu, et al., *Phys. Rev.* **C57**, 3466-3469 (1998).
6. M. Asai, et al., *Phys. Rev.* **C59**, 3060-3065 (1999).
7. P. Möller, et al., *Atomic Data and Nuclear Data Tables*, **59**, 185 (1995).

CALCULATED GROUNDSTATE PROPERTIES OF ERBIUM ISOTOPES IN COMPARISON WITH MEASURED 2+-STATES*

T. CORNELIUS, J. A. MARUHN AND W. GREINER

*Institut für Theoretische Physik, Universität Frankfurt,
D-60054 Frankfurt am Main, Germany
E-mail: corneliu@th.physik.uni-frankfurt.de*

P. FLEISCHER AND P.–G. REINHARD

*Institut für Theoretische Physik II, Universität Erlangen-Nürnberg, Erlangen,
Germany*

T. BÜRVENICH

Theoretical Division, Los Alamos National Laboratory, Los Alamos, NM, USA

M. BENDER

*Service de Physique Nucléaire Théorique, Université Libre de Bruxelles,
Bruxelles, Belgium*

S. SCHRAMM

Nuclear Theory Group, Argonne National Laboratory, Argonne, IL, USA

J. H. HAMILTON

Physics Department, Vanderbilt University, Nashville, TN, USA

The deformation properties of nuclei in the vicinity of Erbium are calculated within the relativistic mean-field and the Skyrme-Hartree-Fock model and compared to experimentally measured 2+–states. The investigation in this region is of special interest since the lowest 2+–energy of an isotope-chain is not located in the midshell region.

---

*This work is supported by the BMBF, GSI and DFG.

1. Models and Parametrisations

For our investigations we use self-consistent mean-field models which are nowadays well developed and provide a pertinent picture of the nuclear properties throughout the whole mass table. We consider two different models, the Skyrme-Hartree-Fock approach (SHF) and the relativistic mean-field model (RMF). We take two typical parametrisations for each model, SkI3 [1] and SLy6 [2] for the SHF and NL-Z2 [3] and NL3 [4] for the RMF model. The numerical calculations consist of solving the coupled SHF and RMF equations on a grid in coordinate space using a Fourier definition of the derivatives and solving them with the gradient iteration method. Pairing correlations are treated in the BCS scheme using a delta pairing force. To calculate the potential energy surface (PES) we constrain the expectation value of the operator $\hat{Q}_{20}$ for the total quadrupole moment of the nucleus. Calculation are performed in axial symmetry assuming reflection-symmetric shapes. All other multipoles can adjust themselves freely to obtain the solution with minimal energy.

2. Deformation Properties and 2^+ Energies

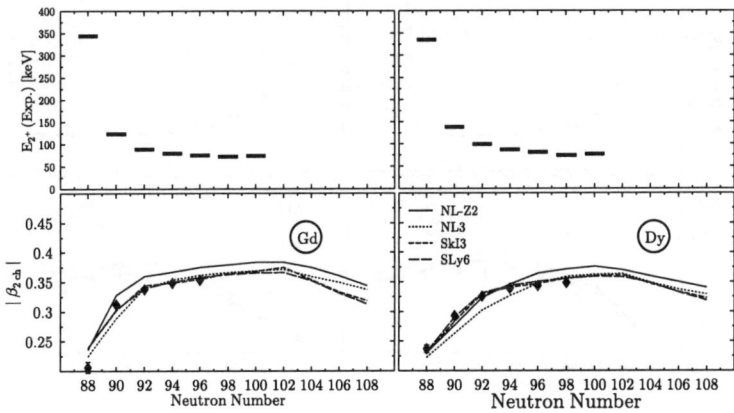

Figure 1.   The experimental measured 2^+-levels (upper panels) are compared with the calculated absolute values of the quadrupole deformation obtained from the charge density (lower panels) for the Gd- (left) and the Dy- (right) isotope chain. As far as known, the quadrupole deformation is compared to experimental data.

The calculation of the potential energy surface under the assumption of axial symmetry gives us the groundstate of the nuclei and their quadrupole deformation. A possible triaxial deformed groundstate was tested for a few nuclei, but could not be found within our models. All calculated nuclei show a well deformed prolate groundstate with a deformation energy of about $E_{\text{def}} = 5 - 20$ MeV. These big energies results from the absence of magic shells.

The rotational–vibrational–model gives us the following relation $E_{2^+} \approx (B\beta_2^2)^{-1}$. Based on it we can compare the global trends of our calculations to the experimental data. Assuming a constant mass-parameter (B) we can conclude that a rising deformation should lead to a drop of the 2^+–states. In Fig. 1 and Fig. 2 the results of our investigation are shown. Comparing the quadrupole deformation with the measured 2^+–levels we see that the global trend is reproduced by our models. The shallow structure of the 2^+–levels in the neutron region of $N = 92 - 100$ for Gd appears in all used parametrisations as well as the fall of the 2^+–states for the Yb-Isotopes which corresponds to an increasing quadrupole moment. Quadrupole deformations derived from experimental data are in agreement with our models even though there are slight differences for higher neutron numbers.

First preliminary results of a self consistent microscopic calculation of ex-

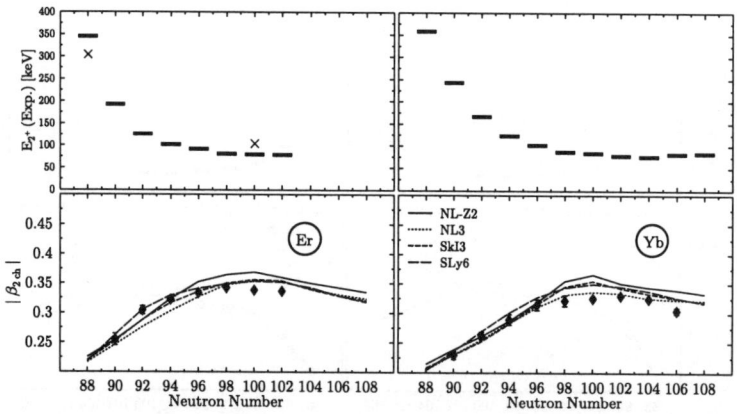

Figure 2. Same as in Fig. 1 only for the Er- (left) and the Yb- (right) isotope chain. In addition first results of the self consistent microscopic calculations of excited states by P. Fleischer are plotted (marked with an ×).

cited states by P. Fleischer are plotted in Fig. 2. These calculations are done with the parameterset SkI3 and give promising resluts [5].

Although the assumption of a constant mass-parameter is fairly rough –

it depends on the mass number and the deformation parameters – we find satisfying results.

References

1. P.–G. Reinhard and H. Flocard, *Nucl. Phys.* **A 584**, 467–488 (1995).
2. E. Chabanat, P. Bonche, P. Haensel, J. Meyer and R. Schaeffer, *Nucl. Phys.* **A 635**, 231–256 (1998).
3. M. Bender, K. Rutz, P.–G. Reinhard, J.A. Maruhn and W. Greiner, *Phys. Rev.* **C 60**, 034304 (1999).
4. G. A. Lalazissis, J. König and P. Ring, *Phys. Rev.* **C 55**, 540–543 (1997).
5. T. Cornelius, M. Bender, T. Bürvenich, P. Fleischer, P.–G. Reinhard, J. A. Maruhn and W. Greiner, *GSI Scientific Report*, (2001).

OBSERVATION OF UNUSUALLY NEUTRON-DEFICIENT FISSION FRAGMENTS IN HEAVY-ION-INDUCED FUSION REACTIONS

N. FOTIADES

Los Alamos National Laboratory, Los Alamos, New Mexico 87545

J. A. CIZEWSKI

Department of Physics and Astronomy, Rutgers University, New Brunswick, New Jersey 08903

R. KRÜCKEN

Physik Department E12, Technische Universität München, 85748 Garching, Germany

D. P. MCNABB, J. A. BECKER, L. A. BERNSTEIN AND W. YOUNES

Lawrence Livermore National Laboratory, Livermore, California 94550

R. M. CLARK, P. FALLON, I. Y. LEE, AND A. O. MACCHIAVELLI

Nuclear Science Division, Lawrence Berkeley National Laboratory, Berkeley, California 94720

Unusually neutron-deficient heavy fragments were observed in the fission of the compound system formed in two different heavy-ion induced reactions and studied with the Gammasphere array. The enhanced production of unusually neutron-deficient fragments cannot be reproduced by statistical calculations and supports very neutron-rich complementary fragments near ^{68}Fe. In conjunction with the observation of neutron-rich light fission fragments in other work, the present results provide support for a new asymmetric fission mode associated with the increased stability of nuclei near ^{70}Ni.

1. Introduction

The study of both symmetric and asymmetric fission is an important source of information about the fission mechanism. One of the highlights in this study has always been the search for new fission modes. The recent observation [1] of enhanced production of unusually neutron-rich fragments with anomalous kinetic energies in the vicinity of ^{70}Ni in the fission of ^{245}Cm induced by thermal neutrons suggests a possible new fission mode. Unfortunately, experimental limitations in Ref. 1 did not allow the observation of the complementary fragments to the unusually neutron-rich fragments. In the present work we report on enhanced production of unusually neutron-deficient heavy fission fragments in two different heavy-ion induced reactions, an observation which could support the need for a new fission mode.

2. Experiments and Results

Fission was studied in the present work following fusion reactions using beams from the 88-Inch Cyclotron Facility at Lawrence Berkeley National Laboratory and establishing coincidences between fission fragment γ-rays with the Gammasphere array. Two fusion-evaporation reactions have been studied: i) a ^{197}Pb compound nucleus (CN) was formed in the ^{24}Mg + ^{173}Yb reaction at 134.5 MeV and ii) a ^{199}Tl CN was formed in the ^{23}Na + ^{176}Yb reaction at a beam energy of 129 MeV. In both cases the 1 mg/cm^2 enriched Yb targets were evaporated on 7-10 mg/cm^2 Au backings. In both measurements there was evidence for fusion reactions between the beam and the ^{197}Au backing forming ^{221}Pa and ^{220}Th CN, respectively. Three-dimensional cubes were constructed to investigate the coincidence relationships between the γ rays. An example of the quality of the data is shown in Figure 1 with a spectrum gated on known [2] transitions of ^{124}Ba.

Relative intensities obtained for the even-mass Te, Xe, Ba and Ce fragments are shown in Figure 2. To determine the relative intensities single gates were set on the known $2_1^+ \rightarrow 0^+$ transitions [2] for each isotope and the intensities (corrected for possible contamination) of the $4_1^+ \rightarrow 2_1^+$ transitions were deduced. Since all of these transitions are stretched quadrupole in character, no corrections to the intensities for directional correlation effects were required.

For Te and Xe isotopes only one maximum in the yield distribution in Figure 2 is observed, while in the Ba and Ce isotopes two maxima are present in the distributions. The maxima at higher mass for Ba and Ce isotopes can be readily understood as symmetric fission of compound nuclei produced by heavy-ion fusion reactions with the target material or target backing. However, the maxima at lower mass for Ba and Ce isotopes cannot be understood in symmetric fission. In the next section we discuss possible

284

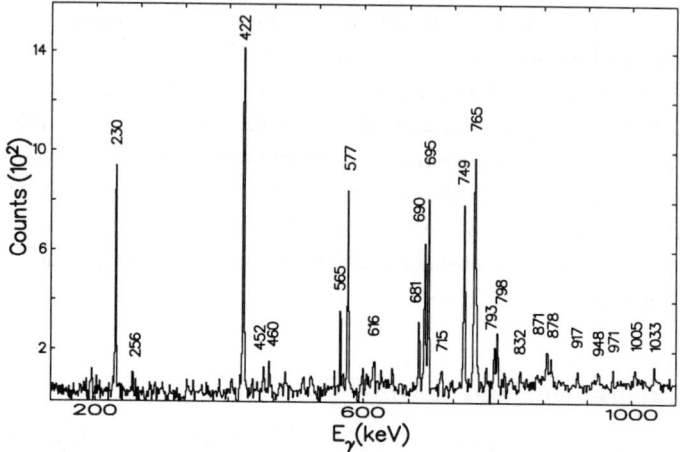

Figure 1. Spectrum obtained from the sum of double gates on the 230- and 577-keV and on the 421- and 695-keV transitions of ^{124}Ba from the ^{24}Mg+^{173}Yb(Au) experiment. The energies of the ^{124}Ba transitions are given in keV.

explanations for the presence of these unusually neutron-deficient isotopes in our data.

3. Discussion

The enhanced production of neutron-deficient even-even Ba and Ce isotopes displayed in Figure 2 is not expected. The observed shapes of the high-mass yields of Ba and Ce isotopes, as well as the lighter Te and Xe isotopes, can be reproduced by the statistical model code GEMINI [3] and the results of these calculations for the fission of ^{197}Pb CN are included in Figure 2a (calculations for the ^{221}Pa CN predict similar distributions within ±2 mass numbers). In contrast, the calculations displayed in Figure 2a cannot reproduce the enhanced intensity of the light-mass Ba and Ce fragments.

Typically, 5 to 8 neutrons are emitted in the fission events associated with our fusion reactions. For example, coincidences between known transitions [2] of ^{83}Br and ^{132}Ba in the first experiment and between known transitions of ^{80,82}Se and ^{132}Ba in the second experiment were clearly established in our data. The ^{132}Ba-^{83}Br pair of fragments originate from the fission of ^{221}Pa CN and 6 neutron emission, while the ^{132}Ba-^{80,82}Se pairs of fragments originate from the fission of ^{220}Th CN and emission of

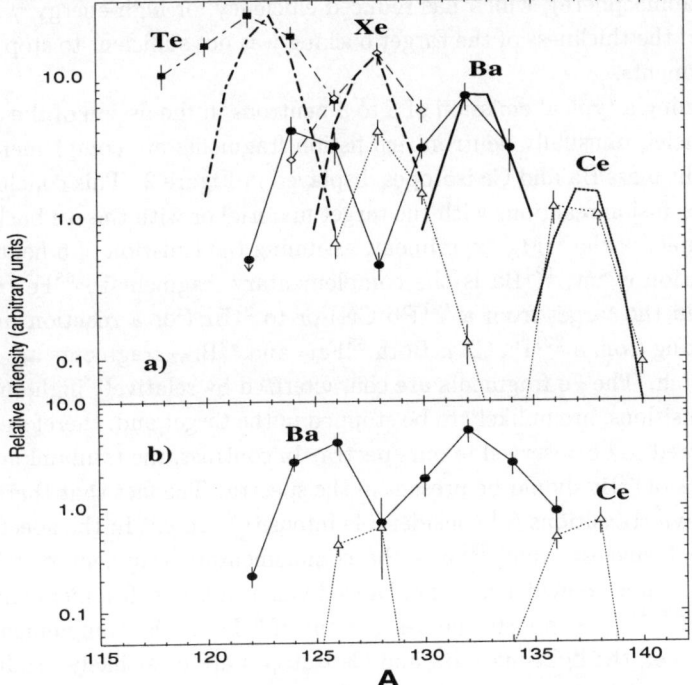

Figure 2.   a) Relative intensities observed in the ^{24}Mg + ^{173}Yb(Au) experiment (symbols - thin lines) compared to statistical calculations for the ^{197}Pb CN using the code GEMINI (thick lines) for Te (solid squares - dot-dashed lines), Xe (open diamonds - long dashed lines), Ba (solid circles - solid lines) and Ce (open triangles - dotted lines) isotopes. Only an upper limit for the intensity of ^{122}Ba could be established. Calculations for the ^{221}Pa CN (not shown in the figure) produce similar distributions (within A±2). b) Relative intensities for Ba (filled circles - solid line) and Ce (open triangles - dotted line) isotopes observed in the ^{23}Na + ^{176}Yb(Au) experiment.

6-8 neutrons, following in both cases the fusion of the beam with the Au target backing. Therefore, at least part of the intensity of the heavier Ba and Ce fragments comes from reactions with the target backings. Reactions with the Yb targets in both experiments would result in Fe or Mn fragments associated with ^{132}Ba. However, the setup used in the present experiments reduces the ability to detect low-mass fragments. In the first experiment the intensity distributions of fission fragments were observed to drop dramatically for $Z < 30$ [4], due mainly to the detection system used (Gammasphere), which has reduced efficiency for high-energy γ rays. Moreover, the thickness of the target backing was not sufficient to stop very light fragments.

Assuming a typical emission of 5 to 8 neutrons in the fission of the compound nuclei, unusually neutron-rich fission fragments are complementary to the light mass Ba and Ce isotopes displayed in Figure 2. This conclusion is valid for fusion reactions with the target material or with the Au backing. For example, in the ^{24}Mg experiment, assuming the emission of 5 neutrons in the fission event, ^{124}Ba is the complementary fragment to ^{68}Fe (for a reaction in the target from a ^{197}Pb CN) or to ^{92}Br (for a reaction in the gold backing from a ^{221}Pa CN). Both ^{68}Fe$_{42}$ and ^{92}Br$_{57}$ fragments are very neutron rich. The Fe fragments are characterized by relatively high-energy γ-ray transitions, are unlikely to be stopped in the target and, therefore, are not expected to be observed in our spectra. In contrast, the (still unknown) transitions of ^{92}Br should be present in the spectra. The fact that there are no unknown transitions (of considerable intensity) present in the spectrum in Figure 1 suggests that ^{68}Fe is the complementary fragment to ^{124}Ba. Moreover, reported evidence [1] of enhanced yields of fission fragments in the vicinity of ^{70}Ni$_{42}$, supports the assignment of ^{68}Fe as the complementary fragment, i.e., the light-mass Ba and Ce isotopes are most likely produced in reactions with the target. Abnormal kinetic energy distributions were measured for the neutron-rich light fragments in the vicinity of ^{70}Ni$_{42}$ [1]. Unfortunately, it is not possible to deduce kinetic energy distributions for any of the fragments observed in the present work. However, the evidence for populating neutron-rich isotopes near ^{70}Ni from two different fission reaction mechanisms (heavy-ion-induced and thermal-neutron-induced fission) suggests that a new asymmetric fission mode might be associated with the increased stability in N~42 Fe and Ni isotopes.

Finally, we discuss here the possibility of production of light Ba and Ce isotopes in evaporation after reactions of the beams with Z ~48 contaminants in the targets. From a spectrographic analysis of our targets [5], the only possible contaminant observed in this mass region was natural silver, with an upper limit of 200 parts per million. A rough calculation suggests that this maximum contamination is not sufficient to produce the number of counts we observed for the isotopes in the light-mass maxima in Figure 1. A more quantitative assessment, obtained by comparing cross section results from a statistical calculation performed with the code EVAP [6], confirmed that possible silver contamination could not reproduce the observed intensities of light-mass Ba and Ce isotopes. Therefore, the original interpretation remains valid: these neutron-deficient isotopes were produced in reactions between the beam and Yb targets which implies complementary fragments which are neutron-rich species near ^{68}Fe.

4. Summary

Intensity distributions of even-even Te, Xe, Ba and Ce fragments following the fission of compound nuclei in two different heavy-ion induced reactions have been studied via in-beam γ-ray spectroscopy. Production of unusually neutron-deficient Ba and Ce fission fragments was observed in both cases. Statistical model calculations performed with the code GEMINI cannot predict the presence of such neutron-deficient fragments. The observation of neutron-deficient Ba and Ce fission fragments imply very neutron-rich complementary fragments near ^{68}Fe, and in conjunction with the observation of neutron-rich light fission fragments [1], support a possible new asymmetric fission mode associated with the increased stability of nuclei near ^{70}Ni.

Acknowledgments

We would like to thank Professor R. J. Charity for providing the computer code GEMINI. This work has been supported in part by the U.S. Department of Energy under Contract Nos. W-7405-ENG-36 (LANL), W-7405-ENG-48 (LLNL) and AC03-76SF00098 (LBNL) and by the National Science Foundation (Rutgers).

References

1. D. Rochman, Thesis, CEA Cadarache SPRC/LEPh-ILL Grenoble, (2001).
2. R. B. Firestone, V. S. Shirley, C. M. Baglin, S. Y. Frank Chu, and J. Zipkin, *Table of Isotopes*, (Wiley, New York, 1996), and references therein.
3. R. J. Charity *et al.*, Nucl. Phys. **A483**, 371 (1988); R. J. Charity, Computer code GEMINI (unpublished).
4. N Fotiades *et al.*, Physica Scripta Vol. **T88**, 127 (2000).
5. Oak Ridge National Laboratories, Isotope Sales Division.
6. Code EVAP, by N. G. Nicolis, D. G. Sarantites and J. R. Beene (unpublished); extension of the code PACE, by A. Gavron, Phys. Rev. **C21**, 230 (1980).

COLLECTIVE NUCLEAR STRUCTURE
OF NEUTRON RICH $^{104\text{-}108}$Mo

J.B. GUPTA
Ramjas College, University of Delhi, Delhi-110 007 India

The complex structure of $^{104\text{-}08}$Mo in having large quadrupole deformation β_2 and the γ-soft vibrational structure reflected in the γ-g B(E2) ratios, were studied in the Interacting Boson Model-1. Through the introduction of L=3 coupling term representing O(5) symmetry these special features were reproduced. Comparison is made with other methods of introducing γ-softness in the Hamiltonian. We also study the $\gamma\gamma$-structure for the K=4 bands in Mo.

1. Introduction

While the shape transition for Sr and Zr is sharp at N=60, reinforced by the Z=38 and Z=40 deformed subshell gaps [1], in Mo the shape transition is smooth. Using the high efficiency 4π counter array of 124 Compton suppressed Ge detectors, Guessous et al [2, 3] studied the level structures of $^{104\text{-}108}$Mo produced in the spontaneous fission of ^{248}Cm. Besides the K^{π}=0+ g-, β and K^{π}=2+ γ-band, the K^{π}=4+ band have been identified in ^{104,106}Mo. They have a low 2_1+, large quadrupole deformation β_2, and the 0_2+ state rises between N=60-64. The γ-soft vibrational structure is reflected in the lowering of the K^{π} =2+ γ-band, and in the γ-g B(E2) ratios. Smith et al. noted a fall in the transition quadrupole moment at 8+ in the g.s. band [4], ascribed to γ-softness.

The energy level structure of $^{104\text{-}108}$Mo are almost similar, so that the illustration of one, say ^{104}Mo, shall apply to the other two as well except when specified. Its level structure does not correspond to any of the three limiting symmetries of IBM. The ratio $R_{4/2}$=2.9 should put Mo isotopes either in the U(5)-S U(3) transition Class-A, with critical phase transition of X(5) symmetry [5] as in ^{152}Sm or in the SU(3)-O(6) transition class B, as in $^{186\text{-}192}$Os. In Mo, the gamma band is lower, indicating γ-soft, while in ^{152}Sm, the beta band is low making it beta-soft. This excludes Class-A for ^{104}Mo. These special features of $^{104\text{-}108}$Mo can serve as a lab. for the nuclear models. The Interacting Boson Model–1 [6] with variable input parameters is well suited for studying their collective structure.

In the MULT version of the IBM-1 Hamiltonian,

$$H = \varepsilon\, n_d + a_0\, P^+.P + a_1\, L.L + a_2\, Q.Q + a_3\, U.U + a_4\, V.V \qquad (1)$$

usually, the first four terms are used for an empirical fit to the low energy spectrum of a nucleus. For the E2 transitions one uses the transition operator

$$T(E2) = e_b Q = \alpha[d^+s + s^+d]^{(2)} + \beta\,[d^+d]^{(2)}. \qquad (2)$$

2. Calculation and Results

2.1. *Level energies*

To test the Class-A, we set the $P^+.P$ term to zero and varied the other three terms. We get two sets, with ($\varepsilon=407.6$, $2a_1=ELL=13.5$, $2a_2=QQ=-13.5$ keV) and (61.8, 34.3 and -26.3 keV respectively). The first set yields a larger $E(2_1)$ and large odd-even staggering in the γ-band. The boson distribution of the 2+ states indicates a O(6) like pattern. The other set, yields a lower $E(2_1)$ and almost no odd-even staggering as in experiment. But the K=0, 2 bands are almost degenerate, as for the exact SU(3) symmetry and $R_{4/2}$ is 3.3. Its boson distribution indicates a double hump for the 2_2 state, thus putting the β-band below the γ-band. Thus SU(3) symmetry perturbed through the boson energy term did not reproduce the energy structure. This excludes transition Class-A.

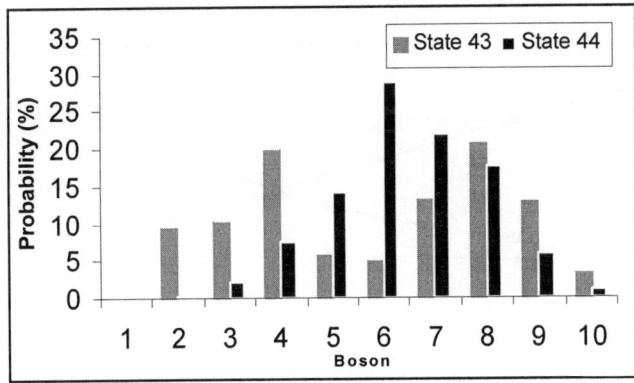

Figure 1. The boson probability distribution for the 4_3 and 4_4 states.

So we set ε=0 and introduced the P⁺.P term. This yields a fairly good fit, but the γ-g B(E2) ratios correspond to a good rotor, unlike the data. Then we introduced the OCT(=a_3/5) term as in Table 1. There is ambiguity in the assignment of the calculated third and fourth 4+, 6+ to the $K^{\pi}=0_2$, 4+ bands in ^{104,106}Mo, since both bands from our calculation almost overlap the $K^{\pi}=4+$ experimental values. The plot of the boson probability distribution exhibits a double peak for 4_3 state (Fig. 1). It may be the K=0+ peaks or a 2γ $K^{\pi}=4+$ state. Data on the $K^{\pi}=0+$ band is required to resolve this.

Table 1. The input parameters for IBM-1 Hamiltonian for $^{104-108}$Mo. χ=-1.32, ε=a_4=0.0, α=0.10, β=-0.12, N_p=4.

Isotope	PAIR	ELL	QQ	OCT
A=104	34.7	31.4	-14.7	10.0
A=106	65.6	35.3	-12.5	7.0
A=108	58.8	35.3	-7.07	10.0

2.2. The B(E2) values and B(E2) ratios

The B(E2, 2+→0+) and the quadrupole moment $Q(2_1+)$ in $^{104-108}$Mo correspond to a deformed nucleus and are well given in IBM-1. But the B(E2) ratio R= B(E2, 2_2→0_1/2_1) is small in the three isotopes which does not correspond to a good axially symmetric rotor. For OCT=0, the B(E2) ratio R=B(E2, 2_2→0_1/2_1) from IBM-1 is closer to the Alaga value of 0.7. The dependence of R for ^{104}Mo on the OCT value is shown in figure 2 [7].

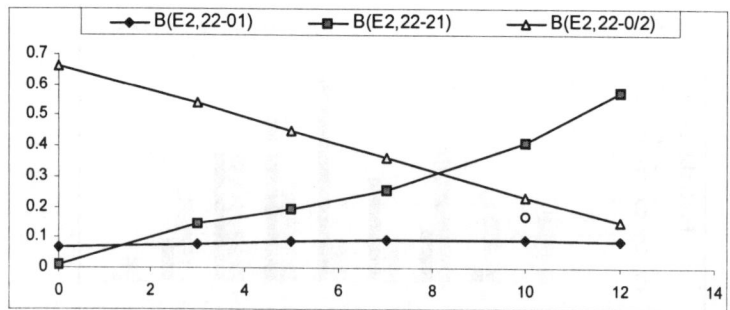

Figure 2. The dependence of calculated γ-g B(E2) values and B(E2) ratios on a_3 (=5 OCT) value for ^{104}Mo. The open circle on the B(E2) ratio curve is its experimental value derived from the γ-ray data in [3]. Scale for B(E2) values is x 1/10.

Table 2. The Interband B(E2) ratios in $^{104\text{-}108}$Mo isotopes for transitions from the levels in the γ-band, calculated in IBM-1, are compared with EXP values derived from the E_γ and I_γ values in Ref. [3] (taken from [7]).

B(E2) ratio I_I	I_f'/I_f''	A= 104 EXP	104 IBM	106 EXP	106 IBM	108 EXP	108 IBM
2_2+	$0_1/2_1$	0.17	0.22	0.25	0.32	0.073	0.12
3_1+	$2_1/4_1$	0.44	0.56	0.39	0.79	0.08	0.34
	$2_1/2_2$	0.044	0.043	0.073	0.043	0.02	0.04
4_2+	$2_1/4_1$	0.057	0.015	0.072	0.044	0.032	0.001
	$2_1/2_2$	0.010	0.004	0.024	0.009	0.027	0.0002
5_1+	$4_1/4_2$		0.032	0.040	0.046	0.044	0.03
	$4_2/3_1$		0.84	1.88	0.76	1.3	0.62
	$4_1+/3_1$	0.06	0.027	0.075	0.035	0.057	0.019
6_2+	$4_1+/4_2+$	0.0075	2×10^{-4}	0.007	0.0006		$0.002\times$
	$4_1/6_1$	0.032	0.001	0.020	0.006		0.01
	$4_2/6_1$	4.0	8.3	2.6	10		6.0

The individual B(E2) values vary smoothly with the OCT value. The B(E2) ratio R drops from a high of about 0.6 to about 0.2 for OCT=10 keV in ^{104}Mo. Similarly, the B(E2, $3\to2_1+/4_1+$) ratio falls from 2.5 (Alaga value) to a low of 0.1 for OCT=10 keV. This corresponds to the variation expected for transition Class-B [6].

A plot of the probability distribution over the boson number for the 2_1+ and 2_2+ states shows a greater overlap of the enhanced amplitude of wave functions of the two states as the OCT value is increased, which explains in part the increase in the E2 transition between them.

In the a_3 term of Eq. (1), U is the generator of O(5) subgroup in the subchains I and III of U(6) in IBM-1. In terms of the quantum number τ of the generator of subgroup O(5), here the $\Delta\tau=1$ transitions are affected most compared to the $\Delta\tau=0$ and 2. For example, the $2_2\to2_1$ transition is $\Delta\tau=1$ and the $2_2\to0_1$ transition is $\Delta\tau=2$. The same is true for B(E2, $3\to2_1$)/ B(E2, $3\to4_1$) ratio.

For the γ-g B(E2) ratios for 2+, 3+ and 4+ of the γ-band, the calculated values come closer to experiment for an OCT value of 5 to 8 keV for ^{106}Mo and about 10 keV for ^{104,108}Mo [7]. The open circle on the B(E2) ratio curve indicates the experimental values of the B(E2) ratio as deduced from the relative γ-ray intensities in Ref. [3]. As a compromise we choose the value of 7 keV for ^{106}Mo and 10 keV for ^{104,108}Mo, and calculated all other B(E2) values for them (Table 2).

The inter-band B(E2) ratios from the states of the K^π =2+ γ-band to the ground band, viz. the γ-g/γ ratios B(E2, $3_1 \rightarrow 2_1 + /2_2 +$), B(E2, $4_2 \rightarrow 2_1 + /2_2 +$), B(E2, $5_1 \rightarrow 4_1 + /4_2 +$) and B(E2, $6_2 \rightarrow 4_1 + /4_2 +$) ranging from 0.044 to 0.007 are well given from IBM for the above choices of the OCT values in the three isotopes. Similarly the γ-γ/γ' ratio B(E2, 5_1-4_2/3_1) is given within a factor of two.

2.3 Study of γ softness

The coefficient χ of the $[d^+.d]^{(2)}$ is zero for O(6) symmetry. So we varied the χ value in H_{IBM-1} to about -0.88, -0.4 and obtained appropriate values of the a_1 and a_2 coefficients (table 3). This led to the lowering of the K^π =2+, 4+ bands and the raising of the K^π =0+ excited band. The corresponding B(E2) values yield the B(E2) ratio R decreasing from 0.7 up to ~0.4. Similarly the B(E2, $3 \rightarrow 2_1/4_1$) ratio decreased from a value of ~2.0 to ~0.9. Further lowering of χ to -0.2 yields the requisite lower B(E2) ratios. Note the large value of the quadrupole term and low 2_2+.

Table 3. Variation of the χ parameter (CHQ=$\sqrt{5}$ χ) and effect of coefficient a_0 in the $P^+.P$ term. All values are in keV. The ε and a_3 values are set to zero for ^{104}Mo .

PAIR	ELL	QQ	CHQ	E(2+)	E(2_2)	E(4_3)	E(4_4)	B(E2, 2_2-0/2)	B(E2, 3-2/4)
21.0	38.3	-24.2	-2.968	140	812	1470	1669	0.66	2.0
14.2	34.4	-30.7	-1.958	139	796	1491	1634	0.59	1.65
-5.5	25.7	-53.8	-0.958	142	748	1510	1549	0.37	0.9
54.2	16.5	-76.2	-0.458	162	644	1436	1844	0.12	0.34

Next, we noted that for EPS larger than some critical value, we get a energy fit with somewhat larger odd-even staggering, towards O(6) γ-soft character. This yields small R=B(E2, $2_2 \rightarrow 0_1/2_1$) and B(E2, $3 \rightarrow 2_1/4_1$), as in experiment. Study of the boson distribution again indicates an O(6) like character, with much less spread out of the wave function for the 0+, 2+ states.

3. The $K^{\pi}=4+$ band

The $K^{\pi}=4+$, I=4 state: In ^{104}Mo, for the B(E2, $4_4 \rightarrow 2_2/4_2$) and B(E2, $4_4 \rightarrow 2_2/3_1$) ratios, the values for decays from the third 4+ (0.14 and 0.16) are closer to experiment. The same is true for ^{106}Mo, so that the third 4+ state should correspond to the KI=44 state in both isotopes.

The $K^{\pi}=4+$, I=5 state: In ^{104}Mo and ^{106}Mo, the B (E2, $5_2 \rightarrow 4_4/4_\gamma$) and B(E2, $5_2 \rightarrow 4_4/5_1$) ratios again favour the third 4+ level to correspond to the KI=44 state in experiment. The values for the third 4+ state as the final state are about a factor of 10 larger than for the fourth 4+ state, and are closer to experiment. For the B(E2, $5_2 \rightarrow 3_1/5_1$) and B(E2, $5_2 \rightarrow 3_1/4_2$) ratios, the IBM-1 values agree with experiment.

The $K^{\pi}=4+$ I=6 state: The B(E2) ratios in the columns of 4_3 state are favoured slightly over the IBM-1 values for 4_4 state. These values are much off the experimental values. The B(E2) values here are rather small and a small variation in the interaction can affect the B(E2) values by a large factor.

Table 5. The B(E2) ratios in ^{104,106}Mo for the transitions from the $K^{\pi} = 4_1+$ band. The results from IBM-1 are given for the third 4+ and 6+ states.

B(E2) ratio	A=104		106	
I_i(K=4) $I_f/I_{f'}$	EXP	IBM-1	EXP	IBM-1
$4_4 \rightarrow 2_\gamma/4_\gamma$	0.075	0.14	0.24	0.37
$4_4 \rightarrow 2_\gamma/3_\gamma$	0.26	0.16	0.30	0.39
$5_2 \rightarrow 3_\gamma/4_\gamma$	0.26	0.17	0.11	0.004
$5_2 \rightarrow 3_\gamma/5_\gamma$	0.064	0.054	0.114	0.138
$5_2 \rightarrow 4_4/4_\gamma$	59.5	15.5	67	12.6
$5_2 \rightarrow 4_4/5_\gamma$	14	4.8	18.3	7.0
$6_4 \rightarrow 4_\gamma/4_4$	0.10	0.0005	0.11	0.004
$6_4 \rightarrow 4_4/5_2$	0.032	7.0	0.026	1.1

4. Discussion

The unique structure of $^{104-108}$Mo, with rotational g-bands and γ-soft features reflected in γ-g interband B(E2) ratios, have been simulated by the use of nonzero a_3 value in Eq, (1). Certain B(E2) values and B(E2) ratios vary smoothly with a_3, which represents the d-boson coupling to L=3. The

Casimir operator of O(5) is related to the U.U term in Eq. (1). Reduced χ value in the quadrupole operator Q of the Hamiltonian also induces the transition in Class B from SU(3) to O(6) as illustrated in table 3 but needs very low χ value. Alternatively, the increase of single boson energy term beyond a critical value also induces the γ-softness, but the odd-even staggering in the K=2+ band is large. We also did IBM study of ^{186}Os. Here the first four term Hamiltonian yields the good energy fit and low B(E2, $2_2 \rightarrow 0_1/2_1$) and B(E2, $3_1 \rightarrow 2_1/4_1$) are obtained.

The IBM-1 calculation results on the energy spectra and the B(E2) values for the γ-g support the 2γ phonon band character of the K^π =4+ bands in $^{104\text{-}106}$Mo.

Acknowledgement

The facility for work at Ramjas College is gratefully acknowledged.

References

[1] J. H. Hamilton, *Reports on Progress in Physics* **48** 631 (1985)

[2] A. Guessous *et al*, *Phys. Rev. Lett.* **75** 2280 (1995)

[3] A. Guessous *et al*, *Phys. Rev.* **C53** 1191 (1996)

[4] A.G. Smith *et al*, J.Phys. **G28**, 2307 (2002)

[5] F. Iachello, *Phys. Rev. Lett.* **87**, 052502 (2001)

[6] F. Iachello and A. Arima, *The Interacting Boson Model* (Cambridge University Press) (1987)

[7] J.B. Gupta, *J. Phys. G 28*, 2376 (2002)

NANOSECOND ISOMERS IN NEUTRON-RICH $N\approx19$ NUCLEI

M. ASAI AND T. ISHII

Advanced Science Research Center,
Japan Atomic Energy Research Institute,
Tokai, Ibaraki 319-1195, Japan
E-mail: asai@tandem.tokai.jaeri.go.jp

A. MAKISHIMA

Department of Liberal Arts and Sciences,
National Defense Medical College,
Tokorozawa, Saitama 359-8513, Japan

M. OGAWA

Research Laboratory for Nuclear Reactors,
Tokyo Institute of Technology,
Meguro, Tokyo 152-8550, Japan

M. MATSUDA

Department of Materials Science,
Japan Atomic Energy Research Institute,
Tokai, Ibaraki 319-1195, Japan

Nanosecond isomers have been found in neutron-rich $N\approx19$ nuclei produced by heavy-ion deep-inelastic collisions. The $\nu f_{7/2}$ excitations were established in ^{32,33}Si and ^{34}P through experimental spin-parity assignments.

1. Introduction

The $N=20$ closed shell is formed by a large energy spacing between the sd-shell orbitals and the next $f_{7/2}$ orbital. However, recent experimental and theoretical investigations[1,2] revealed that the closed shell properties of $N=20$ vanish in the neutron-rich region around ^{32}Mg. Neutron excitations across the $N=20$ shell gap play an important role in this phenomenon. The $\nu f_{7/2}$ excited states in neutron-rich $N=19$ nuclei are known in ^{37}Ar, ^{36}Cl, and ^{35}S at 1611, 2518, and 1991 keV, respectively. These levels are depopulated by M2 transitions between the $\nu f_{7/2}$ and $\nu d_{3/2}$ orbitals with nanosecond lifetimes. In more neutron-rich $N=19$ nuclei, the 1435 and 2305 keV levels in ^{33}Si and ^{34}P are the candidates for the $\nu f_{7/2}$ excitations,[3] although their spin-parities have not been determined experimentally. In ^{32}Si, the $t_{1/2}=27(2)$ ns isomer was reported at 5583 keV[4] which was considered as the $(\nu f_{7/2}\nu d_{3/2})5^-$ state. In this work, the $\nu f_{7/2}$ excitations in neutron-rich $N\approx19$ nuclei are investigated through the observation of nanosecond isomers and experimental spin-parity assignments using an isomer-scope.[5]

2. Experiment

Neutron-rich $N\approx19$ nuclei were produced in the heavy-ion deep-inelastic collisions of ^{198}Pt + ^{37}Cl (9 MeV/nucleon). Projectile-like fragments (PLFs) were stopped in an annular Si detector placed 55 mm downstream from the target, and γ rays emitted from the stopped PLFs were measured with four Ge detectors by PLF-γ(-γ) coincidences. The atomic number of the PLF was identified using Si ΔE detectors. Lifetimes of isomers were deduced from PLF-γ time spectra. Multipolarities of γ transitions were determined through an in-plane to out-of-plane γ-ray anisotropy analysis.[6]

3. Results and Discussions

Figure 1 shows decay curves of γ rays associated with the isomers in ^{32,33}Si and ^{34}P obtained from the PLF-γ coincidence data. The half-lives of the isomers were determined to be 33.1(5) ns, 10.2(3) ns, and $0.3<t_{1/2}<2.5$ ns for ^{32,33}Si and ^{34}P, respectively. The lower limit for ^{34}P was estimated from the detection limit of the isomer-scope. Figure 2 shows the decay schemes of the isomers established through the γ-γ coincidence analysis. The γ-ray anisotropy analysis revealed that the 1435 keV transition in ^{33}Si and the 1876 keV one in ^{34}P are stretched quadrupole transitions. Combined with their lifetimes, the M2 multipolarity was assigned to these transitions, and the spin-parities of the 1435 and 2305 keV isomers were determined to be $7/2^-$ and 4^-, respectively.

The $\nu f_{7/2}$ excitations decrease in energy from ^{36}Cl, ^{35}S to ^{34}P, ^{33}Si, but it is still high at ^{33}Si. On the other hand, level structures in ^{32}Al and ^{31}Mg seem significantly different from those in ^{34}P and ^{33}Si. This fact indicates that the shell structure of $N\approx20$ drastically changes just below the $Z=14$, which is consistent with the Monte Carlo shell model calculations.[2]

Fornal et al.[4] reported the 79 keV transition above the 5504 keV level in ^{32}Si and assigned this transition as the isomeric E1 transition of $5^- \rightarrow 4^+$ with $t_{1/2} = 27(2)$ ns. In the present experiment, however, the 79 keV γ rays were not observed in ^{32m}Si, and the in-plane to out-of-plane ratio of the 3562 keV transition was consistent with that of the stretched octupole transition. Therefore, we conclude that the 5504 keV level is the isomeric 5^- state with $t_{1/2} = 33.1(5)$ ns depopulated by the 3562 keV E3 transition to the 1942 keV first 2^+ state.

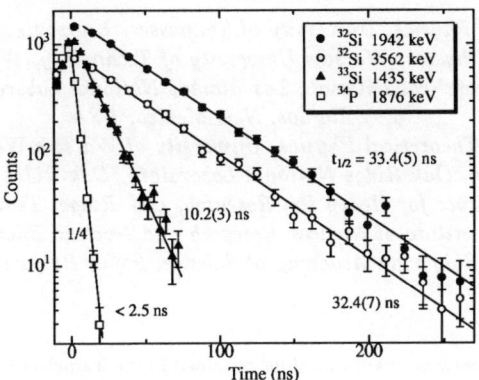

Figure 1. Decay curves of γ rays associated with the isomers in ^{32,33}Si and ^{34}P.

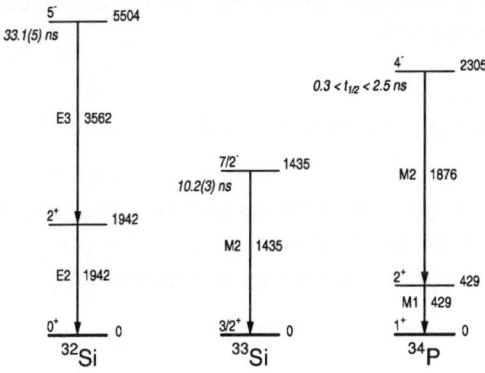

Figure 2. Decay schemes of the isomers in ^{32,33}Si and ^{34}P.

References

1. T. Motobayashi *et al.*, *Phys. Lett.* **B346**, 9 (1995).
2. Y. Utsuno *et al.*, *Phys. Rev.* **C60**, 054315 (1999).
3. B. Fornal *et al.*, *Phys. Rev.* **C49**, 2413 (1994).
4. B. Fornal *et al.*, *Phys. Rev.* **C55**, 762 (1997).
5. T. Ishii *et al.*, *Phys. Rev. Lett.* **84**, 39 (2000).
6. M. Asai *et al.*, *Phys. Rev.* **C62**, 054313 (2000).

NUCLEAR WAVE FUNCTIONS FOR SPIN AND PSEUDOSPIN PARTNERS*

P. J. BORYCKI,[1,2] J. GINOCCHIO,[3] W. NAZAREWICZ,[1,4,5] AND
M. STOITSOV[1,5-7]

[1] *Department of Physics, University of Tennessee, Knoxville, Tennessee, USA*
[2] *Institute of Physics, Warsaw University of Technology, Warsaw, Poland*
[3] *Theoretical Division, Los Alamos National Laboratory,*
Los Alamos, New Mexico, USA
[4] *Institute of Theoretical Physics, University of Warsaw, Warsaw, Poland*
[5] *Physics Division, Oak Ridge National Laboratory, Oak Ridge, Tennessee, USA*
[6] *Joint Institute for Heavy Ion Research, Oak Ridge, Tennessee, USA*
[7] *Institute of Nuclear Research and Nuclear Energy,*
Bulgarian Academy of Science, Sofia, Bulgaria

Using relations between wave functions obtained in the framework of the relativistic mean field theory, we investigate the effects of pseudospin and spin symmetry breaking on the single nucleon wave functions in spherical nuclei. In our analysis, we apply both relativistic and non-relativistic self-consistent models as well as the harmonic oscillator model.

1. Pseudospin/Spin Symmetry and the Dirac Hamiltonian

The Dirac Hamiltonian with external scalar $V_S(\mathbf{r})$ and vector $V_V(\mathbf{r})$ potentials, vanishing space components and non-vanishing time component, is invariant under an SU(2) algebra if the scalar potential $V_S(\mathbf{r})$ and the vector potential $V_V(\mathbf{r})$ are related up to a constant $C_{ps/s}$:

$$V_S(\mathbf{r}) \pm V_V(\mathbf{r}) = C_{ps/s}. \tag{1}$$

In the pseudospin[1,2,3] (spin) symmetry limit[4], the radial wave functions of the lower (upper) components are equal, while the upper (lower)

---

*This work was supported in part by the U.S. Department of Energy under Contract Nos. DE-FG02-96ER40963 (University of Tennessee), DE-AC05-00OR22725 with UT-Battelle, LLC (Oak Ridge National Laboratory), W-7405-ENG-36 (Los Alamos), and by the Polish Comittee for Scientific Research (KBN) under Contract No. 5 P03B 014 21.

298

components $g_{\tilde{n}\tilde{\ell}j=\tilde{\ell}\pm\frac{1}{2}}(r)$ $(f_{n\ell j=\ell\pm\frac{1}{2}}(r))$ satisfy differential relations[5]:

$$D_{\tilde{n}\tilde{\ell}j=\tilde{\ell}-\frac{1}{2}}(r)g_{\tilde{n}\tilde{\ell}j=\tilde{\ell}-\frac{1}{2}}(r) = D_{\tilde{n}\tilde{\ell}j=\tilde{\ell}+\frac{1}{2}}(r)g_{\tilde{n}\tilde{\ell}j=\tilde{\ell}+\frac{1}{2}}(r), \tag{2}$$

$$D_{n\ell j=\ell-\frac{1}{2}}(r)f_{n\ell j=\ell-\frac{1}{2}}(r) = D_{n\ell j=\ell+\frac{1}{2}}(r)f_{n\ell j=\ell+\frac{1}{2}}(r), \tag{3}$$

where $D_{\tilde{n}\tilde{\ell}j=\tilde{\ell}-\frac{1}{2}}(r) = \frac{d}{dr} - \frac{\tilde{\ell}-1}{r}$, $D_{\tilde{n}\tilde{\ell}j=\tilde{\ell}+\frac{1}{2}}(r) = \frac{d}{dr} + \frac{\tilde{\ell}+2}{r}$.

Relations (2,3) are strictly fulfilled only under condition (1). Therefore, comparing the differences between the left and right hand sides of the equations, one can learn about the pseudospin and spin symmetry-breaking effects.

2. Comparison within the Harmonic Oscillator Model

For the spherical harmonic oscillator potential, we take the analytical form of wave functions with an oscillator frequency $\hbar\omega = 41/A^{1/3}$. Then, one can express $D_{\tilde{n}\tilde{\ell}j}(r)g_{\tilde{n}\tilde{\ell}j}(r)$ defined by Eq. (2) as:

$$D_{\tilde{n}\tilde{\ell}j=\tilde{\ell}-\frac{1}{2}}(r)g_{\tilde{n}\tilde{\ell}j=\tilde{\ell}-\frac{1}{2}}(r) = \chi_{\tilde{n}\tilde{\ell}}(x) \sum_{a=0}^{\tilde{n}-1} \frac{(-1)^a(2\tilde{n}+\tilde{\ell}-3/2-a)x^a}{(\tilde{n}-1-a)!a!\Gamma(a+\tilde{\ell}+3/2)}, \tag{4}$$

$$D_{\tilde{n}\tilde{\ell}j=\tilde{\ell}+\frac{1}{2}}(r)g_{\tilde{n}\tilde{\ell}j=\tilde{\ell}+\frac{1}{2}}(r) = \sqrt{\frac{\tilde{n}+\tilde{\ell}-1/2}{\tilde{n}-1}}\chi_{\tilde{n}\tilde{\ell}}(x) \times$$

$$\times \sum_{a=0}^{\tilde{n}-1} \frac{(-1)^a(2\tilde{n}-2-a)x^a}{(\tilde{n}-1-a)!a!\Gamma(a+\tilde{\ell}+3/2)}, \tag{5}$$

where $x = r^2 2\nu$, $\nu = \frac{m\omega}{2\hbar^2}$,

$$\chi_{\tilde{n}\tilde{\ell}}(x) = \sqrt{\left(2(2\nu)^{\frac{6-\tilde{\ell}}{2}}(n-1)!\right)/\Gamma(\tilde{n}+\tilde{\ell}-1/2)^3 x^{\frac{\tilde{\ell}}{2}}e^{-x/2}}. \tag{6}$$

Expressions (4,5) can be expressed as products of the common envelope function $\chi_{\tilde{n}\tilde{\ell}}(r)$ and certain polynomials. These polynomials are of the same order $(\tilde{n}-1)$ independent of j, whereas the original harmonic oscillator eigenfunction with $j = \tilde{\ell} \pm \frac{1}{2}$ involves a polynomial of order $\tilde{n} + 1/2 \pm 1/2$ in x.

Systematic self-consistent calculations of several doubly magic nuclei have shown that Eq. (2) holds better as $\tilde{n}$ increases or $\tilde{\ell}$ decreases, in agreement with a simple harmonic oscillator estimate.

3. Summary and Conclusions

In the pseudospin (spin) symmetry limits the radial wave functions of the upper and lower components of pseudospin (spin) doublets satisfy certain differential relations. We demonstrated that these relations are not only approximately valid for the relativistic mean field eigenfunctions but also for the non-relativistic Hartree-Fock and harmonic oscillator eigenfunctions (see, e.g. Fig. 1). Generally, we expect them to be approximately valid for eigenfunctions of any non-relativistic phenomenological nuclear potential that fits the single-particle levels in nuclei.

Hence we seem to have both spin and pseudospin dynamic symmetry; that is, the energy levels are not degenerate but the eigenfunctions well preserve both symmetries.

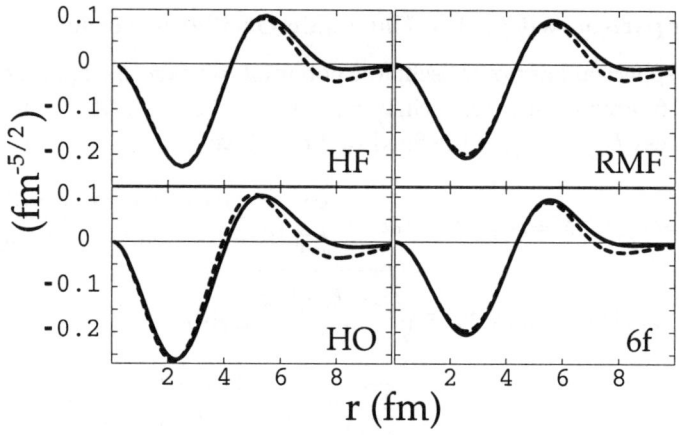

Figure 1. Numerical check of identity (2). Comparison between $D_{\tilde{n}\tilde{\ell}j=\tilde{\ell}-\frac{1}{2}} g_{\tilde{n}\tilde{\ell}j=\tilde{\ell}-\frac{1}{2}}$ (dashed line) and $D_{\tilde{n}\tilde{\ell}j=\tilde{\ell}+\frac{1}{2}} g_{\tilde{n}\tilde{\ell}j=\tilde{\ell}+\frac{1}{2}}$ (solid line) for $2\tilde{d}$ pseudospin doublet in ^{208}Pb obtained in different methods: HF, HO and RMF. The plot labeled '6f' shows 6 times scaled lower components of the RMF wave function (see Ref. [6] for details).

References

1. K.T. Hecht and A. Adler, Nucl. Phys. **A137**, 129 (1969)
2. A. Arima, M. Harvey, and K. Shimizu, Phys. Lett. **30B**, 517 (1969)
3. R.D. Ratna-Raju, J.P. Draayer, and K.T. Hecht, Nucl. Phys. **A202**, 433 (1973)
4. J.N. Ginocchio, Phys. Rev. Lett. **78**, 436 (1997)
5. J.N. Ginocchio, Phys. Rev. **C66**, 064312 (2002)
6. P.J. Borycki, J. Ginocchio, W. Nazarewicz, M. Stoitsov, *to be published*, nucl-th/0301098

GAMMA RAY SPECTROSCOPY OF NUCLEI NEAR N=Z LINE

LAKSHMAN CHATURVEDI

Department of Physics, Banaras Hindu University
Varanasi 221005, India
e-mail: lakshman@banaras.ernet.in

With new innovations in detection technology it has become possible to study nuclei far from stability line with very small cross sections for a deeper understanding of the nuclear structure/behaviour at high angular momentum and exotic shapes. The new generations of highly efficient large gamma detector arrays coupled with recoil mass analyzer and other ancillary detectors (such as neutron detectors/charge particle detectors and/or other type of detectors) have made possible studies of both proton-rich and neutron rich nuclei populated with extremely low cross sections. Some interesting measurements [1] performed at Nuclear Science Centre, New Delhi, India, current results [2,3] and future prospects to probe the N=Z nuclei from Z=26 to Z=50 will be described in terms of its underlying physics of reinforcing proton and neutron shell gaps which will lead to new superdeformed doubly-magic nuclei.

1. Introduction

In beam γ-ray studies of nuclei in the mass region 70 formed by heavy-ion fusion reactions performed in the recent years have revealed a variety of excitation modes in these nuclei. These transitional nuclei are mainly characterized by complicated energy spectra comprising of both single-particle and collective degree of freedom. In the present work, our aim was to investigate the high spin states of ^{70}As and ^{70}Ge nucleus. The problem of studying these nuclei at high-spin is very low and that of nearby channels are comparatively large. With the help of a recoil mass spectrometer like Heavy Ion Recoil Analyzer (HIRA) coupled with Gamma Detector Array (GDA) at the Nuclear Science Centre, it has been possible to identify directly the recoils of such weakly populated nuclei and carry out spectroscopic studies of these nuclei. Fig.1 shows eight Compton-Suppressed High Purity Germanium (CS-HPGe) detectors in conjunction with HIRA used during the course of the present measurements. In recent years there has been considerable theoretical and experimental interest to study N=Z nuclei, in order to investigate shape transition dominated by the T=0 residual interaction. The approach for the experimental studies of these nuclei has become possible only due to the development of the state-of-the-art facilities. A major world-class national facility – a Large Gamma Detector Array (24 Clover detectors) coupled with a high efficiency gas-filled Hybrid Recoil Analyzer and ancillary detectors is going to be available at the Nuclear Science Centre. A systematic study of N=Z nuclei has been planned to understand the neutron-proton correlation effects.

2. Experimental Procedure

The near yrast states in ^{70}Ge nucleus were populated in two different experiments using the single reaction ^{46}Ti(^{28}Si,4p)^{70}Ge at a beam energy of 110 MeV The production cross-section of ^{70}Ge at this beam energy was around 2.5% of total fusion, while that of the mostly produced nucleus – ^{70}As was around 24%. Two self supporting, stacked and enriched targets of thickness ~0.30 and 0.21 mg/cm^2 were used in both of the experiments. Similarly, the near yrast state in ^{70}As nucleus were populated using the two reactions, ^{51}V(^{28}Si,2αn) and ^{46}Ti (^{28}Si,3pn) at beam energies of 125 and 110 MeV, respectively. The first experiment used a Lead-backed target of thickness 2mg/cm^2, while the second one had two self-supporting, stacked and enriched targets of thickness ~300 and 210μgm/cm^2. The heavy-ion beams were provided by the 15UD-16 MV Pelletron facility at Nuclear Science Centre, New Delhi, India.

3. Results and Conclusions

Seventeen new transitions have been found and placed in the level scheme of ^{70}As. The onset of collectivity is observed at high spin with the discovery of three rotational bands, similar to ^{72}As. The kinematic moments of inertia of these bands follow the general systematics of mass 70 region. The $\pi[440]1/2^+ \otimes$ v [431]3/2$^+$ and $\pi[301]3/2^- \otimes$ v [431]3/2$^+$ configurations are proposed for +ve and –ve parity bands, respectively. The finding of quite different signature splitting for +ve parity bands in ^{70}As and ^{72}As is interpreted as due to different shapes of these nuclei from the Total Routhian Surface (TRS) calculations.

However, more detailed calculations are needed to definitely conclude about the cause of the difference in signature splitting between the two isotopes. The level structure of ^{70}As as obtained from the present work is shown in Fig. 2. Width of the arrow indicates the relative intensity of gamma rays.

The proposed level scheme of ^{70}Ge obtained from present study is shown in Fig. 3. All the low-spin states reported earlier are confirmed; in addition, 14 new transitions have also been found and placed in the level scheme. The level scheme is extended up to a tentative spin-parity, $J^\pi = 21^-$ and excitation energy $Ex \sim 13$ MeV. A rotational band like structure is clearly observed to build up on the known 7^- state. The spin-parity of the newly observed states of this band is based on the DCO (Directional Correlation of Oriented states) ratio measurement. Here it should be mentioned that we could not find 1707 keV transition as was reported in earlier works [4,5]. We are trying to understand the observed band structure in the framework of Cranking model with the Woods-Saxon potential, the results of which will be published shortly.

Acknowledgements

The author would like to thank all those who participated in the experiments and all the Pelletron staff of Nuclear Science Centre for providing a good quality of beam during the experiments.

References

1. B. Mukherjee et al., APH N.S., Heavy Ion Phys. 11 (2000) 305
2. D. Bucurescu et al., Phys. Rev. C 56 (1997) 2497
3. S.M. Fischer et al., Phys. Rev. Lett. 87 (2001) 132501-1
4. C. Morand, et al., Phys. Rev. C 13 (1976) 2182
5. L. Cleeman, et al., Nucl. Phys. A386 (1982) 367

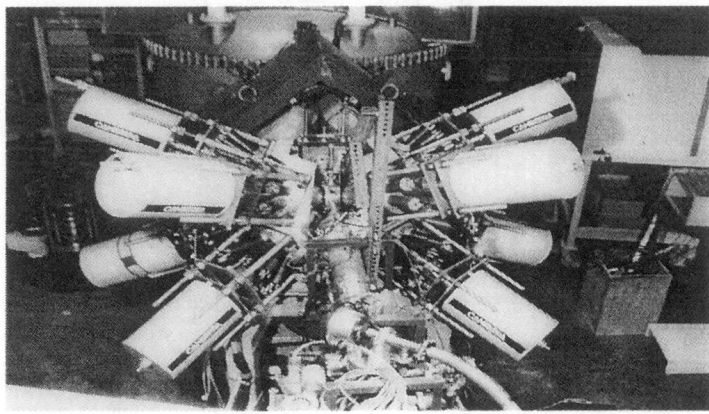

Figure 1. Heavy Ion RecoilAnalyzer (HIRA) and Gamma Detector Array (GDA), coupled facility, for the evaporation residue tagged gamma spectroscopy measurements.

304

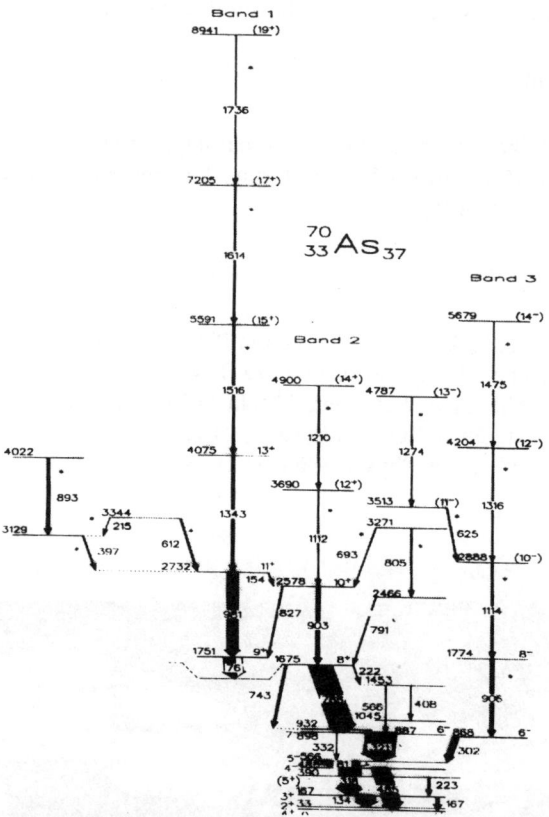

Figure 2. Level Scheme of ^{70}As a obtained from the present work. Width of the arrow indicates the relative intensity of γ-ray and new transitions are marked by '*'.

305

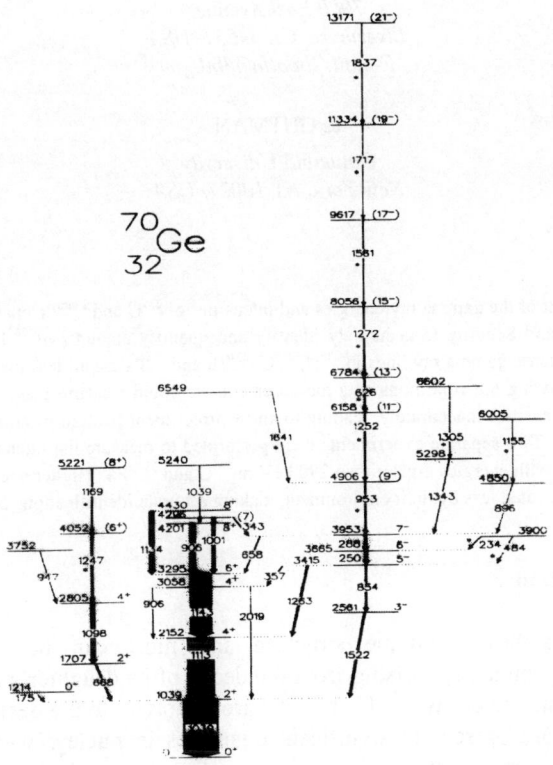

Figure 3. Level scheme of ^{70}Ge nucleus as obtained from present experiments where the width of the arrows represents the relative intensity of transitions and the new transitions are indicated by '*' mark.

GAMMA SPECTROSCOPY OF ^{235}U AND ^{234m}Pa

J. B. PATIN, M. A. STOYER, K. J. MOODY, D. A. SHAUGHNESSY, N. J. STOYER, J. F. WILD AND P.T. WOODDY

Lawrence Livermore National Laboratory
7000 East Avenue,
Livermore, CA 94551, USA
E-mail: jbpatin@llnl.gov

E. GUTMAN

Columbia University
New York, NY 10027, USA

Measurement of the gamma ray energies and intensities of ^{235}U and ^{234m}Pa enable those in Safeguards and Security to accurately identify and quantify amounts of ^{235}U and ^{238}U. The most intense gamma ray lines in ^{235}U, ^{238}U, ^{234}Th and ^{234}Pa are all less then 290 keV. Without knowing the conditions of a measurement, we could measure these gamma ray lines and intensities inaccurately, leading to an incorrect identification of an isotope and its quantity. Two separate experiments were performed to measure the intensities of the gamma rays with energies greater then 290 keV in ^{235}U and ^{234m}Pa. Higher energy gamma rays are attenuated less by their environment, making accurate identifications probable.

1. Introduction

Every nucleus has a unique structure and most can be identified by characteristic gamma rays arising from the decay of its daughter's excited states populated in nuclear decay. ^{235}U and ^{238}U are important to the design of nuclear weapons and are present in significant quantities in nuclear warheads. It is important therefore to have a good understanding of the characteristic gamma rays arising from the decay of ^{235}U and ^{238}U, so that identification of these nuclear components can be simplified. Work trying to identify the amounts of nuclear material from Russian warheads has been done for some time [1].

Gamma rays less than 290 keV arising from the decay of ^{235}U and ^{234m}Pa, the decay granddaughter of ^{238}U, have intensities that are well known. However, the intensities of gamma rays greater than 290 keV are not as well known, and recent measurements have been in disagreement with literature values [2-4]. It is important to know the intensities of these higher energy gamma rays as they are attenuated less by their environment, making identification easier and more accurate.

2. Production and Measurement

2.1. ^{235}U Samples

Three different ^{235}U samples, with isotopic purities of 93.2%, 99.3% and 99.94%, were used in the determination of intensities of gamma rays greater than 290 keV. Uranium metal was dissolved in 10 mL of dilute HCl and placed in a vial of known geometry. This vial was placed in a low background high purity germanium detector for gamma analysis. Each sample was counted close to and far away from the detector to eliminate interference from peaks associated with summing in the detector. After the peaks acquired a significant number of counts, they were fit using the FITEK program to obtain experimental energies and peak areas. To obtain the experimental intensities, the peak areas were corrected for the energy efficiency of the detector and normalized to the intensity of the 187.5 keV gamma ray (57.2% absolute intensity).

2.2. ^{234m}Pa samples

75 g of Q-metal were dissolved in a minimum volume of dilute HCl containing HF. Thorium was isolated using a LaF$_3$ precipitate. The LaF$_3$ precipitate was then dissolved in a minimum volume of boric and nitric acid, warmed and loaded onto an anion exchange column. The column was washed with copious amounts of 8M nitric acid before the thorium was eluted with 9M HCl. The thorium fraction was evaporated to dryness, dissolved in 8M nitric acid, and the anion column procedure was repeated three more times. Counting samples were prepared by volatilization onto platinum, mounted onto a plastic counting disk, and covered with a layer of aluminum foil and scotch tape.

The pure ^{234}Th fractions were counted for a period of 120 days in a variety of low background high purity germanium detectors. The samples were all counted at a distance of 20 cm or more from the face of the detector. The data obtained are the weighted average of at least 20 measurements, all of which agreed within statistics. The procedure was done three times, and half-lives for ^{234}Th of 24.25 ± 0.05 days, 23.99 ± 0.17 days, and 24.42 ± 0.15 days were obtained. The reported half-life of ^{234}Th is 24.10 ± 0.10 days [2]. To obtain the experimental intensities, peak areas from fits to the gamma spectra were corrected for the energy efficiency of the detector and normalized to the 766.36 keV gamma ray of ^{234m}Pa (0.2940% absolute intensity).

3. Results for ^{235}U and ^{234m}Pa

The following tables contain a small sample of the total number of gamma rays observed. Table 1 contains the results of the ^{235}U analysis and Table 2 contains the results of the ^{234m}Pa analysis.

Table 1. Experimental gamma ray intensities (uncertainties in parenthesis) for the three different ^{235}U samples compared to literature values.

Energy(keV)	Intensity(%)			Energy(keV)	Intensity(%)			
Experimental	93.2%	99.3%	99.94%	TOI[4]	TOI[4]	ToRI[2]	ENSDF[3]	
109.04(4)	3.819(15)	2.737(15)	2.245(5)	109.16(2)	1.54(5)	1.54(5)	1.54(5)	
143.777(1)	6.5715(12)	9.938(2)	10.8740(7)	143.76(2)	10.96(8)	10.96(8)	10.96(8)	
163.423(5)	3.9479(8)	4.8297(12)	4.450(2)	163.33(2)	5.08(4)	5.08(4)	5.08(4)	
185.7982(3)	57.2	57.2	57.2	185.715(5)	57.2(5)	57.2(5)	57.2(5)	
202.207(2)	1.2858(5)	1.1098(2)	1.0976(3)	202.11(2)	1.08(2)	1.08(2)	1.08(2)	
205.411(11)	6.0863(14)	5.14(5)	5.03(3)	205.311(10)	5.01(5)	5.01(5)	5.01(5)	
291.89(13)	0.05155(6)	0.02426(2)	0.02078(5)	291.65(3)	0.038(5)	0.038(5)	0.04(5)	
311.24(6)	0.0141(12)	0.00462(2)	0.00369(5)	310.69(6)	0.004	0.004	0.005	
346.096(12)	0.16482(13)	0.06266(6)	0.05584(18)	345.4(1)	0.07	0.07		
356.38(11)	0.01064(7)	0.00420(3)	0.00287(1)	356.03(5)	0.005	0.005	0.005	
388.16(7)	0.03004(13)	0.01080(5)	0.00818(19)	387.82(3)	0.038(5)	0.038(5)	0.04(5)	

Table 2. Experimental gamma ray intensities (uncertainties in parenthesis) for ^{234m}Pa samples compared to literature values.

Energy(keV)	Intensity(%)	Energy(keV)	Intensity(%)		
Experimental	Experimental	TOI[4]	ToRI[2]	ENSDF[3]	
63.29(2)	3.43(15)	63.29(2)	4.8(5)	4.8(6)	
92.38(2)	2.2(9)	92.38(1)	2.81(15)	2.8(3)	
92.80(2)	2.23(9)	92.80(2)	2.77(15)	2.8(3)	
73.92(2)	0.0138(16)	73.92(2)	0.0172(13)	0.0172(19)	
258.22(5)	0.068(3)	258.26(10)	0.0728(4)	0.0728(4)	
766.37(3)	0.294(12)	766.36(2)	0.294(12)	0.294(12)	
1001.05(7)	0.77(3)	1001.03(3)	0.837(10)	0.837(10)	
1737.67(7)	0.0194(9)	1737.73(10)	0.0211(3)	0.0211(3)	
131.31(2)	16.5(9)	131.30(1)	18.0	18.0(1.8)	
883.25(4)	12.5(7)	883.24(4)	9.6(6)	9.6(1.1)	
945.97(4)	20.5(1.1)	946.00(3)	13.4(8)	13.4(1.5)	

4. Conclusions

For gamma ray energies less than 290 keV, as the purity of the ^{235}U sample increases, the experimental intensities more accurately reflect the literature values. For gamma ray energies greater than 290 keV, the experimental intensities do not reflect literature values. The analysis of the ^{234m}Pa data suggests that there are significant differences in the intensities of some of the more prominent gamma lines in the decay of ^{238}U, ^{234}Th, ^{234m}Pa, and ^{234g}Pa. A more thorough chemical separation of ^{234}Th from the Q-metal, and elimination of ^{230}Th which has grown in from the ^{234}U present in the Q-metal will provide a much better understanding of the intensities of these major gamma lines.

Acknowledgments

This work was performed under the auspices of the U.S. Department of Energy by the Lawrence Livermore National Laboratory under contract W-7405-ENG-48.

References

1. S. Fetter, T. B. Cochran, L. Grodzins, H. L. Lynch, and M. S. Zucker, *Science.* **248**, 828 (1990).
2. S. Y. F. Chu, L. P. Ekström and R. B. Firestone, Table of Radioactive Isotopes, http://nucleardata.nuclear.lu.se/nucleardata/toi/
3. ENSDF, http://www.nndc.bnl.gov/nndc/ensdf/
4. Richard B. Firestone, <u>Table of Isotopes</u>, 8th ed., Vol. II, (John Wiley and Sons, NY, 1996) pp. 2759-64.

LEVEL STRUCTURE OF ^{141}BA AND ^{139}XE AND LEVEL SYSTEMATICS OF N=85 EVEN-ODD ISOTONES

Y.X. Luo[1,2], J.O. Rasmussen[2], J.H. Hamilton[1], A.V.Ramayya[1], J.K. Hwang[1], C.J. Beyer[1], S.J. Zhu[3], J. Kormicki[1], X.Q. Zhang[1], E.F. Jones[1], P.M. Gore[1], T.N. Ginter[2], I.Y. Lee[2], K.E. Gregorich[2], A.O. Macchiavelli[2], G.M. Ter-Akopian[4], Yu.Ts. Oganessian[4], A.V. Daniel[4], M.A. Stoyer[5], S.J. Asztalos[5], J.D. Cole[6], R. Donangelo[7], and S.C. Wu[2]

[1]*Vanderbilt University, Nashville, TN 37235*
[2]*Lawence Berkeley National Laboratory, Berkeley, CA 94720*
[3]*Tsinghua University, Beijing, China*
[4]*Joint Institute for Nuclear Research, RU-141980 Dubna, Russia*
[5]*Lawrence Livermore National Laboratory, Livermore, CA 94551*
[6]*Idaho National Environmental and Engineering Laboratory*
[7]Federal University of Rio de Janeiro (UFRJ), 21945-970 Rio de Janeiro, Brasil

From high statistics γ–γ–γ ^{252}Cf coincidence studies, new level schemes of ^{139}Xe and ^{141}Ba are proposed and level systematics of N=85 isotones from Z=64-66 down to Z=56, 54 completed and studied. These studies provide valuable information in the 'intermediate' region between the doubly-magic ^{132}Sn and well-deformed nuclei for the interplay of spherical shell model excitations and the quasi-collective motions. In going above the maximum stretched three-neutron limits and moving to mid-way between Z=50 and Z=64, contributions of valence protons and quadrupole collectivity must be taken into account. Octupole correlations were observed in ^{141}Ba and ^{139}Xe located at the border of the predicted octupole island. The level systematics provides evidences that octupole strength is becoming weaker when approaching Z=50 shell.

1. Introduction

With three valence neutrons outside the N=82 closed shell, the N=85 even-odd isotones with Z ~ 50-64 are located in the 'intermediate' region between the well-deformed nuclei and doubly magic ^{132}Sn. The studies of these 'quasi-$f_{7/2}$' nuclei provide a good opportunity to gain insight into the interplay between single particle excitations and quasi-collective motions [1,2,3]. Furthermore, the N=85 isotones are located at the border of the predicted octupole correlation island centered at N=88 [4]. Studies of octupole correlations and couplings with single particle excitations in these isotones have also drawn much attention.

Low-lying states were first studied by β^- decay [5] and low excitation low spin states were populated by alpha or heavy ion reactions for the N=85 isotones ^{149}Gd (Z=64) [6], ^{147}Sm (Z=62) and ^{145}Nd (Z=60) [7]. Levels observed were interpreted as three valence neutron excitations and that coupled with octupole phonons [6,7]. The combination of multi-gamma detection array with fission source provides good opportunities to explore in the neutron-rich region [8]. Lighter N=85 isotones [139]Xe (Z=54) [9,10] and [141]Ba (Z=56) [10] were studied earlier at Eurogam and Gammasphere, and [137]Te (Z=52) at Eurogam II [11].

The $5.7*10^{11}$ 3- and higher-fold [252]Cf fission coincidence events accumulated at Gammasphere in the 2000 run opened up studies of weakly populated bands and levels to higher spins and excitations. Level structures of [141]Ba and [139]Xe were reinvestigated with level schemes considerably expanded, and level systematics of the N=85 even-odd isotones extended to Z=56 and 54 [2,3].

Following our work [2,3] similar less extended level schemes of [139]Xe and that of [141]Ba with fewer bands and lower spins were reported [12]. A number of spin and parity assignments were made based on measurements of angular correlations and linear polarizations [12].

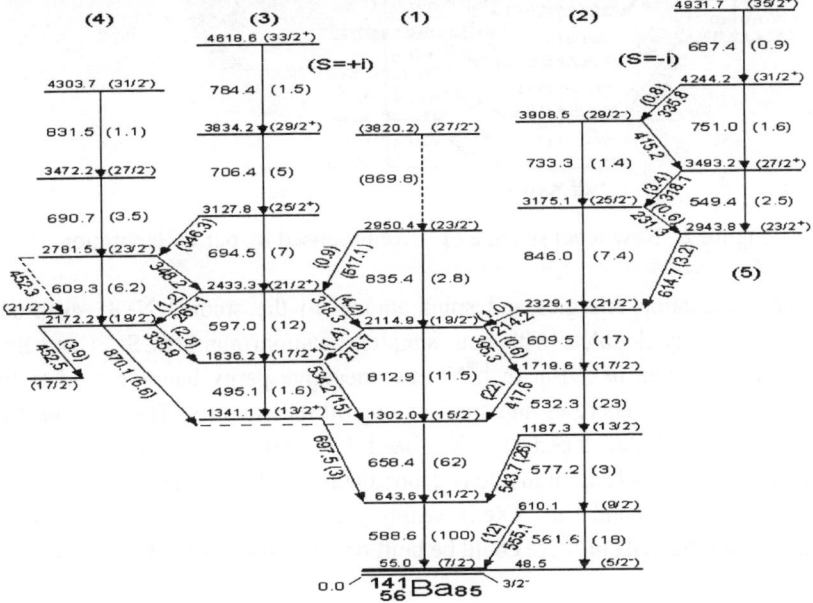

Figure 1 New level scheme of [141]Ba proposed by our collaboration [2,3].

2. New Level schemes of [141]Ba (Z=56) and [139]Xe (Z=54)

The level schemes of [141]Ba and [139]Xe proposed by Zhu et al. [10] are considerably expanded (see Figs. 1,2 for [141]Ba, and [139]Xe) [2,3]. New bands 4 and 5 are observed in [141]Ba. Additional analyses was made on [141]Ba and [139]Xe with less compressed spectra from our [252]Cf 1995 Gammasphere data [2,3].

By analogy with the N=85 even-odd isotones [149]Gd [6], [147]Sm and [145]Nd [7], spins and parities were tentatively assigned to the excited states in [141]Ba and

^{139}Xe. In ^{141}Ba, two positive parity bands were identified and extended to the

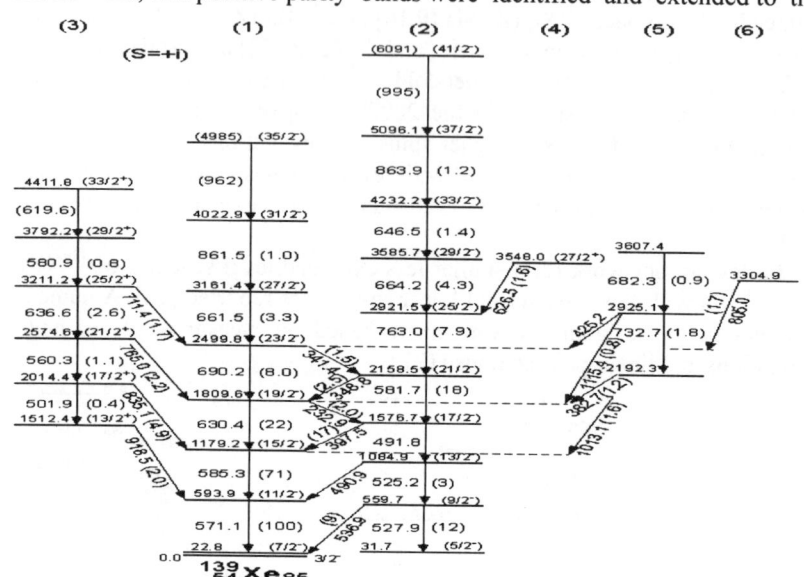

Figure 2   New level scheme of ^{139}Xe proposed by our collaboration.

highest excitation energies and spins among all the studied N=85 even-odd isotones. Parity doublets with both simplex quantum numbers S=+i and S=-i now are observed in ^{141}Ba. In ^{139}Xe, two negative parity bands extend to the highest excitation energies and spins among all the isotones. The only positive parity band well-developed in ^{139}Xe, band 3, is pushed up to high excitation energies so that only E1 transitions depopulating the band 3 and feeding band 1 are possible. The band 3 in ^{139}Xe is weakly populated. Another possible positive parity band (band 4) in ^{139}Xe could be built on the 3548.0 keV (27/2$^+$) level.

3.   Interpretations of level structure of ^{141}Ba and ^{139}Xe, and level systematics of the N=85 even-odd isotones

The earlier studies of N=85 even=odd isotones ^{149}Gd [6], ^{147}Sm and ^{145}Nd [7] generally do not populate as high spins as does SF. Also, these studies for Z=60, 62, and 64, are all close to the 64 subshell with filled g$_{7/2}$ and d$_{5/2}$ proton orbitals. Configurations like $\upsilon(f_{7/2})^3$ or $\upsilon(f_{7/2})^2$ h$_{9/2}$ were assigned, with proton participation entering only indirectly through octupole phonon coupling.

However, for ^{141}Ba (Z=56) and ^{139}Xe (Z=54) the interpretations above only provide dominant components of configurations for the low spin states.

The levels of ^{141}Ba and ^{139}Xe [2,3] differ considerably with those of the N=85 isotones near Z=64 subshell. When going above the maximum spins of the stretched configurations of the three valence neutrons, band 1 and 2 in ^{141}Ba and ^{139}Xe show a gap in spacing and more regular spacing than in isotones of Gd, Sm, and Nd near Z=64 subshell (see Figs 1, 2, and 3). Bands 1 and 2 are more appropriately considered as signature partners of one band. The near equal spacing of the levels at higher excitations is characteristic of harmonic vibrations and is interpreted as mixture of configurations involving $f_{7/2}$ and $h_{9/2}$

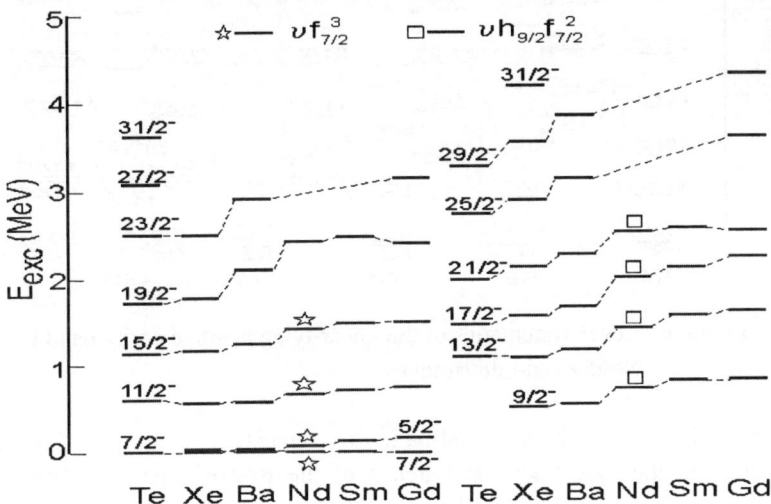

Figure 3 Level systematics of the negative parity bands in N=85 isotones.

neutron plus quadrupole phonons [2,3]. The participation of valence protons gradually increases as one goes up the bands to higher spins. Single neutron excitations $\upsilon f^2_{7/2}i_{13/2}$ and that coupled with octupole phonon excitations $\upsilon f^3_{7/2} * 3^-$ are assigned to the positive parity bandhead $13/2^+$ in ^{141}Ba and ^{139}Xe , with increasing contributions of valence protons at higher spins in this band. Band 4 in ^{141}Ba shows smaller energy spacing than that of band 1, suggesting larger deformation or quadrupole softness. In stead of assigning double octupole phonon coupled with ground single neutrons to the $19/2^-$ of band 4 as in [6,7], which requires considerable anharmonicity, $\pi(g^{-2}_{7/2})_2\upsilon(f^3_{7/2})_{15/2}$ and $\upsilon(f^2_{7/2})_6h_{9/2}$ are assigned to the bandhead, and mixture with quadrupole phonons from valence protons is expected in going up the band. Band 5, the higher-lying positive parity band, is interpreted to involve excited neutrons coupled with octupole phonons, $\nu f^2_{7/2} h_{9/2} * 3^-$, with growing participation of valence protons at the higher spins.

The level systematics of the N=85 isotones from Gd (Z=64) to Te (Z=52) are

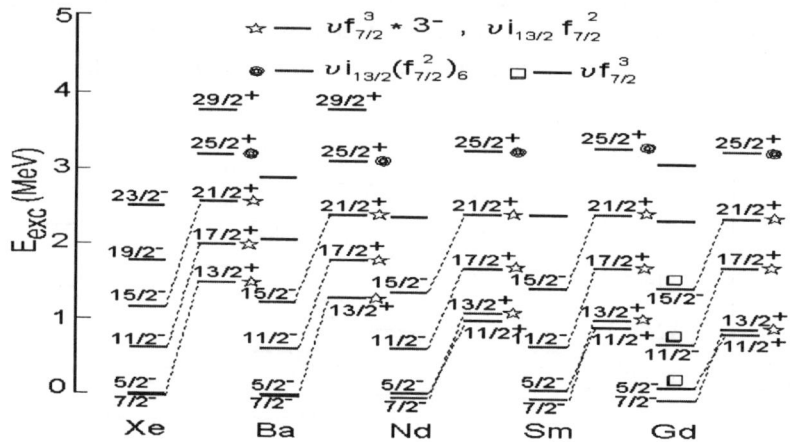

Figure 4   Level systematics of the lower-lying positive parity band in N=85 even-odd isotones.

shown in Figs. 3 - 5. The spin and parity assignments for ^{141}Ba and ^{139}Xe are supported by the systematics and trends of the isotone's level patterns. In moving down from Gd, the irregularities of the two lowest negative parity bands are decreasing, and around Ba and Xe, the bands become more regular as mentioned above (Fig. 3). This changing irregularity may be attributed to moving away from the Z=64 subshell to midway between Z=50 and 64, when proton excitations must be included. The level structures may suggest that quadrupole/ octupole collectivity has come into play in this 'intermediate' region, a region, however, only three valence neutrons away from the N=82 closed shell.

Figure 4 shows the systematics of lower-lying positive parity bands (band 3 in ^{141}Ba) in N=85 even-odd isotones. Of particular interest is the pronounced increasing of the excitation energies of the $13/2^+$ bandhead with decreasing Z. The weakly populated positive parity band 3 in ^{139}Xe is already located at a high excitation energy, and no such band observed in ^{137}Te [11]. On the other hand, the excitation energies of the 13/2⁻ states are decreasing with decreasing Z. The opposite trend of 13/2⁺ and 13/2⁻ levels seen in Fig. 5 makes the 13/2⁺ states from yrast in ^{149}Gd, ^{147}Sm and ^{145}Nd to nonyrast in Ba and Xe with no 13/2⁺ state observed in ^{137}Te. This evolution of the excitations of the 13/2⁺ states may account for the weakening of the population of the positive parity bands as Z decreases, and provides evidences that when approaching the Z=50 proton shell, octupole excitations in the N=85 even-odd isotones are becoming weaker.

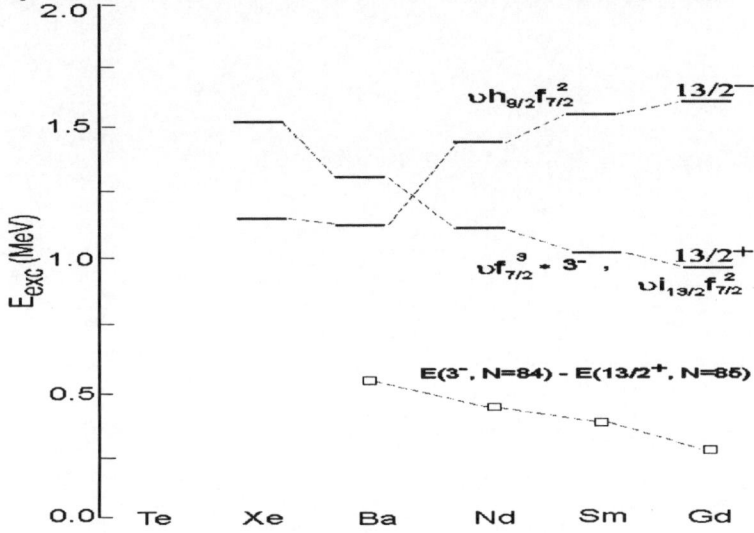

Figure 5 Trends of the excitation energies of the bandhead, 13/2⁺state, of the lower-lying positive parity band of the N=85 isotones.

Works at Vanderbilt University, LBNL, and LLNL were supported by the U.S. DOE under Grant No. DE - FG05 - 88ER40407, DE-AC07-761D1570 and DE-AC03-76SF00098, and W-7405-ENG-48, respectively.

References
1. V. Paar *et al.*, *Nucl. Phys.* **A350**, 139(1980).
2 Y.X. Luo *et al.*, in *Proceedings of INPC2001, Berkeley,* July30, 2001, p.807.
3. Y.X. Luo *et al.*, *Phys. Rev.* **C56**, 014305-1(2002).
4. W. Nazarewicz and P. Olanders, *Nucl. Phys.* **A441**, 420(1985).
5. *Table of Isotopes*, 8th ed., edited by R.B. Firestone and V.S. Shirley, (Wiley, New York, 1996).
6. M. Piiparinen *et al.*, *Z. Phys.* **A300**, 133(1981).

7. W. Urban *et al.*, *Phys. Rev.* **C53**, 2516(1996).
8. J.H. Hamilton *et al.*, *Prog. Part. Nucl. Phys.* **35**, 635(1995).
9. M. Bentaleb *et al.*, *Z. Phys.* **A348**, 245(1994)
10. S.J. Zhu *et al.*, *J. Phys.* **G23**, L77(1997).
11. W. Urban *et al.*, *Phys. Rev.* **C61**, 041301(R)(2000).
12. W. Urban *et al.*, *Phys. Rev.* **C66**, 044302-1(2002)

HIGH SPIN STUDIES OF NEUTRON-RICH ^{166}Er[*]

M. BURNS, R. CHAPMAN, M. LABICHE, X. LIANG, J. OLLIER AND K.M. SPOHR

Electronic Engineering and Physics, University of Paisley, Paisley, PA1 2BE, UK.

M. AXIOTIS, E. FARNEA AND T. MARTINEZ

INFN – Laboratory Nazionali di Legnaro, Via Romea 4, 35020 Legnaro (Padova), Italy.

TH. KRÖLL, D.R. NAPOLI AND C.A. UR

Dipartmento di Fisica, University of Padova, Via F. Marzolo 8, 35131 Padova, Italy.

High spin states in ^{166}Er were populated in the incomplete fusion of 55MeV ^{7}Li ions with a ^{164}Dy target. The GASP γ–detector array and the ISIS charged particle telescope array at the INFN Laboratory, Legnaro, Italy were used for this work. The present data has allowed an extension of the ^{166}Er yrast level scheme to 24ħ. The gamma band has been extended to 19ħ and four new bands have been identified. Work is in progress to identify linking transitions.

1. Introduction

The study of neutron-rich rare earth nuclei is still in its infancy due to the inaccessibility of this region via normal (HI,xn) reactions with stable beams. Incomplete fusion-evaporation experiments have recently been shown to be very successful in populating high-spin states of neutron-rich stable nuclei. The theoretical focus of such work is the study of the interplay between quasiparticle and collective degrees of freedom, within the framework of the cranked shell model. The level structure of ^{166}Er has previously been studied by Coulomb excitation (Fahlander *et al* [1]) and in the (α,2nγ) work of Fields *et al* [2] where the yrast sequence was extended to spin 16ħ.

Here we have used a beam of ^{7}Li to study the yrast and near-yrast states of ^{166}Er populated in the incomplete fusion (ICF) reaction, ^{164}Dy(^{7}Li,p4n)^{166}Er. In the proximity of the target nucleus, the projectile nucleus breaks up and only part of it (^{6}He) fuses forming a highly excited compound nucleus. The proton which does not take part in the fusion process carries a much higher kinetic energy than protons evaporated from a compound system and their coincident detection with γ–rays from the de-excitation of the final product nucleus can be used for a precise off-line selection of the exit channel.

---

[*] This work is supported by the EPSRC.

2. Experiment and Results

The ^{164}Dy(^{7}Li,p4n)^{166}Er reaction was studied at the INFN, Legnaro Laboratory. The Tandem-XTU accelerator was used to accelerate ^{7}Li ions to 55MeV. The isotopically enriched ^{164}Dy target was of thickness 3.5 mg·cm^{-2}. The ICF reaction populates only a few exit channels of which ^{166}Er is predominant via the (p,4n), (d,3n) and (t,2n) channels. The high efficiency and resolving power of the GASP multi-detector γ-ray array coupled with the selectivity provided by the ISIS array of silicon detector telescopes yielded good statistics and channel selection. The trigger conditions for an 'event' corresponded to two or more unsuppressed Ge signals and two or more BGO inner-ball element signals. The data were sorted into a γγγ cube

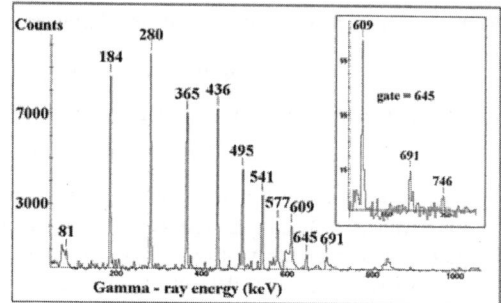

Figure 1. Proton gated γ-ray total projection spectrum (^{166}Er). The inset shows the higher energy transitions. γ-ray energies are in units of keV

with proton detection in ISIS as a condition. The cube was analysed using the RADWARE analysis package. Figure 1 shows the γ-ray total projection spectrum in coincidence with protons. The yrast scheme has been extended to spin 24ℏ and the two signatures of the gamma band to spins 16ℏ and 19ℏ (fig 2a). Four new rotational sequences have also been found (fig 2b).

3. Discussion

Figure 3(a) shows the experimental Routhians and (b) the quasiparticle aligned angular momenta for the yrast and γ-vibrational decay sequences of ^{166}Er as a function of rotational frequency. A reference band has been subtracted with Harris parameters $J^{(0)} = 34$ ℏ^2 MeV^{-1} and $J^{(1)} = 128$ ℏ^4 MeV^{-3}. The alignment of a pair of $i_{13/2}$ quasineutrons takes place at ℏω ~ 0.25MeV, which agrees well with the results of cranked shell model calculations. ^{166}Er, with neutron number 98, is expected to exhibit a strong interaction between the ground state band and the crossing band based on an aligned pair of $i_{13/2}$ quasineutrons. The behaviour of the quasiparticle alignment as a function of rotational frequency (fig 3b) is consistent with this expectation.

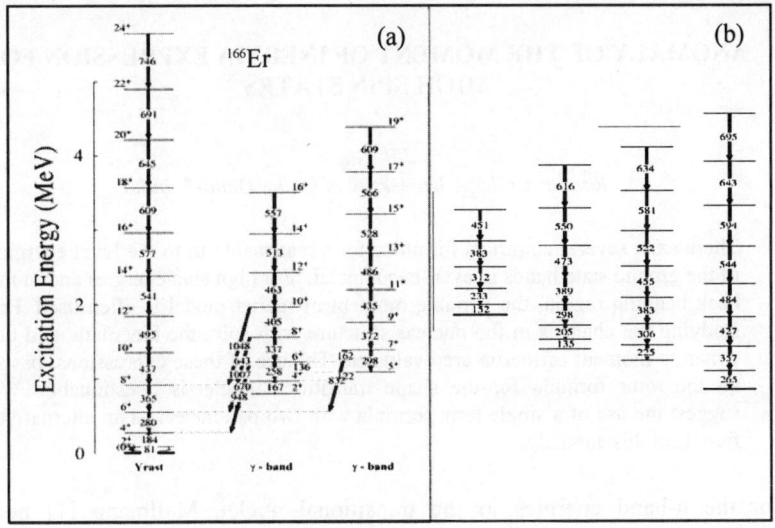

Figure 2(a) Level scheme and (b) four new observed decay sequences in ¹⁶⁶Er.

4. Conclusion

The ICF reaction has been successfully used to populate rotational decay sequences of ¹⁶⁶Er. The AB band crossing has been observed and comparison with results of CSM

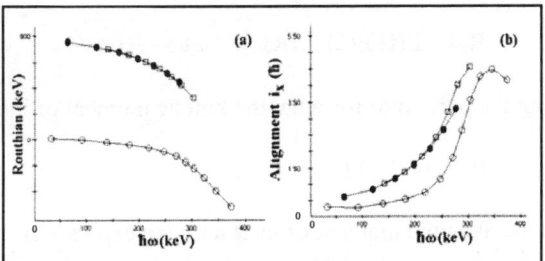

Figure 3(a) Experimental Routhians and (b) quasiparticle alignments for the yrast (○) and the two signatures of the γ−vibrational decay sequences (● □) in ¹⁶⁶Er plotted as a function of rotational frequency.

calculations shows good agreement. Data analysis is in progress.

References

1. C. Fahlander *et al*, *Phys. Lett.*, **B388**, 475 (1996).
2. C.A. Fields *et al*, *Nucl. Phys.*, **A440**, 301 (1985).

ANOMALY OF THE MOMENT OF INERTIA EXPRESSION FOR HIGH SPIN STATES

J.B. Gupta

Ramjas College, University of Delhi, Delhi-7, India

There exist several empirical formulae for a reasonable fit to the level energies of the ground state bands in even-even nuclei. For high spin energies and in the back bending region, the variable moment of inertia model is often used. For studying the changes in the nuclear structure with spin, the kinematic and the dynamic moment of inertia are evaluated. The use of these expressions, based on the rotor formula for the shape transitional nuclei is questionable. We suggest the use of a single term formula with two parameters as an alternative, free from this anomaly.

For the g-band energies in the transitional nuclei, Mallmann [1] noted a relationship of $R_{I/2} = E(I)/E(2)$ with $R_{4/2} = E(4)/E(2)$. For the g-bands, Ejiri et al. [2] suggested a two term expression $E(I) = aI(I+1) + bI$, the sum of the rotational and vibrational energies. On the Mallmann plot of $R_{I/2}$ versus $R_{4/2}$, this yields a linear relation, equivalent to the geometric relation:

$$R_{I/2} = E(I)/E(2) = (R_{4/2})I(I-2)/8 - I(I-4)/4 \qquad (1)$$

Using the pure rotor formula, the kinetic moment of inertia θ_I is defined as

$$\theta_I^{(1)} = (2I-1)/E_\gamma \ (I \to I-2). \qquad (2)$$

and the dynamic moment of inertia by the expression

$$\theta_I^{(2)} = 4/ \ [E_\gamma \ (I+2 \to I) - E_\gamma \ (I \to I-2)] \qquad (3)$$

Based on the linear plot, the variation of $E(I)$, E_γ, $\theta_I^{(1)}$ and $\theta_I^{(2)}$, for a few $R_{4/2}$ values, are listed in Table 1. *Since E_γ increases linearly with I in each case, (hbar) ω also increases linearly with spin I.* A plot of moment of inertia versus I or ω will have the same form. Note the increasing slope of $\theta_I^{(1)}$ with I (or hbar ω) with respect to the horizontal, corresponding to the increasing spherical content and the decreasing $R_{4/2}$. There is phase transition from spherical to E(5), to O(6), to X(5) and to the deformed shape. The dynamic moment of inertia is constant in each case, but increases with decreasing $R_{4/2}$. The deviation from linearity yields varying $\theta_I^{(2)}$, i.e. it indicates the changes in the collective structure or that of the amount of sphericity with increasing spin.

The variable moment of inertia model (VMI) expression [3] is

$$E\,(I) = \{hbar^2/2\theta_1\}I(I+1) + (1/2)C\,(\theta_1 - \theta_0)^2, \qquad (4)$$

Here θ_1 and C are determined from the equilibrium condition $dE/d\theta|_1 = 0$, which yields a cubic equation in θ_1 viz. $2C\theta_1^3 - 2C\theta_0\,\theta_1 - (hbar)I(I+1)=0$. One has to determine the coefficients θ_0 and C, then solve for θ_1 for every spin I. A recursive process is required for the solution. The calculated moment of inertia are generally slightly larger than the kinetic moment of inertia, which form the input.

Table 1. The values of $R_{1/2}$ or E(I) relative to E(2), given by the Eq. 1. Hbar $\omega = E_\gamma/2$. The dynamic M.I. is equal to 4.0, 8.0 and 20.0 for $R_{4/2} =3.0$, 2.5 and 2.2 respectively

$R_{4/2}$	Spin I	2	4	6	8	10	12	14	16	18
3	E(I)	1	3	6	10	15	21	28	36	45
	E_γ	1	2	3	4	5	6	7	8	9
	$\theta_1^{(1)}$	3	3.5	11/3	15/4	19/5	23/6	27/7	31/8	35/9
2.5	E(I)	1	2.5	4.5	7	10	13.5	17.5	22	27
	E_γ	1	1.5	2	2.5	3	3.5	4	4.5	5
	$\theta_1^{(1)}$	3	14/3	11/2	6	19/3	46/7	54/8	62/9	7
2.20	E(I)	1	2.20	3.60	5.20	7.00	9.00	11.20	13.60	16.2
	E_γ	1	1.20	1.40	1.60	1.80	2.00	2.20	2.40	2.60
	$\theta_1^{(1)}$	3	5.8	7.8	9.4	10.6	11.5	12.3	13.0	13.5

The anomaly in the use of $\theta_I^{(1)}$ (Eq. 2) is that the numerator is from the pure rotor expression, while the denominator comes from experiment, so that $\theta_I^{(1)}$ increases linearly, even when there is no change in the structure of the nucleus. The increase is merely arithmetic. The slope of $\theta_1^{(1)}$ measures the sphericity of the core. Let us look at an alternative representation [4]: $E(I) = a\,I^b$. We showed that for ^{148}Ce neither E/I , nor E/I(I+1) was a constant with I, but $E(I)/I^{3/2}$ was indepedent of I. In general, the index b is different from 3/2. For a spherical vibrator the index $b=1$ and for a good rotor, the index b is less than 2 (to compensate for the replacemt of I(I+1) by I^b). If the nuclear structure is constant with spin I, b remains constant. A plot of log E(I) versus log I yields b as the slope and a as the intercept [4]. The value of the index b is a direct measure of the degree of deformation of the nuclear core. For example, the index b is 1.60, 1.58, 1.40 and 1.22 for ^{152}Sm, ^{104}Mo ^{116}Pd, and ^{150}Sm respectively, with $R_{4/2}$ =3.00, 2.92, 2.56 and 2.32.

The variation of b with I directly indicates the change in the nuclear structure with spin, as indicated for ^{116}Pd at I=8+, at which the nucleus back bends. For the soft vibrator ^{150}Sm, b rises with spin but for ^{152}Sm it is constant (see Fig. 1).

The co-efficient a is analogous to the inverse moment of inertia. The plots of a also show similar behavior: very good constancy in ^{152}Sm, a rise in ^{150}Sm, and a sharp variation in ^{116}Pd. For comparison we show in Fig. 1 the variation of the kinetic moment of inertia with spin I. On a plot of $\theta_I^{(1)}$ versus hbar ω for ^{116}Pd, there is back band at 8+. On the latter plot (not shown) the coefficient a also indicates the back bending behavior. The advantage with the power expression is that it does not suffer from the inadequacy of the definition of moment of inertia in (2). Further, as compared to the other energy formulae discussed in above, the advantage with the single term formula is that one can evaluate a and b continuously between the consecutive levels and get a_I and b_I as one does for the kinetic and dynamic moments of inertia.

References

1. C. A. Mallmann, Phys. Rev. Lett. **2**, 501 (1959)
2. H. Ejiri et al., J. Phys. Soc. Jpn **24**, 1189 (1968)
3. MAJ Mariscotti, G. Scharff-Goldhaberand B. Buck, Phys. Rev. **178**, 1964 (1969)
4. J.B. Gupta, A.K. Kavathekar and R. Sharma, Phys. Scripta **51**, 316 (1995)

Fig.1. Index b, Coeff.a of E(I) = a I^b and the kinetic moment of inertia.

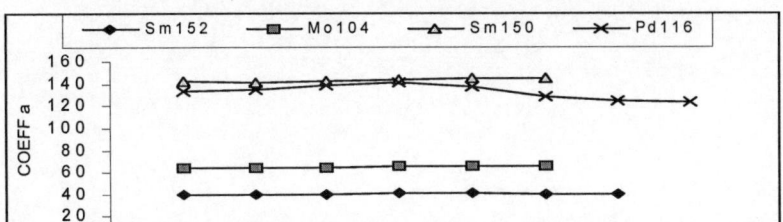

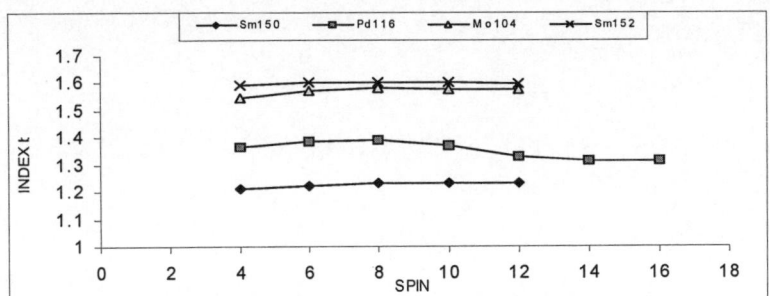

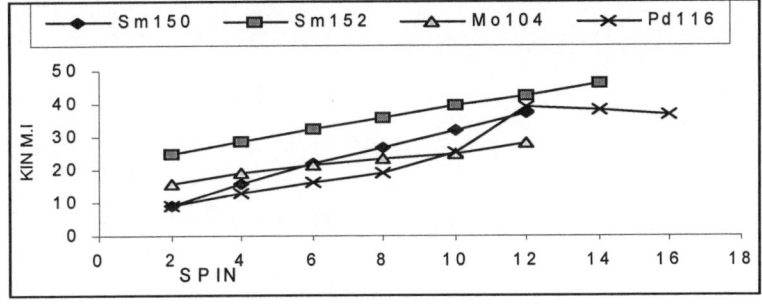

Nuclear
Astrophysics

ASTROPHYSICS OF NEUTRON RICH NUCLEI

M. WIESCHER[†]

Joint Institute for Nuclear Astrophysics
Department of Physics University of Notre Dame
225 Nieuwland Science Hall
Notre Dame, IN 46556, USA
E-mail: Wiescher.1@nd.edu

Neutron induced nucleosynthesis processes are responsible for the formation of heavy elements beyond iron. I present a short overview on the physics of the s- and the r-process and discuss the associated nuclear physics uncertainties. I demonstrate the need for improved nuclear structure and reaction data on neutron rich nuclei along the range of the r-process path. I also discuss possible nucleosynthesis scenarios at the extreme conditions of the early supernova shock and cataclysmic binaries.

1. Introduction

The nuclear astrophysics of neutron rich isotopes represents a multifaceted variety of scenarios ranging from late stellar evolution phenomena to supernova shock and the conditions in accreting or merging neutron stars. These scenarios provide the necessary conditions for the production of sufficiently high neutron flux which is important for the synthesis of heavy elements above iron. The production of heavy element is displayed in the so-called galactic abundance distribution curve which shows the observed abundances of the elements plotted as a function of their mass number A [1], most of the data are based on solar system observations. This abundance distribution curve reflects the nucleosynthesis history of our solar system material, until recently it was the sole observational evidence for the existence of neutron induced nucleosynthesis of the heavy elements.

There are two fundamentally important scenarios for the production of heavy nuclei. The first one is the so-called s-process which operates on the basis of a rather moderate neutron flux. Because of the low neutron flux, the timescale for the process is slow, which gave the process its name, slow (s) neutron capture process. All radioactive neutron-rich isotopes produced in this process have sufficient time to β-decay and the reaction path runs along the line of stability as shown in figure 1. The s-process is responsible for the production of about 50% of the heavy elements above iron up to lead. The site of the s-process has been a

---

[†] Work supported by grant PHY-0140324 of the National Science Foundation.

matter of debate for a long time; the main uncertainty has been the nature of the s-process neutron source. The observational evidence pointed towards two independent sites, named the weak s-process and the strong s-process [2]. Only recently these scenarios were associated with actual stellar sites, the helium core burning in massive red giant (RG) stars ($M>10M_\odot$), and helium shell burning processes in asymptotic giant branch (AGB) stars [3]. Recent efforts in theoretical modeling these scenarios yet still indicate considerable uncertainties associated with both the nucleosynthesis aspects as well as the hydro dynamical aspects of the rather complex convection and mixing processes during late stellar evolution.

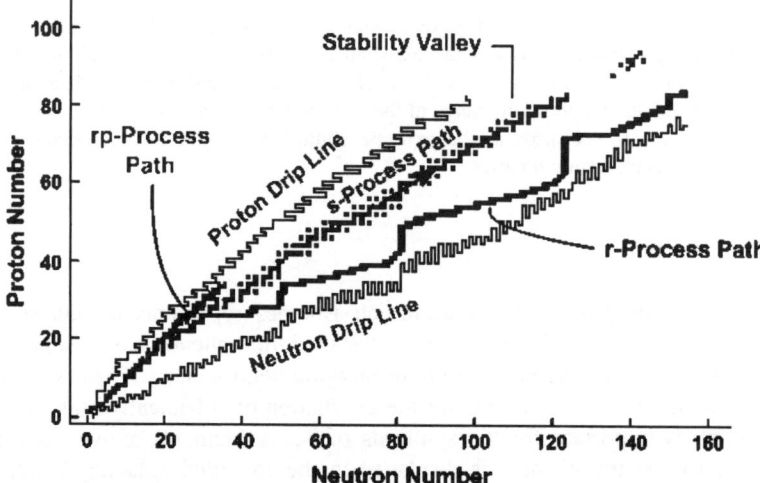

Figure 1. Shown are the reaction paths of the two main neutron induced nucleosynthesis processes, the s-process and the r-process (above Z=26). Also indicated is the rp-process reaction path which drives the thermo-nuclear runaway on accreting neutron stars.

The second neutron induced nucleosynthesis process is the so called rapid neutron capture process or r-process. Based on generic fits of the observed abundance distribution the r-process is expected to take place at conditions which provide a neutron flux of $>10^{23}$ n/s [4]. At such conditions the neutron capture rates are much faster than the β-decay rates and the r-process path is situated far outside from the line of stability as illustrated in figure 1. The actual site of the r-process is still a matter of debate. High entropy conditions in the neutrino driven wind of the supernova shock are presently considered as the favored site for the r-process [5]. Difficulties in modeling the conditions necessary for simulating the observed r-process abundance distributions did however lead to speculation about alternative scenarios such as merging neutron stars [6]. Only limited experimental information is available on r-process nuclei. This represents a major uncertainty for any r-process model. However, the uncertainties are not only limited to site and nuclear structure information for the r-process. Until recently the main signature for the r-process was the r-process abundance distribution which was obtained by subtracting the s-process abundances from the solar abundance distribution [2]. The uncertainties therefore depend strongly on the accuracy of s-process abundance predictions.

Another important role for neutron rich nuclei in nuclear astrophysics is in the physics of neutronization of matter which does take place at the extreme conditions provided in the crust of an accreting neutrons star. At high temperature conditions neutron rich nuclei are produced by fast neutron capture processes, at high density conditions neutron rich nuclei are formed by electron capture processes. Very little is known about the electron capture and electron capture induced neutron emission processes near the neutron drip-line. These reactions are critical for simulating the fate of matter at high density conditions. A second important aspect is the question of pycno-nuclear fusion processes. Pycno-nuclear reactions take place at extreme density conditions and have been proposed for stable nuclei fusion like $^{12}C+^{12}C$ in the core material of white dwarfs [7]. Pycno-nuclear processes in the deep layers of neutron star crusts involve fusion reactions between very neutron rich near-drip line nuclei [8]. The rates for such fusion processes are extremely uncertain and may depend sensitively on the pronounced halo structure of neutron rich isotopes. Pycno-nuclear reactions may generate substantial amount of heat in the deeper layers of the neutron star crust which may affect the thermal and electrical characteristics of the crust material [8].

2. The s-Process and the r-Process

The nucleosynthesis of most of the observed elements above iron is due to neutron induced reaction processes. The analysis of the elemental abundance distribution curve [1] revealed the existence of two fundamental neutron induced nucleosynthesis processes, the s-process and the r-process. Recent observational data indicate that these nucleosynthesis processes take place at different stellar temperature and density conditions. The s-process is characterized by slow neutron capture and β-decay processes along the line of stability as shown in figure 1. The time scale for the production of heavy elements depends strongly on the neutron flux for the s-process. The s-process neutron sources are associated with (α,n) reactions triggered during the helium burning phases in stellar evolution. The weak s-process component takes place towards the end of helium core burning in massive red giant (RG) stars ($M>10M_\odot$) and is triggered by the slightly exothermic $^{22}Ne(\alpha,n)$ reaction. The necessary ^{22}Ne fuel is being built up towards the end of helium core burning from the ^{14}N ashes of the preceding CNO dominated hydrogen burning phase by the reaction sequence ^{14}N$(\alpha,\gamma)^{18}$F$(\beta)^{18}$O$(\alpha,\gamma)^{22}$Ne [9]. The nucleosynthesis towards heavy elements proceeds only up to the A=100 range for the weak s-process component. The strong s-process is associated with helium shell burning in low mass, M<4M$_\odot$, asymptotic giant branch (AGB) stars [3,10]. The scenario is relatively complex and depends on the evolution of the helium burning shell on top of the inert ^{12}C core of the star [3,10]. The neutrons for the s-process in the helium shell are produced by the ^{13}C(α,n) reaction, the ^{13}C fuel is provided by the ^{12}C$(p,\gamma)^{13}$N$(\beta)^{13}$C process which depends on mixing hydrogen from the outer hydrogen rich zone into the helium and carbon -rich layers. For a typical temperature of 10^8 K in this zone a neutron density of 10^7 n/cm^3 is provided. The strong s-process is responsible for nucleosynthesis of heavy elements up to the Pb Bi mass range [10,11]. In the subsequent He-flash the s-process material is mixed with the deep convective outer envelope of the star, transported towards the stellar surface, and ejected into the interstellar space by radiation driven winds. While the complex thermodynamical and mixing aspects for such a scenario need more detailed and self-consistent theoretical treatment, there is also still a multitude of nuclear physics questions to be addressed. The neutron production reactions depend on possible contributions from low energy near-threshold resonances that only have been estimated at best. Other important parameters are neutron capture reaction cross sections along the s-process path. While such cross sections on stable nuclei have been measured extensively for decades [2,10] there still remain critical questions which are associated with n-capture on long-lived radioactive nuclei. These captures open side branches in the reaction path leading to different final abundance distributions. Since such branching reactions depend strongly on temperature, density, and subsequently the neutron-flux conditions, the analysis of the isotope abundance distributions in this mass regions [12] coupled with a detailed knowledge of the associated capture cross sections represent a perfect thermometer and barometer for the s-process sites. Another important aspect is the question about the actual end-point of the s-process in the Pb Bi region. This is closely associated with the neutron capture cross sections on the Pb and Bi isotopes and the reprocessing through subsequent β- and α-decay processes. These reactions are of particular importance for the addressing the question if the origin of Pb isotopes is more associated with s-process or r-process sites [10,11]. For these particular reasons experimental efforts have focused over the last few years on the development of new facilities for measuring neutron capture cross sections on small samples of rare stable and long-lived radioactive isotopes using high neutron flux spallation sources such as n-TOF at CERN and LANSCE at Los Alamos.

The exact determination of neutron capture cross section for the s-process is crucial for deriving reliable s-process abundance distributions which are used to predict the expected r-process abundance distribution. Figure 2 shows the s-process abundance distribution and the r-process abundance distribution which has been obtained by subtraction the first from the observed solar abundance distribution curve [1,2].

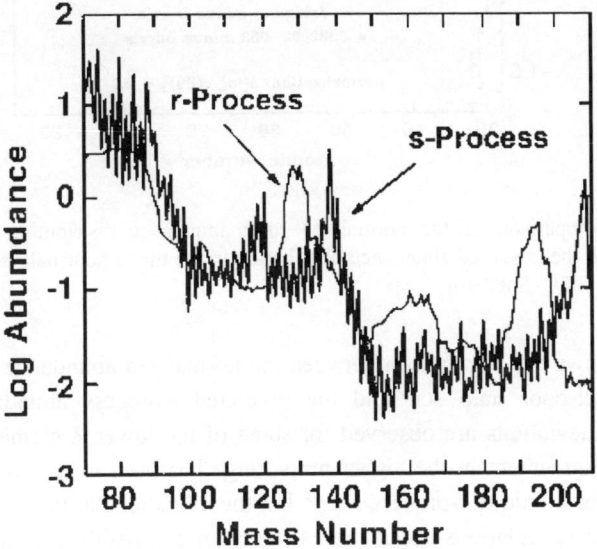

Figure 2. Abundance distributions originated from s-process nucleosynthesis near the line of stability and r-process nucleosynthesis along the neutron rich side of stability.

Recent observations of heavy element abundance distributions in old metal-poor galactic halo stars revealed an abundance distribution that is nearly identical with the one observed for the solar system [13] except that the absolute abundances are significantly lower. The difference in absolute abundance is explained by the fact that the stars originated during a much earlier phase of the universe than the solar system and therefore have a considerably shorter nucleosynthesis history.

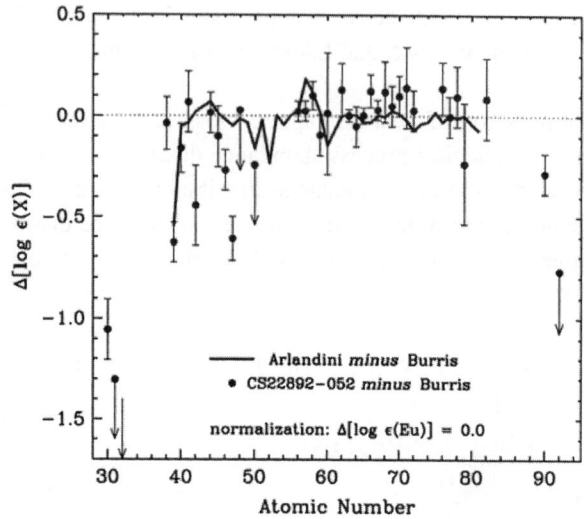

Figure 3. Comparison of the normalized solar abundance distribution of r-process elements with the observed abundance distribution in the metal poor halo star CS22892-052 (courtesy of C. Sneden).

Figure 3 shows the comparison between the normalized abundance distribution of the metal-poor halo star and the predicted r-process abundance curve. Substantial deviations are observed for some of the lower Z elements, namely silver. The agreement in the higher mass range has been interpreted as a clear indication for a unique r-process site [13] but the deviations in the lower Z-range have lead to speculations about the existence of a possible second "weak" r-process site which provide additional nucleosynthesis of low Z isotopes [13]. The possibility of a second weak r-process component is also supported by the analysis of isotope distributions in meteoritic inclusions [14]. While the existence of the "weak" r-process is still a matter of debate, there is also no agreement on the actual site of the "main" r-process. Both, simulations of the neutrino wind driven shock front scenario [5] as well as simulations of r-process nucleosynthesis in merging neutron stars [6] give abundance distributions which are in reasonably good agreement with the observed values. However, the first kind of models requires very high neutron densities (or entropy) to produce the observed r-process abundance distributions. Such neutron densities are difficult to achieve. On the other hand a frequent occurrence of neutron mergers appears too low to account for the overall observed r-process abundances. The observations of r-process abundance distributions in an increasing number of old metal poor halo stars provides a unique source of new observational data to address the question for the actual r-process site [15]. These data represent the nucleosynthesis of a single (or a least only very few) r-process event rather than a multitude of neutron induced nucleosynthesis events over the history of the universe. The analysis of the abundance distribution requires detailed knowledge of the nuclear structure and nuclear reaction conditions along the r-process path [4]. This requires nuclear data such as masses, β-decay properties, and possibly even neutron capture reaction rates for neutron rich r-process nuclei. Particular interesting are also fission probabilities of Z>94 neutron rich isotopes which will determine the "endpoint" of the r-process. Fission and decay properties of these heavy isotopes define the r-process abundances of potential radioactive clocks such as long-lived Thorium [16] and influence the abundance distribution in the A=120-200 range depending on their fission product distributions. Rarely discussed is the need of nuclear data for far of stability neutron rich light nuclei between Z=2 and Z=26 which can be associated with the ignition of the r-process. The actual ignition depends critically on the site and the associated initial abundance distributions. For some cases of the neutrino wind r-process model it has recently been shown that neutron induced nucleosynthesis may well start at much lighter nuclei triggering the r-process by rapid neutron capture processes in the helium to carbon range rather than in the iron range [17].

3. Nuclear Halo Structure and the On-Set of the r-Process

In the neutrino driven wind model nuclei have been dissociated to protons neutrons and alpha particles in the early high temperature and density phase of the shock. With the expansion of the shock the ^{4}He and neutron abundance dominate and three particle reactions such as $^4\text{He}(2\alpha,\gamma)^{12}\text{C}$ and $^4\text{He}(\alpha n,\gamma)^9\text{B}$ bridge the mass 5 and mass 8 gaps and trigger the reassembling of heavy nuclei with Z>6 by alpha capture reactions in statistical equilibrium. This process is called the alpha process [5] and leads to the nearly instantaneous formation of higher Z isotopes until the ^{4}He abundance drops and neutron capture reactions drive the abundance distribution toward the neutron rich side triggering the r-process. However, taking into account neutron capture reactions on light nuclei Z≥6 may introduce an r-process component already at lower masses. This drastically reduces the neutron numbers at an early stage of the shock. To fully understand the consequences for the r-process we need to take fully into account neutron capture reactions at low Z-nuclei. This is particularly interesting in view of the new magic number N=16 emerging on the neutron rich side which may affect the production of ^{23}N, ^{24}O, and ^{25}F. The reaction rates itself are highly uncertain, shell model estimates have been made [18] but they don't include possible enhancement effects due to the pronounced halo structure of near drip line neutron rich nuclei. It has been pointed out before that such a halo structure as observed in ^{6}He, ^{8}He, and ^{11}Li may open new reaction links bridging the mass 5 and mass 8 gaps [19,20] first attempts to include these processes in the simulations did not show significant effects [17]. Recent experiments to explore the ^{8}He, and ^{11}Li halo structure [21], however, indicated that the actual rates may be significantly higher than previously estimated. At the present time the results remain still inconclusive, more detailed nuclear structure studies for the very neutron rich light nuclei are required to reliably explore the role and impact of neutron capture on light nuclei for the r-process.

4. Exotica, Pycno-Nuclear Processes

Accreting neutron stars provide a unique laboratory for probing the fate of matter at extreme density conditions. It has been discussed frequently that accreted matter ignites at highly electron degenerate conditions through the triple alpha process and through break-out processes from the hot CNO cycles such as ^{15}O(α,γ) and ^{18}Ne(α,p). These reactions trigger the rp-process which rapidly converts the accreted abundances towards the mass A=100 range [22]. Such events are observed as x-ray bursts [23]. Due to the gravitational potential most of the processed material remains on the surface of the neutron star. Further accretion causes gradual increase in density for the buried material which triggers slow conversion by electron capture processes towards the neutron drip-line and subsequent electron capture induced neutron emission. This gradually converts the initial mass distribution from the A=100 range towards lower Z nuclei along the neutron drip line at increasing density. At densities higher than $\rho=10^9$g/cm^3 the nuclei freeze out in a bcc lattice structure at further increase of density pycno-nuclear reactions can occur since the nuclear wave functions overlap and effectively reduce the defective Coulomb barrier. The pycno-nuclear reaction rates depend exponentially on the density but are nearly independent of temperature since the nuclei are bound in the lattice structure. Figure 4 shows the pycno-nuclear reaction rates calculated for ^{26}O+^{26}O and ^{34}Mg+^{34}Mg as a function of density which were calculated using the formalism by Koonin and Schramm [24]. A strong Z-dependence in the reaction rate explains the difference of ~25 orders of magnitude between the two rates. Towards high densities pycno-nuclear reaction rates become faster than e$^-$ capture processes at high density conditions [25] and even faster than neutron decay of neutron unbound nuclei such as ^{26}O as indicated in the figure.

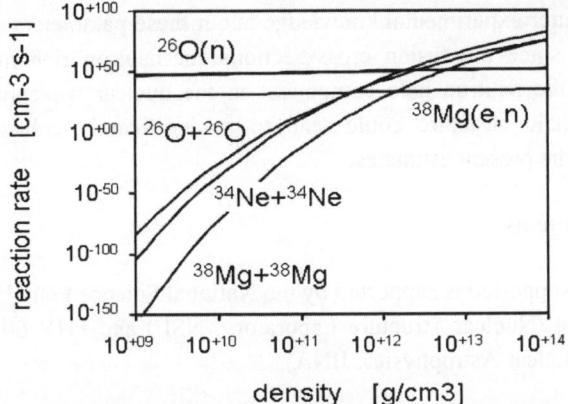

Figure 4. Pycno-nuclear fusion reaction rates for oxygen and magnesium isotopes at the neutron drip line. The rates for $^{26}O+^{26}O$, $^{34}Ne+^{34}Ne$ and $^{38}Mg+^{38}Mg$ are shown in comparison with the estimated rates for neutron decay of neutron unbound ^{26}O and β^- decay plus e$^-$ induced neutron emission from ^{34}Ne and ^{38}Mg.

This indicates that at densities above $\rho=10^{11}g/cm^3$ pycno-nuclear reactions play a dominant role in neutron star crust nucleosynthesis. Depending on the characteristics of heat transport mechanism and direction in the neutron star crust, the energy release may lead to new signatures in the gravitational wave or x-ray emission spectrum of the accreting neutron star. Since the pycno-nuclear reaction rates depend directly on the astrophysical S-factor of the fusion

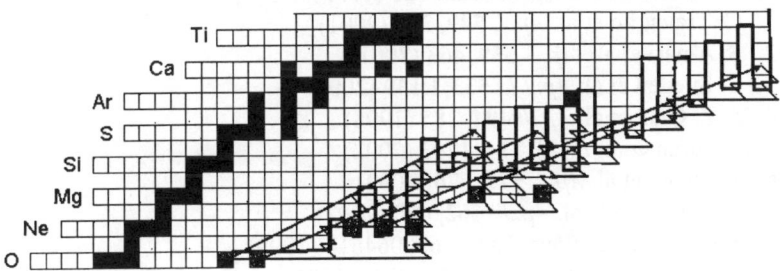

Figure 5. Electron capture induced neutron emission and pycno-nuclear reaction processes in the neutron star crust at densities above $10^{11}g/cm^3$.

Figure 5 demonstrates the pycno-nuclear reaction induced processing path emerging at density conditions above $10^{11}g/cm^3$. Since pycno-nuclear reactions generate an appreciable amount of energy [8], such a reaction path may lead to internal heat release in the deeper layers of the neutron star crust. Depending

process [24], better experimental knowledge about these parameters are necessary in particular since the fusion cross sections for neutron rich nuclei depend critically on the neutron halo component in the nuclear wave function [26]. A pronounced halo structure could lead to a significantly enhanced S-factor compared to the present estimates.

Acknowledgments

This work is supported is supported by the National Science Foundation through PHY 0140324 (Nuclear Structure Laboratory, NSL) and PHY 0072711 (Joint Institute of Nuclear Astrophysics, JINA).

References

1. E. Anders, N. Grevesse, *Geochim. Cosmochim. Acta* **53**, 197 (1989)
2. F. Käppeler et al. *Rep.Prop.Phys.* **52**, 945 (1989)
3. M. Busso et al. *Annu. Rev. Astron. Astophys.* **37**, 239 (1999)
4. K.L. Kratz et al. *Hyperf. Int.* **129**, 185 (2000)
5. S.E. Woosley, R.D. Hoffmann *Ap.J.* **395**, 202 (1992)
6. C. Freiburghaus et al. *Ap.J.* **516**, 381 (1999)
7. E.E. Salpeter, H.M. van Horn *Ap.J.* **155**, 183 (1969)
8. P. Haensel, J.L. Zdunik *Astron.&Astrophys.* **227**, 431 (1990)
9. F. Käppeler et al. *Ap.J.* **437**, 396 (1994)
10. C. Arlandini et al., *Ap.J.* **525**, 886 (1999)
11. C. Travaglio et al. *Ap.J.* **549**, 346 (2001)
12. S. Amari et al., *Ap.J.* **559**, 463 (2001)
13. C. Sneden et al. *Ap.J.L.* **533**, L139 (2000)
14. Y.-Z. Qian et al. *Ap.J.* **494**, 285 (1998)
15. J. Truran et al. *PASP* **114**, 1293 (2002)
16. H. Schatz et al. *Ap.J.* **579**, 626 (2002)
17. M. Terasawa et al. *Ap.J.* **562**, 470 (2001)
18. H. Herndl et al. *Phys. Rev. C* **60**, 064614 (1999)
19. J. Görres et al. *Phys. Rev. C* **52**, 2231 (1995)
20. V. Efros et al. *Z.Phys. A* **355**, 101 (1996)
21. T. Aumann et al. *Phys. Rev. C* **59**, 1252 (1999)
22. H. Schatz et al. *Phys. Rev. Lett.* **86**, 3471 (2001)
23. W.H.G. Lewin et al. *Space Sci.Rev.* **62**, 223 (1993)
24. S. Schramm, S.E. Koonin *Ap.J.* **356**, 296 (1990)
25. J. Bahcall, R.A. Wolf *Ap.J.* **139**, 622 (1964)
26. E. F. Aguilera et al. *Phys. Rev. Lett.* **84**, 5058 (2000)

BETA-DECAY STUDIES OF R-PROCESS NUCLIDES IN THE ^{132}SN REGION

A. WÖHR[1],[*] W.B. WALTERS[1], K.-L. KRATZ[2], O. ARNDT[2], B.A. BROWN[3],
I. DILLMANN[2], D. FEDOROV[4], V. FEDOSEYEV[5], L. FRAILE[5], P. HOFF[6],
U. KÖSTER[5], A.N. OSTROWSKI[2], B. PFEIFFER[2], H.L. RAVN[5],
M.D. SELIVERSTOV[7], D. SEWERYNIAK[8], J. SHERGUR[1],[8] B. TRUETT[8]
AND THE ISOLDE COLLABORATION [5]

[1] Department of Chemistry, University of Maryland, College Park, USA
[2] Institut für Kernchemie, Universität Mainz, Mainz, Germany
[3] NSCL, Michigan State University, East Lansing, USA
[4] CERN, Geneva, Switzerland
[5] Petersburg Nuclear Physics Institute, Gatchina, Russia
[6] Department of Chemistry, University of Oslo, Norway
[7] Institute of Spectroscopy, Russian Academy of Science, Troitzk, Russia
[8] Physics Division, Argonne National Laboratory, Argonne, USA

Very neutron rich Ag, Cd, In and Sn isotopes lying in or near the r-process path have been studied at CERN/ISOLDE using the highest achievable isotopic selectivity. These methods include the use of a resonance-ionization-laser-ion-soure,a two step neutron-converter target in combination with a high-resolution mass separator. During these studies a number of nuclear-physics surprises have been discovered. They andtheir astrophysical consequences on the A$\simeq$130 solar r-process abundance peak are discussed.

1. Introduction

One of the most important challenges for the study of the structure and decay of neutron-rich nuclei has been the investigation of those that lie in and near the path of r-process nucleosynthesis. Once the connection between the closed N=82 neutron shell and the abundance peak at A=130 had been established [1,2] it was possible to infer that the path of rapid neutron capture must pass through nuclides like ^{129}Ag and ^{130}Cd that lie 20 and 14 neutrons away from β-stability. Consequently, a high priority has been conferred on the study of nuclides that lie that distance from stability in the whole A=130 region.

---

*present address: department of physics, university of notre dame, notre dame, usa

The understanding of the synthesis of the heaviest elements in the Universe requires the knowledgeof nuclear-structure and decay properties of the order of 6000 nuclides from β-stability out to the neutron drip line. Clearly, most of these values must be obtained by various theoretical methods. On the other hand, the reliability of the results can be greatly improved by the use of experimental data where available. Moreover, the models can be improved through testing against new experimental data. The recent observation of elemental abundances in metal-poor halo stars that are quite similar to the solar r-process abundance pattern for elements with $Z>56$ (see Cowan, this conference) has placed even greater emphasis on the nuclear properties of nuclides in the $A=130$ "waiting-point" region.

2. Development of Experimental Methods

The study of r-process isotopes around $N=82$ began over a decade ago with the discovery of ^{130}Cd by Kratz et al. in 1986 [4] at the old CERN/ISOLDE facility situated at the Synchrocylotron. At that time, the β-decay half-life $(T_{1/2})$ was measured with great experimental difficulties via multiscaling of so-called β-delayed neutrons to be (195 ± 35) ms. A major step forward had thereafter been the move of ISOLDE to the PS-Booster, with a pulsed proton beam of 1 GeV. With this move, higher yields of very neutron-rich, short-lived isotopes could be achieved with the pulsed proton beam. Simulaneously with the move of ISOLDE, the development of resonance-ionization laser ion sources (RILS) was going on. For each of the elements of interest an efficient ionization scheme had to be developed and the ion source parameters had to be studied carefully. However, a disadvantage of the RILIS system is, that elements, such as In and Cs – with a low ionization potential – are still surface ionized. Therefore, they are the major source of isobaric contamination. Another significant improvement in order to obtain the new results has been the implementation of a "secondary target" at CERN/ISOLDE – proposed by Nolen et. al. [3] for the Rare Isotope Accelerator project. This so-called neutron converter acts as a kind of "mini-spallation-source". In a two step process, the 1 GeV proton-beam of the PS-Booster is impinging on a Ta or W rod in close proximity of the ^{238}U target. The fast neutrons produced by the proton beam then induce fission of uranium. The big advantage of this method is, that no or only very little of the proton-rich spallation products are produced in the target (for details. see e.g. [5]). In addition, the experiments had been carried out at the high-resolution separator (HRS) at a mass resolution of $\frac{M}{\Delta M} = 4300$. Additional selectivity can be achieved using β-delayed neutron counting and multi-parameter γ-coincidence measurements. Without the combination of all of these methodsthe required isotopic selectivity would not have been possible.

3. Experiments on Neutron-rich Ag Isotopes

At the Sanibel conference in 1997 [6], we reported on some improvements of the experimental methods, in particular the hyperfine splitting of the Ag isotopes which finally made the "waiting-point" nucleus ^{129}Ag accessible. Recently, the $T_{1/2}$ of the N=83 isotope ^{130}Ag was measured to be (35 ± 10) ms [7] for the first time. In addition, the quality of the γ-spectroscopic data of ^{124}Ag to ^{128}Ag [8] has improved considerably. These data are presently being analyzed. At this point, we have to note, that one of the first nuclear-structure surprises was the level systematics of the heay even-even Cd isotopes populated in the decay of odd-odd Ag nuclides, where it was found, that the 645-keV 2^+ energy of ^{128}Cd is actually 7 kev lower than the 2^+ energy of ^{126}Cd. Such an observation is counter to most notions of the evolution of collectivity which is usually an increase (or decrease) in 2^+ energy as nucleons are added to or removed from a closed shell. Therefore, we come to the conclusion, that we have observed the first weakening of the N=82 neutron shell closure as Z drops below 50 [8].

4. Experiments on Neutron-rich Cd Isotopes

After the development of a RILIS for Ag, the next development was done for Cd. Again there were surprises to be discovered. With the Cd-RILIS, the $T_{1/2}$ of ^{130}Cd was remeasured with higher precision to 162(7) ms and the P_n value to 3.6(10)%. Furthermore, the measurments were extended to both ^{131}Cd and ^{132}Cd, where $T_{1/2}$ (and P_n values) of 68(3) ms (3.6(10)%) and 97(1) ms (60(15)%) were determined, respectively [11]. The surprise was the short $T_{1/2}$ and the low P_n value for ^{131}Cd to levels of the N=82 isotone ^{131}In. This indicates, that the levels populated via Gamow-Teller decay accross the N=82 closed shell must lie at lower energies than predicted by some shell model calculations. Very recently, the measurements could be extended to ^{133}Cd with a half-life of 57(10) ms [7].

Recently, it has been possible to use the Cd-RILIS to study the β- and γ-decay of ^{130}Cd to levels of ^{130}In [14,5,10]. This particular nuclide is of considerable importance to r-process nucleosynthesis as its half-life of 162 ms provides the ultimate barrier to the matter flow in the A$\simeq$130 region. Moreover, a full understanding of the structure and decay of ^{130}Cd is essential to the calculation of the lower-Z N=82 nuclides that also play a role in the buildup of the rising wing of the A$\simeq$130 $N_{r,\odot}$ peak.

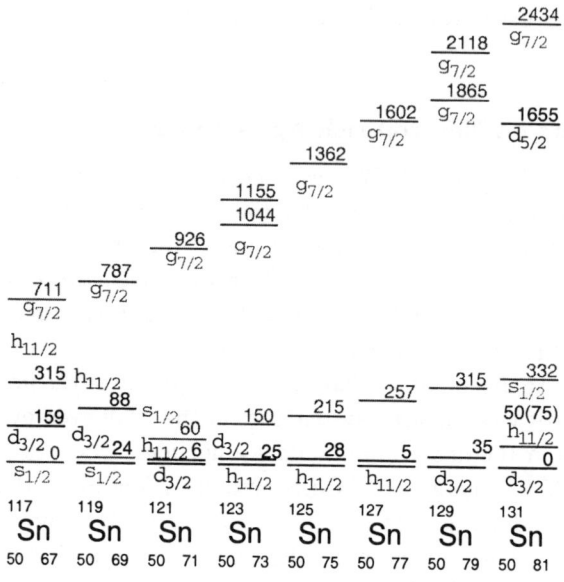

$g_{7/2}$ to $d_{3/2}$ versus 3^+ to 1^+ difference
= 1100 1120 914 ~800 700

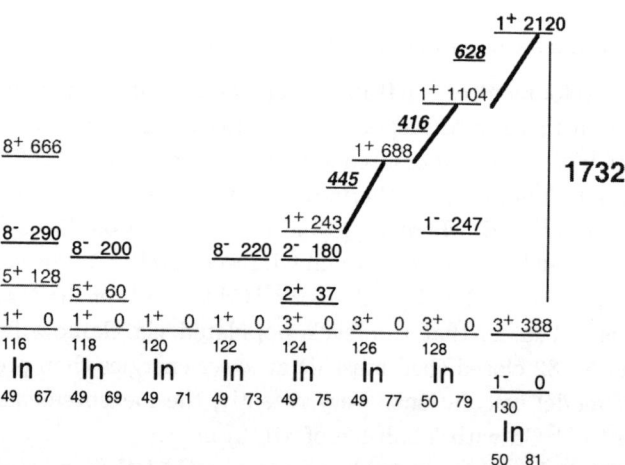

Figure 1. Evolution of nuclear structure upon approaching the N=82 closed shell for the odd-mass Sn nuclides and the odd-odd In nuclides. The structure of ^{130}In is shown graphically relative to the 3^+ levels in the adjacent lighter In nuclides which are the ground states.

In order to observe γ-rays from of ^{130}Cd decay in a region where large backgrounds of surface-ionized Cs and In isobars are present, all of the improvements mentioned in section 2. had been incorporated. Hence, the ^{130}Cd was produced via neutron induced fission, but not by spallation. The use of the neutron converter target thus resulted in a considerable lowering of the ^{130}In contaminant and an almost complete suppression of spallation produced ^{130}Cs.

For ^{130}Cd, it has been possible to determine the 2120 keV energy for the 1^+ level to which the Gamow-Teller β-decay goes. Using $\beta\gamma$-coincidences, we have measured the end-point for β-population of this level which allowed to determine the Q_β value for ^{130}Cd decay. That Q value appears to be considerably higherthan predicted by the mass formulae of Möller et al.[19], Aboussir et al.[21] or Duflo and Zuker[20], but rather lies in the high-energy range of recent "quenched" mass models (HFB/SKP, HFB-2) [22] values listed in the Atomic Mass Evaluation of Audi and Wapstra[15]. The 8 major gamma rays found to follow the β-decay of ^{130}Cd include a γ-ray at 388 keV, reported by Hellström et al.[12], to stem from a microsecond isomer. Scince it seems, that this level is populated in the decay of the 1^+ level at 2120 keV by the 1732 keV γ-ray, we can assume that it is likely to have spin and parity of 3^+.

The relatively (and unanticipatedly) high energy of the 2120 keV 1^+ level is a consequence of two unexpectedly large position shifts in going from ^{128}In via ^{130}In. These levels are shown in Figure 1 where the evolution of the positions of the 1^+ levels in the odd-odd In nuclides are shown as N increases toward the closed shell at N=82. The position of the 1^+ level is seen to increase by 445 and 416 keV in going from ^{124}In via ^{126}In to ^{128}In, but then jumps by 628 keV to ^{130}In. Added to that larger than expected shift is the dramatic inversion of the 3^+ and 1^- levels which lifts the 1+ level to a position 2120 keV above the ground state. Earlier shell-model calculations had shown a position for the 1^+ level at about 1500 keV and near degeneracy for the 1^- and 3^+ levels.

5. Experiments on Neutron-rich In Isotopes

The combination of the RILIS with the neutron converter allowed to further extend the information on extreme neutron rich Indium isotopes to ^{135}In with 86 neutrons[9]. A $T_{1/2}$ of 92(10) ms has been determined and a very large P_n value close to 100% has been estimated. With respect to the r-process, the ^{135}In isotope is a major "waiting point" beyond the N=82 closed neutron shell at neutron densities in the range of $n_n \simeq 10^{25} n/cm^3$.

6. Experiments on Neutron-rich Sn Isotopes

Another focal point of our investigations has been the study of neutron-rich Sn isotopes beyond the N=82 closed neutron shell. So far, it has been possible to identify and measure the β-decay half-lives via delayed neutron emission of ^{135}Sn to ^{138}Sn. The study of ^{138}Sn has been possible in spite of a production rate of only two nuclides per PSB pulse, as compared with 100 per pulse from ^{137}Sn and 3000 per pulse for for ^{136}Sn. With respect to the r-process, ^{136}Sn is a "waiting-point" nucleus for moderate neutron desities of the order of $10^{24} n/cm^3$ needed to drive the r-process the A$\simeq$130 $N_{r,\odot}$ peak.

The important nuclear physics surprise for the study of ^{135}Sn decay has been the discovery of the low 282 keV energy for the first excited state of ^{135}Sb, thought to have a significant $\pi d_{5/2}$ configuration. This surprise arises as the same state lies at 963 keV in ^{133}Sb at the N = 82 closed shell. This dramatic drop in energy has been attributed to neutron skin effects which favor the binding of the lower-l $d_{5/2}$ state relative to the higher-l $g_{7/2}$ proton state. Once this was observed for the Sb nuclides with a single proton beyond the Z=50 closed shell, it was noted that a similar effect is observed in the Bi nuclides with a single proton beyond the Z=82 closed shell. There, the lower-l $f_{7/2}$ level is found to be steadily lowered by the addition of neutron pairs beyond N=126 relative to the higher-l $h_{9/2}$ level. For a detailed description of the results, see ref. [13].

7. Conclusions and Outlook

7.1. *Astrophysical Implications*

As a consequence of the renormalization of the $\pi\nu$ monopole interaction, for ^{130}Cd decay, the half-lives of the so far unknown N=82 "waiting-point" nuclei ^{128}Pd to ^{122}Zr will become longer than those predicted by recent shell models (see, e.g. [16]). This leads to the build-up of the A$\simeq$130 $N_{r,\odot}$ peak. s can be seen from figure 2, each N=82 r-process isotope acts as a "waiting-point" for different neutron-density ranges. In consequence, the classical $N_{r,prog} \times \lambda_\beta \simeq const.$ concept [1,4] has to be modified, including the effect of the different neutron separation energies on the r-abundances when the r-process "climbs up" the N=82 ladder from Z$\simeq$40 to Z$\simeq$50 [18].

In r-proccess abundance calculations, resulting in a better reproduction of the rising wing of this peak compared to calculations with the previous "short" shell-model half-lives. The direct consequence for astrophysics is a better understanding of the r-process matter flow through this "bottle-neck" region, which – to a large extent – also determines the total duration

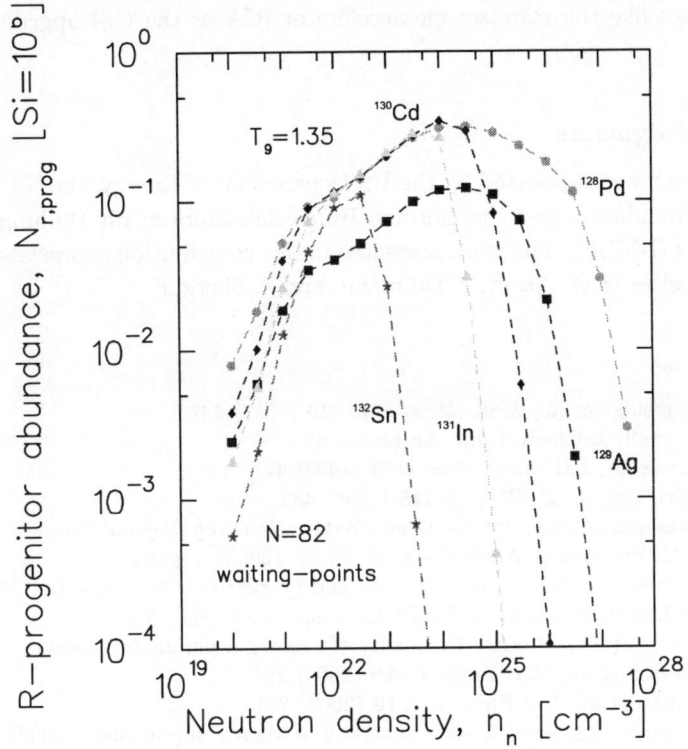

Figure 2. Calculated r-process abundances of the N=82 "waiting-point" isotopes ^{128}Pd to ^{132}Sn as a function of neutron density.

of a *classical* r-process[17].

7.2. Outlook

Presently, we are approaching or have already reached the technical limits (2 ^{138}Sn nuclei per PSB pulse) of studying very neutron-rich nuclei around N≃82 with the currently available ISOL techniques. However, there are still improvements that can be sought to obtain faster diffusion in the target as well as to lower isobaric backgrounds from surface isonization. In the near future the remaining r-process "waiting-points" may be easier reached with fragmentation processes and their cocktail beams (see Schatz, this conference) at fragment separators such as the A1900 at MSU. For further improvements and the accessibility of the N≃126 waiting-point nuclei devices like the rare isotope accelerator RIA or the GSI-upgrade are needed.

Acknowledgments

This research was supported by the US Department of Energy, the National Science Foundation and the German Bundesministerium für Bildung und Forschung (BMBF). The work presented in this contribution comprises part of the thesises of O. Arndt, I. Dillmann, and J. Shergur.

References

1. E.M. Burbidge et al., *Revs. Mod. Phys.* **29** (1957) 547
2. C.D. Coryell, J. Chem. Educ. **38** (1961) 67
3. J. Nolen et al., *AIP Conf. Proc.* **473** (1998) 477
4. K.-L. Kratz et al., *Z. Phys.* **A325** (1986) 489
5. I. Dillmann et al. Proc. Int. Conf. on *Capture Gammay Ray and Related Topics – CGS11*, Pruhonice, World Scientific Press, (2003) in print
6. K.-L. Kratz et al., Int. Conf. on Fission and Properties of Neutron-Rich Nuclei, Sanibel Island, World Scientific Press, Singapore (1998), 586
7. O. Arndt, Diploma Thesis, University of Mainz, unplublisched (2003)
8. T. Kautzsch et al., Eur. Phys. J. **A9** (2000) 201
9. I. Dillmann et al., Eur.Phys. J. **A13** (2002) 281
10. I. Dillmann, Diploma Thesis, Univeristy of Mainz, unpublished (2002)
11. M. Hannawald et al., Phys. Rev. **C62** (2000) 054301
12. M. Hellström et al., Proc. Int. Workshop XXXI on *Gross Properties of Nuclei and Nuclear Exctiations*, Hirschegg, in print
13. J. Shergur et al., Phys. Rev. **C65** (2002) 034313
14. A. Wöhr, et al., *Proc. 11th Workshop on "Nuclear Astrophysics"* MPA/P13 (2002)
15. G. Audi and A.H. Wapstra, *Nucl. Phys.* **A624** (1997) 1
16. G. Martinez-Pinedo and K. Langanke, *Phys. Rev. Lett.* **83** (1999) 4502; and priv. comm. (2001)
17. P. Möller, B. Pfeiffer and K.-L. Kratz, *Phys. Rev C*, in print
18. K.-L. Kratz, Nucl.Phys. **A688** (2001) 308
19. P. Möller et al., ADNT **59** (1995) 1983
20. J. Duflo and A.P. Zuker, Phys. Rev. **C52** (1995) R23
21. Y. Aboussir et al., ADNT **61** (1995) 127
22. M. Samyn et al.. Nucl. Phys. **A700** (2001) 142

WEAK R-PROCESS EXPERIMENTS WITH FAST RADIOACTIVE BEAMS

H. SCHATZ *, P. HOSMER*, R.R.C. CLEMENT*, T. ELLIOT,
A. ESTRADE, M. GOUPELL, P.F. MANTICA † F. MONTES*,
A.C. MORTON, M. OUELLETTE*, P. SANTI, M. STEINER, A. STOLZ

National Superconducting Cyclotron Laboratory,
Michigan State University,
South Shaw Lane
East Lansing, MI 48824

A. WÖHR ‡AND W.B. WALTERS

Department of Chemistry and Biochemistry,
University of Maryland,
College Park, MD 20742

K.-L. KRATZ AND B. PFEIFFER

Institut für Kernchemie
Universität Mainz, F. Strassmann Weg 2
D-55128 Mainz, Germany

P. REEDER

Pacific Northwest National Laboratories
Nuclear Chemistry
P.O. Box 999
902 Battelle Boulevard
Richland, WA 99352

*and Department of Physics and Astronomy, Michigan State University
†and Department of Chemistry, Michigan State University
‡and Physics Division, Argonne National Laboratory

A particularly difficult problem in nuclear astrophysics is the origin of the light r-process elements with $A < 130$ synthesized in the weak r-process. Over the last years new observations have begun to shed some light on the problem. At the same time, the new Coupled Cyclotron Facility at MSU now brings the majority of the nuclei in the path of the weak r-process within reach. To that extent we have begun an experimental program to measure β-decay half-lives and probabilities for β-delayed neutron-emission of neutron rich nuclei in the path of the weak r-process. We show first, preliminary data from measurements near ^{78}Ni.

1. Introduction

The r-process is one of the major nucleosynthesis processes in the universe producing roughly half of all elements heavier than iron. One of the biggest problems in nuclear astrophysics remains the question of the site of the r-process. The proposed scenarios include (i) the neutrino-driven wind in core-collapse supernovae [1,2], (ii) accretion onto and jets from a forming neutron star in core collapse supernovae [3,4], and (iii) neutron star mergers [5,6]. Currently none of the proposed models can synthesize all r-process nuclides in the right proportions, especially in the case of the lighter $A < 130$ nuclei [7].

Recently, observations of r-process elements in ultra metal poor (UMP) halo stars have shed new light on the operation of the r-process during the early history of our Galaxy and on the question of the r-process site (see Truran et al. [8] for a recent review). These UMP halo stars formed very early in galactic history before large scale mixing of elements from various sources occurred. They therefore preserved the specific composition of the interstellar medium at the location of their formation. Some of them formed from material locally polluted by an r-process event, hence the enhancement in r-process elements of up to ≈ 50 times compared to solar found in some of the UMP halo stars. These observations provide a unique glimpse at abundance distributions synthesized by individual (or very few) r-process events unlike the solar system abundance distribution, which represents a mixture of contributions from a large number of events. Abundances of up to 28 r-process elements ranging from strontium to uranium have now been determined in some UMP stars providing accurate information on patterns of abundances. The observation of long lived radioactive nuclei such as thorium and, more recently, uranium [9,10] provides important constraints on the r-process beyond lead. In addition uranium and thorium observations can be used to date the r-process events and to obtain stringent lower limits for the age of the universe [11,12,13].

One of the important conclusions from these observations is the possibility of two distinct r-process components. The so-called strong component seems to produce mostly the heavier mass r-elements with $A > 130$. The abundance pattern produced by the strong componenet resembles the solar system abundance distribution of r-elements and shows no event to event variations. The weak component seems to be restricted to the production of several lighter r-nuclei with $A < 130$. The existence of a weak and a strong r-process has also been proposed based on meteoritic data pointing to different chemical evolution histories of the radioactive r-process isotopes ^{129}I and ^{182}Hf that were present at the time of solar system formation [14].

To summarize, in many aspects the origin of the light ($A < 130$) r-process elements seems to be a particularly complex and difficult problem. First of all, none of the proposed r-process models mentioned above can reproduce the correct light r-process element distribution. Second, unlike for the heavy r-process elements, the light r-process element abundances show a considerable star to star scatter in UMP stars. Third, there might be an additional, unknown weak r-process site contributing to the light r-process abundances. One possible scenario for the weak r-process is explosive helium burning in type II supernovae when the shock wave passes through the outer layers of the exploding star [15]. However, it remains doubtful whether sufficiently high neutron densities can be achieved in such a scenario. More recently, Meyer [16] pointed out that a new type of r-process environments with extreme expansion timescales and entropies but, quite surprisingly, not necessarily with neutron excess, would show the extreme sensitivities to model parameters needed to explain the observed variability in the weak r-process abundances. However, no realistic scenario that would provide such environments has been identified so far.

To address these complex open questions there is clearly a need for more data. Most importantly one needs: (i) more observations of light r-process element abundances in metal poor stars, and (ii) much more information on the nuclear physics underlying the weak r-process. Only with reliable nuclear physics it will be possible to disentangle the contributions from different r-processes and to obtain stringent constraints on the conditions required for the various potential sites. Fig. 1 shows an r-process path calculated with the site independent classical r-process model. The data needed for r-process calculations of the weak r-process are β decay properties, masses, and to some extent level structure and neutron capture rates. Especially important are the neutron rich r-process nuclei in the vicinity of the $N = 50$ and $N = 82$ shell closures. The shell closures serve as bottlenecks that determine the timescale of the process and therefore provide the most stringent constraints for r-process conditions. In addition, it is in the vicinity of the shell structure related r-process abundance peaks where one might find site specific signatures from freezeout timescales [7] or neutrino post processing [17]. At the same time, nuclear properties are particularly uncertain around shell closures owing to the unknown shell structure in extremely neutron rich nuclei. It has been suggested that shell gaps for nuclei in the r-process path might be significantly reduced compared to the ones in stable nuclei [18] with significant consequences for r-process model calculations [19].

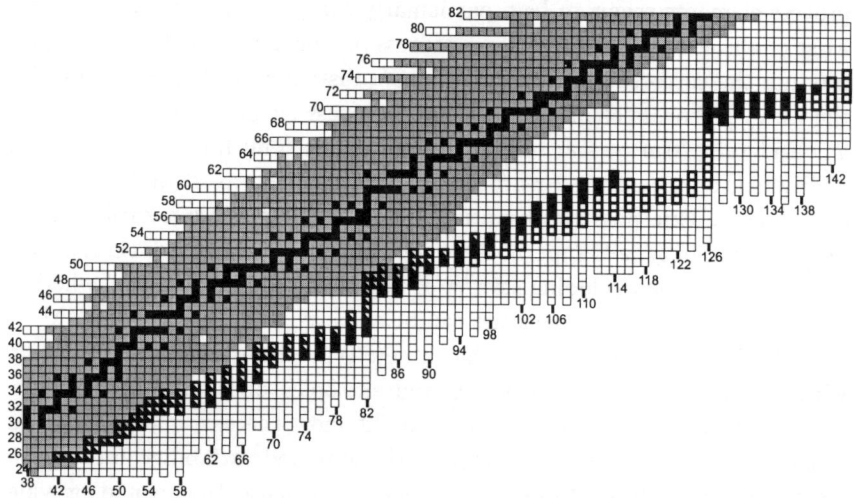

Figure 1. The classical r-process path marked with thick squares (from Schatz et al. 2002 [11]). Half filled and fully filled squares denote nuclides in the r-process path that are within reach for at least a half-life measurement at the new NSCL Coupled Cyclotron Facility and the proposed RIA facility, respectively. Shaded in grey are nuclides with experimentally known half-lives.

The nuclei in the r-process path are extremely neutron rich and short lived. Fig. 1 shows, that to a large extent the r-process path lies beyond the region of nuclei for which experimental data are available. Exceptions are some nuclei around the $N = 50$ and $N = 82$ shell closures, mostly owing to pioneering experiments at ISOLDE (see Pfeiffer et al. 2001 [20] and Kratz et al. 2000 [21] for recent reviews). Recent examples include the spectroscopy of $^{135-137}$Sn [22] and measurements of decay properties of

^{135}In [23] and ^{131,132}Cd [24]. Other experiments with r-process nuclei include measurements in the Ti-Co region at GANIL using projectile fragmentation [25], and measurements on ^{135}Sn at OSIRIS using fast neutron induced fission [26]. Nevertheless, most nuclei in the weak r-process including particularly important isotopes such as the waiting point ^{78}Ni and nuclei in the vicinity of the $N = 82$ shell closure for $Z < 47$ have been out of reach so far. To address this problem we have begun to develop an experimental program at the new Coupled Cyclotron Facility at MSU focusing specifically on the weak r-process.

Our goal is to measure β decay half-lives and branchings for β delayed neutron emission of r-process nuclei between Fe and Sn. Both quantities are critical input parameters in r-process models. In particular, β-delayed neutron emission can mimic abundance characteristics caused by neutrino induced neutron emission [17]. Data on β delayed neutron emission are therefore needed to extract neutrino postprocessing signatures from observed abundances.

2. First Experiment and Preliminary Data

In our first experiment, r-process nuclei in the vicinity of ^{78}Ni were produced via fragmentation of a 140 MeV/nucleon ^{86}Kr beam with an average intensity of 5 pnA impinging on a 376 mg/cm^2 beryllium target. The r-process nuclei were separated using the A1900 fragment separator. Nuclei in the radioactive beam were identified individually by measuring their momentum, charge, and velocity. Momenta were measured by tracking particles at the dispersive intermediate image of the A1900, charge numbers were obtained from energy loss measurements in a Si PIN diode, and velocities were determined from the time-of-flight between two plastic scintillators. The identified nuclei were transported to the NSCL β counting system and implanted into a highly segmented (40x40 strips) double-sided silicon strip detector measuring time and location of the implantation. The same detector measured time and location of subsequent β decays of implanted nuclei. Implantations and decays were measured continuously at total radioactive beam intensities of ≈ 10 particles per second. A more detailed description of the NSCL β counting system and first results on nuclei not related to the r-process can be found in Morton et al. 2002 [27].

Neutrons emitted during β-decay were detected in the new neutron long-counter NERO surrounding the β-endstation. NERO consists of a large volume of polyethylene to moderate emitted neutrons before they are detected in embedded ^{3}He and BF$_3$ gas counters. NERO provides a particularly large inner volume for the beam stop detector systems needed for this type of experiments. At the same time, it has been designed to achieve a neutron efficiency of $\approx$ 45% for low energy neutrons with only a 30% drop at 2 MeV. Experimental efficiency data are currently being analyzed.

β-decay half-lives can be determined from the measured time differences between implantation and decay for each species. Branchings for β-delayed neutron emission are derived by comparing the number of measured β-decays with and without the detection of a neutron. This experimental method based on fast fragment beams has several strengths: (i) because of the high selectivity provided by individual particle identification, experiments can be done with very low beam intensities down to one particle per hour, or even one particle per day for just a half-life measurement. This pushes the experimental reach towards more exotic nuclei and into the path of the r-process. (ii) Measurements can be performed for several nuclides simultaneously using a mixed beam. (iii) There are no limitations imposed by the very short half-lives of exotic nuclei (down to milliseconds) as the time between production and detection of the radioactive nuclei is very short (less than a microsecond).

Fig. 2 shows preliminary results for the particles identified in a 24 h period of beam time. All events shown correspond to successful implantations into the detector. Also shown is the border to the path of the r-process. Clearly, already we have reached the path of the (classical) r-process. The data analysis to extract β-decay half-lives and branchings for β-delayed neutron emission is underway.

3. Conclusions and Outlook

Experimental data on nuclei in the weak r-process path are of great importance to address the many open questions related to the origin of neutron rich heavy nuclei found in the solar system and in stars. With the advent of a new generation of radioactive beam facilities many of the critical nuclei have now come within reach. Fig. 1 shows the estimated reach at the new NSCL Coupled Cyclotron Facility for at least a half-life measurement. Most of the nuclei in the weak r-process up to $A < 130$ will be accessible. As we have demonstrated, the facility and equipment to do the necessary experiments is now in place and first measurements have begun.

As far as the strong r-process is concerned, some of the heavier r-process

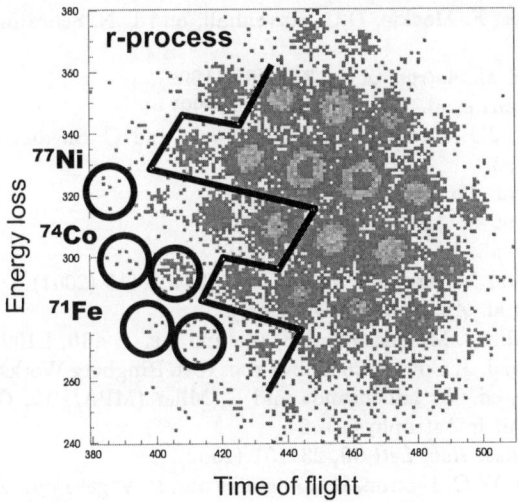

Figure 2. Preliminary particle identification in the energy loss (measure of charge number) and time-of-flight (measure of mass/charge) diagram. Each peak denotes the events corresponding to a specific nuclide. The classical r-process path runs to the left of the solid line. Clearly, data for nuclei in the r-process path can be obtained.

nuclei in the Sn-Xe mass region can be accessed at current facilities, for example with further developments at ISOLDE or using the in-flight fission of uranium at 1 GeV/nucleon at GSI [28]. However, to reach a significant number of the heavier nuclei in the strong r-process path, in particular nuclei in the vicinity of the important $N = 126$ shell closure, higher beam intensities and energies will be required. With the proposed Rare Isotope Accelerator (RIA) even those nuclei might be within reach in the not too distant future (see Fig. 1).

Acknowledgments

This research has been supported by the National Science Foundation under grants PHY 0110253 (NSCL) and PHY 0072636 (Joint institute of Nuclear Astrophysics). H S. is an Alfred P. Sloan Fellow.

References

1. S.E. Woosley and R.D. Hoffman *Ap. J.* **395**, 202 (1992).
2. K. Takahashi, J. Witti and H.-Th. Janka, *A&A* **286**, 857 (1994).
3. J.M. LeBlanc and J.R. Wilson *Ap. J.* **161**, 541 (1970).
4. A.G.W. Cameron, *Ap. J.* **562**, 456 (2001).

352

5. J.M. Lattimer, F. Mackie, D.G. Ravenhall, and D.N. Schramm, *Ap. J.*, **213**, 225 (1977).
6. S. Rosswog et al. *Astron. Astr.* **341**,499 (1999).
7. C. Freiburghaus et al. *Ap. J.* **516**, 381 (1999).
8. J.W. Truran, J.J. Cowan, C.A. Pilachowski and C. Sneden *Publ. Ast. Soc. Pac.* **114**, 1293.
9. R, Cayrel et al. *Nature* **409**, 691 (2001).
10. J.J. Cowan et al. *Ap. J.* **572**, 861 (2002).
11. H. Schatz et al. *Ap. J.* **579**, 626 (2002).
12. S. Goriely and M. Arnould *Astron. Astr* **379**, 1113 (2001).
13. J. Cowan et al. *Ap. J.* **521**, 194 (1999).
14. G.J. Wasserburg, M. Busso and R. Gallino *Ap. J.* **466**, L109 (1996).
15. J.W. Truran, J. J. Cowan, 2000, in Proc. 10th Ringberg Workshop on Nuclear Astrophysics, ed. W. Hillebrandt and E. Mller (MPA/P12; Garching: Max-Planck-Institut fr Astrophysik), 64
16. B. Meyer *Phys. Rev. Lett* **89**, 231101 (2002).
17. Y.-Z. Quian, W.C. Haxton, K. Langanke and P. Vogel *Phys. Rev. C* **55**, 1532 (1997).
18. J. Dobaczewski, I. Hamamoto, W. Nazarewicz and J.A. Sheikh *Phys. Rev. Lett.* **72**, 981 (1994).
19. B. Chen et al. *Phys. Lett. B* **355**, 37 (1995).
20. B. Pfeiffer, K.-L. Kratz, F.-K. Thielemann and W.B. Walters *Nucl. Phys. A* **693**, 282 (2001).
21. K.-L. Kratz et al. *Hyperf. Int.* **129**, 185 (2000).
22. J. Shergur et al. *Phys. Rev. C* **65**, 4313 (2002).
23. I. Dillmann et al. *Eur. Phys. J. A* **13**, 281 (2002).
24. M. Hannawald et al. *Phys. Rev. C* **62**, 4301 (2000).
25. O. Sorlin et al. *Nucl. Phys. A* **669**, 351 (2000).
26. A. Korgul et al. *Phys. Rev. C* **64**, 021302 (2001).
27. A.C. Morton et al. *Phys. Lett. B* **544**, 274 (2002).
28. M. Bernas et al. *Phys. Lett. B* **415**, 111 (1997).

R-PROCESS ABUNDANCE SIGNATURES

JOHN J. COWAN*

Department of Physics and Astronomy,
University of Oklahoma,
Norman, OK 73019, USA
E-mail: cowan@nhn.ou.edu

CHRISTOPHER SNEDEN

Department of Astronomy and McDonald Observatory,
University of Texas,
Austin, TX 78712, USA
E-mail: chris@verdi.as.utexas.edu

Abundance observations indicate the presence of rapid-neutron capture (i.e., r-process) elements in old Galactic halo and globular cluster stars. These observations demonstrate that the earliest generations of stars in the Galaxy, responsible for neutron-capture synthesis and the progenitors of the halo stars, were rapidly evolving. Abundance comparisons among large numbers of stars provide clues about the nature of neutron-capture element synthesis both during the earliest times and throughout the history of the Galaxy. In particular, these comparisons suggest differences in the way the heavier (including Ba and above) and lighter neutron capture elements are synthesized in nature. Understanding these differences will help to identify the astrophysical site (or sites) of and conditions in the r-process. The abundance comparisons also demonstrate a large star-to-star scatter in the neutron-capture/iron ratios at low metallicities- which disappears with increasing [Fe/H]- and suggests an early, chemically unmixed and inhomogeneous Galaxy. The very recent neutron-capture element observations indicate that the early phases of Galactic nucleosynthesis, and the associated chemical evolution, are quite complex, with the yields from different (progenitor) mass-range stars contributing to different chemical mixes. Stellar abundance comparisons suggest a change from the r-process to the slow neutron capture (*i.e.*, s-) process at higher metallicities (and later times)in the Galaxy. Finally, the detection of thorium and uranium in halo and globular cluster stars offers a promising, independent age-dating technique that can put lower limits on the age of the Galaxy and thus the Universe.

---

*Also at: Department of Astronomy and Mcdonald Observatory, University of Texas, Austin, TX 78712

1. Introduction

The heavy solar system abundances (here, Z > 30) are formed in neutron capture (n-capture) processes, either the slow (s-) or rapid (r-) process. We show in Figure 1 the solar system – also thought of as cosmic – abundances with the neutron-capture elements highlighted. Abundance observations of these elements in halo stars contain vital clues to the nucleosynthesis history and chemical evolution of the Galaxy, and abundances of radioactive elements can also be utilized to obtain age determinations for the oldest stars, which in turn put lower limits on age estimates for the Galaxy and the Universe (e.g., [1,2]).

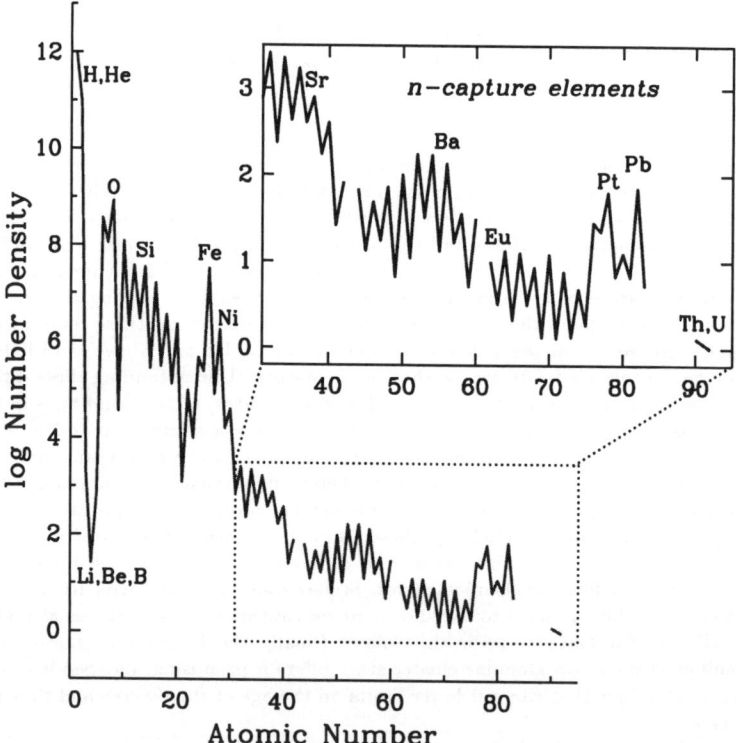

Figure 1. Abundances of elements in the Sun and in undifferentiated solar system material [3]. This abundance set is normalized by convention to log N(H) = 12 in astronomical literature on stars (planetary and meteoritic studies usually normalize the abundances to log N(Si) = 6). The main figure shows the entire set of stable and long-lived radioactive elements, while the inset is restricted to the n-capture elements, defined here as those elements with Z > 30.

2. Neutron-Capture Abundances in Metal-Poor Halo Stars

Extensive abundance studies have been made for metal-poor (*i.e.*, low iron abundance) Galactic halo studies [4–7]. We show in Figure 2 detailed abundances for the heaviest Z > 30 *n*-capture abundances in three stars: CS 22892–052, BD +17°3248 and HD 115444 [6–8]. The abundances in these stars have been compared to a scaled solar system curve, indicated by the solid line.

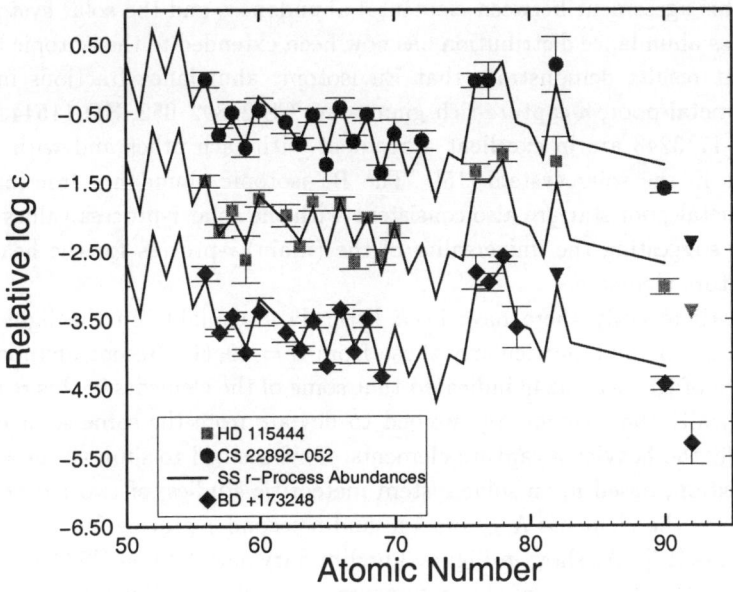

Figure 2. A comparison of observed abundances in the three stars CS 22892-052, HD 115444 and BD +17°3248 with a solar system *r*-process elemental abundances. Upper limits are indicated by inverted triangles.

This solid-line curve was obtained by summing the individual isotopic contributions from the *s*- and the *r*-process in solar system material from *n*-capture cross section measurements [9,10]. These solar elemental *r*-process abundances that were derived are based upon the "classical approach" to the *s*-process, which is empirical and by definition model-independent. Nevertheless, more sophisticated stellar models based upon *s*-process nucleosynthesis in low-mass AGB stars, predict very similar solar abundances [11]. As Figure 2 illustrates there is a remarkable agreement between the solar *r*-process abundances and the abundance patterns of the heaviest *n*-

capture elements in these very old metal-poor halo stars. This agreement implies a robust r-process, one that seems to reproduce the same relative n-capture abundances over many Gyr in the history of the Galaxy. These results further suggest a very narrow range of astrophysical conditions and/or mass ranges for the site(s) for the r-process. It has been suggested, for instance, that only low mass (8–10 $M_\odot$) supernovae may be a likely site for the main r-process and may be responsible for the synthesis of these heavy ($Z \geq 56$) n-capture elements [12–14].

This agreement between elemental abundances and the solar system r-process abundance distribution has now been extended to the isotopic level. Recent results demonstrate that Eu isotopic abundance fractions in the very metal-poor, n-capture-rich giant stars CS 22892–052, HD 115444 and BD +17°3248 are in excellent agreement with each other and with their values in the solar system [15]. The Ba isotopic abundance fractions in one metal-poor star are also consistent with the solar r-process values [16], again suggesting the universality of the (main) r-process for the heaviest n-capture elements.

Until recently there have been little data available on the lighter n-capture elements, particularly those from $Z = 40$–50. Recent abundances studies of BD +17°3248 indicated that some of the elements in this regime, specifically the element Ag, seemed to deviate from the same solar curve that fit the heavier n-capture elements. This seemed to support an earlier suggestion, based upon solar system meteoritic studies, of two r-processes - one for the elements $A \gtrsim 130$-140 and a second r-process for the lighter elements [17]. We show in Figure 3 preliminary new data on CS 22892–052. we note the detection of elements never seen before in this star, including Mo, Lu, Au, Pt and Pb [18]. Also significant new upper limits have been found for Ga, Ge, Cd and Sn. Comparison of the abundances with the solar r-process curve [10] demonstrates the same agreement found previously for the heaviest n-capture elements in this star and other similar stars [6,1,2]. It is clear from Figure 3 that some of the lighter elements between $Z = 40$–50 (e.g., Ag) show significant deviations while others appear to fall near the line. This new result is consistent with earlier studies of this star and with that of BD +17°3248. However, with only a few stars and very limited data available, it is not clear at this point what is the source of the synthesis for these lighter elements. It has been suggested that perhaps, analogously to the s-process, the lighter elements might be synthesized in a "weak" r-process with the heavier elements synthesized in the more robust "strong" or "main" r-process [19]. While a second r-process site, perhaps supernovae of a a different mass range or frequency [20] or perhaps the helium zone of an exploding supernovae [19], might be responsible for the synthesis of nuclei with A ≲ 130–140, there have also been suggestions that the entire abundance distribution could be synthesized in a single core-collapse supernova [6,21].

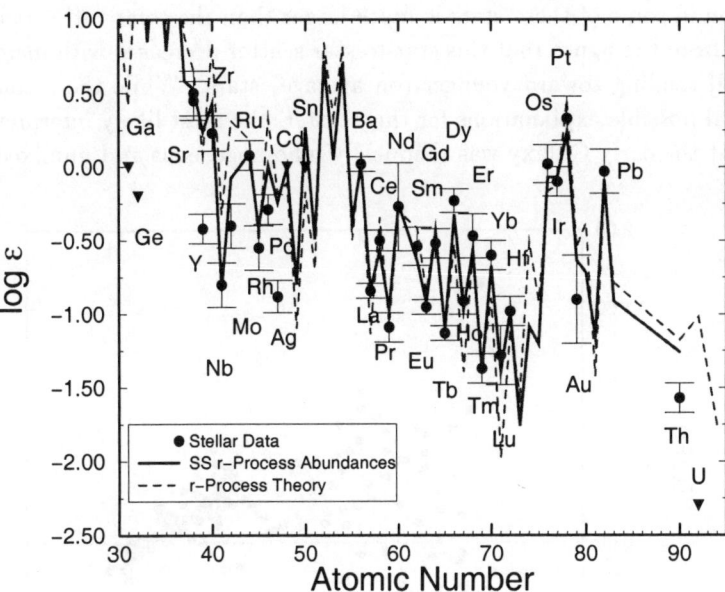

Figure 3. A comparison of new observed abundances from CS 22892-052 with solar system [10] and theoretical [5] r-process elemental abundances.

3. Abundance Scatter and the Chemical Evolution of the Galaxy

A number of studies have demonstrated a dramatic and very large star-to-star scatter in the abundance level of the heavy n-capture elements with respect to the to iron abundances. This star-to-star scatter increases dramatically with decreasing stellar metallicity, as shown in Figure 4, in which we plot [Eu/Fe][a] abundance ratios as a function of [Fe/H] metallicity for a large number of halo and disk stars [1,2]. Eu is employed for such studies

---

[a][A/B] ≡ log[$(N_A/N_B)_{star}/(N_A/N_B)_{Sun}$]

since it is relatively easy to detect in the spectra of metal-poor stars and because it is predominantly produced in the r-process. It is seen in Figure 4 that near [Fe/H] = −3, (which includes some of the oldest stars in the Galaxy), the [Eu/Fe] ratio reaches a peak of ∼ 50 and varies from star-to-star by more than a factor of 100. Thus, even though the [Eu/H] ratios are still less than that of the sun, the relative ratio of this r-process element to iron in some of these stars is much larger than the solar ratio. It is also clear from the figure that this star-to-star scatter decreases with increasing [Fe/H] tending toward younger (on average) stars. While there could be several possible explanations for this scatter, the most likely interpretation is that the early Galaxy was chemically inhomogeneous and unmixed.

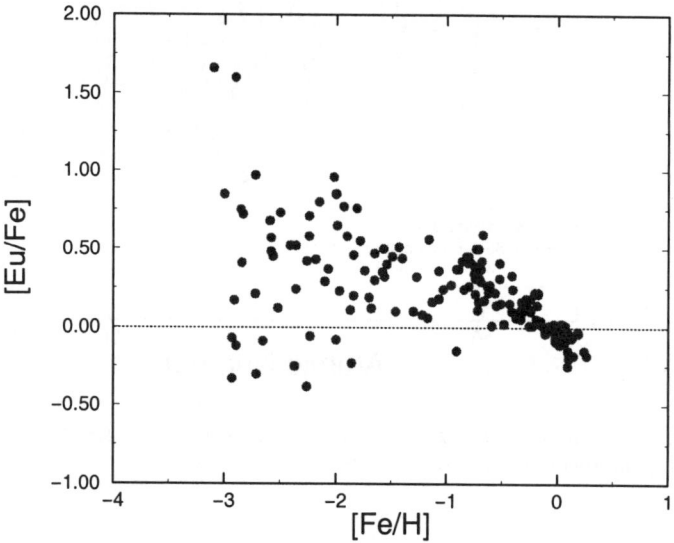

Figure 4. The ratio [Eu/Fe] is plotted as a function of [Fe/H] from various surveys of halo and disk stars [1,2]. The dotted line represents the solar value.

Abundance trends with metallicity also demonstrate the ongoing chemical evolution of the Galaxy. In the past [Ba/Eu] has been employed to study the changing contributions to the r- and s-processes. The element Ba is overwhelmingly synthesize in the s-process and provides a good indication of the s-process n-capture nucleosynthesis history of the Galaxy. There is, however, significant observationally related scatter in the Ba abundance data making it less reliable for such studies. As an alternative, new studies are now employing La, also a predominantly s-process element, to Eu to examine Galactic chemical evolution. Very preliminary results [22] seem to indicate some initial indication of the s-process below [Fe/H] $= -2$ with the major onset of the ejection of this material into the Galaxy occurring near a metallicity of -2. These results lend support to earlier studies with similar findings [10,23].

4. Radioactive Chronometers and the Age of the Galaxy

Abundance detections of certain long-lived radioactive isotopes can be employed as "chronometers" or clocks to determine the ages of the oldest stars. There have been a number of recent detections of the element thorium, with a half-life of 14 Gyr, in the metal-poor halo halo stars [4–8,24,25]. This element, along with uranium, is synthesized solely in the r-process. Comparison of the observed stellar abundance of this radioactive element with its initial (time-zero) abundance in an r-process site leads to a direct radioactive-age estimate of the star. We show in Figure 3 the abundance distribution, including Th, in the ultra-metal-poor star CS 22892–052. While (as noted before) the heavy n-capture elements are consistent with the scaled solar r-process curve, the observed Th abundance lies below this same line. This difference is a clear demonstration that this star is older than the sun. To determine how much older requires knowledge of the initial Th abundance that must be predicted from r-process models. Such a model calculation is illustrated in Figure 3 by the dashed line [5]. The goal of such theoretical calculations is to reproduce the stellar, and hence the solar system r-process, abundance pattern and at the same time predict the abundances of the radioactive elements. Such predictions, to reduce errors, employ the ratio of Th to another r-process element, typically Eu. Utilizing this technique has led to chronometric age estimates ranging from $\simeq 11$–15 ± 4 Gyr [4–8,24,25] that are consistent with globular cluster ages and cosmological age estimates based upon the observed supernova expansion rates. However, the Th/Eu chronometer gives a very different and completely inconsistent age in the star CS 31082–001 [25]. This star was the first with a detection of U and the Th/U chronometer does give an age of 15.5 Gyr [26]. Since Th/Eu and Th/U give similar results in BD +17°3248 [7], it is not clear yet why CS 31082–001 is so different or if it is rare. Clearly additional U detections will be needed to answer this question. Just as importantly reliable nuclear data, experimental where available but mostly theoretical predictions, for the neutron-rich nuclei in the r-process will be necessary to determine more accurately the ages of these old stars and put limits on the age of the Galaxy and the Universe.

Acknowledgments

Partial support for this research was provided by the National Science Foundation (AST-9986974 to JJC and AST-9987162 to CS) and by the Space Telescope Science grant GO-08342. JJC thanks the University of Texas at Austin Department of Astronomy John W. Cox Fund for partial support while this paper was being written.

References

1. C. Sneden and J. J. Cowan, *Science*, in press (2002).
2. J. W. Truran, J. J. Cowan, C. Sneden and C. Pilachowski, *Publ. Astron. Soc. Pac.*, in press (2002).
3. N. Grevesse and A. J. Sauval, *Space Sci. Rev.* **85**, 161 (1998).
4. C. Sneden, et al., *Astrophys. J.* **467**, 819 (1996).
5. J. J. Cowan, et al., *Ap. J.* **521**, 194 (1999).
6. C. Sneden, et al., *Astrophys. J.* **533**, L139 (2000).
7. J. J. Cowan, et al., *Astrophys. J.* **572**, 861 (2002).
8. J. Westin, C. Sneden, B. Gustafsson, and J. J. Cowan, *Astrophys. J.* **530**, 783 (2000).
9. F. Käppeler, H. Beer, and K. Wisshak, *Rep. Prog. Phys.* **52**, 945 (1989).
10. D. L. Burris, C. A. Pilachowski, T. A. Armandroff, C. Sneden, J. J. Cowan, and H. Roe, *Astrophys. J.* **544**, 302 (2000).
11. C. Arlandini, et al., *Astrophys. J.* **525**, 886 (1999).
12. G. J. Mathews, G. Bazan, and J. J. Cowan, *Astrophys. J.* **391**, 719 (1992).
13. J. C. Wheeler, J. J. Cowan, and W. Hillebrandt, *Astrophys. J.* **493**, L101 (1998).
14. Y. Ishimaru and S. Wanajo, *Astrophys. J.* **511**, L33 (1999).
15. C. Sneden, et al., *Astrophys. J.* **566**, L28 (2002).
16. D. Lambert and C. Allende-Prieto, *MNRAS* **335**, 325 (2002).
17. G. J. Wasserburg, M. Busso, and R. Gallino, *Astrophys. J.* **466**, L109 (1996).
18. C. Sneden, et al., in preparation (2002).
19. J. W. Truran and J. J. Cowan, in *Nuclear Astrophysics*, eds. W. Hillebrandt and E. Müller 64 (2000)
20. G. J. Wasserburg and Y.-Z. Qian, *Astrophys. J.* **529**, L21 (2000).
21. A. G. W. Cameron, *Astrophys. J.* **562**, 456 (2001).
22. J. Simmerer, et al., in preparation (2002).
23. K. K. Gilroy, C. Sneden, C. A. Pilachowski, and J. J. Cowan, *Astrophys. J.* **327**, 298 (1988).
24. J. A. Johnson and M. Bolte, *Astrophys. J.* **554**, 888 (2001).
25. R. Cayrel, et al., *Nature* **409**, 691 (2001).
26. H. Schatz, et al., *Astrophys. J.* **579**, 626 (2002).

THE SINGLE-PARTICLE STRUCTURE OF NEUTRON-RICH NUCLEI OF ASTROPHYSICAL INTEREST AT THE ORNL HRIBF*

D.W. BARDAYAN[1], J.C. BATCHELDER[2], J.C. BLACKMON[1], C.R. BRUNE[3],

A.E. CHAMPAGNE[4], J.A. CIZEWSKI[5], T. DAVINSON[6], U. GREIFE[7], A.N. JAMES[8],

M. JOHNSON[5], R.L. KOZUB[9], J.F. LIANG[1], R.J. LIVESAY[7], Z. MA[10],

C.D. NESARAJA[9], D.C. RADFORD[1], D. SHAPIRA[1], M.S. SMITH[1], J. S. THOMAS[5],

P.J. WOODS[6], E. ZGANJAR[11], AND THE UNIRIB COLLABORATION

[1]*Physics Division, Oak Ridge National Laboratory*, Oak Ridge, TN 37831 USA*
[2]*UNIRIB†, Oak Ridge Associated Universities, Oak Ridge, TN 37831 USA*
[3]*Dept. of Physics and Astronomy, Ohio University, Athens, OH 45701 USA*
[4]*Dept. of Physics and Astronomy, Univ. of North Carolina, Chapel Hill, NC 27599 USA*
[5]*Dept. of Physics and Astronomy, Rutgers University, New Brunswick, NJ 08903 USA*
[6]*Dept. of Physics, Univ. of Edinburgh, Edinburgh EH9 3JZ United Kingdom*
[7]*Dept. of Physics, Colorado School of Mines, Golden, CO 80401 USA*
[8]*Physics Dept., University of Liverpool, Liverpool L693BX United Kingdom*
[9]*Dept. of Physics, Tennessee Technological University, Cookeville, TN 38505 USA*
[10]*Dept. of Physics and Astronomy, University of Tennessee, Knoxville, TN 37996 USA*
[11]*Dept. of Physics and Astronomy, Louisiana State Univ., Baton Rouge, LA 70803 USA*

The rapid neutron-capture process (*r* process) produces roughly half of the elements heavier than iron. The path and abundances produced are uncertain, however, because of the lack of nuclear structure information on important neutron-rich nuclei. We are studying nuclei on or near the r-process path via single-nucleon transfer reactions on neutron-rich radioactive beams at ORNL's Holifield Radioactive Ion Beam Facility (HRIBF). Owing to the difficulties in studying these reactions in inverse kinematics, a variety of experimental approaches are being developed. We present the experimental methods and initial results.

1. Introduction

Almost all of the elements heavier than lithium populating the Cosmos were created in stellar burning or explosions [1]. One important source is the astrophysical *r* process which is thought to have created roughly half of the elements heavier than iron [2]. Heavy nuclei are produced in the *r* process by a series of neutron captures and β decays that flows through extremely neutron-

---

* ORNL is managed by UT-Battelle, LLC for the U.S. Dept. of Energy under contract DE-AC05-00OR22725.
† UNIRIB is a consortium of universities, the state of TN, Oak Ridge Associated Universities, and ORNL and is partially supported by them.

rich nuclei ending near uranium. Abundances recently measured in some very old stars are consistent with an *r*-process origin [3]. Comparisons of measured abundances with model calculations can constrain the astrophysical conditions in which the *r* process occurs (e.g., can it occur in supernovae) and even the age of the Galaxy. The *r*-process models are currently very uncertain, however, because the nuclear structure of most *r*-process nuclei is not known. Especially important is the structure of nuclei near closed neutron shells where the *r*-process abundances peak as a result of the small neutron-capture cross sections of these nuclei. A recent study [4] found that order of magnitude variations in the predicted *r*-process abundances result from uncertainties in the estimated neutron-induced reaction cross sections. Better nuclear structure information is needed on these neutron-rich nuclei to better constrain nuclear structure and mass models, improve cross section estimates, and in turn improve the *r*-process abundance calculations.

Accelerated beams of many neutron-rich nuclei have recently become available at ORNL's Holifield Radioactive Ion Beam Facility [5]. These nuclei are produced by proton-induced fission of uranium, which has been deposited as UC on a low-density graphite matrix. The radioactive nuclei are then accelerated to energies up to 4-5 MeV/amu by the 25-MV Tandem accelerator. The available neutron-rich beams with intensities greater than 10^3 ions/second

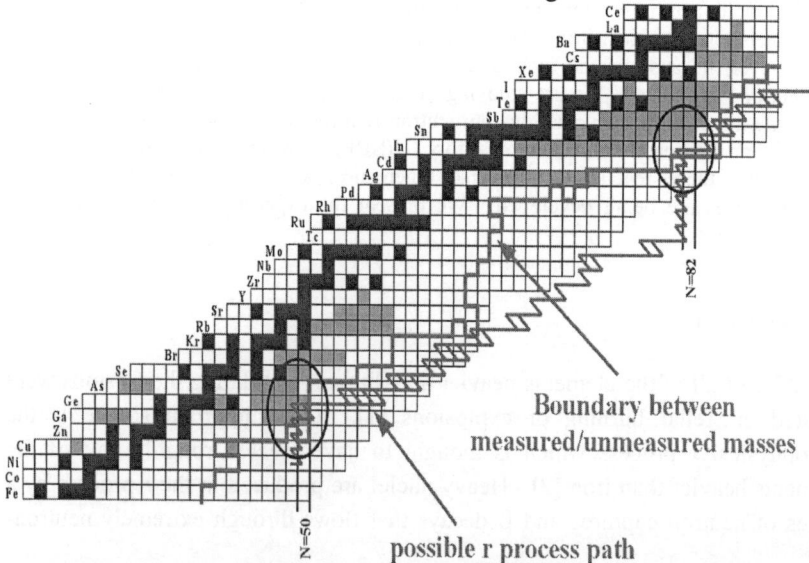

Figure 1. The available neutron-rich beams are shown as gray boxes. Stable isotopes are shown as black boxes. A possible r-process path is also shown. Near the neutron closed shells N=50 and 82, studies of r-process nuclei are possible.

are shown as gray boxes in Fig. 1. Near the neutron closed shells, N = 50 and 82, available beams include nuclei on or near the r-process path. Single-nucleon transfer reactions are being used with these beams to study the structure of these neutron-rich nuclei of astrophysical importance.

2. Neutron-Transfer Reactions

Single-neutron transfer reactions are being studied via (d,p) or (^{9}Be,^{8}Be) reactions in inverse kinematics. The excitation energies of single-particle neutron states are determined either from the proton-energy spectrum at a fixed angle or from the energies of the γ-rays emitted when the excited nucleus decays. Information on spectroscopic factors and the spins of populated states are also obtained from the measured (d,p) cross sections and angular distributions. The kinematics of such reactions, however, present difficult challenges that require multiple experimental approaches to overcome. As an example, we plot in Fig. 2 the emitted proton energies calculated for d(^{132}Sn,p)^{133}Sn at 5 MeV/amu as a function of their laboratory angles. The forward center-of-mass angles, at which the angular distributions are most sensitive to the transferred angular momentum, correspond to backward laboratory angles. At these angles, however, the proton energies for excited states rapidly drop below what is reasonable for particle identification with a silicon-detector telescope. The (d,p) cross section is also much smaller at backward laboratory angles than at forward angles. The best solution typically is to cover a relatively large angular range (θ_{lab}=70°-130°) with

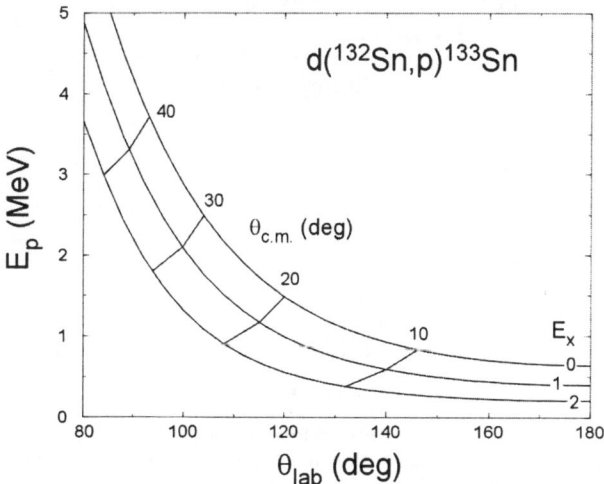

Figure 2. Reaction kinematics for d(^{132}Sn,p)^{133}Sn in inverse kinematics at 5 MeV/amu.

two-dimensional position-sensitive silicon-strip detector telescopes. The two-dimensional position sensitivity is required to determine the reaction angle of the detected protons upon which the proton energy depends sensitively. A typical target-detector configuration is pictured in Fig. 3 [6]. In a recent stable beam test [7], beams of ~4 MeV/amu ^{124}Sn bombarded 200 µg/cm^2 CD$_2$ targets rotated 60° to the beam axis.   Protons from the d(^{124}Sn,p)^{125}Sn [7] reaction were

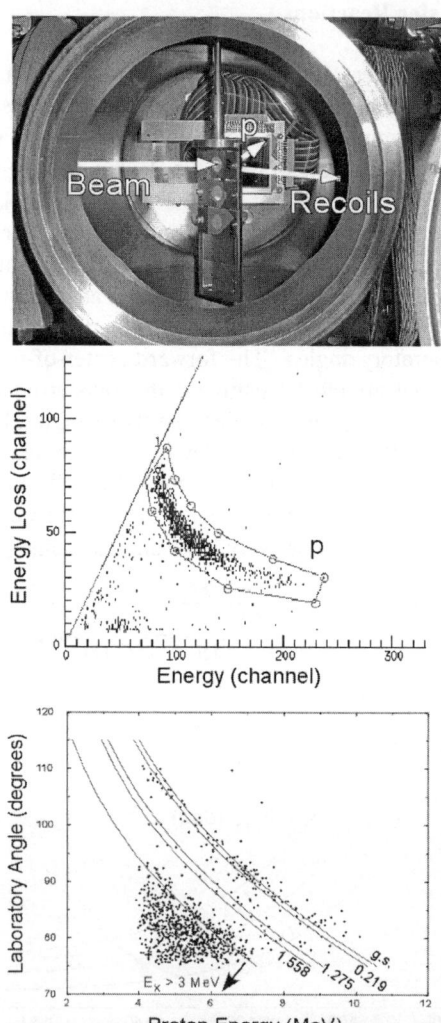

Figure 3. A typical detector setup is shown along with particle identification, and proton energies from the d(^{124}Sn,p)^{125}Sn reaction [7] at 4 MeV/amu. Expected kinematical curves for protons populating ^{125}Sn states (labeled in MeV) are shown. Excitation energies are from Ref. [8].

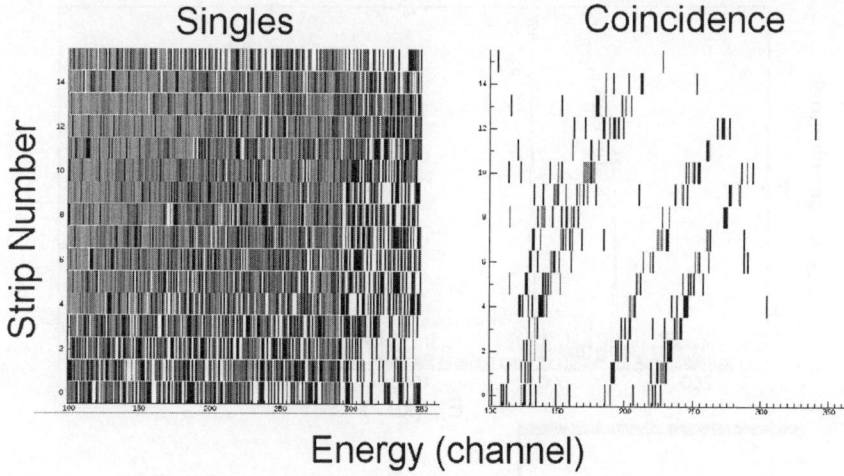

Energy (channel)

Figure 4. The energy spectrum of charged particles observed in a recent study of the $d(^{18}F,p)^{19}F$ reaction at 6 MeV/amu is plotted for each detector strip in singles and in coincidence with ^{19}F ions at the focal plane of the DRS. Strips 0 - 15 correspond to laboratory angles from 157° - 118°, respectively.

distinguished from scattered target ions by standard energy-loss techniques (Fig. 3). The identification of protons was sufficient to identify the events of interest as the yield of protons from other reactions with contaminants in the target was found to be negligible. Protons populating several states in ^{125}Sn were identified from the plot of proton energy versus laboratory angle (Fig. 3).

In other cases to identify the events of interest, it will be necessary to detect the heavy recoil in coincidence with reaction protons. This can be accomplished by placing a microchannel plate [9] downstream of the target or by separating the recoils from beam particles using a mass separator such as the Daresbury Recoil Separator (DRS), which has been installed at the HRIBF for studies of reactions of astrophysical interest with radioactive ion beams [10]. This technique was used recently for a study of the $d(^{18}F,p)^{19}F$ reaction [11]. Protons from the $d(^{18}F,p)^{19}F$ reaction were detected at backward laboratory angles in coincidence with recoil ^{19}F ions detected at the focal plane of the DRS. As shown in Fig. 4, protons populating states in ^{19}F were readily identified when the coincidence requirement was applied.

Because of large kinematical shifts and high level densities, it may not be possible to achieve the necessary energy resolution in the charged-particle spectrum to resolve all the states in the compound nucleus. In these cases, the

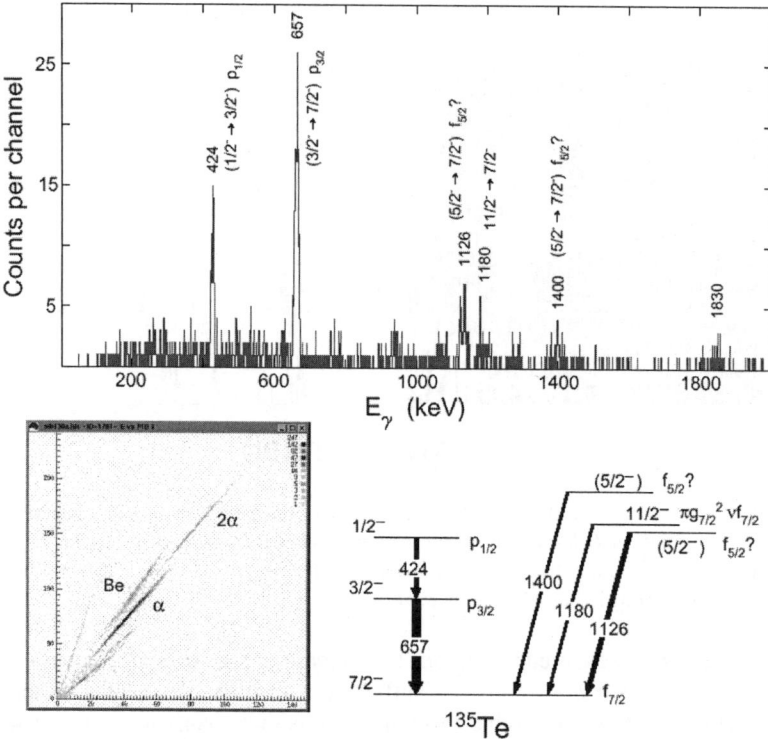

Figure 5. TOP: Gamma-ray spectrum from the ^{9}Be(^{134}Te,^{8}Be)^{135}Te reaction at 4 MeV/amu in coincidence with 2α clusters detected by the HYBALL. BOTTOM LEFT: Particle identification spectrum for one of the CsI detectors. BOTTOM RIGHT: Partial level scheme for ^{135}Te, showing tentative configuration assignments.

excitation energies of single-particle levels can be inferred from the energies of γ-rays observed following a neutron-transfer reaction. This method has been used to probe the level structure of ^{135}Te using a radioactive ^{134}Te beam and a ^{9}Be target. The resulting ^{8}Be ions from the ^{9}Be(^{134}Te,^{8}Be)^{135}Te reaction broke up into two α-particles that were detected in coincidence in single CsI crystals of the ORNL HYBALL detector [12]. As shown in Fig. 5, the detection of coincident α-particle pairs was an extremely clean signature of neutron transfer. The CLARION Ge detector array [12] was triggered on these 2α events finding several γ-ray transitions in ^{135}Te.

3. Proton-Transfer Reactions

Proton-transfer studies on neutron-rich radioactive beams can also be used to probe the structure of nuclei near neutron closed shells. For instance, ^{131}In is believed to be on the r-process path and is one proton away from doubly-magic ^{132}Sn, an available radioactive beam at the HRIBF. For doubly-magic nuclei such as ^{132}Sn, the high energy of core excitations limits the mixing of core-coupled states into the states of adjacent nuclides, and thus the single-particle levels of ^{131}In should be relatively pure.

Knowledge of the energies of the single-particle levels in ^{131}In would provide a much needed test of the shell model in this crucial mass region. The $\pi p_{3/2}$ and $\pi f_{5/2}$ hole states in ^{131}In are important single-particle levels that are not observed in decay studies. They should, however, be observable using single-proton transfer reactions with a ^{132}Sn beam. Similar to the method used for the ^{9}Be(^{134}Te,^{8}Be)^{135}Te study, 2α clusters would be detected in the HYBALL detector from the ^{7}Li(^{132}Sn,^{8}Be)^{131}In reaction [13]. The excitation energies of single-particle levels in ^{131}In would be deduced from the γ-rays detected in the CLARION Ge array in coincidence with the 2α clusters.

4. Conclusions

The availability of neutron-rich radioactive beams at the HRIBF provides exciting opportunities for study of the structure of nuclei far from stability. Such information is critical to our understanding of the astrophysical r process. Especially important is the understanding of structure near neutron closed shells where the r-process abundances tend to peak. We are developing the techniques needed to study these important nuclei via single-nucleon transfer reactions using neutron-rich radioactive beams. The need to study the reactions in inverse kinematics presents many experimental difficulties that are being overcome by using a variety of techniques. The detection of protons from the (d,p) reaction in silicon-strip detectors in coincidence with heavy recoils provides a promising mechanism for the extraction of single-particle strengths and angular-momentum transfers. A complimentary technique is to use a single-particle transfer reaction resulting in the emission of a ^{8}Be nucleus that would, in turn, break up into particle clusters providing a highly-selective trigger. The detection of these particle clusters in coincidence with γ-rays provides a good measure of the energies of single-particle levels in the resulting nucleus. In addition to providing important nuclear structure information, the techniques we are developing will also be useful when the Rare Isotope Accelerator (RIA) [14] comes online, and the study of reactions in inverse-kinematics on nuclei further from stability becomes possible.

Acknowledgments

We thank the staff of the HRIBF whose hard work made these experiments possible.

References

1. W.A. Fowler, *Rev. Mod. Phys.* **56**, 149 (1984).
2. S.E. Woosley and R.D. Hoffman, *Astrophys. J.* **395**, 202 (1992).
3. J.J. Cowan et al., *Astrophys. J.* **572**, 861 (2002).
4. R. Surman and J. Engel, *Phys. Rev. C* **64**, 035801 (2001).
5. http://www.phy.ornl.gov/hribf/
6. J.A. Cizewski *et al.*, HRIBF experiment RIB-092.
7. R.L. Kozub *et al.*, HRIBF experiment RIB-100.
8. A. Strömich *et al.*, *Phys. Rev. C* **16**, 2193 (1977).
9. D. Shapira, T.A. Lewis, and L.D. Hulett, *Nucl. Instrum. Meth. Phys. Res. A* **454**, 409 (2000).
10. http://www.phy.ornl.gov/astrophysics/nuc/rib/drs.html
11. R.L. Kozub *et al.*, HRIBF experiment RIB-044; http://www.phy.ornl.gov/hribf/usersgroup/news/sum-02/sum_02.html
12. C.J. Gross *et al., Nucl. Instrum. Meth. Phys. Res. A* **450**, 12 (2000).
13. E. Zganjar *et al.*, HRIBF experiment RIB-097.
14. http://www.orau.org/ria/

OPEN QUESTIONS IN STELLAR HELIUM BURNING STUDIED WITH REAL PHOTONS *

MOSHE GAI

Laboratory for Nuclear Sciences, Department of Physics, U3046,
University of Connecticut, 2152 Hillside Rd., Storrs, CT 06269-3046, USA
E-mail: gai@uconn.edu URL: http://www.phys.uconn.edu

The outcome of helium burning is the formation of the two elements, carbon and oxygen. The ratio of carbon to oxygen at the end of helium burning is crucial for understanding the final fate of a progenitor star and the nucleosynthesis of heavy elements in Type II supernova, with oxygen rich star predicted to collapse to a black hole, and a carbon rich star to a neutron star. Type Ia supernovae (SNeIa) are used as standard candles for measuring cosmological distances with the use of an empirical light curve-luminosity stretching factor. It is essential to understand helium burning that yields the carbon/oxygen white dwarf and thus the initial stage of SNeIa. Since the triple alpha-particle capture reaction, $^8Be(\alpha, \gamma)^{12}C$, the first burning stage in helium burning, is well understood, one must extract the cross section of the $^{12}C(\alpha, \gamma)^{16}O$ reaction at the Gamow peak (300 keV) with high accuracy of approximately 10% or better. This goal has not been achieved despite repeated strong statements that appeared in the literature. Constraint from the beta-delayed alpha-particle emission of ^{16}N were shown to not sufficiently restrict the p-wave cross section factor; e.g. a low value of $S_{E1}(300)$ can not be ruled out. Measurements at low energies, are thus mandatory for determining the ellusive cross section factor for the $^{12}C(\alpha, \gamma)^{16}O$ reaction. We are constructing a Time Projection Chamber (TPC) for use with high intensity photon beams extracted from the HIγS/TUNL facility at Duke University to study the $^{16}O(\gamma, \alpha)^{12}C$ reaction, and thus the direct reaction at energies as low as 0.7 MeV. This work is in progress.

1. Introduction: Oxygen Formation in Helium Burning and The $^{12}C(\alpha, \gamma)^{16}O$ Reaction

The outcome of helium burning is the formation of the two elements, carbon and oxygen [1,2,3]. The ratio of carbon to oxygen at the end of helium burning is crucial for understanding the fate of Type II supernovae and the nucleosynthesis of heavy elements. While an oxygen rich star is predicted

---

*work supported by usdoe grant no. de-fg02-94er40870.

to end up as a black hole, a carbon rich star leads to a neutron star [2]. At the same time helium burning is also very important for understanding Type Ia supernovae (SNeIa) that are now being used as a standard candle for cosmological distances [4]. All thus far luminosity calibration curves and the stretching factor are based on empirical observations without fundamental understanding the relation between the time characteristics of the light curve and the maximum magnitude of Type Ia supernova. Since the first burning stage in helium burning, the triple alpha-particle capture reaction ($^8Be(\alpha,\gamma)^{12}C$), is well understood [1], one must extract the p-wave [$S_{E1}(300)$] and d-wave [$S_{E2}(300)$] cross section of the $^{12}C(\alpha,\gamma)^{16}O$ reaction at the Gamow peak (300 keV) with high accuracy of approximately 10% or better to completly understand stellar helium burning, and better understand Type II and Type Ia supernova.

1.1. Beta-Delayed Alpha-Particle Emission of ^{16}N

Early hopes for extracting the astrophysical E1 S-factor [$S_{E1}(300) = \sigma_{E1} \times E \times e^{2\pi\eta}$] of the $^{12}C(\alpha,\gamma)^{16}O$ reaction through constraints imposed by new data on the beta-delayed alpha-particle emission of ^{16}N [5,6,7,8] were examined in detail over the last few years and a few observations were made. The original Yale data [5,6] were improved [9,10] in a phase II experiment of the Yale-UConn group, and were found to be in disagreement with the TRIUMF data [8] but consistent with the unpublished data of the Seattle group [11], see Fig. 1. In addition, an independent R-matrix analysis [12] of the world data including the beta decay of ^{16}N data was found to not rule out a small S-factor solution (10-20 keV-b). It is thus doubtful that one can extract the p-wave cross section factor with a reasonable accuracy as stated in Ref. [8]. The confusion in this field mandates a direct measurement of the cross section of the $^{12}C(\alpha,\gamma)^{16}O$ reaction at low energies. A recent measurement at lower energies [13], suggest a d-wave cross section factor that is at least twice larger than "the accepted value", and the their low energy data point(s) measured with low precision can not rule out a small p-wave cross section factor.

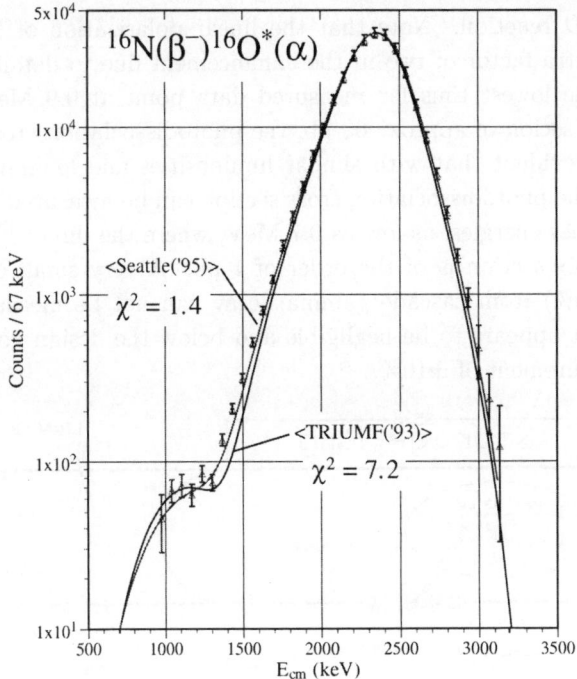

Fig. 1: The Yale-UConn data on the beta-delayed alpha-particle emission of ^{16}N [9,10] compared to the TRIUMF [8] and Seattle results [11]. The TRIUMF and Seattle data are averaged over the energy resolution of the Yale-UConn experiment and are shown by continuous lines. The unpublished Seattle data are listed (by permision) in the appendix of Ref. [10].

2. The Proposed $^{16}O(\gamma, \alpha)^{12}C$ Experiments

For determining the cross section of the $^{12}C(\alpha, \gamma)^{16}O$ at very low energies, as low as $E_{cm} = 700$ KeV, considerably lower than measured till now [13], it is advantageous to have an experimental setup with larger (amplified) cross section, high luminosity and low background. It turns out that the use of the inverse process, the $^{16}O(\gamma, \alpha)^{12}C$ reaction may indeed satisfy all three conditions. The cross section of $^{16}O(\gamma, \alpha)^{12}C$ reaction (with polarized photons) at the kinematical region of interest (photons approx 8-8.5 MeV) is larger by a factor of approximately 100 than the cross section of the direct

$^{12}C(\alpha, \gamma)^{16}O$ reaction. Note that the linear polarization of the photons yields an extra factor of two in the enhancement due to detailed balance. Thus for the lowest thus far measured data point at 0.9 MeV with the direct cross secion of approx. 60 pb, the photodissociation cross section is 6 nb. It is evident that with similar luminosities and lower background, see below, the photodissociation cross section can be measured to yet lower center of mass energies, as low as 0.7 MeV, where the direct $^{12}C(\alpha, \gamma)^{16}O$ reaction cross section is of the order of 1 pb. A very small contribution (less than 5%) from cascade gamma decay can not be measured in this method, but appears to be negligible and below the design goal accuracy of our measurement of ±10%.

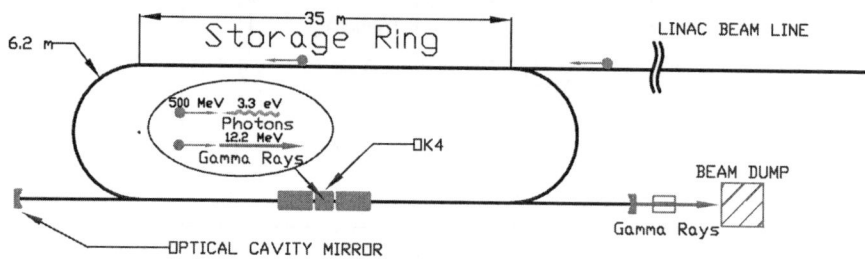

Fig. 2: Schematic diagram of the HIγS facility [14] for the production of intense MeV gamma beams.

The High Intensity Gamma Source (HIγS) [14], shown in Fig. 2, has already achieved many of its design milestones and is rapidly approaching its design goal for 2-200 MeV gammas. For 9 MeV gammas we expect an energy resolution of 0.1% and intensity of order 10^9 /sec. Currently achieved intensities are of the order of 10^8 /sec with energy resolution of approx. 0.5%. The backscattered photons of the HIγS facility are collimated (3 mm diameter) and enter the target/detector TPC setup as we discuss below. With a Q value of -7.162, our experiment will utilize gammas of energies approximately 8 to 10 MeV. Note that the emitted photons are linearly polarized [15] and the emitted particles are primarily in a horizontal plane (parallel to target room floor) with a $sin^2\phi$ azimuthal angular dependence [16], thus simplifying the tracking of particles in this experiment. The pulsed photon beam (0.1 ns every 180 nsec with at most 500 gammas per pulse) provides additional trigger for removing background. The image intensified CCD camera is triggered by light detected in the PMT, see below, and the time projection information from the drift chamber yields the azimuthal angle of the event of interest. The scattering angle is measured with high accuracy using the (8 cm long) alpha tracks and (2 cm long) carbon tracks. Background events from contaminants carbon, oxygen and fluorine isotopes, ar discriminated using the TPC as a calorimeter with a 2% energy resolution. Time of flight techniques, and flushing of the CCD between two events will also be used. To reduce noise, the CCD will be cooled. We note that similar research program with high intensity photon beams and a TPC already exists at the RCNP at Osaka, Japan [17], proving that tracks from low energy light ions can be identified in the TPC with a managable electron background. An $^{16}O(e, e'\alpha)^{12}C$ experiment with virtual photons proposed at the MIT-Bates accelerator [18] is useful to extract only the d-wave astrophysical cross section factor and thus it complements our experiment proposed for the HIγS-TUNL facility [19].

2.1. *Proposed Time Projection Chamber (TPC)*

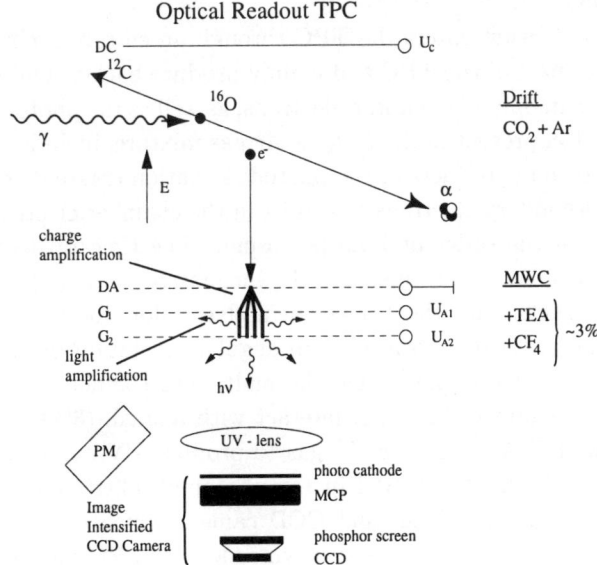

Fig. 3: Schematic diagram of the Optical Readout TPC [20].

We are constructing an Optical Readout Time Projection Chamber (TPC), similar to the TPC constructed in the Physikalisch Technische Bundesanstalt, (PTB) in Braunschweig, Germany and the Weizmann Institute, Rehovot, Israel [20], for the detection of alphas and carbon, the byproduct of the photodissociation of ^{16}O. Since the range of available alphas is approximately 8 cm (at 100 mbars) the TPC is 40 cm wide and up to one meter long. We first construct a 40 cm long TPC for initial use at the HIγS beam line at TUNL/Duke. The TPC is largely insensitive to single Compton electrons, and the large compton electron flux, if a problem, can be blocked using a standard beam blocker placed between the drift chamber volume and the Multi Wire Proportional Counter of the TPC. The TPC allows for tracking of both alphas and carbons emitted almost back to back from the beam position in time correlation. The very different range of alphas and carbons (approx. a factor of 4), and differences in the lateral ionization density, will aid us in particle identification. The TPC also allow us to measure angular distributions with respect to the photon beam thus seperating the E1 and E2 components of the $^{12}C(\alpha,\gamma)^{16}O$ reaction. The excellent energy resolution of the TPC (approx. 2%) allows us to exclude events from the photodissociation of nuclei other than ^{16}O, including isotopes of carbon, oxygen and fluorine, that are present in the gas. In Fig. 3, taken from Titt et al. [20], we show a schematic diagram of the Optical Readout TPC.

The photon beam enters the TPC through an entrance window in the drift chamber part of the TPC and mainly produce background e^+e^- pairs and a smaller amount of Compton electrons, as well as the photodissociation of various nuclei present in the $CO_2 + Ar$ gas mixture, including ^{16}O. The charged particle byproducts of the photodissociation create delta electrons that create secondary electrons that drift in the chamber electric field with a total time of the order of 1 μs per 5 cm. The time projection of the drift electrons allows us to measure the inclination angle (ϕ) of the plane of the byproducts, and the tracks themselves allow for measurement of the scattering angle (θ), both with an angular resolution better than two degrees. The electrons that reach the multi-wire chamber are multiplied (by approx. a factor of 10^5) and interact with a small (3%) admixture of triethylamine (TEA) [20] or CF_4 [21] gas to produce UV or visible photons, respectively. The light detected in the photomultiplier tube, see Fig. 3, triggers the Image Intensifier and CCD camera which takes a picture of the visible tracks. The picture is downloaded to a PC and analyzed for recognition of the two back-to-back alpha-carbon tracks originating from the beam position. The background electrons lose approx. 0.5 KeV/cm in the TPC and are removed by an appropriate threshold in the trigger Photo Multiplier Tube (PMT). Events from the photodissociation of nuclei other than ^{16}O are removed by measuring the total energy (Q-value) of the event with a resolution of 2%.

2.2. Design Goals

The luminosity of our proposed $^{16}O(\gamma, \alpha)^{12}C$ experiment can be very large. For example, with a 30 cm long fiducial length target with 30% CO_2 at a pressure of 76 torr (100 mbar) and a photon beam of 2×10^9 /sec, we obtain a luminosity of 10^{29} $sec^{-1}cm^{-2}$, or a day long integrated luminosity of 10 nb^{-1}. Thus a measurement of the photodissociation of ^{16}O with cross section of 1 nb, yields 10 count per day, leading to a sensitivity for measuring the direct $^{12}C(\alpha, \gamma)^{16}O$ reaction with a cross section as low as 10 pb, corresponding to energies as low as 700 keV. The construction and tests of the TPC is in progress at the Univeristy of Connecticut, at the PTB in Braunschweig and the Weizmann Institute. A mark I experiment to measure coincidences between α-particles and ^{12}C is in progress at the TUNL/HIγS facility.

3. Acknowledgements

The author would like to acknowledge discussions with Amos Breskin and the work of Steve O. Nelson and Joseph A. Dooley on the design and construction of the TPC. This work is in collaboration with Henry Weller of the TUNL/(HIγS-TUNL) facility.

References

1. W.A. Fowler, Rev. Mod. Phys. **56**(1984)149.
2. T.A. Weaver, and.E. Woosley; Physics Report, **227**(1993)65.
3. M. Gai; From the Sun to the Great Attractor, Lecture Notes in Physics, Springer, 2000, p. 49.
4. P. Hoeflich, Ch. Gerardy, E. Linder, and H. Marion; astr-ph/0301334.
5. Zhiping Zhao; Ph.D. thesis, Yale University, 1993.
6. Z. Zhao, R.H. France III, K.S. Lai, S.L. Rugari, M. Gai, and E.L. Wilds; Phys. Rev. Lett. **70**(1993)2066, ER **70**(1993)3524.
7. L. Buchmann; et al.; Phys. Rev. Lett. **70**(1993)726.
8. R.E. Azuma, et al.; Phys. Rev. **C50**(1994)1194.
9. Ralph H. France III; Ph.D. thesis, Yale University, 1996, http://www.phys.uconn.edu/~france/thesis.html.
10. R.H. France III, E.L. Wilds, N.B. Jevtic, J.E. McDonald, and M. Gai; Nucl. Phys. **A621**(1997)165c.
11. Z. Zhao, L. Debrackeleer, and E.G. Adelberger; 1995, *Private Communication*.
12. G.M. Hale; Nucl. Phys. **A621**(1997)177c.
13. R. Kunz et al.; Phys. Rev. Lett. **86**(2001)3244.
14. A Proposal: "A FREE-ELECTRON LASER GENERATED GAMMA-RAY

BEAM FOR NUCLEAR PHYSICS", W. Tornow, R. Walter, H.R. Weller, V. Litvinenko, B. Mueller, P. Kibrough, Duke/TUNL, 1997.

15. V.N. Litvinenko *et al.*, Phys. Rev. Letts. **78**(1997)4569.

16. E.C. Schreiber *et al.*, Phys. Rev. **C61**(2000)061604(R).

17. T. Shima, Y. Nagai, T. Kii, T. Baba, T. Takahashi, and H. Ohgaki; Nucl. Phys. **A629**(1998)475c.

18. I. Tsentalovitch *et al.*, Bates proposal, 1999.

19. Toni Feder, Physics Today, Volume 55, No. 12, p. 26, 2002.

20. U.Titt, A.Breskin, R.Chechik, V.Dangendorf, H.Schmidt-Bocking, H.Schuhmacher; Nucl. Instr. Meth. **A416**(1998)85. U.Titt, V.Dangendorf, H.Schuhmacher; Nucl. Phys. **B Supp.** **78**(1999)444. U.Titt, Dissertation zur Erlangung des Doktorgrades. (Ph.D. thesis), J.W.Goethe Universitat Frankfurt, 1999.

21. Pansky *et al.*, Nucl. Instr. Meth. **354**(1995)262.

NUCLEAR DATA ON NEUTRON-RICH NUCLEI FOR ASTROPHYSICS

M. S. SMITH

Physics Division, Oak Ridge National Lab,* Oak Ridge, TN, 37831-6354

R. A. MEYER

RAME', Inc., Teaticket, MA , 02536, USA, &
Chemistry Department, Clark University, Worcester, MA, 01610, USA

To understand the stellar origins of the heavy elements in supernovae, a knowledge of the structure of, and reactions involving, thousands of neutron-rich nuclei are essential. To ensure that the latest experimental and theoretical information on these nuclei is incorporated in element synthesis models, dedicated efforts in data compilation, evaluation, dissemination, and coordination are needed. Strategies to improve the utilization of nuclear data for astrophysics studies are described, including establishing an Astrophysics Data Coordinator position, a new web site to aid in locating available nuclear data sets, **www.nucastrodata.org**, and a new visualization program with an easy-to-use graphical user interface to over 8000 reaction rates.

More than half of the nuclei heavier than iron are believed to have their cosmic origin in a series of rapid neutron capture reactions - the r process. This process requires high temperatures, and entropies as may be found in the hot, neutrino-driven wind off the surface of a newly-born neutron star in a core-collapse supernova[1]. The development of sophisticated computer programs predicting the synthesis of abundances under these conditions, coupled with new observations of abundances on the surface of extremely metal-deficient stars which have a r-process-like pattern[2], are generating enormous excitement in the study of the origin of the heavy elements.

Calculations of r-process nucleosynthesis require, as input, the structure of and reactions involving thousands of neutron-rich nuclei from Fe to Bi, from stability out to the neutron drip line. Studies have shown the

---

*Managed by UT-Batelle, LLC, for the U.S. D.O.E. under contract DE-AC05-00OR22725.

sensitivity of r-process nucleosynthesis predictions to both the structure of nuclei (e.g., the quenching of shell gaps[3]) and their reactions[4]. Since nuclear physics uncertainties correspond to orders of magnitude uncertainties in the predicted r-process abundances, better determinations of these quantities - both theoretical and experimental - are crucial to improve our knowledge of heavy element origins. The importance of this work is reflected in research programs at present and future radioactive beam facilities[5].

To have an impact on our understanding of the cosmos, it is not enough to perform state-of-the-art measurements or theoretical calculations. This information must be appropriately modified for input into astrophysics simulation codes, requiring *dedicated efforts* in data compilation, evaluation, dissemination, and coordination. There are a number of valuable online nuclear datasets produced out of such efforts. However, because these datasets do not include many of the latest measurements and calculations, recent advances are often not utilized to help solve the very astrophysical puzzles that motivated their generation. The situation is worsening in time as the latest nuclear measurements are not being used by astrophysics modelers, and there is an international recognition that this needs to be rectified[6].

While some efforts in recent years have yielded important progress, the field suffers from subcritical manpower. Strategies are needed to ensure a more effective utilization of nuclear physics information in astrophysics studies. For example, some manpower gains may be realized by exploiting the overlap between the nuclear data and nuclear astrophysics communities[7]. It is also crucial to coordinate plans to evaluate nuclear reactions or structure properties on a national and international basis in order to share expertise and avoid duplication. These, and other data activities, would be greatly facilitated by the establishment of a Nuclear Astrophysics Data Coordinator position[8,9] whose duties would be to: establish, maintain, and update a central WWW archive of relevant datasets; modify archive datasets for compatibility with astrophysical codes; improve dataset accessibility via the creation of indices, search capabilities, graphical user interfaces, bibliographies, error checking, online plotting tools, and other enhancements. Other activities of the coordinator would be to: help coordinate international data activities; establish and maintain a nuclear astrophysics email distribution list; publicize new nuclear astrophysics meetings, experimental results, and publications; and establish and maintain a priority list of important nuclear reactions and properties that require further study. With a modest investment, these data activities can make a significant positive impact on astrophysics research efforts worldwide.

In addition to coordination, the development of web-based tools such as web sites, datasets, and data viewers is essential to ensure that the latest nuclear data is effectively utilized in astrophysics studies. For example, researchers are now able to search, browse, display, and download a wide variety of nuclear physics information via the WWW - including some data sets specialized for nuclear astrophysics studies (e.g., reaction rates).There has, however, been no site linking *all available nuclear astrophysics datasets*, and there has been little synergy between the existing nuclear astrophysics data websites. To rectify this situation, a new website has been launched: **www.nucastrodata.org**. This site features an extensive list of nuclear datasets (over 60 so far) important for nuclear astrophysics studies available from around the world. It is designed to help users navigate through these datasets, as well as to publicize them to the research community. Many of the linked datasets contain extensive information on neutron-rich nuclei needed for supernova studies. The site is independent of any host institution and does not compete with any existing nuclear astrophysics data site. To ease the access to these datasets, a new visualization tool[10] has been developed which provides an easy-to-use, interactive, platform-independent, graphical user interface to the rates of over 8000 nuclear reactions. This program, viewable through a web browser, enables users to plot multiple rates, access rate parameters, add new rates and plot them, and create rate vs. temperature tables, all through a point-and-click graphical user interface based on the chart of the nuclides. With community support, these dissemination tools will grow into valuable resources for astrophysics research.

References

1. B.S. Meyer, *Ann. Rev. Astron. Astrophys.***32**, 153 (1994).
2. C. Sneden *et al.*, . *Astrophys. J.* **467**, 819 (1996).
3. B. Pfeiffer *et al.*, *Z. Phys.* **A357**, 235 (1997).
4. R. Surman and J. Engel, *Phys. Rev. C***64**, 035801 (2001).
5. M.S. Smith and K.E. Rehm, *Ann. Rev. Nucl. Part. Sci.* **51**, 91 (2001).
6. Muir, D.W., and Herman, M., *IAEA Report* **INDC(NDS)-423**, 7 (2001).
7. M.S. Smith *et al.*, unpublished (1995); http://www.phy.ornl.gov/astrophysics/data/task/taskforce_report.html.
8. P.D. Parker *et al.*, Nuclear Astrophysics Data Project White Paper, unpublished (1996); http://ie.lbl.gov/whitepaper.html
9. M.S. Smith and R.A. Meyer, *IAEA Report* **INDC/P(02)-22**, 1 (2002).
10. E. Lingerfelt *et al.*, *Bull. Am. Phys. Soc.* **47**, 36 (2002); http://csep10.phys.utk.edu/RatePlotter/start.html

Fragmentation
and
Fission

THE STRUCTURE OF NEUTRON-RICH NUCLEI INVESTIGATED THROUGH REACTIONS OF HIGH-ENERGY BEAMS

THOMAS AUMANN

Gesellschaft für Schwerionenforschung mbH (GSI).
Planckstraße 1,
D-64291 Darmstadt, Germany
E-mail: t.aumann@gsi.de

The structure of neutron-rich nuclei in the mass range from A=6 to A=22 has been investigated via reactions of secondary beams at energies around 250 to 600 MeV/u. One-neutron knockout reactions as well as electromagnetic dissociation experiments have been utilized to explore the single-particle structure and the evolution of the dipole response with increasing isospin. The coincident detection of fragments, neutrons, and γ-rays allows the measurement of unbound resonant states and correlations, as well as differential cross sections.

1. Introduction

Reactions of high-energy radioactive beams have evolved in the past decade as a major tool for detailed nuclear-structure studies of unstable and even unbound nuclei [1]. Single-nucleon knockout reactions have emerged as a very promising tool to study the single-particle structure of exotic nuclei. Similar to transfer reactions, angular momentum assignments can be made and spectroscopic factors can be extracted from measurements of single-nucleon knockout cross sections. The advantage of knockout reactions at high energy is their applicability even with weak secondary beams [2-5] and the availability of precise reaction theory.

Nuclear and electromagnetic inelastic scattering in inverse kinematics opens up the possibility of studying the multipole response of exotic nuclei. Theory predicts considerable changes of the excitation spectra for asymmetric nuclei in comparison to what is known from stable nuclei. An intense theoretical activity has developed in the past years in order to understand this phenomenon. For neutron-rich nuclei different types of calculations predict pronounced effects, in particular a redistribution of the strength towards lower excitation energies well below the giant resonance region [6-10]. The predicted strength functions depend strongly on the effective forces used in the calculations [8] and thus, in turn, the measurement of the multipole response of exotic nuclei can yield information on the isospin dependent part of the in-medium nucleon-nucleon interaction. A first systematic experimental investigation of the dipole response of neutron-rich nuclei was carried out utilizing electromagnetic excitation at high energies, i.e. at several hundred MeV/u [11]. One aspect here is the threshold strength as observed for halo nuclei, which characteristically depends on the single-particle configurations in the ground state. A systematic investigation of this effect was carried out in order to establish the exclusive measurement of differential one-nucleon removal cross sections for electromagnetic excitation as a spectroscopic tool to investigate the ground-state structure of short-lived nuclei. This aspect is discussed in detail in the contribution by Palit *et al.* [12].

2. Experimental Method

The radioactive beams were produced in fragmentation reactions of ^{40}Ar and ^{18}O primary beams delivered by the synchrotron SIS at GSI, Darmstadt, impinging on a beryllium target. The settings of the Fragment Separator FRS [13] and the beam line to the experimental area were chosen to accept secondary fragments with a magnetic rigidity corresponding to a certain mass-over-charge ratio. This mixed secondary beam contained several isotopes, which were identified uniquely according to their nuclear charge and mass number on an event-by-event basis by utilizing energy-loss and time-of-flight measurements. In a similar manner, the fragments produced in the reaction target are identified by measuring energy loss, time-of-flight and the magnetic rigidity. The latter is accomplished by three position measurements in order to determine the trajectory through a large-gap dipole magnet placed behind the target. Neutrons emitted from the excited projectile or excited projectile-like fragments are kinematically focused in the forward direction and detected with high efficiency ($\approx$90%) in the LAND neutron detector [14], which was placed at zero degrees about 11 m downstream from the target. The gap of the dipole magnet allowed an angular acceptance for the neutrons of about ±80 mrad in both horizontal and vertical planes. The angular range for fragments and neutrons covered by the detectors correspond to a 4π measurement in the rest frame of the projectile for fragment-neutron relative energies below 5 MeV (at 500 MeV/u beam energy). Gamma-rays were detected by the 4π Crystal Ball spectrometer surrounding the target. The excitation energy prior to decay is obtained by reconstructing the invariant mass. In order to extract the electromagnetic excitation cross section from the measurement with the lead target, the nuclear contribution was determined from a measurement with a carbon target and scaled accordingly.

3. One-neutron knockout to continuum states

At high beam energies the internal degrees of freedom can be considered as frozen during the reaction. In this 'sudden approximation' the momentum of the fragment after one-neutron removal is equal (with opposite sign) to the initial momentum of the knocked-out halo neutron in the projectile. For 'Borromean' nuclei like ^{6}He or ^{11}Li, the remaining system is unbound and decays in flight into fragment plus neutron. In this case, the momentum distribution of the unbound intermediate system has to be reconstructed from the coincident measurement of the momenta of fragment and neutron. Figure 1 shows the measured momentum distribution of the unbound ^{10}Li (left panel) [2]. The data are well described by considering two components with about equal weight for $l=0$ and $l=1$ contributions to the ^{11}Li wave function, thus demonstrating the disappearance of the N=8 closed shell. The asymmetric angular correlation (right panel of Figure 1) observed in the ^{10}Li decay has its origin in the interference between paths involving states with different parity. For a pure s-wave configuration an isotropic distribution is expected, while a pure p-wave configuration would yield an anisotropic but symmetric angular correlation as it is the case for ^{6}He [15]. The observation of the asymmetric angular correlation gives independent and direct proof for the admixture of s and p components in the ^{11}Li ground state.

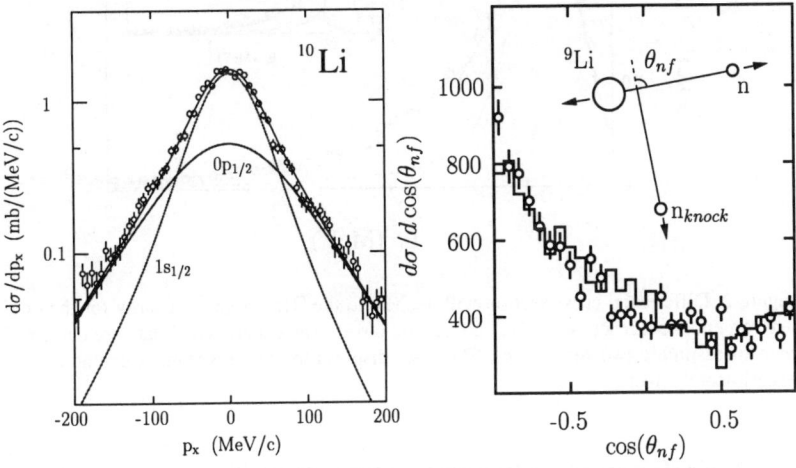

Figure 1. Left panel: Momentum distribution of the unbound ^{10}Li after one-neutron knockout from ^{11}Li [2]. From the fit of the calculated distributions for the s and p states, the occupancy of the two angular-momentum components in the ground state of ^{11}Li are determined. Right panel: Neutron-fragment angular distribution for one-neutron knockout in ^{11}Li and subsequent decay of ^{10}Li [2]. The asymmetric distribution indicates interference between s- and p-wave components.

Figure 2 shows the ^{6}He+n relative energy spectrum after knockout of one neutron from ^{8}He. The 3/2$^-$ ground state of ^{7}He, which is unbound by approximately 0.4 MeV, is clearly visible. In contrast to the breakup of ^{6}He, for which the alpha-n relative-energy spectrum is dominated by the $p_{3/2}$ ground state [15,16], a tail towards higher energies is visible in case of ^{7}He. The spectrum can be described by a fit of two resonances, as displayed in Figure 2 by the solid curve. The second resonance at an energy of 1.0 MeV with a width of 0.75(8) MeV is interpreted as the 1/2⁻ state of ^{7}He [17]. The energy difference between the ground state and the first excited state in ^{7}He amounts to about 0.6 MeV, which should be compared to 1.2 MeV in the case of ^{5}He. Thus, a considerable decrease of the splitting of the two lowest states is observed when adding two more neutrons to ^{5}He. The reduction of the splitting due to a change of the spin-orbit force with diffuseness might be responsible for this observation.

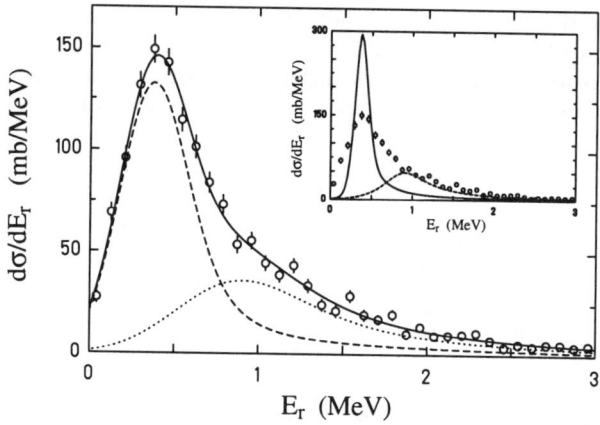

Figure 2. Differential cross section with respect to the ^{6}He-n relative energy for the one-neutron knockout reaction of ^{8}He on a carbon target. The experimental data are described by a fit assuming two resonances. The inset displays the two resonances excluding any experimental effects.

4. Dipole strength of neutron-rich oxygen isotopes

In a first attempt to study giant resonances and lower lying modes in exotic nuclei, we investigated systematically the dipole strength distributions of all neutron-rich oxygen isotopes up to ^{22}O. ^{16}O is a strongly bound doubly magic nucleus. For the heavier isotopes, one may expect a decoupling of the valence neutrons from the inert ^{16}O core. The separation energies of the last neutron lies between 7 and 8 MeV for the even isotopes with A=18 to 22, and around 4 MeV for the odd isotopes, to be compared to a separation energy of 16 MeV for ^{16}O. Thus the neutron-rich oxygen isotopes might be good candidates for a collective soft-dipole excitation.

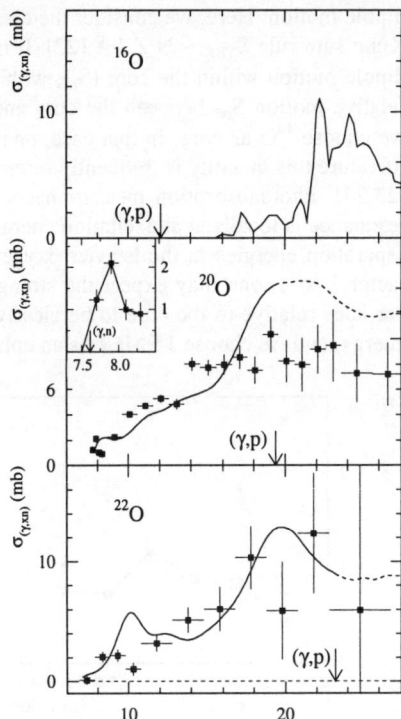

Figure 3. Photo-neutron cross sections $\sigma_{(\gamma,xn)}$ for ^{16}O [18] (upper panel) and for the unstable isotopes ^{20,22}O [19] (lower panels) as extracted from the measured electromagnetic excitation cross section (symbols). The thresholds for decay channels involving protons (which were not observed in the present experiment) are indicated by arrows. For ^{20,22}O, the data are compared to the shell model calculations [10]. Adapted from [19].

From the electromagnetic dissociation cross sections $d\sigma/dE^*$, the photo-neutron cross sections $\sigma_{(\gamma,xn)}$ are extracted by applying semi-classical calculations [20]. Figure 3 shows the deduced $\sigma_{(\gamma,xn)}$ for ^{20,22}O, where neutron multiplicities up to $x=3$ are included [19]. For comparison, the photo-neutron cross sections are shown also for the stable nucleus ^{16}O (upper frame) as measured with real photons [18]. Evidently, the dipole response changes significantly due to the presence of the valence neutrons. Most noticeable is the sizeable dipole absorption cross section below the GDR energy region.

The data are compared to a large-scale shell-model calculation by Sagawa and Suzuki [10] using the Warburton-Brown interaction [21] in Figure 3. The calculated photo-absorption cross sections were convoluted with the experimental resolution. Qualitatively, the shell-model calculations reproduce the experimental observation of a redistribution of the E1 strength towards excitation energies below the GDR compared to that of the doubly magic nucleus ^{16}O.

In the following, the observed low-lying dipole strength is compared to an absolute scale, a measure of which is provided by (model-independent) sum rules. Such a comparison may help to judge on the degree of coherence of the dipole motion. Here, we consider the classical energy-weighted Thomas-Reiche-Kuhn sum rule $S_{TRK} \sim N\,Z\,/\,A$ [22]. It may be decomposed into strength of the dipole motion within the core (S_c), within the valence neutrons S_v, and into the relative motion S_{rel} between the core and the valence nucleons [23]. Naturally, we choose ^{16}O as core. In that case, one obtains $S_{rel} = S_{TRK}\ Z_c/A_c\ N_v/N$. In the literature this quantity is frequently referred to as the cluster sum rule, $S_{Clus} = S_{rel}$ [23,24]. Photoabsorption measurements in ^{16}O [18] show that its giant dipole resonance is localized at excitation energies of 20 - 30 MeV. As the one-neutron separation energies in the heavier oxygen isotopes are smaller than in ^{16}O by a factor 2 to 4, one may expect the strength of the dipole motion of the valence particles relative to the core to be clearly localized below 20 MeV in excitation energy, and we choose 15 MeV as an upper limit.

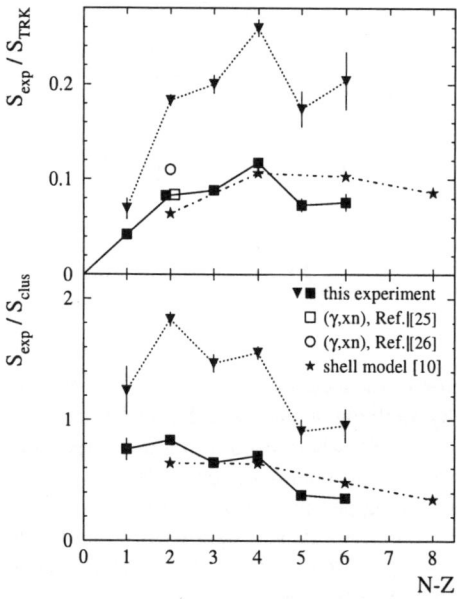

Figure 5. Evolution of integrated (up to 15 MeV excitation energy) strength S_{exp} in units of the TRK sum rule S_{TRK} (upper panel) and of the cluster-sum-rule limit S_{clus} (lower panel) for oxygen isotopes with the neutron excess N-Z. The data (filled squares) are compared to shell-model calculations by Sagawa and Suzuki [10] (stars). The filled triangles display S_{exp} if integrated up to 20 MeV excitation energy. Adapted from [19].

Figure 5 shows the energy-weighted dipole strength integrated over an excitation-energy interval from the one-neutron separation energy up to 15 MeV plotted as a fraction of the classical and the cluster dipole sum rule. We notice, first of all, that this low-lying strength exhausts a sizeable fraction of the classical sum rule, up to 12%. Next, it appears that the limit given by the cluster sum rule is nearly approached in the two lightest isotopes only, with a clear tendency of values decreasing with neutron number.

The missing strength is most likely localized at higher excitation energies, i.e. around the giant resonance domain. This means that a strict separation into core- and valence-neutron-dominated domains, underlying the simplified picture described above, is not realized. This is also indicated if we plot the strength integrated up to 20 MeV, as shown in Figure 5 as filled triangles. In this case, the cluster-sum-rule limit is exhausted for ^{21,22}O, but for $^{18-20}$O the integrated strength clearly exceeds unity. Thus, core excitations already play a role in this energy region, while for ^{16}O the integrated photo-neutron cross section between 15 and 20 MeV corresponds to only 1.6% [18] of S_{TRK}. To which extent the low-lying dipole strength observed in the neutron-rich oxygen isotopes involves coherent excitations or is due to single particle transitions must remain a subject of a detailed theoretical study. In this context, we refer to recent relativistic RPA calculations [9] for neutron-rich nuclei. Therein, in case of oxygen isotopes, low-lying dipole strength is obtained which is comparable in magnitude to our experimental results, but it was concluded that the low-energy transitions can essentially be assigned to single neutron particle-hole excitations rather than to a coherent superposition of many particle-hole configurations.

5. Summary

The kinematically complete measurement of reactions of high-energy secondary beams, such as those discussed here using the combination FRS-Crystal Ball-ALADIN-LAND at GSI, allows the study of the structure of radioactive nuclei at the neutron drip-line and to investigate new phenomena in a systematic way. Break-up reactions give precise information on the spatial extension of the wave function and on the associated quantum numbers and occupation probabilities. The electromagnetic excitation allows the dipole strength distribution of extremely neutron-rich nuclei to be determined. Soft dipole excitations well below the giant dipole resonance are observed, in stark contrast to the dipole response of stable nuclei.

Acknowledgments

The experiments discussed in this article are the result of a collaborative effort. The author likes to thank all members of the S188, S233 and S135 collaborations: D. Aleksandrov, L. Axelsson, K. Boretzky, M.J.G. Borge, L.V. Chulkov, D. Cortina, U. Datta Pramanik, Th.W. Elze, H. Emling, M. Fallot, H. Geissel, A. Grünschloß, M. Hellström, S. Ilievski, N. Iwasa, K.L. Jones, B. Jonson, J.V. Kratz, R. Kulessa, Y. Leifels, A. Leistenschneider, E. Lubkiewicz, G. Münzenberg, F. Nickel, T. Nilsson, C. Nociforo, G. Nyman, R. Palit, P. Reiter, A. Richter, K. Riisager, G. Schrieder, H. Simon, M.H. Smedberg, K.Sümmerer, O. Tengblad, E. Wajda, W. Walús, and M.V. Zhukov.

390

References

1. For a collection of recent review articles we refer to the *Special Issue on Research Opportunities with Accelerated Beams of Radioactive Ions*, edited by I. Tanihata, Nucl. Phys. A **693** (2001).
2. H. Simon *et al.*, Phys. Rev. Lett. **83** (1999) 496.
3. T. Aumann *et al.*, Phys. Rev. Lett. **84** (2000) 35.
4. A. Navin *et al.*, Phys. Rev. Lett. **85** (2000) 266.
5. V. Maddalena *et al.*, Phys. Rev. C **63** (2001) 024613.
6. I. Hamamoto and H. Sagawa, Phys. Rev. C **53** (1996) R1492.
7. F. Ghielmetti, G. Colò, P.F. Bortignon, and R.A. Broglia, Phys. Rev. C **54** (1996) R2143.
8. P.G. Reinhard, Nucl. Phys. A **649** (1999) 305c.
9. D. Vretenar, N. Paar, P. Ring, and G.A. Lalazissis, Phys. Rev. C **63** (2001) 047301; Nucl. Phys. A **692** (2001) 496.
10. H. Sagawa and T. Suzuki, Phys. Rev. C **59** (1999) 3116.
11. T. Aumann *et al.*, Nucl. Phys. A **649** (1999) 297c; Phys. Rev. C **59** (1999) 1252; A. Leistenschneider *et al.*, Phys. Rev. Lett. **86** (2001) 5442.
12. R. Palit *et al.*, these proceedings.
13. H. Geissel *et al.*, Nucl. Instr. and Meth. B **70** (1992) 286.
14. T. Blaich *et al.*, Nucl. Instr. and Meth. A **314** (1992) 136.
15. L. Chulkov *et al.*, Phys. Rev. Lett. **79** (1997) 201.
16. D. Aleksandrov *et al.*, Nucl. Phys. A **633** (1998) 234.
17. M. Meister *et al.*, Phys. Rev. Lett. **88** (2002) 102501.
18. E.G. Fuller, Phys. Rep. **127** (1985) 187; A. Vessière *et al.*, Nucl. Phys. A **227** (1974) 513; B.L. Berman, At. Data Nucl. Data Tables **15** (1975) 319.
19. A. Leistenschneider *et al.*, Phys. Rev. Lett. **86** (2001) 5442.
20. C.A. Bertulani and G. Baur, Phys. Rep. **163** (1988) 299.
21. B.A. Brown, Progr. in Part. and Nucl. Phys. **47** (2001) 517.
22. A. Bohr, B.R. Mottelson, *Nuclear Structure* (Benjamin, London, 1975), Vol. 2.
23. H. Sagawa and M. Honma, Phys. Lett. B **251** (1990) 17.
24. Y. Alhassid, M. Gai, G.F. Bertsch, Phys. Rev. Lett. **49** (1982) 1482.
25. U. Kneissl *et al.*, Nucl. Phys. A **272** (1976) 125.
26. J.G. Woodworth *et al.*, Phys. Rev. C **19** (1979) 1667.

ISOMER SPECTROSCOPY OF HEAVY NEUTRON-RICH NUCLEI PRODUCED IN RELATIVISTIC FRAGMENTATION

M. PFÜTZNER

IEP, Warsaw University, Hoża 69, 00-681 Warszawa, POLAND
E-mail: pfutzner@mimuw.edu.pl

P.H. REGAN, ZS. PODOLYÁK, P.M. WALKER

Dept. of Physics, University of Surrey, Guildford, GU2 7XH, UK

J. GERL, M. HELLSTRÖM, P. MAYET, CH. SCHLEGEL

GSI, Planckstrasse 1, Darmstadt, Germany

FOR THE GSI ISOMER COLLABORATION

We present an overview of results obtained at the Fragment Separator of GSI Darmstadt where beams of ^{238}U and ^{208}Pb at energies of 1 GeV/nucleon were used to study microsecond isomers in neutron-rich heavy nuclei. Selected examples of new isomeric states identified in these experiments will be given. A systematic study of the population probabilities of isomeric states will be discussed. Measured values of the isomeric ratio will be compared with the estimates from a simple model using the sharp cut-off approximation.

1. Introduction

The method of projectile fragmentation followed by a separation of reaction products in a magnetic spectrometer has proven to be an efficient and highly sensitive technique for the production and study of microsecond isomers[1]. The combination of fragmentation with delayed gamma-ray spectroscopy has opened possibilities for obtaining nuclear structure information on nuclei which are virtually inaccessible using other techniques. The unprecedented sensitivity of this method results from the unambiguous in-flight identification of single ions transmitted by the separator. This allows a clear correlation between the detected, delayed γ rays with individually identified ions, resulting in a substantial background reduction. In some cases, the production of as little as a few hundred ions is sufficient to provide useful information on the excited states of some exotic nuclei, which makes this technique an ideal tool for studies of nuclei very far from the line of beta-stability.

The systematic application of isomeric spectroscopy following projectile fragmentation to nuclear structure studies was initiated by Grzywacz *et al.* [2] at GANIL laboratory. Using the LISE spectrometer fragmented beams of ^{112}Sn, ^{92}Mo, ^{78}Kr and ^{86}Kr were studied and a wealth of interesting and important results were obtained in medium-mass nuclei on both the neutron-deficient[3,4] and neutron-rich[5,6] sides of the valley of stability.

In order to apply this technique to heavier nuclei, high-energy beams of heavier projectiles, such as ^{238}U or ^{208}Pb, are required. Presently, only GSI Darmstadt laboratory offers a unique combination of such beams, together with necessary instrumentation to undertake these studies. Results of the first experiment, performed at GSI with the ^{238}U beam at 1000 MeV/nucleon and devoted to microsecond isomers in vicinity of ^{208}Pb, were reported at the first conference of this series[7]. In a following GSI experiment, isomers were searched among fragmentation products of the ^{208}Pb beam at 1000 MeV/nucleon. Some preliminary results were presented at the second conference[8]. Here, we shortly summarize these two experiments. Most of the results presented here have been published in a number of publications and conference proceedings [9,10,11,12,13,14,15,16,17].

2. Experimental Considerations

The ions of interest were produced by the fragmentation reaction of a heavy ion beam, delivered by the GSI synchrotron SIS, impinging on a target located at the entrance of the Fragment Separator (FRS). The FRS was operated in the standard achromatic mode with a wedge-shaped degrader mounted in the central focal plane. The unambiguous separation and identification of heavy fragments suffers from a substantial probability that ions emerging from the target and degrader are not fully stripped of their atomic electrons. In order to maximize the stripping probability, a high energy of primary beam (1000 MeV/nucleon) was selected and relatively thin targets of about 1 g/cm^2 beryllium were used. Additionally, niobium stripping foils were mounted behind (downstream) the target and the intermediate degrader. Typically, the probability for fully stripped ions in the first section of the FRS (between the target and the degrader) was larger then 90 %, while in the second section it was above 70 %.

Ions transmitted to the final focus were identified in-flight in the second section of the FRS. This was achieved by measuring the time-of-flight (TOF) as well as horizontal positions in the intermediate and in the final focal planes by means of position sensitive plastic scintillators. Additional information on the atomic number Z was delivered by an ionization chamber located at the final focus. Combining the values of TOF, position, energy loss and magnetic fields, the mass-over-charge ratio A/q and atomic number Z could be uniquely determined for each ion.

After passing the identification detectors at the final focus, the ions were slowed down in an aluminum degrader of variable thickness and implanted into an aluminum catcher surrounded by an array of germanium γ detectors. Gamma rays which were detected within a period of up to 80 μs after the implantation of an ion were recorded together with the particle identification data as a single event. In the original ^{238}U beam experiment only two germanium detectors were used. The array mounted for the ^{208}Pb beam experiment consisted of three segmented clover detectors from the EXOGAM collaboration[18] and a large-volume GSI 'Super Clover' detector. In both runs the germanium detectors were mounted as close as possible to the catcher but without blocking the heavy-ion trajectories.

A disadvantage of the FRS separator is the large transverse area covered by ions in the final focus. In order to stop all transmitted nuclei, a catcher of dimensions of 10 cm high and 20 cm wide is necessary. As a consequence, the efficiency of the γ-ray array depends on the isotopically varying implantation position. This effect is illustrated in Fig. 1 which shows the full energy peak efficiency for 100 keV and 1000 keV determined for the array used in the ^{208}Pb beam experiment. A further difficulty is posed by a burst of prompt radiation which accompanies the stopping process of each ion in the catcher[1]. Since only the first γ ray is detected by each germanium crystal, and detection of prompt events is allowed (even 'necessary' for ns isomers), the burst reduces the effective sensitivity of the array to delayed radiation. The estimated influence of such prompt blocking is shown in Fig. 1.

3. Selected Results

3.1. *Fragmentation of* ^{238} *U*

The ^{238}U beam experiment represented the first attempt to produce heavy neutron-rich isotopes by fragmentation of uranium. This approach proved to be very successful, in spite of the primary-beam intensity being limited to only 5×10^6 ions per second, 7 previously unreported isotopes in the region

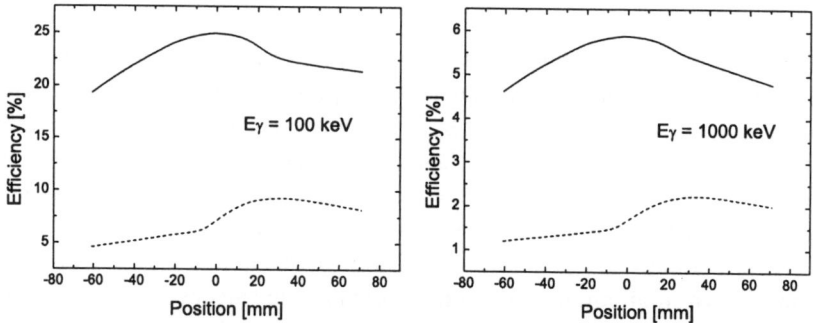

Figure 1. Total efficiency of the γ-ray array used in the ^{208}Pb beam experiment as a function of the horizontal position of the emitting source for gamma-ray energies of 100 keV (left panel) and 1000 keV (right panel). Solid lines show the efficiency measured with the standard calibration sources, the dashed lines represent the effective efficiency, corrected for the prompt blocking (see Ref.[1]for details). Note that the detectors were not mounted symmetrically with respect to the central position.

between mercury and polonium were identified for the first time[9]. Particularly striking was the production of new isotopes of thallium (^{211}Tl, ^{212}Tl) and polonium (^{219}Po, ^{220}Po). The difficulty in producing these heavy, neutron-rich isotopes can be illustrated by noting that the first observation of the isotopes of these elements with one neutron less were first reported[19] more then 90 years ago by authors such as E. Rutherford, O. Hahn and L. Meitner! The isomer spectroscopy aspect of this first experiment resulted in the observation of 9 previously known isomers as well as evidence for 4 new ones. For the first time it was shown that, contrary to expectations, states with spins as high as $25/2\ \hbar$ could be populated in the violent relativistic fragmentation process. Moreover, the population probabilities (see the next section) were found to be rather large, ranging between 25 % and 50 % (see Table 1).

Among the isomers observed for the first time, arguably the most interesting was ^{212m}Pb, with a deduced half-life of $T_{1/2} \approx 5\ \mu$s. Although the gamma spectrum correlated with this nucleus had very low statistics, at least one γ-ray transition at an energy of 315 keV was clearly evident. Our analysis of this nucleus suggests that most probably this line arises from the $4^+ \rightarrow 2^+$ transition following the decay of a predicted $I^\pi = 8^+$ isomer which a predominantly $\nu(g_{9/2}^2)_8$ structure. It is worth pointing out that this information was gained with only approximately 400 implanted ions of ^{212}Pb. A measurement of the delayed γ-ray spectrum of this nucleus with much larger statistics would be certainly desired, and recent work on the use of heavy-ion induced deep-inelastic reactions with large efficiency gamma-ray arrays such as GAMMASPHERE may help in this area[20]. However, it is worth noting that the fragmentation isomer spectroscopy of ^{212}Pb is particularly difficult because the magnetic rigidity of the ^{212}Pb is close to that of two charge states of the primary beam, both of which have very large intensities and thus could not be allowed to enter any of the identification detectors.

3.2. *Fragmentation of* ^{208}Pb

The highly promising results from the first uranium fragmentation experiment motivated the next experiment in which a primary beam of ^{208}Pb was used to study nuclei mainly in two regions: one centered on the deformed $A \sim 180$ nuclei close to stability, and the second covering very neutron-rich nuclei with $A \sim 190$.

The first region, which is well known to be rich in K isomers, was explored in order to obtain new information on the angular momentum population distributions in fragmentation reactions. Indeed, decays of about 20 known isomeric states were observed[1]. Notably, three separate isomers with the spin of $35/2 \, \hbar$ were populated, namely ^{175}Hf, ^{179}W, and ^{181}Re [16]. To dates, these isomers represent the highest discrete spin observed in a fragmentation reaction.

In the second region studied with the ^{208}Pb primary beam, 11 previously unreported neutron-rich isotopes of elements between hafnium and osmium were identified and evidence for 10 new isomeric states was obtained. Additionally, the previously reported isomers in ^{200}Pt and ^{206}Hg, which were transmitted as H-like and He-like ions, respectively, were observed. The former represents a particularly interesting case. The 7^-, $T_{1/2} = 14$ ns state survives the 300 ns flight time through the FRS because its main decay mode (by L and M-shell internal conversion) is blocked in the highly charged ionic state. Moreover, the large statistics in the γ spectrum collected for this isotope allowed a γ-γ coincidence analysis which revealed a previously unknown isomeric state with $T_{1/2} = 10$ ns which is predicted to have a 4 quasi-particle nature[12].

Among the isomers observed for the first time in this experiment, the case of ^{190m}W [11] is of particular note. The observed cascade of 5 γ-ray transitions results from the decay of a proposed $K^\pi = 10^-$ isomer to the 10^+ level of the ground-state rotational band. The deduced E_{4+}/E_{2+} energy ratio, equal to 2.72, suggests a γ-soft nucleus. However, when compared with neighboring nuclei, a striking deviation from systematics is evident. The observed pattern is similar to the one found in vicinity of ^{152}Gd, manifesting a breakdown of the $Z = 64$ sub-shell closure [21]. This may indicate that new and interesting structure phenomena occur close to ^{190}W. Future studies will certainly focus on this region.

Another case worth mentioning is ^{195}Os in which a $T_{1/2} = 25$ ns isomer was observed yielding the first information on γ-ray transitions in this nucleus. This information has been complemented by a recent experiment using a deep-inelastic reaction between a ^{136}Xe beam and a ^{198}Pt target with CHICO detector plus GAMMASPHERE in which much larger statistics were collected[22]. The region of heavy osmium isotopes focuses attention because of predicted rapid change of shape between ^{190}Os and ^{196}Os [23].

Finally, in a setting centered on lighter neutron-rich isotopes, the doubly-mid-shell nucleus $^{170}_{66}$Dy$_{104}$ was tentatively identified for the first time. This is the first step towards studying this exotic nucleus which has the maximal number of valence particles for any nucleus below ^{208}Pb. Among its predicted properties[24] is a low lying $K^{\pi} = 6^+$ state which should be one of the best examples of a good, axially symmetric K-isomeric state.

4. Isomeric ratios

From the number of implanted ions and γ-rays depopulating an isomeric state, the population probability of this isomer (isomeric ratio) can be determined. Because of necessary corrections for the in-flight decay losses and for the finite observation time, the half-life and the detailed decay scheme have to be known. Therefore, reliable information on the angular momentum population in fragmentation reactions can only be derived from data on 'well known' isomers. Five such isomeric ratios were determined from ^{238}U beam data[9] and 26 from the ^{208}Pb beam experiment[1].

In references [1,9] we proposed a simple model of isomers population based on the abrasion-ablation approach to the fragmentation process, which yields the distribution of angular momentum in final fragments after the particle evaporation phase[25]. It is then assumed that in the subsequent statistical γ-ray cascade, all levels with angular momenta larger than that of the isomer (and only those) decay to the isomer. This, so called *sharp cut-off approximation* is expected to be justified only for isomers lying on the yrast line or very close to it. Moreover, it is assumed that there are no very long-lived states which feed into the isomeric state of interest.

The isomeric ratios for the yrast states, measured in both experiments,

Table 1. Previously reported isomers at (or very close to) the yrast line, identified among fragmentation products of ^{238}U (upper part) and of ^{208}Pb (lower part). For each nucleus the spin, parity, excitation energy, and the half-life of the corresponding isomer are given, followed by the experimentally determined isomeric ratio and a value predicted by the sharp cut-off model.

Nuclide	I^π	E^* [keV]	$T_{1/2}$ [μs]	isomeric ratio exp.	th.
^{205}Tl	$25/2^+$	3291	2.6	0.25(5)	0.23
^{205}Pb	$25/2^-$	3196	0.217	0.29(5)	0.23
^{206}Pb	7^-	2200	123	0.52(8)	0.58
	12^+	4027	0.197	0.29(5)	0.25
^{208}Pb	10^+	4895	0.500	0.34(9)	0.34
^{175}Hf	$35/2^-$	3016	1.21	2.5(6)	3.8
^{179}Ta	$21/2^-$	1252	0.32	9(3)	24.9
^{180}Ta	15^-	520	31	10(3)	6.1
^{179}W	$35/2^-$	3348	0.75	2.7(5)	2.4
^{200}Pt	7^-	1617	0.014	30(5)	35.5

are summarized in Table 1, together with the predictions of the sharp cut-off model. A very good agreement between the experiment and the model can be seen, except for the ^{179}Ta case. However, in this nucleus there is a long-lived isomer ($T_{1/2} = 11$ ms, $J^\pi = 25/2^+$) located just above (65 keV) the $21/2^-$ state [26]. A part of the γ-ray cascade may be trapped by this long-lived state, which would have the effect of substantially reducing the feeding and thus the observed decay rate of the lower-lying isomer.

Other, non-yrast isomers identified in the ^{208}Pb experiment were found to be populated mostly with lower probabilities than predicted by our model. This can be explained partly by the fact that the statistical cascade most likely omits levels located far from the yrast line but also by a wealth of long-lived, higher-lying K-isomers occurring in the rare-earth $A \sim 180$ region. That said, a general conclusion clearly emerges from this study that the sharp cut-off model represents un upper limit for the isomeric ratio. This conclusion has also been recently reached for analogous studies of neutron-*deficient* nuclei with uranium beams [27].

5. Summary and Future Directions

In two experiments performed at the GSI Fragment Separator heavy neutron-rich projectile fragments of ^{238}U and ^{208}Pb were investigated. Several new isotopes were identified demonstrating the potential of the fragmentation technique for future studies of the structure of heavy, exotic nuclei. Delayed γ-ray spectroscopy was applied to search for and study microsecond isomers. It was shown that levels with angular momenta up to $35/2$ $\hbar$ are populated with substantial probabilities. A simple model of isomeric ratios, based on the abrasion-ablation picture of fragmentation reactions was found to be generally in good agreement with experimental results, specifically in case of near-yrast isomeric states. In a more general case it was found to represent an upper limit of the isomeric ratio.

The first information on γ transitions in very neutron-rich, heavy nuclei, including ^{212}Pb, ^{195}Os and ^{190}W, was obtained triggering interest for deeper studies of their structure. The success of these isomer spectroscopy studies at the FRS was one of the motivating factors behind the installation of RISING array at GSI. This high efficiency gamma-ray array, composed of EUROBALL detectors, will start operating for in-beam fragmentation studies in 2003, with a dedicated Stopped Beam Campaign aimed at isomeric states envisaged for 2005. By yielding new data with much better statistics, this program should be able to shed new light on the structure of very neutron-rich isotopes of heavy elements.

References

1. M. Pfützner et al., Phys. Rev. C **65**, 064604 (2002), and references therein.
2. R. Grzywacz *et al.*, Phys. Lett. B **355**, 439 (1995).
3. R. Grzywacz *et al.*, Phys. Rev. C **55**, 1126 (1997).
4. C. Chandler *et al.*, Phys. Rev. C **56**, R2924 (1997).
5. J.M. Daugas *et al.*, Phys. Lett. B **476**, 213 (2000).
6. R. Grzywacz *et al.*, a contribution to this volume.
7. M. Pfützner *et al.*, Proc. of the First Int. Conf. on Fission and Properties of Neutron-Rich Nuclei, Sanibel Island, Florida, USA, 1997, edited by J.H. Hamilton and A.V. Ramayya, World Scientific, p.431.
8. Zs. Podolyák *et al.* in Proc. of the Second Int. Conf. on Fission and Properties of Neutron-Rich Nuclei, St. Andrews, Scotland, 1999, edited by J.H. Hamilton, W.R. Phillips and H.K. Carter, World Scientific, London, p. 156.
9. M. Pfützner *et al.*, Phys. Lett. B **444**, 32 (1998).
10. C. Schlegel *et al.*, Phys. Scripta T **88**, 72 (2000).
11. Zs. Podolyák *et al.*, Phys. Lett. B **491**, 225 (2000).
12. M. Caamaño *et al.*, Acta Phys. Pol. B **32**, 763 (2001).
13. M. Caamaño *et al.*, Nucl. Phys. A **682**, 223c (2001).
14. Zs. Podolyák *et al.*, in Proc. of the Int. Conf. on Exotic Nuclei at the Proton Drip Line, ed. by C.M. Petrache and G. Lo Bianco, September 25-28, 2001, Camerino, Italy, p. 189.
15. Zs. Podolyák *et al.*, in Proc. of the Int. Conf. on the Structure of the Nucleus at the Dawn of the New Century: Nuclear Structure, Bologna, Italy (2001) ed. by G.C. Bonsignori, M. Bruno, A. Ventura and D. Vretenar, World Scientific,

London, p. 298.

16. Zs. Podolyak *et al.,* Progress of Theoretical Physics Supplement **146**, 467 (2002).

17. M. Pfützner *et al.,* in Proc. of the Int. Conf. on Experimental Nuclear Physics in Europe, Sevilla, Spain 1999, ed. by B. Rubio, M. Lozano and W. Gelletly, AIP Conf. Proc. **495** p. 113.

18. J. Simpson *et al.,* Heavy Ion Physics **11**, 159 (2000).

19. E. Rutherford, Phil. Trans. Roy. Soc. **204**, 169 (1904); O. Hahn, L. Meitner Phys. Zeit. **10**, 697 (1909).

20. G.J. Lane et al., Nucl. Phys. **A682**, 71c (2001); B. Fornal *et al.,* Phys. Rev. Lett. **87**, 212501 (2001).

21. R.F. Casten *et al.,* Phys. Rev. Lett. **47**, 1433 (1981).

22. P. H. Regan, private communication.

23. C. Weldon *et al.,* Phys. Rev. **C 63**, 011304(R) (2000).

24. P. H. Regan *et al.,* Phys. Rev. **C 65**, 037302 (2002).

25. M. de Jong, A.V. Ignatyuk, K.-H. Schmidt, Nucl. Phys. **A 613**, 435 (1997).

26. T. Saitoh et al., Phys. Scripta **T88**, 67 (2000); F.G. Kondev *et al,* Nucl. Phys. **A617**, 91 (1997)

27. K. Gladnishki *et al.,* Acta Phys. Pol. **B**, in press.

REACTION MECHANISMS INVOLVED IN THE PRODUCTION OF NEUTRON-RICH ISOTOPES

J. BENLLIURE

Universidad de Santiago de Compostela,
15706 Santiago de Compostela,Spain
E-mail: j.benlliure@usc.es

K. HELARIUTTA, K.-H. SCHMIDT, M.V. RICCIARDI

Gesellschaft für Schwerionenforschung, Planckstrasse 1
64291 Darmstadt, Germany
E-mail: k.h.schmidt@gsi.de

The reaction mechanisms best suited for the production of neutron-rich nuclei, fragmentation and fission, are discussed. Measurements of the production cross sections of reaction residues together with model calculations allow to conclude about the expected production rates of neutron-rich isotopes in future facilities.

1. Introduction

The production of nuclei far from stability has provided a new ground to investigate the structure and the dynamics of nuclei. The present experimental facilities are able to produce a large variety of neutron-deficient isotopes, however, the neutron-rich side of the chart of the nuclides is only accessible up to the drip lines for the lightest nuclei. Consequently, the challenge for most of the future planned rare-beam facilities is the production of neutron-rich isotopes.

Together with the technological improvements concerning high-current accelerators, large-acceptance separators and larger extraction and charge-breeding efficiencies, the optimal choice of the reaction mechanisms will play a major role on the final production rates of neutron-rich nuclei. Different reaction mechanisms can be used to produce these nuclei, however, heavy-ion collisions at low energies like fusion, deep inelastic or multinucleon transfer can only be applied with thin targets limiting the final production rates. Better suited seem to be fragmentation or spallation at high energies and fission. In addition, these two reaction mechanisms allow to produce a

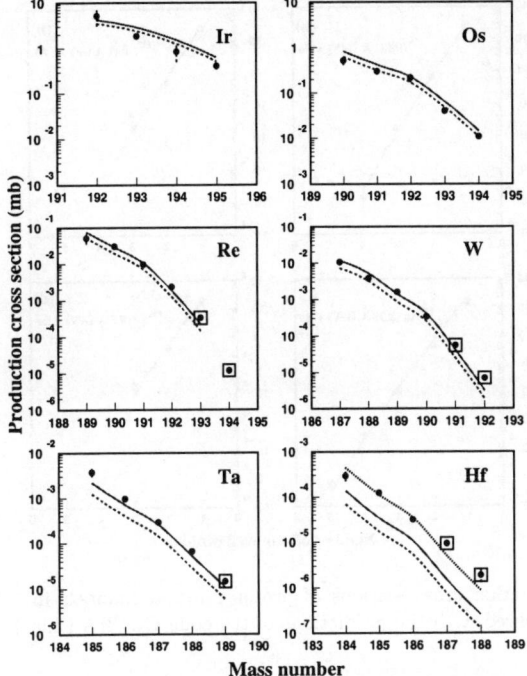

Figure 1. Production cross sections of neutron-rich nuclei in cold-fragmentation reactions of ^{197}Au at 1 A GeV in a beryllium target. The lines represent different model calculations with the code COFRA (see text). The squares mark those isotopes which were observed for the first time in reference [6].

large variety of neutron-rich nuclei.

In this paper we discuss the main issues of these two reaction mechanisms. The large collection of data obtained during the last years has brought new model descriptions of these reactions providing reliable predictions on the expected production rates of neutron-rich isotopes in future planned rare-beams facilities.

2. Production of heavy neutron-rich nuclei

Heavy exotic nuclei ($Z > 70$) can be produced by means of fusion-evaporation reactions or in fragmentation(spallation) of heavy nuclei. Both reaction mechanisms lead mainly to the production of neutron-deficient residues. However, it has recently been shown that fragmentation reactions at relativistic energies present large fluctuations in the N/Z distribution of

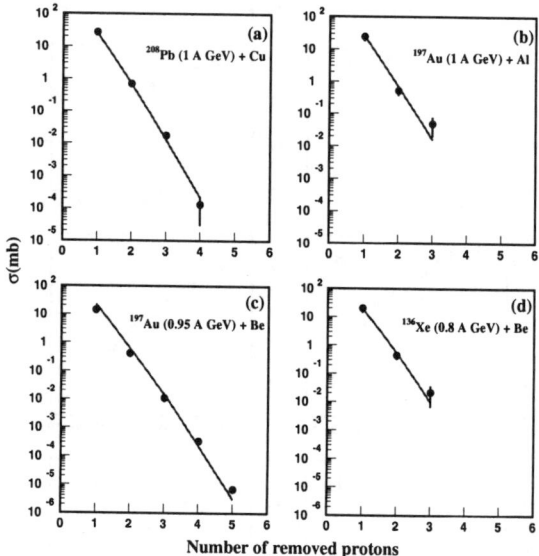

Figure 2. Production cross sections of proton-removal channels in different reactions (dots) [2,3,4] compared with the predictions of the code COFRA (lines).

the final residues and in its excitation-energy distribution. In particular, the proton-removal channels have been investigated in cold-fragmentation reactions[1] where only protons are abraded from the projectile, while the induced excitation energy is below the particle-emission threshold. These reactions can lead to the production of heavy neutron-rich nuclei beyond the present limit of the chart of the nuclides as shown in Fig. 1.

Cold-fragmentation reactions can be described in terms of the abrasion-ablation model as a two-step process. First, the interaction between projectile and target leads to a projectile-like residue with a given excitation energy which statistically de-excitates by particle evaporation or fission. A new analytical formulation of the abrasion-ablation model, the code COFRA, has been developed[1] in order to calculate the expected low production cross sections of extremely neutron-rich nuclei which are not reachable with Monte Carlo codes.

The results of these calculations are shown in Fig. 1. The different lines correspond to calculations with different enhancement factors of the excitation energy induced in the collision due to particle-hole excitations (see reference[1] for details). As can be seen, the production cross sections

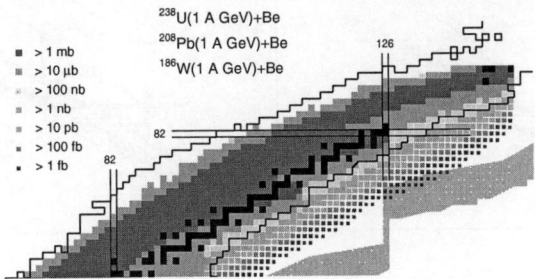

Figure 3.   Estimated production of heavy neutron-rich residues in cold-fragmentation reactions induced by ^{238}U, ^{208}Pb and ^{174}W at 1 A GeV in a beryllium target on top of a chart of the nuclides. The color scale indicates the production cross section.

of very neutron-rich nuclei can nicely be described with the new analytical formulation of the abrasion-ablation model. In Fig. 2 we compare the predictions of the code with the production cross sections of proton-removal channels populated in different reactions [1,2,3].

Since the COFRA code describes correctly all the measured proton-removal channels in relativistic heavy-ion collisions, it was used to determine the expected production of heavy neutron-rich nuclei in future rare-beam facilities. The results of these calculations are shown in Fig. 3. In this figure we report the expected production cross sections of heavy neutron-rich nuclei that can be obtained in the fragmentation of ^{238}U, ^{208}Pb and ^{174}W. According to these calculations, large progress is expected in this region of the chart of the nuclides, where the r-process path may even be reached around the end point N=126.

3.   Production of medium-mass neutron-rich nuclei

3.1. *Fission*

Fission has largely been used to produced medium-mass neutron-rich nuclei up to the present limits of the chart of the nuclides [4]. The isotopic distribution of residues produced in fission can be understood in terms of the potential governing this process. The Coulomb term of the nuclear potential is responsible for the neutron excess of the stable fissile nuclei leading to fission residues with an even larger neutron excess with respect to the valley of beta stability. However, the asymmetry term preserves the same N/Z of the fissioning nucleus in the fission residues. Shell effects and temperature induce a polarization effect which allow to produce even more

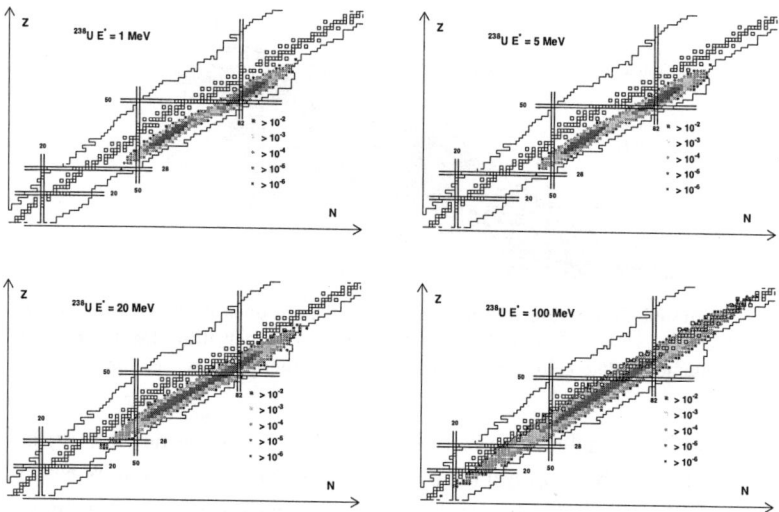

Figure 4. *Residual nuclei produced after the fission of* ^{238}U *at different excitation energies.*

neutron-rich residues.

In order to investigate the fluctuations in N/Z and mass asymmetry induced by the temperature of the fissioning system we have performed several simulations with the fission code of Ref.[5]. In figure 4 we represent on top of a chart of the nuclides the distributions of residues after the fission of ^{238}U at different excitation energies. As can be seen in the figure, when increasing the excitation energy, shell effects (double humped distribution) disappear, and the fluctuations in mass asymmetry and N/Z increase, populating a larger variety of neutron-rich residues. However, at high excitation energies neutron evaporation becomes more important, and the residue distribution moves to the neutron-deficient side. One can define an optimum excitation energy of the fissioning system around 50 MeV to produce the largest variety of neutron-rich nuclei.

Once we know the optimum conditions to produced neutron-rich nuclei in fission reactions, the final production rates will be defined by the evolution of the production cross sections. In Fig.5, we report the production cross sections of several neutron-rich tin and nickel isotopes produced in the fission of ^{238}U projectiles on Be[4], Pd[6] and H_2[7]. These data show a dramatic decrease of the production cross section of around one order of magnitude per additional neutron. Consequently, one can not expect a spectacular expansion of the chart of the nuclides in the region of medium-mass neutron-rich isotopes by using fission reactions with the future rare-beam facilities.

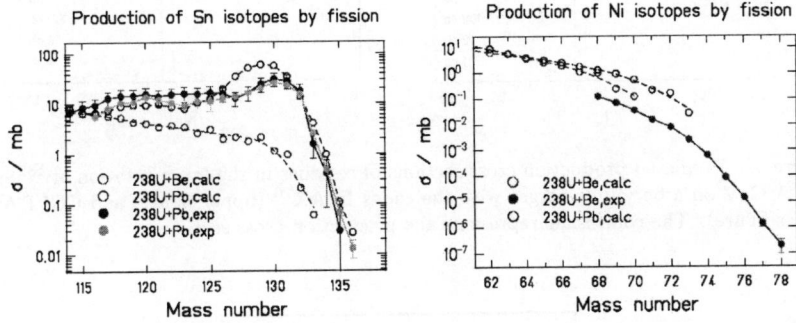

Figure 5. Production cross sections of tin and nickel isotopes produced in the fission of ^{238}U induced by different projectiles

3.2. *Two-step scenario: fission + cold fragmentation*

Recently it has been proposed to use both fission and cold-fragmentation reactions, in a two-step reaction scheme[8], in order to overcome the limitations of fission to produce extremely neutron-rich nuclei in the medium-mass region. However, nowadays it is difficult to make reliable predictions of the final production rates using this idea. As shown in Fig. 6, present fragmentation codes show clear discrepancies in the predicted production rates when neutron-rich projectiles are used. In addition, the energy at which the fragmentation stage takes place plays a major role as shown in Fig. 7. In this figure we report the measured production cross sections of different tin isotopes in the reaction ^{129}Xe on aluminum at 790 A MeV [10] and 50 A MeV [11]. According to these data the higher energies enhance the production of neutron-rich residues.

Nevertheless, we used the cold-fragmentation code COFRA, which is one of the more reliable codes to predict the production of fragmentation residues from neutron-rich projectiles to estimate the production rates in a two-step reaction scheme at energies above 100 A MeV. In these calculations, the primary production cross sections in the fission step were taken from measured data in the reactions ^{238}U(1 A GeV)+p [7] and ^{238}U(750

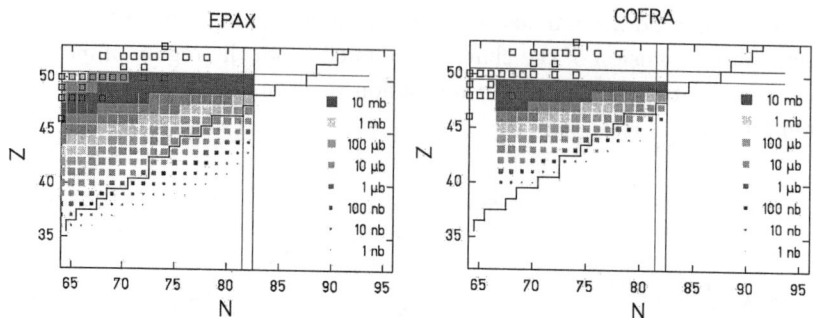

Figure 6. Predicted production cross sections of residues in the fragmentation of ^{132}Sn at 1 A GeV on a beryllium target with the codes EPAX[10] (upper figure) and COFRA[2] (lower figure). The color scale represents the production cross section.

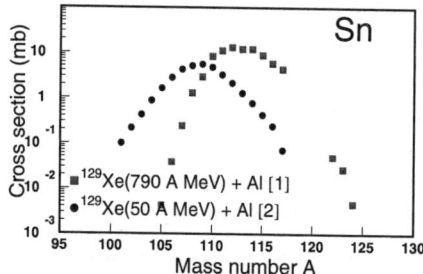

Figure 7. Measured production cross sections of tin isotopes in the reaction ^{129}Xe on aluminum at 790 A MeV (squares)[11] and 50 A MeV (dots)[12].

A MeV)+Be [4]. The most representative results of these calculations are shown in figure 8. In this figure we represent the production of different neutron-rich isotopes along the neutron shells N=50 and N=82. The thick line represents the direct production in the fission of ^{238}U induced by 1 GeV protons, while the thin lines correspond to the two-step production after the cold fragmentation in a Be target of different Ga and Sn isotopes produced by fission. As can be observed in both pictures, the two-step scheme is competitive with the direct production only for the largest neutron excess.

However, if one considers extraction, ionization and re-acceleration efficiencies, the situation could change. In this realistic scenario the two-step scheme could be used to produce by fission an abundant long-lived neutron-rich nuclei like ^{132}Sn and fragmenting it to produce those neutron-rich isotopes that have low extraction efficiencies.

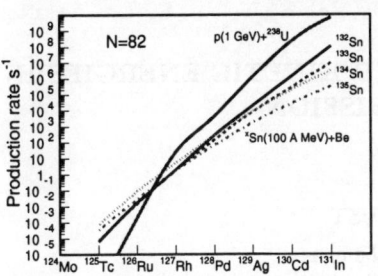 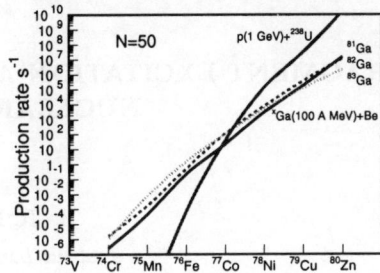

Figure 8. Production of neutron-rich isotopes along the neutron shells N=50 and N=82 after the fission of ^{238}U induced by 1 GeV protons (thick lines), and after the cold fragmentation in a beryllim target of different neutron-rich gallium and tin isotopes produced in by fission (thin lines).

4. Conclusion

In this paper we have reviewed the two reaction mechanisms best suited to extend the present limits of the chart of the nuclides in the neutron-rich side. Recent experiments have shown that the cold-fragmentation process constitutes the appropriate tool to produce heavy neutron-rich residues. In the intermediate-mass region it is well established that fission allows to produce abundantly moderately neutron-rich isotopes. However, it will be difficult to enlarge the present limits of the chart of the nuclides using this reaction mechanism. An alternative would be to use a two-step scenario where the cold fragmentation of neutron-rich isotopes produced by fission is foreseen. Although this new scenario is only competitive with fission for isotopes with a large neutron excess, it would be an optimum solution to produce neutron-rich isotopes with low extraction efficiencies.

References

1. J. Benlliure et al., *Nucl. Phys.* **A660**, 87 (1999).
2. M. de Jong et al., *Nucl. Phys.* **A628**, 479 (1998).
3. K.-H. Schmidt et al., *Nucl. Phys.* **A542**, 699 (1992).
4. M. Bernas et al., *Phys. Lett.* **B415**, 111 (1997).
5. J. Benlliure et al., *Nucl. Phys.* **A628**, 458 (1998).
6. T. Enqvist et al., *Nucl. Phys.* **A658**, 47 (1999).
7. M. Bernas et al., in preparation
8. K. Helariutta et al., submitted to *Eur. Phys. J.* **A**
9. K. Sümmerer, B. Blank, *Phys. Rev.* **C61**, 034607 (2000).
10. J. Reinhold et al., *Phys. Rev.* **C58**, 247 (1998)
11. K.A. Hanold et al., *Phys. Rev.* **C52**, 52 (1995)

FRAGMENT EXCITATION AND KINETIC ENERGIES IN NUCLEAR FISSION

H. R. FAUST

Institut Laue Langevin
6, rue Jules Horowitz
F-38043 Grenoble, FRANCE
E-mail: faust@ill.fr

The model for the statistical excitation of fragments in nuclear fission (Random EXcitation Model, REX-M) is reformulated, and closed expressions are given for the calculation of mean values for fragment excitation, total excitation energy, total kinetic energy and single kinetic energies. We give furthermore the expressions for the distribution functions for the above mentioned observables. The calculations are compared to the results of a new experiment, where in particular the kinetic energy distributions of fragments from thermal neutron induced fission of ^{233}U have been determined with high resolution.

1. The Model

The Ansatz which is made in the model is, that at the end of the process the temperature kT of the fragments in spontaneous fission is proportional to the Q-value which is available for the specific mass and charge split [1]

$$kT = \bar{f}Q. \tag{1}$$

The above equation states that a constant fraction of the Q-value is converted into temperature of the fragments. The constant $\bar{f}$ is taken to be independent of the compound nucleus and the way this nucleus disintegrates. The value of $\bar{f}$ has been fitted to a variety of fissioning systems between Thorium and Fermium [2], and its numerical value is

$$\bar{f} = 0.0045 \tag{2}$$

The Q-value of the process is calculated from the mass excesses of the particles involved

$$Q = \Delta_{CN} - (\Delta_{f1} + \Delta_{f2}), \tag{3}$$

where Δ_{CN} is the mass excess of the compound nucleus, and $\Delta_{1,2}$ are the mass excess values for the fragments.

From the statistical model the relationship between the temperature of a nucleus and its mean excitation is known to be

$$< E^* >= a \cdot kT^2. \tag{4}$$

The constant which connects excitation energy and temperature is the level density parameter a. Its value is taken from tables.

2. Mean values for fragment excitation and kinetic energies

2.1. *Spontaneous fission*

For spontaneous fission the process starts from the ground state of the compound nucleus, and the mean excitation energy of fragment 1 is

$$< E_1^* >= a_1 \cdot (\bar{f}Q)^2. \tag{5}$$

The mean excitation energy of fragment 2 becomes

$$< E_2^* >= a_2 \cdot (\bar{f}Q)^2 \tag{6}$$

and the mean total excitation energy $< TXE >$ is the sum of the mean excitation of both fragments

$$< TXE >= (a_1 + a_2) \cdot (\bar{f}Q)^2. \tag{7}$$

We assume that no other mechanism for the excitation of the fragments is present than the one given by Eq. (1). Therefore the mean value for the total kinetic energy $< TKE >$ of the fragments is the difference between the total available energy Q and their mean total excitation energy

$$< TKE >= Q- < TXE > . \tag{8}$$

In binary reactions momentum and kinetic energy of the outgoing particles are conserved quantities. It follows that the ratio of the kinetic energies of the two fragments is inversely proportional to the ratio of their respective masses m_1 and m_2

$$\frac{E_{kin}^1}{E_{kin}^2} = \frac{m_2}{m_1}, \tag{9}$$

and the mean single kinetic energies $< E_{kin}^1 >$ and $< E_{kin}^2 >$ become

$$< E_{kin1} >= \frac{< TKE >}{1 + \frac{m_2}{m_1}} \tag{10}$$

$$< E_{kin2} >= \frac{< TKE >}{1 + \frac{m_1}{m_2}}. \tag{11}$$

2.2. *Fission induced by a reaction*

If fission is induced by a nuclear reaction, the process starts from an already excited compound nucleus with excitation energy ϵ_c. This compound nucleus energy has to be added to the Q-value of the reaction itself

$$Q_R = Q + \epsilon_c. \tag{12}$$

The temperature of the final fragments will be increased accordingly:

$$kT = \sqrt{(\bar{f}Q)^2 + \frac{\epsilon_c}{a_1 + a_2}} \tag{13}$$

and the mean values for fragment excitation becomes

$$< E_1^* >= a_1(\bar{f}Q)^2 + \frac{a_1 \cdot \epsilon_c}{(a_1 + a_2)} \tag{14}$$

$$< E_2^* >= a_2(\bar{f}Q)^2 + \frac{a_2 \cdot \epsilon_c}{(a_1 + a_2)}. \tag{15}$$

The mean value of the total excitation energy is

$$< TXE >= (a_1 + a_2) \cdot (\bar{f}Q)^2 + \epsilon_c. \tag{16}$$

This signifies that all of the compound nucleus excitation energy goes into excitation energy of the fragments, and nothing of it is converted into kinetic energy. Therefore the mean total kinetic energy of the fragments is the same than in spontaneous fission:

$$< TKE >= Q_R - < TXE > . \tag{17}$$

Also the expressions for the mean kinetic energy of the single fragment remain the same than for spontaneous fission.

3. Distributions functions for fragment excitation and kinetic energies

3.1. *Distribution functions at low excitation energies*

At the end of the fission process the fragments are at a temperature kT. Consequently, excited states in the fragments are populated according to the Boltzmann equation, which for low and medium excitation energies reads

$$\Phi(E_{1,2}^*) = N_0 \cdot exp(-\frac{E_{1,2}^*}{a_{1,2}(\bar{f}Q)^2}). \tag{18}$$

Excited states in fission fragments are therefore populated according to an exponential function. It is possible to give the expression for the distribution function for the total excitation energy, if the level density parameter of the two fragments are equal:

$$\Phi(TXE) = N_0 \cdot TXE \cdot exp(-\frac{TXE}{\frac{a_1+a_2}{2}(\bar{f}Q)^2}) \qquad (19)$$

If $a_1 \neq a_2$ the above expression does not longer lead to independent distributions for the two fragments, but a covariance dependent on $| a_1 - a_2 |$ is induced (see later section). N_0 is a constant, which normalizes the surface below the exponential to 1. With $TKE = Q - TXE$ the distribution function for the total kinetic energy becomes

$$\Phi(TKE) = N_0 \cdot (Q - TKE) \cdot exp\left(-\frac{(Q - TKE)}{\frac{a_1+a_2}{2}(\bar{f}Q)^2}\right), \qquad (20)$$

and the unnormalized distribution functions for the single kinetic energy are

$$f^{(i)}(E_{kin}) = (Q - \frac{E_{kin}}{\tilde{c}}) \cdot exp\left(\frac{-(Q - \frac{E_{kin}}{\tilde{c}})}{\frac{(a_1+a_2)}{2}(\bar{f}Q)^2}\right). \qquad (21)$$

Here

$$\tilde{c}_H = \frac{m_L}{(m_L + m_H)} \qquad (22)$$

for the heavy fragment, and

$$\tilde{c}_L = \frac{m_H}{(m_L + m_H)} \qquad (23)$$

for the light fragment.

If fission is induced by a nuclear reaction, we have to account for the excitation from the compound nucleus, and the equations become

$$\Phi(E_1^*) = N_0 \cdot exp\left(-\frac{E_1^*}{a_1 \cdot (\bar{f}Q)^2 + \frac{a_1 \cdot \epsilon_c}{(a_1+a_2)}}\right) \qquad (24)$$

$$\Phi(E_2^*) = N_0 \cdot exp\left(-\frac{E_2^*}{a_2 \cdot (\bar{f}Q)^2 + \frac{a_2 \cdot \epsilon_c}{(a_1+a_2)}}\right) \qquad (25)$$

$$\Phi(TXE) = N_0 \cdot TXE \cdot exp\left(-\frac{TXE}{\frac{a_1+a_2}{2}(\bar{f}Q)^2 + \frac{\epsilon_c}{2}}\right) \qquad (26)$$

$$f^{(i)}(E_{kin}) = (Q_R - \frac{E_{kin}}{\tilde{c}}) \cdot exp\left(\frac{-(Q_R - \frac{E_{kin}}{\tilde{c}})}{\frac{(a_1+a_2)}{2}[(\bar{f}Q)^2 + \frac{\epsilon_c}{a_1+a_2}]}\right). \qquad (27)$$

Mean values for excitation and kinetic energy from the distributions are calculated using the usual expression

$$<x> = \frac{\int \Phi(x) \cdot x}{\int \Phi(x)} \qquad (28)$$

and performing the integrals. The assumption for writing down the distribution functions for TXE, TKE and the single kinetic energies was $a_1 = a_2$. In order to ensure that these functions are valid for $a_1 \neq a_2$ the covariance may be calculated and compared to the total variance of the distributions:

$$\mu = \frac{1}{2}(\sigma^2_{E_1*} + \sigma^2_{E_2*} - \sigma^2_{TXE}) \qquad (29)$$

which leads to

$$\mu = (\bar{f}Q)^4 \cdot \frac{1}{4}(a_1 - a_2)^2. \qquad (30)$$

In the case of binary fission in general it is a very good approximation to calculate mean values from the distribution functions.

3.2. Distribution function for excitation at higher excitation energy

It is known in nuclear physics that for high excitation energies the Boltzmann function for the probability of population of excited levels in the nucleus changes from the distribution Eq. (18) to the function of the form

$$\Phi(E^*) \sim exp(\frac{-E^*}{kT}). \qquad (31)$$

This function decreases much faster than the function of Eq. (18), which is valid at low and medium excitation energies only. Furthermore all distributions outlined above do not take care of the requirement that the total excitation energy of the fragments must not exceed the Q-value of the reaction. Both requirements lead to a cut-off of the distribution function, Eq. (18) which has to be determined experimentally.

The cut-off value for the function appears to be situated at about 30 MeV excitation energy, dependent on the specific fission product.

The cut-off of the functions for fragment excitation leads to a strong decrease of the variances for all distributions, which is calculated according to

$$\sigma^2(x) = \frac{\int \Phi(x) \cdot (x- <x>)^2}{\int \Phi(x)} \tag{32}$$

Furthermore the cut-off strongly effects the higher moments which are present in the kinetic energy distributions of the fragments, and in particular the strong tailing towards lower kinetic energies.

4. Comparison with experiment

Mean fragment excitation and mean kinetic energies calculated from the foregoing model have been compared to experimental data for many systems in between Actinium and Fermium and for different excitation energies ranging from spontaneous fission up to 12 MeV excitation for the compound nucleus, see [1] and [2]. In all cases the calculated values agree within few percent with the experimental ones. In the following we will present new data on thermal neutron induced fission of ^{233}U, taken at the mass separator $LOHENGRIN$ at the ILL. Particular emphasis was put on the high resolution measurement of the kinetic energy distribution of correlated fragments.

4.1. Experimental set up and measurements

The mass separator $LOHENGRIN$ allows for the measurement of unslowed fission products from a thin source which is placed in a neutron flux of about $5.3 \cdot 10^{14} n/cm^2 \cdot s$. Fragments are separated in a combination of magnetic and electric fields, and detected in a high resolution ionization chamber. The energy resolution of the set-up is about 0.5 MeV. The fission source consisted in a layer of ^{233}U as oxide, with a thickness of about $65\mu g/cm^2$, covered with a tantalum layer of about $151\mu g/cm^2$. The kinetic energy distribution of the following masses were scanned in energy steps of 1 MeV: $A = 86$, $A = 91$, $A = 94$, $A = 101$, $A = 102$, $A = 104$, $A = 106$, $A = 108$, $A = 128$, $A = 132$, $A = 140$, $A = 147$, $A = 148$ and $A = 150$. Most of the masses were scanned for several ionic charge states located around the most probable ionic charge. The data were corrected for the energy dispersion of the spectrometer, and for energy loss in the target and the cover layer.
To perform the calculations we used the tables of Wapstra *et al.* [3] and

414

Møller and Nix [4] for the mass excesses, and the values of Butz-Jørgensen and Knitter [5] for the level density parameters.

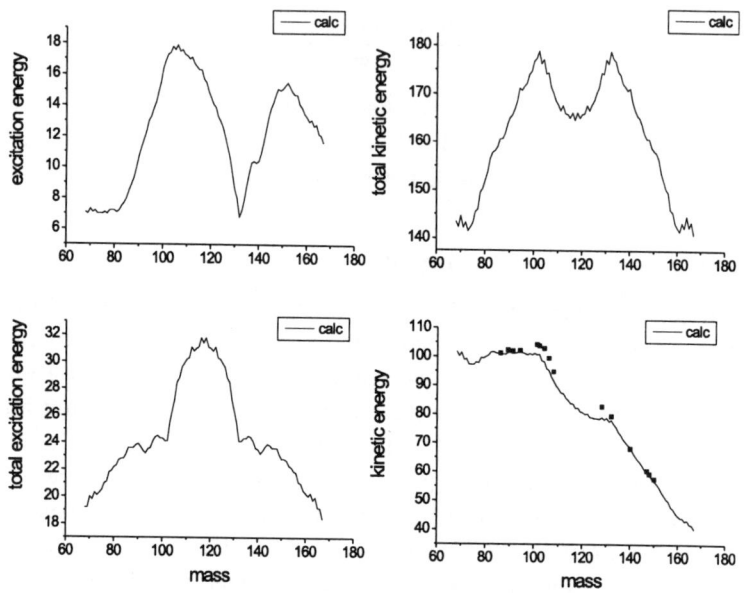

Figure 1. Calculated mean values for fragment excitation energy, total excitation energy, total kinetic energy and single fragment kinetic energy as function of fragment mass for $^{233}U(n,f)$. Values from the experiment are included as data points.

4.2. *Mean values for fragment excitation and kinetic energies*

For the masses cited above the experimental mean kinetic energies were determined from the data by fitting the distributions with a Gaussian. In Fig. 1 the results of the calculations for mean values of fragment excitation, TXE, TKE and kinetic energy of the fragments are shown. Data extracted from the experiment are included as points. It is seen that mean fragment excitation as a function of fragment mass has a saw-tooth behavior, with very low values for excitation around magic numbers, in particular the isotopes ^{78}Ni and ^{132}Sn. Highest fragment excitation is found in mid-shell regions. This behavior reflects the fact that at magic numbers the level density parameter is low, whereas for midshell nuclei level density parameters tend to be high. The values for the mean total excitation energy

$< TXE >$ are high for nuclei in the symmetric mass region, because for the system ^{234}U the sum of the level density parameters is high. In consequence the values for the mean total kinetic energies $< TKE >$ in symmetry are low. Values for the single kinetic energies are constant for the light fission wing, and are strongly decreasing for heavy masses. This is a consequence of momentum conservation in binary reaction processes.

The calculations, as already seen in numerous other systems, do reproduce the measured data within few per cent.

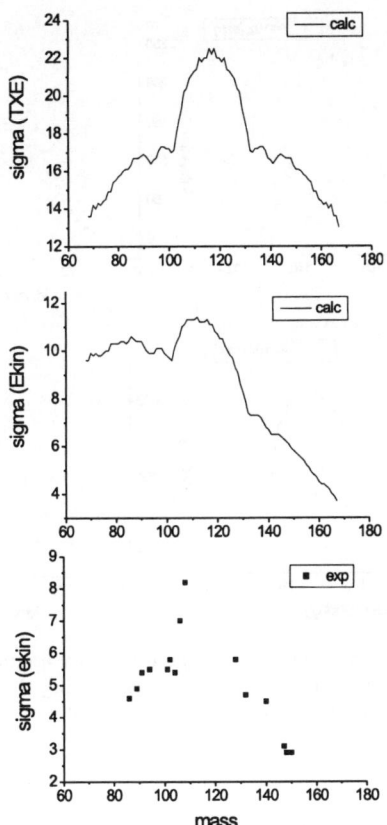

Figure 2. Calculated (solid line) and measured values (points) of sigma for TXE and the kinetic energy as function of fragment mass for $^{233}U(n, f)$.

416

4.3. *Variances of the kinetic energy distributions*

The variances of the kinetic energy distributions can be directly determined from Eq. (32). The value of sigma of the distributions are shown in Fig. 2, together with the measured value for sigma from the experiment. Here sigma has been determined using a Gaussian fit to the kinetic energy distributions. Without a cut-off on the distributions for the excitation energy the calculated variances are too large. However, the general pattern of the calculations is reproduced, in particular a strong increase for the halfwidths for masses around symmetry, from $A = 102$ to $A = 130$.

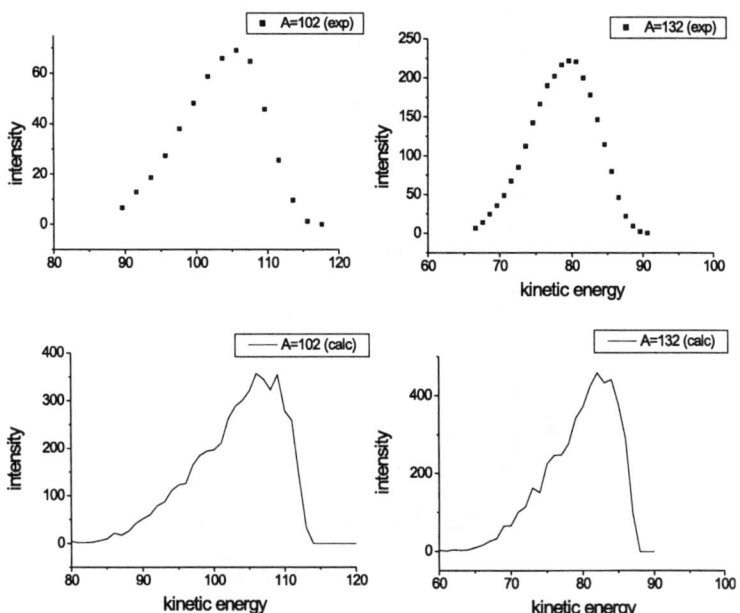

Figure 3. Kinetic energy distributions for the correlated fragments $A = 102$ and $A = 132$ in $^{233}U(n, f)$. The upper part of the figure is showing the data from experiment, the lower part shows the result of the Monte-Carlo calculation.

4.4. *Higher moments of the distributions*

Higher moments in the kinetic energy distributions of the fragments are in particular seen in the persistence of a tail towards low energies for almost all mass splits. In the calculations this tailing arises because of the finite values for the exponential function for high excitation energies. At these energies the Boltzmann function for an atomic nucleus changes from Eq. (18) to Eq. (31), causing the tailing for the kinetic energy distributions to become less pronounced. To calculate the lineshape for the kinetic energy distributions for single fragments including the cut-off at high excitation energies, a Monte-Carlo code was written which selects events in the allowed region for the excitation function of Eq. (18), and constructs then the functions for TXE, TKE and the single fragment kinetic energies.

In Fig. (3) an example of the lineshape for the kinetic energy distributions for the correlated masses $A = 102$ and $A = 132$ is shown. Here it is demonstrated that a cut-off of the excitation energy at 35 MeV reproduces quantitatively the lineshapes of the kinetic energy distributions.

5. Conclusions

We have shown that the model for statistical excitation of fission fragments $(REX - M)$ leads to the calculation of mean values for fragment excitation and kinetic energies, which are in agreement with the experimental values. Furthermore it was demonstrated that by cutting off the excitation function of fission fragments at higher values, very good descriptions of the kinetic energy function are obtained.

References

1. H.R. Faust, *Eur. Phys. J.* **A14**, 459-468 (2002).
2. H.R. Faust in: Proceedings 'Dynamical Aspects of Nuclear Fission', Casta-Papiernicka (2001), *World Scientific*.
3. A.H. Wapstra, G. Audi and R. Hoekstra, *At. Data Nucl. Data Tables* **39**, no. 2, 281 (1988).
4. P. Møller and J.R. Nix, *At. Data Nucl. Data Tables* **39**, no. 2, 213 (1988).
5. C. Butz-Jørgensen and K.H. Knitter, *Nucl. Phys.* **A 490**, 307 (1988).

EXPERIMENTAL STUDIES OF PROMPT NEUTRON AND PHOTON EMISSION IN INTERMEDIATE ENERGY NEUTRON-INDUCED FISSION

T. ETHVIGNOT,* T. GRANIER, P. CASOLI

Commissariat à l'Énergie Atomique,
Service de Physique Nucléaire,
BP 12, 91680 Bruyères-le-Châtel, France
E-mail: Thierry.Ethvignot@cea.fr

R. O. NELSON, R. C. HAIGHT, M. DEVLIN, R. DROSG, N. FOTIADES, J. M. O'DONNELL

Los Alamos National Laboratory,
Los Alamos, NM 87545, USA

W. YOUNES, P. E. GARRETT, J.A. BECKER

Lawrence Livermore National Laboratory,
Livermore, CA 94551, USA

An experimental program for studying the properties of fission induced by intermediate energy neutrons, i.e. several hundreds of keV to several hundreds of MeV is in progress at the Los Alamos Neutron Science CEnter (LANSCE). Experiments using the GEANIE high-resolution gamma-ray and x-ray spectrometer and the FIGARO neutron detector array were performed with uranium 238 targets. Information on the fragment yields, i.e. excitation functions and isotopic distributions, were extracted from GEANIE inclusive γ and $\gamma - \gamma$ data. Charge yields were extracted by measuring prompt x-rays using a thin, fission-sensitive target specially designed. On the FIGARO beam line, prompt neutron spectra as a function of incident neutron energy were measured with neutron detectors and a multiple-plate fission chamber, with a double time-of-flight method. The experimental results are presented and compared to model calculations .

*partially supported by Délégation Générale pour l'Armement of the French Ministery of Defense

1. Introduction

Within the last century, neutron-induced fission has been extensively studied with the fairly low energy neutrons available at nuclear reactors and at mono-energetic sources below 15 MeV. There is presently a need for experimental studies at higher incident neutron energies to improve our understanding of fission mechanisms and fission dynamics. This knowledge is particularly crucial for the future industrial applications of spallation, such as destruction of nuclear waste and concurrent power generation.

In this perspective, a collaborative experimental program is being carried out by Bruyères-le-Châtel and Los Alamos. This program takes advantage of the WNR facility at LANSCE, a spallation neutron source which provides a relatively large flux of fast neutrons in a broad energy range extending up to more than 400 MeV.

Experiments utilizing two existing detector systems, GEANIE (GErmanium Array for Neutron-Induced Excitations) and FIGARO (Fast neutron-Induced Gamma-Ray Observer), have provided new data on prompt neutron, gamma and x-ray emission.

Fission is accompanied by prompt emission of neutrons and photons which originate from several stages of the phenomenon. Depending on the incident neutron energy, neutrons and photons may be emitted by the fissioning nucleus itself. After scission, the prompt de-excitation of the fission fragments occurs through evaporation of neutrons and electromagnetic processes which may be coupled to the atomic system. Most of the prompt electro-magnetic de-excitation occurs mainly once the neutron evaporation is exhausted. Two processes participate in the electro-magnetic de-excitation: gamma-ray emission and internal conversion. The latter is accompanied by the emission of K x-rays characteristic of the primary fragment's atomic number.

2. Experiments

2.1. *Prompt photon studies*

GEANIE is an array of 26 high-resolution germanium (Ge) gamma-ray detectors with 20 BGO (bismuth germinate) background suppression shields. Eleven of the Ge detectors are dedicated to low energy photon spectroscopy, from 10 keV up to 300 keV. The other 15 detectors are more adapted to higher energy gamma-rays, typically from 300 to 2000 keV. GEANIE is located 20 meters from the WNR neutron production target at the end of a collimated flight path [1].

Two types of experiments were done at GEANIE with uranium 238 targets. In both cases the time-of-flight technique was used to measure the energy of the incident neutrons.

The first class of experiments consisted of inclusive gamma-ray measurements. They were performed with a ~400 mg/cm^2 target of uranium 238. GEANIE's many detectors gave the opportunity for analysis of the recorded gamma-rays in both single-gamma and double-gamma-coincidence modes. The target's relative thickness provided high gamma counting rates. However, the recorded gamma-rays originate not only from the de-exciting post-neutron fragments but also from other neutron-induced reactions occuring in the target, such as (n,xn) and (n,n') on uranium.

This problem of separating fission product γ rays from those of other reactions is minimized in the second class of experiments, with a thin, fission-sensitive target of uranium 238. This target consists of eight x 1 mg/cm^2-layers deposited onto photo-voltaic cells used as fission fragment detectors [2]. A coincidence with the detection of a fission fragment in the target was required in the gamma-ray recording. Moreover, both the uranium samples and the photo-voltaic cells were thin enough for the measurement of prompt fission K x-rays. This is the first time fission x-rays have been measured at a broad spectrum neutron source.

2.2. *Prompt neutron studies*

An experiment was performed at the FIGARO flight path[3] in order to measure the energy spectrum of prompt neutrons emitted in fission induced on uranium 238. The target, situated 20 meters from the WNR neutron production target, was a multi-plate gas ionization fission chamber containing 380 mg of uranium 238. Neutrons were detected in coincidence with fission by using six liquid scintillator detectors located at one to two meters from the fission chamber. Neutrons and gamma-rays were discriminated off-line by means of pulse shape analysis. The double time-of-flight technique was used to measure the kinetic energy of both incident and emitted neutrons.

3. Results

3.1. *Fragment Excitation Functions*

From the inclusive gamma-ray measurements, more than one hundred post-neutron fission fragments were identified from Ge to Nd. Among them, the even-even nuclei are particularly interesting. Indeed, it is known that the

2+ to 0+ (G.S.) transition of the ground state rotational band of an even-even fragment is usually fed at $>= 90$ percent in the de-excitation process[4]. Measuring the corresponding gamma-ray intensities gives a measurement of the fragment's production. In this way, the fragment production cross section has been extracted for about thirty even-even post-neutron-emission fragments as a function of incident neutron energy. The evolution of this cross section as a function of neutron energy (excitation function) exhibits correlations with the thresholds of fission chances. However, for a given element, the behavior differs strongly from one isotope to another.

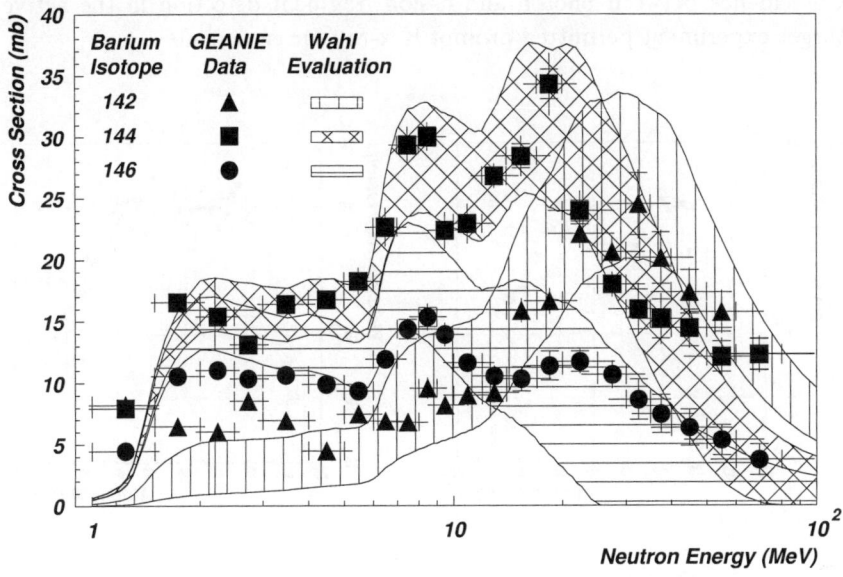

Figure 1. Excitation functions for three barium isotopes for the system ^{238}U(n,f) measured with the GEANIE setup. The data are compared to Wahl systematics. The data are not normalized to the calculations, but are absolute cross section values.

As an example, Fig. 1 displays the fragment production excitation functions for three barium isotopes. These excitation functions have been obtained from the 2+ to 0+ ground-state-band transitions. They are compared to the systematics by Wahl [5]. The systematics values are calculated from a five Gaussian fit to mass and charge distributions of library data. The energy dependence is interpolated between data sets. At energies above 14 MeV, Wahl uses data from ^{238}U(p,f) and ^{232}Th(p,f). The calculated values fit the data reasonably well, and illustrate the strong dependence of the excitation functions on successive fission chances. One can see that the lighter barium isotopes are produced more at high energy.

Also, gamma spectroscopy of some fragments may provide information on the fissioning nucleus. Differences in the intensity of some gamma-ray transitions are observed at high neutron energies. They may reflect a change in the orbital momentum and excitation energy in the fissioning system [6]. These results are currently under investigation.

3.2. Charge Yields

Coincidence between photon and fission fragment detection in the active target experiment permitted prompt K x-ray measurements.

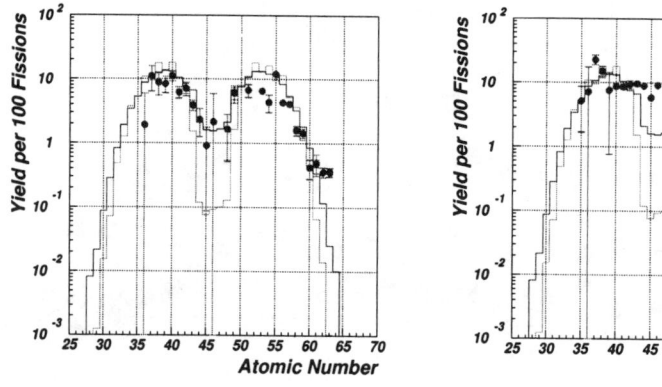
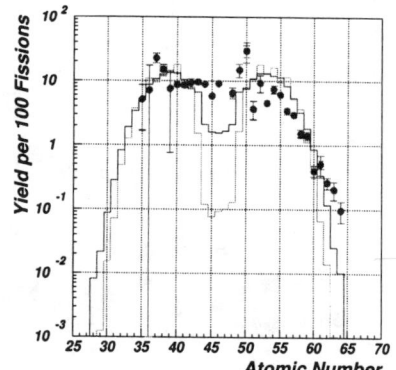

Figure 2. Charge yield distributions measured with GEANIE for energies around 14 MeV (left) and 50 MeV and above (right) for the system ^{238}U(n,f). The histograms represent the Wahl systematics for 14 MeV and fast neutrons. The data are not normalized to the calculations, but are the absolute values deduced from this work.

Due to the large number of K x-ray lines, five per element, the x-ray energy spectrum is quite complex. Nevertheless, the energies and relative intensities of these lines are known to a high precision [7]. It has been possible, within the limits of the recorded statistics, to deconvolute the contribution of the different elements produced for different groups of incident neutron energy. This was done by fitting the data with a parametric function with one free parameter per element.

In the seventies, Reisdorf and collaborators measured the x-ray yields per fission fragment Y(Z) in spontaneous and low energy neutron-induced fission [9]. It appears that Y(Z) depends only weakly on the nature of the fissioning system. By applying Reisdorf's prescription to our x-ray yields, one is able to extract an elemental distribution of the primary fission fragments. Two such distributions are displayed in fig. 2. They correspond respectively to incident neutron energies of 14 MeV and 50 MeV and up (190 MeV average energy). The x-ray data have been corrected for self absorption and detection efficiency with point source calibrations and a MCNP [8] simulation. The preliminary results exhibit the expected increase in the probability for symmetric fission as one goes towards high incident neutron energies. Moreover, they agree quite well with the systematics from Wahl for 14 MeV neutrons [5]. The agreement with the systematics is particularly striking at 14 MeV. At 50 MeV and up the symmetric region gap is filled up. These results agree with the work by Zoller, et al. [10] on mass distribution measurements for the same system.

3.3. Prompt Neutron Spectra

From the neutron data, prompt neutron spectra have been obtained for different groups of incident neutron energies. The detection efficiency was obtained by dividing the measured spectrum around 14 MeV with the precision measurement by Boykov, et al. [11] at the same energy. The neutron detectors were situated at 114 cm from the fission target. The overall time resolution for the emitted neutrons was about 8 ns. In fig. 3 is plotted the mean energy of the emitted neutron spectrum versus the incident neutron energy. There is an experimental threshold of 0.8 MeV. The mean energy slowly rises from 1 to 5 MeV with a sudden decrease of 5% above this value. The rise is constant again from 6 to 12 MeV and some structure is seen above 12 MeV. The data are compared to calculations with the improved Los Alamos Model [12]. With a similar threshold of 0.8 MeV, the calculated values agree well with the data up to 20 MeV. In particular, the dip at 6 MeV is reproduced. The dip corresponds to opening of the second chance fission channel. It is interesting to see that the transition is well described with a rather crude observable, i.e. the mean energy. The discrepancy between the data and the calculation above 20 MeV is being investigated.

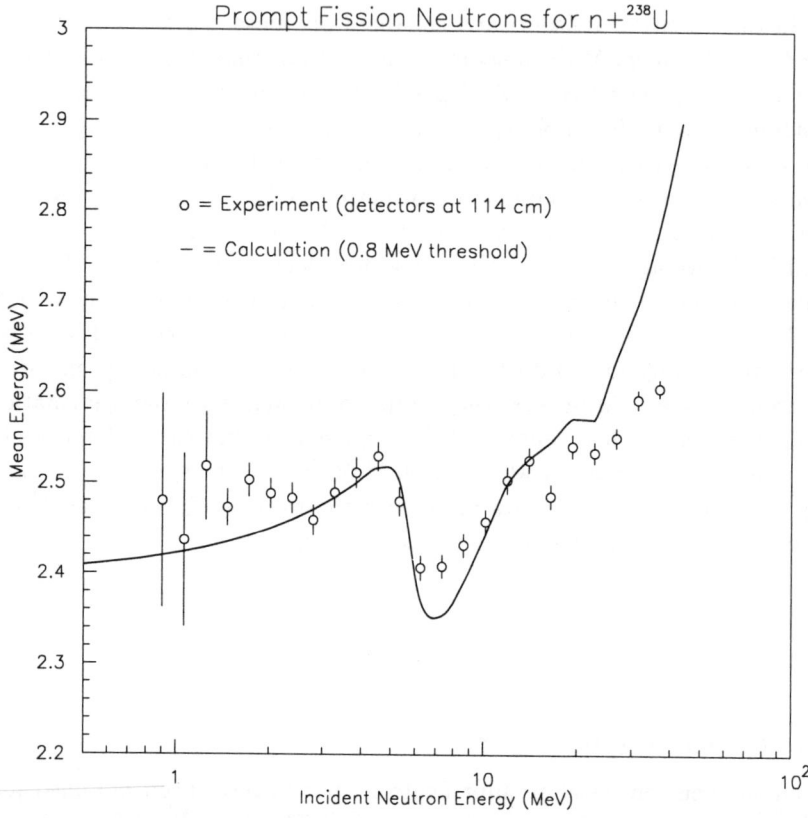

Figure 3. Mean energy of neutrons in coincidence with fission for the system ^{238}U(n,f) measured with the FIGARO setup. The data points are compared to calculations with the improved Los Alamos Model (line).

4. Conclusion

New data on prompt neutron and photon emission in intermediate energy neutron-induced fission of uranium 238 have been obtained. Gamma ray measurements have enabled the extraction of individual post-neutron-emission fragment production excitation functions. From x-ray measurements, the primary fragments' Z distribution has been extracted for different incident neutron energies. Although these last results have to be finalized, they agree quite well with evaluations and other measurements.

Moreover, prompt-fission neutron-energy measurements have been performed for the first time at a broad range intermediate energy neutron source. The results, still preliminary, show interesting features intimately correlated to fission mechanisms.

The final results, hopefully available soon, will provide new constraints on theory and models of fission.

References

1. R. Nelson *et al.* Italian Physical Society Conference Proceedings, **59**, 445 (1997).
2. T. Ethvignot, T. Granier, L. Giot, P. Casoli and R.O. Nelson, Nucl. Instrum. Meth. **A490**, 559 (2002).
3. R.C. Haight *et al.* in Int. Conf. on Nucl. Data for Science and Tech., Tsukuba, Japan, Oct 2001, Journal of Nuclear Science and Technology, supplement 2, 634 (August 2002).
4. E. Cheifetz, J.B. Wilhelmy, R.C. Jared, S.G. Thompson, Phys. Rev **C4**, 1913 (1971).
5. A. Wahl, Contribution to "Compilation and Evaluation of Fission Yields Nuclear Data" IAEA-TECDOC-1168 (2000) 58 and Los Alamos Report LA-13928 (May 2002).
6. T. Ethvignot *et al.* in Int. Conf. on Nucl. Data in Science and Tech., Tsukuba, Japan, Oct 2001, Journal of Nuclear Science and Technology, supplement 2, 254 (August 2002).
7. J.A. Bearden, Rev. of Modern Physics **39**, 78 (1967).
8. J. Breismeister, ed., "MCNP - A Monte-Carlo N-Particule Transport Code", LANL Report, LA-12625-M (1997).
9. W. Reisdorf, J.P. Unik, H.C. Griffin and L.E. Glendenin, Nuclear Physics **A177** (1971) 379-392.
10. C.M. Zoller, A Gavron, J.P. Lestone *et al.* Proc. Seminar on Fission, Pont d'Oye III, (1995) 56 and Proc. VII School on Neutron Physics, Dubna (1995) vol. 1, 130.
11. G.S.Boykov, V.D.Dmitriev, G.A.Kudyaev, Yu.B.Ostapenko, M.I.Svirin, G.N.Smirenkin, Yad. Fiz. **53**, 3,628 (1993).
12. B. Morillon, private commnunication and Vladuca and Tudora, Annals of Nuclear Energy **28**, 1643 (2001).

SEARCH FOR LIMITS OF STABILITY: ^{16}Be

T. BAUMANN, N. FRANK, B. A. LUTHER*, D.J. MORRISSEY, J. P. SEITZ,
B. M. SHERRILL, M. STEINER, J. STETSON, A. STOLZ,
I. WIEDENHÖVER†, M. THOENNESSEN

*National Superconducting Cyclotron Laboratory and
Department of Physics & Astronomy
Michigan State University, East Lansing, MI 48824, USA
E-mail: thoennessen@nscl.msu.edu*

In order to search for the heaviest possibly bound beryllium isotope, ^{16}Be, a primary beam of ^{40}Ar was accelerated to 140 MeV/nucleon using the newly completed Coupled Cyclotron Facility at the National Superconducting Cyclotron Laboratory at Michigan State University. Neutron-rich fragmentation products emerging from a beryllium production target were separated with the A1900 fragment separator set to its maximum magnetic field of 6 Tm. The isotopes ^{6,8}He, ^{9,11}Li, ^{12,14}Be, ^{17,19}B, and ^{20}C were identified using time-of-flight and energy-loss information, but no events of ^{16}Be were recorded.

1. Introduction

The recently reported emergence of new shell structures in very neutron-rich nuclei initiated a re-evaluation of binding energies along the neutron dripline for light nuclei.

The last bound Be isotope is ^{14}Be. The next heavier Be isotope, ^{15}Be, is unbound with respect to one-neutron emission. ^{16}Be is predicted to be bound regarding one-neutron emission, with a one-neutron separation energy of 1.828 MeV (see Fig. 1a). However, ^{16}Be becomes unbound regarding two-neutron decay.

Previous measurements of light neutron-rich nuclei [4,5] failed to observe ^{16}Be, although they did not specifically aim at measuring it.

According to mean-field and shell-model configuration mixing calculations that recently have been undertaken by B. A. Brown [2] the two-neutron separation energy of ^{16}Be is only about −628 keV [3].

The same calculation predicts ^{19}B and ^{22}C to be unbound as well (see Fig. 1b). These nuclei, however, already have been measured and are confirmed to be bound [4–6]. This comparison opens the question if ^{16}Be could be bound as well, despite having been predicted to be unbound.

---

* Permanent address: Dept. of Physics, Concordia College, Moorhead, MN 56562
† Permanent address: Dept. of Physics, Florida State University, Tallahassee, FL 32306

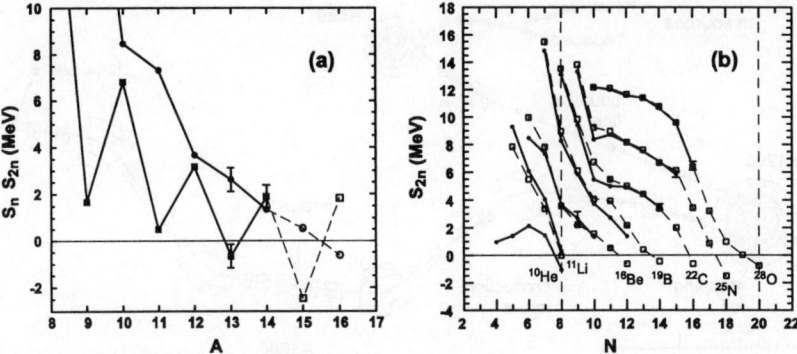

Figure 1: Neutron separation energies for selected nuclei: (a) one-neutron (squares) and two-neutron (circles) separation energies for Be isotopes. Filled symbols are according to measured values [1], open symbols are from calculations [2,3]. (b) comparison of two-neutron separation energies from experiments (filled symbols) [1] and from a mean-field and shell-model configuration mixing model (open symbols) [2,3].

2. Experimental Setup

The experiment was performed utilizing the Coupled Cyclotron Facility (CCF) at the National Superconducting Cyclotron Laboratory at Michigan State University (see Fig. 2). The superconducting ECR ion source (SC-ECR) delivered a beam of $^{40}Ar^{7+}$ with an intensity of $4.4 \cdot 10^{12}$ particles per second, which was attenuated by a factor of 10 and then accelerated to an energy of 12.3 MeV/nucleon in the K500 cyclotron. After extraction from the K500, the beam was transported through the coupling line to the K1200 cyclotron, where it was radially injected with an intensity of $4.2 \cdot 10^{10}$ s^{-1}. A 0.2 mg/cm^2 carbon foil in the K1200 injection line fully stripped the ^{40}Ar ions which could finally be accelerated to the full beam energy of 140 MeV/nucleon. The beam was extracted from the K1200 with an intensity of $9.7 \cdot 10^{9}$ s^{-1} and directed onto a 1455 mg/cm^2 beryllium target at the entrance of the A1900 fragment separator.

The first half of the separator up to image 2 was set to the maximum rigidity of 6.00 Tm. At the second intermediate image, the dispersive plane, a thin (327 mg/cm^2) plastic start detector as well as an achromatic wedge-shaped plastic degrader (960 mg/cm^2) were installed. The momentum acceptance was ±2.5%. The magnetic rigidity of the dipole stages after the intermediate image was set to 5.6719 Tm. The detector set at the focal plane comprised a 5x5 cm^2 500-μm-thick silicon detector for the energy loss measurement and a 10-cm-thick plastic scintillation detector for the total particle energy.

428

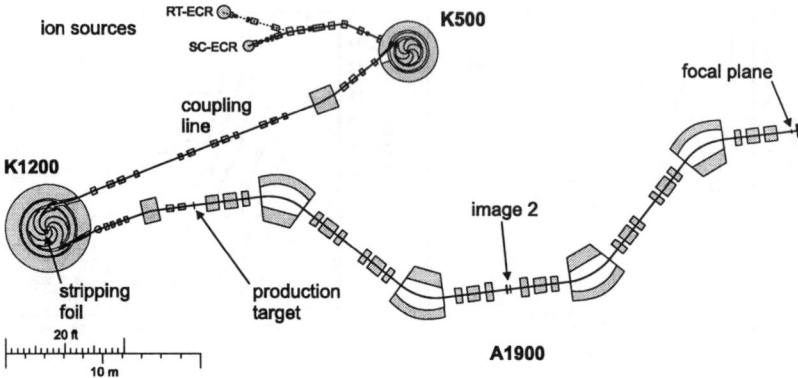

Figure 2: Schematic layout of the coupled cyclotron facility at the NSCL and the experimental setup.

3. Data Analysis

Complete particle identification was achieved by determining the mass-to-charge ratio with a time-of-flight measurement and the particle charge with the energy-loss measurement. The particle velocity at the focal plane was calculated from the time of flight between the image 2 plastic scintillator and the thick plastic scintillator at the focal plane. The charge was deduced from the energy-loss signal in the silicon detector. The focal-plane particle identification plot is shown in Fig. 3. We can clearly identify ^{6,8}He, ^{9,11}Li, ^{12,14}Be, ^{17,19}B,and ^{20}C.

We used the simulation code LISE [7] to calculate event rates taking into account extrapolated production cross sections (EPAX 2.15 [8]) and ion optical transmission. Table 1 lists production cross sections for ^{19}B and ^{16}Be from ^{40}Ar on a beryllium target at relativistic energies. Cross sections from an earlier version of EPAX (1.0) [9] are also included.

Table 1. Comparison of production cross sections for ^{19}B and ^{16}Be from ^{40}Ar on a beryllium target.

Isotope	EPAX 1.0 [9]	EPAX 2.15 [8]	Experiment [6]
	(mb)	(mb)	(mb)
^{19}B	$2.35 \cdot 10^{-6}$	$2.13 \cdot 10^{-8}$	$4.86 \cdot 10^{-7}$
^{16}Be	$1.40 \cdot 10^{-5}$	$2.41 \cdot 10^{-7}$	—

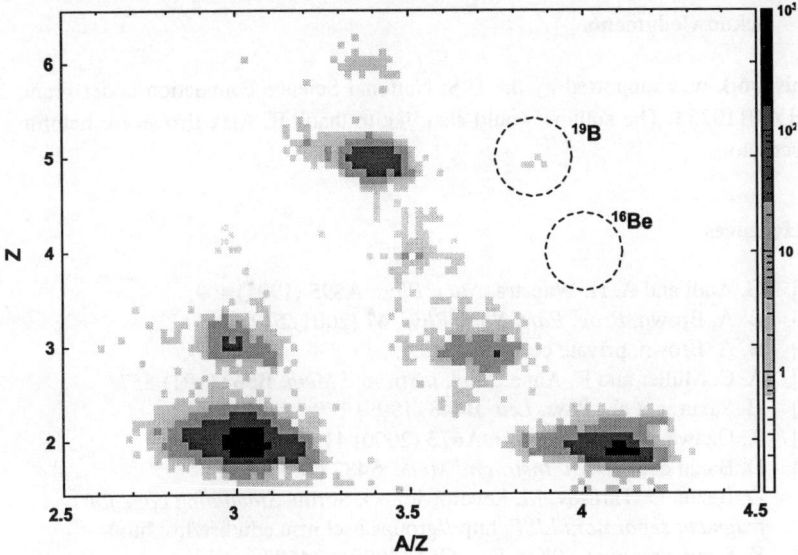

Figure 3: Measured particle identification at the focal plane. The isotopes ^{6,8}He, ^{9,11}Li, ^{12,14}Be, ^{17,19}B, and ^{20}C are clearly identified.

From these cross sections one would expect to detect ^{16}Be at a higher rate than ^{19}B if the beryllium isotope was bound. Both isotopes have a very similar magnetic rigidity and transmission through the separator. However, while we detect five counts of ^{19}B, no counts for ^{16}Be were recorded.

4. Conclusions

The fact that ^{16}Be was not detected leads to the conclusion that it is unbound. The location of the neutron dripline has been unambiguously determined for Z = 4. A recent unsuccessful search for ^{21}B [11] also establishes ^{19}B as the last bound Z =5 isotope. Thus, the change of the shell structure for neutron-rich isotopes has not lead to any new previously unexpected observations of new isotopes.

430

5. Acknowledgments

This work was supported by the U.S. National Science Foundation under grant PHY-0110253. The authors would also like to thank B. Alex Brown for helpful discussions.

References

[1]   G. Audi and A. H. Wapstra, *Nucl. Phys.* **A595** (1995) 409.
[2]   B. A. Brown, *Prog. Part. Nucl. Phys.* **47** (2001) 517.
[3]   B. A. Brown, private communication.
[4]   A. C. Müller and R. Anne, *Nucl. Instr. and Meth.* **B56** (1991) 557.
[5]   H. Sakurai et al., *Phys. Lett.* **B448** (1999) 180.
[6]   A. Ozawa et al., *Nucl. Phys.* **A673** (2000) 411.
[7]   D. Bazin et al., *Nucl. Instr. and Meth.* **A482** (2000) 314.
[8]   D. Bazin, O. Tarasov, M. Lewitowicz, O. Sorlin, *Simulation code for fragment separators LISE*, http://groups.nscl.msu.edu/lise/lise.html
[9]   K. Sümmerer et al., *Phys. Rev.* **C61** (2000) 034607.
[10] K. Sümmerer et al., *Phys. Rev.* **C42** (1990) 2546.
[11] Y. Yamaguchi et al., *Bull. Am. Phys. Soc.* **46** (2001) 117.

^{239}PU(N,F) AT RESONANCE ENERGIES AND ITS MULTI-MODAL INTERPRETATION

F.-J. HAMBSCH, AND H. BAX

EC-JRC-Institute for Reference Materials and Measurements.
Retiseweg,
B-2440 Geel, Belgium
E-mail: hambsch@irmm.jrc.be

I. RUSKOV

Institute for Nuclear Research and Nuclear Energy-BAS
72 Tzarigradsko Chaussee Blvd.
BG-1784 Sofia, Bulgaria
E-mail: ivanruskov@yahoo.com

L. DEMATTÈ

CINECA, I-40033 Casalecchio di Remo, Italy

A measurement of fission fragment total kinetic energy (TKE) and mass yield distributions Y(A,TKE) in the ^{239}Pu(n,f) resolved resonance region has been performed applying the twin Frisch gridded ionization chamber technique. Special emphasis was devoted to cope with the strong α-activity of this isotope by an improved pile-up rejection system. Up to about 200 eV all fission resonances could be resolved and their two-dimensional mass yield and TKE distribution, Y(A,TKE), measured. Compared to the results on ^{235}U(n,f), much smaller fluctuations of the fission fragment mass and TKE have been observed in the case of ^{239}Pu. From a physical point of view such fluctuations have been expected for the fission fragment properties, because the only possible low-energy spin states ($J^{\pi}=0^{+},1^{+}$) belong to well separated (about 1.25 MeV) transition state bands. Hence, it was expected to observe differences in the fission fragment mass and TKE distributions between spin 0^{+} and 1^{+} resonances. However, no spin dependence and only a slight anti-correlation of the TKE with the prompt neutron multiplicity, v_p, has been found in the resolved resonance energy region above 1 eV.
Within the multi-modal random neck-rupture (MM-RNR) model the Y(A,TKE) distributions have been fitted assuming three fission modes, two asymmetric and one symmetric one. The branching ratio of the two asymmetric modes shows similar fluctuations as the experimental TKE. Recently, a new theoretical approach has given a solution to the absence of pronounced fluctuations of the fission properties in the case of ^{239}Pu. Since only one transition state is involved in the fission of 0^{+} and 1^{+} resonances with a given fission fragment distribution, no fluctuations are expected.

1. Introduction

Fission is described by a continuous change of the shape of the nucleus, accompanied by considerable changes of the nuclear potential energy. Theoretical models are still unable to solve the longstanding question where in the fission path the fission fragment mass and kinetic energy distributions are determined. Fluctuations in the fission fragment parameters from resonance to resonance have in the past been discussed in terms of scission point models. Based on the Bohr model, Wheeler [1] suggested that, due to symmetry properties of the fissioning nucleus, the symmetrical fission yield should vary from resonance to resonance.

This has triggered the search for correlations between the quantum numbers of the compound nuclear states and the experimentally observable fission fragment properties with questionable results.

Recent measurements [2,3] of the fission reaction of ^{235}U with resonance neutrons, although showing fluctuations of the fission fragment properties from resonance to resonance, did not show any spin dependence.

Evaluations of the fission cross-section are now also concerned by the inclusion of the prompt neutron multiplicity, ν_p, as a function of resonance energy. This quantity, however, shows fluctuations which are very dependent on changes in the mass distribution, due to the sawtooth behaviour of $\nu_p(A)$. With high-resolution measurements of mass and TKE of the fission fragments, fluctuations in ν_p can be related to fluctuations observed in the mass distribution, as it has already been shown in case of ^{235}U [2].

The understanding of the observed fluctuations in ν_p is also very important for reactor physics. These fluctuations have a significant impact, e.g. on the reactivity coefficient of advanced water reactors.

2. Experiment

A sample of about 500μg of ^{239}Pu (enrichment: 99.98%), evaporated on a very thin gold coated polyimide backing, was prepared by the Sample Preparation Group of IRMM and was placed on the common cathode of a twin Frisch gridded ionization chamber. As working gas, pure CH_4 at $\sim 1.1 \cdot 10^5 Pa$ was used with a constant flow rate of ~0.1l/min. The chamber was placed at a distance of ~9.4m from the rotating U target of the white-spectrum neutron time-of-flight spectrometer GELINA of the IRMM. In this way, all the neutron energies were being measured simultaneously. Due to the high intrinsic α-activity of the Pu-sample (~1MBq) a special pile-up rejection system was applied. More

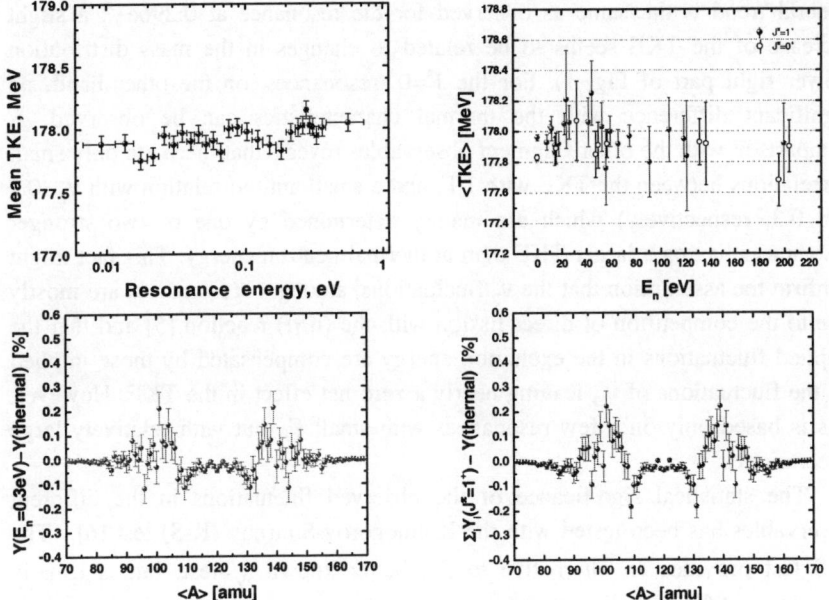

Figure 1. Left part: mean TKE as a function of neutron energy below 1eV and right part above 1eV. Lower part: difference in the mass distributions between thermal and: right E_n=0.296eV and left the sum of all 1^+ resonances.

information on the experimental details and the specialized pile-up rejection scheme can be found in Ref. [4].

3. Results and discussion

In the left part of Fig. 1 the mass and TKE characteristics are shown for the neutron energy region E_n=0.01 to 1eV. In the upper part, the average TKE is plotted as a function of E_n, showing a slight increase to the position of the first resonance at 0.296eV ($J^\pi=1^+$). By calculating the corresponding mass distribution and comparing it with the thermal one (lower left part of Fig. 1), a change of up to 0.3% is visible (equals 5% of the total cumulative yield of ~6%). Since a change of the mass yield distribution yields immediately a change in the TKE distribution, the observed increase in TKE is correlated with a lower yield in the mass-symmetric region and a higher yield around A_H~136.

In the upper right part of Fig. 1 the resulting mean TKE for 31 resolved resonances (22 with $J^\pi=1^+$ and 9 with $J^\pi=0^+$), lying in the region up to E_n~200eV, are shown. The magnitude of the TKE fluctuations is in the range of the experimental errors and only a few strong 1^+ resonances show significant deviations from the thermal characteristics. For most of the $J^\pi=1^+$ resonances the

434

general trend is the same as observed for the resonance at 0.296eV: a slight increase of the TKE seems to be related to changes in the mass distribution (lower right part of Fig. 1). For the $J^\pi=0^+$ resonances, on the other hand, no significant difference with the thermal characteristics can be observed. A comparison with the other fragment observables reveals that there are only small correlations between the TKE with $1/\Gamma_f$ and a small anti-correlation with ν_p (0.2 and 0.3, respectively) which are mainly determined by one or two stronger resonances having a larger TKE than at thermal neutron energy. This fact might confirm the assumption that the ν_p fluctuations, among 1^+ resonances are mostly due to the competition of direct fission with the (n,γf) reaction [5] and that the implied fluctuations in the excitation energy are compensated by those implied by the fluctuations of ν_γ, leaving nearly a zero net effect in the TKE. However, this is based only on a few resonances with small Γ_f, but with relatively large uncertainties.

The statistical significance of the observed fluctuations in the different observables has been tested with the Kolmogorov-Smirnov (K-S) test [6]. The K-S test provides an alternative to the better known χ^2-test, but is usually superior to it for the following reasons:

• it does not require the assumption that the population is normally distributed;

• it does not require a minimum number of events per bin, so it works well with data having low statistics;

• it can be used to compare both unbinned or binned data. In the second case, the bin width should be less than or comparable with the physical quantity of interest (for example, the experimental resolution);

• it takes into account not only the difference between corresponding bins, but also the sign of this difference.

Details about this test will be presented in Ref. [7]. In conclusion it has shown that for all $J^\pi=1^+$ resonances the mass yield differs compared to the mass yield at thermal fission.

The kinetic energy distribution for all 1^+ resonances is also different compared to the thermal one, with one exception being the 0.296eV resonance, which shows a heavy fragment energy distribution similar to the one at thermal energy. The Y(A,TKE) distributions for 0^+ resonances differ compared to the Y(A,TKE) distributions for 1^+ resonances, but insignificantly from the Y(A,TKE) distribution at thermal energy.

The mass distributions deduced from between resonance energies are closer to the mass distribution of the 0.296eV resonance, but the energy distributions are different from the one at thermal energy with 95% confidence. This can be

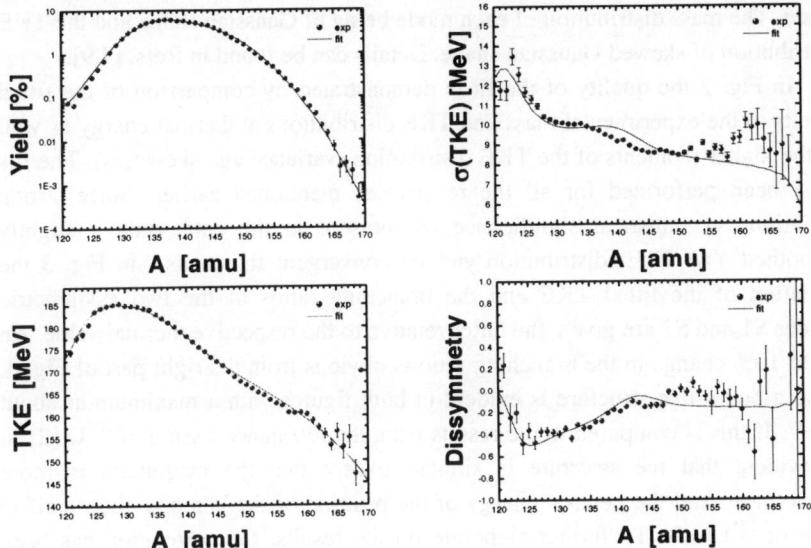

Figure 2. Comparison between fit results (full line) and experimental points at thermal energy for the mass distribution (upper left) the TKE distribution (lower left) and the higher moments of the TKE distribution, variance (upper right) and dissymmetry (lower right).

understood as a result of the influence of the high-energy tail of the 0.296eV resonance, which is quite a broad resonance.

4.   Comparison to theory

Several theoretical models have been proposed in the past to explain the fission process with, mainly only qualitative explanations of the experimental observables. The only quantitative approach, so far, is based on the MM-RNR model of Brosa *et al.* [8]. In this model the shape of the undergoing fission compound nucleus (CN) is parameterized. Fission is described as a continuous evolution of this shape towards two touching ellipsoids at the moment of scission. The potential energy landscape shows a number of valleys and saddles, the path of the CN in this complex surface is marked by bifurcation points, forming in this way different modes of fission. Within this model, a mathematical description was proposed [9] to describe the fission fragment yields as a function of the fragment mass A and pre-neutron emission TKE.

In the frame of this approach the experimental two dimensional Y(A,TKE) distributions have been decomposed into the three main modes, two asymmetric modes, called standard I and II (S1 and S2), and one symmetric superlong (SL) mode. The mass distribution of each mode being of Gaussian shape and the TKE distribution of skewed Gaussian shape. Details can be found in Refs. [7,9].

In Fig. 2 the quality of the fit is demonstrated by comparison of the fitted results to the experimental mass and TKE distributions at thermal energy as well as the higher moments of the TKE distribution (variance and skewness). The fits have been performed for all the resonances mentioned earlier. Since strong variations in statistical significance are present in the data, only a slightly smoothed Y(A,TKE) distribution yielded convergent fit results. In Fig. 3 the variation of the fitted TKE and the branching ratios of the two asymmetric modes S1 and S2 are given, the latter relative to the respective thermal value. An up to 10% change in the branching ratio is obvious from the right part of Fig. 3. Also a bump like structure is evident in both figures with a maximum at about 40eV. If this is compared to the results from the resonance fission of ^{235}U [2], it is evident that the structure is similar, except that the magnitude is more pronounced and the neutron energy of the position of the bump is about half in case of ^{235}U(n,f). To further elaborate on the results, the correlation has been investigated between the different observables available for the resonance fission of ^{239}Pu. In Fig. 4 the correlation between the branching ratio and the relative changes in TKE is shown. A strong correlation is obvious, which is also understood from energetics point of view. If the mass distribution is changed, an immediate consequence is that also the mean TKE averaged over all masses is changing.

The prompt neutron multiplicity, as mentioned already in the introduction, is also subject to changes, if the mass and TKE distributions are varying. Again this is obvious from the energy balance. In Fig. 5 the available v_p data from the evaluated library JEF2.2 are given. Except two resonances around 40eV, the variation is about 3%. Nevertheless, it is just those resonances were also a larger than average change in the branching ratio and in TKE is observed. However,

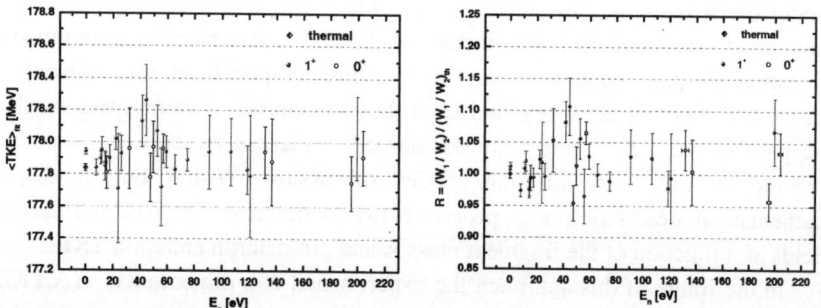

Figure 3. Left part: Resulting TKE from the fit and right part resulting relative branching ratio of the two asymmetric fission modes, both as a function of resonance energy.

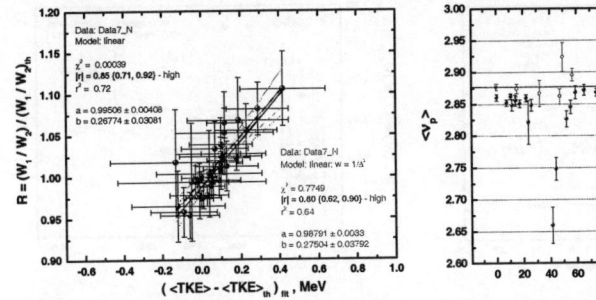

Figure 4. Correlation between branching ratio and relative TKE.

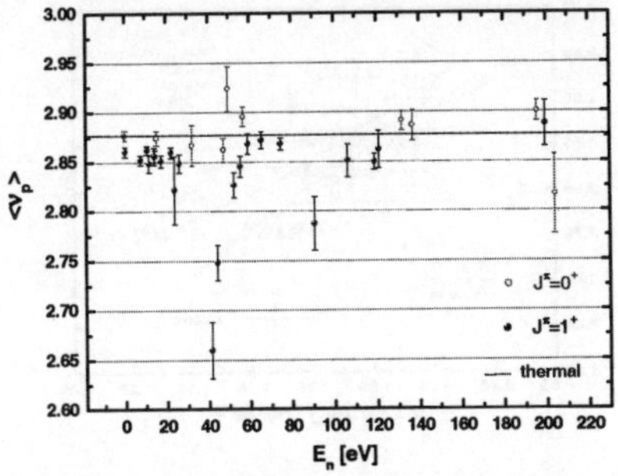

Figure 5. Variation in ν_p as a function of E_n. Data points are from JEF 2.2.

the corresponding anti-correlation with the TKE shows only a moderate correlation, if all the resonances are included. If the two 1^+ resonances at 41.66eV and 44.53eV are omitted, the correlation becomes insignificant (see Fig. 6). The same holds for the correlation with $1/\Gamma_f$ (shown in Fig. 7). Also here the correlation is based on only weak resonances with large uncertainties. Omitting them makes the correlation again insignificant.

As it was already mentioned, in contrast to resonance neutron induced fission of ^{235}U [2], less pronounced fluctuations in the mass yield and TKE-distributions for ^{239}Pu(n,f) were observed A possible explanation for the rather weak fluctuations of the fission fragment parameters is to be found in a new theoretical approach by Furman [10]. In ^{239}Pu, for each spin state J^π, only one possible K-channel, above the outer fission barrier, can be open, whereas for ^{235}U a mixture of two to three K-channels can take place [2,10]. If the quantum number K is considered to be a "good" quantum number, i.e. if it is conserved from saddle to scission, the scission configuration should have the same K. In this way, the fission fragment properties, for a given fission mode and K quantum number, should be "fixed". Hence, the superposition of different transition states with different K-quantum numbers (as in ^{235}U) and thus, different fission fragment property distributions can make them fluctuating from resonance to resonance. In case of ^{239}Pu, with only a single transition state, such fluctuations should be absent or be less pronounced, as it was observed.

5. Conclusions

In contrast to the ^{235}U(n,f) case, fluctuations in mass yield and TKE distributions as a function of resonance energy appear much weaker for ^{239}Pu. Only a slight anti-correlation (r=~0.3) with ν_p and an even smaller (r=~0.2) correlation with $1/\Gamma_f$ has been found. Within the MM-RNR model a fit to the

438

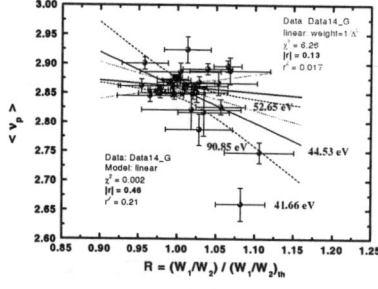

Figure 6. Correlation of the TKE with ν_p.

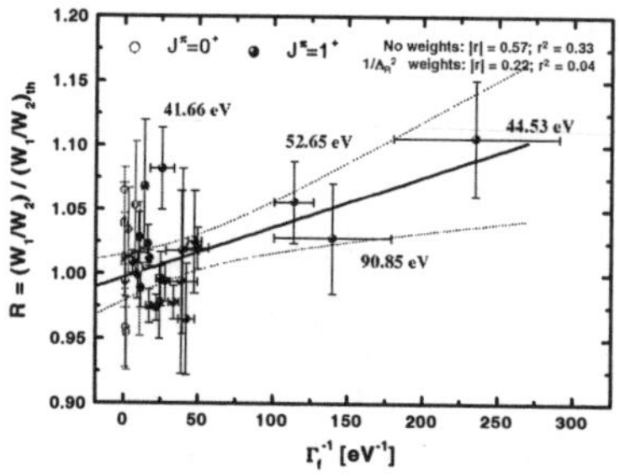

Figure 7. Correlation of the branching ratio with $1/\Gamma_f$.

Y(A,TKE) distribution has been performed, showing branching ratio fluctuations of up to 10% for the two asymmetric fission modes.

The absence of larger fluctuations of the fission fragment properties is understood within the model of Furman [10], supposing that only one transition state with given J, K quantum numbers is responsible for the fission path and the fission fragment properties are therefore "fixed" for each fission mode.

References

1. J.A. Wheeler, Physica **22**, 1103 (1956).
2. F.-J. Hambsch, H.-H. Knitter, C. Budtz-Jørgensen and J.P. Theobald, Nucl. Phys. **A491**, 56 (1989).
3. Sh. S. Zeinalov, M. Florek, W.I. Furman, V.A. Kriatchkov and Yu.S. Zamyatnin, Proc. 7th Int. Seminar on Interaction of Neutrons with Nuclei (ISINN-7), Dubna, Russia, May 25-28, 1999, p. 258.
4. L. Demattè, F.-J. Hambsch, H. Bax, Nucl. Inst. and Methods **A480**, 706 (2002).
5. E. Fort, J. Fréhaut, H. Tellier and P. Long, Nucl. Sci. Eng. **99,** 375 (1988).
6. W.T. Eadie, D. Drijard, F. James, M. Roos, B. Sadoulet, "Statistical Methods in Experimental Physics", North-Holland, 1971.
7. I. Ruskov, F.-J. Hambsch, L. Demattè, to be submitted to Nucl. Phys. A.
8. U. Brosa, S. Grossmann, A. Müller, Phys. Reports **197**, 167 (1990).
9. U. Brosa, H.-H. Knitter, Proc. of the XVIIIth Int. Symp. on Nuclear Physics, Physics and Chemistry of Fission, Gaussig, GDR, 1988, eds. H. Märten and D. Seeliger, ZfK-732, 145 (1988).
10. W. Furman, Proc. FJ/OH Spring Session 99, May 17-21, 1999, Geel Belgium p. 124.

FISSION BARRIER PHYSICS, RESONANCE FLUCTUATIONS AND ISOMER FISSION CROSS-SECTIONS

J. ERIC LYNN AND ANNA C. HAYES

Group T-16, Los Alamos National Laboratory, New Mexico, USA, 87545

Although the main picture of fission barrier physics was established some time ago many of the details still have to be settled. Consequently, the application to evaluation of cross-sections of unmeasurable or exotic nuclides and their excited states is still in its early stages. In this paper I consider some of these details and explore the possibility of quantitative estimation of fission cross-sections. Special examples include the isomer of ^{235}U and ^{232}U.

1. Introduction

Our present understanding of the early stages of the fission process (i.e. the passage across the barrier) is largely based on the intensive experimental and theoretical work of the 1960s and 70s (for a review, see [1]). This gave us an appreciation of the role of nucleon shells in the deformed nucleus and their effect on the topography of the potential energy surface and on the inertial parameter. Of course earlier work, especially that of A. Bohr [2] on the role of saddle point channels with special properties such as that of carrying a good K quantum number as well as total angular momentum and parity, remained as very valuable insights into the fission process and could be carried over into the new picture of the multi-humped barrier. All this information and understanding offers a good prospect of estimating with some confidence, at least at lower neutron energies, the cross-sections of very heavy exotic nuclides in which fission is a significant reaction mode. In recent years the demand for such cross-section data for applications ranging from nuclear astrophysics to nuclear criticality safety has become considerable. In this paper we review the current work at Los Alamos on models and calculations of neutron-induced fission cross-sections and those of competing cross-sections.

2. Intermediate structure

A major feature in the topography of the fission barrier of the actinides is the existence of a secondary well between inner and outer barrier peaks [3]. This gives rise to shape meta-stable states (class-II states) that are very complicated at the excitation energies for which fission becomes a significant competing reaction in compound nucleus decay from the primary well. The statistical model for dealing with this feature in the overall barrier postulates a transmission coefficient T_A for the deforming nucleus to cross the inner peak, A, whereupon the system equilibrates into the complicated class-II states of the secondary well. The

439

probability of these states to decay by fission rather than to return to the primary well is $T_B / (T_A + T_B)$ where T_B is the transmission coefficient for making the transition across the outer peak B. The overall transmission coefficient for the class-I states of the primary well to decay by fission is thus $T_F = T_A T_B / (T_A + T_B)$. This is the transmission coefficient for fission that would be used in standard Hauser-Feshbach theory.

Near and below the barrier peaks the class-II states manifest themselves as intermediate structure in the fission cross-section. This has very important implications for estimating fission cross-sections. Because the fission strength is clustered into intermediate resonances, the average fission cross-section over many intermediate resonances is lower than the value given by the statistical model. This reduction effect can be modelled [4] by a uniform "picket-fence" of class-II states with equal spacing D_{II} with the same values of coupling width $\Gamma_{II\odot} = T_A D_{II} /2\pi$ and the fission width $\Gamma_{II(F)} = T_B D_{II} /2\pi$. Below the barrier peaks this reduction can be as much as an order of magnitude.

There are also important effects from quantum chaos in both the class-II states and the class-I states. In the class-II states the factors to be considered are:
a) the Porter-Thomas fluctuations of the partial fission widths through the saddle-point channels of the outer peak,
b) the Porter-Thomas fluctuations of the coupling width to the class-I states through the inner barrier,
c) the Wigner-type fluctuations of the class-II level spacings.
In the class-I states we must consider:
a) the Porter-Thomas fluctuations of the squared coupling matrix elements of the individual class-I states to the closest class-II states,
b) the Porter-Thomas fluctuations of reduced neutron widths for elastic and inelastic scattering,
c) class-I level spacing fluctuations; when intermediate structure is sufficiently sharp to be describable by perturbation theory the fluctuation of the position of the class-I levels close to a class-II level is particularly important.
An analytical approach to the problem of averaging over all these fluctuations seems intractable. We have therefore resorted to a Monte Carlo approach, in which individual values of the parameters listed above are selected, using pseudo-random numbers, from the appropriate distribution functions with specified mean values. The eigenvalue problem is solved to give R-matrix resonance parameters across the whole range of an intermediate resonance. From these the cross-section is calculated and averaged. A large number of such calculations for the specified mean values yields an ensemble of σ_F values from which the mean value of the cross-section and the variance, skew and maximum likelihood estimator of the local cross-section averaged over individual class-II states are obtained.

The averaging factor S is defined as the ratio of the average cross-section $<\sigma_F>$ calculated by this procedure to the cross-section calculated from the uniform picket-fence model, $<\sigma_F> = S<\sigma_{UPF}>$. Some calculated results are shown in Fig. 1

as contours of S in the plane of inner and outer barrier transmission coefficients. The particular case shown is for neutron energy 10keV, mean class-I level spacing of 1eV and mean class-II level spacing of 50eV. These and other parameters are similar to those encountered in actinide nuclei. The low values of S over large parts of the plane demonstrate the importance of taking this factor into account in analyzing and estimating fission cross-section data.

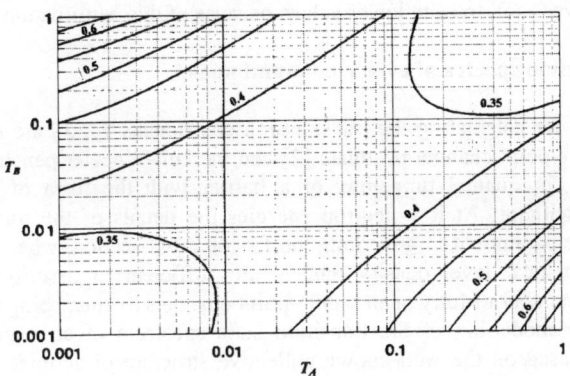

Figure 1. Contours of the averaging factor S as a function of T_A and T_B.

Towards the bottom left-hand corner of the plane shown in Fig.1 (low T_A and T_B values) the distribution of σ_F values becomes increasingly extreme. In Figure 2 we show the ratio of the variance of σ_F to the squared mean $<\sigma_F>^2$ as a function of T_B for a few T_A values. A ratio of unity is equivalent to an exponential

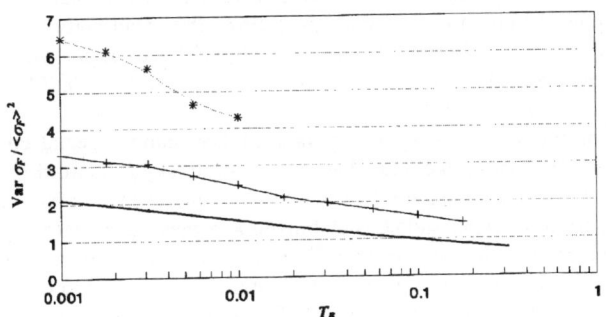

Figure 2. Variance of fission cross-section. T_A: 1.0—, 0.1—+—, 0.01—*—

distribution and a ratio of two to a Porter-Thomas distribution. The reason for much higher values of varσ_F/$<\sigma_F>^2$ is that the class-II levels become very narrow and no longer encompass many class-I levels within a Lorentzian envelope. Perturbation theory becomes applicable and the magnitude of the individual value of σ_F depends on the degree of accidental degeneracy between the class-II levels and the nearest class-I levels. For much lower values of the transmission coefficients than those shown in Figure 1 we expect the averaging factor S as defined above to fall to even lower values because of this perturbation regime.

3. Transmission spectra of even compound nuclei

The method for calculating the transmission coefficients T_A, T_B are based on A. Bohr's saddle-point channel postulate [2] and the Hill-Wheeler penetration factor formula [5] for quantal tunnelling of a barrier with the form of an inverted harmonic oscillator. At low neutron energies the details of the transition state spectrum are paramount in governing the fission cross-section, whereas at higher energies statistical level density models are sufficient to describe the cross-section, the number of fully open saddle-point channels by then being very large.

The modelling of the transition state spectrum of an even fissioning nucleus is based on the well-known collective structure of actinide nuclei with normal deformation. Thus we postulate a "ground" state (at the saddle) carrying a rotational band $J^{\pi} = 0^+, 2^+, 4^+, 6^+, ...$; a mass-asymmetry vibration with spin-projection along the symmetry axis $K^{\pi} = 0^-$, and associated rotational band $J^{\pi} = 1^-, 3^-, 5^-, ...$; a bending vibration with $K^{\pi} = 1^-$, and associated rotational band $J^{\pi} = 1^-, 2^-, 3^-, ...$; a gamma vibration (or rotation - see below) with $K^{\pi} = 2^+$, and associated rotational band $J^{\pi} = 2^+, 3^+, 4^+, ...$; and combinations of these.

The energies of these various state are dictated by the potential energy surface in deformation space at the relevant barrier. The moments of inertia for the rotational bands are assumed to follow the rigid-body form and are extrapolated from the observed values of rotational bands at normal deformation. At the inner barrier the bending vibration is assumed to be at 0.8MeV and at the outer barrier to be at 0.6MeV.

The mass asymmetry and gamma modes require special consideration. Calculations [6] using methods based on Strutinsky's liquid-drop-plus-shell-correction theory demonstrate that while the inner saddle point has a mass symmetric form the outer saddle is stabilized at a mass-asymmetric shape for the main actinides, and this feature is well supported by experiment in the thorium isotopes. Therefore, the mass asymmetry vibration is assumed to be at 0.7MeV at the inner barrier but to have a low value (0.1MeV) at the outer barrier because of the low tunnelling frequency between the mirror-image saddle point shapes through the mass symmetric hill. On the other hand, theoretical calculations indicate that the inner-saddle is axially asymmetric [7]. The energy difference between the axially symmetric and asymmetric shapes is least in the uranium isotopes, so it is necessary to explore a range of models for this feature in calculating fission cross-sections for these. In Model 1 we assume normal axial symmetry with the transition state energy as 0.8MeV for both barriers. In Model 2 we assume the axial asymmetry parameter γ to be 11°; this gives a gamma rotation with spin $J^{\pi} = 2^{+}$ and energy about 250keV and a second member of the gamma rotational band with spin $J^{\pi} = 4^{+}$ and energy about 1MeV. In Model 3 the degree of axial asymmetry is maximized at $\gamma = 30°$, and within the energy gap there are at least three members of the gamma rotational band with spin $J^{\pi} = 2^{+}$, energy ≈60keV, $J^{\pi} = 4^{+}$, energy ≈200keV and $J^{\pi} = 6^{+}$, energy ≈400keV. In Model 4 we assume an axially asymmetric form that is soft to axial deformation with a basic gamma vibration energy of 0.25MeV.

As in normally deformed nuclei we expect that these collective states and their low energy combinations will lie within the energy gap due to pairing forces. Above this energy gap a large number of 2 quasi-particle states will arise. These will act as heads for vibration-rotation bands. In this region we assume therefore that the transition state spectrum can be described by a statistical level density form. The transition state density above the energy gap $E_{G(A,B)}$ is assumed to have the simple temperature form:

$$\rho_{A,B}(I^{\pi}) = C_{A,B}(2J+1)\exp[-(J+1/2)^{2}/2\sigma_{A,B}^{2}]\exp(U/\Theta_{A,B}) \tag{1}$$

where U is the energy above the barrier. Because these denser states above the energy gap will affect significantly the cross-section at lower energies it is necessary to take them into account through integrals convoluting the level density with the Hill-Wheeler formula.

4. Application to neutron-induced fission of odd-mass uranium targets

The discrete transition state spectra within the energy gap are essentially sufficient to allow the determination of the barrier heights of the even mass uranium nuclides from particle transfer induced fission reactions, such as (d,pf) and (t,pf). With Model 1 the inner barrier height is found to be considerably lower than the outer barrier, but with models 2 to 3 the two barriers are essentially equal and approximately 1 MeV below the neutron separation energy of ^{234}U and ^{236}U. The difference is closer to 0.5 MeV for ^{238}U. We expect therefore that the top of the energy gaps at the barriers will appear a few hundred keV or less above the neutron separation energy of ^{234}U and ^{236}U. The residual states for inelastic neutron scattering out of these compound nuclei appear to be rather completely known (at least for the residual nucleus ^{235}U, which can also be used as a guide for completing an assessment of the likely level scheme for ^{233}U). Therefore, from the fission cross-sections and calculated inelastic neutron competition of these target nuclides we can fit the density of transition states above the energy gaps.

In fitting the data for the target nucleus ^{235}U the spin dispersion coefficient is assumed to have the value 6. The energy gaps and transition density coefficients are given in Table 1. The energy gap for the inner barrier seems very high in Model 1, and for this reason we feel that this model is less viable than the others.

Table 1. Deduced level density parameters at barriers and for residual nucleus. All energies are in MeV or reciprocal MeV units.

	Model 1		Model 2		Model 3		Model 4	
	A	B	A	B	A	B	A	B
Barrier	5.2	5.7	5.53	5.53	5.53	5.53	5.53	5.53
$E_{G(A,B)}$	1.65	1.03	1.25	1.15	1.32	1.22	1.1	1.15
$C_{A,B}$	0.20	0.05	0.34	0.07	0.34	0.07	0.34	0.07
$\Theta_{A,B}$	0.42	0.42	0.47	0.47	0.46	0.46	0.47	0.47
C_R	0.19		0.21		0.22		0.19	
Θ_R	0.54		0.57		0.50		0.55	

Above about 1.2 MeV neutron energy we must describe the level density of the residual nucleus by a simple statistical formula. We use the temperature form of Eq.1. Extrapolating the barrier level density parameters of Table 1 into this energy region we find the residual nucleus parameters listed in the last 2 lines of the Table.

The transition state spectra and barrier and level density parameters of Table 1 enable us to make good fits to the fission cross-section of ^{235}U. With inner and outer barrier heights of 5.83MeV for ^{234}U we can also fit the cross-section of ^{233}U very well using the other parameters of Table 1.

5. The fission cross-section of the isomer of ^{235}U

These models of the transition spectra can be extended to other nuclides for which no or little experimental data on the fission cross-sections are available. A particularly interesting case is that of the 77eV isomer of ^{235}U with half-life 26m. Measurements of the thermal neutron cross-section have shown that this is about twice as large as the cross-section of the ground state. This has led to the speculation that the fast neutron fission cross-section will also be higher causing the calculated fission yield in a very intense neutron flux to be increased, and conversely that the capture yield (to the product ^{236}U) will be lowered. The argument is not strong because of the highly stochastic nature of thermal neutron cross-sections. We have therefore used our models to calculate the isomer cross-section over the energy region from 1keV to 2MeV to test this speculation.

At neutron energies up to a few tens of keV s-wave neutron absorption to form the compound nucleus is dominant. Above that, p-wave neutron absorption becomes dominant and remains so until well above 0.5MeV. It follows that only a few saddle-point channels out of the total transition state spectrum are important for the fission cross-section in the lower part of the neutron energy range. Because the target spins I^π of the ground state and isomer of ^{235}U are very different ($7/2^-$ and $1/2^+$ respectively) the important saddle-point channels for the relevant compound nucleus spins J^π are quite different for the two states. These are shown in Table 2; only the lowest states in the transition state spectra are listed.

Table 2. Saddle-point channels for the ground state and isomer of ^{235}U (Model 3).

I^π	l	J^π	CN weight	Trans.state	$V_A(J^\pi) - S_n$	$V_B(J^\pi) - S_n$
Ground:						
$7/2^-$	0	3^-	7/16	mass asym.vib.	0.3	0.9
		4^-	9/16	bend.vib.+rot.	0.2	0.5
	1	2^+	5/16	rotation	1.0	1.0
		3^+	7/16	γ-vib.+rot.	0.9	0.1
		4^+	9/16	rotation	0.9	0.9
		5^+	11/16	γ-vib.+rot.	0.9	0.0
Isomer:						
$1/2^+$	0	0^+	1/4	zero excn.	1.0	1.0
		1^+	3/4	bend.vib.+m.a.	-0.5	0.3
	1	0^-	1/4	2q.p.(> en.gap)	-ve	-ve
		1^-	3/2	bend.vib.	0.3	0.9
		2^-	5/4	bend.vib.+ rot.	0.2	0.5

Table 2 shows that s-wave neutron-induced fission will be considerably suppressed and to a much smaller extent p-wave induced fission will be reduced as well. Detailed calculations verify this expectation. In Figure 3 we show the experimental fission cross-section data of ^{235}U and the Model 3 calculated cross-section for both the ground state and the isomer. Contrary to the expectation from the high value of the thermal neutron cross-section, the isomer cross-section is only about half the value of the ground state cross-section at low neutron energies and the two cross-sections do not attain near-equality until about 0.5MeV neutron energy. The capture cross-section of the isomer is correspondingly enhanced.

6. Neutron-induced fission of even U nuclides

Most of the long-lived nuclides have neutron separation energies below the barrier. Therefore, information on their barrier height parameters can be determined from the neutron-induced fission cross-section. Transition states are taken from calculations [8,6] of neutron single-particle orbitals at deformations

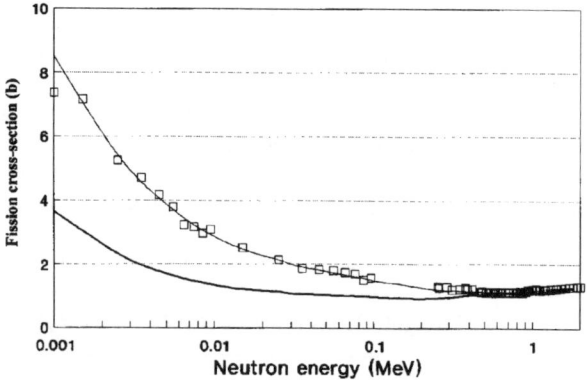

Figure 3. The fission cross-section of ^{235}U and calculated cross-section for the ground state (light line) and the isomer (bold line).

corresponding to the inner and outer barriers; to these are added vibration-rotation bands as prescribed for the even compound nuclei (Section 3). We find that the single-particle level density has to be about double that of normal odd-neutron nuclei in order to reproduce the magnitude of the fission cross-sections. The determined barrier heights are $V_A = 5.45$ MeV, $V_B = 5.65$ MeV for the compound nucleus ^{235}U, $V_A = 5.9$ MeV, $V_B = 5.7$ MeV for the compound nucleus ^{237}U, and $V_A = 5.95$ MeV, $V_B = 5.85$ MeV for the compound nucleus ^{239}U. Above the energy gap of the target nucleus the density of states for inelastic scattering to the residual nucleus are found to be consistent with the parameters $C_R \approx 0.1$-0.15MeV^{-1} (considerably lower than C_A for the corresponding even fissioning nucleus), $\Theta_R \approx 0.5\text{MeV}$.

The fission cross-section of ^{232}U is of considerable interest, especially in nuclear criticality safety. There are few experimental measurements and these are incomplete and inconsistent. The downward trend of barrier heights with decreasing mass number in the odd uranium nuclei that we observe above suggests that $V_A = 5.55$ MeV, $V_B = 5.55$ MeV might be a reasonable choice of barriers for ^{233}U. This gives the theoretical curve shown in Fig.4, where it is compared with the three available sets of experimental data [9]. The theory helps eliminate one experimental set and agrees quite well with the others.

7. Fission cross-sections of other actinide isomers

As an even-odd nuclide ^{235}U is unusual in having an isomer. Most isomers occur in odd-odd nuclides because the lowest 2quasi-particle state can occur quite close in energy with the spins of the odd particles either aligned or anti-parallel, giving, often, a very large K and spin difference. In ^{235}U, the long

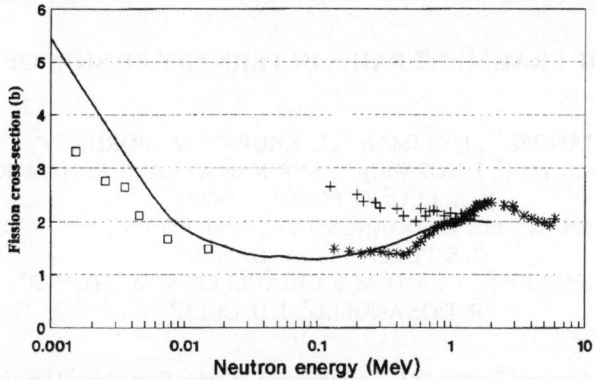

Figure 4. Fission cross-section of ^{232}U.

half-life of the isomer appears to be due primarily to the extremely small energy difference between it and the ground state.

We have not studied the fission cross-sections of the odd-odd actinides. However, a general remark is in order. The neutron separation energy of the compound nucleus is generally significantly higher than the fission barrier. The transition state spectrum of the compound nucleus will be quite dense, with vibration-rotation bands built on single proton quasi-particle states. It follows that there will be many more saddle-point channels available at low neutron energies for compound nucleus spins derived from a high-spin target nucleus than from a low-spin target. The high-spin target can thus be expected to have a considerably higher fission cross-section than its low spin counterpart.

References

1. S.Bjornholm and J.E.Lynn, *Rev.Mod.Phys.* **52**, 725 (1980)
2. A.Bohr, *PeacefulUses of Atomic Energy* **2**, 220 (United Nations, 1955)
3. V.M.Strutinsky, *Nucl.Phys.A,* **95**, 420 (1967)
4. J.E.Lynn and B.B.Back, *J.Phys.A* **7,** 395 (1974)
5. D.L.Hill and J.A.Wheeler, *Phys.Rev.* **89**, 1102 (1953)
6. P.Moller and J.R.Nix, *Phys.&Chem.of Fission* **1**, 103 (1974)
7. S.E.Larsson and G.Leander, *Phys.&Chem.of Fission* **1**, 177 (1974)
8. S.G.Nilsson *et al., Nucl.Phys.*A **131**, 1 (1969)
9. B.I.Fursov *et al.At.En.***61**, 383 (1986)

YIELDS OF FRAGMENT PAIRS IN TERNARY FISSION OF ^{252}Cf

M. JANDEL[1,3], J. KLIMAN[1,3], L. KRUPA[1,3], M. MORHÁČ[1,3],
J. H. HAMILTON[2], J. KORMICKI[2], A. V. RAMAYYA[2], J. K. HWANG[2],
Y. X. LUO[2], D. FONG[2], P. GORE[2],
G. M. TER AKOPIAN[3], YU. TS. OGANESSIAN[3], A. M. RODIN[3], A. S. FOMICHEV[3],
G. S. POPEKO[3], A. V. DANIEL[3],
J. O. RASMUSSEN[4], A. O. MACCHIAVELLI[4], M. A. STOYER[4],
R. DONANGELO[5], J. D. COLE[6]

[1]Department of Nuclear Physics, SASc., Dubravská cesta 6, Bratislava, Slovak Republic
[2]Department of Physics, Vanderbilt University, Nashville, Tennessee 37235
[3]Flerov Laboratory for Nuclear Reactions, JINR, Dubna, Russia
[4]Lawrence Berkeley National Laboratory, Berkeley, CA 94720
[5]Instituto de Fisica, UFRJ, 21945-970 Rio de Janeiro, Brazil
[6]Idaho National Engineering and Environmental Laboratory, Idaho Falls, Idaho 83415

The yields of correlated fragment pairs in He ternary fission for Zr/Ba and Mo/Xe charge splits were obtained. Neutron multiplicity distributions for these splits was determined with $<v_n>$=2.9(1) for both charge splits Zr/Ba and Mo/Xe. With the use of minimization method, the primary fragments' mass and excitation energy distributions were obtained, with the mean masses primary fragment 102.7/145.3 for Zr/Ba and 106.4/141.6 for Mo/Xe.

Introduction

It has been shown in previous works [1-3] that with the use of large detector arrays such as Gammasphere, the relative yields of correlated fragment pairs from fission can be extracted using prompt gamma-ray spectroscopy methods. In our experiment, we measured prompt gamma-rays from spontaneous fission of ^{252}Cf in coincidence with ternary particles. Gamma-rays were detected by Gammasphere array and eight Si ΔE-E telescopes provided the light charged particle [LCP] detection [4]. The total intensities of the lowest $2^+ \rightarrow 0^+$ ground-state band transitions observed in the de-excitation of even-even fission products reflect the total independent yields of these isotopes [5,6]. The relative yields of correlated fragment pairs have been extracted In works [1,3] using the analysis of γ-γ and γ-γ-γ coincidences measured in binary fission of ^{252}Cf. The primary fragments characteristics: mass distribution and excitation energies of primary fragments, has been determined using minimization method.

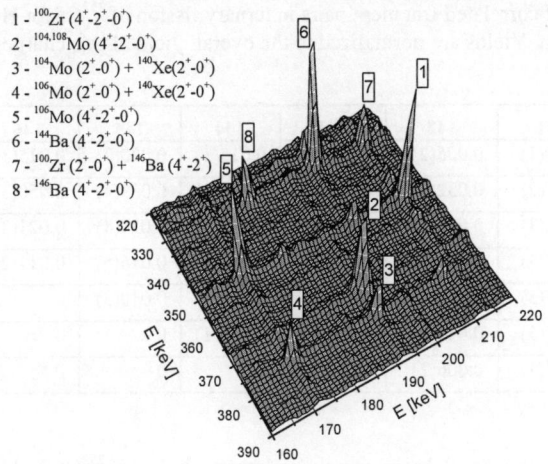

Figure 1. The cut from matrix of γ-γ coincidences for He particles. The peaks of transitions in correlated fragment pairs are shown together with the peaks of coincidences of the intraband transitions in the fragments.

We will follow these procedures and extract the primary fragments' characteristics for ternary fission of ^{252}Cf. The results are very important for our understanding of the fission process and the peculiarities in ternary fission.

Measured yields of correlated fragments' pairs

The matrix of γ-γ coincidences was built for the He particles. When analyzing the matrix it is very necessary to have a background properly subtracted. The background subtraction was carried out in the matrix of γ-γ coincidences by a two-dimensional subtraction method published in [7]. The cut from the matrix of coincidences after the background subtraction is shown in Figure 1. The peaks of transitions in correlated fragments are clearly seen. Fitting the peaks corresponding to the lowest ground state transitions in both of the fragments determine the yield of correlated fragment pair. The fitting procedure was carried out using the two-dimensional fit [8]. The area of peaks has been then corrected to internal conversion and detector efficiency according to:

$$Y\left(A'_{LF}, Z'_{LF}, A'_{HF}, Z'_{HF}\right) = \sum_{\gamma_1, \gamma_2} \left(1 + \alpha_1\right)\left(1 + \alpha_2\right) V\left(\gamma_1, \gamma_2\right) / \varepsilon_1 \varepsilon_2 p_1 p_2 ,$$

Table 1 Yields of correlated fragment pairs in ternary fission of ^{252}Cf for He particles and charge split Zr/Ba. Yields are normalized to the overall yield of this charge split.

Zr/Ba	141	142	143	144	145	146	147
99	0.001(1)	0.006(2)	0.007(2)	0.015(3)	0.020(3)	0.015(3)	0.007(3)
100	0.007(2)	0.032(3)	0.037(3)	0.099(2)	0.039(4)	0.050(3)	0.029(3)
101	0.012(3)	0.019(3)	0.054(6)	0.130(3)	0.037(4)	0.021(3)	0.004(1)
102	0.015(3)	0.030(4)	0.039(5)	0.106(3)	0.046(5)	0.012(2)	0
103	0.012(3)	0.038(4)	0.005(2)	0.012(3)	0.010(3)	0	0
104	0.006(3)	0.012(2)	0.002(1)	0.005(1)	0	0	0
105	0.003(1)	0.006(2)	0.001(1)	0	0	0	0

Table 2 Yields of correlated fragment pairs in ternary fission of ^{252}Cf for He particles and charge split Zr/Ba. Yields are normalized to the overall yield of this charge split.

Mo/Xe	137	138	139	140	141	142	143
102	0	0	0	0	0	0.012(2)	0.012(2)
103	0	0	0.005(1)	0.022(3)	0.010(3)	0.029(3)	0.019(3)
104	0.006(2)	0.010(2)	0.013(3)	0.079(3)	0.037(4)	0.072(4)	0.012(3)
105	0.009(3)	0.024(3)	0.024(4)	0.058(3)	0.026(3)	0.019(4)	0.003(1)
106	0.025(4)	0.053(4)	0.074(5)	0.091(3)	0.031(3)	0.019(4)	0
107	0.008(2)	0.043(4)	0.030(4)	0.023(3)	0.004(1)	0	0
108	0.012(3)	0.037(3)	0.011(2)	0.011(2)	0	0	0

where α, ε, and p are the internal conversion coefficient, detection efficiency and probability of emission for γ-ray with given energy, respectively. Relative yields of correlated fragment pairs are given in Tables 1,2 for charge splits Mo/Xe and Zr/Ba in He ternary fission of ^{252}Cf. The mass distribution of fragments is shown in Figure 2. By summing the yields for the same number of evaporated neutrons we built a neutron multiplicity distribution for both of the charge splits. The neutron multiplicity distributions for both splits are shown in Figure 3 For Zr/Ba charge split the mean neutron multiplicity determined by the fit is $<v_n>=2.9$ with $\sigma=1.4$. For Mo/Xe charge split the fit determined the $<v_n>=2.9$ with $\sigma=1.25$. When comparing to the values obtained for binary fission [1] the difference in neutron emission makes approximately 0.8 n which when assuming 8 MeV of excitation energy represents 6.4 MeV difference between binary and ternary fission. Similarly, by summing the yields of the same heavy

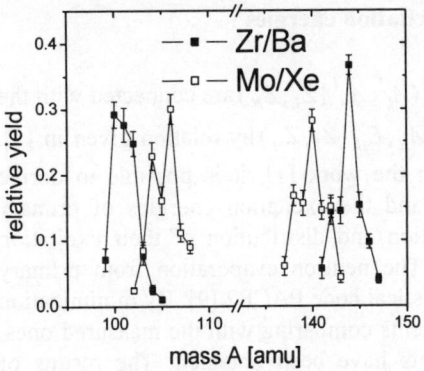

Figure 2 The mass distribution of fragments in He ternary fission of ^{252}Cf

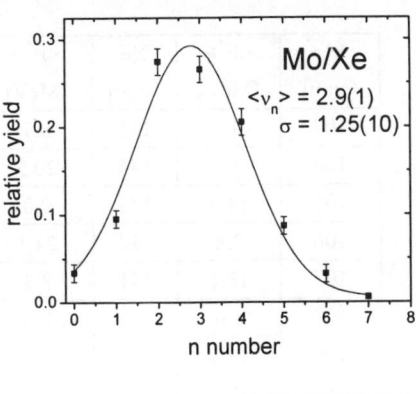

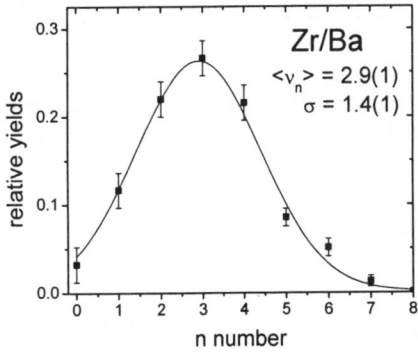

Figure 3 The neutron multiplicity distribution, in the upper part for Mo/Xe and in the lower part for Zr/Ba.

and light fragment mass, we obtained the mass distribution of the fragments, as they appear after a neutron evaporation.

We see an enhanced yields for pairs ^{100}Zr+^{144}Ba, ^{101}Zr+^{144}Ba and ^{102}Zr+^{144}Ba, making almost 30% of the yield of Zr/Ba charge split. The even-odd neutron number effects are clearly observable in the Mo/Xe charge split and also the highest yields representing approximately 30% of the total yield of Mo/Xe were recorded for pairs ^{104}Mo+^{140,142}Xe, ^{106}Mo+^{139,140}Xe. The pattern of the yields has an even-odd neutron number structure. However, here we would like to stress, that the statistics of the data was not very high (in the peaks of highest yield we recorded approximately 1200 events) and for low yield pairs the peaks were close to the detection limit. This was mainly connected with the odd mass isotopes and determination of their yields. The ratio of the total yields of the charge splits is approximately Y(Mo/Xe)/ Y(Zr/Ba) $\simeq 16230/9894$ $\simeq 1.64$.

In the next section we will shortly mention the model [1] needed for determination of primary mass distribution and excitation energies of primary fragments and the results of this procedure applied on our data.

3. Primary fragments' masses and excitation energies

The yields of secondary fragments $Y^{calc}(A_L', A_H' | Z_L, Z_H)$ are connected with the yields of primary fragments $Y(A_L, E_L^*, A_H, E_H^* | Z_L, Z_H)$ by relation given in [1]. By unfolding procedure, given also in the work [1], it is possible to derive primary fragments' mass distribution and the excitation energies of primary fragments. The primary mass distribution and distribution of their excitation energy is considered to be gaussian. The neutron evaporation from primary fragments was calculated using the statistical code PACE2 [9]. By minimization of calculated yields of secondary fragments comparing with the measured ones, the characteristics of primary fragments have been obtained. The results of minimization procedure are shown in Tables 3,4 and 5.

Table 3,4 Mean excitation energies obtained by minimization procedure for primary fragments in charge splits Zr/Ba and Mo/Xe.

Zr [amu]	$<E^*>$ [MeV]	Ba [amu]	$<E^*>$ [MeV]
100	3.9	148	18.6
101	11.8	147	10.6
102	6.4	146	18.0
103	15.6	144	8.6
104	24.4	143	2.0
105	10.6	142	13.5
106	23.7	141	2.0

Mo [amu]	$<E^*>$ [MeV]	Xe [amu]	$<E^*>$ [MeV]
103	9.1	145	8.7
104	2.0	144	20.5
105	14.9	143	7.8
106	2.8	142	24.1
107	18.1	141	8.3
108	5.6	140	23.8
109	19.3	139	8.9
110	2.0	138	28.8

Table 5 Characteristics of mass distribution of primary fragments and total kinetic energy distribution obtained from the minimization and values from binary fission are shown for comparison [1].

		$<A_{LF}>/<A_{HF}>$	$\sigma(A)$	$<TKE>$	σ_{TKE}
ternary	Zr/Ba	102.7/145.3	1.8	186	9.4
	Mo/Xe	106.4/141.6	1.8	189	9.4
binary	Zr/Ce	101.9/150.1	1.9	183.3	9.9
	Mo/Ba	106.3/145.7	2.0	189.3	9.5
	Ru/Xe	112.3/139.7	2.6	193.3	9.7

The widths of primary fragments' mass distribution are for both charge splits $\sigma(A)=1.8(2)$ mass units with the mean values of mass distributions $<A_{Zr}>/<A_{Ba}>$ $=102.7/145$ and $<A_{Mo}>/<A_{Xe}>$ $=106.4/141.6$. The structure is observed in Mo/Xe excitation energies. The Mo fragments with even mass have low excitation energy comparing to the odd mass Mo primary fragments. Here, this effect can be explained when assuming, that 10-20% of He particles are emitted as ^5He isotopes [10] and we were performing the unfolding procedure only for the primary fragments' configuration assuming the emission of ^4He. Apparently, the emission of ^5He corresponds to different characteristics of primary mass distribution and it might be incorporated into unfolding procedure as an additional mode. However, the number of free parameters of minimization model increase twice, that is why we didn't attempt to involve this mode. This may be the reason of such oscillating effects in excitation energies of primary odd – even mass fragments.

Acknowledgments

The work was partially supported by the Grant Agency of Slovak republic through contract GAV 2/5121/98. Research at Vanderbilt University is supported in part by the U.S. Department of Energy under Grant No. DE-FG05-88ER40407. Work at Idaho National Engineering Laboratory is supported by the U.S. Department of Energy under Contract No. DE-AC07-76ID01570. The work at LLNL was performed under the auspices of the U.S. Department of Energy under contract No. W-7405-ENG-48, and that of LBNL under contract No. DE-AC03-76SF00098. The Joint Institute for Heavy Ion Research is supported by the member institutes, the University of Tennessee, Vanderbilt University and the US Department of Energy.

References

1. G. M. Ter-Akopian et al., *Phys. Rev.* **C55**, 1146 (1997).
2. J. H. Hamilton et al., *J. Phys. Rev.* **20**, L85 (1994).
3. G. M. Ter-Akopian et al., *Phys. Rev. Lett.* **77**, 32 (1996).
4. M. Jandel et al., *J. Phys. G* **28**, 2893 (2002).
5. E. Cheifetz et al., *Phys. Rev.* **C4**, 1913 (1971).
6. R. Aryaeinejad et al., *Phys. Rev.* **C48**, 566 (1993).
7. M. Morhac et al., *NIM* **A401**, 113 (1997).
8. M. Morhac et al., *Appl. Spectr.*, accepted.
9. A. Gavron, *Phys. Rev.* **C21**, 230 (1980).
10. J. K. Hwang et al., *Phys. Rev.* **C61**, 047601 (2000)

HOT BIMODAL TERNARY FISSION IN ^{252}Cf

D. FONG[1], M. JANDEL[1,2,3,4], J.H. HAMILTON[1], A.V. RAMAYYA[1], J.K. HWANG[1], P. GORE[1], J. KORMICKI[1], Y.X. LUO[1], J. KLIMAN[2,3], L. KRUPA[2,3,4], A.V. DANIEL[3], A. FOMICHEV[3,4], G.S. POPEKO[3], A. RODIN[3], G. M. TER-AKOPIAN[3], Y. T. OGANESSIAN[3], J.O. RASMUSSEN[5], A. O. MACCHIAVELLI[5], I.Y. LEE[5], M.A. STOYER[6], R. DONANGELO[7], J.D. COLE[8], W. GREINER[9], D. POENARU[10]

[1] Physics Department, Vanderbilt University, Nashville, Tennessee 37235

[2] Department of Nuclear Physics, Slovak Academy of Sciences, Bratislave, Slovak Republic

[3] Flerov Laboratory for Nuclear Reactions, JINR, Dubna, Russia

[4] Joint Institute of Heavy Ion Research, Oak Ridge, TN 37831-6374

[5] Lawrence Berkeley National Laboratory, Berkeley, CA 94720

[6] Lawrence Livermore National Laboratory, Livermore, CA 94550

[7] Instituto de Física, Universidade Federal de Rio de Janeiro, 21945-970 Rio de Janeiro, Brazil

[8] Idaho National Engineering and Environmental Laboratory, Idaho Falls, Idaho 83415

[9] Johann Wolfgang von-Goethe Universitat, Frankfurt am Main 11 60054, Germany

[10] National Institute of Physics and Nuclear Engineering, Bucharest-Magurele 76900, Romania

Spontaneous fission of ^{252}Cf was studied with 8 light charged particle detectors in Gammasphere. 9×10^5 alpha coincident events were recorded. In coincidence with LCP, the γ-γ matrix was analyzed. We obtained the relative yields of fission fragment pairs as a function of the neutron number. In the Ba-α-Zr and Mo-α-Xe splits, the 6n-10n channels show enhanced neutron emissions, just as previously found for the hot bimodal Ba-Mo binary yield. Two Gaussian curves describe the data, centered on 3.5 and 7 neutrons. For Mo-α-Xe, the average neutron numbers are approximately 3 and 6, and for Ba-α-Zr, approximately 3 and 8.

Introduction

A recent experiment has yielded evidence for a hot bimodal process in ternary fission of ^{252}Cf. By analyzing the light charged particle (LCP) from the ternary fission as well as the γ-γ coincidence data, the fission fragments were identified. Selecting alpha particles as our LCP and determining the fission yields from the γ spectra, the distribution of neutrons emitted was calculated. This distribution is not well-described by a single Gaussian in two splits : Ba-α-Zr and Mo-α-Xe. Instead, there is a second Gaussian at a high number of emitted neutrons. This is intriguing, as previous works have found a similar hot bimodal process in binary fission to Ba-Mo.

By forming fragments with low average kinetic energy, the fragments carry away more internal excitation energy. This results in greater deformation of the primary fission fragments. The hyperdeformed nucleus then boils off a greater than normal number of neutrons. This process was first observed in binary fission of ^{252}Cf to Ba-Mo by Ter-Akopian et al. [1]. Here, we have a second Gaussian centered around 7-8 neutrons in addition to the stronger Gaussian centered around n =3. A more recent study by Wu et al. [2] lowered the strength of the high neutron distribution to a few percent of the main distribution. This leads us to examine the ternary fission counterparts of Ba-Mo binary fission, which are Ba-α-Zr and Mo-α-Xe. A study of the energetics of the system shows that a ternary fission process is energetically favorable for a large number of neutrons emitted. The Q-value decreases from ~200 MeV for zero neutrons released to ~160 MeV for eight neutrons, but there is certainly sufficient energy released.

Experimental Details

A source of ^{252}Cf with an intensity of $3x10^4$ spontaneous fission events per second was placed in Gammasphere in 2001. The 3 mm diameter active area was deposited on a 1.5 μm Ti foil. Both sides were covered with a thin gold foil (4.5 μm on the active layer, 3.8 μm on the Ti backing). In addition, a 2.5 μm Mylar foil was added to both sides to provide full absorption of the fragment. This eliminated Doppler shift problems and limited damage to the Si detectors. These Si detectors were 8 identical ΔE-E detectors surrounding the source to provide LCP identification. The ΔE detectors were 10 μm thick and 10x10 mm^2 in area. The E detectors were 400 μm in thickness. With the LCP identification, a γ-γ coincidence matrix was constructed in coincidence with observed α particles.

456

Data Analysis and Conclusions

From the γ-γ matrix in coincidence with α, we find the relative yields of each fission split. This is done by gating on the dominant transition in one secondary fragment and looking for the strongest transitions in the corresponding partner fragments. A sample spectrum is shown here gating on the α and the 2^+ - 0^+ transition in ^{142}Xe in Figure 1.

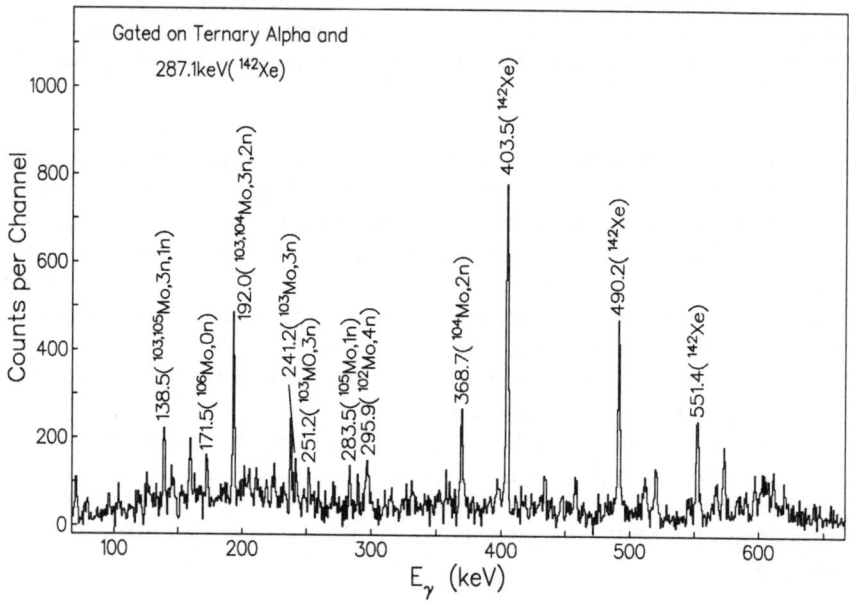

Figure 1: Gamma ray spectrum gating on α and γ

Each split is associated with a certain number of neutrons emitted by the primary fragments. By combining all possible splits, we construct the total neutron multiplicity. For most fission processes, this is fit by a single Gaussian centered around 3-4 neutrons emitted.

Tables 1 and 2 show calculated yields of α ternary fission fragment pairs for Ba-α-Zr and Mo-α-Xe. These yields are corrected for internal conversion, detector efficiency, and probability of γ emission. N/A refers to splits that are not possible due to total number of neutrons in the fragments being greater than the original number of neutrons.

Table 1: Ba-α-Zr yield table

	139Ba	140Ba	141Ba	142Ba	143Ba	144Ba	145Ba	146Ba	147Ba	148Ba
97Zr	<0.03	0.11(4)	<0.03	<0.03	0.06(4)	<0.03	<0.03	<0.03	<0.03	<0.03
98Zr	<0.03	0.30(6)	0.09(4)	0.11(4)	<0.03	0.19(5)	0.09(5)	0.22(5)	0.08(6)	0.07(5)
99Zr	0.05(3)	0.59(7)	0.28(6)	0.49(6)	0.24(7)	0.47(6)	0.28(6)	0.38(6)	0.21(8)	0.56(7)
100Zr	0.04(3)	0.65(6)	0.06(5)	0.64(5)	0.43(7)	1.63(7)	0.62(6)	1.09(6)	0.39(9)	0.16(6)
101Zr	0.04(3)	0.24(5)	<0.03	0.23(5)	0.48(6)	0.65(6)	0.27(5)	0.24(5)	0.21(7)	N/A
102Zr	0.05(3)	0.54(6)	0.32(6)	1.16(6)	0.54(7)	1.66(7)	0.47(6)	0.24(6)	N/A	N/A
103Zr	<0.03	0.20(5)	0.10(4)	0.29(4)	0.21(5)	0.15(4)	0.13(4)	N/A	N/A	N/A
104Zr	0.06(3)	0.60(5)	0.16(5)	0.35(7)	0.32(7)	0.29(6)	N/A	N/A	N/A	N/A

Table 2 : Mo-α-Xe yield table

	102Mo	103Mo	104Mo	105Mo	106Mo	107Mo	108Mo	109Mo
136Xe	<0.03	<0.03	<0.03	<0.03	<0.03	0.08(3)	0.08(5)	<0.03
137Xe	0.13(3)	0.05(3)	0.6(1)	0.04(3)	0.35(5)	0.45(5)	0.5(1)	<0.03
138Xe	0.16(5)	<0.03	0.5(1)	0.25(5)	1.69(8)	0.42(6)	0.7(1)	0.19(5)
139Xe	0.11(5)	0.06(5)	0.6(1)	0.26(5)	1.28(8)	0.63(6)	0.3(2)	0.19(5)
140Xe	0.11(5)	0.06(4)	1.5(1)	0.72(5)	1.74(7)	0.41(6)	0.6(1)	N/A
141Xe	0.12(4)	0.11(5)	6.1(2)	0.22(5)	0.43(6)	<0.03	N/A	N/A
142Xe	0.41(5)	0.23(4)	1.0(1)	0.12(3)	0.25(5)	N/A	N/A	N/A

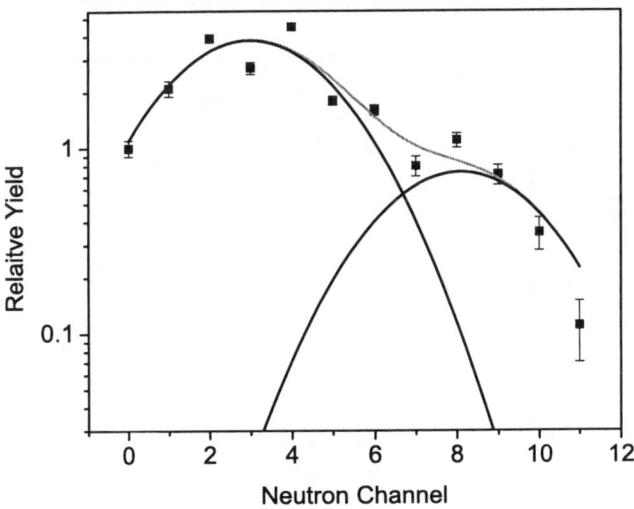

Figure 2 : Ba-α-Zr neutron yields and double Gaussian fit

458

Figures 2 and 3 show the neutron multiplicities from these tables. The features in this distribution are the main peak centered at n=3, and additional strength described by a second Gaussian centered around 6-8 neutrons. Similar to the binary Ba-Mo fission, there appear to be two competing fission modes. To contrast this with a non-bimodal α ternary pairing, Figure 4 shows the distribution for Ru-α-Te with no additional strength above the single Gaussian.

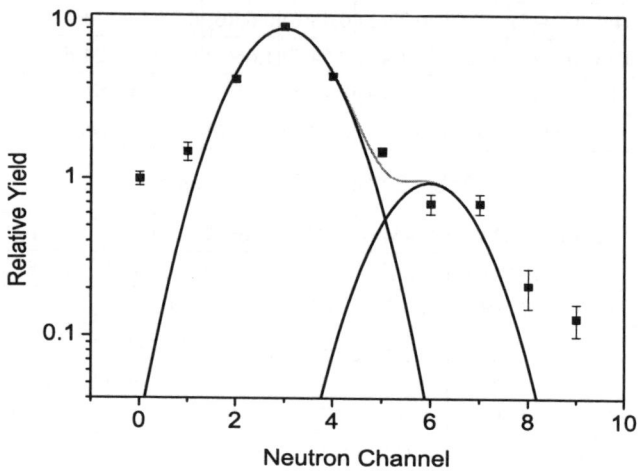

Figure 3 : Mo-α-Xe neutron yields and double Gaussian fit

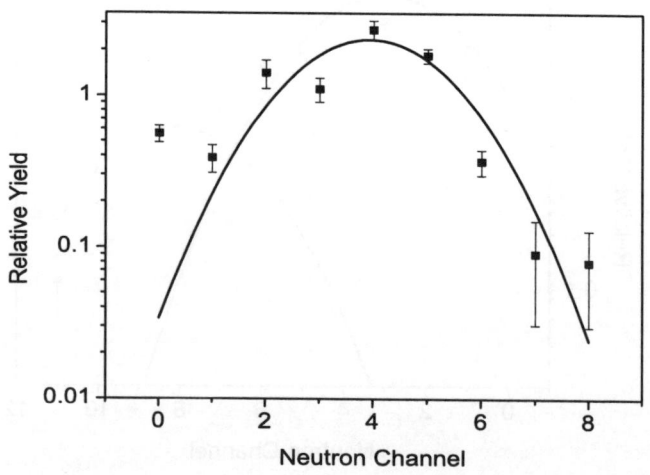

Figure 4 : Ru-α-Te neutron yields and single Gaussian fit

In the binary Ba-Mo split, it is believed that this hot fission mode is because of hyperdeformation of the soft Ba nucleus, with a smaller contribution from hyperdeformation of the Mo nucleus. This is consistent with these ternary results, as the strength of the hot mode is stronger in the Ba associated split (~20% of the main peak) than in the Mo associated split (~10% of the main peak). These are both enhanced compared to the binary fission. This is plausible as the LCP involved in ternary fission limits the kinetic energy each fragment can carry away, leaving more energy for internal excitation.

Summary

This experiment extends the phenomenon of hot bimodal fission to ternary processes. The Ba and Mo fragments appear to have a normal and a hyperdeformed mode in α ternary fission as well as binary fission. It appears enhanced in strength for the ternary process, but the statistics are poor, leading to larger error bars.

Acknowledgements

Work at Vanderbilt University was supported in part by U.S. DOE under grant No. DE-FG05-88ER40407. Work at Lawrence Berkeley National Laboratory was supported in part by the U.S. DOE under grant No. W-7405-ENG48.

References

1. Ter-Akopian et al. *Phys. Rev. Lett.* **73**, 1477 (1996)
2. Wu et al. *Phys. Rev.* **C62**, 041601 (2000)

INVESTIGATION OF NEUTRON MULTIPLICITY DISTRIBUTIONS IN SEPARATE REACTION CHANNELS USING GADOLINIUM LOADED LARGE LIQUID SCINTILLATION COUNTERS[*]

A. S. VOROBYEV, A. B. LAPTEV, O. A. SHCHERBAKOV

Petersburg Nuclear Physics Institute.
188300, Gatchina, Leningrad district, Russia
E-mail: alexander.vorobyev@pnpi.spb.ru

V. A. KALININ, V. N. DUSHIN, V. A. JAKOVLEV, B. F. PETROV

V.G. Khlopin Radium Institute.
2-nd Murinski Ave. 28, 194021, St. Petersburg, Russia
E-mail: kalinin@atom.nw.ru

F.-J. HAMBSCH

EC-JRC-Institute for Reference Materials and Measurements.

B-2440 Geel, Belgium
E-mail: hambsch@irmm.jrc.be

Measurements of the neutron multiplicity distributions in fission are related with some experimental and data processing difficulties and never have been described in a comprehensive manner. Therefore there are still some doubts about the results. To clarify the situation, a subsequent experiment has been performed. The description of the experimental set-up and data unfolding procedure, as well as some results of this research, is presented in this report.

1. Experimental Set-up

An ionization chamber with a ^{252}Cf-source was used for the measurement, because the fission fragment and neutron emission properties are comparatively well known for ^{252}Cf [1-7]. The ^{252}Cf sample with an activity of ~300 fission/s was prepared by cluster transposition of the mother-source onto a ~ 20 μg/cm^2 thick carbon backing. The sample was placed on a common cathode with a system of cylindrical holes, a so-called pin-hole collimator.

The fission chamber was placed between two large gadolinium loaded liquid scintillation neutron counters. Such a composition allows to carry out measurements either in a 2×2π-geometry – to measure the number of neutrons emitted by each of the complementary fragments in every fission event or in a 4π-geometry – for total neutron multiplicity measurements. Our experimental set-up has been described elsewhere [8,9].

---

[*] This work was carried out under the support of the ISTC Project 554-97.

The neutron events were measured in a 20-μs time window, opened ~0.3 μs after the fission event. Fission events were defined by registration of coincident fission fragment pulses from both halves of the ionization chamber. Similarly, background events were measured for counting intervals started randomly. The fission event was considered as "useful" and stored when only one fission event was registered in the 20 μs interval. A total of 3×10^6 and 0.5×10^6 "useful" events have been accumulated for the $2\times2\pi$-geometry and 4π-geometry set-ups, accordingly. The list-mode data included pulse-height, the number of emitted neutrons for both fragments and the neutron background multiplicity.

2. Data Processing and Results

2.1. *General*

The data processing may be divided into several steps: 1) conversion of the obtained pulse-height distributions to the provisional mass and "post-neutron" total kinetic energy (TKE) distributions; 2) calculation of the mean values and variances of the neutron distributions as a function of provisional fission fragment mass and "post-neutron" TKE; 3) calculation of the "pre-neutron" mass and TKE using the obtained provisional mass and "post-neutron" TKE dependence of the mean neutron multiplicity; 4) unfolding of the initial neutron multiplicity distributions and its moments for a fixed value of the fission fragment pre-neutron mass TKE. The three first steps of the data processing were described sufficiently well in refs. [8, 9]. Further on we shall concentrate only on the unfolding of the initial neutron multiplicity distributions from the measured ones.

2.2. *Unfolding of the Neutron Multiplicity Distribution*

2.2.1 *Main idea*

An initial neutron multiplicity distribution is distorted in the measurement due to the detection efficiency, background and neutron pulse pile-up.

The transformation of the initial neutron multiplicity distribution, P_{init}, into the measured one corrected for the neutron pulse pile-up, P_{cp}, can be presented in matrix form:

$$K \cdot P_{init} = P_{cp}, \tag{1}$$

where K is a convolution of two cores describing distortions due to background and the detector efficiency. A direct solution of such a system is successful only for high efficiency detectors ($\varepsilon > 0.6$) and large statistics. Because of the last condition is not always fulfilled especially when one builds up neutron multiplicity distributions for fixed mass and TKE, the solution leads to oscillations in the final values of the probability. In this connection the special regularization method should be used. The solution obtained using the method of statistical regularization [10] may result in a strong distortion of moments of the initial neutron multiplicity distribution. Therefore the method proposed by Alkhazov et al. [6] was used.

The main idea of the unfolding procedure consists in the following successive operations. First, an initial model distribution, P_0 , is constructed on the basis of the reconstructed moments of the neutron distribution (see ref. [9]) $<v_T>$, $\sigma^2(v_T)$ and $<v_1>$, $<v_2>$, $\sigma^2(v_1)$, $\sigma^2(v_2)$, $cov(v_1, v_2)$ for the 4π- and 2×2π-geometry, respectively. Second, the model distribution is convoluted with the distorting core, K (see Eq. (1)). The neutron multiplicity distribution obtained in this way corresponds to the measured one after pile-up correction (see ref. [9]) but differs by a small value ΔP_{cp}. Third, the small correction ΔP_{init} in the model distribution, P_0, is reconstructed using the regularization equation:

$$(K^T \cdot K + s^2 \cdot \alpha \cdot \Omega) \cdot \Delta P_{init} = K^T \cdot \Delta P_{cp} , \qquad (2)$$

where Ω is the functional of the norm of second derivatives, α is the regularization parameter, s^2 is the measurement uncertainty. The corrected distribution, P_{init}:

$$P_{init} = P_0 + \Delta P_{init} , \qquad (3)$$

is used as a new model one in the second iteration step and so on. The iteration process is completed when the difference ΔP_{cp} becomes less than the statistical uncertainty of the measurement.

2.2.2 Neutron detection efficiency

The detection efficiency of a large liquid scintillation counter in 4π-geometry is practically constant and can be determined using a fission source with well known mean total number of neutrons.

In 2×2π-geometry the detection efficiency depends on the reaction kinematics, on the angular distributions and energy spectra of neutrons entering into the scintillation counters, on the detector shape and on the shielding characteristics. The detection efficiency can be defined as a matrix:

$$\varepsilon_{2\pi} = \begin{pmatrix} \varepsilon_{11} & \varepsilon_{12} \\ \varepsilon_{21} & \varepsilon_{22} \end{pmatrix}, \qquad (4)$$

where ε_{11}, ε_{22} are the registration efficiencies of "direct" neutrons, i.e. the registration efficiencies of the counters 1 or 2 for neutrons emitted by the fragments flying towards the respective counters. The non-diagonal terms ε_{12} and ε_{21} are the registration efficiencies of "indirect" neutrons, i.e. the registration efficiencies of the counters 1 or 2 for neutrons emitted by the fragment flying in the opposite direction. The matrix elements have been calculated using the Monte-Carlo code [11].

2.2.3 *Total neutron multiplicity distribution*

For the 4π-geometry the measured neutron multiplicity distribution corrected for the neutron pulse pile-up, $P^{4\pi}_{cp}$, is related with the initial one, $P^{4\pi}_{init}$, by the expression:

$$P^{4\pi}_{cp}(i) = \sum_{k=0}^{i} \sum_{v=i-k}^{v_{max}} \frac{v!}{(i-k)!(v-i-k)!} \cdot \varepsilon_{4\pi}^{i-k} \cdot (1-\varepsilon_{4\pi})^{v-i-k} \times$$

$$\times P^{4\pi}_{init}(v) \cdot B^{4\pi}_{cp}(k) \tag{5}$$

where $P^{4\pi}_{cp}(i)$ is the probability of observing i events per fission after pile-up correction, $P^{4\pi}_{init}(v)$ is the probability that v neutrons are emitted per fission, $B^{4\pi}_{cp}(k)$ is the probability of k background events after pile-up correction in $20\mu s$. The maximum number of neutrons considered, v_{max}, was 9.

To demonstrate the quality of the unfolding procedure, the mass distribution formed by the selection of fission events with total fitted excitation energy, Q_{fit} − TKE, not higher than ~ 8 MeV is shown with that obtained for $v_{tot} = 0$ in Figure. 1. The values of Q_{fit} were obtained by smoothing of the mass odd-even effect at the maximum total available energy taken from Audi et al [12]. The good agreement between the two data sets can be seen. The errors shown here are due to the unfolding procedure and the asymmetry of the fission fragment detector as well as of the neutron counters.

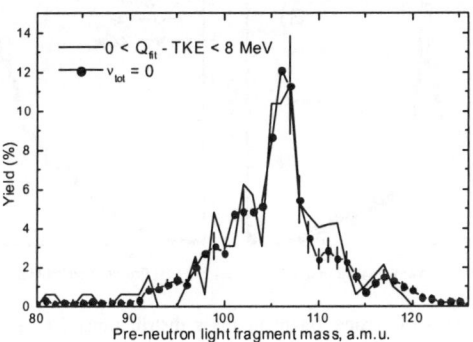

Figure 1. Comparison of cold fission fragment mass distributions obtained by two different methods for ^{252}Cf (see text).

2.2.4 *Two-dimensional neutron multiplicity distribution*

The measured neutron multiplicity distribution corrected for the neutron pulse pile-up, $P^{2\pi}_{cp}$ is related with the initial one, $P^{2\pi}_{init}$, by an expression analogous to Eq. (1). Columns and rows of this matrix correspond to the neutrons registered by the first and second scintillator counters, respectively. In this case the core K of the equation is also a convolution of two cores describing distortions due to

464

the background $B^{2\pi}_{cp}$ and the detector efficiency R^ε. The relation for R^ε can be written as:

$$R^\varepsilon (n^{cp}{}_1, n^{cp}{}_2, v_1, v_2) = \sum_{r=0}^{v_1} \sum_{r'=0}^{v_1-r} C_{v_1}^r \cdot \varepsilon_{11}^r \cdot C_{v_1-r}^{r'} \cdot \varepsilon_{12}^{r'} \times$$

$$\times (1 - \varepsilon_{11} - \varepsilon_{12})^{v_1 - r - r'} \cdot C_{v_2}^{n^{cp}{}_1 - r} \cdot \varepsilon_{21}^{n^{cp}{}_1 - r} \times \qquad (6)$$

$$\times C_{v_2 - n^{cp}{}_1 + r}^{n^{cp}{}_2 - r'} \cdot \varepsilon_{22}^{n^{cp}{}_2 - r'} \cdot (1 - \varepsilon_{22} - \varepsilon_{21})^{v_2 - n^{cp}{}_1 - n^{cp}{}_2 + r + r'}$$

where $C^n{}_v$ are the binomial coefficients, v_1 is the number of neutrons emitted by the fragment flying towards the first scintillator tank, out of v_1 neutrons r and r' neutrons may be registered by the first and second neutron counters, respectively. As this takes place the following conditions has to be satisfied:

$$r + r' \le v_1 ; \quad n^{cp}{}_1 + n^{cp}{}_2 - (r + r') \le v_2 ; \quad n^{cp}{}_1 > r ; \quad n^{cp}{}_2 > r'. \qquad (7)$$

The basic criteria of the correctness of the unfolded distributions for fixed numbers of prompt neutrons was the equality of the summed up total unfolded distribution to that known in literature and a similarity of the partial distributions obtained in $2 \times 2\pi$- and 4π-geometry.

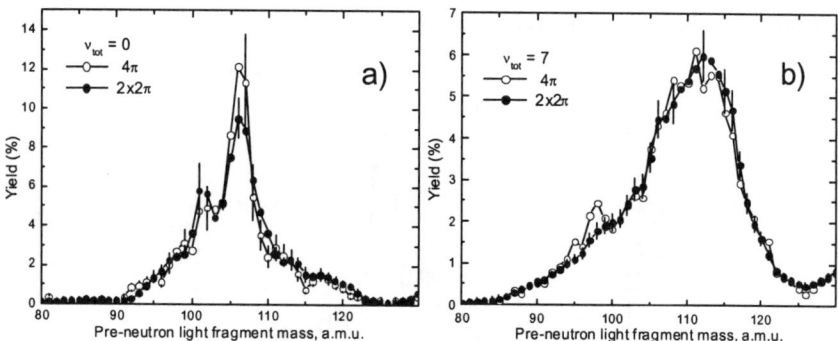

Figure 2. The comparison of fission fragment mass distributions for (a) $v_{tot}=0$ and (b) $v_{tot}=7$, obtained by the 4π-geometry (solid line with hollow circles) and $2 \times 2\pi$-geometry (solid line with full circles) for ^{252}Cf.

In Figure 2 the comparison of the unfolded results for the $2 \times 2\pi$- and 4π-geometry set-ups is presented. The ^{252}Cf fragment mass distribution for fixed numbers of emitted neutrons from complementary fragments are given in Figure 3 together with the data of Alkhazov et. al. [6] and Düring et. al. [7]. The data of refs. [6, 7] and those obtained in the present paper are quite consistent. Also a good agreement is seen for our data in $2 \times 2\pi$- and 4π-geometry. A general coincidence of the data obtained by different groups (refs. [6,7] and ours) and in various geometries indicate a correctness of the method used for data processing.

465

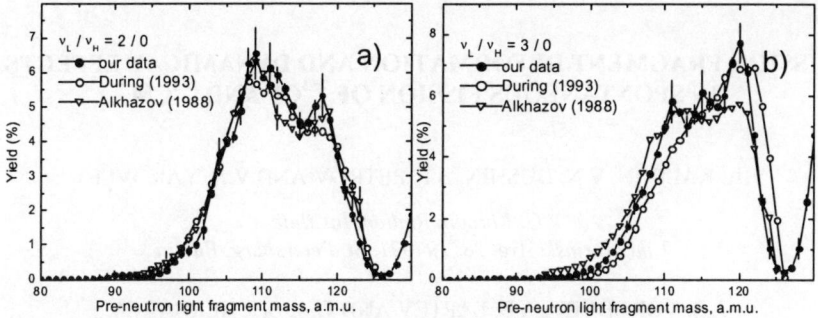

Figure 3. Fission fragment mass distributions of ^{252}Cf for (a) $\nu_L / \nu_H = 2 / 0$ and (b) $\nu_L / \nu_H = 3 / 0$, obtained in this work (full symbols) and compared to literature [6,7] (open symbols).

References

1. H. R. Bowman, J. C. D. Milton, S. G. Thompson and W. J. Swiatecki, *Phys Rev.* **129**, 2133 (1963).
2. R. L. Walsh, J. W. Boldeman, *Nucl. Phys.* **A276**, 189 (1977).
3. C. Budtz-Jørgensen, H.-H. Knitter, *Nucl. Phys.* **A490**, 307 (1988).
4. A. Gavron, Z. Fraenkel, *Phys. Rev.* **C9**, 632 (1974).
5. Signarbieux C., Poitou J., Ribrag M., Matuszek J., *Phys. Lett.* **39B**, 503 (1972).
6. I. D. Alkhazov, A. V. Kuznetsov, S. S. Kovalenko, B. F. Petrov, V. I. Shpakov, *Proc. Int. Conf. "Nuclear Data for Science and Technology"*, Mito, Japan, 24-27 May, 1988, p.991 (1988).
7. Düring I., Adler M., Märten A., Ruben A., Cramer B., Jahnke U., *In Proc. Second Int. Conf. on Dynamical Aspects of Nuclear Fission*, Smolenice, Slovakia, June 14-18, 1993, ed. Kristiak J., Pustylnik B. I. JINR Report E7-94-19. Dubna, , pp.133-141 (1994).
8. V. A. Kalinin, V. N. Dushin B. F. Petrov, V. A. Jakovlev, A. S. Vorobyev, A. B. Laptev, G. A. Petrov, Y. S. Pleva, O. A. Shcherbakov, V. E. Sokolov, F.-J. Hambsch, *Proc. Seminar on Fission Pont d' Oye IV*, Castle Pont d'Oye, Habay-la-Neuve, Belgium, 5 - 8 October 1999 ed. C. Wagemans et. al., p. 33 (1999).
9. A. Vorobyev, V. Dushin, F.-J. Hambsch, V. Jakovlev, V. Kalinin, I. Kraev, A. Laptev, B. Petrov, G. Petrov, Y. Pleva, O. Shcherbakov, V. Sokolov, *J. Nucl. Sci. Techol., Supplement 2*, 630 (2002).
10. M. Dakowski, Yu. A. Lazarev, V. F. Turchin, L. S. Turovtseva, *Nucl. Instr. and Meth.* **113**, 195 (1973).
11. A. V. Daniel, I. D. Alkhazov, V. N. Dushin, S. S. Kovalenko, V. I. Shpakov, *Preprint RI-170* of Khlopin Radiun Institute, Leningrad (1983).
12. G. Audi, A. H. Wapstra, *Nucl. Phys.* **A595**, 409 (1995).

FISSION FRAGMENT DEFORMATION AND DYNAMICAL EFFECTS IN SPONTANEOUS FISSION OF ^{252}CF AND ^{248}CM [*]

V.A. KALININ, V.N. DUSHIN, B.F. PETROV AND V.A. YAKOVLEV

V.G. Khlopin Radium Institute
2-nd Murinski Ave. 28, 194021, St. Petersburg, Russia

A.S. VOROBYEV, A.B. LAPTEV AND O.A. SHCHERBAKOV

Petersburg Nuclear Physics Institute
188350, Gatchina, Leningrad district, Russia

F.-J. HAMBSCH

EC-JRC-Institute for Reference Materials and Measurements
Retieseweg B-2440, Geel, Belgium

The number of prompt neutrons emitted in the fission event have been measured separately for each complementary fragment in coincidence with fragment mass and kinetic energies in spontaneous fission of ^{252}Cf and ^{248}Cm. Two high efficient Gd-loaded liquid scintillator tanks were used for the neutron registration. Approximately $3 \cdot 10^6$ fission events coincident with prompt neutron emission have been accumulated for each isotope. Neutron multiplicity distributions corrected for efficiency, background and pile-up have been obtained as a function of fission fragment mass and kinetic energy. Based on these unfolded multiplicity distributions fragment mass distributions for specific numbers of emitted neutrons have been obtained. These mass spectra demonstrates the presence of cold compact and deformed fission and the strong influence of deformed neutron shells in fragments at the descent to scission. Possible explanations for deformed cold fission are given on the basis of an analysis of the behavior of the moments of the multiplicity distributions on the fragment total kinetic energy.

1. Introduction

Information about neutron multiplicities is crucial for the better understanding of the fission process. This quantity contains information on the deformation of both fragments at the vicinity of the scission point, on the energy partition released in the fission process between different degrees of freedom, the distribution of the excitation energy between the fragments, and on the neutron-γ competition at the fission fragment de-excitation. Generally, these detailed data can help to investigate the dynamics of the strongly deformed fissioning system at the descent from saddle to scission.

---

[*] This work is supported by ISTC, http://www.istc.ru.

The first multi-parameter measurements of neutron multiplicities combined with the direction-selected spectroscopy of scission fragments for spontaneous fission of ^{252}Cf have been done at KRI (Russia) [1] more than a decade ago, followed by analogous measurements carried out by the TUD-HMI collaboration (Germany) [2]. These experiments revealed some unusual effects, such as deformed cold fission. Recently, a new series of such experiments was started by the KRI-PNPI-IRMM collaboration within the framework of an ISTC Project. In the present report, some results of the measurements carried out for spontaneous fission of ^{252}Cf and ^{248}Cm are presented.

2. Experiment

A thin fission source was placed between two large 200 l Gd-loaded liquid scintillator tanks, which were used for neutron detection in a 2x2π-geometry - to separate contributions from complementary fragments. The efficiency of the neutron registration was about 55% for each detector. A 16 cm thick iron shielding inserted between the tanks was used to decrease both fission neutron and capture γ-ray scattering from one tank into the other (cross talk). The fission fragments were collimated towards the neutron detectors by means of a pin-hole collimator combined with a common cathode of the twin parallel plate gas-flow ionization chamber with Frisch grids. At the counting rate of "useful" fission events corresponding to registration of collimated fragments of 1-3 s^{-1} approximately 3·10^6 fission events have been accumulated for each nuclide.

3. Results and Discussion

Mass and kinetic energy distributions of fission fragments have been obtained on the basis of reference values of the most probable mass and energies of fission fragments. The first moment (mean multiplicity) and second moments (variance and co-variance) of the one-dimensional multiplicity distributions for both neutron detectors have been calculated for each fragment mass and total kinetic energy (TKE). These data have been corrected for neutron pile-up, background and detector efficiency, including cross talk effects. An iterative procedure was applied to introduce the correction for neutron emission based on the measured mass and TKE dependence of the mean neutron multiplicity.

On the basis of the obtained moments approximate multiplicity distributions have been modeled for each value of the fragment mass and TKE. The true shape of the distributions was found using an iterative procedure of minimization of the difference between the model distributions distorted by operators of efficiency, background and pile-up and experimentally measured. The unfolding procedure of the two-dimensional multiplicity distributions of neutrons emitted from the complementary fragments is described more in detail in the report of A.Vorobyev et. al. in the proceedings of this conference. The multiplicity distributions for total numbers of neutrons emitted from both fragments were reconstructed for each fragment mass using the method of minimum divergence proposed by Tarasko [3], based on an iterative renormalization of the experimental distribution. On the basis of the obtained multiplicity distributions the fission fragment mass distributions for fixed number of emitted neutrons have been formed.

In Fig.1 the fragment mass distribution for fixed total number of emitted neutrons v_{tot} is presented for spontaneous fission of ^{248}Cm and ^{252}Cf. The distributions demonstrate the overlapping of the dominating yields, which may be explained in the framework of the scission point model of fission [4]. Also visible is a 3-5 mass unit fine structure, which may be related to cold fission. At $v_{tot} = 0,1$ the deformed neutron shell N=88 corresponding to the heavy fragment mass M~145 for ^{252}Cf and M~143 for ^{248}Cm is playing a dominating role. In case of ^{252}Cf the yield of M~106 is additionally enhanced due to the deformed neutron shell N=66. At $v_{tot} = 2-4$ the N=66 shell is dominating and the light fragment yield is stabilized in both nuclides. It can also be seen that fine structures are almost vanished. At $v_{tot} = 5-7$ the spherical neutron shell N~82 (M~132-134) plays the stabilizing role in the formation of the mass distribution . Hence, the light fragment takes the largest part of the deformation of the fissioning system and fine structures denoting that the system is cold at the scission point are re-appearing. It may be noted that in both cases of cold fission (neutronless and deformed) the heavy fragment is playing the stabilizing role in the formation of the mass distribution. The fine structures for ^{252}Cf have an approximate period of 5 mass units that may point to an enhanced yield of even charges. For ^{248}Cm the period is smaller. The reasons for that are not clear, but these irregularities may be due to the unfolding procedure at low statistics, especially for $v_{tot} = 7$.

In Fig. 2 the partial fission fragment mass distributions for fixed numbers of emitted neutrons in spontaneous fission of ^{248}Cm and ^{252}Cf are shown. v_L/v_H denotes the number of neutrons emitted from light and heavy fragments, respectively. The figure demonstrates that only configurations with significantly different number of neutrons emitted or with significant asymmetry in fragment deformation are responsible for structures in the mass distributions shown in Fig.1.

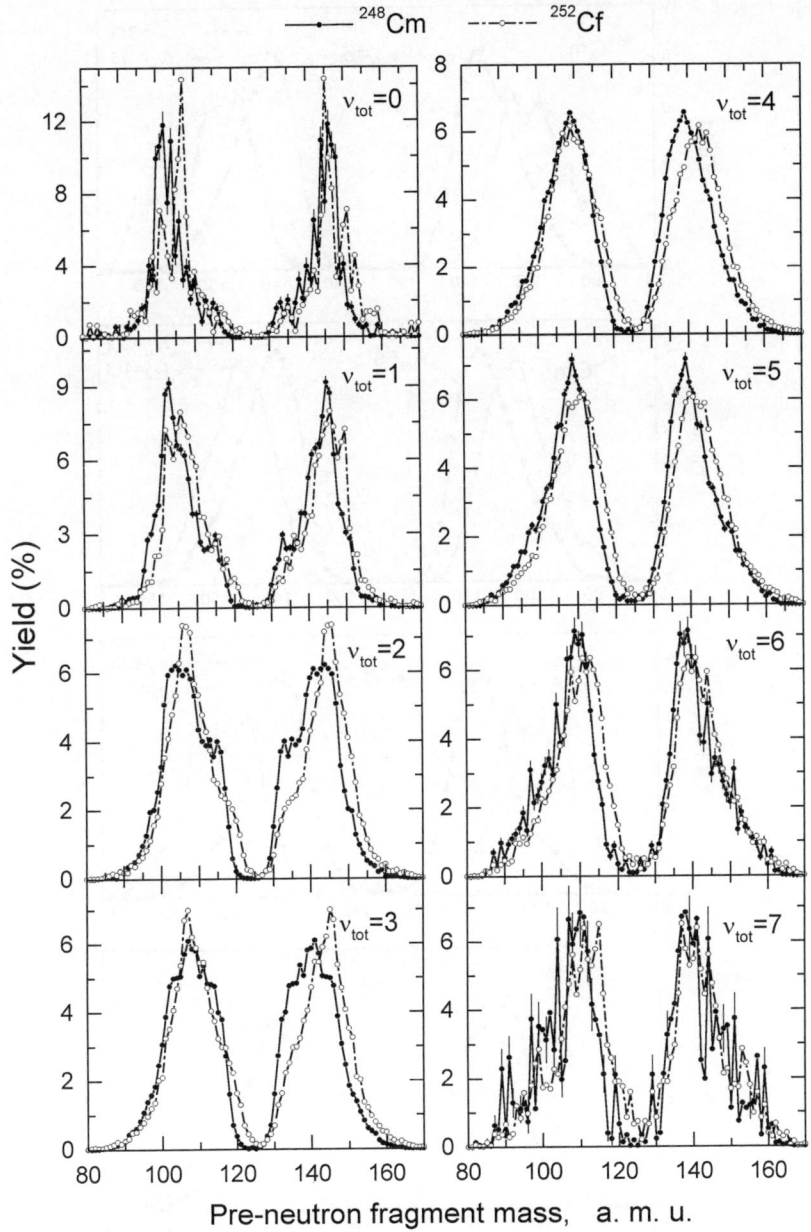

Figure 1. Fission fragment mass distributions for fixed total number of emitted neutrons.

470

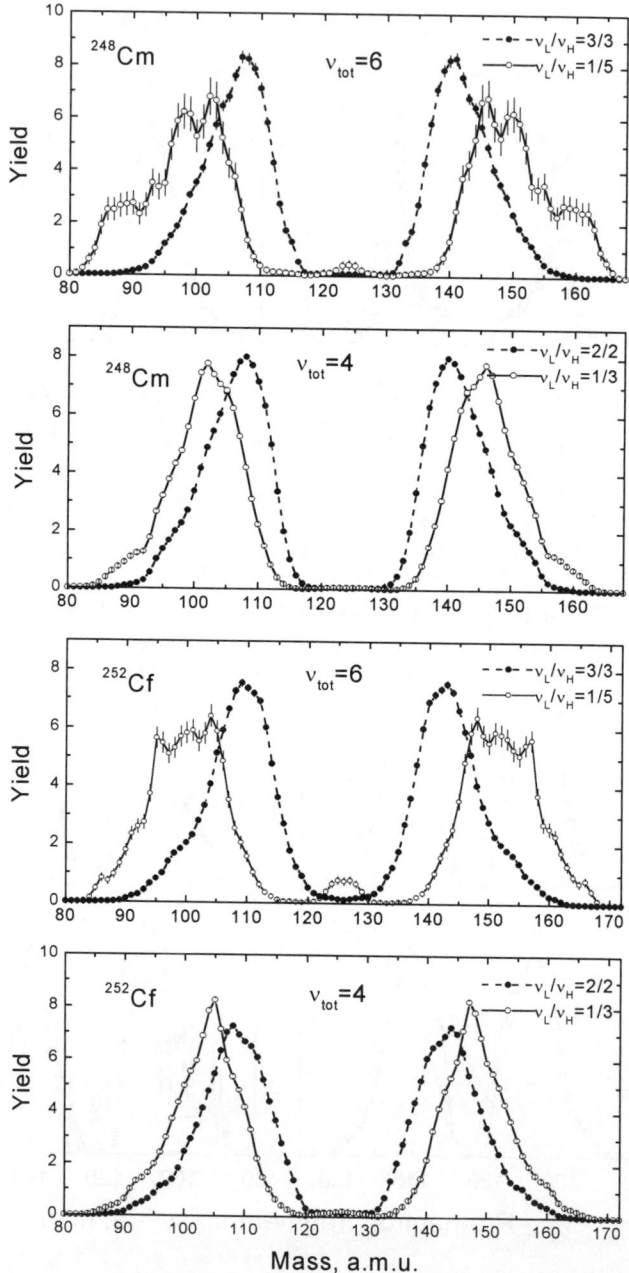

Figure 2. Partial fission fragment mass distributions for fixed number of emitted neutrons.

The mechanism of formation of cold deformed configurations at the scission point is not clear up to now. But it is obvious that it strongly depends on the thermodynamics of the descent from saddle to scission. The following two principal possibilities for formation of cold configuration at the scission point can be assumed − superfluid descent without any energy dissipation or partial system de-excitation in the vicinity of the scission point. In ref. [5] the latter was discussed. A rather strong decrease of the mean neutron multiplicity $<\nu_l>$ emitted from light fragments was found in the TKE dependence of $<\nu_l>$ below TKE=160-170 MeV. Taking into account the specific features of the experimental set-up used a decreasing number of registered neutrons could mean isotropic emission of scission neutrons in the laboratory system before full fragment acceleration. Such neutrons may be emitted due to single particle excitation of the fragment, caused by energy dissipation in a non-adiabatic transition from saddle to scission.

4. Conclusion

The data obtained in the present experiment with high efficient neutron detectors have been successfully used for the reconstruction of neutron multiplicity distributions. Partial fission fragment mass distributions obtained for fixed numbers of emitted neutrons shows good agreement with the scission point model [4] and appearance of cold fission structures at smallest and largest fragment deformations. As it was demonstrated, strong asymmetry in fragment deformation is an important condition of appearance of cold deformed fission.

References

1. I.D. Alkhazov, A.V. Kuznetsov, S.S. Kovalenko, *et al.*, *Proc. of Int. Conf. on Nucl. Data for Sci. and Technology,* Mito, Japan, May 30 − June 3, 1988, p.991(1988).
2. I. Düring, M. Adler, A. Märten, *et al.*, *Proc. of Second Int. Conf. on Dynamical Aspects of Nuclear Fission,* Smolenice, Slovakiya, June 14-18, 1993, p.133 JINR Report E7-94-19 (1994).
3. M.Z.Tarasko, *Voprosy Atomnoy Nauki I Techniki, Issue Metodiki I Programmy Chislennogo Resheniya Zadach Matematicheskoy Fisiki* 2, M., p. 86 (1987) (Russian).
4. Wilkins B.D. et al., *Phys. Rev,* **C14**, 1832 (1976).
5. V.Kalinin, V. Dushin, B.Petrov, *et al.*, *Proc. of Int. Conf. on Nucl. Data for Sci. and Technology,* Tsukuba, Japan, October 7-12, 2001, J. Nucl. Sci. Technol., Supplement 2, 250 (2002).

METAL CLUSTER FISSION *

A. G. LYALIN,† A. V. SOLOV'YOV‡ AND W. GREINER

Institut für Theoretische Physik der Universität Frankfurt am Main,
Robert-Mayer 8-10, D-60054 Frankfurt am Main, Germany

O. I. OBOLENSKY

A. F. Ioffe Physical-Technical Institute of the Russian Academy of Sciences,
Polytechnicheskaya 26, 194021 St. Petersburg, Russia

Fission of doubly charged metal clusters is studied using the open-shell two-center deformed jellium Hartree-Fock and local density approximations as well as within *ab initio* molecular dynamic approach accounting for all electrons in the system. Results of calculations of the fission barriers for the symmetric and asymmetric fission processes $Na_{10}^{2+} \to Na_7^+ + Na_3^+$, $Na_{18}^{2+} \to Na_{15}^+ + Na_3^+$ and $Na_{18}^{2+} \to 2Na_9^+$ are presented. The role of exchange and many-electron correlation effects in metal clusters fission is analyzed. It is demonstrated that the influence of many-electron correlations on the height of the fission barrier is more profound if the barrier arises nearby or beyond the scission point. The importance of cluster deformations in the fission process is elucidated with the use of the overlapping-spheroids shape parameterization allowing one an independent variation of deformations in the parent and daughter clusters.

1. Introduction

Fission of charged atomic clusters occurs when repulsive Coulomb forces, arising due to the excessive charge, overcome the electronic binding energy of the cluster[1,2,3]. This mechanism of the cluster fission is in a great deal similar to the nuclear fission phenomena. Experimentally, multiply charged metal clusters can be observed in the mass spectra when their size exceeds

---

*This work is supported by the Alexander von Humboldt Foundation and the Russian Academy of Sciences (grant 44).

†Permanent address: Institute of Physics, St Petersburg State University, Ulianovskaja 1, 198504 St Petersburg, Petrodvorez, Russia; E-mail: lyalin@th.physik.uni-frankfurt.de

‡Permanent address: A. F. Ioffe Physical-Technical Institute of the Russian Academy of Sciences, Polytechnicheskaya 26, 194021 St. Petersburg, Russia; E-mail: solovyov@th.physik.uni-frankfurt.de

the critical size of stability, which depends on the metal species and cluster charge[4,5].

The microscopic description of energetics and dynamics of metal cluster fission process based on molecular dynamic (MD) simulations has been performed[6,7,8] using the local-spin-density-functional method. This method is, however, strongly restricted by the cluster size, because of computational difficulties, and thus is usually applied to the small metal clusters with the number of atoms $N \leq 20$.

Fission process of metal clusters can be also simulated on the basis of the jellium model, which does not take into account the detailed ionic structure of the cluster core. Jellium model considers the electrons in the usual quantum mechanical way, but approximates the cluster core potential by the potential of the homogeneous positively charged background and, therefore, is better applicable for lager cluster sizes[9].

One of the effective approaches for metal cluster exploration is the shell-correction method (SCM), originally developed[10,11] in nuclear physics. This method was adapted for metal clusters[12,13,14].

Most of the electronic structure calculations of the jellium metal clusters have been performed using self-consistent Kohn-Sham local density approximation (LDA)[15,16,17]. It has been shown that the cluster shape parameterization must be flexible enough to account for the effects generated by the shell structure of the parent and daughter clusters, which in general have not spherical but deformed shapes[14].

Alternatively, one can develop direct *ab initio* methods for the description of electronic properties of metal clusters. It can be achieved by using the Hartree-Fock (HF) approximation and by the construction on its basis the systematic many-body theory. Following this line, the open-shell two-center jellium Hartree-Fock approximation valid for metal clusters with arbitrary number of the valence electrons has been developed[18,19]. This method has been generalized and adopted to study of the metal clusters fission process[20,21].

The atomic system of units, $|e| = m_e = \hbar = 1$, has been used throughout the paper, unless other units are indicated.

2. Theoretical methods

According to the main postulate of the jellium model, the electron motion in a metallic cluster takes place in the field of the uniform positive charge distribution of the ionic background. For the parameterization of the ionic background we consider the model in which the initial parent cluster having the form of the ellipsoid of revolution (spheroid) splits into two independently deformed spheroids of smaller size[20,21]. The two principal diameters a_k and b_k of the spheroids can be expressed via the deformation parameter δ_k as

$$a_k = \left(\frac{2 + \delta_k}{2 - \delta_k}\right)^{2/3} R_k, \quad b_k = \left(\frac{2 - \delta_k}{2 + \delta_k}\right)^{1/3} R_k. \tag{1}$$

Here partial indexes $k = 0, 1, 2$ correspond to the parent cluster ($k = 0$) and the two daughter fragments ($k = 1, 2$), R_k ($k = 0, 1, 2$) are the radii of the corresponding undeformed spherical cluster, which are equal to $R_k = r_s N_k^{1/3}$, where N_k is the number of atoms in the k-th cluster, and r_s is the Wigner-Seitz radius. For sodium clusters, $r_s = 4.0$, which corresponds to the density of the bulk sodium. The deformation parameters δ_k characterize the families of the prolate ($\delta_k > 0$) and the oblate ($\delta_k < 0$) spheroids of equal volume $V_k = 4\pi a_k b_k^2 / 3 = 4\pi R_k^3 / 3$. For overlapping region the radii $R_1(d)$ and $R_2(d)$ are functions of the distance d between the centers of mass of the two fragments. They are determined so that the total volume inside the two spheroids is equal to the volume of the parent cluster $4\pi R_0^3 / 3$.

The ions charge density $\rho(\mathbf{r})$ is kept uniform within the clusters volume including the overlapping-spheroids region,

$$\rho(\mathbf{r}) = \begin{cases} \rho_c, & (x^2 + y^2)/b_1^2 + (z + d/2)^2/a_1^2 \leq 1 \\ \rho_c, & (x^2 + y^2)/b_2^2 + (z - d/2)^2/a_2^2 \leq 1 \\ 0, & \text{otherwise.} \end{cases} \tag{2}$$

Here $\rho_c = Z_0/V_0$ is the ionic charge density inside the cluster and Z_0 is the total charge of the ionic core, d is the fragment separation distance.

The Hartree-Fock and LDA equations have been solved in the system of the prolate spheroidal coordinates as a system of coupled two-dimensional second order partial differential equations. In the present work we use the Gunnarsson and Lundqvist parameterization of the density of electron exchange-correlation energy[22].

The MD calculations have been performed with the use of the Gaussian 98 software package[23]. We have utilized the $6 - 311G(d)$ basis set of primitive Gaussian functions to expand the cluster orbitals and the density-functional theory based on the hybrid Becke-type three-parameter exchange functional paired with the gradient-corrected Lee, Yang and Parr correlation functional ($B3LYP$)[23].

3. Numerical results

Let us present and discuss the results of calculations performed within the model described above. Figure 1 shows fission barriers for the asymmetric

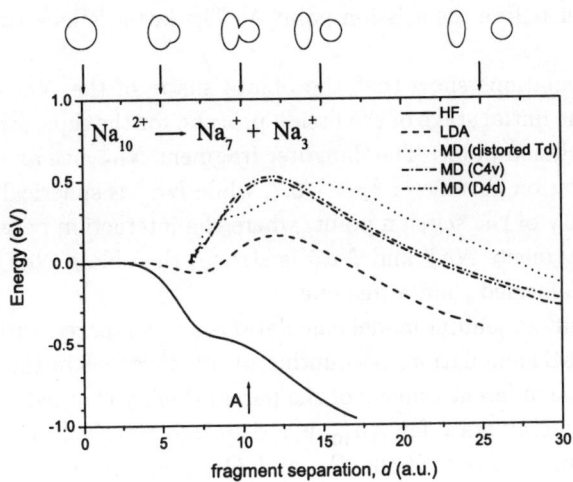

Figure 1. Fission barriers in the two-center deformed jellium Hartree-Fock (solid lines) and LDA (dashed lines) approaches calculated in this work for the asymmetric channel $Na_{10}^{2+} \rightarrow Na_7^+ + Na_3^+$ as a function of fragments separation distance d. The evolution of jellium cluster shape during the fission process is shown on top of figure. Dotted, dashed-dotted and dashed-double-dotted lines show the fission barriers derived from MD simulations for the Na_{10}^{2+} parent cluster having initial geometry of distorted tetrahedron or characterized by the C_{4v} and D_{4d} point symmetry groups respectively.

channel $Na_{10}^{2+} \rightarrow Na_7^+ + Na_3^+$ as a function of the fragments separation distance d. We have minimized the total energy of the system over the parent and daughter fragments spheroidal deformations with the aim to find the fission pathway corresponding to the minimum of the fission barrier. We have also used the assumption of continuous shape deformation during the fission process. The evolution of cluster shape during the fission process is shown for the HF and LDA models on top of figure 1 and the subsequent figures. Solid and dashed lines in figure 1 are the result of the two-center jellium HF and LDA calculations respectively. The zero of energy put at $d = 0$.

Within the framework of the two-center deformed jellium Hartree-Fock approximation, the parent cluster Na_{10}^{2+} is unstable towards the asymmetric channel $Na_{10}^{2+} \rightarrow Na_7^+ + Na_3^+$. Accounting for many-electron correlations within the LDA theory leads to the formation of the fission barrier and the appearance of a local minimum on the energy curve at $d = 7.2$ a.u., corresponding to the super deformed asymmetric prolate state of the parent Na_{10}^{2+} cluster before the scission point A. The latter is located at $d = 10.4$ a.u.

Our calculations show that the oblate shape of the Na_7^+ fragment is formed at the initial stage of the fission process, for the separation distances below the scission point. The daughter fragment Na_7^+ has an oblate shape with deformation parameter $\delta_1 = -0.7$, while Na_3^+ is spherical, i.e. $\delta_2 = 0$. In the vicinity of the scission point, where the interaction between the two daughter fragments Na_7^+ and Na_3^+ is strong, the oblate Na_7^+ fragment is even more deformed than a free one.

The results of jellium model calculations are compared with the results of *ab initio* MD simulations accounting for all electrons in the system. We consider three different isomers of the parent doubly-charged Na_{10}^{2+} cluster. The lowest energy state for Na_{10}^{2+} is a distorted tetrahedron. The isomers with the ionic structure of the C_{4v} and D_{4d} point symmetry groups have the total energy that exceeds the lowest energy state by 0.42 eV and 0.53 eV respectively. The fission barriers derived from MD simulations for the Na_{10}^{2+} parent cluster with the initial geometry of distorted tetrahedron, or characterized by the C_{4v} and D_{4d} point symmetry groups are shown in figure 1 by dotted, dashed-dotted and dashed-double-dotted lines respectively. The fission barriers for all channels considered are close. The heights of the barriers obtained are equal to $\Delta_{MD}^{Td} = 0.49$ eV, $\Delta_{MD}^{C_{4v}} = 0.50$ eV and $\Delta_{MD}^{D_{4d}} = 0.53$ eV. The weak sensitivity of the fission barrier to the choice of the isomeric form of the reactants can be understood if one notices that the barrier top point is located at rather large inter-fragment distance. At such distances the interaction between the fragments, apart from the Coulombic repulsion, is mainly determined by electronic properties of the system rather than by details of the ionic structure of the fragments. Comparison of the asymmetric $Na_{18}^{2+} \rightarrow Na_{15}^+ + Na_3^+$ and symmetric $Na_{18}^{2+} \rightarrow 2Na_9^+$ fission channels of the parent Na_{18}^{2+} cluster is of particular interest, because there has to be a competition between these two channels involving the magic cluster ions Na_3^+ and Na_9^+. It was noticed[6,13] that namely in fission of the Na_{18}^{2+} cluster a magic fragment other than Na_3^+ determines the favoured channel. Figure 2 shows fission barriers for the asymmetric channel $Na_{18}^{2+} \rightarrow Na_{15}^+ + Na_3^+$. We have started from the initial configuration

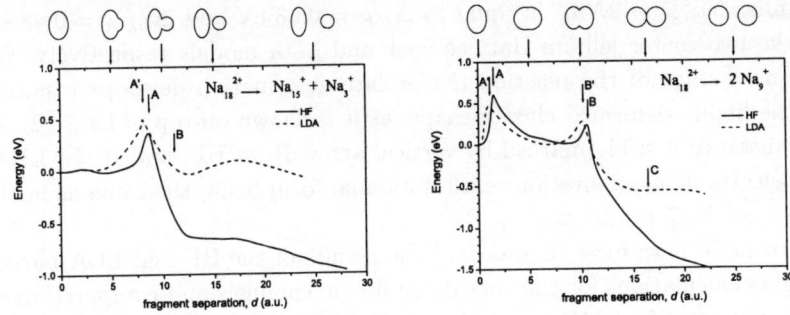

Figure 2. The same as in figure 1 but for the asymmetric channel $Na_{18}^{2+} \rightarrow Na_{15}^{+} + Na_{3}^{+}$ and for the symmetric channel $Na_{18}^{2+} \rightarrow 2Na_{9}^{+}$

corresponding to oblate shape of the parent Na_{18}^{2+} cluster with the deformation parameter $\delta_0 = -0.35$. The daughter fragment Na_{15}^{+} has an oblate shape with deformation parameter $\delta_1 = -0.6$, while Na_3^{+} is spherical, i.e. $\delta_2 = 0$.

The total energy as a function of the fragment separation distance has a maximum (marked by vertical arrow A for HF, and A$'$ for LDA), arising due to the alteration of the electronic configuration $1\sigma^2 2\sigma^2 1\pi^4 2\pi^2 3\sigma^2 1\delta^4 \rightarrow 1\sigma^2 2\sigma^2 1\pi^4 3\sigma^2 4\sigma^2 1\delta^4$. These maxima on the energy curves define the fission barrier hights, being equal to $\Delta_{HF} = 0.36$ eV for Hartree-Fock and $\Delta_{LDA} = 0.50$ eV for LDA. It is interesting to notice that the LDA total energy curve has a pronounced minimum at $d = 12.5$ a.u., located beyond the scission point $d = 11.1$ a.u. We have marked the scission point by vertical arrow B, both for HF and LDA. This minimum means that a quasistable state of the supermolecule $Na_{15}^{+} + Na_3^{+}$ can be created during the fission process. However, the appearance of the minimum and thus the stability of the super molecule is rather sensitive to the model chosen for the description of exchange and correlation inter-electron interaction. This is already clear from the fact that such a minimum does not appear in the HF simulations (see figure 2).

Figure 2 also shows the dependence of total energy E_{tot} on separation distance d for the symmetric channel $Na_{18}^{2+} \rightarrow 2Na_{9}^{+}$. The parent cluster changes its shape from oblate to prolate one on the initial stage of the fission process ($d \approx 1$ a.u.). This transition is accompanied by the first re-arrangement of the electronic configuration (marked by vertical arrow A for HF and A$'$ for LDA). The total fission barrier for the symmetric channel $Na_{18}^{2+} \rightarrow 2Na_9^+$ is equal to $\Delta_{HF} = 0.63$ eV and $\Delta_{LDA} = 0.48$ eV in the two-center jellium Hartree-Fock and LDA models respectively. On the next stage of the reaction the prolate deformation develops resulting in the highly deformed cluster shape, as it is shown on top of figure 2. At the distance $d \approx 11$ (marked by vertical arrow B for HF, and B$'$ for LDA) the electronic configuration reaches its final form being the same as in the spherical Na_9^+ products.

In table 1 we have summarized the results of the HF and LDA barrier heights calculations for the considered fission channels and compared them with the results of MD simulations. The height of the fission barrier for

Table 1. Summary of the fission barrier heights in (eV).

Channel	HF	LDA	MD (this work) T_d C_{4v} D_{4d}	MD[7]
$Na_{10}^{2+} \rightarrow Na_7^+ + Na_3^+$	0	0.16	0.49 0.50 0.53	0.67
$Na_{18}^{2+} \rightarrow Na_{15}^+ + Na_3^+$	0.36	0.50	–	0.50
$Na_{18}^{2+} \rightarrow 2Na_9^+$	0.63	0.48	–	0.52

Na_{10}^{2+} cluster in the two-center deformed jellium LDA model is 0.51 eV lower than its value following from the MD simulations[7]. Since both methods apply the same form of the density functional[22], the discrepancy in the fission barrier heights can be attributed to the manifestation of dynamics of ions during the fission process. One can expect that the influence of the detailed ionic structure on the fission process has to diminish with the growth of the cluster size, making the jellium model approach more and more accurate. MD calculations performed in our work utilize the $B3LYP$ gradient-corrected exchange-correlation functional. As it follows from table 1, the gradient corrections to the exchange-correlation functional lower the Na_{10}^{2+} fission barrier. With increasing cluster size the role of gradient corrections decreases. For the Na_{18}^{2+} cluster, we report a very good agreement of the heights of fission barriers derived from the jellium LDA model with those obtained from the axially averaged MD simulations[7].

4. Summary

We have developed the open-shell two-center deformed jellium Hartree-Fock and LDA method for the description of the metal clusters fission process. The proposed two overlapping sheroids shape parameterization allows one to consider independently a wide variety of shape deformations of parent and daughter clusters, and to investigate the role of deformation effects on the cluster fission process. The role of many-electron correlation effects in metal clusters fission is analysed. The described Hartree-Fock model forms the basis for further systematic development of more advanced *ab initio* many-body theories for the process of metal clusters fission.

References

1. K. Sattler, J. Mühlbach, O. Echt *et al*, Phys. Rev. Lett. **47**, 160 (1981).
2. U. Näher, S. Bjornholm, F. Frauendorf *et al*, Phys. Rep. **285**, 245 (1997).
3. C. Yannouleas, U. Landman, and R.N. Barnett, in *Metal Clusters*, p.145, edited by W. Ekardt (Wiley, New York, 1999).
4. C. Bréchignac, Ph. Cahuzac, F. Carlier, and J. Leygnier, Phys. Rev. Lett. **63**, 1368 (1989); C. Bréchignac, Ph. Cahuzac, F. Carlier, and M. de Frutos, Phys. Rev. B **49**, 2825 (1994).
5. T.P. Martin, J. Chem. Phys. **81**, 4426 (1984).
6. R.N. Barnett, U. Landman *et al*, Phys. Rev. Lett. **67**, 3058 (1991).
7. B. Montag and P.-G. Reinhard, Phys. Rev. B **52**, 16365 (1995).
8. P. Blaise, S.A. Blundell, C. Guet *et al*, Phys. Rev. Lett. **87**, 063401 (2001).
9. W. Ekard, W.D. Schöne, and J.M. Pacheco, in *Metal Clusters*, p.1, edited by W. Ekardt (Wiley, New York, 1999).
10. V.M. Strutinsky, Nucl. Phys. A **95**, 420 (1967); *ibid* **122**, 1 (1968).
11. J.M. Eisenberg, and W. Greiner, *Nuclear Theory. Vol.1. Collective and Particle Models*, (North Holland, Amsterdam, 1985).
12. M. Nakamura, Y. Ishii, A. Tamura *et al*, Phys. Rev. A **42**, 2267 (1990).
13. C. Yannouleas and U. Landman, Phys. Rev. B **51**, 1902 (1995).
14. C. Yannouleas and U. Landman, J. Phys. Chem. **99**, 14577 (1995).
15. S. Saito and S. Ohnishi, Phys. Rev. Lett. **59**, 190 (1987).
16. F. Garcias, A. Mañanes, J.M. Lópes *et al*, Phys. Rev. B **51**, 1897 (1995).
17. H. Koizumi and S. Sugano, Phys. Rev. A **51**, R886 (1995).
18. A.G. Lyalin, S.K. Semenov, A.V. Solov'yov, N.A. Cherepkov, and W. Greiner, J. Phys. B: At. Mol. Opt. Phys. **33**, 3653 (2000)
19. A.G. Lyalin, S.K. Semenov, A.V. Solov'yov, N.A. Cherepkov, J.-P. Connerade, and W. Greiner, J. Chin. Chem. Soc. (Taipei) **48**, 419 (2001).
20. A. Lyalin, A. Solov'yov, W. Greiner *et al*, Phys. Rev. A **65**, 023201 (2002).
21. A. Lyalin, A. Solov'yov and W. Greiner Phys. Rev. A **65**, 043202 (2002).
22. O. Gunnarsson and B.I.Lundqvist, Phys. Rev. B, **13**, 4274, (1976).
23. M.J. Frisch *et al*, computer code GAUSSIAN 98, Rev. A. 9, Gaussian Inc., Pittsburgh, PA, 1998; James B. Foresman and Æleen Frisch *Exploring Chemistry with Electronic Structure Methods* (Pittsburgh, PA: Gaussian Inc, 1996)

NEW BEHAVIOR OF HEAVY-ION FUSION REACTIONS AT EXTREME SUB-BARRIER ENERGIES

C.L. JIANG, H. ESBENSEN, K.E. REHM, B.B. BACK, R.V.F. JANSSENS, J.A. CAGGIANO, P. COLLON, J. GREENE, A.M. HEINZ, D.J. HENDERSON, I. NISHINAKA, T.O. PENNINGTON, D. SEWERYNIAK

Physics Division, Argonne National Laboratory, Argonne, Illinois 60439

The excitation function for fusion-evaporation in the ^{60}Ni + ^{89}Y system was measured over a range in cross section covering six orders of magnitude. The cross section exhibits an abrupt decrease at extreme sub-barrier energies. This behavior, which is also present in a few other systems found in the literature, cannot be reproduced with present models, including those based on a coupled-channels approach. Possible causes are discussed, including a dependence on the intrinsic structure of the participants.

Fusion cross sections between heavy ions at sub-barrier energies have been studied extensively since it was realized that, near and below the Coulomb barrier, the measured yields are strongly enhanced compared to one-dimensional tunneling calculations [1,2]. This enhancement has been explained in terms of channel-coupling effects which create a multi-dimensional potential barrier and result in increased fusion probabilities. In most experiments carried out thus far, fusion cross sections have been measured down to the 0.1 - 1 mb level, and coupled-channels calculations have proven quite successful in reproducing the general trends of the measured yields [3].

The present work deals with fusion cross sections well below 0.1 mb. Naively, one would expect coupled-channels effects to saturate at low bombarding energies and the product σE to exhibit a simple exponential fall-off with decreasing energy [1,4]. Here, we show that the low-energy data, near the low-energy-edge of the barrier distribution, often exhibit a decrease much steeper than anticipated, and much steeper than predicted by a simple exponential fall-off. It is worth pointing out that this phenomenon is of particular interest for fusion reactions of astrophysical interest, as these usually take place at extreme sub-barrier energies.

Two examples of fusion evaporation excitation functions which have been measured to very small cross sections are shown in Fig. 1a, where they are plotted as σE versus E/V_b with the Coulomb barrier V_b calculated from the systematics of L.C. Vaz [5]. The solid line is the result of coupled-channels calculations in the case of ^{58}Ni + ^{58}Ni. Such calculations are, however, not yet available for reactions between heavier nuclei such as the ^{90}Zr + ^{92}Zr system shown in Fig. 1a. In these instances the Wong formula [6] is used as a representation of the data (dashed lines). The latter formula has been introduced successfully in many analyses of heavy-ion induced fusion data [1,8]. In the case of ^{58}Ni + ^{58}Ni, it can be seen that this description compares well with the results of the coupled-channels calculations. It is clear from Fig. 1a that the exponential slope of the data at the low energies exhibits a continuous change with decreasing energy rather than a constant slope. Most importantly, at the lowest energies, the fall-off is steeper than predicted by the theoretical models.

To amplify the low-energy behavior of the cross sections, we present in Fig. 1b the ratio of the experimental cross sections and either the results of coupled-channels calculations, where available, or, again, of the parameterization based on the Wong formula. While these descriptions give good agreement with the data at higher energies, they are unable to reproduce the yields at the lowest energies, where one observes a drop-off in the cross section ratio starting at an energy E_0 indicated by the arrows. Four systems exhibiting this unexpected behavior have been found in the literature. They are: ^{58}Ni + ^{58}Ni [9], and ^{90}Zr + ^{90}Zr, ^{92}Zr and ^{89}Y [10].

It should be noted that precise measurements of very small fusion cross sections are experimentally challenging. Reactions on small amounts of heavier isotopic target contaminants can dominate the low-energy yields due to their higher center-of-mass energies. In addition, even rare isobaric contaminants in the beam (e.g., a ^{58}Fe contamination in a ^{58}Ni beam) can sharply affect the fusion cross sections at the lowest energies because of the strong dependence of these cross sections on the respective Coulomb barriers. For these reasons some systems have not been included in the compilation mentioned above.

The energy E_0 at which the drop-off occurs can be extracted from the data with a relative error of about 2%. Because of the limited amount of data, especially between V_b=100 - 178 MeV, we have chosen the system ^{60}Ni + ^{89}Y for a new measurement of fusion cross sections at very low energies. An additional reason for choosing this system is that neither the beam nor the target has isobaric or isotopic contaminants. The experiment

482

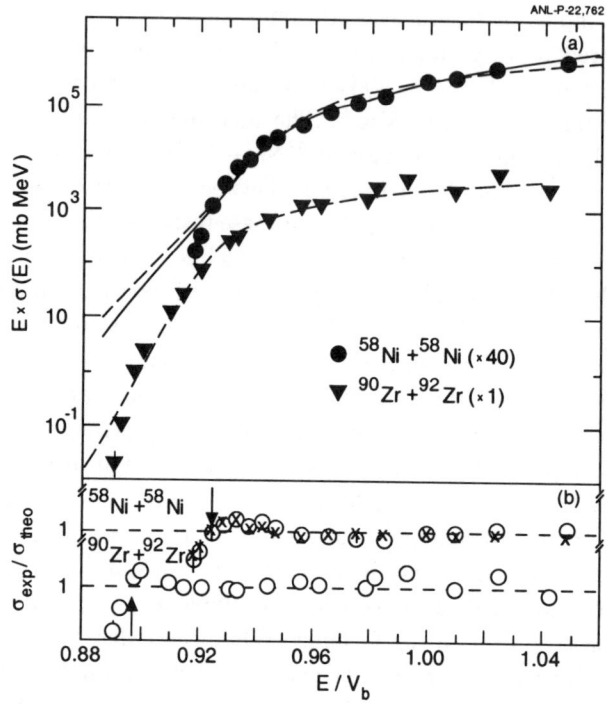

Figure 1. (a) plots of σE versus E/V_b for the systems ^{58}Ni + ^{58}Ni and ^{90}Zr + ^{92}Zr. (b) $\sigma_{exp}/\sigma_{theo}$ vs. E/V_b. Circles: σ_{theo} calculated using the Wong formula; Crosses: σ_{theo} calculated within the coupled-channels formalism. See text for details.

was performed at the Argonne superconducting linear accelerator ATLAS with the Fragment Mass Analyzer (FMA) [11].

The excitation function for evaporation residue production in the system ^{60}Ni + ^{89}Y, covering cross sections from about 100 mb to less than 100 nb, is presented in Fig. 2a. Also indicated in the figure are the results of a calculation using the Wong parameterization. At the lowest energies the experimental cross sections again fall below the calculated values, as is best seen from the ratio $\sigma_{exp}/\sigma_{theo}$ given in Fig. 2b, and an energy E_0 (see arrow in Fig. 2b) can be derived. Remarkably, the data for the five available systems indicate that the onset of the steeper than expected decrease in the fusion cross section can be parameterized by the relation $E_0 \sim 0.91V_b$. This observation may well suggest that this reduction in cross section is an entrance channel effect.

Another way to illustrate the steep fall-off in the σE product is to exam-

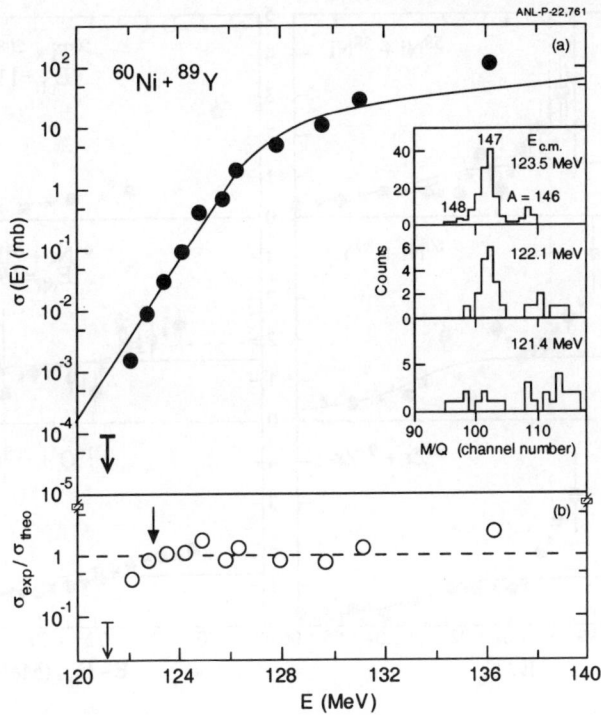

Figure 2. Experimental evaporation residue cross sections $\sigma(E)$, (a) and $\sigma_{exp}/\sigma_{theo}$, (b) ($\sigma_{theo}$ calculated from the Wong formula), plotted as function of the center-of-mass energy for the system ^{60}Ni + ^{89}Y. The incident energies have been corrected for the finite target thickness (including the influence of sharp changes in the excitation function with energy). The uncertainties in the cross sections are smaller than the points. Only an upper limit to the cross section was obtained at the lowest energy. The insert presents M/Q spectra obtained at three center-of-mass energies.

ine the exponential slopes defined as $L(E) = d(ln(\sigma E))/dE$. A large value of $L(E)$ implies a steeper drop-off of the cross sections. These are plotted as a function of E/E_0 for three systems in Fig. 3. The dashed and solid lines in the figure represent the results of coupled-channels calculations and of the Wong formula, respectively. While the calculated slopes approach a constant value of ~ 1.5 MeV^{-1}, the experimental data exhibit a continuous increase with decreasing energies. Thus, at least two representations of the experimental effect are possible. It would clearly be of great interest to investigate whether the behavior observed here persists at even lower energies, and whether it is a common behavior of all heavy-ion systems at extreme sub-barrier energies.

484

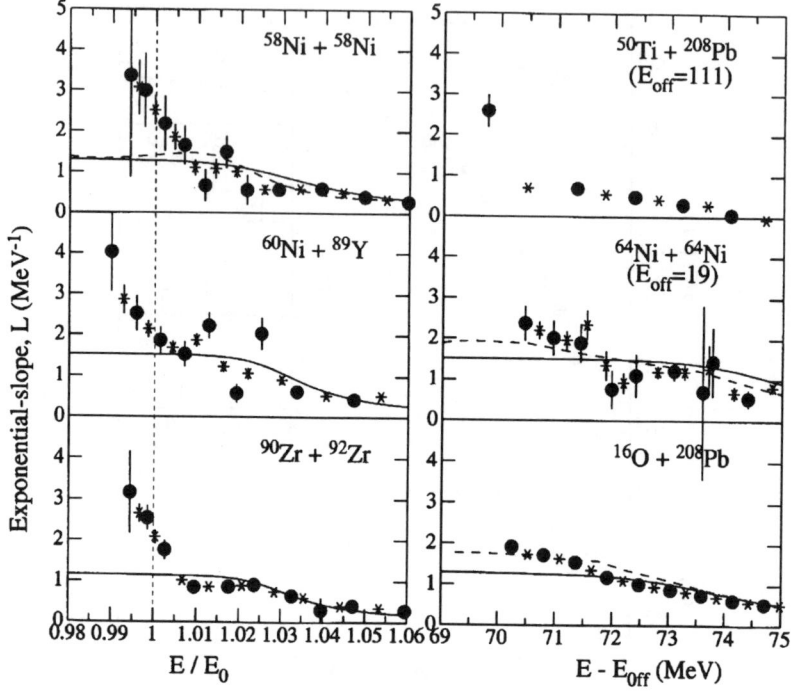

Figure 3. Exponential slopes $L(E)$, $d(ln(\sigma E))/dE$, plotted as function of E/E_0 for systems ^{58}Ni + ^{58}Ni, ^{60}Ni + ^{89}Y and ^{90}Zr + ^{92}Zr, and as fuction of $E - E_{off}$ for systems ^{50}Ti + ^{208}Pb, ^{64}Ni + ^{64}Ni and ^{16}O + ^{208}Pb. Solid circles and stars correspond to slope determinations from consecutive data points and from least-squares fits to three data points, respectively, at the corresponding average energies. Solid and dashed lines are the results of theoretical calculations with the Wong formula and the coupled-channels formalism, respectively.

The values of E_0, ratio $E_0/0.91V_b$, and the compound nucleus excitation energy E_{ex} and cross sections at the energy E_0 are summarized in Table 1 for the five systems. There are also two systems, ^{50}Ti + ^{208}Pb [12] and ^{64}Ni + ^{64}Ni [13], for which a steep fall-off in σE is observed, although it is only apparent at the lowest data point. These systems are included in the second part of Table 1. It should also be noted that many systems can be found in the literature which show an increase in the exponential slope with decreasing energy, although in these instances the measurements do not extend to sufficiently low energies to exhibit the fall-off in σE explicitly. Some of these systems are included in Table 1 and Fig. 3 as well, especially those for which the cross sections have been measured to rather low values.

In these cases, the quoted values of E_0 correspond to the lowest energies measured in the respective experiments, and should be regarded as an upper limit only.

Table 1. Drop-off energy E_0 and other relevant information for various systems. See text for details.

System	E_0 MeV	$E_0/0.91V_b$	$E_{ex}(E_0)$ MeV	$\sigma(E_0)$ mb	Ref.
^{58}Ni + ^{58}Ni	93.9	1.02±.02	27.8	0.40	9
^{60}Ni + ^{89}Y	123.0	1.00±.02	32.5	0.009	present
^{90}Zr + ^{89}Y	170.8	1.00±.02	19.3	0.20	10
^{90}Zr + ^{92}Zr	170.8	0.99±.02	17.1	0.006	10
^{90}Zr + ^{90}Zr	175.2	1.00±.02	17.9	0.020	10
^{50}Ti + ^{208}Pb	181.6	0.98	12.0	0.003	12
^{64}Ni + ^{64}Ni	89.2	1.00	40.4	0.02	13
^{16}O + ^{208}Pb	70.1	1.01	23.6	0.39	14
^{40}Ca + ^{90}Zr	93.4	1.01	36.1	0.84	15
^{16}O + ^{144}Sm	56.6	1.01	28.1	0.15	16
^{19}F + ^{208}Pb	75.1	0.96	24.9	0.023	17
^{16}O + ^{154}Sm	52.4	0.95	36.0	0.18	16
^{40}Ar + ^{144}Sm	116.6	0.96	25.8	0.0016	18
^{40}Ar + ^{154}Sm	108.8	0.91	33.5	0.0016	18
^{40}Ar + ^{112}Sn	96.6	0.96	33.5	0.0084	18
^{40}Ar + ^{122}Sn	94.0	0.95	35.4	0.0018	18
^{86}Kr + ^{92}Mo	160.4	0.98	22.3	0.007	19
^{86}Kr + ^{104}Ru	162.4	0.96	23.9	0.0049	19
^{36}S + ^{110}Pd	79.7	0.97	41.7	0.0021	20

As can be seen from the third column in Table 1, those systems involving closed-shell ("stiff") nuclei in the entrance channel are usually well described by the systematics $E_0 \sim 0.91V_b$. For reactions involving open-shell ("soft") nuclei in the entrance channel, like ^{40}Ar +^{154}Sm, firm E_0 values are not yet available, and the upper limits given in Table 1 suggest that there is a nuclear structure dependence to this phenomenon. It is likely that the larger channel-coupling effects for these softer participants are pushing the appearance of the phenomenon towards lower E_0/V_b values.

Some aspects of the behavior of the cross sections at low energies can be understood on the basis of the underlying Q-values, while others remain unexplained. In fusion reactions between two heavy ions, the reaction Q-values are always negative. As a result, once the excitation energy in the compound system E_{ex} reaches zero, which occurs at a finite bombarding energy, the cross section must be zero, and thus the exponential slope of σE should approach infinity. Hence, for a heavy-ion system, the slope $L(E)$ should increase towards infinity with decreasing energy. Any theoretical model attempting to describe the fusion behavior at extreme sub-barrier energies should include the properties of the fused system as expressed through the Q-value. The Q-value alone is, however, not sufficient to explain the phenomenon explored in this study. Present fusion models assume that the fused system can be populated at the excitation energy and spin corresponding to the entrance channel kinetic energy and angular momentum. This assumption is valid only as long as the total width of the compound state remains larger than the level spacing. When this condition is not satisfied a reduction of the fusion probability ensues, as is well known from capture reactions with thermal neutrons. For most of the systems under study here, however, the fall-off of the cross sections occurs at excitation energies $E_{ex} \sim$ 20-30 MeV where the level densities are still very high. These observations argue for another physical reason for reduced penetration or another hindrance implied by the negative Q-value; the most likely being an entrance channel phenomenon. More low-energy cross section measurements for both fusion-evaporation and fusion-fission are needed to fully understand this interesting behavior, in particular for soft systems.

This work was supported by the U. S. Department of Energy, Nuclear Physics Division, under Contract Nos. W-31-109-ENG-38.

References

1. R. Vandenbosch, Annu. Rev. Nuc. Part. Sci. **42**, 447 (1992).
2. M. Dasgupta et al., Annu. Rev. Nucl. Part. Sci. **48**, 401 (1998).
3. K. Hagino and N. Takigawa, Phys. Rev. C **55**, 276 (1997).
4. J. Fernández-Niello et al., Comput. Phy. Commun. **54**, 409 (1989).
5. L.C. Vaz and J.M. Alexander, Phy. Rep. **69**, 373 (1981).
6. C. Y. Wong, Phys. Rev. Lett. **31**, 766 (1973).
7. C.R. Morton et al.,Phys. Rev. C **62**, 024607 (2000).
8. U. Jahnke et al., Phys. Rev. Lett. **48**, 17 (1982).
9. M. Beckerman et al., Phy. Rev. C **25**, 837 (1982).
10. J. G. Keller et al., Nucl. Phys. **A452**, 173 (1986).
11. C.N. Davids and J. Larson, Nucl. Instr. and Meth. **A345**, 528 (1994).
12. S. Hofmann et al., Z. Phys. **A358**, 377 (1997).
13. D. Ackermann et al., Nucl. Phys. **A609**, 91 (1996).
14. C.R. Morton et al.,Phys. Rev. C **60**, 044608 (1999).
15. H. Timmers et al., Nucl. Phys. **A633**, 421 (1998).
16. J.R. Leigh et al.,Phys. Rev. C **52**, 3151 (1995).
17. D.J. Hinde et al.,Phys. Rev. C **60**, 054602 (1999).
18. W. Reisdorf et al., Nucl. Phys. **A438**, 212 (1985).
19. W. Reisdorf et al., Nucl. Phys. **A444**, 154 (1985).
20. A.M. Stefanini et al.,Phys. Rev. C **52**, R1727 (1995).

WHAT COULD WE LEARN FROM PARITY VIOLATING EFFECTS IN NEUTRON INDUCED FISSION?

VLADIMIR GUDKOV

Department of Physics and Astronomy
University of South Carolina
Columbia, SC 29208
E-mail: gudkov@sc.edu

Parity violation in nuclear fission was observed about 25 years ago. From that time, many experiments have been done and different theoretical models have been proposed. One can ask what we have learned new about fission from parity violating effects and what could be done with the new generation neutron sources.

1. Introduction

The parity violating effects in nuclear fission looks promising for the study of the fission process because these effects, being of tiny differences in the angular distributions (asymmetries) of fission fragments with different masses, could be very sensitive to some properties of nuclear fission models. Parity violation in neutron induced fission is a process of transformation of parity mixing of complex compound (totally statistical) nuclear states into a strongly correlated macroscopic motion of fission fragments. This is possible due to the existence of so called fission mode, which was a source of many surprises in the nuclear fission process[2,3,4,5,6]. The discovery of large parity violating asymmetries[1] in resonance fission made it necessary to add new features to the 'conventional' fission models, and aroused new questions about the Bohr's hypothesis in order to explain non-vanishing effects in the interference of channels with opposite parities (see, for example references[7,8,9]).

Let us recollect the main features of parity violation in nuclear fission. P-odd correlations in angular distributions of fission fragments in spontaneous fission of polarized nuclei have been predicted[10] a long time ago with the assumption that weak interactions mix wave functions of opposite parities of the deformed nucleus. The pear-like shape of nuclear deformation, with the axis being coincided to the direction of fission fragments momenta

$\vec{p}_f$, oriented along the spin $\vec{I}$ of the nucleus, leads to the P-odd correlation $(\vec{I} \cdot \vec{p}_f)$ in the angular distribution of the fragments. Thus the P-odd correlation is proportional[10,11] to the ratio of the weak matrix element between the deformed states with opposite parities and the distance between these states. Also, it is proportional to the ratio of barrier penetrabilities for the states with opposite parities. The barrier penetrabilities play the main role in the enhancement of the P-odd effects in spontaneous fission, since their ratio could be as much as several orders of magnitude due to the difference in penetrabilities for states with opposite parities. (This factor could be also important for a neutron induced sub-barrier fission if one mixing state/both mixing states, is/are under the fission barrier.) Also, in the case of a double structure of the fission barrier, it could be an additional barrier resonance enhancement[11].

The parity violation in nuclear fission has been observed for neutron induced fission but not for predicted effects in spontaneous fission. The experimental effect was extremely large ($\sim 10^{-4}$) if one assumes the simple statistical averaging over the number of final states for fission $N_f \sim 10^8 -$ -10^{10}. Therefore, to explain such large effects the coherence nature of the fission process (such as the Bohr's hypothesis) has to be applied to fission transition states with opposite parities.

2. P-odd and P-even Correlations

To show the coherent nature of the parity violating effects in fission and to obtain a description of the dependency of the effects on neutron energy, masses of fragments, etc, we use the approach based on the theory of nuclear reaction[8] (for other approaches and new developments see, for example [7,9] and references therein). Let us consider P-odd correlations between neutron spin $\vec{\sigma}$ and the light fragment momentum $\vec{p}_f$: $(\vec{\sigma} \cdot \vec{p}_f)$ in neutron induced fission with resonance neutrons. Assuming that parity is mixing on the compound nuclear stage, and taking into account only the two nearest resonances with opposite parities (s- and p-wave neutron resonances), one can obtain the asymmetry of the momentum distribution of light fragment in respect to the neutron spin, as

$$\alpha_{n,fis} \simeq \frac{\Re e \sum_{l,f} \{f_{ss} f_{ps}^* + f_{pp} f_{sp}^*\}}{2\sigma_{fis}}, \qquad (1)$$

where σ_{fis} is a total fission cross section, f_{ss} and f_{pp} are parity-conserving amplitudes, f_{ps} and f_{sp} are parity-violating amplitudes:

$$f_{ss} \sim (\Gamma_l^f)^{1/2} e^{(i\delta_l^f)} \frac{1}{(E - E_s + i\Gamma_s/2)} (\Gamma_s^n)^{1/2} e^{(i\delta_s^n)}, \qquad (2)$$

$$f_{pp} \sim (\Gamma_{l'}^f)^{1/2} e^{(i\delta_{l'}^f)} \frac{1}{(E - E_p + i\Gamma_p/2)} (\Gamma_p^n)^{1/2} e^{(i\delta_p^n)}, \qquad (3)$$

$$f_{ps} \sim (\Gamma_{l'}^f)^{1/2} e^{(i\delta_{l'}^f)} \frac{-W}{(E - E_p + i\Gamma_p/2)(E - E_s + i\Gamma_s/2)} (\Gamma_s^n)^{1/2} e^{(i\delta_s^n)}, \quad (4)$$

$$f_{sp} \sim (\Gamma_l^f)^{1/2} e^{(i\delta_l^f)} \frac{-W}{(E - E_s + i\Gamma_s/2)(E - E_p + i\Gamma_p/2)} (\Gamma_p^n)^{1/2} e^{(i\delta_p^n)}. \quad (5)$$

Here Γ_l^f is the partial width of the compound resonance of the decay into fragments in state f with relative angular momentum l, and δ_l^f is the potential phase shift in this channel; $l' = l+1$. $\Gamma_{s,p}^n$ and $\delta_{s,p}^n$ are neutron width and potential phase for the s- and p-wave compound resonances with the resonance energies $E_{s,p}$ and the total widths $\Gamma_{s,p}$. E is neutron energy and W is parity violating mixing matrix element for s and p compound resonances. These amplitudes can be represented by diagrams on Figure 1, where symbols s and p correspond to s- and p-wave propagators $1/(E - E_{s,p} + i\Gamma_{s,p})$, vertexes correspond to decay amplitudes $(\Gamma_i^k)^{1/2} \exp(\delta_i^k)$ for a channel k with parity i, and W is the parity mixing matrix element. From eqs.(1)-

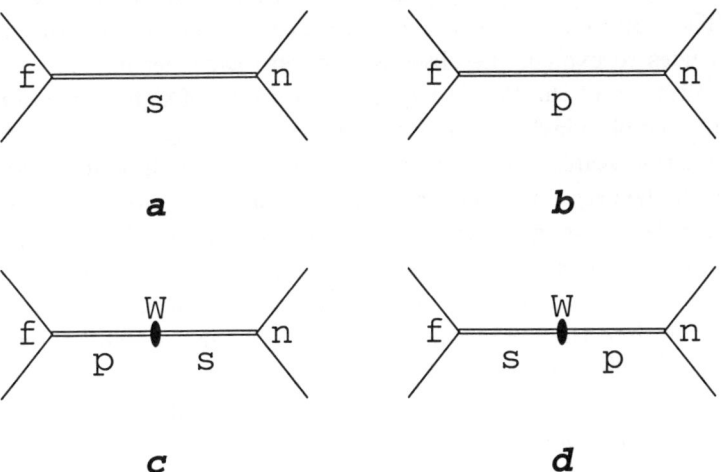

Figure 1.   Parity conserving f_{ss} (diagram a) and f_{pp} (diagram b) and parity violating f_{ps} (diagram c) and f_{sp} (diagram d) partial fission amplitudes.

(5) one can see that for low energy neutrons the first term in eq.(1) is dominant since it is proportional to Γ_s^n ($f_{ss}f_{ps}^* \sim \Gamma_s^n$) but the second one is proportional to Γ_p^n ($f_{pp}f_{sp}^* \sim \Gamma_p^n$)), which is smaller by orders of magnitude.

One can compare the expression (1) for P-odd asymmetry with corresponding expressions for P-even asymmetries: the right-left asymmetry due to the correlation $(\vec{\sigma} \cdot [\vec{p_n} \times \vec{p_f}])$ (where $\vec{p_n}$ is the neutron momentum)

$$\alpha_{RL} \simeq \Im m \sum_{l,f}\{f_{ss}f_{pp}^*\}/(2\sigma_{fis}), \tag{6}$$

and forward-back asymmetry due to the correlation $(\vec{p_n} \cdot \vec{p_f})$

$$\alpha_{FB} \simeq \Re e \sum_{l,f}\{f_{ss}f_{pp}^*\}/(2\sigma_{fis}). \tag{7}$$

It can be shown that if parity is mixing on the compound nuclear stage then the parity violating part in eq.(1) is factorized from the fission process. In that case, the mechanism of production of P-odd asymmetries, which is caused by the interference of fission decay channels with opposite parities (relative orbital momenta of fragments), should be similar to the mechanism for the P-even asymmetries. (For spontaneous or sub-barrier fission, this statement could be incorrect, since the barrier penetrabilities may strongly depend on parity.)

The sum over all states of the fission fragments in the eqs.(1), (6) and (7) results in the term $\sim \sum_{f,l(l')}^{N_f}(\Gamma_l^f\Gamma_{l'}^f)^{1/2}e^{i(\delta_{l'}^f - \delta_l^f)}$, which vanishes if the fission widths of the fragments with opposite parities are uncorrelated. Therefore, one needs to introduce new correlations for the fission amplitudes to explain the observed non-zero asymmetries. It could be a generalization of the Bohr's hypothesis in terms of sign correlations of fragment's amplitudes[8], or a model of fission involving a pear-like deformed transition states[7], or a coherent summation over fragment's amplitudes in the helicity representation[9]. The common feature of these approaches is a very strong coherence of the process of the fission as opposed to almost pure statistical behavior of nuclei on the compound nuclear stage. As a result of a transition of a quantum system from a random statistical state into the 'self-organized' highly coherent system, one gets $\sum_{f,l(l')}^{N_f}(\Gamma_l^f\Gamma_{l'}^f)^{1/2}e^{i(\delta_{l'}^f - \delta_l^f)} \simeq (\Gamma_s^{fis}\Gamma_p^{fis})^{1/2}$, where $\Gamma_{s,p}^{fis}$ is the total fission amplitude. Then, taking into account only the product of two amplitudes f_{ss} and f_{ps} (diagrams a and c) in eq.(1), one obtains[8]

$$\alpha_{n,fis} \simeq \frac{2W(E - E_p)}{(E - E_p)^2 + \Gamma_p^2/4}\left(\frac{\Gamma_p^{fis}}{\Gamma_s^{fis}}\right)^{1/2}. \tag{8}$$

From this equation one can see that due to the mixing on the compound nuclear stage, the P-odd effects are enhanced in the vicinity of p-wave resonance.

It should be noted that values of P-odd and P-even asymmetries have the same order of magnitude in the eV neutron energy region. This is mostly because of a coincidence in the numerical values of neutron barrier penetration factors for P-even asymmetries and the parity mixing parameter for P-odd asymmetries for the heavy fission nuclei (for details see, for example references[8,12]). Therefore, one can see that both P-odd and P-even asymmetries could provide similar information about the fission process in most cases. Moreover, sometimes P-even asymmetries could be even more preferable than P-odd ones because they increase with the neutron energy faster, do not contain unknown weak matrix element, and, what's more, may not require polarized neutrons (for example, $(\vec{p_n} \cdot \vec{p_f})$ correlation). However, the similarities between P-odd and P-even effects (from the point of view of the fission process) are not so straightforward and simple even in the two-level approximation. One example would be the case of a very small p-wave neutron resonance ($\Gamma_n^p \simeq 0$), where the P-even asymmetries became zeros but P-odd $\alpha_{n,fis}$ is not affected by the value of Γ_n^p at all.

3. Fission Process and Asymmetries

Parity mixing on the compound nuclear stage results in the very strong energy dependance of the P-odd effects on the scale of the average distance $\bar{D}$ between neutron resonances. The excitation energy of compound nuclei has a scale of about $\sim 6\ MeV$. Therefore, a small change in neutron energy ($\bar{D} \sim 1\ eV$) could enhance a contribution from different parities (p- or s-wave neutron resonances) to the asymmetry, but it does not change essential nuclear parameters. In the case of over-barrier fission, this will affect only the value of P-odd effects due to resonance enhancement[8]. For sub-barrier fission, however, the process can be more complicated due to dependency of barrier penetration factors on parity. In such cases, the measurement of the energy dependencies of P-odd and P-even effects could be a sensitive tool to clarify details of the fission process and new possible mechanisms of the parity mixing.

Due to experimental difficulties, the P-odd and P-even asymmetries in most experiments have been measured with a crude separation of fragments into two groups: the light and the heavy ones. To understand the fission process in details, it is important to study more detailed properties, such as a multi differential cross sections with dependencies of the asymmetries on neuron energy, masses of fragments, fragment charges, etc. These parameters strongly depend on theoretical models and they are very sensitive to the mechanisms of nuclear fission. To illustrate this statement, let us consider a couple of examples.

The first one is related to mechanism of the fragment formation. Statistical models assume the formation of mass distributions at the scission point, but angular distributions are formed before it. Therefore, for that class of models, one cannot expect differences in P-odd (P-even) asymmetries for fragments with different masses. Some measurements (see, for example[13,14,15]) of these mass dependencies have been done. However, more data and better accuracy are needed for an unambiguous analysis.

Another example is the measurement of P-odd asymmetry in the vicinity of s-wave resonance with zero spin. The energy dependence of this asymmetry will be different for the cases if the parity of transition states are good quantum numbers or not (for more details see, for example[8,12]).

The measurements of P-odd and P-even asymmetries for sub-barrier and partially sub-barrier fission would provide even more opportunities to test fission models because, in these cases, the asymmetries strongly depend on the structure of fission barriers.

4. Summary

The given examples show that P-odd (and P-even) correlations in fission are very sensitive to details of the fission mechanism both for over-barrier and sub-barrier fission (and for binary and ternary[16] fission). However, one needs to measure not only 'averaged' asymmetries but rather multiple differential fission cross sections with high accuracy and in a wide energy range. The high flux neutron beam with a reasonable energy resolution is the key factor for these future experiments. It is very important to increase the current accuracy, at least by an order of magnitude, to be able to answer important questions related to the fission dynamics. The obtained information would be of great interest to the nuclear theory, and it will lead to improvements of nuclear fission models. It would also contribute to our general understanding of theory of transition of quantum system from a pure statistical state to a highly coherent one.

References

1. G.V. Danilyan et al, JETP Pisma **26**, 298 (1977).

2. A. Bohr in *Proc. Int. Conf. on peaceful uses of atomic energy*, vol. 4, p.220, (Geneva, 1955).
3. J.E. Lynn, *The Theory of Neutron Resonance Reactions*, (Clarendon Press, Oxford, 1968).
4. A. Bohr and B.R. Mottelson *Nuclear Structure*, (W.A. Benjamin, New York, 1969).
5. V.L. Sailor, see ref.[2] p.192.
6. V.M. Strutinsky , *JETP* **39**, 781 (1960).
7. O.P. Sushkov and V.V. Flambaum, *Usp. Fiz. Nauk.* **136**, 3 (1982).
8. V.E. Bunakov and V.P. Gudkov, *Nucl. Phys.* A**403**, 93 (1983).
9. A. Barabanov, W. Furman and A. Popov in *Astrophysics, Symmetries, and Applied Physics at Spallation Neutron Sources*, eds. P. E. Koehler, C. R. Gould, R. Haight and T. E. Valentine, p.194, World Scientific, Singapore, New Jersey, London, Hong Kong, 2002.
10. V.V. Vladimirsky and A.N. Andreev, *JETP* **41**, 663 (1961).
11. A.P. Budnik and N.S. Rabotnov, *Phys. Lett.* B**46**, 155 1973.
12. V.P. Gudkov, *KEK Report 91-2*, (KEK, Japan, 1991).
13. A.K. Petukhov *et al*, *JETP Pisma* **30**, 470 (1979).
14. A.Ya. Alexandrovich *et al*, *Nucl. Phys.* A**576**, 541 (1994).
15. A. Kötzle *et al*, *Nucl. Instrum. Methods* A**440**, 750 (2000).
16. V.E. Bunakov and V.P. Gudkov, *Z. Phys.* A**321**, 271 (1985).

FISSION MODES STUDIED WITH MULTI-DIMENSIONAL LANGEVIN EQUATION

T. ICHIKAWA, T. ASANO, T. WADA, M. OHTA

Department of Physics, Konan University,
Okamoto 8-9-1,
Kobe 658-8501, Japan
E-mail: dn021001@center.konan-u.ac.jp

S. YAMAJI

The Institut of Physical and Chemical Research,
Wako-shi,
Saitama 351-0198, Japan
E-mail: yamajis@rarfaxp.riken.go.jp

H. NAKAHARA

Emeritus, Tokyo Metropolitan University,
Minami-Ohsawa,
Hachioji 192-0397, Japan
E-mail: nakahara-hiromichi@c.metro-u.ac.jp

We investigate fission modes of ^{249}Bk with 3-dimensional Langevin equation. The shell correction energy is included in the potential energy surface. The mass distribution and the total kinetic energy (TKE) distribution of fission fragments are calculated and are compared with the experimental results. There are at least two kinds of the fission modes in the calculation results: the symmetric mode having a elongated shape and a lower TKE, and the asymmetric mode having a compact shape and a higher TKE. These results are in agreement with the experimental result. The dynamics of each mode from saddle to scission is also discussed.

1. Introduction

For the fission of a low excited compound nucleus in the region of light and medium actinides, the existence of multi fission modes has been reported [1]. The aim of this study is the investigation of these modes by calculating the dynamical fission paths numerically.

From the systematic study for various nuclides, the distribution of fission

494

fragments was found to be dominant in asymmetric divisions for almost all nuclides. However, in the fission of the compound nucleus with the high excitation energy, these characteristics vanish and the distribution of fission fragments can be explained with the liquid-drop model. Thus, we can expect that shell energy plays an important role in the fission process in the case of the low excitation.

It was also found that the distribution of the total kinetic energy (TKE) consist of several components [2]. In the region of light actinides below Fermium, the existence of mainly two kinds of dynamical fission paths have been suggested [3]. For example, from the analysis of the experiment results of ^{249}Bk [4], it is pointed that at least the two modes exists, i.e., the asymmetric mode with the TKE of about 200 MeV having the compact shape and the symmetric one with the TKE of about 180 MeV having the elongated shape.

The study on the fission mode has been proceeded so far by the static estimation investigating the landscape of the potential energy surface [5,6]. Möllor et. al. have calculated the energy surface in the 5-dimensional parameter space and have investigated the various nuclides [7]. They pointed out that there are two kinds of potential valley leading the asymmetric fission and the symmetric fission in the region of the light actinides, and these valleys clearly separate at the second saddle.

As another way of these static analyses, we introduce the dynamical method by using the Langevin equation on the basis of the fluctuation-dissipation theorem. In the case of the high excitation, the distribution of the fission fragments and the TKE calculated with this method is in good agreements with the experiment results [8]. Thus, taking account of the potential energy including the shell energy, we extend this method to the case of low excitations.

In this paper, we investigate the fission mode of the ^{249}Bk at the excitation energy E^*=19.8 MeV by using the 3-dimensional Langevin equation. We calculate the mass distribution and the TKE distribution, and these results are compared with the experimental one. We also investigate dynamical fission paths and discuss the influence of the shell effect on the dynamics.

2. Framework

The shape of nucleus is described by the two-center parameterization. Z_0 denotes the distance between the harmonic oscillators in the unit of the radius of the spherical compound nuclei $\left(R_0 = r_0 A^{1/3}\right)$, δ denotes the deformation of the fragments with the constraint that both fragments have same deformation ($\delta_1 = \delta_2$) and α denotes the mass asymmetry parameter ($\alpha = (A_1 - A_2)/(A_1 + A_2)$), where A_1 and A_2 are the mass number of the fragments. The liquid drop energy, the surface energy and the coulomb energy are also calculated with this parameterization.

We describe the fission process by the following equation, called as the Langevin equation,

$$\frac{dq_i}{dt} = \left(m^{-1}\right)_{ij} p_j,$$
$$\frac{dp_i}{dt} = -\frac{\partial V}{\partial q_i} - \frac{1}{2}\frac{\partial}{\partial q_i}\left(m^{-1}\right)_{jk} p_j p_k - \gamma_{ij}\left(m^{-1}\right)_{jk} p_k + g_{ij} R_j(t), \quad (1)$$

where the suffix stands for Z_0, δ or α. Summation over repeated indices is implied. $V(q)$ is the potential energy taken account of shell effect, $m_{ij}(q)$ and $\gamma_{ij}(q)$ are the shape dependent collective inertia and dissipation tensors. We assume the random forces as the white noise type of which the normalized random force $R(t)$ is to satisfy $\langle R(t)\rangle=0$, $\langle R_i(t_1)R_j(t_2)\rangle = 2\delta_{ij}\delta(t_1 - t_2)$. The strength of random force g_{ij} is calculated from $\gamma_{ij}T = g_{ik}g_{kj}$ that is given by the fluctuation-dissipative theorem. T denotes the temperature of the compound nucleus that is defined as $E^* = aT^2$ with the excitation energy of compound nucleus and the level density parameter a of Töke and Swiatecki [9]. The inertia tensor is calculated using the hydrodynamical model with Werner-Wheeler approximation [10] for the velocity field, and wall-and-window one-body dissipation [11] is adopted for the dissipation tensor.

The shell correction energy of the two-center shell model is calculated with the code TWOCTR [12,13]. The shell correction energy depends on the temperature of the nucleus. The temperature dependent factor on the shell correction energy is assumed as $\exp\left(-E^*/E_d\right)$, where E^* is the excitation energy and E_d is the shell damping energy that is taken to be 20 MeV [14].

3. Numerical Results

In this paper, we investigate the fission modes of ^{249}Bk at the excitation energy E*=19.8MeV. Figure 1 shows the mass distribution of fission fragments. The open circle denotes the experimental result [4], and the square denotes the mass distribution in the case of the shell damping energy of 20MeV. In this case, this shell damping energy reduces the shell correction energy at zero temperature by 65%. The peek position of this distribu-

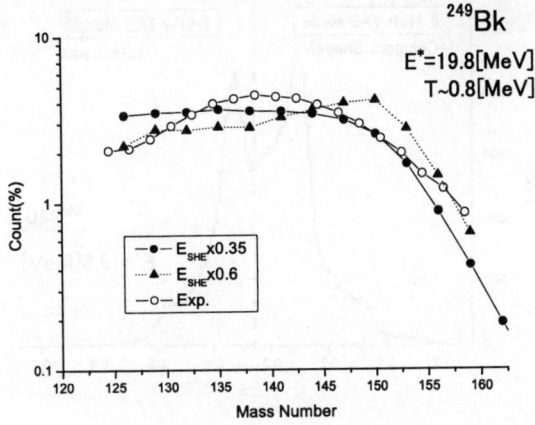

Figure 1. The mass distribution of fission fragments. The open circles denote the experimental results. The solid circles and triangles denote the calculation results of the shell energy multiplied by 0.35 and 0.6 respectively.

tion is around 140 and is in good agreement with the experimental result. However, we overestimate the number of symmetric fission fragments.

Next, we emphasize the shell correction energy to see the influence of the shell effect. The triangle denotes the case of the shell damping energy of 40MeV corresponding to the reduction of the shell energy by 40%. Then, the ratio of the number of the symmetric and the asymmetric fragments is in good agreement with experimental results. However, the peak position of this distribution is around 150. Thus, we cannot understand the mass distribution of fission fragments from only the change of the shell correction energy.

Figure 2 shows the deformation distribution of the fission fragments. In this figure, two peeks can be seen clearly at $\delta=0.26$ and $\delta=0.36$. From this result, we expect the existence of characteristic structures. We also investigated fission modes in the liquid drop model, where we obtained the deformation distribution having a single peak around 0.36. Thus, we expect that the peek at $\delta=0.36$ in Figure 2 corresponds to the liquid drop like mode.

According to the result shown in Figure 2, we divide the events of the fission fragments into two parts; one is the events with the deformation of fragments being $\delta>0.33$, which corresponds to the liquid drop like mode,

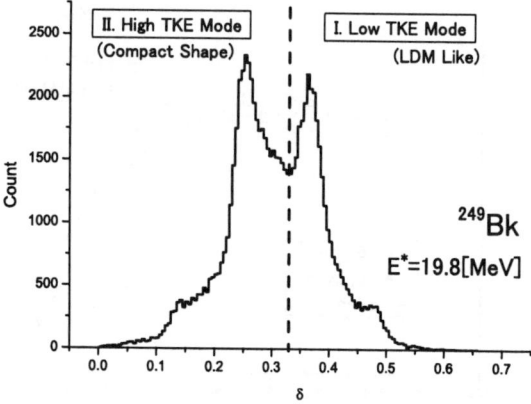

Figure 2. The deformation distribution of fission fragments at scission.

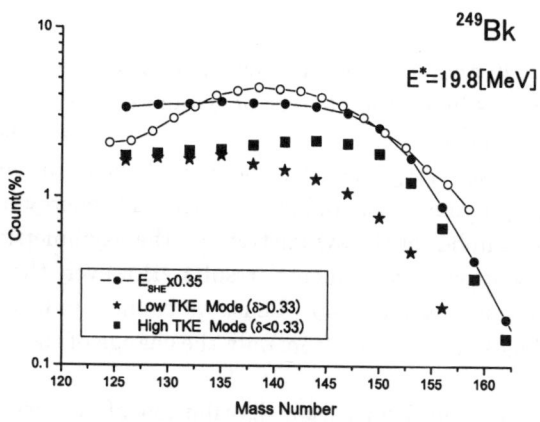

Figure 3. The mass distribution of fission fragments with $\delta>0.33$ and $\delta<0.33$. The solid star and square denotes the mass distribution of the low and the high TKE mode respectively.

and the other is one with $\delta<0.33$, which corresponds to the compact mode.

Figure 3 shows the mass distribution of each mode. The stars denote the mass distribution of the fission fragments with $\delta>0.33$ corresponding to the elongated modes. The mass distribution of this mode has the symmetric one such as the liquid drop like mode, and the average TKE of this mode is 180MeV. Thus, we expect that this mode corresponds to the symmetric mode.

The squares denote the mass distribution of fission fragments with $\delta < 0.33$ corresponding to the compact mode. The mass distribution of this mode has the asymmetric one, and the peak position is at 145. The average TKE of this mode is 200MeV. Thus, we expect that this mode corresponds to the asymmetric mode. However, we overestimate this asymmetric mode in the mass symmetric region compared with experimental results. That is to say the asymmetric mode distributes too widely. This overestimation also can be seen in the average TKE distribution in the next figure.

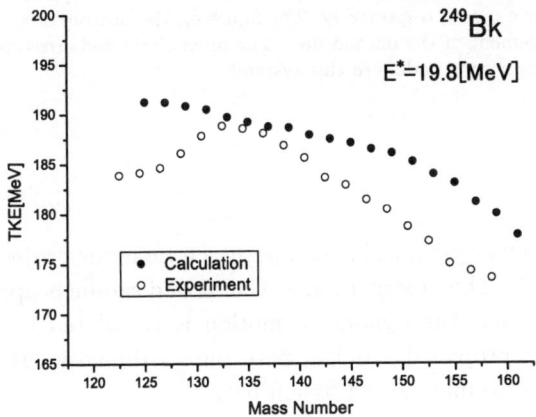

Figure 4.   The average TKE distribution of fission fragments. The open and solid circles denote the experimental and the calculation results respectively.

Figure 4 shows the distribution of average TKE of each fragment. The open circle denotes the experimental result and the solid circle denotes the calculation one. In the calculation result, we could not reproduce the peek in the distribution such as the experimental results. The reason why the average TKE of symmetric fragments increases is coming from the large yield of mass distribution of the aymmetric mode (the solid squares in Figure 3) in the mass symmetric region. It may also be said that this widely spread mass distribution is due to the constraint of $\delta = \delta_1 = \delta_2$ in the present calculation.

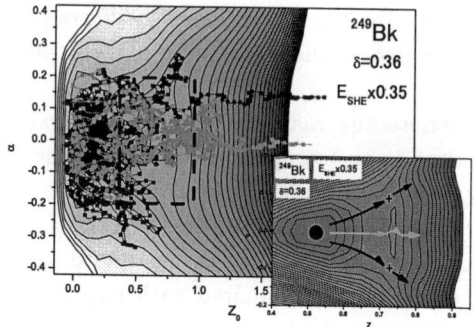

Figure 5. The energy landscape of ^{249}Bk projected onto Z_0-α plane at δ=0.36. The horizontal line denotes the elongation Z_0, and the vertical line denotes the mass asymmetry α. The black and the gray line denote the dynamical fission path of the asymmetric and the symmetric mode, respectively. The figure on the bottom of the right hand side shows the enlargement of the dashed line. The open circle and cross denote the second minimum and the second saddle in this system.

Figure 5 shows dynamical fission path of each mode projected onto Z_0-α plane at δ=0.36. The deformation at the second saddle is appeared around δ=0.36. Note that the dynamical motion is calculated in 3-dimensional spaces, but we project dynamical path onto 2-dimensional map. This is very instructive to understand fission path.

From the analysis of the calculation result, the each mode is separated at the second saddle. The area around the second saddle is enlarged and pasted in the bottom of the right hand side. The fission fragments with asymmetric mode go along the black arrow indicated in the figure and it can be seen the valley leading the asymmetric division. The fission fragments with the symmetric mode go along the gray arrow.

There is the fission barrier to symmetric division, and this barrier height is lager than the second saddle by 0.1MeV. We also calculated the second saddle in 4-dimensional space. Then, the each fragment has the different deformation ($\delta_1 \neq \delta_2$) and the second saddle becomes lower than the 3-dimensional case. Thus, in the 4-dimensianl calculation, we may expect that the wide distribution of the asymmetric mode will be improved.

4. Summary

We investigated the fission modes of ^{249}Bk by calculating the 3-dimensinal Langevin equation numerically and discussed the mass and the TKE distribution. In the fission mode of the low excited compound nucleus, the shell effects has very important role for the dynamics from the saddle to scission. Thus, taking account of the shell correction energy calculated with the two-center shell model, we extended the analysis by using the Langevin equation to the low excitation case.

From the calculation results, the fission fragments were distribution in the δ-space clearly shown the distinct two peaks. This fact shows that at least two fission modes can be appeared in the present dynamical calculation. One is the symmetric mode having the elongated shape ($\delta > 0.33$) and the lower TKE. The other is the asymmetric mode having the compact shape ($\delta < 0.33$) and the higher TKE. We can also find that this separation is coming from the shell correction energy in the potential energy surface. The problem is the insufficient the reproduction of the experimental mass distribution and the average TKE.

The further improvement of this method will be done in the expansion of the deformation space ($\delta_1 \neq \delta_2$) and the temperature dependence due to the neutron evaporation between the saddle and scission.

References

1. P. Armbruster, *Rep. Prog. Phys.* **62**, 465 (1999).
2. E. K. Hulet *et al.*, *Phys. Rev. Lett.* **56**, 313 (1989).
3. A. Turkevich and J. B. Niday, *Phys. Rev.* **84**, 52 (1951).
4. Y. L. Zhao, Ph.D. thesis, Tokyo Metropolitan University, 1999.
5. V. V. Pashkevich, *Nucl. Phys.* **A169**, 275 (1971).
6. P. Möller, R. Nix, and W. J. Swiatecki, *Nucl. Phys.* **A469**, 1 (1987).
7. P. Möller, D. G. Madland, A. J. Sierk, and A. Iwamoto, *Nature* **409**, 785 (2001).
8. T. Wada, Y. Abe, and N. Carjan, *Phys. Rev. Lett.* **70**, 3538 (1993).
9. J. Tōke and W. J. Światecki, *Nucl. Phys.* **A372**, 141 (1981).
10. H. J. Krappe, J. R. Nix, and A. J. Sierk, *Phys. Rev.* **C20**, 992 (1979).
11. J. R. Nix and A. J. Sierk, *Phys. Rev.* **C15**, 2072 (1977).
12. S. Suekane, A. Iwamoto, S. Yamaji, and K. Harada, *JAERI-memo.* 5948 (1974).
13. A. Iwamoto, S. Yamaji, S. Suekane, and K. Harada, *Prog. Theor. Phys.* **55**, 115 (1976).
14. A. V. Ignatyuk, G. N. Smirenkin, and A. S. Tishin, *Sov. J. Nucl. Phys.* **21**, 255 (1975).

THE FISSION TIME SCALE MEASURED WITH AN ATOMIC CLOCK

V.L. KRAVCHUK, H.W. WILSCHUT, M. HUNYADI, S. KOPECKY,
H. LÖHNER, A. ROGACHEVSKIY AND R.H. SIEMSSEN

Kernfysisch Versneller Instituut
Zernikelaan 25,
Groningen, 9747AA, The Netherlands
E-mail: kravchuk@kvi.nl

A. KRASZNAHORKAY

Institute of Nuclear Research of the Hungarian Academy of Science
P.O. Box 51,
Debrecen, H-4001, Hungary
E-mail: kraszna@moon.atomki.hu

We present a new direct method of measuring the fission absolute time scale using an atomic clock based on the lifetime of a vacancy in the atomic K-shell. We studied the reaction ^{20}Ne + ^{232}Th $\rightarrow$ ^{16}O + ^{236}U* at 30 MeV/u. The excitation energy of about 115 MeV in such a reaction is in the range where a fission lifetime of the order of 10^{-18} s has been suggested on basis of blocking experiments. The lifetime of the K-shell hole for U is about 6×10^{-18} s. The reaction channel is detected by a forward hodoscope, operated in coincidence with detectors for fission fragments and with Germanium detectors. The latter detectors allow one to detect the characteristic K x rays. The K-shell ionization probability due to the projectile and the projectile-like fragment is about 2%. Depending on the fission lifetime of the U nuclei, either a continuum (short lifetimes) or an explicit characteristic K x-ray spectrum of U (long lifetimes) can be observed. First analysis indicates a x-ray yield in the U energy region.

1. Introduction

During the last decade theoretical studies have been performed showing an important role of friction in nuclear fission. The determination of friction will give access to the viscosity of hot nuclear matter. It is impossible to measure it directly, however by measuring the time needed for fission, one can estimate the value of the nuclear friction coefficient in an appropriate model.

Many experiments were performed already to measure fission lifetimes. In most cases lifetimes were extracted from the comparison of fusion-evaporation model calculations with the experimental multiplicities of neutrons [1-4] or of γ rays from the giant dipole resonance [5]. Recently the crystal blocking technique was used at GANIL [6]. Fission lifetimes larger than 10^{-18} s were observed in the excitation energy region from 50 to 200 MeV. This directly measured lifetime is at least one order longer than in the previous indirect experiments. However, it was not clear which fraction of the fissioning nuclei in fact has this lifetime.

Here we report on an alternative direct method to measure the fission time scale using an atomic clock. The main idea is to find the probability to observe the characteristic K x-ray spectrum of the pre-scission nucleus in coincidence with fission fragments. We consider alpha-transfer reaction 30 MeV/u ^{20}Ne + ^{232}Th → ^{16}O + ^{236}U*. The excitation energy in this case is about 115 MeV or 2.2 MeV in terms of nuclear temperature. The transfer reaction is chosen to create with a large probability (≈ 2 %) a K-shell hole, while keeping the excitation of the target moderate. The time needed for filling a U K-shell vacancy is τ_K = 6×10^{-18} s, determined from the natural line widths [7]. Thus giving access to fission lifetimes in the 10^{-18} s range.

2. The atomic clock

The shape of the K x-ray spectra for 20 and 80 MeV/u heavy ion induced reactions was studied. K x-ray production cross sections and the effect of the additional L-shell ionization are reported in [8]. The distortion of the spectra due to additional L-shell was found to be minor and can be ignored in the fission experiment.

We also performed an experiment to determine the K-shell ionization probability in 30 MeV/u ^{20}Ne + ^{232}Th reaction. Characteristic x rays were measured in coincidence with elastically scattered particles. The x rays were measured using two large 20 cm^2 area high purity Ge planar x-ray detectors. The elastic channel was identified using ΔE-E detectors positioned in the focal plane of the Big-Bite Spectrometer (BBS) [9]. The BBS was set at 4°, considerably smaller than the grazing angle 11°. An Al absorber with thickness of 3 mm effectively removed the intense contribution of L x rays. The absolute K-shell ionization probability was determined as the ratio between the number of true coincident K x rays and the number of elastically scattered particles taking into account the Ge x-ray detectors efficiency, which was typically 1% for the Th K x-ray region. The K-shell ionization probability was determined as $(2.0 \pm 0.2) \times 10^{-2}$. In this experiment we also used three other targets with Z=50, 65 and 82 in order to find direct K-shell ionization probabilities for different reduced velocities $\xi_K=(2/\theta)(v_P/v_K)$. ξ_K is a measure of the projectile velocity with respect to that of the K-shell electron (for definitions see [8]). We also compiled other data for direct K-shell ionization probabilities [10-12]. It appears that experimental data are in reasonable agreement with recently performed SCA calculations [14] (see figure 1).

504

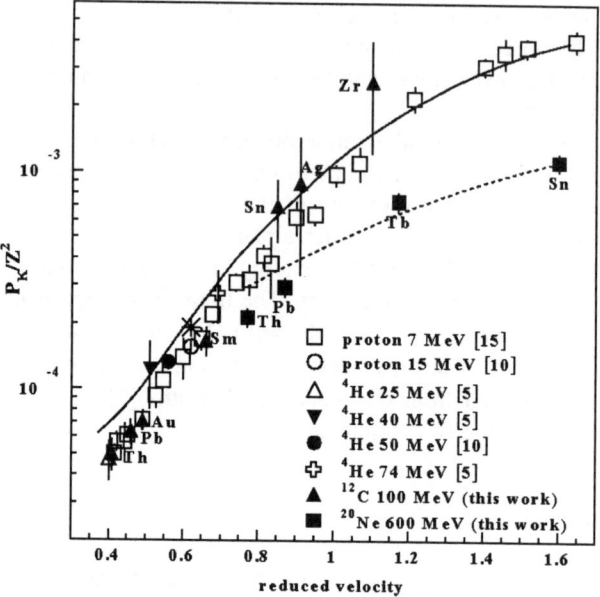

Figure 1. Direct K-shell ionization probabilities. Solid line presents SCA
calculations for proton [13] and dashed line presents SCA calculations for Ne [14]
induced reactions.

Using our experimental data and scaling we can estimate K-shell ionization
probability in case of alpha-transfer reaction 30 MeV/u ^{20}Ne + ^{232}Th → ^{16}O +
^{236}U* by assuming two half trajectory collisions, one for Ne and one for O.
Within this assumption $P_{K\ Ne→O} ≈ (0.5(P_{K\ Ne})^{1/2} + 0.5(Z_O/Z_{Ne})\ (P_{K\ Ne})^{1/2})^2 = 0.81×$
$P_{K\ Ne} ≈ 0.016$. This value is sufficiently large for a meaningful coincidence
experiment with fission fragments.

Anholt [15] has modeled the x ray spectra of symmetric systems, where
colliding atomic nuclei stick for a certain time in the presence of a K-shell hole.
Quite independent from the actual system and the collision energy he found that
for $E_K τ ≥ 20ħ$ (where E_K is the transition energy and $τ$ the lifetime of the
binary system) the K x-ray spectrum keeps it characteristic shape to some
extent. It would be discernable in our characteristic fingerprint fitting procedure
described elsewhere (see e.g. [8]). Therefore, lifetimes down to 10^{-19} s can be
detected in the fission of Uranium. The lifetime for fission in these cases is
given by $τ_{fission} ≈ (P_{K\text{-fission}}/\ (P_{K\ Ne→O} - P_{K\text{-fission}})) × τ_K$, where $P_{K\text{-fission}}$ is the
probability to observe a Uranium K x-ray in coincidence with fission and is, of
course, limited by $P_{K\ Ne→O}$. For times shorter than $3×10^{-20}$ s x-ray emission adds
to the continuum background.

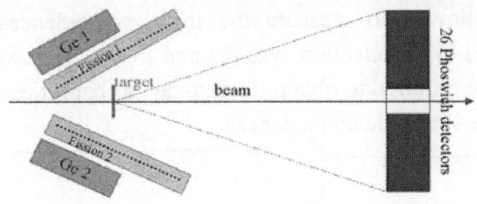

Figure 2. Schematic view of the experimental setup to measure the
fission time scale.

3. Fission time scale experiment

The fission time scale experiment was performed recently. The same
reaction 30 MeV/u ^{20}Ne + ^{232}Th was studied. Experimental setup is shown in
figure 2. The typical beam current was 16 enA. The Th target had a thickness
of 0.35 mg/cm^2. K x rays were registered with 2 large area (20 cm^2)
high purity Ge detectors. The absolute efficiency was about 1% and
typical count rates were about 25 kHz. Fission fragments were detected
with two multiwire gaseous fission detectors, operated with low-pressure
(5 Torr) isobutane gas. Fission detectors were placed inside of the vacuum
chamber with a solid angle Ω=22.6% each. The intrinsic efficiency for the
fission fragments detection was about100%. Typical count rate for each
fission detector was about 25 kHz. Reaction channels were determined
with a Forward Wall detection system consisted of 26 phoswich detectors,
covered sufficiently large area of the forward hemisphere. Angular range was
from -20° to +27° in the vertical and from -29° to +29° in the horizontal
direction. The opening for the beam ranged ±11° in both directions. The
Forward Wall detection system was placed outside of the vacuum chamber. The
exit window of the vacuum chamber consisted of the two foils (50 μm Aramica
and 70 μm Kevlar) combination in order to detect reaction products outside of
the chamber. The geometry was chosen to cover the solid angle just outside of
the grazing angle (11°) in order to reject contribution of the elastically scattered
particles in the forward wall. The average count rate for the Forward Wall was
17 kHz. The phoswich detectors consisted of a 1mm thick fast scintillator acting
as a ΔE detector and of a 5 cm thick slow scintillator acting as an E detector.
The isotope separation was sufficient to distinguish reaction channels. The

electronics setup allowed to measure the triple coincidences between Ge, Fission and Forward Wall detection systems and to obtain downscaled singles events. The latter are required to obtain the K x-ray probabilities associated with the fission fragments – ejectile coincidences.

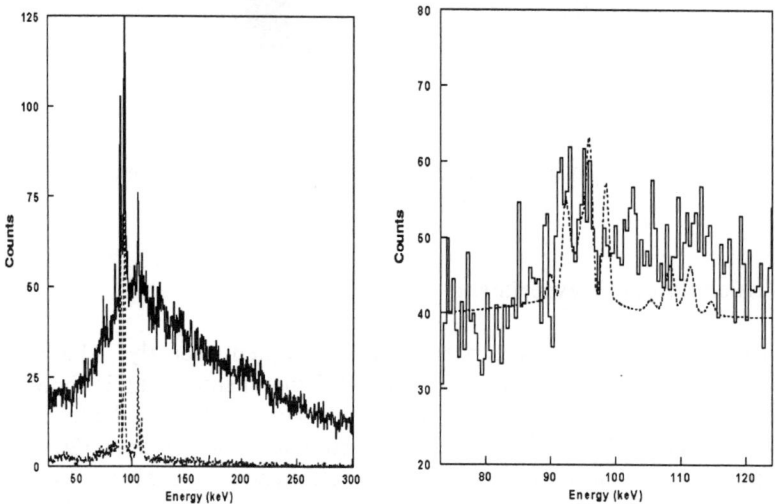

Figure 3. Left figure: prompt and random spectrum gated on Oxygen and on two fission fragments. The sharp peaks are target K x rays. Right figure: The true x-ray spectrum analyzed for Pa and U characteristic x rays. For details see text.

4. First results and conclusions

The left panel in figure 3 shows the prompt and random x-ray spectrum requiring Oxygen in the Forward Wall and two fission fragments. The two spectra are dominated by target K x-rays which reflects the extremely large cross section for target ionization (136 barn). Precise random subtraction, therefore, is difficult and is complicated by the fact that we require a fourfold coincidence. Instead, we normalize the two spectra requiring that no target (Th) x-ray yield is found in the true spectrum. This procedure may remove too much from the prompt spectrum, but was found to affects little final result we discuss next.

The right-hand side panel in figure presents the true K x ray spectrum obtained in the manner discussed above, in addition it was normalized to correct for the energy dependence in the efficiency. The spectrum was analyzed with the fingerprint procedure outlined in [8]. The O gate will also contain reaction channels where a projectile-like fragment breaks-up but where only the O is detected by the Forward Wall. Therefore, we expect yield for all channels with target-like elements $Z \leq 92$ (U). However the yield for the target X-rays is eliminated in the random subtraction procedure. The dashed line shows the characteristic K x-ray spectrum extracted for Pa ($Z_{target} + 1$) and U ($Z_{target} + 2$). We clearly observe yield above Th, but also above U. The extra yield in the latter region may indicate that the characteristic shape does not apply i.e. $\tau_{fission} \leq 10^{-19}$ s. Taking the sum of the excess counts in this region is about 750 which translates in $P_{K-fission} = 0.017 \times P_{K\ Ne \to O}$. Thus we can estimate the fission lifetime as $\tau_f = 0.017 \times 6 \times 10^{-18}$ s $\approx 10^{-19}$ s. If we would take into account all extra yield in the x-ray region of interest an upper limit of $\tau_f \leq 3 \times 10^{-19}$ s is obtained. Our analysis is still preliminary, but at this moment we conclude that the lifetime is much smaller than what has been observed in the blocking experiment at GANIL. In contrast the lifetime is more in agreement with the analysis of neutron multiplicities in the Uranium region lifetimes were extracted around 5×10^{-19} or 5×10^{-20} s, depending on the theoretical assumptions (see [3]). Further analysis, which considers the dependence on reaction channel in more detail, is required to substantiate the present findings. The atomic clock method presented here may proof to be a new versatile tool in fission studies of heavy nuclides.

Acknowledgments

We would like to thank Prof. D. Trautmann for the SCA calculations and Prof. M. Polasik for the calculations of the energies and intensities of the x rays. This work has been performed as part of the research program of the "Stichting voor Fundamenteel Onderzoek der Materie" (FOM), which is financially supported by the "Nederlandse Organisatie voor Wetenschappelijk Onderzoek" (NWO). This work is partly supported by the Hungarian OTKA Foundation No. N32570.

References

1. D. Hilscher and H. Rossner, *Annales-de-Physique* **17**, 471 (1992).
2. D.J. Hinde et al., *Phys. Rev.* **C45**, 1229 (1992).
3. K. Siwek-Wilczynska, J. Wilczynski, R.H. Siemssen and H.W. Wilschut, *Phys. Rev.* **C51**, 2054 (1995).
4. V.A. Rubchenya et al., *Phys. Rev.* **C58**, 1587 (1998).

508

5. P. Paul and M. Thoennessen, *Annu. Rev. Nucl. Part. Sci.* **44**, 65 (1994).
6. F. Goldenbaum, M. Morjean et al., *Phys. Rev. Lett.* **82**, 5012 (1999).
7. S.I. Salem and L.E. Lee, *Atomic Data and Nucl. Data Tables* **18**, 233 (1976).
8. V.L. Kravchuk et al., *Phys. Rev.* **A64**, 062710 (2001).
9. A.M. van den Berg, *Nucl. Instr. and Meth.* **B99**, 637 (1995).
10. R.J. Vader et al., *Phys. Rev.* **A14**, 62 (1976).
11. J.H. van Dijk et al., *Z. Phys.* **A314**, 1 (1983).
12. M. Dost et al., *Phys. Rev.* **A24**, 693 (1976).
13. D. Trautmann and F. Rösel, *Nucl. Instr. and Meth.* **169**, 259 (1980).
14. D. Trautmann and M. Polasik, *Private communication* 2002.
15. R. Anholt, *Rev. of Mod. Phys.* **57**, 995 (1985).

FISSION STUDIES on ^{238}U (1 AGeV) + Pb

T. ENQVIST*, J. BENLLIURE†, F. REJMUND‡, K.-H. SCHMIDT,
P. ARMBRUSTER

GSI, Planckstrasse 1, D-64291 Darmstadt, Germany

M. BERNAS, C. STÉPHAN, L. TASSAN-GOT

IPN Orsay, IN2P3, F-91406 Orsay, France

A. BOUDARD, R. LEGRAIN, C. VOLANT

DAPNIA/SPhN CEA/Saclay, F-91191 Gif sur Yvette, France

Projectile fission, induced in collisions of ^{238}U (1 AGeV) with lead, has systematically been studied. About 600 isotopes of elements between gallium (Z=31) and praseodymium (Z=59) produced in the fission process were identified in nuclear charge and mass, and the production cross sections of these isotopes were determined. In addition, the velocity distributions of all the reaction products have been mapped. By having the access to the reaction kinematics allows resolving different reaction mechanisms.

1. Introduction

Peripheral collisions of heavy ions at relativistic energies have become more important during the last years due to their usefulness in different applications. For example, the preparation of secondary beams using projectile fragmentation and fission with recoil separators, e.g. refs. [1,2,3] has become an important tool for nuclear structure investigations (see e.g. refs. [4,5]). On the one hand, the basic research on the reaction mechanism has revealed new aspects of the fragmentation process [6,7,8,9,10,11,12,13] as well as of nuclear fission from low [14,15,16,17,18] and high excitation energies [19,20,21].

The present work reports on a complete survey on isotopic production cross sections down to 0.1 mb and velocity distributions of all elements

---

*Present address: University of Jyväskylä, Finland
†Permanent address: Universidad de Santiago de Compostela, Spain
‡Permanent address: IPN Orsay, IN2P3, France

produced by fission of ^{238}U induced in collisions at 1 AGeV with lead. It completes previous investigations at 750 AMeV [17,21] reaching smaller lower limit in cross section, but being kinematically less comprehensive and restricted to Z $\leq$ 52.

Electromagnetic interactions predominantly excite the giant dipole resonance (maximum at $\sim$12 MeV for ^{238}U), leading to the evaporation of a few neutrons or to low-energy fission. In nuclear collisions, depending on the impact parameter, a number of nucleons is removed from the projectile, and the average excitation energy of the system can be expressed as 27·ΔA MeV [7,13], ΔA being the abraded mass.

2. Experiment

The experiment was carried out at the GSI fragment separator FRS [2]. The primary beam of ^{238}U, delivered by the heavy-ion synchrotron SIS, with an energy of 1 AGeV, impinged on a 50.5 mg/cm^2 lead target. The total number of ^{238}U ions on the target was approximately 2.2 · 10^{10}. The relatively thin target was used in order not to disturb the kinematical properties of the reaction products.

The FRS was equipped with essentially two 5 mm thick scintillation detectors [22] which measured the horizontal positions at the dispersive intermediate-image-plane (S$_2$) and at the achromatic final-image-plane (S$_4$) as well as the time-of-flight in the second half of the separator. In addition, the energy loss of the reaction products at S$_4$ was determined with a MUSIC detector [23]. The primary-beam intensity was permanently recorded [6].

3. Data analysis

The identification of the isotopes produced in the target was based on the relation between the magnetic rigidity $B\rho$ and the nuclear charge Z and mass number A by the equation

$$B\rho = \beta\gamma \cdot \frac{A}{Z} \cdot c \cdot \frac{m_0}{e}, \tag{1}$$

where γ is the Lorentz parameter, c the velocity of light, m_0 the nuclear mass unit, and e the charge of an electron.

The magnetic rigidity $B\rho$ was determined at the first stage of the FRS from the horizontal position at S$_2$. This information was combined with the time-of-flight in order to determine the A/Z ratio of the reaction products. The nuclear-charge values of the reaction products were deduced from the energy loss in the MUSIC detector.

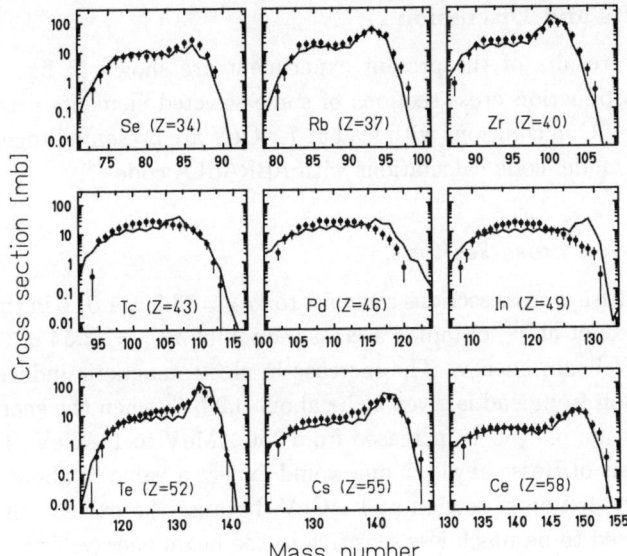

Figure 1. Comparison of the isotopic production cross sections for fission from the present work (full circles) in the reaction ^{238}U + lead at 1 AGeV with calculations performed by the ABRABLA code for some selected elements. Statistical error bars are shown if they exceed the size of the symbol. Additional systematic uncertainty amounts to ±30%.

For an unambigious isotopic identification of the reaction products, the analysis was restricted to the fully stripped ions only. The losses in the counting rate due to the fraction of incompletely stripped ions were corrected for. Another correction to be applied considered the losses due to secondary reactions in the layers of matter in the beam line.

The measured counting rates were normalised to the number of primary fragments. In addition, the data were corrected for the dead time of the data-acquisition system, and for the angular transmission of the fragment separator using the empirically determined angular acceptance of ±(10.4 ± 0.1) mrad. The full momentum distribution of each isotope was measured, and it needs not to be corrected for. The accuracy of the production cross sections is limited mostly by the different corrections on the measured counting rates. We estimated a systematic uncertainty of 30%.

For more details of the data analysis as well as experimental details, see ref. [12].

4. Results and Discussion

The basic results of the present experiment are shown in fig. 1, where isotopic production cross sections of some selected elements produced by fission of ^{238}U in collisions with lead at 1 AGeV are presented together with results of model-code calculations with ABRABLA code [24].

4.1. *Fission cross sections*

The total fission cross sections amounts to $\sigma_{\text{fiss}} = (3.5 \pm 1.0)$ b in the present work. Hesse et al. [15] obtained a cross section of $\sigma_{\text{fiss}} = (3.54 \pm 0.21)$ b at 750 AMeV beam energy. The increase in electromagnetic-induced fission cross section from lead is given to be about 0.2 b [14] when the energy of the primary beam of ^{238}U is increased from 750 AMeV to 1 AGeV. Therefore, on the basis of Hesse et al. [15] one would expect a value of about 3.7 b for the total fission cross section at 1 AGeV, because the nuclear interactions are supposed to be much less sensitive to the beam energy [19,25].

4.2. *Velocities of fission fragments*

The mean velocity values induced in the fission process are shown in fig. 2 for zirconium and tellurium isotopes. The values are corrected for the finite angular acceptance of the spectrometer.

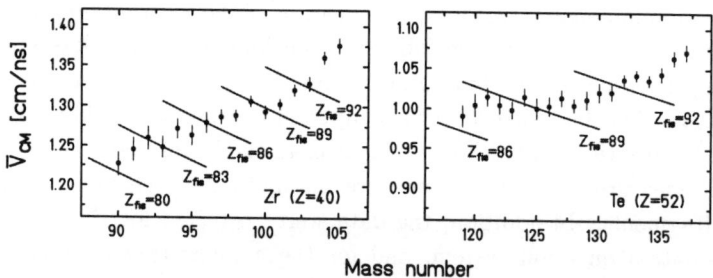

Figure 2. Mean values of the velocities (full dots) induced in the fission process as a function of mass number for zirconium and tellurium isotopes. In addition, the full lines indicate the calculated velocity values. See the text for more details.

The nuclear charge of the fissioning nucleus couldn't be directly determined in the present experiment. However, this information can be deduced from the measured mean velocities of the fission fragments. The total kinetic energy TKE, induced in the fission of sufficiently highly excited nuclei, can well be described by the liquid-drop description [26,27]

$$\text{TKE} = \frac{Z_1 Z_2 e^2}{D}, \tag{2}$$

where

$$D = r_0 A_1^{1/3}(1 + \frac{2\beta_1}{3}) + r_0 A_2^{1/3}(1 + \frac{2\beta_2}{3}) + d. \tag{3}$$

A_1, A_2, Z_1, and Z_2 denote the mass and nuclear charge numbers of the fission products. The values of parameters were: $r_0 = 1.16$ fm, $d = 2.0$ fm, and $\beta_1 = \beta_2 = 0.625$ [27]. By considering momentum conservation, these equations determine the velocities of the fission fragments.

For the velocity calculations, in fig. 2, it is assumed that the fissioning nucleus varies from uranium to mercury having the same A/Z ratio as the observed fission fragment. It is also assumed that no neutrons are emitted from the fission products since they affect to the fission velocities only slightly. One may also argue that the neutrons are predominantly evaporated before fission, since fission from high excitation energies is hindered due to dissipation [19,28,29,30,31].

4.3. Charge polarisation

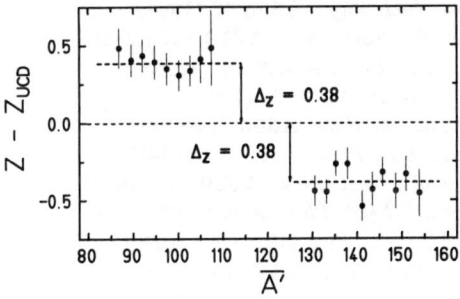

Figure 3. The charge polarisation in the low-energy fission of ^{238}U, determined in the present work.

The charge polarisation, i.e. the difference of the mean N/Z ratio of the fission fragments before the neutron evaporation to the N/Z ratio of the fissioning system, in the low-energy fission of ^{238}U in the present work was determined to be $\Delta_Z = 0.38 \pm 0.06$.

The charge polarisation in the present work is demonstrated in fig. 3.

5. Summary

The secondary-beam facility of GSI has been used to determine the production cross sections and the momentum distributions of about 900 secondary projectiles from a ^{238}U beam at 1 AGeV in lead. The data clearly show the characteristics of the different reaction mechanisms. From the velocities of the fission fragments the average nuclear charge of the fissioning nucleus, responsible for their production, could be deduced. An estimation for the charge polarisation of the low-energy fission of ^{238}U was given.

References

1. R. Anne et al., *Nucl. Instrum. Methods* **A257** 215 (1987).
2. H. Geissel et al., *Nucl. Instrum. Methods* **B70**, 286 (1992).
3. B.M. Sherrill et al., *Nucl. Instrum. Methods* **B56/57** 1106 (1993).
4. Z. Podolyák et al., *Physics Letters* **B491** 225 (2000).
5. M. Pfützner et al., *Phys. Rev.* **C64** 064604 (2002).
6. K.-H. Schmidt et al., *Nucl. Phys.* **A542** 699 (1992).
7. K.-H. Schmidt et al., *Physics Letters* **B300** 313 (1993).
8. E. Hanelt et al., *Z. Phys.* **A346** 43 (1993).
9. M. de Jong et al., *Nucl. Phys.* **A628** 479 (1998).
10. J. Benlliure et al., *Nucl. Phys.* **A628** 458 (1998).
11. J. Benlliure et al., *Eur. Phys. J.***A2** 193 (1998).
12. T. Enqvist et al., *Nucl. Phys.* **A658** 47 (1999).
13. K.-H. Schmidt et al., *Nucl. Phys.* **A710** 157 (2002).
14. Th. Rubehn et al., *Z. Phys.* **A353** 197 (1995).
15. M. Hesse et al., *Z. Phys.* **A355** 69 (1996).
16. P. Armbruster et al., *Z. Phys.* **A355** 191 (1996).
17. C. Donzaud et al., *Eur. Phys. J.***A1** 407 (1998).
18. A.R. Junghans et al., *Nucl. Phys.* **A629** 635 (1998).
19. A.V. Ignatyuk et al., *Nucl. Phys.* **A593** 519 (1995).
20. Th. Rubehn et al., *Phys. Rev.* **C53** 3143 (1996).
21. W. Schwab et al., *Eur. Phys. J.***A2** 179 (1998).
22. B. Voss et al., *Nucl. Instrum. Methods***A364**, 150 (1995).
23. M. Pfützner et al., *Nucl. Instrum. Methods* **B86** 213 (1994).
24. J. Benlliure et al., *Nucl. Phys.* **A628** 458 (1998).
25. C.J. Benesh et al., *Phys. Rev.* **C40** 1198 (1989).
26. B.D. Wilkins et al., *Phys. Rev.* **C14** 1832 (1976).
27. C. Böckstiegel et al., *Phys. Lett.* **B398** 259 (1997).
28. D.J. Hinde et al., *Phys. Rev.* **C45** 1229 (1992).
29. D. Hilscher et al., *Ann. Phys. Fr.* **17** 471 (1992).
30. B. Jurado et al., Proc. Winter Meeting on Nucl. Phys., Bormio (2001).
31. J. Benlliure et al., *Nucl. Phys.* **A700** 469 (2002).

NEUTRON-INDUCED FISSION OF ^{233}U, ^{238}U, ^{232}Th, ^{239}Pu, ^{237}Np, ^{nat}Pb AND ^{209}Bi RELATIVE TO ^{235}U IN THE ENERGY RANGE 1-200 MeV [*]

O. A. SHCHERBAKOV, A. B. LAPTEV, G. A. PETROV AND A. S. VOROBYEV

Petersburg Nuclear Physics Institute
Gatchina, Leningrad district, 188300, Russia
E-mail: laptev@pnpi.spb.ru

A. V. FOMICHEV, A. Y. DONETS, A. V. EVDOKIMOV AND S. M. SOLOVIEV

V.G. Khlopin Radium Institute
2-nd Murinski Ave. 28, St. Petersburg, 194021, Russia

Y. V. TUBOLTSEV

A.F. Ioffe Physico-Technical Institute
Polytekhnicheskaya 26, St. Petersburg, 194021, Russia

T. FUKAHORI AND A. HASEGAWA

Nuclear Data Center/JAERI
Tokai-mura, Naka-gun, Ibaraki-ken, 319-1195, Japan

V. M. MASLOV

Radiation Physics and Chemistry Institute
Minsk-Sosny, 220109, Belarus

Fission cross section ratios of ^{233}U, ^{238}U, ^{232}Th, ^{239}Pu, ^{237}Np, natural Pb and ^{209}Bi to ^{235}U have been measured in a wide energy range of incident neutrons from 1 MeV to 200 MeV using time-of-flight technique at the neutron spectrometer GNEIS based on the 1-GeV proton synchrocyclotron of PNPI.

1 Introduction

Among the applications of the fission data for actinides above 20 MeV, the most important are accelerator-driven transmutation of waste reactor materials and energy production, accelerator-based conversion of weapon plutonium and production of tritium, accelerator and spaceship shielding, radiation therapy, *etc*. A need in information about neutron-induced fission of lead and bismuth at intermediate energies also exists mainly due to probable use of these metals as structural materials in the neutron-producing targets of high-current proton accelerators of new generation. Besides, neutron-induced fission cross sections

---

[*] This work is supported by ISTC grant # 609-97.

of Pb and Bi are very convenient as standards in the intermediate energy range because they have thresholds at 25-40 MeV which eliminates the influence of low energy neutrons.

During the last decade, the measurements of neutron-induced fission cross sections in the energy range 1-200 MeV with continuous spectrum neutrons have been systematically performed at the GNEIS facility (v. refs. in [1]). The target nuclei ^{233}U, ^{238}U, ^{232}Th, ^{239}Pu, ^{237}Np, ^{nat}Pb and ^{209}Bi have been chosen for present investigation according to the Nuclear Data Request List for new JENDL High Energy File (JENDL/HE-3.3). On the moment of the beginning of present measurements, besides our earlier data there was only one experimental data set obtained by P. Lisowski et al. in Los Alamos [2,3] for ^{233}U, ^{234}U, ^{236}U, ^{238}U, ^{239}Pu, ^{237}Np, ^{232}Th relative to ^{235}U in the energy range 1-400 MeV. These data have been published as graphs only. Later on, an experimental data set for ^{239}Pu, ^{240}Pu, ^{242}Pu and ^{244}Pu obtained by P. Staples and K. Morley [4] at LANL in the energy range 0.5-400 MeV (up to 62 MeV for ^{239}Pu) has been published.

As for Pb and Bi, besides our data, the experimental data base on neutron-induced fission cross sections is very scarce and can be roughly divided into two groups: a few results obtained in isolated energy points (e.g., refs. in [5]) and the only one data set obtained by P. Staples et al. [6,7] in Los Alamos using a continuous neutron spectrum up to 400 MeV.

The description of experimental technique and data analysis are presented in this report together with the final results, which are shown in comparison with the data of other authors and theoretical calculations.

2   Experiment

The fission cross section ratios relative to ^{235}U have been measured using the neutron time-of-flight spectrometer GNEIS [8] based on the 1 GeV proton synchrocyclotron of PNPI (Gatchina). A water-cooled lead target situated inside the vacuum chamber of the accelerator is used as a pulsed spallation neutron source with average intensity $3\cdot10^{14}$ n/s, a burst duration 10 ns and repetition rate up to 50 Hz. The measurements were performed using the bare source target, a 48.5-m flight path and neutron beam contained in the evacuated flight tubes (Figure 1). A system of iron, brass and lead collimators gives the beam diameter of 18 cm at the fission chambers location. The measurements were carried out with the use of the "clearing" magnet to remove charged particles from the neutron beam.

The fission reaction rate was measured using simultaneously two fast parallel plate ionization chambers filled with methane (94-100 %) and CF_4 (6-0 %) mixture working gas at the absolute pressure of 2.5 – 3.5 atm. The first (actinide) chamber consisted of 6 sections, every one containing a pair of cathode and anode plates spaced by 5 mm. The second chamber contained 10 foils with lead or bismuth targets. A painting technique has been used for actinide targets

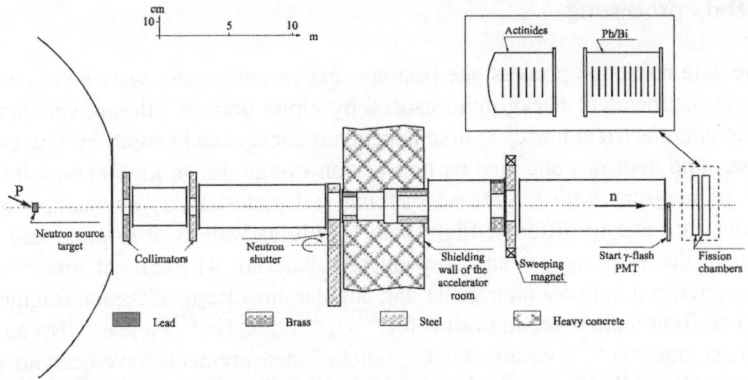

Figure 1. The scheme of the 48.5-m flight path used for fission cross section measurements.

production. The Pb and Bi targets have been produced from high purity metal using evaporation in vacuum. All targets were 180 mm in diameter, deposited on one side of the 0.05-mm-thick Al backings. The actinide and Pb/Bi targets were 150-560 µg/cm² and 260-520 µg/cm² thick, respectively. Both chambers contained 1 or 2 targets of 99.992 % enriched ^{235}U 150-260 µg/cm² thick. A weak ^{252}Cf deposit was applied on actinide targets to match the gains of electronics. The foils were oriented perpendicular to the neutron beam, all target deposits facing away from the neutron source.

The time-of-flight and pulse height spectra were accumulated in a "multi-stop" mode with zero dead time using the data acquisition system based on a 100-MHz FLASH-ADC in each measuring channel. The start signal was provided by gamma flash detector – the bare PMT placed in the neutron beam 2 m upstream from the fission chambers. The "time-of-flight vs neutron energy" calibration with an accuracy (0.025-0.1) % has been made for every target foil using a position of the carbon and lead total cross section resonances (v. insert B to Figure 2) and weak gamma-flash peak observed in the TOF-spectra and used as a true time-zero (v. insert A to Figure 2).

The initial stage of the raw data reduction included the digital processing of the data from FLASH-ADCs and re-arranging of the data into pulse height and TOF-spectra (Figure 2).

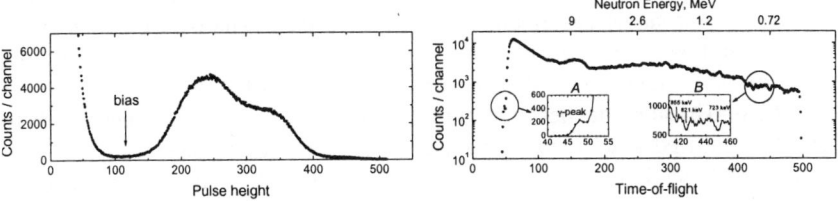

Figure 2. Pulse height spectra observed for ^{235}U target in neutron energy range 0.5-200 MeV (left) and time-of-flight spectra (right) observed with a chosen pulse height bias.

3   Data processing

In the data reduction process, the fission event counting rates were corrected for: 1) time independent background caused by alpha particle pile-up, spontaneous fission and electrical noise; 2) time dependent background caused by low-energy wraparound neutrons and also by non-fission events due to high-energy interactions of neutrons with target and backing foil materials; 3) attenuation of the neutron flux due to different flight path length, as well as absorption and scattering in the working gas and backing foil material; 4) fragment losses in the targets, neutron momentum transfer and angular anisotropy of fission fragments.

The fission cross section ratios for ^{233}U, ^{238}U, ^{232}Th, ^{239}Pu and ^{237}Np and the reference nucleus ^{235}U obtained in the "shape" measurements have been normalized using the following methods: 1) calculation of the target thickness and detection efficiency ratios (for Pb and Bi only this method has been used); 2) normalization to recommended data; 3) threshold cross section method [9,10]. To use the last method, a part of the measurements has been done using "^{235}U-mixed" targets for threshold isotopes ^{238}U, ^{232}Th, ^{237}Np, and "^{238}U-mixed" targets for fissile isotopes ^{239}Pu and ^{233}U.

Finally, the normalized fission cross section ratios have been converted to the cross sections using fission cross section of ^{235}U from JENDL-3.2 [11] below 20 MeV and the recommended data of A. Carlson et al. [12] above 20 MeV.

4   Results and discussion

The results of present measurements are shown in Figures 3,4 in comparison with experimental data of P. Lisowski et al. [2,3,13] (for actinides), P. Staples and K. Morley [6,7] and some others [5] (for Pb and Bi), systematics [14,15] and theoretical calculations. The error bars represent the statistical errors only (one standard deviation). These uncertainties are less than 1% above ~ 1 MeV for actinides, and for lead and bismuth are as much as 4-7 %, 2-3 % and 1.2-1.6 % at 60 MeV, 100 MeV and 200 MeV, respectively.

There is a large disagreement between present data and that of P. Lisowski et al. for ^{233}U, ^{238}U, ^{232}Th and ^{239}Pu above 20 MeV while for ^{237}Np both data sets are not in contradiction. In general, the comparison of our data with results of other authors shows that in the overlapping energy regions (below 20 MeV) our data are in reasonable agreement with evaluated data. The data of systematics [14,15] could be compared with experimental data only above 50 MeV and show a very strong disagreement with experiment in the case of ^{239}Pu.

The statistical model description of the measured fission cross sections up to 150 MeV showed that fission dynamics effects are not evidenced in the total fission cross section of ^{233}U, ^{238}U, ^{232}Th, ^{239}Pu and ^{237}Np. It was made a supposition that collective mode damping is independent on the deformation of nuclei. The most challenging feature of the experimental data is a dependence of the observed fission cross section on the fissility of a target nuclide up to 150 MeV that needs some other modification of the statistical model approach.

Above 50 MeV the present data for ^{nat}Pb are in a good agreement with of Staples *et al.* [6,7] and with the parameterization of Prokofiev *et al.* [5] based on the measurements with ^{nat}Pb and ^{208}Pb, including the data of Staples *et al.* and other known experimental data carried out mainly in separate energy points. The

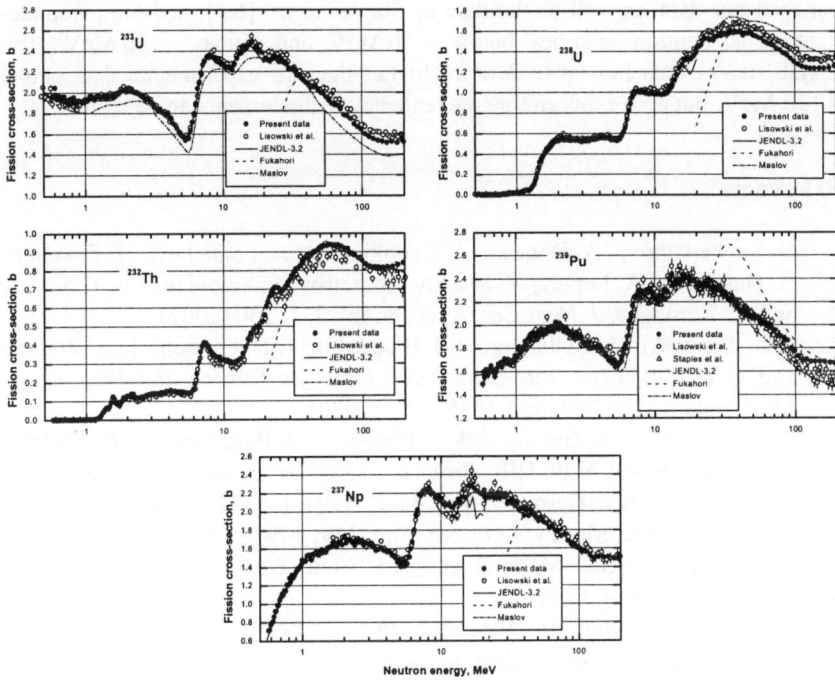

Figure 3. Fission cross section of ^{233}U, ^{238}U, ^{232}Th, ^{239}Pu and ^{237}Np in the energy range 1- 200 MeV.

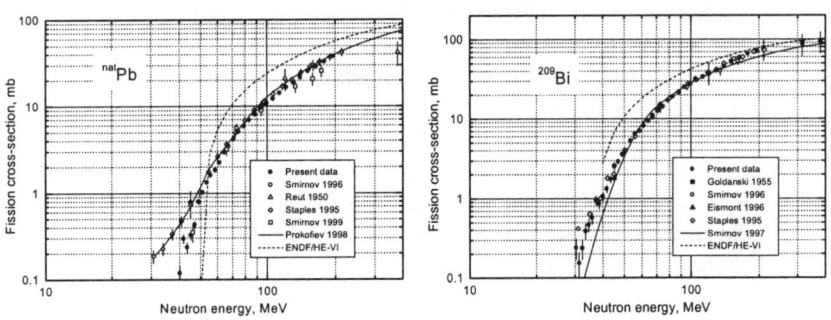

Figure 4. Fission cross section ^{nat}Pb and ^{209}Bi in the energy range up to 200 MeV.

ENDF/HE-VI data based on the systematics of Fukahori and Perlstein [14,15] for ^{208}Pb lies much higher than all experimental data. It should be noted that for comparison of ^{nat}Pb and ^{208}Pb data, the latter were multiplied by 1.43, as it was done in [5]. This coefficient was verified experimentally with an accuracy $\pm$ 10 % in the energy range 45-160 MeV [5].

In the case of ^{209}Bi, an agreement of the present data with that of other authors is good within the stated uncertainties. There is a noticeable discrepancy between our data, as well as the data of Staples et al. [6,7], and the parameterization at neutron energies below $\sim$ 50 MeV and above $\sim$ 120 MeV. The ENDF/HE-VI data lies up to 2 times higher than all experimental data below $\sim$ 150 MeV, but above this energy the tendency to agreement can be seen.

References

1. O. Shcherbakov, A. Donets, A. Evdokimov, A. Fomichev, T. Fukahori, A. Hasegawa, A. Laptev, V. Maslov, G. Petrov, S. Soloviev, Yu. Tuboltsev and A. Vorobyev, J. Nucl. Sci. Tech., Suppl. 2, 1, 230 (2002).
2. P.W. Lisowski, J.L. Ullmann, S.J. Balestrini, A.D. Carlson, O.A. Wasson and N.W. Hill, Proc. Int. Conf. Nucl. Data for Science and Technology, Mito, Japan, May 30-June 3 1988, p.97 (1988).
3. P.W. Lisowski, A. Gavron, W.E. Parker, S.J. Balestrini, A.D. Carlson, O.A. Wasson and N.W. Hill, Proc. Int. Conf. Nucl. Data for Science and Technology, Julich, Germany, May 13-17 1991, p.732 (1992).
4. P. Staples and K. Morley, Nucl. Sci. Eng. 129, 149 (1998).
5. A.V. Prokofiev, S.G. Mashnik and A.J. Sierk, LANL Report LA-UR-98-0418, Los Alamos, 1998.
6. P. Staples et al., Bull. Am. Phys. Soc. 40, 962 (1995).
7. P. Staples, private communication (2000).
8. N.K. Abrosimov, G.Z. Borukhovich, A.B. Laptev, G.A. Petrov, V.V. Marchenkov, O.A. Shcherbakov, Yu.V. Tuboltsev and V.I. Yurchenko, Nucl. Instr. Meth. A242, 121 (1985).
9. J.W. Behrens, Report UCRL-51478, Lawrence Livermore Laboratory,1973.
10. J.W. Behrens and J.C. Browne, Nucl. Sci. Eng. 77, 444 (1981).
11. T. Nakagawa et al., J. Nucl. Sci. Tech. 32(12), 1259 (1995).
12. A.D. Carlson, S. Chiba, F.-J. Hambsch, N. Olsson and A.N. Smirnov, Summary Report of a Consultants' Meeting, Austria, December 2-6, 1996. INDC(NDC)-368, IAEA, Vienna, p.23 (1997).
13. J.L. Ullmann, private communication (1999).
14. T. Fukahori and S. Pearlstein, Report BNL-45200, 1991.
15. T. Fukahori, private communication, 1999.

SYMMETRIC/ASYMMETRIC URANIUM NEUTRON-INDUCED FISSION UP TO 200 MEV*

V.M. MASLOV
JOINT INSTITUTE FOR NUCLEAR AND ENERGY RESEARCH,
220109, MINSK-SOSNY, BELARUS

The symmetric SL-mode and asymmetric lumped (S1+S2)-mode fission cross sections of 238-U(n,f) and 235-U(n,f) reactions are calculated up to 200 MeV within a statistica l model. For each U nuclide, emerging in U(n,xnf) reaction separate triaxial outer fission barrier is assumed for the SL-mode. To fit branching ratio of symmetric and asymmetric fission more fissions coming from higher fission chances, or more fissions from N-deficient nuclei are predicted. Damping of triaxial collective modes contribution to the level density at SL-mode outer saddle seems to be essential for the 238-U(n,f) branching ratio description.

1. Introduction

Below emissive fission threshold steeply rising superlong (SL) mode fission yields[1,2] in reactions ^{235}U(n,f)^{sym} and ^{238}U(n,f)^{sym}, related to symmetric scission, were reproduced recently within a Hauser-Feshbach model[3]. Double-humped fission barrier model[4] was employed previously for Ac and Ra, assuming that symmetric fission "involves a separate outer barrier that is axially asymmetric"[5,6]. A separate outer barrier B was assumed for the SL-mode, while common inner barrier A was assumed for the symmetric SL- and lumped asymmetric (S1+S2)-mode. This is supported by fission mode calculations[7,8] using the multimodal random-neck rapture model[9]. The SL-mode cross sections are controlled by high outer fission barrier E_{fBSL} with significant transparency $\hbar\omega_{fBSL}$ as well as outer saddle point B triaxiality and mass symmetry. Outer saddles for S1- and S2-modes were assumed axially symmetric and mass asymmetric. Above emissive fission threshold a number of nuclei contribute to U fission observables, about $\sim$20 for E_n $\sim$200 MeV. Branching ratio of symmetric to the observed fission

---

*This work is supported by International Science and Technology Center under Project B-404, Funding Party Japan, and International Atomic Energy Agency under Research Contract 9837.

events $r^{sym} = \sigma_{nFSL}/(\sigma_{nFSL} + \sigma_{nF(S1+S2)})$ was obtained for ^{238}U(n,F) by Zoller et al.[10] for E_n up to $\sim$500 MeV. Observed ^{238}U(n,F) data[11,12] were reproduced[13] in fission/neutron evaporation approximation up to E_n $\sim$200 MeV with first-chance pre-equilibrium (PE) neutron emission. Damping of collective modes contribution to the level density at excitations $U \gtrsim 20$ MeV for axial symmetric saddle and equilibrium deformations[14] is essential for the ^{235}U(n,F) and ^{238}U(n,F) data fit. Competition of the symmetric and asymmetric fission of U nuclei[15] would depend essentially on the excitation energies of nuclei, emerging after (n,$x$nf) neutron evaporation. We anticipate that r^{sym} would be much dependent on the contribution to the observed cross section of emissive fission chances and damping of triaxial deformations at outer SL-mode saddles.

2. Statistical model

We assume that emissive fissions come from a chain of U nuclei, i.e. only PE-emission/evaporation of neutrons is allowed[16]. A coupled channel model, fitting ^{238}U total data[17] up to E_n $\sim$200 MeV is employed. SL-mode contribution to the observed fission cross section, coming from (n,$x$nf) fission reactions could be calculated using fission probability $P_{fSLx}^{J\pi}(U)$ of symmetric scission of fissioning $x-$th nucleus[3],

$$\sigma_{nFSL}(E_n) = \sigma_{nfSL}(E_n) + \sum_{x=1}^{X}\sum_{J\pi}\int_0^{U_{max}} W_{x+1}^{J\pi}(U)P_{fSL(x+1)}^{J\pi}(U)dU, \quad (1)$$

$W_x^{J\pi}(U)$ is the population of $(x+1)-$ th nucleus at excitation energy U after emission of x neutrons, excitation energy U^{max} is defined by E_n and energy, removed by evaporated neutrons. Modeling of first chance σ_{nfSL} is described elsewhere[3].

Nuclear level density $\rho(U, J, \pi)$ is represented as the factorized contribution of quasiparticle and collective states[18], quasiparticle level densities $\rho_{qp}(U, J, \pi)$ were calculated with a phenomenological model by Ignatyuk et al.[19] as

$$\rho(U, J, \pi) = K_{rot}(U)K_{vib}(U)\rho_{qp}(U, J, \pi), \quad (2)$$

where $K_{rot}(U, J)$ and $K_{vib}(U)$ are factors of rotational and vibrational enhancement. At saddle and ground state deformations $K_{rot}(U)$ is defined by the deformation order of symmetry, adopted from shell correction model calculations[20]. At excitations $U \geq U_r$ damping of rotational modes was anticipated[21]. It might be different for axially symmetric and triaxial nuclei, i.e.

$$K_{rot}^{sym}(U) = (\sigma_\perp^2 - 1)F(U) + 1, \tag{3}$$

$$K_{rot}^{asym}(U) = K_{rot}^{ax}(U)((2\sqrt{2\pi}\sigma_\parallel - 1)F(U) + 1), \tag{4}$$

$$F(U) = (1 + \exp(U - U_r)/d_r)^{-1}. \tag{5}$$

Here, $\sigma_\parallel^2$ and $\sigma_\perp^2$ are spin distribution parameters. The mass asymmetry for S1(S2)-modes at outer saddles doubles the rotational enhancement factors as defined by Eqs. (3, 4). Shell effects in level density are modelled with the shell correction δW dependence of $a-$parameter as recommended by Ignatyuk et al.[19] $a(U) = \tilde{a}(1 + \delta W f(\widetilde{U})/(\widetilde{U}))$. The value of the main $a-$parameter is defined by fitting neutron resonance spacing $\langle D_{obs} \rangle$ or systematics. We assume $\tilde{a}_n = \tilde{a}_f$, then a_f/a_n ratio depends on $\delta W_{f(n)}$, taken from [22] (δW_n) and [23] (δW_f).

3. Analysis

Observed fission cross section $\sigma_{nF} = \sigma_{nFSL} + \sigma_{nF(S1+S2)}$ depends on contributions of emissive fission to both terms. These contributions are strongly dependent on the asymptotic value $\tilde{a}_f(A)$ of a_f parameter, while branching ratio r^{sym} depends both on the contributions of fission chances and damping of triaxial collective modes for SL- mode. Contributions of fission chances to σ_{nF} are affected by decreasing with energy the asymptotic value of a_f parameter as

$$\tilde{a}_f(U, A) = \tilde{a}_f(A) \left(1 - 0.1 \left(\frac{U - 20}{U}\right)^{1/4}\right). \tag{6}$$

This decrease leads to major redistribution of contributions of chance fission reactions to the observed fission cross sections of ^{238}U(n,F) and ^{235}U(n,F). SL-mode B fission barrier parameters for the ^{239}U and ^{236}U fissioning nuclei for SL fission modes were derived[3] to be higher than those of asymmetric modes by $\sim$3.5 MeV, while $\hbar\omega_{BSL}$ =2.25 MeV. The contributions of lower mass U nuclides via (n,xnf) reaction to the observed ^{239}U symmetric fission might be obtained assuming for them the same difference of the outer barriers for symmetric SL and asymmetric S1(S2) fission modes, shell correction values being defined as $(\delta W_{fBSL} - \delta W_{fBS1(S2)}) \sim$3.5 MeV, $\delta W_{fS1(S2)} \sim$0.6 MeV is assumed[23]. Uranium inner and outer fission barrier parameters, relevant for saddle asymmetries, predicted by SCM calculations[20], were defined in[24,25].

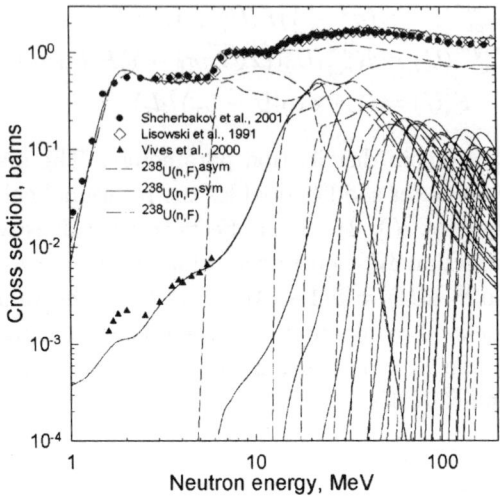

Figure 1. ^{238}U(n,F) reaction cross section.

Figures 1and 2 show calculated symmetric $^{235(238)}$U(n,F)^{sym}, $^{235(238)}$U(n,F)^{asym} asymmetric and symmetric+asymmetric $^{235(238)}$U(n,F) fission cross sections for damping of triaxial deformations with $U_r = 30$ MeV and $d_r = 10$ MeV (Eqs. (4,5)) at outer SL-mode saddles and energy dependent asymptotic $\tilde{a}_f(U, A)$ of a_f parameter. Relative contributions of σ_{nF}^{sym} and σ_{nF}^{asym} to the observed fission cross section σ_{nF} could be controlled by comparing calculated branching ratio r^{sym} with measured data by Zoller et al.[10](see Fig. 3). Relative contributions of fission chances with low and high number of pre-fission neutrons make a major influence on energy dependence of r^{sym} at $E_n \gtrsim 25$ MeV. When energy independent asymptotic $\tilde{a}_f(A)$ of a_f parameter is employed, lower chances make predominant contribution to the observed fission cross section, hence in "no triaxial damping" case calculated branching ratio r^{sym} would be systematically much higher, than data by Zoller et al.[10] predict, triaxial damping in this case produces completely wrong r^{sym} shape (dot-dashed curve on Fig. 3). There is a step-like structure in measured r^{sym} data around $E_n \sim$ 30 MeV, it is due to the contribution of symmetric (second-chance) fission of ^{238}U nuclide after emission of PE pre-fission neutron. Different shape of r^{sym} for ^{235}U(n,F) reaction for $E_n \lesssim 90$ MeV might be attributed to higher contribution of lower fission chances.

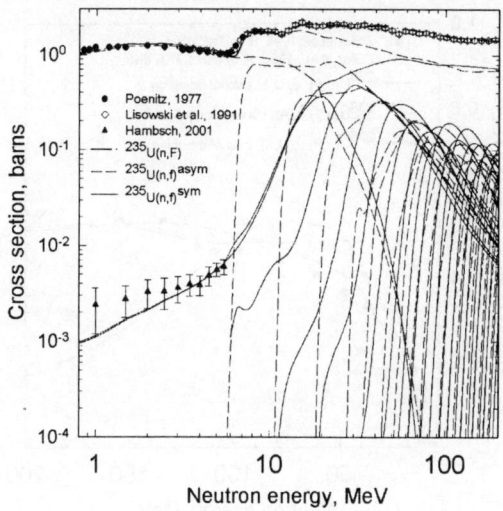

Figure 2. ^{235}U(n,F) reaction cross section.

4. Conclusions

Axial collective modes damping both for inner and outer saddle and equilibrium deformations as well as triaxial damping at SL-mode outer saddle deformations produce reasonable symmetric/asymmetric emissive fission partitioning of ^{235}U(n,F) and ^{238}U(n,F). Estimate of fission intrinsic level densities is equivalent to more fission events at lower intrinsic excitation energies, or more fissions from neutron-deficient U nuclei, i.e. from higher fission chances. Damping of triaxial deformations at outer saddle for SKL-mode is essential for the description of r^{sym} ratio[10] for ^{238}U(n,f). Triaxial damping at outer SL-mode saddle is equivalent to less symmetric fission events at higher excitation energy and more symmetric fission from neutron-deficient isotopes, as compared with "no triaxial damping" case.

References

1. F.-J. Hambsch, private communication (2001).
2. F. Vives, F.-J. Hambsch, H. Bax, S. Oberstedt, Nucl. Phys. A662 (2000) 63.
3. V.M. Maslov. and F.-J. Hambsch, Nucl. Phys. A705, 352 (2002).
4. V.M. Strutinsky, Nucl. Phys. A95 (1967) 420.
5. E. Konecny, H.J. Specht, J. Weber, Phys. Lett., 45B (1973) 329.
6. J. Weber, Britt H.C., Gavron A., et al., Phys. Rev., 13 (1976) 2413.

526

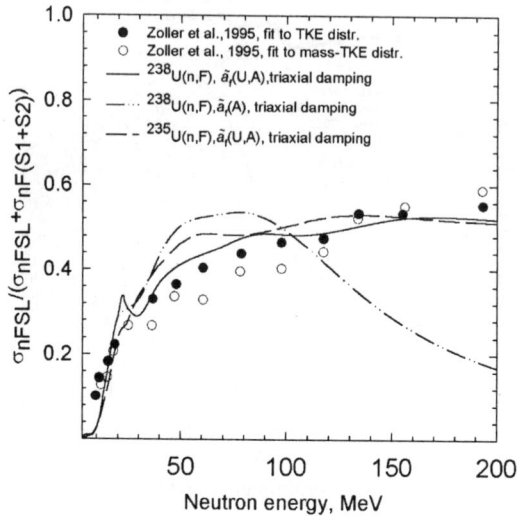

Figure 3. Branching ratio r^{sym} for ^{238}U(n,F) and ^{235}U(n,F) reactions.

7. S. Oberstedt, F.-J. Hambsch, F. Vives, Nucl. Phys. A644 (1998) 289.
8. F.-J. Hambsch, F. Vives, P. Ziegler, et al., Nucl. Phys. A679 (2000) 3.
9. U. Brosa, S. Grossmann, A. Müller, Phys. Rev. 197 (1990) 167.
10. C.M. Zoller, A. Gavron, J.P. Lestone et al., IKDA 95/25, Darmstadt
11. P. Lisowski, A.D. Carlson, O.A. Wasson et al., Proc. Specialists' Meeting on Neutron Cross Section Standards for the Energy Region above 20 MeV, Uppsala, Sweden, 21-23 May, 1991, p. 177, OECD, Paris (1991).
12. O.A. Shcherbakov et al.,Proc. Int. Conf. Nucl. Data for Sci. and Techn., Tsukuba, Japan, October, 7-12, 2001, p. 230 (2002).
13. V.M. Maslov and Yu.V. Porodzinskij, A.Hasegawa, M. Baba, ibid, p. 80.
14. A.R. Junghans, M. de Jong, H.-G. Clerc et al. Nucl. Phys. A629 (1998) 635.
15. P. Möller, D.G. Madland, A.J. Sierk et al., Nature, 409 (2001) 785.
16. M. Uhl, B. Strohmaier, IRK-76/01, IRK, Vienna (1976).
17. W.P. Abfalterer et al., Phys. Rev. C, 63, 044608 (2001).
18. A. Bohr and B. Mottelson, Nuclear Structure, v. 2, (Benjamin, NY, 1975).
19. A.V. Ignatjuk, K.Istekov, G.N. Smirenkin, Sov. J. Nucl. Phys.,29,450(1979).
20. W.M. Howard, P. Möller, At. Data and Nucl. Data Tables,25, 219 (1980).
21. G. Hansen, A.S. Jensen, Nucl. Phys. A406, 236, (1983).
22. W.O. Myers, W.J. Swiatecky, Ark. Fyzik, 36, 243 (1967).
23. S. Bjørnholm and J.E. Lynn, Rev. Mod. Phys. 52 (1980) 725.
24. Handbook for Calculations of Nuclear Reaction Data: Reference input parameter library,IAEA-TECDOC-1034, p.81, 1998, Vienna.
25. V.M. Maslov, INDC(BLR)-013, IAEA, Vienna,1998.

HEAVY ION RADIOACTIVITIES

D. N. POENARU[1,2,3], W. GREINER[2], R. A. GHERGHESCU[1,2],

J. H. HAMILTON[3] AND A. V. RAMAYYA[3]

[1] *National Institute of Physics and Nuclear Engineering,*
PO Box MG-6, RO-76900 Bucharest, Romania
E-mail: poenaru@ifin.nipne.ro
[2] *Institut für Theoretische Physik der Universität,*
Pf 111932, D-60054 Frankfurt am Main, Germany
[3] *Department of Physics, Vanderbilt University, Nashville, TN 37235, USA*

Several heavy nuclei from Fr to Cm were experimentally identified as heavy ion emitters. The measured half-lives against ^{14}C, ^{18,20}O, ^{23}F, $^{22,24-26}$Ne, ^{28,30}Mg and ^{32,34}Si spontaneous emission are in good agreement with calculations within analytical superasymmetric fission model. Only some of the most favourable cases were investigated, leading to magic or almost magic proton and neutron numbers of daughter nuclei. A systematics of experimental data compared to calculations shows other possible candidates for future experiments. Universal curves may be used to estimate the expected half-lives.

1. Introduction

Spontaneous emission of heavy ions (intermediate between ^{4}He and fission fragments) from nuclei was predicted [1] in 1980, four years before the first experimental confirmation [2]. Further studies were rapidly boosted (see [3,4] and the references therein). The obtained until now data on half-lives and branching ratios relative to α-decay of ^{14}C, ^{18,20}O, ^{23}F, $^{22,24-26}$Ne, ^{28,30}Mg and ^{32,34}Si radioactivities are in good agreement with predicted values within the analytical superasymmetric fission (ASAF) model, as we will show below.

Four theoretical models with predicting power were reviewed in 1980: fragmentation theory; penetrability calculations like in traditional theory of α-decay; numerical (NuSAF)- and analytical (ASAF) superasymmetric fission models. A new superasymmetric peak, experimentally confirmed as a "sholder", has been obtained in the ^{252}No fission fragment mass distribution calculation, based on the fragmentation theory and the two center shell model developed by the Frankfurt school. One of the eight decay modes by cluster emission, predicted in 1980 by calculating the penetrability, from sixteen even-even parents, has been ^{14}C decay of ^{222,224}Ra. Three variants of the numerical superasymmetric fission models were developed since 1979 by adding to the macroscopic deformation energy of binary systems with different charge densities a phenomenological shell correction term, and by performing numerical calculations within WKB approximation.

A very large number of combinations parent – emitted cluster had to be considered in a systematic search for new decay modes. In order to check the metastability of more than 2000 nuclides with measured masses tabulated by Wapstra and Audi, against about 200 isotopes of the elements with $Z_e = 2$–28, this number is of the order of 10^5. The numerical calculation of three-fold integrals involved in the models mentioned above were too time-consuming. The large amount of computations can be performed in a reasonable time by using an analytical relationship for the halflife. Since 1980 the ASAF model was developed to fulfil this requirement.

Besides the half-life predictions in several papers since 1984, there are three large tables, two [5,6] published and one [7] unpublished. In the last one cold fission is included too and the systematics is extended in the region of heavier emitted clusters (mass numbers $A_e > 24$), and of parent nuclei far from stability and superheavies.

The unified approach of cold binary fission, cluster radioactivity, and α-decay was extended to cold ternary fission [8] and to multicluster fission [9]. Microscopic approaches [10,11] were published as well. In the region of highly excited states above the barrier, Moretto's model [12] describes in a unified way the light particle emission (evaporation) and fission. In this work we present a systematics of experimental data compared to calculations, from which one may suggest [13] possible new candidates for future experiments. Universal curves can be used to estimate the expected half-lives.

2. Systematics and New Candidates

The experimentally determined quantities are the partial half-life, T, and the kinetic energy of the emitted cluster, $E_k = Q A_d / A$, where Q is the released energy, and A_d and A are the mass numbers of the daughter and parent nuclei.

A parent nucleus with atomic number Z and mass number A decays into an emitted cluster A_e, Z_e and a daughter A_d, Z_d, i.e. $^A Z \to {}^{A_e} Z_e + {}^{A_d} Z_d$, with conservation of hadron numbers (neutrons and protons): $N = N_e + N_d$ and $Z = Z_e + Z_d$, where $A = N + Z$, and similar relationships hold for A_e

and A_d. The process is energetically allowed if the released energy,

$$Q = M(A, Z) - [M_e(A_e, Z_e) + M_d(A_d, Z_d)] \qquad (1)$$

is a positive quantity, $Q > 0$. The atomic masses M, M_e, and M_d in units of energy, are taken from tables of experimental values.

With the SOLENO spectrometer it was possible to discover [14] the fine structure in cluster decay [15] and to perform [16] the most accurate experiment by using high quality implanted sources [17] of Ra isotopes made at the ISOLDE mass separator, CERN, Geneva. Similarly, an Enge split-pole magnetic spectrometer with a gas filled detector in its focal plane has been used at Argonne [18] to confirm the mass number of the emitted ^{14}C fragment from ^{223}Ra. Another method extensively used [19] is based on solid state nuclear track detectors (SSNTD) [20,21].

The data obtained until now (see the reviews [19,18,17,21], the references therein, and the recently published papers [22,23,24,25]) on half-lives and branching ratios against spontaneous emission of ^{14}C, ^{18,20}O, ^{23}F, $^{22,24-26}$Ne, ^{28,30}Mg and ^{32,34}Si from trans-Francium nuclei are in good agreement with predicted values from the ASAF model [13].

In Figure 1 we selected four cluster radioactivities to illustrate the shell effects and to observe that further experiments have a good chance to be performed. More than one cluster decay mode were detected for some isotopes of Pa, U, and Pu. The cluster emitters ^{221}Fr, $^{221-224,226}$Ra, ^{225}Ac, ^{228,230}Th, ^{231}Pa, $^{230,232-236}$U, ^{236,238}Pu, and ^{242}Cm are either β-stable or not far from stability nuclei. The Green approximation for the line of β-stability crosses the following Z, N pairs: 87,133; 88,135; 89,136; 90,138; 91,140; 92,142; 93,144; 94,146; 95,148, and 96,150. The α-decay half-lives are taken from tables of experimental data when available, or otherwise calculated with a semi-empirical formula as in Ref.[6].

There is a strong competition of α-decay. While $10^{11} < T < 10^{30}$ and $10^{1.5} < T_\alpha < 10^{18}$, the branching ratio $10^{-16} < b_\alpha < 10^{-8}$. Spontaneous fission [26] starts to be important in the region of heavy nuclei with $10^{14} < T_f < 10^{29}$. For Pa, U, Np, Am, and Pu isotopes the branching ratio $b_f = T_f/T$ is in the range $(10^{-7}, 10^2)$, but for ^{242}Cm it approaches 10^{-9}, making very difficult the measurement of ^{34}Si radioactivity [24,25].

Data for the *fine structure* of ^{14}C radioactivity of ^{223}Ra (see the reviews [17,3]) were not included. Unlike in α-decay, where the initial and final states of the parent and daughter are not so different from one another, in cluster radioactivities of odd-mass nuclides, one has a unique possibility to study a transition from a well deformed parent nucleus with complex configuration mixing, to a spherical nucleus with a pure shell model wave function. It can be used as a spectroscopic tool to obtain direct information on spherical components of deformed states.

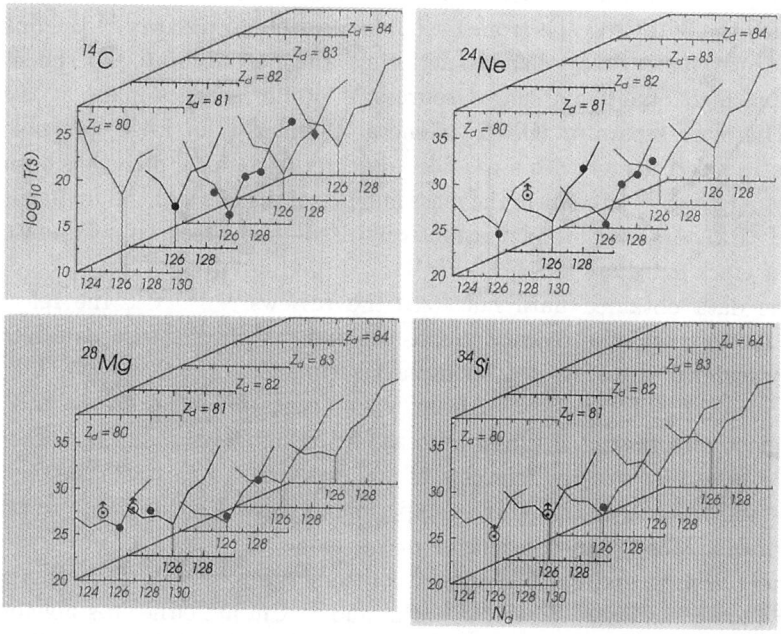

Figure 1. Predicted half-lives within ASAF model (lines) and measurements (points) for four kinds of cluster decay modes versus neutron number of the daughter nucleus. The calculations belonging to the same atomic number of the daughter are joined with a line for $Z_d = 80 - 84$. The experimentally determined upper limits are marked with a vertical arrow. They correspond to: $Z_d = 80$ for ^{24}Ne and ^{28}Mg radioactivity. The two limits for ^{34}Si radioactivity refer to $Z_d = 80$ and to $Z_d = 81$.

A strong shell effect can be seen in Figure 1: as a rule the shortest value of the half-life is obtained when the daughter nucleus has a magic number of neutrons ($N_d = 126$) and protons ($Z_d = 82$). There are few measurements for neighbouring daughters with $Z_d = 80$ or $Z_d = 81$, and only one for $Z_d = 83$.

Possible candidates for future experiments [13] are: ^{220,222,223}Fr, ^{223,224}Ac, and ^{225}Th as ^{14}C emitters; ^{229}Th for ^{20}O radioactivity; ^{229}Pa for ^{22}Ne decay mode; ^{230,232}Pa, ^{231}U, and ^{233}Np for ^{24}Ne radioactivity;

^{234}Pu for ^{26}Mg decay mode; ^{234,235}Np and ^{235,237}Pu as ^{28}Mg emitters, as well as ^{238,239}Am and $^{239-241}$Cm for ^{32}Si radioactivity. Also ^{33}Si decay of ^{241}Cm could be observed.

3. Universal Curves

The (measurable) decay constant $\lambda = \ln 2/T$, can be expressed as a product of three (model dependent) quantities

$$\lambda = \nu S P_s \tag{2}$$

where ν is the frequency of assaults on the barrier per second, S is the preformation probability of the cluster at the nuclear surface, and P_s is the quantum penetrability of the external potential barrier. For α-decay and cluster radioactivities the penetrabilty dominates the half-life variation with A. The frequency ν remains practically constant, the preformation differs from one decay mode to another but it is not changed very much for a given radioactivity, while the general trend of penetrability follows closely that of the half-life. The external part of the barrier (for separated fragments), essentially of Coulomb nature, is much wider than the internal one (still overlapping fragments). The scission configuration is also very important for spontaneous and induced fission phenomena [27].

Consequently, both fission-like and α-like models which take into consideration the external part of the barrier in the same manner, can provide a successful explanation for the measured half-lives.

According to [28], the preformation probability can be calculated within a fission model as a penetrabilty of the internal part of the barrier, which corresponds to still overlapping fragments

$$S = exp(-K_{ov}) \ ; \ K_{ov} = \frac{2}{\hbar} \int_{R_a}^{R_t} \sqrt{2B(R)E(R)} dR \tag{3}$$

where R_a is the internal turning point $(E(R_a) = 0)$, $R_t = R_1 + R_2$ is the separation distance of two fragments at the touching point configuration (scission), $B(R)$ is the nuclear inertia, and $E(R)$ is the deformation energy from which the Q-value was subtracted out.

By taking into account the above mentioned arguments, one may assume as a first approximation, that preformation probability only depends on the mass number of the emitted cluster, A_e, in the following manner:

$$\log S = \frac{(A_e - 1)}{3} \log S_\alpha \tag{4}$$

532

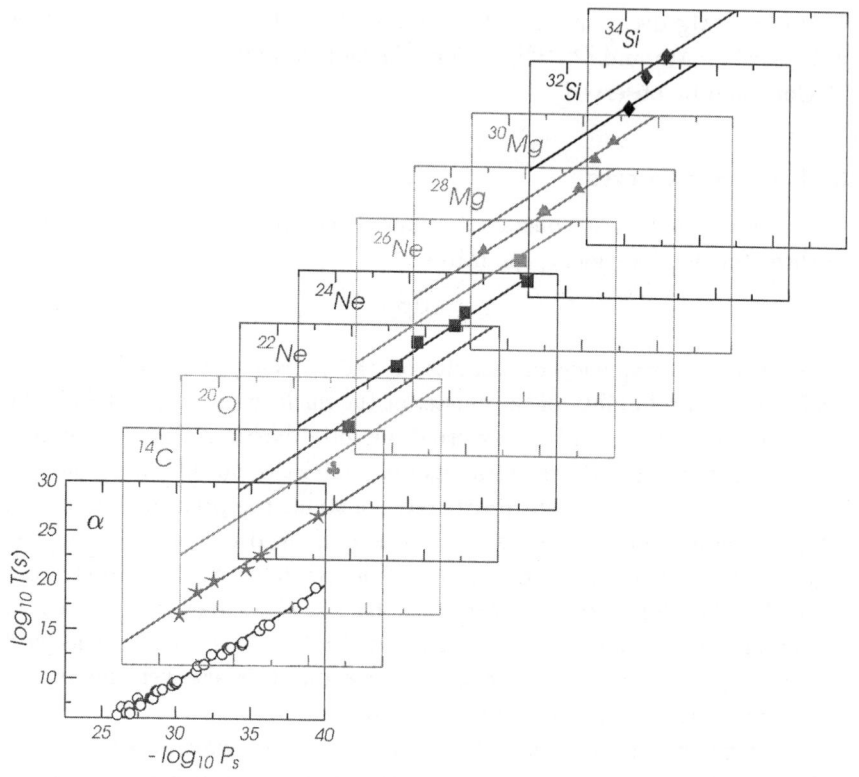

Figure 2. Universal curves and experimental points for heavy ion radioactivities and α-decay.

where the preformation probability of the α particle S_α may be determined by fit to experimental data. The next assumption is that $\nu(A_e, Z_e, A_d, Z_d) = $ constant. From the fit one obtains: $S_\alpha = 0.0160694$ and $\nu = 10^{22.01} \text{ s}^{-1}$.

In this way one arrives at a single straight line *universal curve* on a double logarithmic scale

$$\log T = -\log P_s - 22.169 + 0.598(A_e - 1) \tag{5}$$

where

$$-\log P_s = c_{AZ}\left[\arccos\sqrt{r} - \sqrt{r(1-r)}\right] \tag{6}$$

with $c_{AZ} = 0.22873(\mu_A Z_d Z_e R_b)^{1/2}$, $r = R_t/R_b$, $R_t = 1.2249(A_d^{1/3} + A_e^{1/3})$, $R_b = 1.43998 Z_d Z_e/Q$, and $\mu_A = A_d A_e/A$. For all measurements performed until now the agreement is good, as can be seen in Figure 2. Usually a smooth behaviour can only be obtained for even-even nuclei. Nevertheless, with one exception (^{14}C radioactivity of ^{223}Ra for which the fine structure was observed), the measured parent nuclei with odd neutron or proton numbers were included on this plot, and they behave like the even-even ones.

Sometimes this universal curve is misinterpreted as being a Geiger-Nuttal plot. Geiger and Nuttal found in 1911 an empirical dependence of the α-decay partial half-life on the α-particle range in air. Nowadays this would correspond to a diagram of $\log T$ versus $Q^{-1/2}$. In this kind of systematics the experimental points are considerably scattered.

In *conclusion* up to now the ASAF model predictions have been confirmed. The strong shell effects of the daughter ^{208}Pb were not fully exploited. New searches for cluster decay modes can be made. Possible candidates are ^{220,222,223}Fr, ^{223,224}Ac, and ^{225}Th as ^{14}C emitters; ^{229}Th for ^{20}O radioactivity; ^{229}Pa for ^{22}Ne decay mode; ^{230,232}Pa, ^{231}U, and ^{233}Np for ^{24}Ne radioactivity; ^{234}Pu for ^{26}Mg decay mode; ^{234,235}Np and ^{235,237}Pu as ^{28}Mg emitters, as well as ^{238,239}Am and $^{239-241}$Cm for ^{32}Si radioactivity. Also ^{33}Si decay of ^{241}Cm could be observed. Universal curves provide a means to obtain rapidly estimates of the expected half-lives.

Acknowledgments

This work was partly supported by UNESCO (UVE-ROSTE Contract 875.737.2), by the Centre of Excellence IDRANAP under contract ICA1-CT-2000-70023 with European Commission, Brussels, by Bundesministerium für Bildung und Forschung (BMBF), Bonn, Gesellschaft für Schwerionenforschung (GSI), Darmstadt, by Ministry of Education and Research, Bucharest. The work at Vanderbilt University was supported by U.S. Department of Energy under Grant No. DE-FG05-88ER40407.

References

1. A. Săndulescu, A., D. N. Poenaru and W. Greiner, *Sov. J. Part. Nucl.* **11**, 528 (1980). A. Săndulescu, A., D. N. Poenaru, W. Greiner and J. H. Hamilton, *Phys. Rev. Lett.*, **54**, 490 (1985).
2. H. J. Rose and G. A. Jones, *Nature*, **307**, 245 (1984).
3. D. N. Poenaru and W. Greiner, Eds, *Nuclear Decay Modes*, (Institute of Physics, Bristol, 1996).
4. D. N. Poenaru and W. Greiner, Eds, *Handbook of Nuclear Properties*, (Clarendon Press, Oxford, 1996); *Experimental Techniques in Nuclear Physics*, (Walter de Gruyter, Berlin, 1997).

534

5. D. N. Poenaru, W. Greiner, K. Depta, M. Ivaşcu, D. Mazilu, and A. Săndulescu, *Atomic Data Nucl. Data Tables*, **34**, 423 (1986).
6. D. N. Poenaru, D. Schnabel, W. Greiner, D. Mazilu, and R. Gherghescu, *Atomic Data Nucl. Data Tables*, **48**, 231 (1991).
7. D. N. Poenaru, M. Ivaşcu, D. Mazilu, R. Gherghescu, K. Depta, and W. Greiner, Report NP-54, Central Inst. Phys., Bucharest (1986).
8. D. N. Poenaru, B. Dobrescu, W. Greiner, J. H. Hamilton and A. V Ramayya, *J. Phys. G: Nucl. Part. Phys.*, **26**, L97 (2000).
9. D. N. Poenaru, W. Greiner, J. H. Hamilton, A. V Ramayya, E. Hourany and R. Gherghescu, *Phys. Rev.*, **C59**, 3457 (1999).
10. R. Blendowske, T. Fliessbach and H. Walliser, chap. 7 in Ref.[3] p. 337.
11. R. G. Lovas, R. J. Liotta, A. Insolia, K. Varga, and D. S. Delion, *Phys. Rep.*, **294**, 265 (1998).
12. L. G. Moretto, *Nucl. Phys.*, **A247**, 211 (1975).
13. D. N. Poenaru, Y. Nagame, R. A. Gherghescu and W. Greiner, *Phys. Rev.*, **C65**, 054308 (2002); Proc. of the 4th Catania Relativistic Ion Studies on Exotic Clustering (American Institute of Physics, New York, 2002) Eds. S. Costa, A. Insolia and C. Tuve, in print.
14. L. Brillard, A. G. Elayi, E. Hourani, M. Hussonnois, J. F. Le Du, L. H. Rosier and L. Stab, *C. R. Acad. Sci. Paris*, **309**, 1105 (1989).
15. M. Greiner and W. Scheid, *J. Phys. G: Nucl. Phys.*, **12**, L229 (1986).
16. E. Hourany et al. *Phys. Rev.*, *C*, **52**, 267 (1995).
17. E. Hourany, chap. 8 in Ref.[3], p. 350.
18. W. Henning and W. Kutschera, chap. 7 in *Particle Emission from Nuclei*, Vol. II, (CRC Press, Boca Raton, 1989) p. 188.
19. R. L. Fleischer, P. B. Price, and R. M. Walker, *Nuclear Tracks in Solids. Principles and Applications*, (Univ. of California, Berkeley, 1975); P. B. Price and S. W. Barwick, Ch. 8 in *Particle Emission from Nuclei*, Vol. II, (CRC Press, Boca Raton, 1989), p. 205.
20. S. P. Tretyakova, *Sov. J. Part. Nucl.*, **23**, 156 (1992).
21. R. Bonetti and A. Guglielmetti, chap. 9 in Ref.[3] p. 370.
22. Qiangyan Pan et al., *Phys. Rev.*, **C62**, 044612 (2000).
23. R. Bonetti, C. Carbonini, A. Guglielmetti, M. Hussonnois, D. Trubert and C. Le Naour, *Nucl. Phys.*, **A686**, 64 (2001).
24. S. P. Tretyakova, R. Bonetti, A. Golovchenko et al., *Radiat. Meas.*, **34**, 241 (2001).
25. A. A. Ogloblin et al., *Phys. Rev.*, **C61**, 034301 (2000).
26. D. C. Hoffman, T. M. Hamilton and M. R. Lane, chap. 10 in Ref.[3] p. 393.
27. Nagame, Y. et al. *Phys. Lett.*, **B387**, 26 (1996).
28. D. N. Poenaru and W. Greiner, *Physica Scripta*, **44**, 427 (1991).

SPONTANEOUS FISSION OF ^{252}CF IN THE LIGHT OF PROMPT GAMMA RAYS

G.M. TER-AKOPIAN[1,2], J.H. HAMILTON[3], A.V. RAMAYYA[3], A.V. DANIEL[1],
G.S. POPEKO[1], A.S. FOMICHEV[1,2], A.M. RODIN[1], YU.TS. OGANESSIAN[1],
J.D. COLE[4], J. KORMICKI[3], J.K. HWANG[3], D. FONG[3], P. GORE[3],
J.O. RASMUSSEN[5], A.O. MACCHIAVELLI[5], I.Y. LEE[5], M. A.STOYER[6],
W. GREINER[7], R. DONANGELO[8], M. JANDEL[1,2], L. KRUPA[1,2], J. KLIMAN[1]

[1]*Flerov Laboratory of Nuclear Reaction, JINR, Dubna, 141980, Russia*

[2]*Joint Institute for Heavy Ion Research, Oak Ridge, TN 37835, USA*

[3]*Department of Physics, Vanderbilt University Nashville, TN 37235, USA*

[4]*Idaho National Engineering and Environmental Laboratory, Idaho Falls, ID 83415, USA*

[5]*Lawrence Berkeley National Laboratory, Berkeley, CA 94720, USA*

[6]*Lawrence Livermore National Laboratory Livermore, CA 94550, USA*

[7]*Institut für Theoretische Physik der Universität Frankfurt, Postfach 11 19 12, Germany*

[8]*Instituto de Fisica Unversidade Federal do Rio de Janeiro, Rio de Janeiro, 21945-970, Brazil*

In this paper we report on a study made at Gammasphere for the ternary fission of ^{252}Cf with the use of triggering signals from ternary light charged particles (LCP's).

1. Introduction

Since early 60's the spectroscopic measurements of prompt γ rays have been used for the study of the nuclear fission process. The first systematical data on the intensities of γ-ray cascades and the fragment independent yields and angular momenta measured for the spontaneous fission of ^{252}Cf were reported by the Berkeley group[1,2]. Later on transition intensities between yrast levels have been used by other authors to study the fragment angular momentum in fission[3-5].

Possibilities offered by the modern 4B Ge detector spectrometers and knowledge of nuclear levels populated in fission for a large number of fragments allow one to conduct the study of nuclear fission by analyzing

536

multiple characteristic γ-rays emitted by pairs of secondary fission fragments. We used this method to obtain the independent yields of fragment pairs emitted in the spontaneous fission of ^{252}Cf[6-17]. Independent yields of fragment pairs of different charge split and fragment angular momentum values were obtained from data collected in experiments carried out by our collaboration. By unfolding the data on the yields of fragment pairs we obtained the mass and excitation energy distributions of primary fission fragments. Rare processes, cold (neutronless) binary and ternary fission modes of ^{252}Cf were experimentally observed by the triple γ coincidence technique[10,12,15,16].

In this paper we report on our study made at Gammasphere for the ternary fission of ^{252}Cf with the use of triggering signals from ternary LCP's. Results obtained in the first experiment of this type were reported earlier[16,17]. Here we present some results of the second experiment, which was carried out in Berkeley in 2001. To further investigate the triggering conditions from fission fragments and ternary LCP's we performed a supplementary, test experiment in Dubna. Results deduced from the last Berkeley experiment are also discussed in other papers[18,19] presented at this conference.

2. Experiments

Paper[18] presented at this conference describes well the experimental setup used.

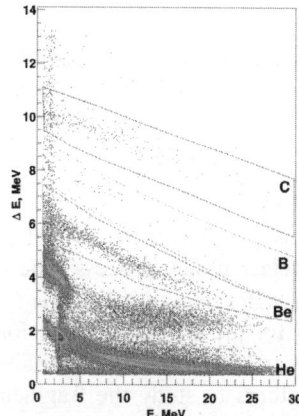

Figure 1: An example of the)E×E plot. Presented are data collected by one detector telescope during the whole measurement time.

In the two-dimensional)E×E plot shown in Fig. 1, one can see the loci of ternary helium and heavier LCP's. The numbers of detected LCP's and their observation conditions are specified in Ref.[18].

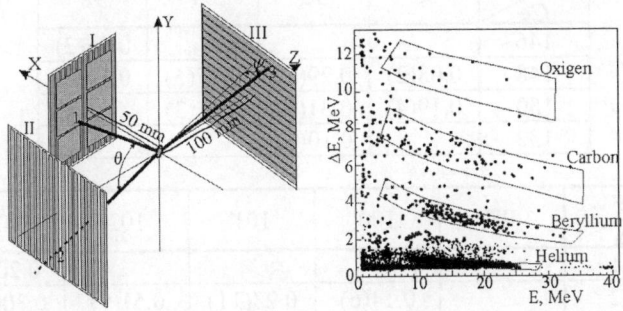

Figure 2: Left panel: Set-up used in the test experiment. The Cf source is in the origin of the coordinate system. Detector telescopes I consist of six ΔE 10 μ thick detectors and one common Si strip detector. Silicon strip detectors II and III produce trigger signals from fission fragments emitted in coincidence with ternary LCP's. Right panel: the)E×E plot obtained with one of the detector telescopes shown in the left panel.

Schematic diagram in Fig. 2 (left panel) outlines the set-up used in the Dubna test experiment. The intensity of the [252]Cf source used in this case was 45,000 spontaneous fission events per second. Except the 1.8μ Ti source backing side, there were no other foils slowing down fission fragments and ternary LCP's. An example of the)E×E plot obtained with one of the detector telescopes of this set-up is shown in Fig. 2 (right panel).

3. New results on the [252]Cf ternary fission from the Berkeley, Gammasphere experiment

The two-dimensional $E_{\gamma1} \times E_{\gamma2}$ spectrum built for the He ternary fission events revealed more than 150 peaks coincidence peaks. From peak values obtained for the ground state γ transitions we obtained the independent yields of fragment pairs. The yields of fragment pairs obtained for five charge splits of [252]Cf are presented in Tables 1–4.

By summing the data of Tables 1–4 over the rows and columns one obtains the independent yields of individual fragment emitted in the He ternary fission. Such yields are shown in Fig. 3 (left panel) as a function of the fragment mass.

We identified 72 peaks recorded in coincidence with Be LCP's. Independent yields obtained for the first time for 38 fragments are presented in

Tables 1 – 4: Independent yields of fragment pairs obtained in four charge splits of ^{252}Cf occurring in the He ternary fission.

Ce \ Sr	95	96	97	98
146				0.06(2)
148	0.08(2)	0.29(5)	0.17(5)	0.19(6)
150	0.19(4)	0.31(3)	0.12(3)	
152		0.10(3)		

Ba \ Zr	98	100	101	102	103
141					0.20(8)
142		0.14(6)	0.23(11)	0.51(9)	0.30(10)
143			0.86(40)	0.94(22)	0.30(17)
144		0.96(6)	1.33(45)	1.17(22)	0.18(7)
145	0.13(8)	0.70(17)	0.81(40)	0.65(17)	
146	0.22(4)	0.60(5)	0.43(20)		
147	0.16(8)	0.30(9)			
148	0.07(5)				

Xe \ Mo	104	105	106	107	108
136					0.05(2)
137			0.20(3)	0.08(3)	0.29(3)
138		0.63(30)	1.00(7)	0.31(9)	0.76(9)
139		0.66(23)	0.75(6)	0.31(5)	0.37(4)
140	0.77(5)	1.50(60)	1.24(7)	0.07(3)	
141	0.18(5)	0.45(18)	0.21(4)		
142	0.39(4)	0.20(10)			

Te \ Ru	109	110	111	112
134	0.11(4)	0.43(4)	0.14(2)	0.12(2)
135		0.13(2)		
136	0.21(8	0.68(11)		

Fig. 3 (right panel). The number of γ transitions identified in the carbon ternary fission was 70. Independent yields were obtained for the first time for 35 fission fragments. These results are presented in Fig. 4 (left panel). The right panel in this figure shows the charge distributions for the first time obtained for ternary fission fragments of ^{252}Cf.

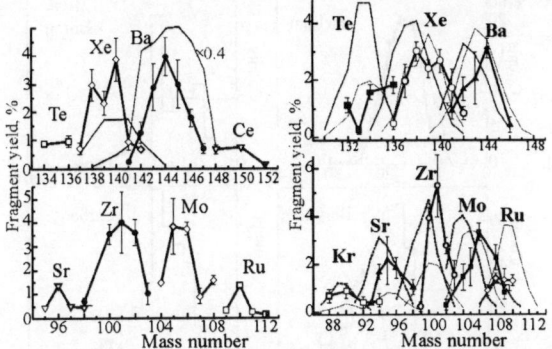

Figure 3: Independent yields of fission fragments in ^{252}Cf ternary fission. The data points are connected with thick solid lines. Left panel: α ternary fission. Thin dashed lines show literature data[4,5] for the binary fission of ^{248}Cm. Right panel: Be ternary fission. The thin solid and dashed lines show literary data[201] for the binary fission of ^{252}Cf and ^{240}Pu*, respectively.

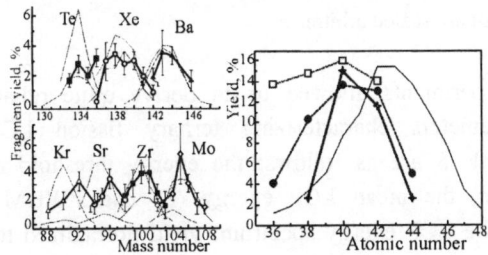

Figure 4: Left panel: Independent yields of fission fragments in ^{252}Cf C ternary fission. The data points are connected with thick solid lines. The thin solid and dashed lines show literature data[20] for the binary fission of ^{252}Cf and ^{236}U*, respectively. Right panel: The yields of fission fragments of different atomic number obtained for the ^{252}Cf spontaneous fission. The thin solid line shows the binary fission data[20]. The thick solid, dashed and dotted lines are for the He, Be and C ternary fission.

4. Characteristics of ternary LCP's

Combined data obtained in the two experiments (in Berkeley and in Dubna) give information on the energy spectra and yields of LCP's. Complementary data on the energy spectra obtained are presented in Fig. 5.

540

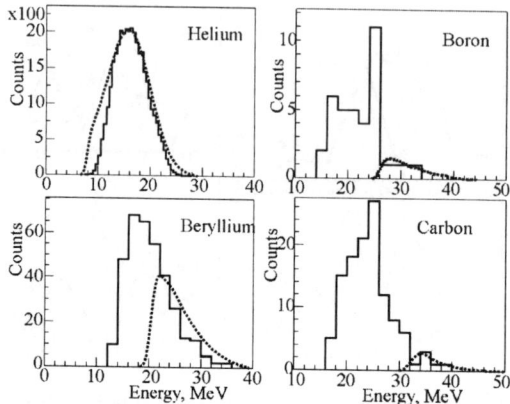

Figure 5: Kinetic energy spectra of ternary He, Be, B and C LCP's obtained in two experiments. The data of experiments made in Berkeley and Dubna are shown, respectively, by dotted curves and solid line histograms. Vertical scales show the count numbers obtained in the Dubna experiment. The much more abundant data of the Berkeley experiment are scaled arbitrarily.

The two experiments allowed us to derive quite reliable data on the kinematical parameters characterizing ternary fission LCP's. Parameters presented in Table 5 are as follows: the energy threshold achieved for the particle detection, the mean LCP energy $\langle E \rangle$ and FWHM obtained for a Gaussian fit of the LCP energy spectrum, yield normalized to 10^4 He LCP's, mean angle $\langle \theta \rangle$ between the trajectories of the LCP and light fragment, and mean angle $\langle \Psi \rangle$ defining the fold originated between the trajectories of the light and heavy fragments due to the LCP emission.

Table 5: Some LCP characteristics.

LCP	He	Be	C	Experiment
Threshold	9	20	32	Berkeley
energy, MeV	9	13	16	Dubna
$\langle E \rangle$, MeV				Berkeley
	15.7	18.0	23.4	Dubna
FWHM, MeV	10.0	17.0	18.7	Berkeley
	10.5	15.2	19.1	Dubna
Yield	10^4	166	103	Berkeley
		171	70	Dubna
$\langle \theta \rangle$, degrees	83.7	83.7	83	Dubna
$\langle \Psi \rangle$, degrees	4.9	8.7	11.3	

5. Conclusion and outlook

Results of the Gammasphere experiment presented above are essentially new and deserve a detailed comparison with other experimental data and theory predictions made. Of special interest are data on the LCP yields in excited states obtained in this experiment for the first time[18].

Future spontaneous fission experiments made with the detection of prompt γ rays are highly desirable. The present results assure us that, working at Gammasphere with an improved setup of the type shown in Fig. 2, one can acquire γ ray spectra specified by fixed fragment masses and total kinetic energies obtained in binary fission and similar spectra sorted according the types and energies of ternary fission LCP's. Within a two week experiment one can collect data for up to 2×10^9 binary fission events and 1×10^5 ternary events associated with the emission of Be LCP's. The use of the folding angle Ψ (see its notation in Fig. 2 and in Table 5) as another marker separating LCP's will considerably enhance the power of the experimental method.

Acknowledgments

Research at Vanderbilt University is supported in part by the U.S, Department of Energy under grant No. DE-FG05-88ER40407. Work at Idaho National Laboratory is supported by the U.S. Department of Energy under Contract No. DE-AC07-76ID01570.The work at LLNL was performed under the auspices the U.S. Department of Energy under Contract No. W-7405-Eng-48, and that of LBNL under contract No. DE-AC03-76SF00098. The Joint Institute for Heavy Ion Research is supported by the member institutes, the University of Tennessee, Vanderbilt University and the U.S. Department of Energy. Work at Joint Institute for Nuclear Research is supported in part by the U.S. Department of Energy, contract #DE-AC011-00NN4125, BBWI Agreement No. 3498 (CRDF grant RPO-10301-INEEL) and by the joint RFBR – DFG grant (RFBR No. 02-02-04004, DFG No. 436RUS 113/673/0-1(R)).

References

1. E. Cheifetz et al., *Phys. Rev. C*, **4**, 1913 (1971).
2. J.B. Wilhelmy et al., *Phys, Rev. C*, **6**, 2041 (1972).
3. Y Abdelrahman et al., *Phys. Lett. B*, **199**, 504 (1987).
4. J.L. Durell, *Proc. 3rd Int. Conf. DANF'96, Casta-Papiernicka, Slovakia 1996*, eds. J. Kliman and B. Pustylnik (1996, Dubna, JINR) p. 270.

542

5. J.L. Durell, *Proc. Int. Conf. "Fission and Properties of Neutron-Rich Nuclei"*, J.H. Hamilton, W.R. Phillips, and H.K. Carter, Eds. (World Scientific, Singapore, 2000) p. 106.

6. J.H. Hamilton et al., *Prog. Part. Nucl. Phys.*, **35** (1995) 635.

7. G.M. Ter-Akopian et al., *Phys. Rev. C*, **55** (1997) 1146.

8. J.H. Hamilton et al., *Prog. Part. Nucl. Phys.*, **38** (1997) 273.

9. G.M. Ter-Akopian et al., *Proc. Int. Conf. "Fission and Properties of Neutron-Rich Nuclei"*, Sanibel Island, Florida, USA, 1997. Eds. J.H. Hamilton and A.V. Ramayya (1998, Singapore, World Scientific) p. 165.

10. G.S. Popeko et. al., *Proc. Int. Conf. "Fission and Properties of Neutron-Rich Nuclei"* Sanibel Island, Florida, USA, 1997. Eds. J.H. Hamilton and A.V. Ramayya (1998, Singapore, World Scientific) p. 645.

11. A.V. Ramayya et al., In: *Heavy Elements and Related Phenomena*, D.K. Gupta and W. Greiner, Eds. World Scientific, Singapore, 1999, p. 477.

12. G.M. Ter-Akopian et al., *Proc. Int. Conf. "Fission and Properties of Neutron-Rich Nuclei"*, J.H. Hamilton, W.R. Phillips, and H.K. Carter, Eds. (World Scientific, Singapore, 2000) p. 98.

13. A.V. Ramayya et. al., *Fission and Properties of Neutron-Rich Nuclei*, J.H. Hamilton, W.R. Phillips, and H.K. Carter, Eds. (World Scientific, Singapore, 2000) p. 246.

14. G.M. Ter-Akopian et al., *Proc. Int. Symp. "Advances in Nuclear Physics"*, Bucharest, Romania, 1999, eds. D. Poenaru and S. Stoica (2000, World Scientific, Singapore) p. 73.

15. E.F. Jones et al., *Yadernaya Fizika (Nucl. Phys.)*, **64** (2001) 1235.

16. A.V. Ramayya et al., *Prog. Part. Nucl. Phys.*, **46** (2001) 221

17. J.H. Hamilton et al., *Yadernaya Fizika (Nucl. Phys.)*, **65** (2002) 1235.

18. A.V. Daniel et al, paper published in these Proceedings.

19. M. Jandel et al., paper published in these Proceedings.

20. A. Wahl, *At. Data Nucl. Data Tables*, **39** (1988) 1.

INFLUENCE OF THE EXCITATION ENERGY ON THE TERNARY TRITON EMISSION PROBABILITY OF THE ^{248}Cm FISSIONING SYSTEM

O. SEROT

CEA-Cadarache, DEN/DER/SPRC/LEPh, Bat. 230, F-13108 Saint Paul lez Durance, France
E-mail: olivier.serot@cea.fr

C. WAGEMANS

Dept. of Subatomic and Radiation Physics, University of Gent, B-9000 Gent, Belgium

J. WAGEMANS*

EC-JRC-Institute for Reference Materials and Measurements, Retieseweg, B-2440 Geel, Belgium

P. GELTENBORT

Institut Laue Langevin, BP 156, F-38042 Grenoble Cedex, France

An interesting characteristic of the ternary fission process is the influence of the excitation energy (E_{exc}) on the ternary particle emission probability. Up to now, this effect was studied only for ternary alpha particles (also called Long Range Alpha (LRA) particles), comparing the LRA emission probability (noted LRA/B) for the same compound nucleus at zero excitation energy ($E_{exc}=0$, for spontaneous fission) and at an excitation energy equal to the neutron binding energy ($E_{exc}=B_n$, for thermal neutron induced fission). It was observed on the three compound nuclei investigated that the increase of the excitation energy leads to an unexpected decrease (by about 20%) of the LRA/B ratio. The aim of the present paper is to study for the first time the influence of the excitation energy on the ternary triton emission probability (noted t/B). This investigation has been carried out on the ^{248}Cm compound nucleus. In a previous measurement, a t/B-value of $(1.79\pm0.11)\times10^{-4}$ for ^{248}Cm(sf) was obtained. We report here the determination of the same quantity for ^{247}Cm(n_{th},f). The measurement was performed at the high flux reactor of the Laue Langevin Institute in Grenoble (France), where both the energy distributions of the ternary particles (tritons and alpha's) as well as their counting rates could be determined. In particular, a t/B-value of $(1.84\pm0.11)\times10^{-4}$ was found, revealing a slight increase of the t/B-value with the excitation energy.

---

*Present address: SCK-CEN, Boeretang 200, B-2400 Mol, Belgium

1. Introduction

During fission, there is a probability of producing a light charged particle which accompanies the two main fragments. This process called ternary fission is important in some nuclear industry applications as it is the main source of tritium production in reactors. In addition, there are several experimental evidences that ternary particles are emitted very close in time to the scission of the compound nucleus. Therefore, information on the fissioning nucleus at the scission point could in principle be extracted from ternary fission studies.

An interesting characteristic of this process is related to the influence of the excitation energy of the fissioning nucleus on the ternary particle emission probability. Up to now, this has been studied only for ternary alpha particles (also called Long Range Alpha (LRA) particles). This effect was measured by comparing the LRA emission probability (noted LRA/B) for the same compound nucleus at zero excitation energy (in the case of spontaneous fission) and at an excitation energy corresponding to the neutron binding energy (in the case of thermal neutron induced fission). In previous work (see [1, 2] and references therein), we obtained such data for three compound nuclei: ^{240}Pu, ^{242}Pu and ^{248}Cm. In each case a surprising effect was observed: the LRA emission probability decreases with the compound nucleus excitation energy.

The aim of this paper is to investigate the influence of the excitation energy of fissioning nucleus on the ternary triton emission probability (noted t/B). For that purpose, a measurement of LRA's and tritons emitted in the thermal neutron induced fission of ^{247}Cm was performed. These experimental data were compared to the ones obtained in the ^{248}Cm spontaneous fission [2].

2. Experimental setup

The ^{247}Cm(n_{th},f) measurement was carried out at the PF1 cold neutron guide installed at the *Institut Laue Langevin* (Grenoble, France). The neutron flux at the sample position was about 5×10^9 neutrons cm^{-2} s^{-1} with an average energy of 11 meV.

Sample

The Cm sample was prepared at the Russian Federal Nuclear Center in Arzamas. A spot curium oxide with a diameter of 15 mm and a thickness of 15μg/cm^2 was deposited on a 30 μm thick Al-foil. The isotopic composi-

Table 1. Isotopic composition of the sample.

Isotope	^{244}Cm	^{245}Cm	^{246}Cm	^{247}Cm	^{248}Cm
Abundance (%)	1.86	0.085	13.17	72.51	12.37

tion of the sample is given in Table 1. Due to the very low ^{245}Cm content, the ternary particles coming from the ^{245}Cm(n_{th},f) reaction were neglected. Those emitted from spontaneous fission decays were measured after closing the neutron beam; their contributions were also negligible.

Detection of the ternary particles

In order to increase the number of ternary particles detected, two ΔE-E telescopes were used. They were placed inside a vacuum chamber on both sides of the sample. Each telescope consisted of two Silicon Surface Barrier (SSB) detectors. The active area was 300 mm^2 for the ΔE-detectors and 450 mm^2 for the E-detectors. The depletion depth was 1500 μm for both E-detectors, and was 50 and 71 μm for the left and right ΔE-detector, respectively. These characteristics were chosen in order to permit a simultaneous detection of tritons and LRA particles. In order to stop both fission fragments and α-particles originating from the radioactive decay of the sample, an Al-foil was put between the sample and the telescope. On the right hand side, this Al-foil (25 μm thick) was placed in front of the telescope. On the left hand side, we took advantage of the Al-foil (30 μm thick) on which the sample was deposited, and therefore the left telescope was not covered. A coincidence between the ΔE-signal (proportional to the energy deposited by the ternary particle traversing the silicon) and the E-signal (proportional to the remaining ternary particle energy) was imposed for both telescopes. These coincident signals were hence digitised and stored in a PC. Note the slightly different experimental conditions on both sides of the target (characteristics of the telescopes, detection geometry and thickness of the Al-foils). Hence, a comparison of the data obtained from the two telescopes constitutes a good test to validate our results. The energy calibration of the four detectors was done using the ^{10}B(n_{th},α)^{7}Li, ^{6}Li(n_{th},α)^{3}H and ^{143}Nd(n_{th},α)^{140}Ce reactions.

Data reduction

The procedure used to identify ternary particles and to separate them from the background was the one proposed by Goulding [3]. This method is based on the difference in energy loss of different particles (characterised by a specific constant a) in the same material leading to the simple relation:

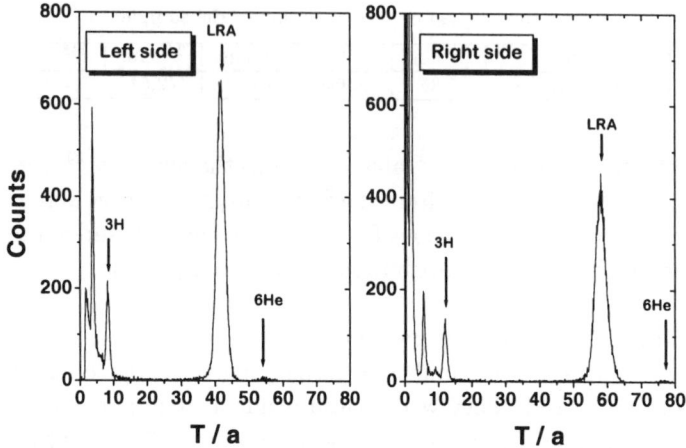

Figure 1.   T/a-spectra obtained from both telescopes. The ternary particles are nicely separated from the background.

$T/a = (E + \Delta E)^{1.73} - E^{1.73}$, where T is the effective thickness of the ΔE-detector. The T/a-spectra are plotted in Fig.1. It shows that a clear separation between ternary particles and background could be achieved. Note that the T/a-positions of the LRA and triton ternary particles are not the same on both sides, which reflects the different thickness of the ΔE-detectors used. The selection of the ternary particles was realised by putting a window on the region of interest of the T/a-spectrum. This was done for LRA and triton particles (due to the poor statistics, the ^{6}He-ternary particles were not investigated). Then, ΔE and E spectra corresponding to a given ternary particle could be obtained and the total energy distribution ($\Delta E + E + \delta E$, where δE is the calculated energy loss by the particle in the sample itself and the covering Al-foil) could be deduced. These energy distributions are plotted in Fig.2. The thresholds observed for each ternary particle are due to the thickness of the ΔE-detector as well as to the presence of the Al-foil. The average energy and the full width at half maximum of the energy distributions were obtained from a Gaussian fit performed on the experimental data.

3. Results and discussion

LRA and triton energy distributions

In order to reduce the contribution of ^{4}He-particles coming from the dis-

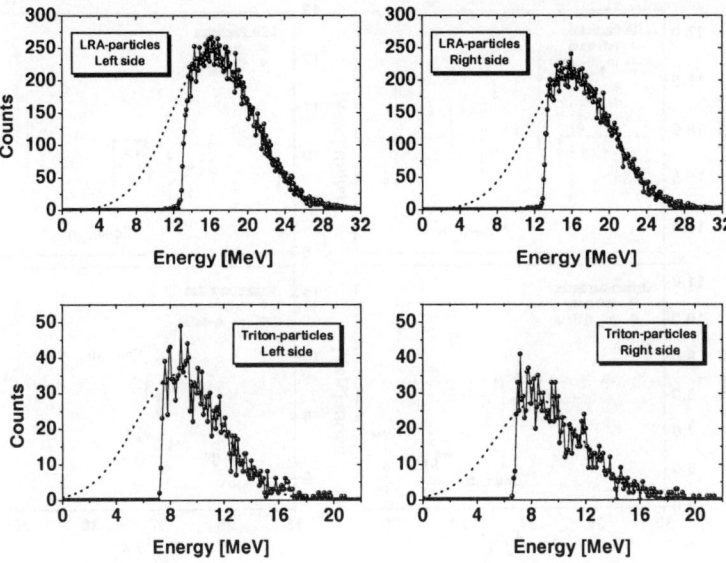

Figure 2. LRA and triton energy distributions obtained from the two telescopes. The Gaussian fits performed on the experimental data are also shown (dotted lines).

integration of ternary ^{5}He-particles (see [4, 5] for a detailed discussion on this subject), the Gaussian fits were performed from 14 MeV up to 40 MeV for each telescope. Their characteristics are compared in Table 2, showing that a good agreement between the two telescopes could be reached. The average values (weighted by the uncertainties) of the data from both sides yields an LRA average energy of 16.01±0.13 MeV and a full width at half maximum of 10.37±0.24 MeV. These results are compared (see top of Fig.3) with those obtained for other fissioning nuclei, where data from spontaneous fission decays and thermal neutron induced fission reactions are plotted. This figure confirms that within the uncertainties, the most probable LRA-energy remains remarkably constant (around 15.9 MeV) for all fissioning systems, while the full width at half maximum clearly increases with the fissility parameter (Z_{CN}^2/A_{CN}) of the compound nucleus. This behavior (which was already observed in the past) is quite well explained from trajectory calculations [1].

For the tritons, the Gaussian fits were performed on the right telescope data from 7 MeV up to 25 MeV, and on the left telescope data from 7.4 MeV up to 25 MeV. Unfortunately, on the right side the fit could be done

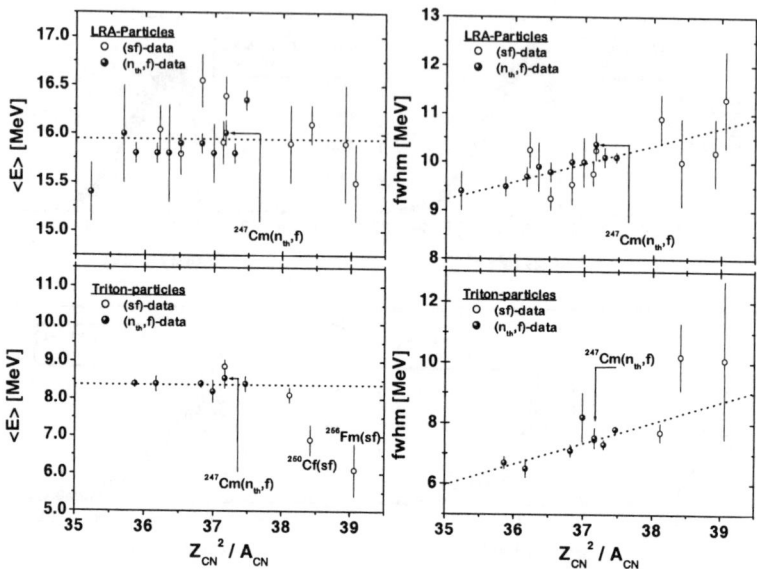

Figure 3. Most probable energy (<E>, left part) and full width at half maximum (fwhm, right part) of the energy distibutions related to the LRA (top) and tritons (bottom) as a function of Z^2_{CN}/A_{CN}. The linear fits (dotted lines) performed on the experimental data are also shown.

only by fixing the average energy, which was taken accordingly with the value deduced from the left side Gaussian fit (<E>=8.55 MeV). Despite this difficulty, both fits lead to FWHM values which are in good agreement (see Table 2). The most probable energy and the full width at half maximum for all fissioning nuclei are plotted on the bottom of Fig.3. Except for ^{250}Cf(sf) and ^{256}Fm(sf), the average energy of the tritons remains constant around 8.4 MeV for all nuclei, while their FWHM seems to increase with the fissility parameter (Z^2_{CN}/A_{CN}) as LRA-particles do.

Emission probabilities
The ratio between triton and LRA emission probabilities (noted t/LRA) was calculated for the ^{247}Cm(n_{th},f) reaction from their counting rates (see Table 2). We found (10.44±0.50)% and (9.78±0.28)% for the left and right telescope, respectively, leading to a weighted average value of t/LRA=(9.94±0.24)%. Since triton and LRA particles were detected simultaneously, only statistical errors were considered. Recently, Serot and Wagemans [2] have measured the LRA emission probability for the same re-

Table 2. Characteristics of the LRA and triton energy distributions. The uncertainties are the sum of the statistical and systematic uncertainties. The number of ternary particles detected (N_{det}) and the counting rate (CR) (deduced from the ratio between the gaussian fit integration and the measuring time) are mentioned.

		LRA	triton
Left	N_{det}	20325	1868
Left	**CR** [h^{-1}]	436.36±16.14	45.56±5.74
Telescope	**<E> [MeV]**	16.16±0.17	8.55±0.27
	fwhm [MeV]	10.37±0.34	7.44±0.52
Right	N_{det}	16855	1630
Right	**CR** [h^{-1}]	373.81±14.58	36.56±3.84
Telescope	**<E> [MeV]**	15.81±0.20	8.55
	fwhm [MeV]	10.38±0.35	7.58±0.42
Average	**<E> [MeV]**	16.01±0.13	8.55±0.27
value	**fwhm [MeV]**	10.37±0.24	7.52±0.33

action resulting in LRA/B=(1.85±0.10)×10^{-3}. Using this value, the triton emission probability could be deduced:

$$t/B = (t/LRA) \times (LRA/B) = (1.84 \pm 0.11) \times 10^{-4} \qquad (1)$$

It should be noted that this t/LRA-value shows that the constant t/LRA-value of 6.55% which has been adopted for all fissioning nuclei in the JEF2.2 fission yield evaluation file [6] is not very realistic.

Influence of the excitation energy
It is generally believed that the energy required to release ternary particles is mainly taken from the available deformation energy of the fissioning nucleus. Since the average prompt neutron multiplicity $< \nu >$ is a measure of the deformation energy, it is expected that the t/B-values are correlated with the $< \nu >$-values. This correlation can indeed be observed in Fig.4, where t/B-data from thermal neutron induced fission reactions are plotted as a function of $< \nu >$. Nevertheless, it was already pointed out that this correlation is valid only if (n_{th},f)-data and (sf)-data are considered separately. In order to overcome this difficulty, Wagemans and Serot [4] have proposed a new correlation between t/B and $< \nu >$, which is valid for spontaneous fission data as well as for thermal neutron induced fission data, if the excitation energy of the fissioning system is taken into account. In particular, they have shown that for a given fissioning nucleus, the variation

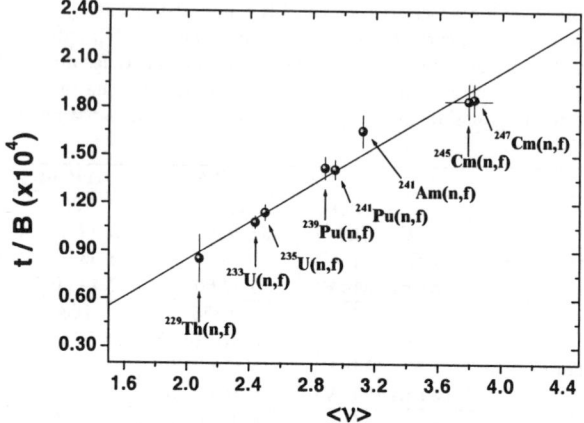

Figure 4. Triton emission probability for (n_{th},f)-data as a function of the neutron multiplicity $<\nu>$. The straight line is a fit of the experimental data.

of t/B with the excitation energy of the compound nucleus is given by:

$$(t/B)(E_{exc}) = (t/B)(E_{exc} = 0) + \frac{\partial(t/B)}{\partial E_{exc}} E_{exc} \qquad (2)$$

where $\frac{\partial(t/B)}{\partial E_{exc}}$ was found to be constant for all fissioning nuclei: $\frac{\partial(t/B)}{\partial E_{exc}} = (1.46 \pm 0.38) \times 10^{-6}$ (t/B)/MeV. Applying Eq.2 with $t/B(E_{exc} = 0) = 1.79 \times 10^{-4}$ (spontaneous fission of ^{248}Cm) and $E_{exc} = 6.213$ MeV (neutron binding energy) leads to: 1.88×10^{-4}. This calculated value is in very good agreement with our experimental result for ^{247}Cm(n_{th},f) ($(1.84 \pm 0.11) \times 10^{-4}$). As mentioned in the introduction, the LRA emission probability decreases with increasing excitation energy. Our result thus indicates that tritons behave in the opposite way. We propose the following explanation for the different behavior of LRA and triton emission:

(1) The LRA emission process seems to be mainly governed by the strong α-cluster preformation probability (S_α) [7, 8, 9]. When the fissioning nucleus is formed after capture of a neutron, the S_α-factor could be decreased, explaining the decrease of LRA/B with the excitation energy.

(2) In the triton emission process, no cluster preformation seems to be involved. After capture of a neutron, it is believed that only a small part of the excitation energy of the compound nucleus goes into the deformation energy (the other part going into intrinsic excitation

energy) [10]. So, the increase of t/B with E_{exc} is expected to be small, which is indeed observed in the present work.

4. Conclusion

The energy distribution as well as the emission probability of the LRA and triton ternary particles emitted from thermal neutron induced fission of ^{247}Cm have been measured. It has been shown that the average energy and the full width at half maximum for both ternary particles (LRA and tritons) follow the trend already observed for other fissioning systems: the average value remains constant (around 15.9 MeV for LRA-particles and 8.4 MeV for tritons) while the FWHM increases with the fissility parameter of the compound nucleus.

From a comparison of the triton emission probability obtained for ^{248}Cm(sf) and for ^{247}Cm(n_{th},f), we could observe for the first time that t/B increases slightly with the excitation energy of the fissioning nucleus. Therefore, tritons do not behave as LRA-particles. In order to confirm this result, an experiment on ^{251}Cf(n_{th},f) is planned in the near future. Data will be compared with available data for ^{252}Cf(sf).

References

1. C. Wagemans, The Nuclear Fission Process, CRC Press, Boca Raton, USA, 1991.
2. O. Serot and C. Wagemans, *Proc. Seminar on Fission Pont d'Oye IV*, Habay-la-Neuve, Belgium, 1999, eds. Wagemans C. et al., p.45.
3. F. Goulding et al., *Nucl. Instr. Meth.* **31**, 223 (1964).
4. C. Wagemans and O. Serot, *Proc. 5th Int. Conf. on Dynamical Aspects of Nuclear Fission*, Casta-Papiernicka, Slovak Republic, 2001, in press.
5. M. Mutterer et al., *Proc. 5th Int. Conf. on Dynamical Aspects of Nuclear Fission*, Casta-Papiernicka, Slovak Republic, 2001, in press.
6. R.W. Mills, in Compilation and evaluation of fission yield nuclear data, IAEA-TECDOC-1168 (2000), p.87.
7. I. Halpern, *Ann. Rev. of Nucl. Sc.* **vol 21**, 245 (1971).
8. N. Carjan, *J. Phys.* **37**, 1279 (1976).
9. O. Serot, N. Carjan and C. Wagemans, *Eur. Phys. J.* **A8**, 187 (2000).
10. A. Ruben, H. Märten and D. Seeliger, *Z. Phys.* **A338**, 67 (1991).

THEORY OF BINARY AND TERNARY COLD FISSION

D. S. DELION[1], A. SĂNDULESCU[2], Ş. MIŞICU[1,3] , AND W. GREINER[3]

[1] National Institute of Physics and Nuclear Engineering
Bucharest-Măgurele, POB MG-6, Romania
[2] Center for Advanced Studies in Physics
Romanian Academy, Calea Victoriei, 125, Bucharest, Romania
[3] Institut für Theoretische Physik, J.W.v.-Goethe
Universität, Robert Mayer str. 8-10, Frankfurt, Germany

We give a description of cold fision processes from ^{252}Cf within the stationary scattering formalism. A strong dependence of binary decay widths upon the internal structure of the considered resonant state is predicted. We describe the angular distribution of ^{4}He, ^{10}Be and ^{14}C in the spontaneous ternary cold fission.

1. Introduction

The use of the Gammasphere and Eurogam facilities in the last years offered the possibility to explore the cold (neutronless) binary and ternary fission of ^{252}Cf [1,2,3,4]. It confirmed the theoretical prediction based on the idea of the cold rearrangements of large clusters of nucleons from the initial to the final ground states, similar to the cluster radioactivity phenomenon [5,6].

In the binary fission the knowledge of the spin distribution for emitted fragments is very important because it provides an important information concerning the fissioning system in the region of the scision point [7,8]. The first aim is to describe the binary cold fission process of ^{252}Cf as a decay of a resonant state. We calculate the yields of the rotational states in coincidence within a stationary scattering theory.

Recently a new experiment was performed by a joint collaboration (Vanderbilt, Berkeley and Dubna) [9] allowing the measurement of the averaged energy and the angular distribution of light clusters in the ternary fission of ^{252}Cf. Several theoretical approaches were devoted to this subject [10,11,12,13,14]. Our second goal is to explain why the averaged values of the light cluster energy spectra are close to each other. We also explain the angular distributions of emitted light clusters.

2. Theoretical background

2.1. *Binary fission*

We describe the decay process by a stationary Schrödinger equation

$$H\Psi(\mathbf{R}, \Omega_1, \Omega_2) = E\Psi(\mathbf{R}, \Omega_1, \Omega_2) , \qquad (2.1)$$

where $\mathbf{R} = (R, \hat{R})$ is the distance between the centers of two deformed nuclei and the orientation of their major axes is given by Euler angles $\Omega_k = (\varphi_k, \theta_k, 0)$, $k = 1, 2$. The Hamiltonian operator can be written in two different ways, depending on the distance $\mathbf{R}$. In the external region, $R > R_c$, each nucleus is left in some excited rotational state, i.e.

$$H = -\frac{\hbar^2}{2\mu}\nabla^2_\mathbf{R} + H_1(\Omega_1) + H_2(\Omega_2) + V(\mathbf{R}, \Omega_1, \Omega_2) , \qquad (2.2)$$

where μ is the reduced mass and $H_k(\Omega_k)$ are the Hamiltonians describing the rotation of the fragments. We estimate the interaction between nuclei using the double folding between nuclear densities [15], where as a nucleon-nucleon force we use M3Y interaction [16]. For smaller distances $R < R_c$ the two nuclear densities overlap each other. The two nuclei oscillate around the most favourable fissioning pole-to-pole configuration, i.e.

$$H = -\frac{\hbar^2}{2\mu}\frac{1}{R}\frac{\partial^2}{\partial R^2}R + H_1(R, \theta_1) + H_2(R, \theta_2) + H_{12}(R, \theta_1, \theta_2) + V_{p-p}(2.3)$$

where $H_k(R, \theta_k)$ describe harmonic oscillations, $H_{12}(R, \theta_1, \theta_2)$ is the coupling term, as given in Ref. [17] and $V_{p-p}(R)$ is the pole-pole potential. The wave functions satisfying the Schrödinger equation in the two regions are given respectivelly by the rotational and vibrational bases [18]. By using the orthonormality of the angular functions one obtains in a standard way a coupled system of differential equations for the radial components

$$\frac{d^2 F_l}{d\rho_l^2} = \sum_{l'=1}^{N} A_{l;l'} F_{l'} , \quad l = 1, ..., N , \qquad (2.4)$$

$$l = (l, I_1, I_2), \quad R > R_c , \quad = (n_1, n_2), \quad R < R_c ,$$

where $\rho_l = k_l R$. The coupling matrix is given in Ref. [18]. We find the resonant states using the standard technique.

The decay width for some channel l can be derived from the continuity equation in a straightforwad way

$$\Gamma_l = \hbar v_l \lim_{R\to\infty} |F_l(R)|^2 = \hbar v_l |S_l|^2 , \qquad (2.5)$$

where S_l are the coefficients multiplying the outgoing Coulomb waves and v_l is the center of mass velocity at infinity in the channel l.

2.2. *Ternary fission*

The Hamiltonian of the ternary fission process can be written as follows

$$H = -\frac{\hbar^2}{2\mu_{12}}\nabla_R^2 - \frac{\hbar^2}{2\mu_3}\nabla_r^2 + V^{(12)}(R) + V^{(3)}(R,\mathbf{r}) \ . \qquad (2.6)$$

The total wave function

$$\psi_K(\mathbf{R},\mathbf{r}) = \sum_n \psi_{nK}^{(12)}(\mathbf{R})\psi_{n-K}^{(3)}(R,\mathbf{r}) \ , \qquad (2.7)$$

where n denotes the eigenvalue resonance number and K the angular momentum projection on the intrinsic axis, satisfies the Schrödinger equation

$$H\psi_K(\mathbf{R},\mathbf{r}) = E_K\psi_K(\mathbf{R},\mathbf{r}) \ . \qquad (2.8)$$

For some distance between heavy fragment R the motion of the light particle is given by

$$\left[-\frac{\hbar^2}{2\mu_3}\nabla_r^2 + V^{(3)}(R,\mathbf{r}) - E_{nK}^{(3)}(R)\right]\psi_{nK}^{(3)}(R,\mathbf{r}) = 0 \ . \qquad (2.9)$$

Therefore the motion of the two heavy fragments is governed by

$$\left[-\frac{\hbar^2}{2\mu_{12}}\nabla_R^2 + V^{(12)}(R) + E_{nK}^{(3)}(R) - E_K\right]\psi_{nK}^{(12)} = \sum_{n'} B_{n,n'}\psi_{n'K}^{(12)} \qquad (2.10)$$

where $B_{n,n'}(R)$ is called the adiabatic matrix [19].

The potential $V^{(12)}$ acting between heavy clusters is given by a double folding procedure using Coulomb+M3Y nuclear interaction. The position of the first resonant state is adjusted by a repulsive core [18]. The potential acting on the light cluster is a sum of the Coulomb and nuclear parts. The nuclear potential in which the light cluster is born is given by a deformed harmonic oscillator (ho) shape

$$V_N^{(3)} = v\frac{\hbar^2\beta^2}{2\mu_3}\left[(1+\frac{4}{3}\delta)(x-x_0)^2 + (1-\frac{2}{3}\delta)(z-z_0)^2\right] - V_0. \qquad (2.11)$$

The ho parameter of the light cluster is defined in a standard way [20]. We estimate z_0 as the half distance between nuclear surfaces at $x_0=0$ [19]. The factor v is a scaling parameter.

The probability that the light particle is emitted in some direction θ is given by the continuity equation

$$\Gamma_{nK}^{(3)}(R,\theta) = \hbar v_{nK}\lim_{r\to\infty}|r\psi_{nK}^{(3)}(R,\mathbf{r})|^2 = \Gamma_{nK}^{(3)}(R)W_{nK}^{(3)}(R,\theta) \qquad (2.12)$$

$$W_{nK}^{(3)}(R,\theta) = \sum_L A_{nLK}Y_{L0}(\theta) \ ,$$

in terms of A_{nLK} coefficients derived in Ref. [19].

3. Numerical application

3.1. *Binary fission*

We studied the following splitting occuring in a binary cold fragmentation

$$^{252}Cf \rightarrow ^{104}Mo + ^{148}Ba \ , \tag{3.1}$$

with the decay energy Q=214.41 MeV. The ground state deformations of the fragments are $\beta_2 = 0.35$, $\beta_4 = 0.03$ for ^{104}Mo and $\beta_2 = 0.24$, $\beta_4 = 0.13$ for ^{148}Ba [18,21]. In the double folding expansion we considered M3Y + Coulomb interaction. For shorter distances we simulated the Pauli effect by using a repulsion with the strength V_{comp}. We considered that the vibrational is separated from the rotational region at $R_c = 15.6$ fm, where the spherical component of total interaction becomes smaller than the pole-to-pole interaction. The vibrational and rotational matrix elements were connected by using a spline procedure.

By changing the value of the repulsive strength V_{comp} we were able to find several resonant states with an energy equal to the experimental Q-value. Then we computed the decay widths to the final rotational states according to Eq. (2.5). In Figure 1 are given the total widths per nucleus versus the fragment angular momentum for the above considered resonances. They are normalized to the first transition connecting the gound states. By solid lines we connected the values for ^{104}Mo and by dashed lines for ^{148}Ba. The experimental yields for ^{104}Mo are $0.21[2^+], 0.74(14)[4^+], 0.05(5)[6^+]$ and for ^{148}Ba $0.8[2^+], 0.2(2)[4^+]$ [2]. They are normalized to the total $2^+ + 4^+ + 6^+$ yield. We denoted in round brackets the error and in squared brackets the angular momentum.

From Figure 1 we conclude that the third resonant state, with the number of nodes and maximal component (0,3), is the closest to the experimental situation. We obtained a good agreement for the normalised yields of ^{104}Mo: $0.22[2^+], 0.67[4^+], 0.11[6^+]$. Concerning ^{148}Ba we obtained the values:$0.26[2^+], 0.46[4^+]$. We mention that for this nucleus the experimental errors are larger. The last state (1,2) has a similar distribution of the decay widths: $0.30[2^+], 0.60[4^+], 0.10[6^+]$ for ^{104}Mo and $0.11[2^+], 0.61[4^+]$ for ^{148}Ba.

In order to discriminate between the two distributions it is necessary to have more information. To this aim we also predicted the widths considering the pairs of fragment angular momenta. Therefore these data are extremely important in understanding the dynamics of the cold fission process.

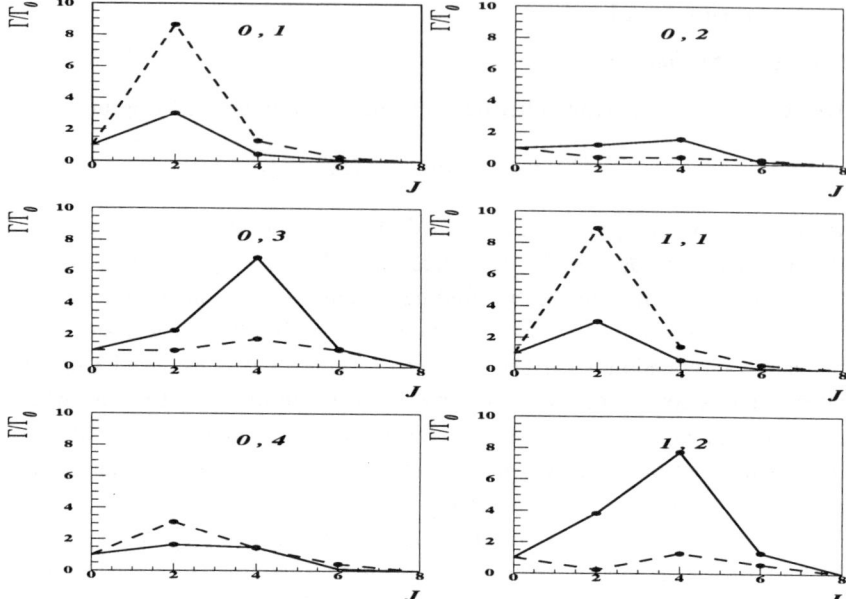

Figure 1. Total widths per nucleus versus the final spin for ^{104}Mo (solid lines) and ^{148}Ba (dashed lines).

3.2. *Ternary fission*

We analyzed the following three ternary splittings of ^{252}Cf

$$(A) \qquad ^{252}\text{Cf} \rightarrow ^{142}\text{Xe} +^{106}\text{Mo} +^4\text{He} , \qquad\qquad (3.2)$$
$$(B) \qquad ^{252}\text{Cf} \rightarrow ^{146}\text{Ba} +^{96}\text{Sr} +^{10}\text{Be} ,$$
$$(C) \qquad ^{252}\text{Cf} \rightarrow ^{142}\text{Xe} +^{96}\text{Sr} +^{14}\text{C} .$$

We considered only the true cold (neutronless) fission process, i.e. the fragments are emitted in their ground states [19]. The energy spectra of emitted light fragments are centered around 16, 18 and 24 MeV for the above three processes [9]. Moreover the light particles are emitted towards 84° with respect to the heavy fragment axis. It is also known that ^{10}Be yields are by two orders of magnitudes less than for ^{4}He [9]. This situation is in evident contrast with the cluster decay phenomenon because Coulomb barriers in the polar direction are very different. A simple explanation can be given by the nuclear potential (2.11) with parameters given in Table 1, having its minimum at $x_0=6$ fm for $R=12$ fm, as predicted in Ref. [14]. This is the potential in which the light cluster is born.

Table 1. The harmonic oscillator parameter $b = 1/\beta^2$ for different emitted light fragments [20] (the second column), $\beta = \mu_3 \omega_0 / \hbar$ (the third column) and the harmonic oscillator strength (the last column).

Nucleus	b (fm)	β (fm^{-2})	$\frac{\hbar^2}{2\mu_3}\beta^2$ (MeV fm^{-2})
^{4}He	1.19	0.706	2.603
^{10}Be	1.44	0.480	0.481
^{14}C	1.65	0.367	0.201

The Coulomb barrier is strongly lowered in the case (B) and becomes comparable with the case (A). We estimated the first eigenvalue of this state versus the inter-fragment distance R. The depth of the h.o. potential, V_0=27.7 MeV, reproduces the experimental value $E^{(3)}$=16 MeV for R=15 fm. The predicted half-lives, corresponding to these dependencies for the eccentricity x_0= 6 fm, are of 2-3 order of magnitude above the characteristic nuclear time $T_N = 10^{-22}$ s. This is consistent with the experimentally known fact that cold binary and α-ternary yields are comparable and the corresponding half-lives are 10^{-18} s [22]. Our numerical calculations showed that the adiabatic matrix is very small. Therefore the only coupling effect is the enhancement of the two-body Coulomb barrier between heavy fragments $V^{(12)}(R) + E_{nK}^{(3)}(R)$. Therefore the half-life of the whole ternary system is given by the shortest light particle half-life. It turns out that the angular distribution practically does not depend upon the inter-fragment distance R, the eccentricity x_0, the angular momentum projection K and the scaling parameter v. In Fig. 2 we see that a small deviation from the equatorial parameter z_0, strongly changes the position of the maximum in the angular distribution. We stress on the fact that the choosen z_0 gives a result very close to the experimental value in Table 1.

It is known that ^{10}Be has a structure in terms of two α-particles plus a neutron pair, and therefore should be very deformed. It is interesting to stress that by increasing deformation δ the maximum approaches experimental value. In this way we conclude that angular distribution is a tool to probe nuclear deformation [23]. The deformation of the light cluster potential has also an important effect on the half-life, because by increasing δ ho potential becomes flatter in the equatorial direction. In the case of the process (C) an increasing of the parameter δ changes the maximum, but the experimental value corresponds to δ=0, which confirms the fact that ^{14}C is not deformed.

558

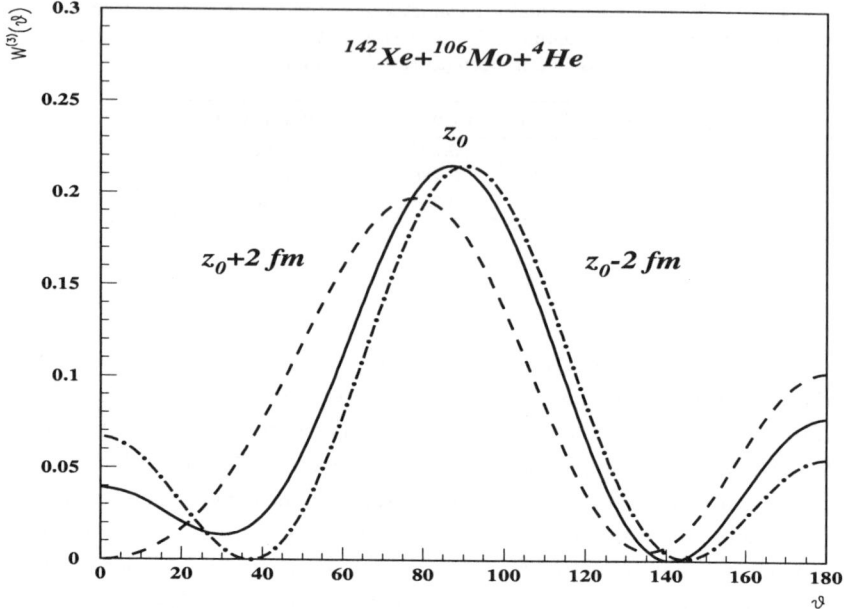

Figure 2. Normalized angular distribution of the process (A) for z_0 (solid line), $z_0 + 2$ fm (dashed line) and $z_0 - 2$ fm (dot-dashed line). The parameters of the nuclear potential are V_0=27.7 MeV, R=15 fm, x_0=6 fm.

4. Conclusions

We computed the decay probabilities to the final rotational states from the fissioning nucleus ^{252}Cf. within the stationary scattering theory. The mutual interaction of the two fragments was computed by using the double folding procedure with M3Y two-body plus Coulomb forces. For short distances between the fragments we simulated the Pauli principle by a repulsive delta-force core. By changing the repulsive strength it was possible to obtain different resonant states inside the resulting molecular potential. Among the computed resonant states we found two possible candidates reproducing the available experimental yields. In order to discriminate between them we give a more detailed picture of the widths versus all possible pairs of the final angular momenta.

We analyzed new experimental data concerning the ternary emission of ^{4}He, ^{10}Be and ^{14}C from ^{252}Cf. We supposed that the light cluster is born in the ho well with a size parameter given by its known nuclear density. Coulomb barriers in the equatorial direction become similar in the three cases, giving comparable eigenvalues of the first resonant states. From the position of the maximum in the angular distribution we concluded that the light cluster is born in the neck region. For ^{10}Be we found out that a significant deformation improves the agreement with the experiment.

References

1. G.M. Ter-Akopian, et al., *Phys. Rev. Lett.* **73**, 1477 (1994).
2. A. Săndulescu, et al. *Phys. Rev.* **C54**, 258 (1996).
3. A.V. Ramayya, et al., in Proc. of the third Int. Conf. *"Dynamical Aspects of Nuclear Fission"*, Eds. J. Kliman and B.I. Pustylnik Častá-Papiernička, Slovak Republic, August 30-Sept 4, 1996, (JINR, Dubna, 1996) p307.
4. J.H. Hamilton, et al., in Proc. of the Symposium on *"Fundamental Issues in Elementary Matter"* Ed. W. Greiner, Bad Honnef, Germany, September 25-29, 2000 (Ep Systema, Debrecen Hungary, 2000).
5. A. Săndulescu and W. Greiner, *J. Phys.* **G3**, L189 (1977).
6. A. Săndulescu and W. Greiner, *Rep. Prog. Phys.* **55**, 1423 (1992).
7. J.O. Rasmussen, et al., in Proc. of the third Int. Conf. *"Dynamical Aspects of Nuclear Fission"*, Částá-Papiernička, Slovak Republic, August 30-Sept 4, 1996, (JINR, Dubna, 1996) p289.
8. J.R.Nix and W.J.Swiatecki, *Nucl.Phys.* **71**, 1 (1965).
9. G.M. Ter-Akopian, seminar May 6, 2002, Frankfurt/Main.
10. A. Săndulescu, Ş Mişicu, F. Carstoiu, and W. Greiner, *Fiz. Elem. Chastits At. Yadra* **30**, 908 (1999); *Phys. Part. Nucl.* **30**, 386 (1999).
11. A. Săndulescu, F. Carstoiu, I. Bulboacă, and W. Greiner, *Phys. Rev.* **C60**, 044613 (1999).
12. F. Carstoiu, I. Bulboacă, A. Săndulescu, and W. Greiner, *Phys. Rev.* **C61**, 044606 (2000).
13. A. Florescu, A. Săndulescu, D.S. Delion, J.H. Hamilton, A. Ramayya, and W. Greiner, *Phys. Rev.* **C61**, 051602(R) (2000).
14. D.S. Delion, A. Florescu and A. Săndulescu, *Phys. Rev.* **C63**, 044312 (2001).
15. F. Cârstoiu and R.J. Lombard, *Ann. Phys. (N.Y.)* **217**, 279 (1992).
16. G. Bertsch, J. Borysowicz, H. McManus, and W.G. Love, *Nucl. Phys.* **A284**, 399 (1977).
17. Ş. Mişicu, A. Săndulescu, G.M. Ter-Akopian, and W. Greiner, *Phys. Rev.* **C60**, 034613-6 (1999).
18. D.S. Delion, A. Săndulescu, Ş Mişicu F. Carstoiu, and W. Greiner, *J. Phys.* **G28**, 289 (2002).
19. D.S. Delion, A. Săndulescu, and W. Greiner, *J. Phys.* **G** (in press).
20. Dao T. Khoa, *Phys. Rev.* **C63**, 034007 (2000).
21. P. Möller, J.R. Nix, W.D. Myers, and W.J. Swyatecki, *At. Data Nucl. Data Tables* **59**, 185 (1995).
22. A. Săndulescu, Ş. Mişicu, F. Cârstoiu, A. Florescu, and W. Greiner, *Phys. Rev.* **C58**, 2321 (1998).
23. D.S. Delion, A. Insolia, and R.J. Liotta, *Phys. Rev.* **C46**, 884 (1992).

SEMICLASSICAL APPROXIMATIONS FOR THE BREAKUP OF WEAKLY BOUND NUCLEI

R. DONANGELO, L.F. CANTO

Instituto de Física, Universidade Federal do Rio de Janeiro (UFRJ)
21941-972 Rio de Janeiro, Brazil
E-mail: donangel@if.ufrj.br

H.D. MARTA

Instituto de Física, Facultad de Ingeniería
CC 30, Montevideo, 11000 Uruguay

We investigate the validity of the semiclassical approximation for the breakup of weakly bound nuclei. For this purpose, we calculate angular distributions in breakup reactions and compare the results with data and quantum mechanical calculations available in the literature. We then discuss how semiclassical calculations could be extended to the case of complete and incomplete fusion reactions induced by unstable nuclear beams.

1. Introduction

With the availability of unstable beams, the interest in nuclei far from stability has grown considerably. For example, measurements of the ^{8}B breakup cross section in collisions with ^{58}Ni were carried out at Notre Dame [1]. These results were compared with semiclassical predictions for the Coulomb dissociation of ^{8}B and great discrepancies were found (theory overestimates experimental results by factors between 2 and 3) [2-4].

To investigate the origin of these discrepancies, Nunes and Thompson performed quantum mechanics calculations with the Continuum Discretization Coupled Channel method (CDCC). In a first paper [5], they included Coulomb and nuclear couplings but did not include multi-step processes or continuum-continuum coupling. In a subsequent paper [6], the importance of multi-step processes was investigated, extending the calculations to include continuum-continuum coupling. They found that the inclusion of such processes did not significantly change the Coulomb dissociation but it dramatically reduced the contribution from nuclear breakup to the cross section.
In this way, refs. [5] and [6] showed that nuclear effects and multistep processes play a fundamental role in the theoretical description of the data of ref. [1].

560

However, they do not investigate the validity of the semiclassical approximation. Since such approximations greatly simplify the numerical calculations involved in a quantum mechanical approach, this may be an important issue. This is the aim of the present work. We perform a semiclassical calculation for the same reaction that was quantum mechanically evaluated in ref. [6], and compare our results with those found in that work. Our calculation is based on the method of Alder and Winther [7] with continuum discretization [8-11]. While nuclear breakup and the effects of the nuclear interaction on the classical trajectories were disregarded in refs. [8-11], they are included here. We use the Coulomb and nuclear form factors of ref. [4] and include the nuclear monopole field in the calculation of classical trajectories. Since the potentials and states used in our work are the same as those in the quantum mechanical calculation of ref. [6], a comparison of the corresponding cross sections provides a meaningful test of the validity of the semiclassical approximation to this problem.

This work is organized as follows. In section 2 we describe the semiclassical model employed in our treatment of breakup reactions. In section 3 we compare the results of our calculations to data, and to other semiclassical and quantum mechanical calculations. Our conclusions are given in section 4, where we also discuss how to extend these semiclassical calculations to the case of complete and incomplete fusion processes.

2. Description of the semiclassical method

The collision is described in terms of the vector joining the centers of the projectile and the target, $\mathbf{R}$ and the intrinsic vector, $\mathbf{r}$ joining the centers of the two breakup fragments. To fix ideas, we consider the case of the breakup of ^{8}B into a proton and a ^{7}Be core. For a given impact parameter $\mathbf{b}$ the projectile-target relative motion is given by a classical trajectory $\mathbf{R}(t)$, and the intrinsic dynamics is treated as a time-dependent quantum mechanics problem. Following refs. [5,6], the interaction is

$$V(\mathbf{R},\mathbf{r})=V_p(r_p)+V_{Be}(r_{Be}), \qquad (1)$$

where V_p is the interaction between the proton and the target and V_{Be} is the interaction of the target with the ^{7}Be core. These potentials contain a nuclear part and also a Coulomb part. They are chosen as in ref. [6]. For the semiclassical calculation, the interaction should be split into an "optical potential", V_0, and a coupling interaction, $U(\mathbf{R},\mathbf{r})$, which leads to breakup. The real part of V_0 only affects the classical trajectory of the projectile-target system, while its imaginary part produces absorption along this motion

In order to derive the semiclassical coupled-channel equations, we first introduce the set of intrinsic eigenstates of ^{8}B. To do that, we use the cluster model, in which this nucleus is described as an inert ^{7}Be core plus a proton. In the present calculation, we treat ^{7}Be as an inert core with spin 3/2.

We now derive the semiclassical coupled-channel equations. The time-dependent wave function describing the proton-core motion in the ^{8}B-frame is expanded as

$$\Psi(t) = c_0(\mathbf{b},t)\, \Psi_0 e^{iBt/\hbar}(t) + \Psi_C(t) \qquad (2)$$

where Ψ_C is the component of $\Psi(t)$ in the continuum, c_0 is the amplitude of the ground state wavefunction Ψ_0 and B the binding energy of the proton in the ^{8}B projectile. We now expand Ψ_C in a set of energy eigenfunctions $\Psi_C = c_\alpha(\mathbf{b},t)\Psi_\alpha$ and replace it into eq. (2). The amplitudes $c_0(\mathbf{b},t)$ and $c_\alpha(\mathbf{b},t)$ are functions of time and also of the classical trajectory, which is specified by the collision energy and the impact parameter $\mathbf{b}$. Since the collision is iniciated with ^{8}B in its ground state, the amplitudes have the inicial values $c_0=1$ and $c_\alpha=0$. These amplitudes evolve as the collision proceeds and their final values contain the relevant information on the breakup cross section. The time evolution of these coefficients is obtained from the Schrödinger equation with the Hamiltonian

$$H = h(p_r,\mathbf{r}) + U(t,\mathbf{r}), \qquad (3)$$

where h is the kinetic energy operator associated with the proton-core motion. The time-dependent interaction $U(t,\mathbf{r}) = U(\mathbf{R}(t),\mathbf{r})$. Note that, in contrast to refs. [8-10], the classical trajectories of the present calculation take into account the effects of both the Coulomb and the nuclear parts of the optical potential. We refer to [12] for details on the semiclassical equations.

For the numerical calculations, we have used 14 bins; ten bins with a width of 100 keV, centered at 0.05, 0.15, 0.25, ..., 0.95 MeV, and four bins with a width of 500 keV, centered at 1.25, 1.75, 2.25, and 2.75 MeV. This set of bins is essentially the same considered by Nunes and Thompson [5,6]. The asymptotic amplitudes lead to the breakup probability, from which, and the classical deflection function, the breakup angular distribution can be evaluated .

3. Results and discussion

In figure 1 we show the ingredients entering our semiclassical calculation of the breakup cross section for the case of ^{8}B. The upper panel shows shows the elastic cross section, the center panel shows the breakup probability, and the lower panel illustrates the semiclassical result, i.e. the product of the quantities in the two other panels.

The semiclassical cross sections thus obtained are compared with those of the quantum mechanical calculation of Nunes and Thompson [6] in figure 2. The differential cross section for the breakup of ^{8}B in the ^{8}B + ^{58}Ni collision at 26 MeV is plotted as a function of the scattering angle of the p+^{7}Be system mass center. Both calculations include continuum-continuum coupling, which was shown in ref. [6] to play a very important role in the angular range studied. We see that the semiclassical calculation reproduces the quantum mechanical one with great accuracy. It gives a good description of the coulomb breakup maximum around 20 degrees and also shows the vanishing of the nuclear peak by multi-step effects [6].

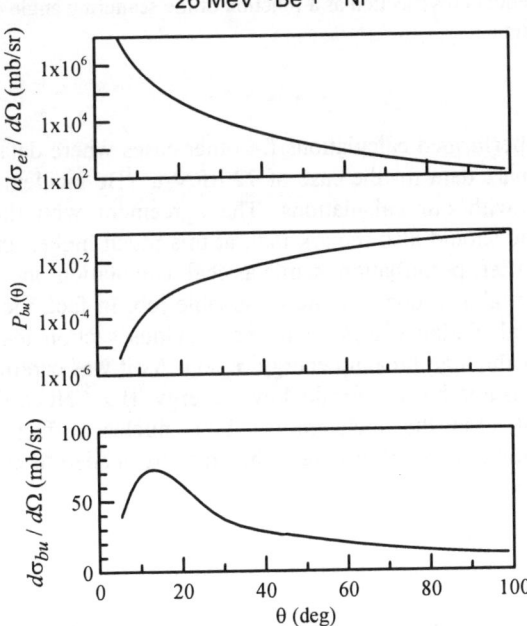

Figure 1. The ingredients entering our semiclassical calculation. For details, see the text.

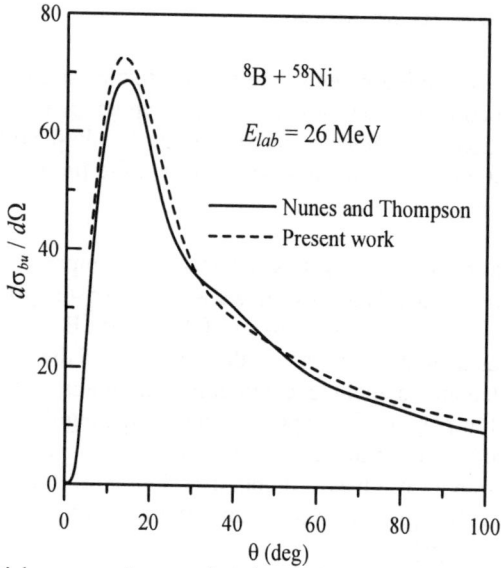

Figure 2. Differential cross section as a function of the scattering angle of the mass center of the p-^{7}Be system.

We have also performed calculations for other cases where data is available. In fig. 3 we compare data for the case of 72 MeV/u ^{11}Be incident on ^{208}Pb, taken from ref.[13], with our calculations. The agreement with the data is quite reasonable. One should also remark that, at this much higher energy, the much simpler first-order perturbation semiclassical calculation including only the Coulomb potential also does a quite reasonable job. In fact, the more complete semiclassical calculation discussed in the previous section leads to improved results only for the high breakup energy region. As it was carefully discussed in refs. [5,6], this is not the case for the lower energy ^{8}B + ^{58}Ni collision mentioned above. In that case the inclusion of both nuclear effects and multistep continuum-continuum couplings were shown to be needed to correctly describe the breakup process.

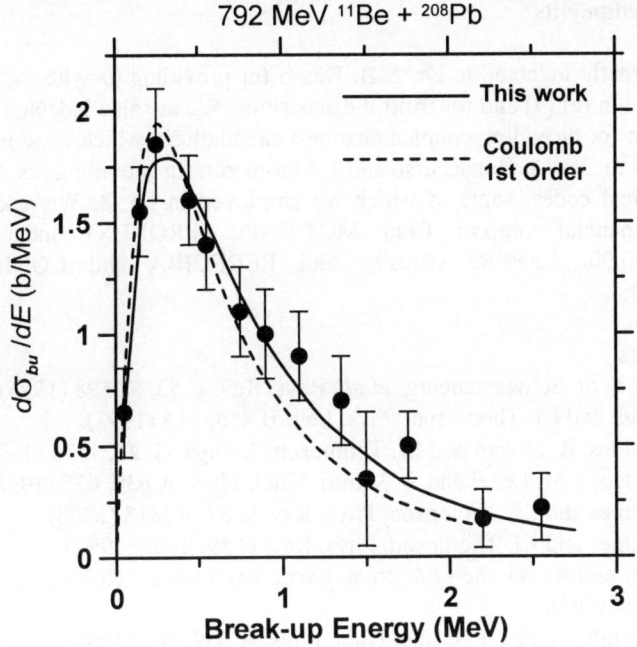

Figure 3. Distribution of breakup energies calculated using the method described in the present work, and compared to the data of ref.[13], and to a first order perturbation calculation where only the Coulomb interaction has been included.

4. Conclusions

We have shown that a semiclassical calculation including nuclear effects and continuum-continuum coupling is able to correctly reproduce the breakup cross section for different collisions induced by weakly bound projectiles. Since the main new ingredient in fusion reactions induced by unstable projectiles is the strong influence of the breakup channel, we believe that these results open up the possibility of performing detailed studies of complete and incomplete fusion reactions. To do that one needs to combine the breakup energy and angular distribution obtained from the semiclassical calculations shown above, so as to obtain the initial conditions for the velocities and impact parameters of the resulting fragments. With this information one then may estimate the fusion probability of the breakup fragments with the target, and thus obtain predictions for the complete and incomplete fusion processes. The development of this procedure is currently under way.

566

Acknowledgments

We are greatly indebted to Dr. C.H. Dasso for providing us with the computer codes used in ref.[4] and for fruitful discussions. We are also indebted to Dr. I.J. Thompson for providing coupled channels calculations, which were included in fig. 2, and to Drs. A. Bonaccorso and J. Margueron for providing results of their semiclassical codes, some of which we employed in fig. 3. We acknowledge partial financial support from MCT/CNPq (PRONEX) under contract 41.96.0886.00, FAPERJ (Brasil), and PEDECIBA and CONICYT-BID (Uruguay).

References

1. Johannes von Schwarzenberg *et al.*, Phys. Rev. C 53, R2598 (1996).
2. R. Shyam and I.J. Thompson, Phys. Lett. B **415**, 315 (1997).
3. F.M. Nunes, R. Shyam and I.J. Thompson, J. Phys. G **24**, 1575 (1998).
4. C.H.Dasso, S.M. Lenzi and A. Vitturi, Nucl. Phys. A **639**, 635 (1998).
5. F.M. Nunes and I.J. Thompson, Phys. Rev. C **57**, R2818 (1998).
6. F.M. Nunes and I.J. Thompson, Phys. Rev. C **59**, 2652 (1999).
7. K. Alder and A. Winther, *Electromagnetic Excitations*, North-Holland, Amsterdam, (1975).
8. C.A. Bertulani and L.F. Canto, Nucl. Phys. A **539**, 163 (1992).
9. A. Romanelli, L.F. Canto, R. Donangelo and P. Lotti, Nucl. Phys. A **558**, 71c (1995), L.F. Canto, R. Donangelo, A. Romanelli and H.Schulz, Phys. Lett. B **318**, 415 (1993).
10. H. Esbensen, G.F.Bertsch and C.A.Bertulani, Nucl. Phys. A **581**, 107 (1995).
11. L.F. Canto, R. Donangelo, A. Romanelli, M.S. Hussein and A.F.R. de Toledo Piza, Phys. Rev. C **55**, R570 (1997).
12. H.D. Marta, L.F. Canto, R. Donangelo and P. Lotti, Phys. Rev C **66**, 024605 (2002).
13. T. Nakamura *et al.*, Phys. Lett. B 331, 296, (1994).

FIRST DIRECT MEASUREMENTS OF THE ^{233}PA FISSION CROSS-SECTION UP TO 8.5 MEV

F. -J. HAMBSCH, F. TOVESSON AND S. OBERSTEDT

EC-JRC-Institute for Reference Materials and Measurements (IRMM)
Retieseweg,
B-2440 Geel, Belgium
E-mail: franz-josef.hambsch@irmm.jrc.be

A. OBERSTEDT

Örebro University
Department of Natural Sciences
SE-70182 Örebro, Sweden
E-mail: andreas.oberstedt@nat.oru.se

B. FOGELBERG AND E. RAMSTRÖM

Uppsala University
Department of Radiation Science
SE-61182 Nyköping, Sweden
E-mail: birger.fogelberg@studsvik.uu.se

The energy dependent neutron induced fission cross-section of ^{233}Pa has been measured for the first time with quasi-monoenergetic neutrons. This isotope is formed as an intermediate between the ^{232}Th source material and the ^{233}U fuel in the thorium-based nuclear-fuel cycle. Its fission cross-section is a key parameter in the modeling of future advanced fuel and reactor concepts. Two experiments have been conducted resulting in cross-section values in the neutron energy range between E_n=1.0-3.0 MeV and E_n=5.0-8.5 MeV. Significant discrepancies were found with model estimates and indirect measurements.

1. Introduction

In recent years there has been a renewed and increased interest in the thorium fuel cycle as an alternative to the traditional uranium-plutonium fuel. Reasons for this are new innovative concepts of accelerator driven systems (ADS) and transmutation of nuclear waste, such as the Energy Amplifier, proposed by Rubbia *et al.* [1].

Any development and implementation of such concepts relies strongly on accurate physics parameters such as neutron reaction data. The ^{233}Pa(n,f) cross-section is a key parameter in the thorium fuel cycle, since this isotope is an intermediate nuclide between the ^{232}Th source material and the ^{233}U-fuel. The reactions involving ^{233}Pa are responsible for the balance of nuclei as well as the average number of prompt neutrons in a contemplated reactor scenario. Due to the relatively short half-life and corresponding high β-activity of ^{233}Pa, only one direct measurement of it fission cross-section, dating back to 1967, exists yielding an average first-chance fission cross-section of 775±190 mb [2]. Fission probability measurements (e.g. [3]), however, indicate a lower value. This has led to the situation that the evaluated nuclear data libraries, such as ENDF/B-VI and JENDL-3.2, differ by approximately a factor of two for this reaction cross-section.

This initiated the work to measure the ^{233}Pa fission cross-section directly in the first and second chance fission region.

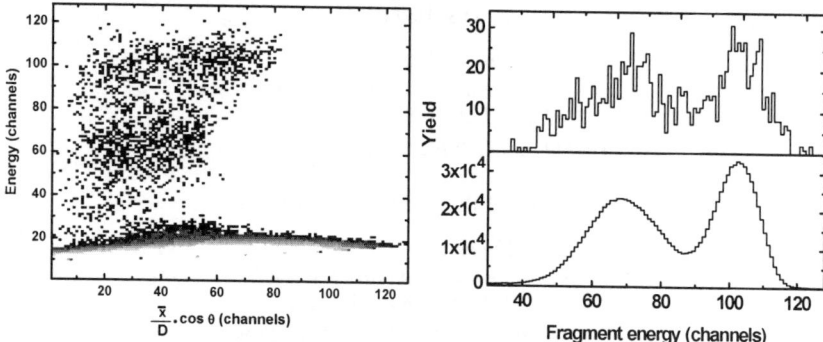

Figure 1. Left: Representation used for discriminating between fission fragments and piled up β-particles. The particle energy is shown versus the sum signal proportional to the center-of-gravity of the particle track. Right: the ^{233}Pa (top) and ^{237}Np fission fragment energy distribution after background rejection.

2. Experiment and results

The ^{233}Pa samples were prepared at the Studsvik Neutron Research Laboratory in Nyköping, Sweden. An elaborated separation procedure was then employed to wash out other metallic elements than Pa from the sample, including the daughter isotope ^{233}U, resulting in a suppression of 10^{-6} compared to Pa.

The cross-section measurement was carried out at the 7 MeV Van de Graaff accelerator at IRMM in Geel, Belgium. A twin ionization chamber was used for the fission detection, with the ^{233}Pa sample and a ^{237}Np sample placed in a back-to-back geometry. The ^{237}Np sample acted as the reference for the neutron flux normalization. Details of the experimental set-up can be found in [4].

Since ^{233}U is fissile, its fission cross-section is strongly effected in the presence of low energy neutrons from scattering on materials surrounding the detector. Hence, a separate measurement with a pure ^{233}U sample relative to ^{237}Np was carried out and the result used as input when correcting the ^{233}Pa data.

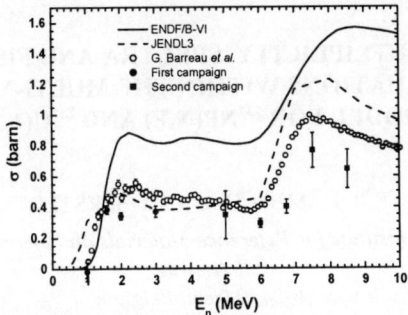

Figure 2. The ^{233}Pa cross-section from this work compared to the ENDF/B-VI and JENDL-3.3 evaluations as well the indirect measurement by Barreau *et al* [4]. The data obtained during the second measurement campaign (squares) are preliminary.

With the high β-activity from the Pa sample, special measures were needed to separate the fission fragments from piled-up electrons. In order to do so, a 2D-plot of the anode energy signal vs. the anode-grid sum signal, proportional to the range of the ionizing particle, was used for the background separation (see Fig. 1).

Once the fission fragment had been identified, the next step in the analysis was the correction for the fission event originating from the ^{233}U decay product in the ^{233}Pa sample. Since the exact time of purification of the Pa sample is known, the amount of ^{233}U in the sample for a given measurement run could be calculated from the well-known decay law. The ^{233}Pa(n,f) cross-section was then calculated relative to that of ^{237}Np(n,f).

The corresponding cross-section plot is shown in Fig. 2. The data in the energy region E_n=5.0-8.5 MeV are preliminary and subject to verification by a third measurement planed for 2002/2003.

As seen in the plot, the measured cross-section is lower than both the evaluations and the indirect measurement by Barreau *et al*. Furthermore, the measured above threshold cross-section favors the JENDL-3.2 evaluation, whereas the threshold energy is closer to that of the ENDF/B-VI evaluation.

References

1. F. Cariminati *et al.*, CERN Report No. CERN/AT/93-47(ET), 1993.
2. H. R. von Gunten *et al.*, *Nucl. Sci. Eng.* **27**, 85 (1967)
3. G. Barreau *et al.* (private communication).
4. F. Tovesson, F. –J. Hambsch, A. Oberstedt, B. Fogelberg, E. Ramström and S. Oberstedt, *Phys. Rev. Lett.* **88**, 062502 (2002).

NEUTRON MUTLIPLICITY, SPECTRA AND FISSION CROSS SECTION EVALUATIONS WITHIN THE MULTI-MODAL FISSION MODEL FOR ^{237}NP(N,F) AND ^{235}U(N,F)

F.-J. HAMBSCH AND S. OBERSTEDT

EC-JRC-Institute for Reference Materials and Measurements.
Retiseweg,
B-2440 Geel, Belgium
E-mail: franz-josef.hambsch@irmm.jrc.be

G. VLADUCA AND A. TUDORA

Bucharest University, Faculty of Physics.
R-76900 Bucharest, Romania
E-mail: anabella@Olimp.fiz.infim.ro

The statistical model for fission cross section evaluation is extended including the concept of multi-modality of the fission process. The three most dominant fission modes are considered. Based on calculations in the frame of the multi-modal random neck-rupture model (MM-RNR), separate outer fission barriers are assumed for each mode, while the inner barriers and isomeric wells are the same. Deconvoluted modal fission cross-sections, in addition to the other reaction cross-sections, for ^{237}Np(n,f) and ^{235}U(n,f), based on experimental branching ratios, were calculated for the first time in the incident energy range 0.01 to 5.5 MeV providing good agreement with the experimental fission cross section data. Also the prompt fission neutron multiplicity and spectra have been calculated for both isotopes. Here, the Los Alamos model was improved too introducing also the multi-modality of the fission process as well as a broader fission fragment mass range. In this way a better agreement with experimental data is achieved.

1. Cross section calculations

The compound nucleus (CN) mechanism is treated in the frame of the statistical model with sub-barrier effects using the statistical code STATIS which was modified to take into account the multi-modal fission concept.

For the neutron induced reactions of actinides in the incident neutron energy range where only the first fission chance is involved, the competitive processes are the elastic and inelastic scattering, the radiative capture and fission. The three most dominant fission modes, S1, S2 and SL are taken into account. Since space in this paper is very limited only results can be given, for more detail see Ref. [1]. In Figure 1 the resulting cross-section calculations for

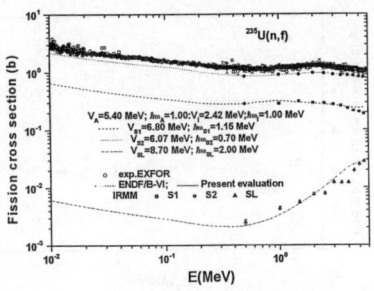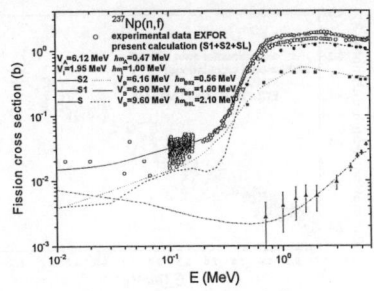

Figure 1. Left part: calculated sum and individual modal fission cross-sections for ^{235}U(n,f), right part: same for ^{237}Np(n,f).

both ^{235}U(n,f) (left part) and ^{237}Np(n,f) (right part) are given and compared to the respective ENDF/B-VI evaluation and EXFOR data. Also the separated fission mode cross-sections and respective calculations are given. The quality of the agreement is obvious from the Figure. Since the calculation extend also to unmeasured neutron energy regions it is possible to calculate the branching ratios at those energies and hence predict the corresponding mass distribution.

2. Neutron multiplicities and spectra calculations

The total prompt fission neutron multiplicity and spectrum for an incident energy E_n in the frame of the multi-modal fission model is calculated as a superposition of the prompt fission neutron multiplicites and spectra associated with a fission mode. The contributions of each mode are calculated using the Los Alamos model [2].

The model parameters needed for the present evaluation are based on the experimental fission fragment yield Y(A) and the total kinetic energy TKE(A) for the neutron induced fission of ^{237}Np (at 11 incident neutron energies) and of ^{235}U (at 12 incident energies) as measured at IRMM [3,4]. From the experimental data the following parameters are used: the branching ratios of the fission modes, the corresponding average fission fragment mass, the standard deviation of the fragment mass distribution and the average TKE. The input Los Alamos model parameters are calculated using the entire fission fragment range.

In Figure 2 an example of the calculation for the prompt neutron multiplicities for ^{235}U(n,f) (left part) and ^{237}Np(n,f) (right part) is given. Figure 3 shows examples for the corresponding spectra, which are compared to available experimental data. More details about the method are given in Ref. [5].

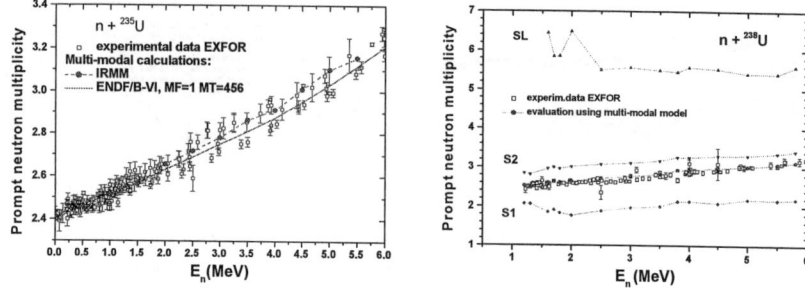

Figure 2. Left part: calculated prompt neutron multiplicities for ^{235}U(n,f), right part: same for ^{237}Np(n,f). The calculations are compared to exiting experimental data (symbols).

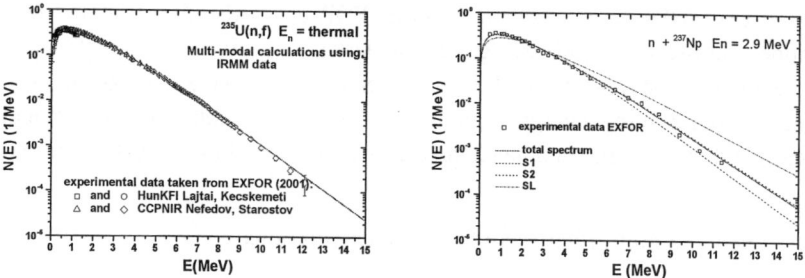

Figure 3. Left part: calculated neutron spectra for ^{235}U(n,f), right part: same for ^{237}Np(n,f). The calculation (full line) is compared to existing experimental data (symbols).

References

1. G. Vladuca, A. Tudora, F.-J. Hambsch, S. Oberstedt, *Nucl. Phys.* **A707**, 32 (2002) and G. Vladuca, A. Tudora, F.-J. Hambsch, S. Oberstedt, submitted to *Nucl. Phys.* A.

2. D.G. Madland, J.R. Nix, *Nucl. Sci. Eng.* **81**, 213 (1982).

3. F.-J. Hambsch, F. Vivès, P. Siegler, S. Oberstedt, *Nucl. Phys.* **A679**, 3 (2000).

4. Ch. Straede, C. Budtz-Jørgensen, H.-H. Knitter, *Nucl. Phys.* **A462**, 85 (1987) and I. Ruskov, private communication (2002).

5. F.-J. Hambsch, S. Oberstedt, G. Vladuca, A. Tudora, *Nucl. Phys.* **A709**, 85 (2002) and F.-J. Hambsch, S. Oberstedt, G. Vladuca, A. Tudora, I. Ruskov, submitted to *Nucl. Phys.* A.

STUDY OF COMPOUND NUCLEI PRODUCED IN LOW ENERGY NUCLEAR REACTIONS IN SOLIDS

G H. MILEY[1] AND H. HORA[2]

[1]Dept.t of Nuclear, Plasma and Radiological Engineering,
Univ. of Illinois at U-C, 103 S. Goodwin, Urbana, IL 61801, USA,
E-mail: g-miley@uiuc.edu
[2]Dept. of Theoretical Physics, Univ. of NSW, Sydney, 2052 Australia

The array of reaction products from experimental studies of Low Energy Nuclear Reactions (LENRs) suggest that such reactions occur through the formation of the intermediate heavy compound nucleus which can then fission or decay into the reaction products. These experiments used an electrolytic technique to obtain a very high loading of protons in thin (2000-8000 Å) metallic films of palladium, nickel, and titanium.

1. Introduction

The debated Pons-Fleischmann "cold fusion" experiment in 1989 was intended to create D-D reactions by loading deuterium (D) into a palladium lattice by electrolytic techniques [1]. Later Miley, Hora, et al. introduced the concept of enhancing reactions via swimming electron layer (SEL) created at interfaces between multi-layer thin films composed of metals with differing Fermi levels [2]. At that time, scattered experiments by various researchers suggested host electrode material might be involved in transmutation reactions [3]. Thereafter the authors and collaborators reported quantitative transmutation data based on multi-layer thin-film electrode technology and high precision analysis techniques, including secondary ion mass spectrometry (SIMS) and neutron activation analysis (NAA)[4]. Additional data from recent in-situ measurements with CR-39 foils and alumina dosimeters show low levels of 2-MeV proton, 16-MeV alpha particle, and soft x-ray emission from the thin films [5].

2. Experimental Apparatus and Results

A flowing packed-bed type electrolytic cell with 1-Molar $LiSO_4$ in light water was employed, the cell being packed with plastic beads (1-mm diameter) coated with a thin film (~500-1000 Å) of metal, (e.g., Ni, Pd or Ti) [see ref. 4 for details]. Beads were removed and analyzed for isotopic composition using NAA, and SIMS before and after multiple week runs. The key finding is the appearance of an array of "new elements" with masses lying above and below that of the base metal, typically having deviations from natural isotopic composition. The mass band peaks occur in an ordered fashion consistent with a magic number sequence [4, 7, 8, 9, 10]. In addition to the prior vs. post-run, great care has been made to distinguish products from impurities by use of passive "reference" electrode runs, a "clean cell" design and high purity components/electrolyte. The key elements observed occurs in quantities significantly exceeding impurity levels, and with statistically significant deviations from natural abundances.

3. Reaction Pathways

Based on these observations, a semi-empirical theory, RIFEX has been formulated. RIFEX assumes multiple body reactions of virtual neutrons (created by proton orbital mixing in the lattice structure) and metallic lattice atoms form a "complex" intermediate nucleus that then fissions, producing the observed product array (1, 3). The small excess energy involved in this reaction explains the observed formation of nearly stable products with little hard radiation emission.

The run with a thin film Ni coating serves as an illustration [rzun #8, ref. 1]. Explaining the formation of products with masses well above or below the base Ni represents the key theoretical challenge. With nickel's high binding energy per nucleon, successive proton-Ni reactions to build up mass are blocked energetically. However, an array of products originating from fission of a single "complex" nucleus is allowed energetically, in effect "sharing" the formation energy of the complex. Consider, the complex nucleus (X*-313) of mass A ~ 313, from the shape of the reaction products yield curve from the Ni Run #8 [1, 8]. The candidate reaction 23n* + 5 Ni-58 = X*-313 corresponds to 8.2 MeV/nucleon for reactants (n* represents a virtual neutron). If the measured yields of all corresponding products are multiplied by their binding energies, 8.1 MeV/nucleon is obtained, giving a net energy release of ~ 0.1 MeV/nucleon. An alternate breakup pathway involves splitting of low mass complexes into pairs or triplets. Tertiary breakup gives; X*-313 = X*-170+ X*-104 + X*-39 which then fission into the low mass products observed in the experiment. In addition to the "virtual neutron" RIFEX model outlined above, G. Stoppini [11] has proposed an e-proton capture mechanism and J. Fisher [12} proposed a polyneutron model to explain the observed product yield curves.

In summary, these results illustrate how the formation and subsequent fission of complexes allows a "pooling" of reaction energy allowing products covering a wide range of masses. These calculations are also reasonably consistent with the measured heat release from the runs. For example, if all of the ~70 reaction products for the Ni run are multiplied by their binding energies, and the same is done for the n*-Ni reactants, an average excess power of 0.9 W is found vs. an experimental value of 0.6 ± 0.4 W. In view of the many uncertainties in the experimental measurement, this very rough agreement is encouraging.

4. Related Studies by Others

Over 15 laboratories worldwide have reported transmutation products using highly loaded p/D in metals like Pd. Recently Y. Iwamura, et al., reported a seminal experiment [13]. Using in-situ XPS with deuterium flow through a multiplayer thin film, they found transmutation of a surface layer of Cs into Pr (alternately Sr into Mo). To explain this, a multi-body reaction of 4 deuterium nuclei and Cs (or Sr) is required. This represents a special case of the compound nucleus phenomenon described here.

5. Conclusion

In closing, the product characteristics observed in the thin-film electrolysis experiments provide important benchmarks for prospective theories. While several theories have been proposed, further development and benchmarking of theories against the experiments is essential to solidify our understanding

Acknowledgement

This research was partially supported by Lattice Energy, LLC.

References

1. Fleischmann, M. and S. Pons. "*J. Electroanal. Chem.*, **261**, 301, 1989.
2. Hora, H., et al," *Physics Letters A*, **175**, 138, (1993).
3. Bockris, J. O'M. and Lin, G. H., *Journal of New Energy*, **1**, 1, 111, (1996).
4. Miley, G. H., and Patterson, J. A., *Journal of New Energy* , **1**, 3, 5, (1996).
5. Lipson, A.G., et al., *Proc. of ICCF-9*, Beijing, China, (2002).
6. Miley, G. H., *Proc. of 4th Asti Workshop on Anomalies in Hydrogen/ Deuterium Loaded Metals*, **64**, 77 (1999).
7. Hora, H. and Miley, G. H., *Czech. J. of Physics*, **48**, 9, 111 (1998).
8. Hora, H., Miley, G. H., *Czech.k J. of Physics,* **50**, 3, 433 (2000).
9. Miley, G. H. et al., *Progress in New Hydrogen Energy*, **2**, 629, (1997).
10. Miley, G. H., and Hora, H., *Blt. of the APS-DNP Annual Meeting*, Lansing, MI, October (2002).
11. Stoppini, G., *Fusion Technology*, **34**, 1, 81 (1998).
12. Fisher, J, *Fusion Technology*, **34**, 1, 66, (1998).
13. Iwamura, Y., et al., *Japanese J. Applied Physics*, **41**, 1, 7A, 4642, (2002).

DEVELOPMENT OF A ^{235m}U GENERATOR FOR MEASUREMENT OF NEUTRON CROSS SECTIONS

D. A. SHAUGHNESSY, Y. M. X. M. DARDENNE, J. M. KENNEALLY, C. A. LAUE,
K. J. MOODY, J. B. PATIN, M. A. STOYER, N. J. STOYER AND J. F. WILD

Analytical and Nuclear Chemistry Division
Lawrence Livermore National Laboratory
P. O. Box 808, L-231, Livermore, CA 94551

We are currently developing an online ^{235m}U generator. Initial tests were performed with ^{232}U, which was electroplated on the inside of a 50-cm long, 1-cm diameter Ni tube at a current density of 2 mA/cm^2 with a plating efficiency of 92%. He or Ar was swept through the tube to collect the recoil activity and deposit it on a Pt plate for counting. Ar did not deposit any activity, but collections made with He were approximately 1% efficient, and the amount of activity collected increased with flow rate. Future experiments will hopefully increase the amount of activity collected from the generator.

1. Introduction

The defense and nuclear technology field is interested in measuring cross sections from ^{235m}U(n,f) reactions at neutron energies of 0 – 20 MeV. ^{235m}U is a 26-min. radionuclide that decays via internal conversion to the ^{235}U ground state.[1,2] One way ^{235m}U is produced is via the alpha decay of ^{239}Pu, which populates the ^{235m}U isomer and members of its rotational band almost exclusively (approximately 99.9%.[3]) To measure ^{235m}U(n,f) cross sections, a sufficient amount of ^{235m}U must be prepared as a target for neutron irradiation.

Following the alpha decay, ^{235m}U can be chemically separated from ^{239}Pu, or the recoiling ^{235m}U can be collected if the ^{239}Pu is thin enough such that the recoils can escape. The disadvantage of a chemical separation is that when freshly purified ^{239}Pu is chemically "milked" to remove ^{235m}U, 10 minutes after purification the milking results in a mixture of 88% isomer and 12% ^{235}U ground state. This ratio becomes more unfavorable as the ingrowth time increases. If the separation is done after short ingrowth times, a favorable isomer ratio is achieved but the overall yield of ^{235m}U is low.

By collecting ^{235m}U recoils directly, a favorable isomer ratio is maintained without the disadvantage of low yield. A layer of ^{239}Pu deposited with a thickness comparable to the daughter recoil range (approximately 0.5 µg/cm^2) should efficiently emit ^{235m}U recoils.

An online generator would provide a constant supply of ^{235m}U that could be quickly transported and deposited for neutron irradiation. We are developing a generator that will consist of ^{239}Pu electroplated on the inside of a Ni tube. After recoiling out of the ^{239}Pu, ^{235m}U is collected in a carrier gas that flows through the tube. Once the generator is developed, we will incorporate it in experiments designed to measure ^{235m}U(n,f) cross sections.

2. Experimental Methods

2.1. *Electrodeposition*

We are first testing the generator with ^{232}U. Electrodeposition was performed according to the methods of Aumann and Müllen.[4,5] 14 mL of an isopropyl alcohol solution containing 0.1 ng/mL ^{232}U was poured into a 50-cm long, 1-cm diameter Ni tube. Pt wire was pulled through the center of the tube such that it did not touch the sides. A high voltage source was connected to the Pt wire and the tube itself was connected to ground. Current density was maintained at 2 mA/cm^2 (50 V.) After one hour, 92% of the initial activity was electroplated on the inside of the tube.

2.2. *Recoil Collection*

A Pt plate was placed underneath the electroplated tube and either He or Ar was passed through the tube to collect the recoiling activity and deposit it on the Pt plate. The plate was then counted in a fission – alpha preset counter operated in Ar with a detection efficiency of approximately 35%. Gross alpha activity was corrected for background, which had been measured prior to the collections.

3. Initial Results

After successfully electroplating ^{232}U inside the tube, one-hour collections were made. The collection efficiency with He was approximately 1% and the amount of activity collected appeared to be dependent on the He flow rate as shown in Figure 1. When Ar was used, no activity was observed on the Pt plate.

578

4. Future Experiments

Future experiments will attempt to increase the amount of activity collected. Different gases will be used and dependence on flow rate, time, and temperature will be examined. The use of an aerosol gas-jet will also be explored.[6] Once the generator is optimized with ^{232}U, ^{239}Pu will be electroplated and the production of ^{235m}U determined. The final configuration will be used for neutron irradiation experiments.

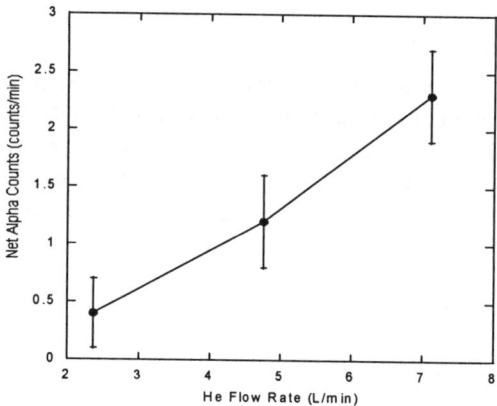

Figure 1. Amount of alpha activity collected versus He flow rate. Each collection was one hour.

Acknowledgments

This work was performed under the auspices of the U. S. Department of Energy by the University of California, Lawrence Livermore National Laboratory under Contract No. W-7405-Eng-48.

References

1. F. Asaro and I. Perlman, *Phys. Rev.* **107**, 318 (1957).
2. J. R. Huizenga, C. L. Rao and D. W. Engelkemeir, *Phys. Rev.* **107**, 319 (1957).
3. R. B. Firestone and V. S. Shirley, Editors, *Table of Isotopes*, 8th ed. (Wiley Interscience, New York, 1996).

4. D. C. Aumann and G. Müllen, *Nucl. Instr. Meth.* **115**, 75 (1974).
5. G. Müllen and D. C. Aumann, *Nucl. Instr. Meth.* **128**, 425 (1975).
6. H. Wollnik, H. G. Wilhelm, G. Röbig and H. Jungclas, *Nucl. Instr. Meth.* **127**, 539 (1975).

MANY-PHONON STATES AS DOORWAY TO FISSION WITHIN THE COMPLEX TRAJECTORY APPROACH

FEODOR F. KARPESHIN
V.A.Fock Institute of Physics
St. Petersburg State University

Shifting the real fission trajectory into the complex plane allows one to regard the fission story as a superposition of contributions to the fission amplitude, arising from various irregular points. In the case of emission of alphas and gamma rays, this is the scission point. In the case of muon-induced prompt fission, or fission of Lambda hypernuclei an avoided crossing of the terms also gives a contribution to the final fate of the particle.

Analytical continuation into the complex plane allows one to calculate suppression of the fission mode, arising from the halo states of one or two neutrons arising in the compound nucleus [1]. Questions of reversibility of cluster emission from fissile nuclei are also considered.

References

1. V.I. Serov, S.N. Abramovich and F.F. Karpeshin. Fission Dynamics of Atomic Clusters and Nuclei, Proc. of the Int. Workshop. Edited by J. da Providencia, David M. Brink, F. Karpechine and F.B. Malik. World Scientific: 2002, P. 94.

TERNARY FISSION OF ^{252}CF: IDENTIFICATION OF ^{10}BE CLUSTERS

A. V. DANIEL[1], G. M. TER-AKOPIAN[1,2], J. H. HAMILTON[3], A. V. RAMAYYA[3],
J. KORMICKI[3], G. S. POPEKO[1], A. S. FOMICHEV[1,2], A. M. RODIN[1],
YU. TS. OGANESSIAN[1], J. D. COLE[4], J. K. HWANG[3], D. FONG[3], P. GORE[3],
J. O. RASMUSSEN[5], A. O. MACCHIAVELLI[5], I. Y. LEE[5], M. A. STOYER[6],
W. GREINER[7], R. DONANGELO[8], M. JANDEL[1,2], L. KRUPA[1,2], J. KLIMAN[1]

[1]*Flerov Laboratory of Nuclear Reaction, JINR, Dubna, 141980, Russia*

[2]*Joint Institute for Heavy Ion Research, Oak Ridge, TN 37835, USA*

[3]*Department of Physics, Vanderbilt University Nashville, TN 37235, USA*

[4]*Idaho National Engineering and Environmental Laboratory, Idaho Falls, ID 83415, USA*

[5]*Lawrence Berkeley National Laboratory, Berkeley, CA 94720, USA*

[6]*Lawrence Livermore National Laboratory Livermore, CA 94550, USA*

[7]*Institut für Theoretische Physik der Universität Frankfurt, Postfach 11 19 12, Germany*

[8]*Instituto de Fisica Unversidade Federal do Rio de Janeiro, Rio de Janeiro, 21945-970, Brazil*

Coincidences between γ rays and light charged particles (LCP) emitted in the ternary fission of ^{252}Cf have been measured in an experiment carried out at the Gammasphere facility. LCPs were detected by)E×E technique. The population ratio was estimated for the ^{10}Be LCPs emitted in the first 2^+ and in the ground states. Supposing that a thermal equilibrium exists at the scission point, a temperature parameter θ=1.84±0.16 MeV was deduced from this ratio.

1. Introduction

The spontaneous fission of ^{252}Cf accompanied by the light charged particle (LCP) emission (often called ternary fission) was widely investigated, see reviews[1,2] and papers cited there. Many experiments were concentrated around the LCP energy and angular distributions, and relative probability of emitting different LCPs[3-6]. The interest to this rare process follows from the possibility to get additional data about the fission mechanism. Indeed, originating from the neck region, between the two fission fragments, the LCPs should supply information about the scission point configuration.

It was supposed[7,8] that LCPs could be sometime emitted in their excited states. If thermal equilibrium is set near the scission point, this could open a direct way to the estimation of nuclear temperature. The authors of Ref.[9,10] reported on the observation of the γ line corresponding to the $2^+ \rightarrow 0^+$ transition in ^{10}Be in the γ spectrum measured in coincidence with the beryllium accompanied fission of ^{252}Cf. There were some evidences that the γ-peak of ^{10}Be was seen without Doppler broadening. More recent experiments[11,12] made at Gammasphere with a much higher energy resolution apparently confirm this result. To be true, the γ rays must be emitted from the ^{10}Be nucleus mainly in the orthogonal direction to its momentum or the ^{10}Be nucleus must stay[13] in the nuclear potential between two fragments an extremely long time ($\sim 10^{-13}$ s) commensurable with the lifetime of the ^{10}Be 2^+ state.

The present experiment had an aim to study the ternary fission of ^{252}Cf by detecting the LCPs in coincidence with the prompt γ rays recorded by Gammasphere.

2. Experiments

A source of ^{252}Cf giving $\sim 4 \times 10^4$ spontaneous fissions per second was installed in the center of the reaction chamber, which was placed in the Gammasphere. The ^{252}Cf specimen was deposited in a 5-mm spot on a 1.8-micron titanium foil and was tightly covered on both sides by gold foils. The thickness of these gold foils was chosen to stop the ternary fission fragments. Only light fragments with energy greater then 110 MeV could penetrate the foils.

Eight $\Delta E \times E$ Si detector telescopes were used to measure LCPs emitted in the ternary fission. Four telescopes were centered at the polar angle $\theta = 30°$ (azimuth angles $\varphi = 45°$, $135°$, $225°$ and $315°$), whereas other four were at $\theta = 150°$ (azimuth angles were the same). The telescopes were inclined so that the lines connecting the centers of two opposite ΔE detectors and going through the center of source were orthogonal to the detector surfaces. Each ΔE detector had an area 10×10 mm^2 and thickness slightly varying between 9μ and 10.5μ. Each E detector was 400μ thick and was 20×20 mm^2 in area. All ΔE detectors were at a distance of 27 mm from the source. The distance between ΔE and E detectors was 13 mm in each telescope.

Data acquisition system was triggered by any of ΔE or E signals having amplitudes above threshold values, which were set to prevent the detection of events occurring from two-fold pile-ups of α particles emitted at the ^{252}Cf radioactive decay. A ternary fission event was stored at a condition that at least one γ ray was detected by Gammasphere within the time interval allocated for this event. Approximately 1.6×10^7 events were recorded during the measuring time which lasted for two weeks almost continuously.

3. Results and Discussion

A resolution of $\Delta E \times E$ telescopes was enough to well identify helium, beryllium, boron and carbon nuclei on the eight two-dimensional plots created separately from the data collected by the telescopes. Only lithium region was shadowed by the random coincidence of the ternary helium LCPs with the ^{252}Cf α-decay particles. Energy calibration of the E detectors was made using the well known energy distribution of the ^{4}He LCPs emitted in the ternary fission of ^{252}Cf. In case of the ΔE detectors, energy calibration was made using a number of mean ΔE values calculated by averaging data obtained in narrow energy (E) intervals selected in the loci of He, Be and C LCPs.

In fact, we could clear marked the regions corresponding to He, Be, B and C LCPs on the $\Delta E \times E$ plots when the energy deposited in the E detector was more then 5 MeV. For these LCPs Table 1 summarizes the threshold energy (E_{th}) and the portion (P) of LCPs above this threshold. The portions were calculated using the known parameters of the LCP energy distributions. The last column in Table 1 gives the numbers of LCPs detected in the experiment.

Table 1. Detection conditions obtained for LCPs.

LCP	E_{th}, MeV	P, %	Counts
He	9	93	4905767
Be	20	39	30960
B	26	26	1940
C	32	14	6445

3.1. Population of the first exited state in ^{10}Be

The detection of prompt fission γ rays triggered by the beryllium LCPs offered a possibility to estimate the probability for the emission of ^{10}Be in its first 2^+ state. The dotted line in Fig. 1 shows the high energy part of the γ ray spectrum corresponding to the beryllium accompanied ^{252}Cf spontaneous fission. Supposing that the main part of the ^{10}Be nuclei decayed in flight, on the way from the source to the detector telescopes (a half-life time of 125 fs is known for the first excited state of ^{10}Be), Doppler correction was applied to the primary data. The resulting spectrum is presented in Fig. 1 by solid line. A distinct peak is seen in this spectrum near the γ ray energy 3368 keV. Fitting this peak with a Gaussian gives a line width FWHM=65.7±4.5 keV. This width is compatible with the value FWHM=62.1 keV obtained by the Monte Carlo simulation of this γ peak. The peak width is determined mainly by the limited angular resolution of the particle telescopes and the Gammasphere detectors.

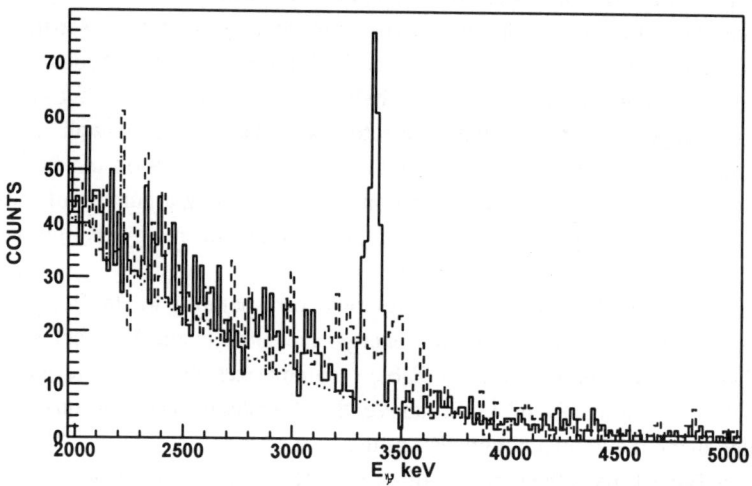

Figure 1. The solid and dashed lines show the spectra of γ rays coinciding with Be LCPs, respectively, after and before the Doppler shift correction. The dotted line shows the spectrum of γ rays coinciding with the He LCPs. The last spectrum was normalized to the total number of counts obtained in the whole spectrum associated with Be LCPs (E_γ=90 – 5000 keV).

Dotted line in Fig. 1 shows the γ ray spectrum recorded in coincidence with helium LCPs. The lack of a bump in this spectrum centered around 3368 keV supports the assumption that the presence of the bump seen in the Be spectrum in the vicinity of this energy (see the Be spectrum shown in Fig. 1 with the dashed line) proves our general conclusion that the 3368 keV line that rises in the Be spectrum after the Doppler shift correction shows up the $2^+ \rightarrow 0^+$ transition in ^{10}Be.

The number of counts obtained in the 3368 keV γ line is 263±22. Taking into consideration the Gammasphere efficiency (~4.6% for this energy), one obtains that the total number 3368 keV γ rays emitted by ^{10}Be nuclei was I_γ=5700±480. After correction of the number given in Table 1 for those Be LCPs, which escaped the detection due the lack of recorded γ ray signals we estimated the total number of Be clusters recorded by the ΔE×E telescopes N_{Be}=51800±5000. Following the experimental results presented in Ref.[14] we assumed that the part of ^{10}Be makes 80% of the Be ternary fission of ^{252}Cf. Estimations based on the comparison of our experimental ΔE×E spectrum with the Monte Carlo simulation is in accord with this result. Finally we estimated the population ratio N_{2+}/N_{0+}=0.160±0.025. Assuming that the thermal equilibrium is set at the scission point, one could take it appropriate to apply the Boltzmann factor to this ratio: N_{2+}/N_{0+}=exp(−3368/θ), to estimate a temperature parameter θ. This gives θ=1.84±0.16 MeV. Most likely the temperature parameter θ is a characteristic of the neck region at the scission point.

3.2. Gamma rays emitted by ^{10}Be in rest

In the spectrum of γ rays coinciding with Be LCPs one does not see any distinct peak near the energy 3368 keV which could be attributed to the γ rays emitted by ^{10}Be in rest (see Fig. 2). A wide bump in the spectrum is explained by the Doppler broadening of the 3368 keV γ line of moving ^{10}Be nuclei. Assuming a probability of about 1% for the γ emission by unmoved ^{10}Be one could see a clear narrow peak (with ~26 counts) rising above the solid line at 3368 keV or nearby.

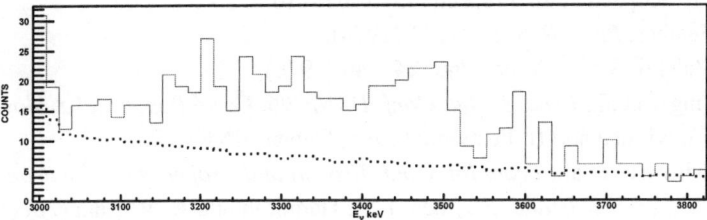

Figure 2. The spectra of γ-rays coinciding with beryllium – solid line, and helium – dotted line. The helium spectrum was normalized to the count number obtained between 90 and 5000 keV in the beryllium spectrum.

Thus, we conclude that the probability to obtain in the ^{252}Cf spontaneous fission ternary ^{10}Be nuclei emitting in rest 3368 γ rays is less than 1%. We notice that this conclusion is valid only for ^{10}Be LCPs with energy more than 20 MeV. But one cannot exclude that the hypothetical triple molecule decays in a way, which could lead to the ^{10}Be LCPs emitted with a low energy. In fact the 3368 keV γ-line was observed in the fission experiment at Gammasphere[11,12] without any bias to the LCP energy.

Acknowledgments

Work at Joint Institute for Nuclear Research is supported in part by the DOE contract #DE-AC011-00NN4125, BBWI Agreement No. 3498 (CRDF grant RPO-10301-INEEL) and by the joint RFBR – DFG grant (RFBR No. 02-02-04004, DFG No. 436RUS 113/673/0-1(R)).

References

1. I. Halpern, *Ann. Rev. Nucl. Sci.*, **21**, 245 (1971).
2. C. Wagemans, *The Nuclear Fission process*, ed. C. Wagemans, (CRS Press, Boca Raton, FL, USA, 1989), Chap. 12.
3. I. A. Kondurov et al., Bull. LINP Data Center, **6**, (1977).
4. F. Gönnenwein et al., *Proc. Int. Conf. NFFS/AMFC* (1993, Bristol, IOP) p. 299.
5. M. Mutterer et al., *Proc. 3rd Int. Conf. DANF'96, Častá Papiernička, Slovakia,* eds. J. Kliman and B. Pustylnik (1996, Dubna, JINR) p.250.
6. Yu. N. Kopach et al., Preprint GSI 2002 – 10.
7. N. Feather, *Proc. R. S. E.* **71**, 21 (1974).
8. G. Valskii, Sov. *J. Nucl. Phys.* **24**, 140 (1976).
9. P. Singer et al., *Proc. 3rd Int. Conf. DANF'96, Častá Papiernička, Slovakia,* eds. J. Kliman and B. Pustylnik (1996, Dubna, JINR) p.262.
10. M. Mutterer et al., *Proc. Int. Conf. Fission and Properties of Neutron-Rich Nuclei, Sanibel Island, FL.,* eds. J. H. Hamilton and A. V. Ramayya (1998, Singapore, World Scientific) p. 119.
11. A. V. Ramayya et al., *Phys. Rev. Lett.* **81**, 947 (1998).
12. J. H. Hamilton et al., *Acta Physica Slovaca* **49**, 31 (1999).
13. W. Greiner, *Acta Physica Slovacis* **49**, 9 (1999).
14. V. A. Rubchenya and S. G. Yavshits, *Z.Phys. A* **329**, 217 (1988).

COLD DEFORMED VALLEYS OF HEAVY AND SUPERHEAVY ELEMENTS

ŞERBAN MIŞICU*AND WALTER GREINER

Institut für Theoretische Physik, J W Goethe Universität
Frankfurt am Main
E-mail : misicu@th.physik.uni-frankfurt.de

Due to quadrupole and hexdecupole ground state deformations the cold synthesis of heavy and super-heavy elements is dependent on the mutual orientation of target and projectile. The cold fission is also strongly dependent on the deformation but the orientation before scission is singled-out by the trajectory which minimizes the action. The measured fission yields of ^{252}Cf are providing a strong support that the dominant orientation window is the one in which the fragments are nose-to-nose. We therefore take the endeavor to interpret the recent mass distributions of quasi-fission and fusion-fission of some superheavy nuclei, recently measured in FLNR-Dubna, based on the orientation windows of the driving potential.

1. Introduction

The role of nuclear deformations was frequently investigated in the study of fusion processes. Many authors devoted their efforts to the case when the target is deformed and the projectile spherical [1]. Naturally, in the case when the nuclear deformations are switched-on, it is necessary to account for the relative orientation of the projectile and the target in the calculation of the fusion barrier. As concluded in [2] the orientation of the colliding nuclei has a significant effect not only on the fusion barrier height but also on the compactness of the touching configuration.

The double-folding procedure of calculating the heavy-ion potential offers an excellent tool for expliciting in a very easy way the orientation of incomming nuclei [3]. Thus, the interaction can be written as a multipolar sum of terms in which the radial dependence, given by the distance R between the centres of the two nuclei, is separated from the orientational one,

---

*Supported by the European Comission through a Marie Curie Fellowship

specified by the angles ϕ_i, θ_i $(i = 1, 2)$ [4]:

$$V(R, \theta_1, \phi_1, \theta_2, \phi_2) = \sum_{\lambda_i \mu} V_{\lambda_1 \lambda_2 \lambda_3}^{\mu - \mu \ 0}(R) \cos \mu(\phi_2 - \phi_1) d_{\mu 0}^{\lambda_1}(\theta_1) d_{-\mu 0}^{\lambda_2}(\theta_2) \quad (1)$$

In the above expression the nuclear part is calculated using M3Y $N - N$ effective interactions. Then driving potential is defined as the difference between the interaction potential (1) and the decay energy $Q = B_{CN} - B_1 - B_2$ of the fusion or fission reaction

$$V_{\text{driv}} = V(R, \theta_1, \phi_1, \theta_2, \phi_2) + B_1 + B_2 - B_{CN} \quad (2)$$

B_1, B_2 and B_{CN} are the binding energies of the projectile, target and compound superheavy nucleus. Although not specified in the above formulas V_{driv} is also depending on the charges Z_i and masses A_i of the participating nuclei.

According to the Fragmentation Theory [5], the driving potential of a given compound nucleus is calculated for all possible projectile-target combinations as a function of the mass and charge asymmetries, $\eta = (A_1 - A_2)/(A_1 + A_2)$ and $\eta_Z = (Z_1 - Z_2)/(Z_1 + Z_2)$ at the barrier distance R_{bar}. The charges of the target and the projectile are determined by requiring that for a fixed η the driving potential $V_{\text{driv}}(R, \eta, \eta_Z)$ attains a minimum in the η_Z direction, i.e. for every fixed mass pair (A_1, A_2) a single pair of charges is determined among all possible combinations. Next, minimas of the potential in the two-dimensional (R, η) landscape are searched. In ref. [5] it was inferred a criterion for cold fusion, i.e. the deep minima of the driving potential are corresponding to the projectile-target combinations where the compound nucleus has a minimum of excitation and will de-excite to the ground-state with the emission of a couple of neutrons. The occurence of the mass-asymmetry valleys is due to the shell effects. It was advocated in ref. [6], using the frame of the fragmentation theory, that due to the existence of different mass-asymmetry valleys for the same compound system, a new, highly asymmetric fission mode appears in which one of the fragments is close to the double magic nucleus ^{208}Pb. This new type of decay was later on confirmed experimentally and is known in the literature as cluster radioactivity [7].

After reviewing the experimental state of the art in the next section, we present calculations for ^{252}Cf and the recently investigated superheavy nuclei ^{286}112 and ^{296}116.

2. The Experimental Status

Recently the the possible formation of superheavy nuclei with $Z = 102-122$ in reactions with ^{22}Ne, ^{26}Mg, ^{48}Ca, ^{58}Fe and ^{86}Kr ions at energies near and below the Coulomb barrier has been studied at FLNR Dubna [8]. Along with other relevant quantities the mass distributions of fission fragments was determined. The nuclei ^{256}No, ^{286}112, ^{292}114 and ^{296}116 have been produced in reactions with ^{48}Ca at the same excitation energy $E^* \approx 33$ MeV. The main pecularity of the data is that the quasi-fission mechanism is dominating in the case of the decay of the nuclei ^{286}112, ^{292}114 and ^{296}116. The mass distribution of quasi-fission products was found to be asymmetric with the peak of the lighter fragment located at $A_1 \approx 80$ for the superheavy nucleus ^{286}112, ≈ 84 for the superheavy nucleus ^{292}114 and ≈ 86 for ^{296}116. The authors also claimed that despite the dominating contribution of the quasi-fission process, in the symmetric region of fission fragment masses $(A_{CN}/2 \pm 20)$, the process of CN fission is prevailing. Indeed, in the light fragment mass-distribution a small and weakly asymmetric bump occurs for all three superheavy nuclei and the light fragment mass is constantly centered around $A_1 = 132 - 134$(see Fig.1 of [8]). For the the nucleus ^{296}116 the data presents another pecularity : an additional small bump centered on $A_1 = A_{CN}/2$, i.e. a totally symmetric distribution of decay products.

3. The pole-pole orientation window in cold fission of ^{252}Cf

In Fig.1 a cut along the mass asymmetry coordinate of the driving potential of ^{252}Cf is represented. The solid curve corresponds to the $p - p$ scenario whereas the dashed one to the $e - e$ one. The differences are tremendous. The Ca-valley (Pb is inside this valley) is very pronounced for the $e - e$ orientation and very shallow for the $p - p$ one. The deepest valley in the $e - e$ configuration is the quasi-symmetric one centered on a Sn isotope (other than the double magic ^{132}Sn) whereas for the $p - p$ configuration the quasi-symmetric valley, this time centered on ^{132}Sn, is overtaken by a broader, deeper and more asymmetric valley which contains the splittings Sr+Nd, Zr+Ce and Mo+Ba.

A couple of years ago we calculated in a $p - p$ scenario the mass-yields of ^{252}Cf [9] and compared them to those observed in experiment [10]. According to the experiment there is a region of small yields where the fragments are spherical or nearly spherical (with a small prolate deformation) and have high-Q values. This region is apparently centered on Sn isotopes and was claimed to be associated with the double magic ^{132}Sn, a fact which should

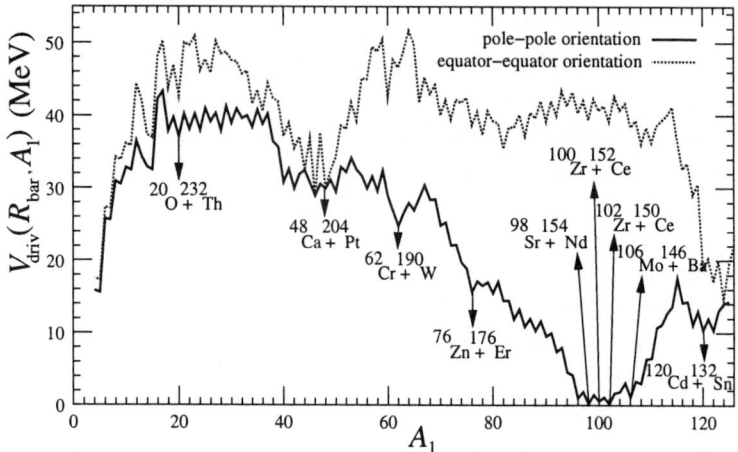

Figure 1. Driving potential of ^{252}Cf as a function of the projectile mass A_1 for the pole-pole orientation (continous line) and for the equator-equator orientation (dashed lines) at the barriers value of the R coordinate.

be interpreted in the light of the cluster radioactivity[7] at weaker asymmetries. In the region of higher yields, extending from $A_2=138$ to 156 the mechanism of fission was claimed to be different. In fact, according to our calculations, the quadrupole and hexadecupole deformations are determining a drastic lowering of the barrier in this region and thus a cold valley, deeper than in the Sn one, occurs. These calculations are consequently excluding the contribution to the cold spontaneous fission of other orientations, such as the $e - e$ for example. Our believe is that in the both regions of yields we deal with the same mechanism of fission, and not with two different mechanisms as adocated in [10]. This unique mechanism is explained by the occurence of pronounced cold valleys beyond the saddle, for strong or weak asymmetric fragmentations, a fact which triggered at the end of 1970's the prediction of cluster radioactivity.

Very recently new experimental studies on the spontaneous fission of ^{252}Cf [11] are confirming the existence of the regions with high yields in the region 134-158 of the heavy fragment. They are also providing much smaller yields in the region of Sn, a fact which is also in agreement to our calculations [12].

4. Orientation Windows in the Cold Fission and Quasifission of $^{286}112$ and $^{296}116$

In Fig.2 a cut along the mass-asymmetry coordinate of the driving potential of the superheavy nucleus $^{112}286$ is given. From the inspection of this figure we remark the differences in the driving potential when the target and the projectile are colliding with different orientations. The Ca-valley (with heavy partner U) is more pronounced when the fragments are coming in contact in $e - e$ or $e - c$ orientations. The valleys corresponding to the standard cluster radioactivity (superasymmetric valleys) are occuring with some differences: for $p - p$, the most pronounced are centered on ^{14}C, ^{22}Ne and ^{28}Mg, whereas for $e - c$ we remark ^{24}Ne, ^{30}Mg and ^{34}Si

For $p - p$ oriented fragments the valley centered on ^{62}Cr is separated by a high barrier from what we call the Pb-valley centered on ^{76}Zn and ^{80}Ge, and obviously for such configurations the tendency of the initial cold strong asymmetric system to move in the symmetric direction, before undergoing quasi-fission, is hindered.

The $p - e$ orientation presents features common to the $p - p$ but also to the $e - e$ and $e - c$ orientations. Similar to the $p - p$ case, the valley for Ca is less pronounced and the Mo-valley is broader and the target ^{96}Sr will give a new minimum in the potential for this orientation.

For the $e - e$ orientation we deal with three important valleys: the Ca-valley, the Pb-valley,centered on ^{80}Ge and the Sn-valley centered mainly on ^{134}Te. As can be seen in Fig.2 this last valley is common also to the $p - p$ orientation and is assigned in the FLNR-Dubna experiment to events emerging from the fusion-fission process. On the other hand between the Pb-valley and the Sn-valley a huge barrier is showing up which determines an even stronger hindrance to symmetric quasi-fission compared to $p - p$ orientation. The Pb-valley, which is very pronounced, is centered on the same light fragment mass numbers as the mass distribution reported in [8] and assigned to quasi-fission.

Making the association $p - p$ *deepest valley* $\rightarrow$ *maximum cold fission yield*, which seems to be valid in the case of ^{252}Cf, we infer that the occurence of light fragments with masses around A_1=132-134 can be due also to the fission of the compound superheavy nucleus $^{286}112$. On the other hand the quasi-fission mass distribution seems to be explained by means of the prevailance of an orientation close to the $e - e$ one. In the entrance channel we have a stable Ca-valley in the $e - e$ scenario. Since the closest valley from this one is the Pb- valley, then we expect the highest

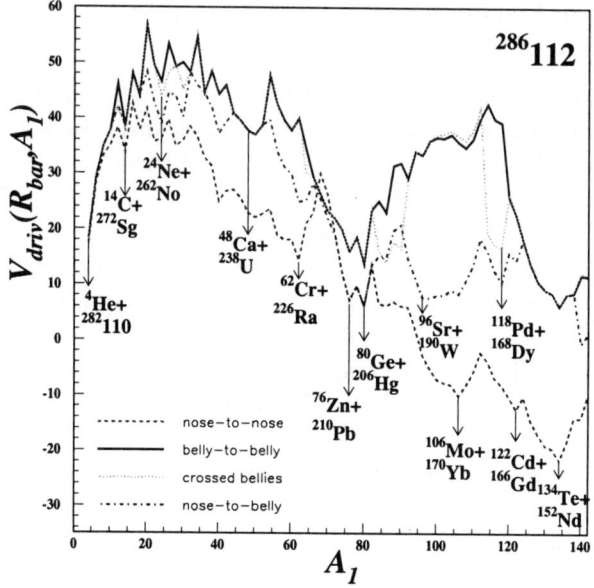

Figure 2. Driving potential as a function of the projectile mass A_1 for the superheavy nucleus $^{286}112$. The driving potential along the A_1-coordinate at $R = R_{\text{barier}}$ is represented for spherical fragments (full line), for deformed fragments oriented in nose-to-nose configuration (full line), belly-to-belly (full line), crossed bellies (dotted lines) and nose-to-belly(dot-dashed).

quasi-fission yields to correspond to a region around $A_1 \approx 80$. Since the excitation energy produced in the reaction ($E^* \approx 33$ MeV) is large enough such that a part of the flux moving in the symmetric region reaches the weak asymmetric valley of Sn and subsequently decay, we do not exclude that a part of the bump recorded in experiment, and assigned to the fusion-fission, is contaminated to a certain extent with quasi-fission events.

In the case of the superheavy nucleus $^{296}116$ the driving potential is qualitatively very similar to the one corresponding to $^{286}112$ (see Fig.3). There are however some differences. the minimas of the valleys are shifted towards higher values of the mass. In the $e - e$ orientation the Pb-valley is centered around $A_1 \approx 86$ whereas the Sn-valley remains almost unchanged. Instead the deepest $p - p$ valley is now centered on the symmetric region and the fragmentations with light mass number 132-134 are contributing,

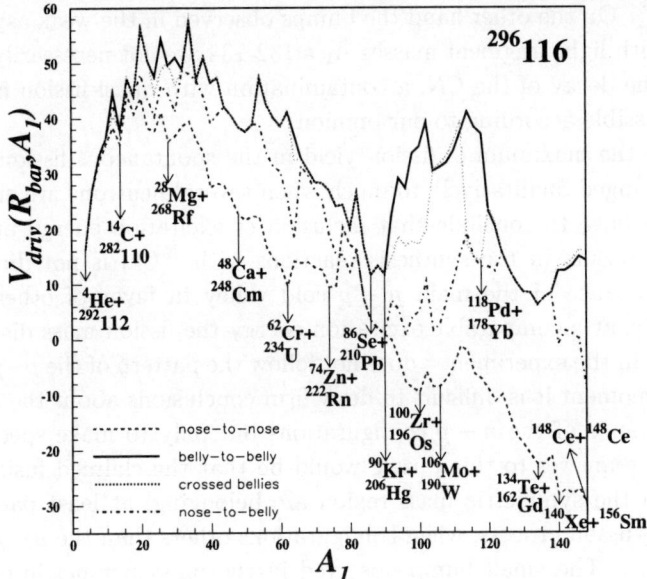

Figure 3.   Same as in previous figure for the superheavy nucleus ^{296}116.

at least for low excitation energy, with yields smaller than in the preceeding case.

The additional bump recorded in experiment for ^{296}116 and centered on the symmetric region is, in the light of our calculations, a good candidate for fusion-fission signature. However the weak asymmetric yields centered on $A_1 \approx 132\text{-}134$ could be , like in the case of ^{286}112, an admixture of quasi-fission and fusion-fission events.

5. Conclusions

Binary cold spontaneous fission is a rearrangement process of the mother nucleus in two fragments with a minimum of energy dissipation. The decaying system tends to choose an optimum path in which the emerging fragments are oriented pole to pole. Only for such a configuration the mass distribution observed in experiment can be understand.

The $p - p$ orientation produces no valley in the mass region where high quasi-fission yields are observed, confirming thus the conclusion of the au-

594

thors of [8]. On the other hand the bumps observed in the weak asymmetric region with light fragment masses $A_1 \approx 132$-134 are not necessarily entirely due to the decay of the CN, a contamination with quasi-fission fragments being possible according to our opinion.

Since the maximum of fission yield in the spontaneous fission of ^{252}Cf is not changed qualitatively to much when several neutrons are emitted [11] one then have to conclude that inclusion of excitation energy around 33 MeV, as occurs in the synthesis reactions with ^{48}Ca, is not diminishing the importance of the main $p - p$ cold valley in favor of other valleys. Obviously, at a comparable excitation energy the fission mass distribution obtained in the experiment [8] does not follow the pattern of the $p-p$ valleys. For the moment it is difficult to draw firm conclusions about the apparent unimportance of the $p - p$ configurations but only to make speculations. A possible answer to this puzzle would be that the claimed fusion-fission events in the symmetric mass region are belonging, at least partially, to the quasi-fisson process, when configurations others than the $p - p$ one are important. The small bump observed in the mass-symmetric region for $^{296}116$ is likely to be due to the fission of the compound nucleus in our opinion.

References

1. A.Iwamoto and P.Möller, *Nucl.Phys.* **A605** (1996) 334.
2. A.Iwamoto, P.Möller, J.Rayford Nix and H.Sagawa, *Nucl.Phys.* **A596** (1996) 329.
3. F.Carstoiu and R.J.Lombard, *Ann.Phys. (N.Y.)* **217**, 279 (1992).
4. Ş. Mişicu and W.Greiner, *J.Phys.G* **28** (2002) 2861.
5. A.Săndulescu, R.K.Gupta, W.Scheid and W.Greiner, *Phys.Lett.*60B, 225 (1976).
6. A.Săndulescu and W.Greiner, *J.Phys.G* **3** L189 (1977).
7. A. Săndulescu and W.Greiner, *Rep.Prog.Phys* **55** (1992) 1423.
8. M.G.Itkis et al.,in International Workshop on *Fusion Dynamics at the Extremes*, edited by.Yu.Ts.Oganessian and V.I.Zagrebaev, World Scientific, Singapore, p.93, 2001.
9. A.Săndulescu, Ş.Mişicu, F.Cârstoiu, A.Florescu and W.Greiner, *Phys. Rev.* **C 57** (1998) 2321.
10. M. Crönni et al., in *Fission and Properties of Neutron-Rich Nuclei*, eds.J.H.Hamilton and A.V.Ramayya, World Scientific, Singapore, 1998, p.109.
11. V.A.Kalinin et al., contribution published in this volume (1981) 281.
12. Ş. Mişicu and W.Greiner, *J.Phys.G* **28** (2002) 2861.

New Experimental Techniques and Directions

SASSYER: CURRENT STATUS AND FUTURE PROSPECTS

C.W. BEAUSANG[†]

Wright Nuclear Structure Laboratory.
Yale University,
New Haven, CT 06520-8124, USA
E-mail: Cornelius.beausang@yale.edu

The Small Angle Separator Array at Yale for Evaporation Residues (SASSYER), is a large-acceptance gas-filled recoil-separator recently commissioned at the Wright Nuclear Structure Laboratory where it is coupled to the Yale Rochester Array for SpecTroscopy (YRAST) Ball gamma-ray spectrometer. Some of the features of SASSYER and YRAST Ball are briefly described and some of the early physics results reported.

1. Introduction

Over the last several years the coupling of powerful germanium detector arrays to recoil separators has lead to tremendous advances in nuclear structure physics, enabling exquisitely sensitive experiments to be carried out at, and perhaps beyond, the limits of nuclear existence. For example the proton drip line has been delineated essentially up to $Z \sim 82$ resulting in the observation of many new interesting phenomena, among them the observation of proton decay and fine structure in proton decay. It is the extremely clean channel selectivity of recoil separators, particularly when combined with the additional selectivity of alpha decay tagging that made such experiments possible. Examples of such coupled devices include the RITU/Jurosphere array at the University of Jyvaskyla, Finland [1] and the Argonne FMA/Gammasphere combination in the U.S. [2].

The Wright Nuclear Structure Laboratory at Yale University has recently commissioned a new gas filled separator, the Small Angle Separator System at Yale for Evaporation Residues, or SASSYER. For gamma-ray detection, the device is coupled to the YRAST Ball (Yale Rochester Array for SpecTroscopy) array [3], the largest university based spectrometer in the U.S. In this paper the design and performance of the SASSYER/YRAST Ball combination will be briefly described and some of the early physics program outlined.

[†] Work supported by the U.S. D.O.E. under grant number DE-FG02-91ER-40609.

598

2. SASSYER and YRAST Ball

SASSYER consists of three magnets arranged in order as a vertically focusing gradient field dipole, a horizontally focusing quadrupole lens and finally a second gradient field vertically focusing dipole. The overall length of the separator is 2.4 m while the acceptance is approximately 5 msr. The maximum $B\rho$ is 2.2 Tm. These numbers are similar to those of other comparable separators. When used in conjunction with YRAST Ball the target is located approximately 15 cm upstream from the entrance to the first dipole, at the center of the field of view of the YRAST Ball clover Ge detectors. The entire magnet volume and the target chamber are filled with ~ 1 Torr of He gas which is isolated from the beam line vacuum by means of a 50 $\mu g/cm^2$ carbon window. The use of gas offers several advantages, primarily a large increase in the transmission efficiency for fusion evaporation residues relative to vacuum devices. Due to charge exchange interactions with the gas the recoils acquire an average charge state so that can be focused onto a sensible sized focal plane detector. The transmission efficiency of SASSYER was measured to be ~ 16% in our commissioning experiments, although clearly this changes from experiment to experiment. A second advantage is the convective cooling of the target by the gas allows the use of much larger beam currents with a consequent increase in reaction rate.

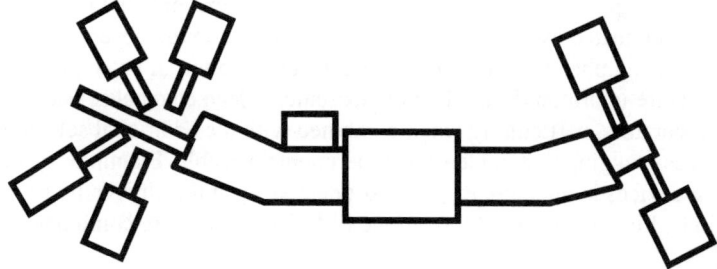

Figure 1: Schematic illustration of SASSYER and YRAST Ball. The beam enters from the left and is dumped into the beam dump located on the side of the first dipole magnet. The detectors of YRAST Ball are schematically illustrated at the target and focal plane positions.

SASSYER has a long and distinguished history. Designed and built by Al Giorso at Lawrence Berkeley National Laboratory, SASSY I and SASSY II was used in the synthesis of element 110 at the HILAC in Berkeley [4]. We gratefully acknowledge the generosity and help of our many friends and

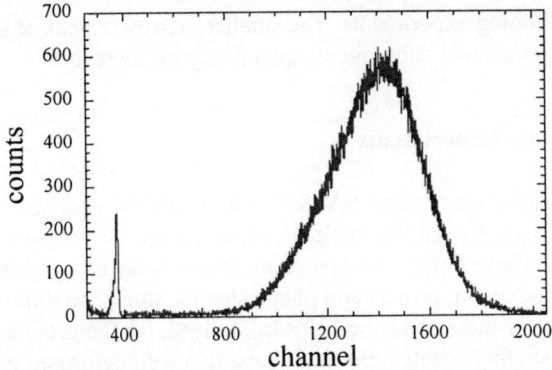

Figure 2: Typical energy spectrum recorded in one of the solar cell detectors. The broad peak corresponds to an implantation of a residue while the narrower peak at lower energies to a subsequent alpha decay. The recoils in this case were ^{150}Dy nuclei produced following the ^{120}Sn(^{34}S,4n) reaction.

colleagues at Berkley who made the separator available to us and whose advice and expertise was invaluable in commissioning the device.

The YRAST Ball multi-germanium detector array [3] is accommodated at the target position of SASSYER. In this location, two mechanical configurations of YRAST Ball are possible. The first consists of a maximum of 13 clover Ge detectors arranged at 41° (4 detectors) and 90° (9 detectors) with respect to the beam axis, giving a total photopeak efficiency of ~4%. A second configuration accommodates 9 clover Ge detectors at 90° and 11 smaller (~25% relative efficiency) Ge detectors arranged in two rings at 160° (3 detectors) and 126° (8 detectors) to the beam axis giving a total photopeak efficiency of ~3.1%. Currently 4 clover detectors are mounted at 41° and 5 and 90°. In addition several LEPS detectors are typically also mounted at 90°.

The SASSYER focal plane instrumentation currently consists of a 30-element solar cell array to detect fusion evaporation recoils and their subsequent alpha decay. These are arranged in a 3 cell high by 10 cell wide configuration. Each solar cell has an active area of 1 cm^2. In addition, a DSSD detector and a PPAC are also available. To search for isomeric gamma decays six or more ~25% coaxial Ge detectors from the YRSAT Ball array are arranged in a close geometry around the focal plane chamber. The overall setup is illustrated schematically in figure 1.

Figure 2 shows a typical spectrum in one of the solar cell detectors following an implantation of a recoiling fusion product. The large broad peak corresponds to the residue implantation, in this case a ~30 MeV ^{150}Dy nucleus produced in one of our commissioning experiments. The smaller, narrower peak at lower energies corresponds to a subsequent alpha decay of the recoil.

3. Early Physics Experiments

Early physics experiments using SASSYER and YRAST Ball have focused on the structure of light Rn and Ra nuclei in particular on the structure of ^{202,203}Rn and ^{209,210,211}Ra. These nuclei are interesting for a variety of reasons, including predictions of coexisting prolate and oblate shapes, similar to those found in Pt/Hg nuclei below the Z = 82 shell closure. Indeed, for the ground state, a change from a slightly oblate deformed shape to a well deformed prolate shape has long been predicted to occur for neutron number ~ 110. In addition, more exotic phenomena, including shears bands and superdeformation may also be found in this region.

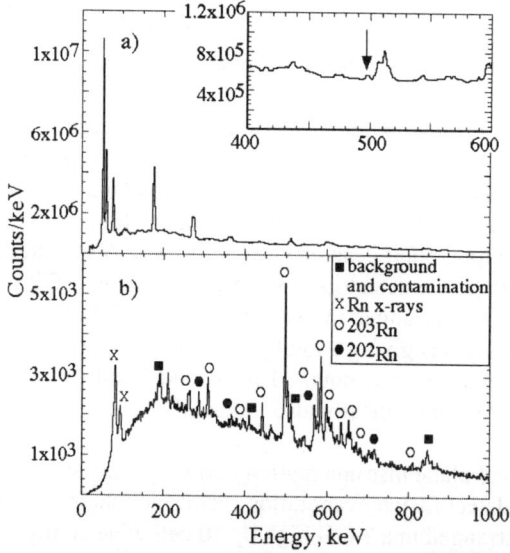

Figure 3: Top: YRAST Ball spectrum following the ^{32}S + ^{174}Yb reaction. The insert shows the region around the 511 keV annihilation peak. The arrow points to the 498 keV transition, the most intense transition in ^{203}Rn. Bottom. The same spectrum, but now gated by a recoil residue at the SASSYER focal plane

Little experimental information is available for these nuclei, particularly for the odd Rn and Ra isotopes, primarily due to the low fusion cross section and very large fission background encountered in such heavy systems. Such systems provide an ideal testing ground for SASSYER as the fusion cross section, while low is still on the order of several hundred microbarns. In addition, we have previously studied the Rn nuclei at Yale using YRAST Ball and we can directly compare our previous raw gamma-gamma data to that selected by SASSYER.

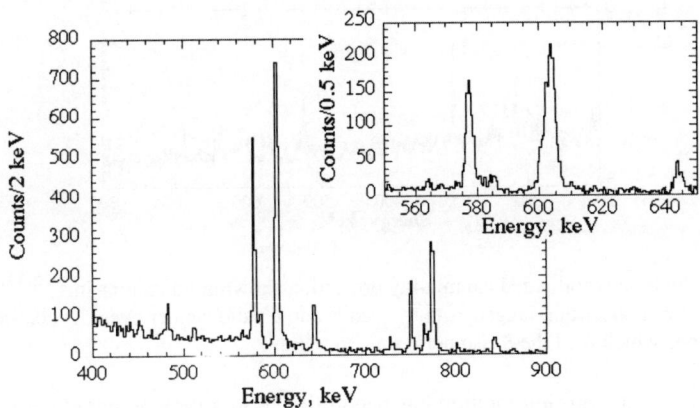

Figure 4: Gamma-ray transitions recorded at the focal plane of SASSYER within 1 µs of a recoil implantation. The spectrum is dominated by transitions below the 8$^+$ isomeric state in ^{210}Ra. The insert shows in more detail the region around 600 keV.

That the comparison is indeed dramatic is illustrated in figure 3. The top portion of which shows a total projection of a gamma-gamma matrix measured with YRAST Ball following the ^{32}S + ^{176}Yb $\rightarrow$ ^{203}Rn + 5n reaction. The spectrum is dominated by a very large, essentially smooth, background originating from fission fragment gamma-rays and by several very intense Coulomb excitation gamma-rays. The largest fusion evaporation peak, the 498 keV (17/2$^+$) $\rightarrow$ (13/2$^+$) transition in ^{203}Rn is just barely visible to the left of the 511 keV annihilation peak.

On the other hand, the lower portion of figure 3 shows a similar projection of a gamma-gamma matrix, but now selected by Rn recoils detected at the SASSYER focal plane. The difference is dramatic, the spectrum is now dominated by transitions from fusion-evaporation residues (mostly ^{202,203}Rn), the 498 keV peak being the largest in the spectrum. A total of ~6 x 10^6 recoil gated gamma-gamma coincidence events were recorded in around 1 week of beam time, enough to go recoil gated gamma-gamma coincidences. Preliminary analysis of these data confirm most of our previous results [5] while adding several new transitions to both ^{202}Rn and ^{203}Rn.

602

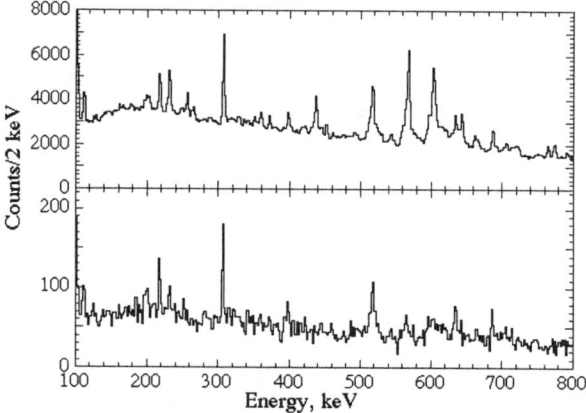

Figure 5. The total recoil gated gamma-ray projection showing transitions in ^{209,210,211}Ra (top). The lower spectrum, tagged both by recoils and a valid isomer event, is dominated by transitions which feed the 8^+ isomeric state in ^{210}Ra.

The isomer decay tagging method has been utilized in a very recent experiment to investigate the structure of the $Z = 88$ Ra nuclei ^{209,210,211}Ra. Prior to this experiment, no information was available on these odd mass Ra nuclei, while the level structure (without transition energies) of ^{210}Ra has been reported only up to the 8^+ state [6]. This state is isomeric with a lifetime of 2.2 µs. Figure 4 shows the gamma-ray spectrum recorded at the focal plane of SASSYER within 1 µs of a recoil implantation. As can be seen, the spectrum is dominated by transitions in ^{210}Ra that lie below the 8^+ isomeric state. The isomer tagged spectrum of prompt gamma-rays, i.e. those which feed the 8^+ isomeric state, is shown in the lower part of figure 5 where is can be compared to the total recoil gated gamma-ray spectrum (figure 5, top).

While analysis of these data is still underway it is clear that a wealth of new information will be available on both higher lying states in ^{210}Ra and for the first time, on the odd ^{209,211}Ra nuclei.

Looking to the future, we anticipate placing the New Yale Plunger Device (NYPD), a recoil distance plunger device at the target position of SASSYER. By use of a retardation foil instead of a stopper foil, recoils can still be focused through SASSYER. Hence, with the extremely clean channel selection provided by SASSYER this should allow the measurement of lifetimes in extremely weak reaction channels or in regions dominated by fission events.

4. Conclusions

In conclusion, the SASSYER gas filled recoil separator has been commissioned at the Wright Nuclear Structure Laboratory at Yale University. Coupled to the YRAST Ball array, the separator provides a new and powerful tool for gamma-ray spectroscopy at the extremes, especially in regions of the chart of the nuclei where fission processes dominate fusion. Early experiments using SASSYER/YRAST Ball have concentrated on Rn/Ra nuclei with $N \sim 115$. Data analysis is underway, but early results look promising. In the near future we anticipate placing the NYPD plunger device at the target position of SASSYER/YRAST Ball allowing us to carry out sensitive lifetime measurements on weakly populated nuclei in regions dominated by fission processes.

Acknowledgments

Many people are responsible for the successful commissioning of SASSYER. In particular I wish to acknowledge the efforts of Jo Ressler and Reiner Kruecken and the technical staff at WNSL. This work is supported by the U.S. D.O.E. under grant number DE-FG02-91ER-40609.

References

1. M. Leino et al., *Nucl. Inst. and Meth.* **B99,** 653 (1995).
2. C.N. Davids et al., *Nucl. Inst. and Meth.* **B70,** 358 (1992), and C.N. Davids and J.D. Larson, *Nucl. Instrum. and Meth.* **B40/41,** 1224 (1989), and see http:// www.phy.anl.gov/fma for information regarding the FMA and its focal plane detectors.
3. C. W. Beausang et al., *Nucl. Inst. and Meth.* **A452/3,** (2000) 431.
4. A. Ghiorso et al., *Phys. Rev.* **C51,** (1995) R2293.
5. H. Newman et al. *Phys. Rev.* **C64,** 2001 027304.
6. J.F.C. Cocks et al., *J. Phys.* **G25,** (1999) 839

CROSS-SECTION MEASUREMENTS WITH THE RADIOACTIVE ISOTOPE ACCELERATOR (RIA)[*]

M.A. STOYER, K.J. MOODY, J.F. WILD, J.B. PATIN, D.A. SHAUGHNESSY,
N.J. STOYER, L.J. HARRIS

Lawrence Livermore National Laboratory, Livermore, CA 94550
E-mail: mastoyer@llnl.gov

RIA will produce beams of exotic nuclei of unprecedented luminosity. Preliminary studies of the feasibility of measuring cross-sections of interest to the science based stockpile stewardship (SBSS) program will be presented, and several experimental techniques will be discussed. Cross-section modeling attempts for the A = 95 mass region will be shown. In addition, several radioactive isotopes could be collected for target production or medical isotope purposes while the main in-beam experiments are running. The inclusion of a broad range mass analyzer (BRAMA) capability at RIA will enable more effective utilization of the facility, enabling the performance of multiple experiments at the same time. This option will be briefly discussed.

1. Introduction

Effective stewardship of the U.S. nuclear weapons stockpile requires a program based on deeper scientific understanding of the technical issues rather than the "trial-and-error" program prevalent during the underground testing era. This requirement is forcing us to examine all aspects of the Stockpile Stewardship Program (SSP) with renewed vigor and with the aim to improve our scientific understanding of the operation of a weapon. The nuclear cross-section networks used to model device performance are being re-examined to include "state-of-the-art" evaluated cross-sections; a combination of improved theoretically modeled cross-sections and evaluated experimental data, which wasn't available when the cross-section sets were originally constructed. Many of the required cross-sections are for difficult or impossible to measure radioactive isotopes or isomers.

This paper focuses on several examples of cross-sections that could be measured, albeit in complex and difficult experiments, at the radioactive beam facility currently being discussed in the U.S., the Rare Isotope Accelerator (RIA).[1] RIA is envisioned as a luminous radioactive beam facility that is a hybrid of the projectile fragmentation (PF) method and the on-line isotope

---

[*] This work was performed under the auspices of the U.S. Department of Energy by the Lawrence Livermore National Laboratory under contract No. W-7405-Eng-48.

separator (ISOL) method. A simplified schematic of RIA is shown in Figure 1. The flexibility provided by having a source of higher energy projectile-like fragments from the PF method and lower energy target-like fragments from the ISOL method will enable such a facility to address a wide variety of physics issues and deliver beams spanning the extent of the chart of nuclides over a wide energy range. While the scientific basis for such a radioactive beam facility has been enunciated for decades[2], only recently has such a facility been considered as useful for SSP.

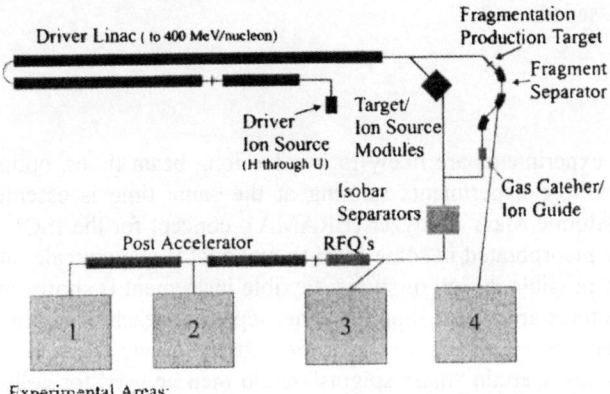

Experimental Areas:
1: < 12 MeV/u   2: < 1.5 MeV/u   3: Nonaccelerated   4: In-flight fragments

Figure 1. Simplified schematic layout of the RIA facility. Note that the isobar separator could include something like the BRAMA concept discussed in the text.

Several classes of experiments are possible at RIA. Since production of radioactive isotopes, regardless of the method employed, is not limited to just one nuclide at a time, on obvious type of experiment involves the harvesting of radioactive isotopes, and subsequent chemical or isotopic purification for use as target material.   Relevant SSP neutron-induced cross-sections could be measured with this harvested target material at a DT or DD neutron source located at RIA, a "white neutron" source such as WNR at LANSCE in Los Alamos, or following an irradiation at a reactor facility. Harvested isotopes would not be limited to those of interest to SSP, but would also include medically, astrophysically, environmentally, or chemically important isotopes.

A second type of experiment would involve the use of beams of isotopes of interest on surrogate neutron targets such as hydrogen, deuterium, helium, etc. Such cross-section measurements would help quantify parameters such as level densities and strength functions for input into models used to calculate (n,2n) and (n.γ) cross-sections. These inverse reactions on surrogate materials would be an indirect method of obtaining cross-sections of interest to SSP radiochemistry.

The choice of performing an in-beam experiment versus harvesting target material for subsequent experiments is determined by the half-life of the isotope to be studied, resultant activities of the material following production, and the cross-section to be measured. The design of the experiment would vary depending on the target isotope, reaction to be studied, and produced activity. Determination of the cross-section might involve mass spectrometry and/or radioactive decay counting, and would rely on the decay properties of the final isotope to be assayed.

2. BRAMA

Because these experiments are likely to involve long beam times, optimization of RIA for multiple experiments running at the same time is essential. The Broad Range Atomic Mass Analyzer (BRAMA)[3] concept for the ISOL method will need to be incorporated in some form to make RIA economically attractive and flexible. A possible design of such a flexible instrument is shown in Figure 2. The main features are a large dipole magnet separating each mass onto a focal plane in which individual masses can be extracted and sent to different experimental areas. Certain "mass spigots" could then be used for collection of target material, while others are used for in-beam experiments.

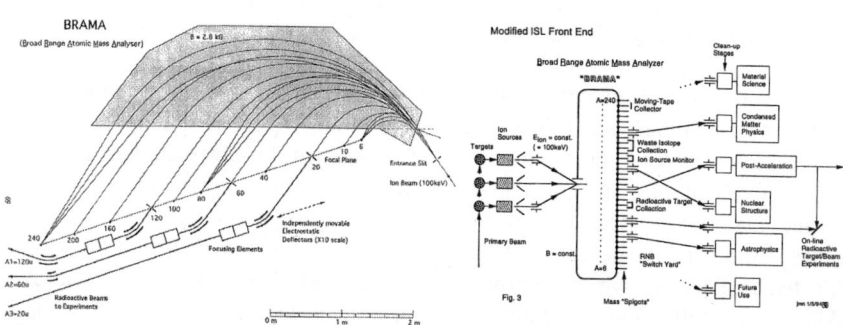

Figure 2. Schematic of the BRAMA separator. Note the multitasking capabilities of such an instrument.

3. Y/Zr Experiments

Y and Zr were sometimes positioned in a nuclear device to provide diagnostic information on its fusion performance in an underground test. Because core samples from a test would take days to be retrieved, the nuclides typically measured were ^{87,88}Y and ^{88,89}Zr. A large cross-section network (as shown in Figure 3 for Y) was used in conjunction with a sophisticated computer model of the test to calculate ^{87}Y/^{88}Y, ^{88}Y/^{89}Y, ^{88}Zr/^{89}Zr, and ^{89}Zr/^{90}Zr ratios to compare with the measured data. Much of the cross-section network is unmeasured and only calculated based on a reaction model and nuclear structure information. Historically, nuclear structure information for radioactive isotopes or isomers might be poorly known or completely unknown. Any measurement of a cross-section within these networks would be an improvement. Some estimates of important cross-sections, employing a $\Delta R/\Delta\sigma$ figure-of-merit, where ΔR is the change in a calculated ratio and $\Delta\sigma$ is the perturbation to the cross-section, were made. For the Y cross-section network, some obvious important cross-sections involving radioactive targets are ^{88}Y(n,2n)^{87m3}Y, ^{87m3}Y(n,γ)^{88}Y, ^{88}Y(n,2n)^{87}Y and the ^{88}Y(n,γ)^{89m3}Y. Similarly, for Zr the ^{89}Zr(n,2n)^{88}Zr, ^{89}Zr(n,γ)^{90m4}Zr, ^{89}Zr(n,γ)^{90}Zr and ^{88}Zr(n,γ)^{89m3}Zr are important. Note the importance of isomeric levels.

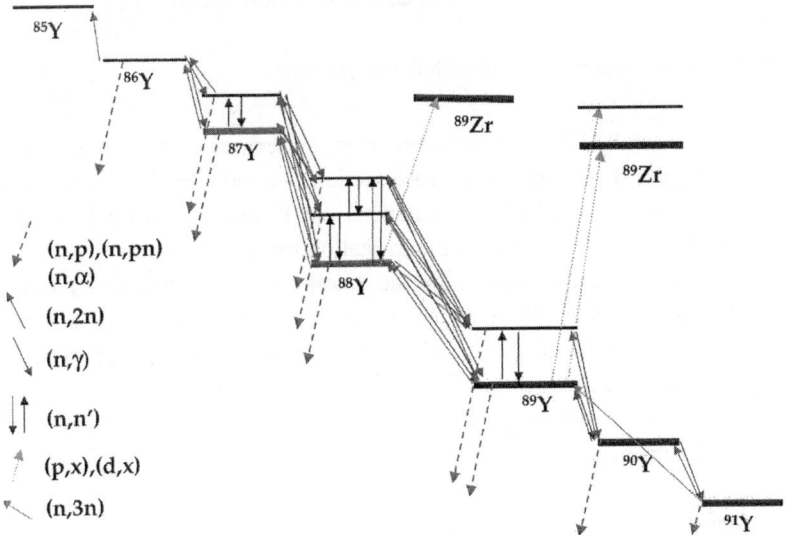

Figure 3. The Y cross-section network including charged particle reactions. The arrows indicate reaction cross-sections included over all relevant energies up to 20 MeV.

608

4. Fission Product Burnup

One of the isotopes measured which indicated the amount of fission that occurred during a nuclear test was [95]Zr. However, [95]Zr was not the nuclide present during the explosion—the long-lived precursor nuclides within the A=95 mass chain were. Any perturbation to the distribution of nuclides within the A=95 mass region (see Figure 4) due to subsequent neutron-induced reactions, especially those modifying the mass 95 chain yield, need to be determined.

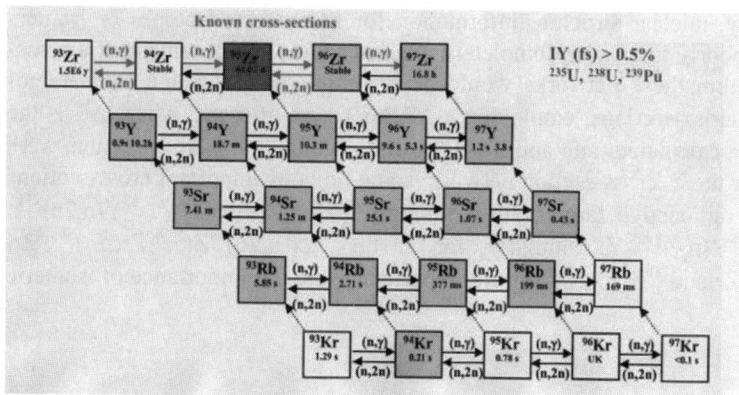

Figure 4. Mass 95 region of neutron-rich fission products.

STAPRE[4] and ALICE[5] calculations were performed on the A=95 nuclides shown in Figure 4 to generate a cross-section network for these neutron-rich nuclides. The (n,2n) STAPRE calculations are shown in Figure 5. Note the fluctuating threshold and shape of the excitation function with this region. Since the relative cross-sections determine the amount of A=96 nuclides producing A=95 nuclides, and A=95 nuclides producing A=94, and since the cross-sections are not identical, it is likely that an improved cross-section set will improve the interpretation of the [95]Zr measured. Neutron capture cross-sections, as well as any other important neutron-induced cross-section (such as (n,p) or (n,α)) will be included in the cross-section set.

5. Conclusions

Identification of important unknown cross-sections for SSP has begun. Because neutron-induced reactions on radioactive isotopes and isomers are important, a radioactive beam facility such as RIA will play a crucial role in measuring these cross-sections. Such measurements will improve the prediction of isotope ratios and the estimation of fission product burnup. BRAMA may provide RIA an important multi-tasking capability because most experiments will require long beam times.

(n,2n)

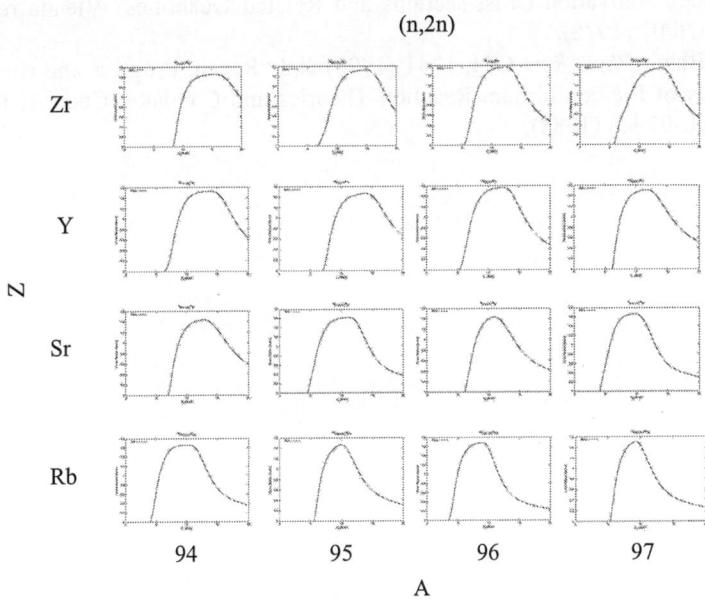

Figure 5. Results of STAPRE calculations for the A=95 region. All scales are the same with the y-axis from 0-1.8 barns and the x-axis from 0-20 MeV.

Acknowledgments

The authors would like to acknowledge Dr. Robert Hoffman for substantial help with the STAPRE calculations and Dr. Lee Bernstein for useful discussions. This work was performed under the auspices of the U.S. Department of Energy by the Lawrence Livermore National Laboratory under contract No. W-7405-Eng-48.

References

1. ISOL Task Force Report to NSAC on Nov. 22, 1999 found at URL = http://srfsrv.jlab.org/isol/ISOLTaskForceReport.pdf.
2. Proceedings of 1st International Conference on Radioactive Nuclear Beams held in Berkeley, CA (1989), "ISL Research Opportunities with Radioactive Nuclear Beams" LALP 91-51 (1991), NUPECC Report on European Radioactive Beam Facilities (May 1993).
3. J. M. Nitschke Proceedings of the Workshop on Post Accelerator Issues at the ISL held in Berkeley, CA Oct. 27-29 (1993) 64.
4. M. Uhl and B. Strohmaier "STAPRE: A Computer Code for Particle Induced Activation Cross-sections and Related Quantities" Vienna report IRK 76/01 (1976).
5. M. Blann *Phys. Rev.* **C54**, 1341 (1996) and "Recent Progress and Current Status of Pre-equilibrium Reaction Theories and Computer Code ALICE" UCRL-97948 (1988).

CHALLENGES OF FISSION RESEARCH AT THE IMPROVED IGISOL FACILITY

HEIKKI PENTTILÄ, TOMMI ERONEN, PETER DENDOOVEN, JANI HAKALA,
WENXUE HUANG, JUSSI HUIKARI, ARI JOKINEN, ANU KANKAINEN,
VELI KOLHINEN, STEFAN KOPECKY, ARTO NIEMINEN, ANDREY POPOV,
SAMI RINTA-ANTILA, YOUBAO WANG AND JUHA ÄYSTÖ

University of Jyväskylä, Department of Physics.
P.O.Box 35,
FIN-40014 University of Jyväskylä, Finland
E-mail: Heikki.Penttila@phys.jyu.fi

The new instrumentation for radioactive ion manipulation at the IGISOL facility has risen up a demand to improve the performance of the ion guide itself. The original gas cell technique IGISOL is shortly described and the current weaknesses pointed out. The program in progress to improve the ion guide performance is described. These improvements are expected to give the highest gain in the studies of fission fragments.

1. The ion guide technique

The ion guide technique for fast production of atomic ions, developed in Jyväskylä in early 1980s [1,2], is based on thermalization of nuclear reaction products recoiling out of the target in noble buffer gas, usually helium, where the thermalized reaction products after some charge exchange in the beginning of the slowing down process remain as 1^+ ions due to high ionization potential of the buffer gas. The ions are flushed out of the ion guide with the buffer gas flow and guided through a differential pumping section to the extraction and acceleration stages of an isotope separator.

Since the primary ionization mechanism is stripping when atoms are recoiling out of the target, it is universal and independent on the Z of the reaction product. When combined with proton-induced fission it provides a unique possibility to study nuclear structure of neutron rich nuclei in the mass range $A \approx 70 - 170$. In addition, the method is fast; for fission products, the evacuation time from the gas cell is of the order of a few milliseconds. [3] The proton-induced fission has indeed proven very successful in the research of neutron-rich nuclei and the fission process [2].

Figure 1 illustrates the present possibilities of the IGISOL facility. The yields given are based on theoretical proton-induced fission cross sections [4,5].

612

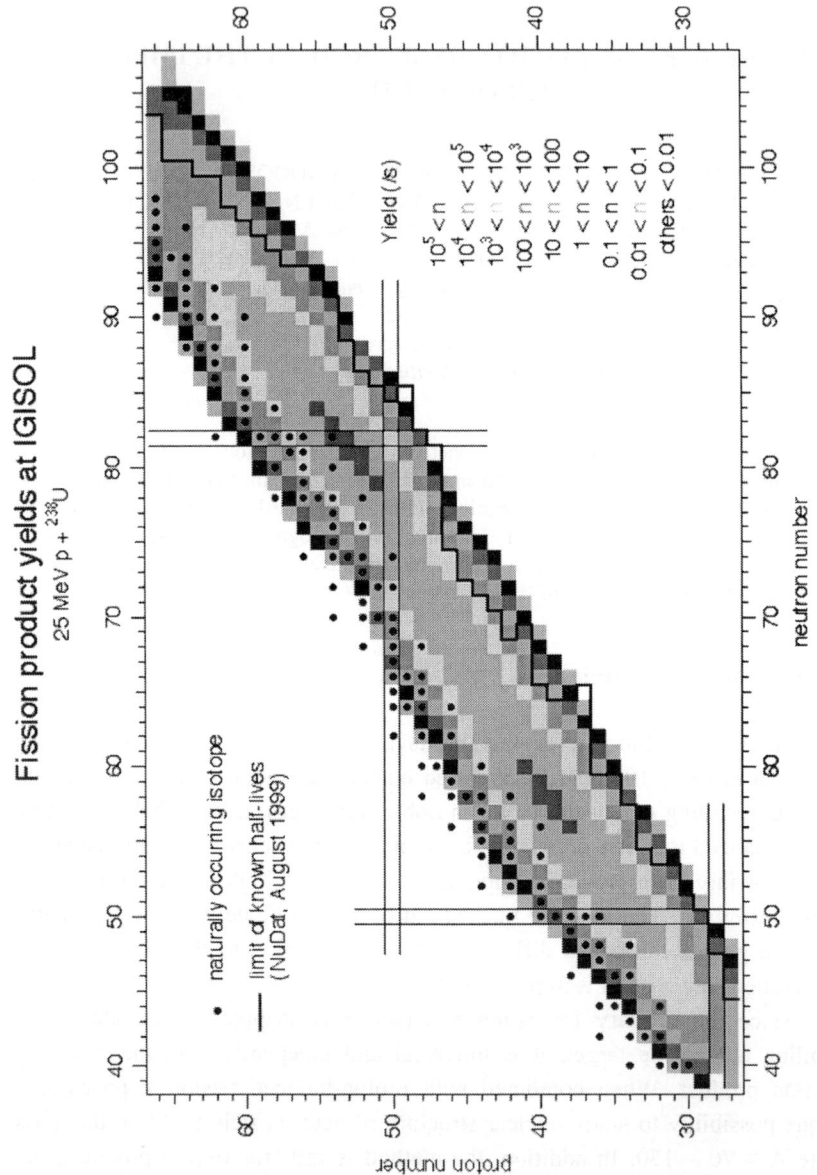

Figure 1. The estimated 25 MeV proton induced fission yield as mass separated ions at the present IGISOL. The limit of known half-lives is marked on the figure.

They are converted to independent mass separated yields by scaling with the known yield of ^{112}Rh with 15 µA proton beam on 15 mg/cm^2 target tilted at   7 degrees [6].   It is immediately seen that not even the half-lives of some nuclei with predicted yields of several hundreds per second are known. This is mostly due to the background from more copiously produced nuclei with the same mass number.   The mass resolving power of the dipole magnet is not adequate to separate different elements within an isobar.

2.   Ion cooler and trap project at JYFL

Isobaric separation of fission products was one initiative of the ion cooler and trap project [7,8] at JYFL. The separator beam is purified from the isobaric contaminants by a cleaning Penning trap whose designed mass resolving power $R = 10^5$ is adequate to this purpose.

Reduction of background is done partly on the cost of speed and efficiency of the system. Since the energy spread of the initial ion beam is too large that the beam could be directly shot into the Penning trap, it needs to be cooled and bunched first. An RF-cooler is build for this purpose. Cooling itself takes place in tens of microseconds and an efficiency of 60% in the transmission of the cooler has been reached. The improved ion beam quality and the possibility of bunching the ions have already proven very useful in collinear laser spectroscopy [9,10]. However, for the most short-lived species the radioactive decay losses in the ion beam bunching have to be taken into account.   The bunching period is dictated by the purification time in the Penning trap that depends of the required mass resolving power. In a recent test run a mass resolving power $R = 3 \times 10^4$ was reached in 120 ms. Thus, the isobaric separation time will stay well below half a second. [11].

3.   IGISOL upgrade

After all this development of sophisticated ion handling techniques to improve ion beam quality, the front end has become the weakest point of the IGISOL facility. Especially the system is handicapped by the limitation of pumping efficiency that is most pronounced in the extraction region of the isotope separator.

The efficiency of an ion guide depends on the initial energy of the stopped ions. For low energy ions from an α-recoil source an efficiency of 50% has been reached [12]. For light ion induced fusion evaporation reaction products a typical ion guide efficiency varies from 1 to 10 %. The reason for lower efficiency is the ionization of helium caused by the projectile beam.

The initial energy of fission fragments is of the order of 1 MeV/nucleon, a few hundred times of that of fusion or α-decay recoils. The typical efficiency for fission ion guide is less than 10^{-4}, the main reason of low efficiency being the inadequate stopping of fission fragments in the buffer gas. The pressure is limited to a few hundred millibars due to limited capacity of differential pumping.

There is place for improvement already in the first differential pumping section, where the bulk of helium gas is pumped away with Roots blowers [13]. The current design works satisfactorily for the fusion ion guide, but it leads to unsatisfying compromises for the physically larger fission ion guide (figure 2). The narrow pumping channels leave higher pressure in the crucial region near the skimmer hole. This result in higher gas load in the second stage of differential pumping that is originally designed for a hot cathode ion source, that is, for gas load way less than those from an ion guide of any kind. A narrow and long pumping channel between the extraction region and oil diffusion pump reduces the pumping efficiency.

In addition, the ion optical simulations propose that the ion transmission through the current extraction electrode may in some conditions be as low as 40%. This result is supported by the measured emittance of the isotope separator [14] that is smaller than expected. This is readily explained if the emittance of the separator is mainly dictated by the size of extraction electrode aperture. A wider extraction aperture would also have required increased pumping capacity in the extraction region. The design of the improved front end of the IGISOL facility is displayed in figure 2. The target chamber, enclosing the ion guide, is made larger to allow more efficient pumping of the skimmer region. In the new design there is also sufficient space for a sextupole RF ion guide or a RF funnel system between the ion guide and the extraction chamber.

The extraction chamber will be evacuated by one 8000 l/s oil diffusion pump, instead of the current one with 2000 l/s nominal efficiency. The pumping channel is widened and shortened. This should allow considerably increase the helium gas pressure inside the ion guide.

Other improvements will be better radiation shielding to allow higher beam intensities in the target, and remote control of valves and movable electrodes in the front end. A system to remotely change the ion guide has also been planned. These improvements are aimed to diminish the need to work in the high radiation area, also allowing for more flexible scheduling of experiments.

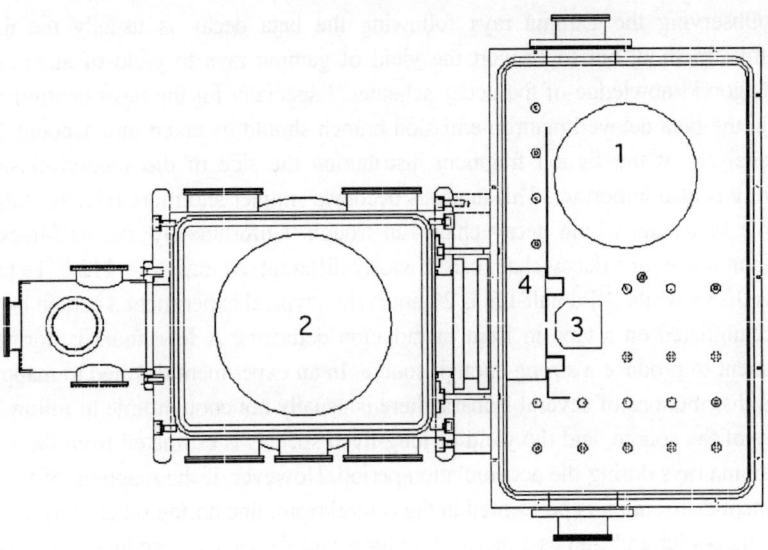

Figure 2. View of the upgraded IGISOL facility front end. 1. Target chamber, evacuated by Roots pumps. 2. Extraction chamber, evacuated by an oil diffusion pump. Extraction electrode is not displayed. 3. The ion guide. Dimensions are those of the fission ion guide. 4. Skimmer plate. The thicker line shows the location of skimmer system and target chamber wall in the present arrangement.

4. Fission studies

After these improvements, the IGISOL facility will provide an extraordinary opportunity to study the properties of more than 60 neutron rich nuclei that are not currently available anywhere else in the world. The spectroscopic studies of those include measurements of masses, moments and radii as well as decay properties and excited states.

In addition to this, it is possible perform detailed studies of the fission process itself.

It has to be stressed that all the ions from the ion guide are primary ions from the reaction. The distribution of separated ions is the same as the distribution of reaction products. Since the ion guide is fast, the observed distribution does not suffer from decay losses either. The distribution can be mapped utilizing the purification Penning trap, or via decay studies.

Observing the gamma rays following the beta decay is usually the most sensitive method, but to convert the yield of gamma rays to yield of atoms one needs good knowledge of the decay schemes. Especially for the most neutron-rich nuclei the beta delayed neutron emission branch should be taken into account. For the analysis of the fission fragment distribution the side of distribution close to stability is also important. This analysis becomes simpler and more reliable if there is time to let the whole decay chain run trough. Unfortunately, the half-lives of different nuclei in a decay chain can be vastly different. At mass A = 112 ^{112}Tc half-life is 0.28 s while ^{112}Pd half-life is 21 hours. In a typical experiment a certain isobar is accumulated on a tape in front of radiation detectors. A few hours is normally sufficient to produce a strong enough source. In an experiment devoted to mapping the Z distributions of several isobars, there is usually not enough time to follow the decay of the source, and the yield of long-lived species is estimated from the yield of gamma rays during the accumulation period. However, if the mapping of fission fragment distribution is performed in the central beam line during other experiments utilizing cooler and trap as experimental tools and fission as a production reaction, no beam time would be wasted in the source decay period.

References

1. Dendooven, P., *Nuclear Instr. and Methods B* **126**, 182 (1997).
2. Äystö, J., *Nuclear Physics A* **693**, 477 (2001).
3. Taskinen, P., et al, *Nuclear Instr. And Methods A* **281**, 539 (1988).
4. Huhta, M. et al, *Phys. Lett. B405*, 230 (1997).
5. Rubchenya, V., *Nuclear Physics A,* **701**, 127 (2002), and private communication.
6. Dendooven,P., et al, in *Proceedings of the 2nd Int. Workshop on Nuclear Fission and Fission-Product Spectroscopy, April 22-25, 1998*, Seyssins, France.
7. Jokinen, A., Huikari,J., Nieminen, A., and Äystö J, *Nuclear Physics A*, **701**, 557 (2002).
8. Szerypo, J., Jokinen, A., Kolhinen, V. S., Nieminen, A., Rinta-Antila, S. and Äystö J., *Nuclear Physics A*, **701**, 588 (2002).
9. Nieminen, A, et al, *Physical Review Lett.* **88**, 094801 (2002).
10. Campbell, P., et al. *Physical Review Lett.* **89**, 082501 (2002).
11. Kolhinen, V.S., private communication.
12. Peräjärvi, K., Huikari, J., Rinta-Antila, S., Dendooven, P., *Nuclear Physics A*, **701**, 570 (2002).
13. Penttilä, H., et al, *Nuclear Instr. and Methods B* **126**, 182 (1997).
14. JYFL Annual report 2001, p. 10, University of Jyväskylä (2002).

NEW REACTIONS FOR DETERMINING γ-RAY LINEAR POLARIZATION SENSITIVITIES WITH COMPTON POLARIMETERS

G. D. JONES

Oliver Lodge Laboratory, University of Liverpool, Liverpool, L69 7ZE, UK

Compton polarimeters have proved to be extremely useful in determining the electric or magnetic character of γ-radiation for transitions of 500 keV energy upwards. The use of Compton polarimeters (using a Ge scattering crystal) at γ-ray energies below a few hundred keV has been compromised in the past because the photo-electric cross section becomes greater than the Compton scattering cross section at energies just below $E_\gamma = 200$ keV. Therefore, the need to have a small scattering crystal (to allow Compton scattering events to escape) reduces the solid angle of a polarimeter to an unacceptable degree. The new generation of multi-pixel double-sided strip detectors and γ-ray tracking technology significantly increases the effective solid angle particularly for low energy Compton scattered events. However, since low energy γ-rays of high γ-ray linear polarization are not known from the literature, linear polarization sensitivities over a wide range of energies are difficult to determine.

It will be shown that the formula for the angular distribution of γ-ray linear polarization can be written in a more transparent form than in terms of the more usual Legendre and associated Legendre polynomials. This form emphasises the special nature of pure dipole $1 \rightarrow 0$ and pure quadrupole $2 \rightarrow 0$, γ-ray transitions, which produce full linear polarization ($|P| = 1$) over the complete angular range (except at exactly 0^0 or 180^0 to the quantization axis) when the initial states are fully aligned in their m = 0 substates.

This discovery gives rise to several new ways of producing γ-rays of high linear polarization over a wide range of energies (from below 200 keV upwards), which will allow linear polarization sensitivities to be determined for polarimeter calibration purposes. Several methods of production and important related problems will be discussed.

1. Introduction

The theoretical linear polarization angular distribution of γ-rays can be denoted [1] by $W(\theta,\xi)$ where as shown in Figure 1, θ is the angle between the quantization axis of the system (i.e. along Oz) and an emitted γ-ray. The angle ξ lies between the electric vector of the γ-ray and the reaction plane. We can resolve the amplitude of the electric vector into any two components at right angles, clearly the simplest choice is to choose one component in the reaction plane where $\xi = 0^0$ and the other perpendicular to it where $\xi = 90^0$. Squaring amplitudes we then have two components of intensity $W(\theta,\xi = 0^0)$ and $W(\theta,\xi = 90^0)$ perpendicular to each other. The degree of linear polarization may then be written as

$P(\theta) = \dfrac{W(\theta,\xi = 0) - W(\theta,\xi = 90)}{W(\theta,\xi = 0) + W(\theta,\xi = 90)}$ where $-1 \leq P(\theta) \leq 1$ and $P(\theta) = 0$ for unpolarized γ-rays.

However, γ-rays preferentially Compton scatter in the plane normal to the direction of the electric vector of the incident photon and the experimental observables are usually $N(\theta, \phi = 0)$ and $N(\theta, \phi = 90)$ representing the full energy peaks of the sum coincidence spectra of energy deposited in the region of the scattering centre plus energy deposited in the analyser region (see Fig. 1).

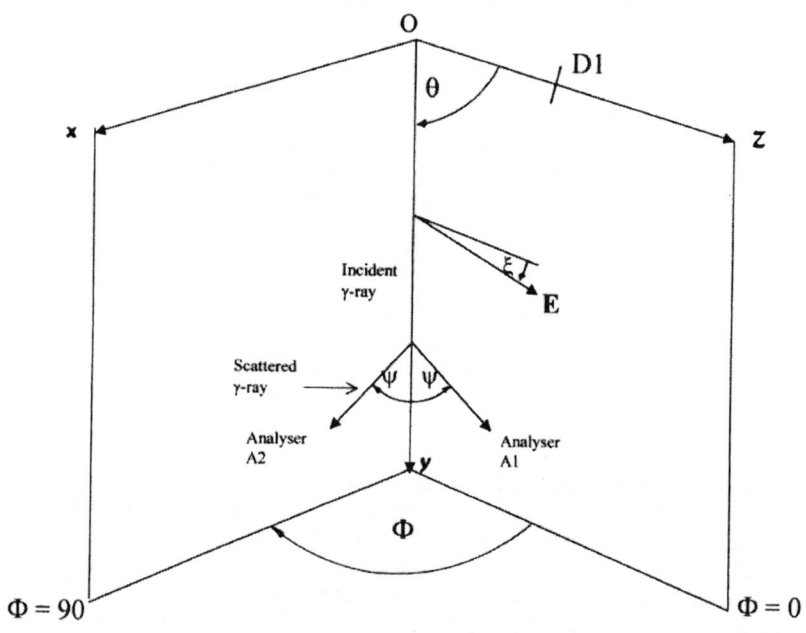

Fig. 1. Schematic representation of a Compton polarimeter, in a Cartesian coordinate system Oxyz, for two cases. Case (a): A radioactive source is placed at O and α-particles are detected in a planar detector Dl positioned along Oz (which then defines the quantization axis). Case (b): An even—even target is placed at O and a beam of spinless particles is incident in the negative Oz direction passing through an annular detector Dl (again Oz is the quantization axis). The Compton polarimeter S + (Al or A2) is at an angle θ, defined by the angle of OS relative to Oz, where S lies in the reaction plane Oyz, although $\theta = 90°$ is sketched for simplicity in both cases.

If we assume that these observables arise from the scattering of the two resolved components of the photon's electric vector $\mathbf{E}$ (in the reaction plane and perpendicular to it) then $N(\theta, \phi = 0)$ arises from $W(\theta, \xi = 0)$ $d\sigma_0$ (the component in the reaction plane being scattered but remaining in the reaction plane) plus $W(\theta, \xi = 90)$ $d\sigma_{90}$ (where the component at 90^0 to the reaction plane is scattered through 90^0 into the reaction plane) and similarly $N(\theta, \phi = 90)$ arises from $W(\theta, \xi = 0)$ $d\sigma_{90} + W(\theta, \xi = 90)$ $d\sigma_0$ where $d\sigma_0$ and $d\sigma_{90}$ are the differential cross sections for Compton scattering of an incident photon through an angle ψ, and into an analysing crystal such that the plane of the component of the electric vector of the incident photon is co-planar with the plane into which it scatters $(d\sigma_0)$ or perpendicular to it $(d\sigma_{90})$.

The experimental measurement of linear polarization may then be written [2] as $P_E(\theta) = \dfrac{(d\sigma_{90} + d\sigma_0) [N(\theta, \phi = 90) - N(\theta, \phi = 0)]}{(d\sigma_{90} - d\sigma_0) [N(\theta, \phi = 90) + N(\theta, \phi = 0)]}$

or $P_E(\theta) = \dfrac{A}{Q(E_\gamma)}$ where A is known as the experimental asymmetry

with $A = \dfrac{[N(\theta, \phi = 90) - N(\theta, \phi = 0)]}{[N(\theta, \phi = 90) + N(\theta, \phi = 0)]}$ and

$Q(E_\gamma) = \dfrac{(d\sigma_{90} - d\sigma_0)}{(d\sigma_{90} + d\sigma_0)}$ is known as the polarization sensitivity of the polarimeter at γ-ray

energy E_γ and $0 \leq Q \leq 1$. Estimates of Q can be obtained using the Klein-Nishina formula [1] where we find that $Q = \dfrac{\sin^2\psi}{k/k_0 + k_0/k - \sin^2\psi}$ k_0, k are wave numbers for incident and scattered photons. This shows that Q is a function of the angle ψ (see Fig. 1), through which a γ-ray is scattered following a Compton event. $Q(E_\gamma) \rightarrow 1$ as $E_\gamma \rightarrow 1$ and $\psi \rightarrow 90^0$. When ψ is restricted to angles around 90^0 by the thickness of material in a strip detector then clearly Q decreases with energy.

2. Statistical Accuracy

Experimentally $P(90^0)$ can be determined easily for pure dipole or pure quadrupole transitions [2].
e.g. For pure E2 : $P(90^0) = \dfrac{3a_2 + 1.25a_4}{2 - a_2 + 0.75a_4}$

where a_2 and a_4 are Legendre polynomial coefficients determined from an angular distribution measurement of the γ-rays in question. For example, the previous measurement of 197 keV(E2) γ-ray in ^{19}F to calibrate a Eurogam "clover" detector with an experimentally determined linear polarization of $P(90^0) = 0.23 \pm 0.06$ (from a_2 and a_4 measurements) and a measured asymmetry $A = 0.077 \pm 0.008$.

From this A we find that $N(\theta = 90, \phi = 0) = 0.857 \times N(\theta = 90, \phi = 90)$ with a polarization sensitivity of $Q = 0.34 \pm 0.09$. Let us assume $N(\theta = 90, \phi = 90) =$

1000 counts and so $N(\theta = 90, \phi = 0) = 857$ counts (same A) but with these statistics we would only determine A to an accuracy of $\sim 30\%$ [then we would have to fold in the uncertainty in $P(90^0)$]. However, for $P(90^0) = 1$ assuming the same $Q = 0.34$ then $A = 0.34$ and if we assume $N(\theta = 90, \phi = 90) = 1000$ counts as before then $N(\theta = 90, \phi = 0) = 493$ counts (for this A) with A and Q known to $\sim 7\%$ for these statistical errors. It is clear that high values of accurately known linear polarization are needed to obtain Q to high accuracy. We should then ask if it is possible to have $P(\theta) = 1$ for all θ ?

3. Linear polarization, angular distribution formula

The general formula for $W(\theta, \xi)$ is given in an article by Twin et al [2] who transcribed linear polarization angular distribution formulae [1] into the notation of Rose and Brink [3]. From this $W(\theta, \xi)$, we can deduce $P(\theta)$ for the case of pure multipoles.

For pure dipoles we can write $P(\theta) = \pm \frac{1}{2} a_2 P_2^2(\cos \theta)$ with the positive sign for M1 and

$$1 + a_2 P_2(\cos \theta)$$

the negative sign for E1 transitions

$$\text{or} \quad P(\theta) = \pm \frac{\frac{1}{2} a_2 (3 - 3 \cos^2 \theta)}{1 + a_2(3 \cos^2 \theta - 1)/2} \qquad = \pm \frac{3 a_2 \sin^2 \theta}{2 + 2a_2 - 3a_2 \sin^2 \theta}$$

It is clear that if $2 + 2 a_2 = 0$ then $P(\theta) = \pm 1$, where now the positive sign is for E1 and the negative sign for M1 transitions, for all θ [**N.B.** Not at $\theta = 0^0$ or 180^0 where we have a line and not a plane]. Therefore, the condition is satisfied if $a_2 = -1$ which occurs for 100% population of the $m = 0$ substate in a $J = 1$ state followed by a pure dipole transition to a $J = 0$ state.

For pure E2 radiation the general polarization formula is

$$P(\theta) = \frac{\frac{1}{2} a_2 P_2^2 (\cos \theta) - 1/12 a_4 P_4^2 (\cos \theta)}{1 + a_2 P_2 (\cos \theta) + a_4 P_4 (\cos \theta)}$$

$$= \frac{\frac{1}{2} a_2 (3 - 3\cos^2 \theta) - 1/12 a_4 15/2 (-1 + 8 \cos^2 \theta - 7 \cos^4 \theta)}{1 + a_2 (3\cos^2 \theta - 1)/2 + a_4 (35 \cos^4 \theta - 30 \cos^2 \theta + 3)/8}$$

$$= \frac{3 a_2 \sin^2 \theta + 5/4 a_4 \sin^2 \theta - 35/16 a_4 \sin^2 2\theta}{2 + 2a_2 + 2a_4 - 3a_2 \sin^2 \theta - 5/4 a_4 \sin^2 \theta - 35/16 a_4 \sin^2 2\theta}$$

If $2 + 2a_2 + 2a_4 = 0$ and $3 a_2 + 5/4 a_4 = 0$ then $P(\theta) = 1$ for all θ (except $\theta = 0^0$ and 180^0 again)

The solution is $a_2 = 5/7$ and $a_4 = -12/7$, which occurs for 100% population of the $m = 0$ substate in a $J = 2^+$ state followed by pure E2 transition to a $J = 0^+$ state. How do we achieve this?

(a) α-decay from an even-even nucleus

Detection of an α-particle emitted from a radioactive source defines a quantization axis. The orbital angular momentum vector (of magnitude J_1) is by definition perpendicular to the direction of linear momentum and hence perpendicular to the quantization axis. Hence, the projection of angular momentum on to the quantization axis is always zero. Since α-particles have zero intrinsic spin, there is no projection of intrinsic spin onto the quantization axis. Therefore, since we start from the ground state of an even-even nucleus with $J = 0$ only the $m = 0$ substates of natural parity states are populated.

Branching ratios to $J_1 = 1$ states following α-decay are usually very small. However, branching ratios to $J_1 = 2^+$ states are sometimes reasonably large although the subsequent γ-decays may be too low in energy in many cases. A suitable case could perhaps be the α-decay of ^{226}Ra to the 186 keV, 2^+ state in ^{222}Rn.

(b) α-decay from ^{227}Th ($J^\pi = \frac{1}{2}^+$)

Following the above argument, if we start from a $J^\pi = \frac{1}{2}^+$ ground state i.e. $m = \pm 1/2$ substates populated then following α-decay we must populate $m = \pm 1/2$ substates in the states of the daughter nucleus ^{223}Ra. This condition of maximum alignment has been shown to hold for all the states observed in this α-decay by Jones et al [4] and was also shown to lead to high linear polarizations [5].

(c) Method 2 of Litherland and Ferguson with heavy ion Coulomb excitation of 2^+ states

With this technique the beam axis is defined as the quantization axis. For even-even beams, i.e. spinless particles in their $m = 0$ substates ($J^\pi = 0^+$ g.state) then $\mathbf{L} = \mathbf{r} \times \mathbf{p}$ is perpendicular to the linear momentum $\mathbf{p}$. Therefore the projection of $\mathbf{L}$ onto the quantization axis is zero. Hence, if the inelastic particles are detected at 0^0 or 180^0 we get $m = 0$ population of Coulomb excited 2^+ states. This technique has been frequently employed for other spectroscopy measurements with an annular detector placed at 180^0. Since $P(\theta) = 1$ at all angles, no solid angle correction factors are needed and one can determine $Q(E_\gamma)$ directly. The main advantage of this technique is that Coulomb excitation cross sections with heavy ions are very large under the appropriate conditions and many 2^+ states of widely different excitation energies can be studied in a relatively short time

4. Solid angle correction factors

Techniques (a) and (c) above produce $P(\theta) = 1$ for all angles. Method (b) could be used to investigate solid angle correction factors since some measure of these would need to be known for actual experimental measurements of polarization. However, a second method is possible which could be used in conjunction with method (c) One could use a beam of ^{19}F ($J^{\pi} = \frac{1}{2}^{+}$ g. state) incident on even-even targets to perform Coulomb excitation measurements as outlined in (c). In this case the spin projection m_s is $\frac{1}{2}$ for both incident and scattered particles. So we find $m = 0$ and $m = \pm 1$ substates populated in the $J^{\pi} = 2^{+}$ state of the even-even target. We can write a_2 and a_4 in terms of the $m = 0$ and $m = \pm 1$ populations, $w(0)$ and $w(1)$ as $a_2 = w(0) \rho_2(2,0) R_2(2\,2\,2\,0) + w(1) \rho_2(2,1) R_2(2\,2\,2\,0)$ and $a_4 = w(0) \rho_4(2,0) R_4(2\,2\,2\,0) + w(1) \rho_4(2,1) R_4(2\,2\,2\,0)$ where ρ_2 and ρ_4 are statistical tensor coefficients and R_2 and R_4 are angular distribution coefficients defined by Rose and Brink [3]

Substituting numerical values we find $a_2 = [w(0) + w(1)] \times -\sqrt{(10/7)} \times -\sqrt{(5/14)}$ and

$a_4 = w(0) \times \sqrt{(18/7)} \times -\sqrt{(8/7)} + w(1) \times -\sqrt{(32/7)} \times -\sqrt{(8/7)}$

or $a_2 = 5w(0)/7 + 5w(1)/7$ and $a_4 = -12w(0)/7 + 16w(1)/7$

Now $w(1) = w(-1) = \dfrac{1 - w(0)}{2}$

Thus $a_2 = 5/14 \, [w(0) + 1]$ and $a_4 = -20/7 \, w(0) + 8/7$, eliminating $w(0)$, we obtain

$a_2 + a_4/8 - \frac{1}{2} = 0$

Surprisingly when we examine the locus of a_2 and a_4 for E2 transitions when $P(90^0) = 1$ we obtain

$P(90^0) = 1 = \dfrac{3\,a_2 + 1.25a_4}{2 - a_2 + 0.75a_4}$ which also gives the result $a_2 + a_4/8 - \frac{1}{2} = 0$

So $P(90^0) = 1$ irrespective of $w(0)$ and $w(\pm 1)$ but $P(\theta) \neq 1$ at $\theta \neq 90$. Therefore, measurements of polarizations using ^{19}F projectiles enables one to investigate solid angle correction factors for extended detectors. It is perhaps worth pointing out that $(p, p'\gamma)$ resonance reactions which have been widely used for polarization sensitivity measurements are appropriate for similar reasons. In those cases a $3/2^{+}$ resonance is formed in the compound system by capture of $\ell = 2$ protons. The 2^{+} excited state is then populated in the target nucleus by emission of $\ell = 0$ protons. Taking the beam axis as quantization axis the only angular momentum projection is from the spin of incident and outgoing protons. Hence, only $m = 0$ and $m = \pm 1$ substates are populated giving the same result as above

It was noted by Butler et al [6] that solid angle correction factors reduced the theoretical polarization sensitivity Q, given by the Klein-Nishina formula earlier, by a fixed amount independent of energy. It will be interesting to note what happens when Q is obtained experimentally using methods (a) or (c).

5. Relative efficiencies

For most reactions used to determine polarization efficiencies any differences in efficiencies or response of analyser detectors were measured by rotating the polarimeter from $\theta = 90^0$ to 0^0 where the theoretical polarization is expected to be zero. With the methods proposed here it is not immediately apparent how one should do this if $P(\theta) = 1$ for all θ.

One suggestion would be to do two experiments

In (i)   measure $\underline{N(\theta, \phi = 0)}$ [system D1 - S - A1]
$\qquad\qquad N(\theta, \phi = 90)$ [system D1 - S - A2]

In (ii)   measure $\underline{N(\theta, \phi = 90)}$ [system D1 - S - A1]
$\qquad\qquad N(\theta, \phi = 0)$ [system D1 - S - A2]

experiment (i) $\div$ (ii) gives $\underline{N(\theta, \phi = 0)}^{\,2}$ which eliminates systematic differences in
$$N(\theta, \phi = 90)^{\,2}$$
geometry and/or electronic set up.

References

[1] L W Fagg and S S Hanna, Rev Mod Phys 31 (1959) 711
[2] P J Twin et al, Nucl Phys A 143 (1970) 481
[3] H J Rose and D M Brink, Rev Mod Phys 39 (1967) 306
[4] G D Jones et al, Eur Phys J A 2 (1998) 129
[5] G D Jones, NIM A 491 (2002) 452
[6] P A Butler et al, NIM 108 (1973) 497

DIGITAL SIGNAL PROCESSING IN DECAY STUDIES

K.P. RYKACZEWSKI

ORNL, Physics Division, Oak Ridge, TN 37830, USA

The development of a data acqusition system based on digital processing of detector signals is described. This system, initiated within the UNIRIB collaboration, was developed originally for proton radioactivity studies at the Holifield Radioactive Ion Beam Facility. It was applied successfully for γ- and β-spectroscopy, as well as in decay studies of the relativistic heavy-ion fragmentation products. A short summary of the results achieved with this system and the possible future applications are given.

1. Introduction

Hybrid systems using shaping amplifiers and peak sensing Analog-to-Digital Converters (ADC) have been successfully used in nuclear spectroscopy for many years. It is still the most common method of data acqusition. With recent advances in technology this type of electronics can be compacted and serve large detector systems. However, already in early nineties, there were attempts[1,2] to digitize the detector signals much earlier, at the detector preamplifier level.

Some advantages of applying fully digital systems in nuclear spectroscopy are presented in the recent reviews[3,4,5,6]. The first profit is obvious - once digitized, the signal will not be disturbed by electronic noise. The digitized signal can be numerically filtered in real time, thus providing equivalent functionality to the analog system. Digital pulse processing provides several options not available within traditional methods. The digitized preamplifier waveform can be recorded, awaiting for a more sophisticated off-line signal shape analysis. It translates to the analog language of having a chance to change and test different settings of the amplifier/shaper module, in order to find the best processing of an incoming signal. Besides obtaining global parameters like the arrival time and amplitude of single signals, one can also process overlapping pulses. One can choose pile-up rejection within a preset range and measure individual signal properties beyond this limit. More sophisticated analysis uses the shape of the signal's rising edge to trace charge collection effects in the detector. The most ambitious projects are aiming now in tracking the absorption of γ-rays in a detector volume, compare[7,8,9].

2. Analog and digital detection systems

Proton radioactivity studies started at the HRIBF[10,11,12,13] with a hybrid system of analog shapers and peak sensing ADCs[14,15]. Fusion-evaporation reaction products recoiling from the target were separated by the Recoil Mass Separator[16] according to their mass-over-charge ratio, with a $\pm$ 10% kinetic energy acceptance. The selected recoils triggered a position sensitive gas avalanche counter (PSAC) and were slowed down by a degrader foil before implantation into a double-sided silicon strip detector (DSSD). Ion implantation signals as well as the proton and alpha decays were preamplified and shaped with the analog electronics[14,15]. The signals were finally recorded by a system of CAMAC 8-channel peak sensing ADCs (Silena 4108). With a DSSD having 40 front strips and 40 back strips, i.e., 80 DSSD detector channels, there were 160 shaping amplifiers and 160 ADC channels used for the recoil implantation and decay signals. The presence or absence of a PSAC signal selected the implantation signal (after a low gain amplifier) from the decay signal (after a high gain amplifier). The charged particle signals were recorded together with a continuosly running recoil- or decay-gated clock. The high gain amplifiers were overloaded by the implantation signal blocking the detection of the decay signal occuring within first $\approx$ 10 μs after the ion arrival. It was preventing the detection or at least strongly reducing the rate of recorded signals from short-lived μs activities.

Experiments on proton radioactivity usually were seen as an identification of single-particle proton orbitals beyond the limit of proton drip-line and as a measurement of the nuclear mass differences for exotic nuclei[17,18]. In the late nineties, there was a progress made in the theoretical analysis of the emission process[19]. Also, experimental studies reached several new deformed proton emitters[20,11]. Understanding of the wave function of these unbound narrow resonance states became a main goal of proton radioactivity studies[21,22]. Mapping the wave function of the emitter requires precise data on the partial decay width, free from the inaccuracies resulting from unknown partial beta half-life. This pointed to the proton emitters with μs half-lives, where the competing β decay width can be neglected.

However, to achieve efficient counting of proton signals in the μs-range, the data acquisition system[14,15] developed at the late eighties had to be modified. At the HRIBF, it was proposed in 1998 to change the system to fully digital processing of the detectors signals, in particular for the DSSD. This decision was supported by the university collaboration UNIRIB as well as by the ORNL Physics Division. The four-channel Digital Gamma Finder (DGF-4C) modules[7], designed by the X-ray Instrumentation Associates (California), were selected for the new data acquisition system[5,6]. These are single-slot CAMAC modules digitizing the incoming signal within 25-nanosecond steps (40 MHz clock). There are 4096 channels (12-bit) available for the amplitude analysis. The incoming signal is adjusted by the analog part before the digitization, to fit a range of a flash ADC. There is no need for separating the data acquisition for the low and high gain signals, i.e., the number of channels for the DSSD electronics was reduced to 80. The time and amplitude of a signal can be analyzed on-board, using an adjustable trapezoidal filter algorithm. This filtering scheme has a built-in option of detecting and analyzing (or rejecting) the overlapping signals in real time.

DGF-4C can act as an *intelligent* digitizer. It can record the waveform of the incoming signal up to 100 μs range. Moreover, it can *select* on-board the signals of interest, like. e.g., the pile-up's occuring within 10 μs interval. Only the signals fullfilling the latter condition are accepted for further readout, which dramatically reduces the data stream and respective deadtime of the system. Such an operating mode, nicknamed "proton-catcher"[23] proved to be extremely useful for the studies of μs proton emitters like a 3-μs activity of ^{145}Tm[24].

The DGF-4C has adjustable fit parameters for the algorithm identifying and analysing on-board the valid signals. It helps to achieve and keep good energy resolution by reducing the effects of noisy experimental conditions (e.g., by requiring a minimum signal length for a fast trigger). The observation of a fine structure in the proton emission from 4-millisecond activity of ^{141gs}Ho profitted from this feature. In a one-week long irradiation with energetic recoils resulting in a high ion dose in the DSSD, a good energy resolution was kept by the DGF-based acqustion. It enabled the measurement of a very weak (0.7%) fine structure branching[25]. The digital rejection of high frequency noise was also crucial for achieving 1 keV energy resolution and a low energy threshold for conversion electron counting in on-line conditions at the mass separator[26].

DGF-4C modules store the valid signals in internal memory. The simultaneous readout of time-synchronized modules occurs, when the memory of one of the modules in the system is filled. This feature allows us to record several subsequent signals, without blocking the detection for the readout time after the first valid trigger. It turned out to be very important for the discovery of a new type of radioactivity. Two-proton radioactivity of ^{45}Fe, with a few millisecond half-life, was identified among the fragmentation products of 650 MeV/u ^{58}Ni beam at the Fragment Separator (FRS) at GSI Darmstadt with the DGF-based data acquisition[6,27,28].

Each event recorded by the DGF-4C gets an absolute time stamp, with the 25-nanosecond clock precision. It allowed us to select and correlate the signals of isomeric recoils with the following γ-radiation, e.g., for an exotic nucleus ^{96}Ag produced and studied at the HRIBF[29]. Moreover, within the same data set, one can inspect isomeric decays with very different lifetimes. In the HRIBF experiment on mass A=140 nuclei, the decay of drip-line radioactivity of 7μs ^{140m}Dy[30] and of 300-ns ^{140m}Eu[31,32] were discovered and studied simultaneously. The relative position of the two isomers present in ^{140}Eu could be experimentally proven for the first time. The decay of 300-ns $\pi h_{11/2} \otimes \nu h_{11/2}$ configuration is feeding the 125-millisecond $\pi h_{11/2} \otimes \nu s_{1/2}$ excited state at lower energy[31].

The processing of overlapping signals was used also in studies of short-lived isomers produced at GSI in relativistic heavy-ion fragmentation at the FRS[5]. The detection scheme requires slowing down and stopping of the energetic isomeric fragments in the catcher material surrounded by the γ-detectors[33]. These studies often offer the only way to access excited states in very exotic nuclei, in particular in very neutron rich isotopes[34,35,36]. The process of slowing down and stopping of very heavy fragments is accompanied by a few hundred nanosecond long burst of radiation (delta-electrons, bremsstrahlung) blocking the detection of subsequent isomeric γ-rays. Analog acqusition systems for γ-detectors record the first burst event, which stops the acqusition until the end of the readout. It results in a strong reduction of the detection rate for isomeric decays, up to 90% for Pb-like fragments. The application of a DGF-based system helps to solve this problem, in two ways. The DGF can analyze properly the overlapping signals and extract the time and amplitude, up to an interval of a few microseconds. This information is stored in the DGF memory, so the observation of photons is still possible after the first burst signal. Also, the pulse shape recording mode of acqusition can be selected, with an external trigger reducing the total data rate. These recorded waveforms contain the burst radiation and overlapping isomeric γ-rays. They can be analyzed off-line,

628

and information on the decay properties of produced metastable states is restored.

3. Future applications

The development of a novel data acquisition system based on digital signal processing was made following the funding decision made at the UNIRIB meeting in November 1998. It helped to achieve a number of unique experimental results. Further decay spectroscopy studies, planned at the HRIBF and expected to profit from the DSP system, are aiming at the structure of exotic particle emitters. Among them are nuclei near the doubly-magic ^{100}Sn, important for testing the predicted rp-process path and its termination region[37,38].

New applications of the HRIBF DSP system are under development, in a close collaboration with XIA. They include the separation of isobars in the radioactive "cocktail" beam by using a "Bragg-curve" detector[39,40,41]. This separation can be achieved through the analysis of the properties of the charge collection[6].

There are projects to apply digital signal processing in the studies of cold neutron decay[42]. The electron and proton emitted in the neutron decay process will be detected in a pair of segmented silicon detectors placed opposite to each other in an electromagnetic field. The reconstruction of the origin of the electron signal should help to obtain the electron energy spectrum corrected for effects related to the electron backscattering at the detector surface.

Help of R.Grzywacz is gratefully acknowledged. The described work has been supported by a number of institutions collaborating within the UNIRIB consortium. ORNL is managed by UT-Battelle, LLC, for the U.S. Department of Energy under Contract DE-AC05-00OR22725.

References

1. A. Georgiev and W. Gast, *IEEE Trans. Nucl. Sci.* **40** (1993) 770.
2. A. Georgiev *et al.*, *IEEE Trans. Nucl. Sci.* **41** (1994) 1116.
3. J. Basilio Simoes *et al.*, *Nucl. Instr. Meth. Phys. Res.* **A422**, (1999) 405.
4. W.K. Warburton *et al.*, contr. to the IRRMA-99 Conf., *Applied Radiation and Isotopes* (2000), see *www.xia.com*.
5. R.K. Grzywacz *et al.*, in contr. to 3rd Int. Conf. on Exotic Nuclei and Atomic Masses, Hämeenlinna, Finland, July 2001, in press.
6. R.K. Grzywacz, in contr. to 14th EMIS Conf. *Nucl. Instr. Meth. Phys.* **B**, (2003), in press.

7. B. Hubbard-Nelson, M. Momayezi and W.K. Warburton, *Nucl. Instr. Meth. Phys. Res.* **A422**, (1999) 411.
8. K. Vetter *et al.*, *Nucl. Instr. Meth. Phys. Res.* **A452**, (2000) 223.
9. K. Vetter, *Nucl. Phys. News Int.* **12**, vol.2, (2002) 15 and earlier refs therein.
10. J.C. Batchelder *et al.*, *Phys. Rev.* **C57**, (1998) R1042.
11. K. Rykaczewski *et al.*, *Phys. Rev.* **C60**, (1999) 011301(R).
12. C.R. Bingham *et al.*, *Phys. Rev.* **C59**, (1999) R2984.
13. T.N. Ginter *et al.*, *Phys. Rev.* **C61**, (1999) 014308.
14. S.L. Thomas *et al.*, *Nucl. Instr. Meth. Phys. Res.* **A288**, (1990) 212.
15. P.J. Sellin *et al.*, *Nucl. Instr. Meth. Phys. Res.* **A311**, (1992) 217.
16. C.J. Gross *et al.*, *Nucl. Instr. Meth. Phys. Res.* **A450**, (2000) 12.
17. S. Hofmann, *Radiochimica Acta* **70/71**, (1995) 93.
18. P.J. Woods and C.N. Davids, *Annu. Rev. Nucl. Part. Sci.* **47**, (1997) 541.
19. S. Åberg, P.B. Semmes,W.Nazarewicz, Phys.Rev. **C56**,(1997) 1762, and Phys. Rev. **C58**, (1998) 3011.
20. C.N. Davids *et al.*, *Phys. Rev. Lett.* **80** (1998) 1849.
21. E. Maglione *et al.*, *Phys. Rev. Lett.* **81** (1998) 538.
22. A.T. Kruppa *et al.*, *Phys. Rev. Lett.* **84** (2000) 4549.
23. M.Momayezi *et al.*, in contr. First Int. Symp. Proton-Emitting Nuclei, AIP Proc. **518**, J.C. Batchelder (ed.), (2000) p. 307.
24. M. Karny *et al.*, *Phys. Rev. Lett.* **90** (2003) 012502.
25. K.P. Rykaczewski *et al.*, in contr. to Int. Conf. "Mapping the Triangle", AIP Proc. **638**, (2002) p. 149.
26. J.C. Batchelder *et al.*, in contr. to this conference
27. M. Pfützner *et al.*, *Eur. Phys. Jour.* **A14** (2002) 279.
28. M. Pfützner *et al.*, *Nucl. Instr. Meth. Phys. Res.* **A493**, (2002) 155.
29. R.K. Grzywacz , priv. comm. (2002)
30. W. Królas *et al.*, *Phys. Rev.* **C65**, (2002) 031303(R).
31. M.N. Tantawy *et al.*, BAPS **47**,No.2,(2002) p. 70, and to be published.
32. K.P. Rykaczewski, *Eur. Phys. Jour.* **A15** (2002) 81.
33. R. Grzywacz *et al.*, *Phys. Lett.* **B355** (1995) 439.
34. R. Grzywacz *et al.*, *Phys. Rev. Lett.* **81** (1998) 766.
35. M. Mineva *et al.*, *Eur. Phys. Jour.* **A11** (2001) 9.
36. M. Pfützner *et al.*, in contr. to this conf.
37. H. Schatz *et al*, Phys. Rev. Lett. **86**, 3471 (2001).
38. R.Page and R.Grzywacz (spokepersons) "Search for New Alpha Emitters above ^{100}Sn" HRIBF experiment RIB-101, (2002).
39. Ch.Schiessl *et al.*, *Nucl. Instr. Meth.* **192**, (1982) 291.
40. J.M. Asselineau *et al.*, *Nucl. Instr. Meth. Phys. Res.* **204**, (1982) 109.
41. A. Galindo-Uribarri *et al.*, in contr. to AMS-9 Conf., Nagoya, Japan, September 9-13, 2002.
42. G. Greene, in contr. to this conf.

THE FUNDAMENTAL PROPERTIES OF THE NEUTRON

G. GREENE

Physics Division, Oak Ridge National Laboratory, Oak Ridge, TN 37831
Department of Physics and Astronomy, University of Tennessee, Knoxville, TN 37996

Precise determination of the properties of the free neutron can shed light on important issues in nuclear physics, particle physics, astrophysics, and cosmology. An overview of the significance of some these will be given. Of particular interest is the accurate measurement of the parameters that describe the beta decay of the neutron. Because of the simplicity of the neutron, measurements of free neutron decay can provide a robust value for the weak vector coupling constant which in turn provides information about the unitarity of the CKM matrix, yields a sensitive test for non-standard couplings, and can address the issue of physics beyond the standard model. The physics of free neutron decay, as well as the prospects for a new series of measurements, employing advanced neutron sources such as the Spallation Neutron Source at Oak Ridge, will be discussed.

INDUCED FISSION STUDIES: EXPERIMENTAL CAPABILITIES, DATA CORRELATIONS, AND APPLICATIONS

J.D. COLE, M.W. DRIGERT, R. ARYAEINEJAD, E.L. REBER, J.K. JEWELL

INEEL, Bechtel BWXT Idaho, LLC. Idaho Falls, Idaho 83415, USA
E-mails: jdc@inel.gov, mwd@inel.gov, rax@inel.gov, reber@inel.gov, jewejk@inel.gov

J.H. HAMILTON, A.V. RAMAYYA

Vanderbilt University, Nashville, TN 37215, USA
E-mails: j.h.hamilton@vanderbilt.edu, a.v.ramayya@vanderbilt.edu

G.M. TER-AKOPIAN, TS.YU. OGANESSIAN, A.V. DANIEL

Joint Institute for Nuclear Research, Dubna, Russia
E-mails: Gurgen.TerAkipian@jinr.ru, oganessian@flnr.jinr.ru, dainel@jinr.ru,

R.V.F. JANSSENS, I.WIEDENHOVER**, RAYMOND KLANN

Argonne National Laboratory, Argonne, Illinois, IL 60439, USA
E-mails: janssens@anl.gov, ingo@nucmar.physics.fsu.edu, klann@anl.gov,

The current experimental configuration and its capabilities, as well as planned improvements and additional capabilities will be presented. The status of appropriate actinide target samples will be presented. This will include the samples currently available as well as future sample plans. The importance and use of the correlated event-by-event parameter information for extracting information about the fundamental fission process will be discussed. The importance of both the experimental techniques and the information extracted for specific applications will be outlined.

1. Introduction

A complement to the ^{252}Cf fission studies is neutron-induced fission of uranium, neptunium, plutonium and higher actinides. These data from neutron induced fission are important for particular applications as well as for the fundamental information about the fission process that they provide. An experimental program started at the Argonne National Laboratory/Intense Pulsed Neutron Source (ANL/IPNS) to support these studies uses an array of 12 Compton-suppressed high-purity germanium (CSHPGe) detectors. Multi-parameter data like those taken in the ^{252}Cf experiments are acquired with some additional parameters beyond the normal γ-γ coincidence data. Time-of-flight

---

* Current address: Florida State University, Department of Physics, Tallahassee, FL 32306

measurement on the incident neutron energy is made for each event. Future plans include the installation of a fission chamber to monitor the incident neutron flux.

2. The ANL Intense Pulsed Neutron Source

IPNS is a spallation neutron source operated primarily for material science and solid state physics research. The IPNS accelerator system consists of an H⁻ ion source, a Cockcroft-Walton preaccelerator, a 50 MeV Alvarez linac, and a 450 MeV Rapid Cycling Synchrotron (RCS). The accelerator normally operates at an average beam current of 14 to 15 μA, delivering 30 pulses per second of

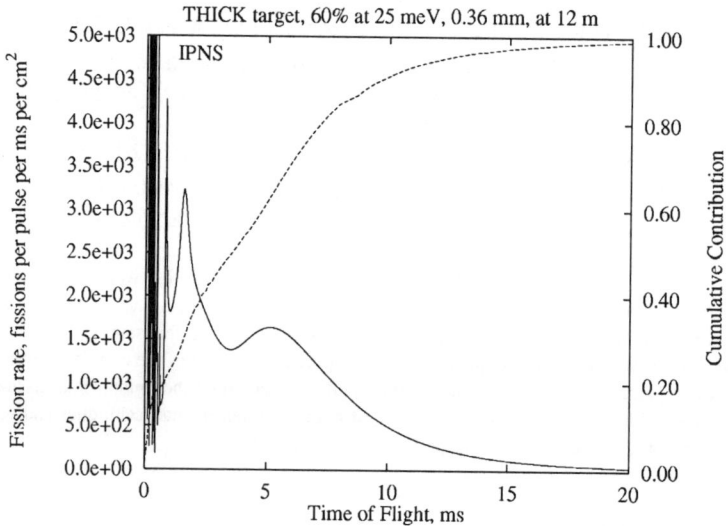

Figure 1 Fission intensity as a funtion of neutron time-of-flight.

approximately 3×10^{12} protons at 450 MeV to the target. The proton pulse width is 70-80 ns, giving good timing resolution for the neutrons. The target is uranium with a frozen methane moderator at $\sim 30°$K. The intensity and energy spread of the neutron pulse depend on the distance between the neutron-producing target and the experimental target position. For the fission experiments, this distance is 12 meters with a neutron energy range from ~ 0.01 eV to ~ 300.0 eV. With this distance and moderation, the neutron spectrum relates to the fission intensity as shown in Figure 1. The figure shows that, with a pulse rate of 30 Hz, all neutron events are observable, even to energies less than thermal and considerable high-energy neutrons are present. These high-energy neutrons are usable with a longer flight path. The instantaneous fission rate is such that the detectors do not become saturated, with a detected rate of ~1300 fission events per second observed with a 93% ^{235}U target.

IPNS is operated on a 24-hour, seven-day-a-week basis for 27 weeks of the year. This results in the operation of the accelerators for a period of three weeks, followed by one week down. There are two longer down periods of three to four weeks scheduled at roughly six-month intervals. This allows the experimental equipment to be maintained and calibrated on a regular basis without loss of data collection time. The longer shutdowns allow annealing of the HPGe detectors and more extensive repairs on the equipment.

3. Detector Array at IPNS

The CSHPGe detector design is based on the design used in the original ANL-Notre Dame array at ATLAS. The primary reason for this is that the bismuth germinate (BGO) shield design for the "Golf Club" style germanium detector put 2.8 inches of BGO behind the germanium crystal so that the high energy γ–rays

Figure 2. View from above of the HPGe detectors around the central target position. The horizontal tube is the neutron beamine with the neutron coming from the right to the left. The central cylinder is the taget chamber.

(>1 MeV) that are forward scattered have a high probability of rejection in the shield. Figure 2 shows the top six CSHPGe detectors mounted around the target location. The detectors are mounted in two rings of six detectors each, with one ring above the plane of the target and the other below the target plane. Lead-copper collimators and shields can be seen in Figure 2. The germanium crystal is irradiated from the side, using only about 1.4 inches of the 2.5 inch crystal diameter. Since the length of the germanium crystal is slightly less than its diameter, there is some slight increase in high-energy sensitivity.

An array of eight neutron detectors is located above the target location allowing γ-n and γ-γ-n data to be collected is shown in Figure 3, placed above the target location. These are BC501 scintillators from Bicron that use pulse-

Figure 3. The neutron detectors mounted above the target location.

shape discrimination to reject the γ-rays. Each of these detectors is shielded by 3 mm of tungsten to reduce the γ-ray signal. The neutron detectors have an operational energy range from about 40 keV to 2 MeV for the neutrons. The lower limit of the range is determined by the neutron-γ-ray separation, while the size of the detector is the limit for the upper end of the range.

4. Calculated and Observed Count Rates

If the following parameters are used in equation $[1]^{1,2}$ which is the probability of detecting a p-fold event in N detectors:
- Fission Rate 10^5 s^{-1}

- Total Efficiency Ω = 9 x 10^{-4},per detector,
- Gamma Multiplicity M = 15,
- p-fold = 2,

$$P_{N_p}^{M} = \binom{N_p}{p}\binom{M}{p} p!\Omega^P[1-(N-p)\frac{1}{2}\Omega]^{M-p} \tag{1}$$

the 2-fold rate is 1100 cps and the 3-fold rate is 45 cps. The random rate is 130 cps using $R_r = \tau S_s^2$ (N-1) where S_s is the singles rate in each detector. The actual observed rate in the array is ~1300 for 2fold events. The discrepancy of ~70 cps is a geometry effect and comes from scattering.

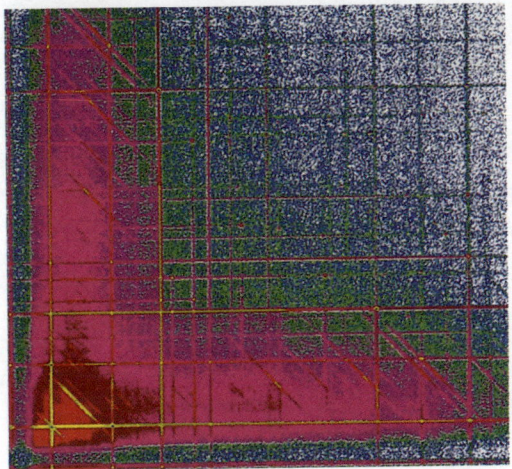

Figure 4. A two-dimensional E_γ-E_γ display of a ^{152}Eu spectrum for a coincidence measurement with an array of eight detectors.

To understand this effect, measurements were made at INEEL using an array of eight germanium detectors. Figure 4 shows the 2-D coincidence spectrum from this array with a ^{152}Eu source. The axes correspond to energy but are not shown. The lines that are parallel to the axes in Figure 4 are from the discrete γ–rays. The last two on the right are for the 1085.9 and 1112.1 keV γ–rays. The lines at ~45° to the γ–ray lines are due to scattering between the detectors. The scattering between adjacent or near detectors can generally be controlled by the Compton shields, but the 180° scattering between opposing detectors is not stopped by the shields. This scattering will always produce a "true" coincidence by the timing circuit logic. The 45° lines can be extrapolated to the axis to determine the γ–ray that is being scattered. For low-energy γ–rays, the energy is partitioned on a continuous basis between the two detectors, and the lines at 45° are continuous. For the higher energy γ–rays, a small energy is deposited in one detector, and the remaining energy is observed in the opposite detector. That is the reason the 45° lines are close to the x- and y-axes.

For a large array like Gammasphere, several detectors can participate in this process, and the scattering to opposite sides of the array can produce significant contributions to the background and extended tails on the low-energy side of a γ–ray peak.

5.   Correlated Fission-Product Pairs

The significant information that the fission studies using arrays of CSHPGe detectors provide is the correlated fission-pair information.   With sufficient

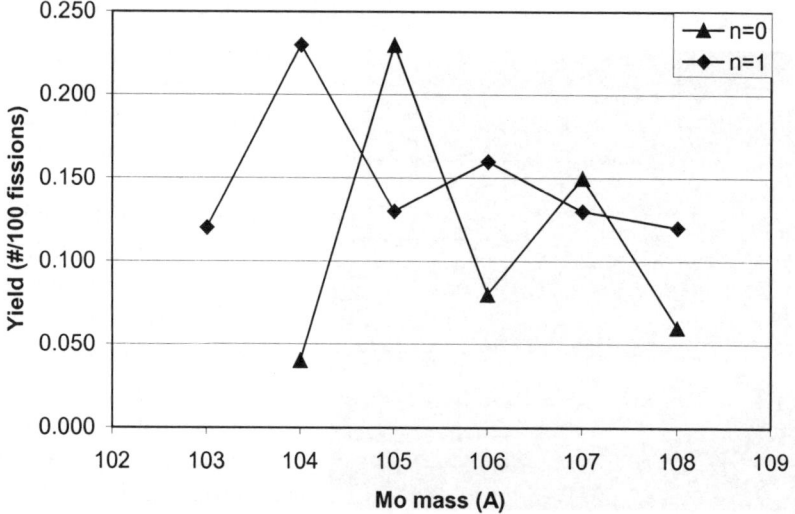

Figure 5. Odd-Even staggering of Mo/Ba pair yields for n=0 and n=1 neutron emitted versus the molybdenum mass number.

statistics, explicit data can be be obtained, *e.g.*, neutron and γ–ray multiplicities, and new information can be determined, *e.g.*, fragment excitation energy and fission-fragment pair yield. Particular effects can be observed in better detail. Figure 5 shows the staggering of the fission-pair yields for n=0 and n=1 neutron channels of the Mo/Ba charge split in ^{252}Cf[3]. The staggering in these two channels is much more pronounced than in other channels up to n=6, although it is present. This cannot be observed without the multi-parameter correlated pair data.

6.   Yields versus Incident Neutron Energy

The filling of the valley between the two mass peaks in the fission yield versus mass number plots for increasing incident neutron number is well known[4]. This has always been presented as yield summed over mass number (A), and has not been measured for the promptly produced isotopic pairs. With an array of detectors, the prompt isotopic yield as a function of incident neutron energy can be measured. This specific pair yield has importance in the performance of nuclear reactors when the fuel burn-up is five to 10 times that of current reactors. Although the incident neutron energy range at IPNS is too small for the needed measurements, an increase in the neutron flight path will allow measurements for part of the needed energy range.

7. Applications of Fission Studies

For some areas of physics, the immediate or applied use of experimental results may not be an important aspect of the work. For studies using low-energy neutrons (<20 MeV incident neutron energy). this is not currently the case. The high-energy, easily observed γ-rays from fission fragments are now a tool for the identification of fissile material in environmental programs. These same γ-rays have use in nuclear weapons verification and monitoring efforts, and may be a new method for use in special inspections for the smuggled of fissile material. A continued effort to identify and correlate high-energy γ-rays with prompt fission products is needed in these areas of study.

Neutron-induced fission cross sections and fission yields are needed in the design of new types of nuclear power reactors[5]. Much of the information and models used in the design of currently operating reactors were measured and developed prior to 1970. Vandenbosh and Huizenga's[6] book on the fission process is the standard reference for reactor physics and description of the fission process. The statistical model as presented in the Los Alamos model[7] is widely used for calculations in reactor design.

The planned new reactor designs require better and more extensive nuclear data on induced fission cross sections, n-γ, and fission product yields. The isotopes of interest are: most thoriums, ^{231}Pa, ^{233}U, ^{235}U, ^{236}U, ^{238}U, ^{235}Np, ^{236}Np, ^{237}Np, ^{238}Pu, ^{239}Pu, ^{240}Pu, ^{241}Pu, ^{242}Pu, ^{241}Am, ^{242}Am, ^{243}Am, ^{242}Cm, ^{243}Cm, ^{244}Cm, ^{245}Cm, ^{246}Cm, ^{247}Cm, ^{248}Cm, ^{250}Cm, ^{247}Bk, ^{251}Cf, ^{252}Cf. Neutron capture cross sections are needed for many of the fission products. All data are needed over an energy range from thermal to 20 MeV, with detailed information needed below 10 MeV.

Multi-parameter measurements made with arrays of detectors, and list mode data systems are needed for the detailed design and modeling of nuclear reactors. The renewed emphasis on nuclear power development in the United States increases the need for measurements of parameters describing neutron interactions with both fissile and non-fissile material. The average values of neutron emission, fragment-excitation energy based on β-decay, limited neutron absorption and scattering data only at energies less than 1 MeV, and yields measured by mass number are inadequate. New design codes are planned with "what if" capabilities to reduce the need to build full-scale test reactors. New requirements for reactor fuel and its performance are also drivers for new measurements and models.

8. Conclusion

The nuclear engineering and reactor physics community now acknowledges the need to perform new experiments using multi-parameter methods and to integrate new results into a theoretical framework that can support calculations. The type of fission and neutron interaction data that existed prior to multi-parameter measurements has limited the development of fission models. The needs described for new nuclear data can be extended to a fission model that better describes specific characteristics of the fission process and not just an average behavior or average value.

Acknowledgments

The U.S. Department of Energy supports this work under contracts DE-AC07-99ID13727, DE-FG05-88ER40407, DE-FG05-87ER40311, and W-31-109-ENG-38.

References

1. S. Y. Van Der Werf, Nucl. Instrum. Methods **153**, 221 (1978).
2. W. J. Ockels, Z. Phys. **A286**, 181 (1978).
3. G. M. Ter-Akopian, et al., Phys. Rev. **C55**, 1146 (1997).
4. L.E. Glendenin, et al., Phys. Rev. **C24**, 2600 (1981).
5. A Technology Roadmap for Generation IV Nuclear Energy Systems: Technical Roadmap Report, to be released.
6. Robert Vandenbosh and John Huizenga, Nuclear Fission, (Academic Press, New York, 1973).
7. D.G. Madland, J.R. Nix, Nucl. Sci. Eng. **81**, 213 (1982).

FIRST EXPERIMENTS WITH ACCELERATED RADIOACTIVE ION-BEAMS AT REX-ISOLDE

J. CEDERKALL, F. AMES, U. BERGMANN, V. FEDOSSEEV, L. FRAILE,
U. KÖSTER, T. NILSSON, T.SIEBER, B.WOLF AND F. WENANDER

CERN, 1211. Geneva 23, Switzerland
E-mail: Joakim.Cederkall@cern.ch

S. EMHOFER, D. HABS AND O. KESTER

Ludwig Maximilians University, Munich ,Germany

R. VON HAHN, M. LAUER, O. NIEDERMAIER, H. SCHEIT AND D. SCHWALM

Max Planck Institute for Nuclear Physics, Heidelberg, Germany

P. VAN DUPPEN, M. HUYSE AND P. MAYET

Catholic University, Leuven, Belgium

J. EBERTH, P. REITER, N. WARR AND D. WEISSHAAR

University of Cologne, Cologne, Germany

G. SCHRIEDER, H. SIMON AND M. PANTEA

Technical University, Darmstadt, Germany

REX-ISOLDE and MINIBALL collaborations

The first experiments using post-accelerated radioactive ion beams from REX-ISOLDE have taken place during 2002 after approximately six years of development. The novel techniques used for the production of the radioactive ion beam as well as at the experimental set-up are described in this paper. It is attempted to show that REX-ISOLDE is a unique machine for experiments in nuclear physics as it involves methods for beam production that lie closer to front-line physics research than to traditional accelerator construction.

1. The Proton Driver

The physics program at ISOLDE [1] covers not only different aspects of nuclear physics but involves vigorous research programs in atomic physics, solid state physics and the life sciences. For this reason the Radioactive Beam Experiment (REX) at ISOLDE [2] shares the secondary ISOL beams with a number of other users. In its current location at the PS-booster at CERN, ISOLDE is driven by a pulsed proton beam with a repetition rate of one pulse every 1.2 s. The maximum pulse intensity is $3.2 \cdot 10^{13}$ protons per pulse and the beam energy is

1.0 or 1.4 GeV. Just as REX shares the secondary beam with a number of other ISOLDE experiments so does ISOLDE share the driver beam with other fixed target experiments. The proton beam is passed from the PS-booster to the different experiments according to a proton-beam super cycle typically containing 12 pulses. On average the ISOLDE target receives half of the pulses in the super cycle during an experiment which corresponds to a driver beam current of 2 μA. The pulsed structure of the driver beam results in a quasi-continuous secondary beam where the decay and release times of the radioactive species are superimposed on each other to give the final time structure. A consequence of the quasi-continuous beam is that the ion pulses accelerated through REX-ISOLDE will have a macroscopic time structure superimposed on the micro- and nanostructures. This time structure will be of the order of the decay and/or release time of the specific isotope/element.

1.1 Targets

Since beginning operation at CERN in 1967 the experience with radioactive beam production at ISOLDE has continuously grown. Up to now more than 700 isotopes of approximately 70 elements have been produced [3]. Common target materials include UC_x/graphite and ThC_x/graphite for fission products but targets for proton rich or very light isotopes are used as well. In the case of fission products one of the latest developments is to let the proton beam impinge on a cylindrical neutron converter situated close to the normal target chamber. Similar converter targets are in use or being studied at several other laboratories as well [4].

1.2 Ion sources

The radioactive species produced in the primary target is transferred from the target cylinder into an ionization cavity. The cavity can typically be a surface ionizer made of tungsten or tantalum. Other types of ion sources, such as various versions of FEBIAD ion sources, are in use as well. A 1^+ ECR ion source is currently also being tested. Another rather recently developed ion source, which continuously gains importance at ISOLDE, is the Resonant Ionization Laser Ion Source (RILIS). The experiments using Mg isotopes have under 2002 relied on this type of source. The RILIS is therefore described briefly in the following. Laser ionization is particularly useful when the element of interest has an ionization potential (IP) that is lower than the work function of the material

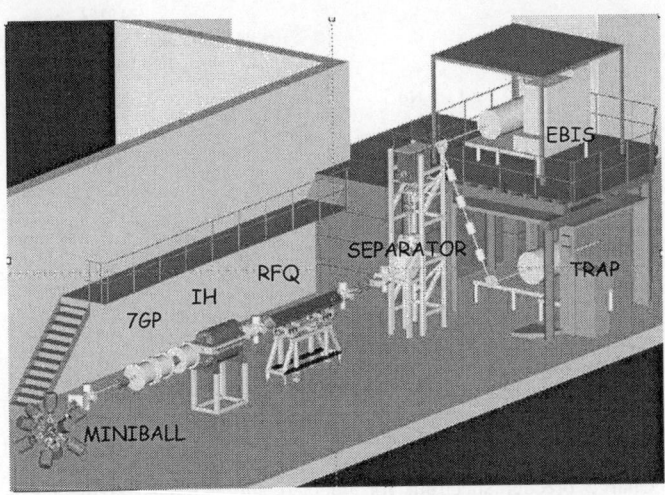

Fig. 1 Overview of REX-ISOLDE. To the right REX-TRAP (lower level) and REX-EBIS (upper level) connected via the beam transport line. REX-EBIS is followed by the mass separator and the REX-LINAC. The MINIBALL is also shown schematically to the far left in the picture.

normally used for surface ionization [5]. Most elements can be ionized in a plasma ion source. This causes, however, a low selectivity and thus the outcoming beam may be an isobaric mix. To obtain element selectivity the laser excitation is performed in intermediate steps given by selected excited atomic states of the element of interest. The ISOLDE RILIS consists of two copper vapor lasers that drive tunable dye lasers. Frequency doubling and tripling in non-linear crystals makes it possible to cover wavelengths from 210 nm to 850 nm. The repetition frequency of the lasers is 11 kHz which roughly corresponds to the time the radioactive atom spends within the cavity. The typical efficiency of the ISOLDE RILIS is 10% [5]. The gain in ionization efficiency over that for surface ionization varies but can be very high. In a recent experiment with the neutron deficient isotope ^{62}Ga this ratio was ~20. For the particular case of Mg, with an IP of 7.65 eV, ionization is accomplished in three steps. The first excitation is given by a laser beam at 285.2 nm and the second by a beam at 552.8 nm. The final step into the continuum is accomplished by the 578 nm and 511 nm beam of one of the copper vapor lasers non-resonantly. The production

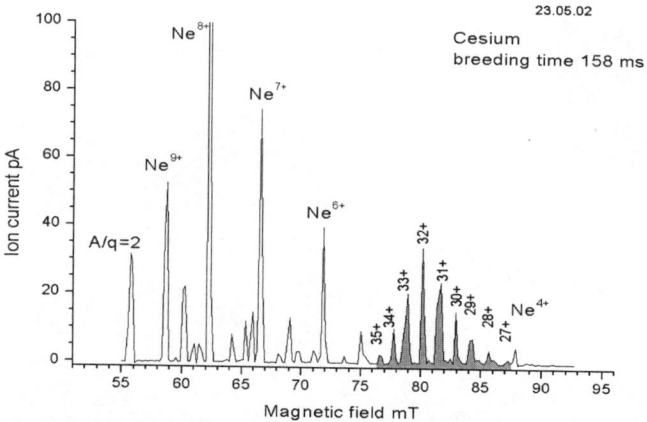

Fig. 2. Charge breeder spectrum for injected and extracted stable Cs. Note the multitude of charge states (marked in grey) for Cs to the right in the picture. Ne was used as buffer gas in REX-TRAP during this test.

rate of heavy Mg isotopes ranges from 10^7 ions/μC for ^{27}Mg to 10^2 ions/μC for ^{34}Mg.

2. REX-ISOLDE

It was realized during the design phase of REX-ISOLDE that a compact post-accelerator would be advantageous taking into account the limited space available and the total cost of the project. The natural way to reduce the size of an accelerator is to boost the charge state of the beam. In REX this is accomplished by introducing a beam preparation stage.

2.1 Beam Preparation and Post-acceleration

The traditional method to increase the charge state of an ion beam is to strip the ions after a first acceleration step. Following the boundary conditions for REX this avenue is not open. It was instead decided to inject the 60 kV 1^+ beam coming from ISOLDE into a charge breeder. At the time when the design was finalized the best option appeared to be to use an Electron Beam Ion Source (EBIS) for this purpose. It should be noted that a parallel development of ECR sources has taken place in the mean time and that this has proven to be a successful approach as well [6].

ISOLDE beams have a typical transversal emittance of 35 πmm mrad or larger [7]. Furthermore, the emittance varies depending on the type of ion source used and on its setting. REX-EBIS [8], on the other hand, has an acceptance of 11 πmm mrad at the 95% level. A cooling stage is therefore necessary. This stage should also be capable of bunching the beam since cyclic injection schemes have proven to be more effective than continuous ones for sources of this type [8]. A completely new approach for beam preparation was therefore chosen for REX where the beam is injected into a cylindrical Penning trap (REX-TRAP) for bunching and cooling. REX-TRAP and its operational principle has been described extensively in previous papers [9,10]. Below we therefore focus on some new results for this part of the system.

According to the single particle theory for ion cooling in a buffer gas filled Penning trap one can expect a centering of the ions in both the transversal and longitudinal directions. The reader is reminded that the transversal cooling is attained by applying an electric rf quadrupole field to couple the oscillations of the magnetron and cyclotron motions to each other. Damping of the stable cyclotron motion then leads to an overall transversal cooling of the ion cloud. This effect is indeed present in REX-TRAP but space charge effects have, as a consequence of the sometimes large number of ions that has to be injected, been observed as well. In a first attempt to understand this effect a number of simulations were carried out by D. Beck et al.[11]. The general feature that can be observed in measurements and simulations is that the centering frequency, which in the single particle case is equal to the cyclotron frequency, is shifted to higher values. One way to qualitatively understand this is to recognize that when space charge repulsion becomes important it is necessary for the cloud to rotate in the magnetic field in order to counteract the repulsion. A single ion that follows in this motion will therefore see a lower effective centering frequency than the one applied, which accounts for the observed effect. With this in mind it becomes logical to try to increase the ion density by exciting the ion cloud to even higher rotational frequencies. Tests with this so-called rotating wall technique [12] have been carried out and are planned to continue.

One of the more challenging aspects of the REX scheme lies in obtaining a high injection (see Fig. 2) efficiency into the EBIS. Typical values for this parameter

Fig. 3 The MINIBALL detector system mounted at the REX beam line.

are today around 50%. Some improvements in the beam diagnostic system as well as in the electrostatic beam optics have been carried out during the past year which explains the new improved efficiency. In the first successful injections during 2001 and early 2002 only a few per cent of the ion beam could be extracted after charge breeding. It should also be mentioned that the electrostatic settings for transferring a maximum of the beam from REX-TRAP to REX-EBIS are in accordance with the simulated beam settings. Further improvements might be found by realignments of the beam axes to the field axes of REX-TRAP and REX-EBIS. Another possibility of improvement lies in gathering further information on the emittance of the beam leaving REX-TRAP. Two major improvements have been made on REX-EBIS during the past year. The first REX-EBIS runs suffered from beam contaminations that originated from the cathode. A new barrier at the electron gun side has therefore been introduced. It effectively stops ions produced in or close to the cathode from entering the central trapping region of REX-EBIS and consequently the contamination observed in the first tests is now gone. The second improvement in the EBIS was aimed at reducing the level of rest gas components in the beam. To this end two new inner structures were prepared. The one currently in use has 2000 holes distributed uniformly over the surfaces of the trapping tubes. As a result of this the pressure inside the EBIS is now in the 10^{-11} mbar region. For completeness it is mentioned that the mass separator [12] after REX-EBIS, which is needed to remove remaining rest-gas components in the beam has not seen any major changes during the past year. A general re-alignment has been carried out during

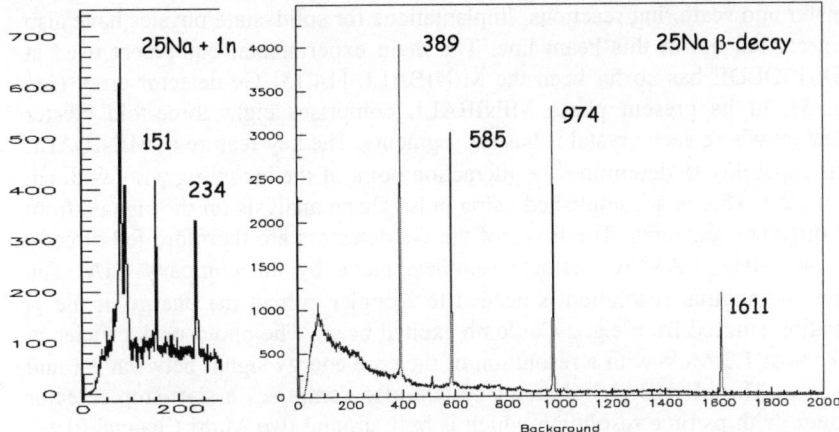

Fig. 4. Proof-of-principle spectra taken with MINIBALL with a ^{25}Na beam impinging on a PE target. The insert on the left-hand side shows two known lines in ^{26}Na after 1n pick-up. The indicated transitions on the right are from the de-excitation of excited states in ^{25}Mg following the decay of ^{25}Na and constitutes a background measurement. Energies in keV on the abscissa.

the shutdown period in order to increase the transmission which previously was approximately 80 %.

The LINAC [2] of REX-ISOLDE is based on conventional accelerator structures. To reiterate the layout the first step of REX-LINAC consists of an RFQ which is a conventional choice for a LINAC fed by a beam without microstructure. After leaving the RFQ with an energy of 300 keV/u the beam passes through a matching section and a buncher before it enters an Interdigital H-structure. The IH consists of a 0° part which is followed by a triplet lens for refocusing and a final accelerating stage. The beam leaves the structure with an energy of 1.1 MeV/u. The three following 7 gap resonators can either be used for increasing the beam energy to 2.2 MeV/u or to decelerate to 0.8 MeV/u. The latter mode has still to be tested. The resonators run at 101.28 MHz and with a duty factor of 10%. All structures have been tested with radioactive beam.

3. Target stations

A final bending magnet steers the beam to one of two target stations. One of these is used for general purposes and has no fixed detector system. Typical experiments at this beam line employ charged particle detectors for studies of transfer and scattering reactions. Implantations for solid–state physics have also been carried out at this beam line. The main experimental equipment used at REX-ISOLDE has so far been the MINIBALL [14,15] Ge-detector array (see Fig. 3). In its present phase MINIBALL comprises eight three-fold cluster detectors where each crystal is has six segments. The key feature of MINIBALL is its capability to determine the interaction point of the incoming γ-ray with cm resolution. This is accomplished using pulse shape analysis on the signals from the different segments. The FETs of the Ge-detectors are therefore fed directly into 40 MHz, CAMAC resident samplers made by the company XIA. The improved angular resolution is needed to Doppler correct the energy of the γ-radiation emitted from e.g. a Coulomb excited beam. The photo-peak efficiency is 9.5% at 1.3 MeV with a resolution of the core energy signal between 2.1 and 2.3 keV. The MINIBALL detector system also comprises a start-stop detector system, with ps time resolution which is built around two Multi Channel Plates and a charged particle detector. The latter is a segmented 0.5 mm thick Si-detector comprising 4 sectors with 24 radial and 16 concentric strips [16]. The radial and angular pitches are 2 mm and 3.5° degrees respectively. The beam can also be monitored after the target position using a PPAC. A BaF_2 detector will also be used to identify heavy ions downstream the target. Proof of principle spectra from the first experiments on heavy Na and Mg isotopes are shown in Fig. 3.

References

1. E. Kugler, Hyperfine Interactions 129 (2000) 23
2. D. Habs et al., Nuclear Instrum. and Methods B 126 (1997) 298
3. M. Huyse, Nuc. Phys. A 701 (2002) 265c
4. U. Köster, Eur. Phys. J. A 15 (2002) 255
5. U. Köster et al., Nuc. Phys. A 701 (2002) 441c
6. C. Tamburella et al. Rev. Sci. Instr. 68 (1997) 2319
7. J. Lettry, Private Communication
8. F. Wenander, Ph.D. thesis, CTH, Gothenburg (2001)
9. P. Schmidt, Ph.D. thesis, University of Mainz (2001)
10. J. Cederkall et al., AIP conf. proc. 610 (2001) 987
11. D. Beck et al., Hyperfine Interactions 132 (2001) 469
12. K. Reisinger, Diploma thesis, LMU Munich (2002)
13. R. Rao et al., Nuclear Instrum. and Methods A 427 (1999) 427
14. P. Reiter et al., Nuclear Physics A 701 (2001) 209c
15. C. Gund, Ph.D thesis, MPI-K, Heidelberg (2002)
16. A. N. Ostrowski et al., Nuclear Instrum. and Methods A 480 (2002) 448

PRODUCTION OF NEUTRON-RICH NUCLIDES AT HRIBF*

E.H. SPEJEWSKI[1], J.C. BATCHELDER[1], H.K. CARTER[1], J.-C. BILHEUX[2],
R. GRZYWACZ[2], D.J. HARTLEY[3], W. KROLAS[3,4], Y. LAROCHELLE[3],
A. PIECHACZEK[5], K.P. RYKACZEWSKI[2], D.W. STRACENER[2], M.N. TANTAWY[3]

1. UNIRIB/Oak Ridge Associated Universities, Oak Ridge TN 37831
2. Physics Division, Oak Ridge National Laboratory, Oak Ridge TN 37831
3. University of Tennessee, Knoxville TN 37996
4. Lawrence Livermore National Laboratory, Livermore CA 94550
5. Louisiana State University, Baton Rouge, LA 70803

A series of experiments has been performed to investigate the production of neutron-rich nuclides for decay-spectroscopy purposes at UNISOR Data obtained for the elements Zn through Kr were well fit by a model designed to predict production rates.

1. Introduction

Development of neutron-rich radioactive beams at the Holifield Radioactive Ion Beam Facility offers the possibility of decay-spectroscopy measurements. In order to propose and plan such experiments, an approximate knowledge of the production rates is necessary. A simple production model has been developed to predict the rates. A series of measurements on the elements Zn through Kr has been performed to determine the applicability of the model.

2. Procedure

A 40-MeV proton beam was incident on a UC_x target [1] installed on the UNISOR isotope separator. The induced-fission products were collected on a metallic tape at and transported to the measurement location. Three Ge clover detectors and a Si electron detector were located around this position. Transition energies and absolute branching ratios were obtained from the literature.

A simple production model, based in part on that by Rudstam [2], has been developed. Employing a phenomenological model [3] of proton-induced fission, the production model requires two parameters, effectively a chemical efficiency (ε_Z) and a system hold-up time (t_Z). The system efficiency for a given isotope is then a function only of the half-life ($t_{1/2}$):

$$\varepsilon_{Z,N} = \varepsilon_Z/[1 + t_Z/t_{1/2}]$$

648

3. Results and Conclusions

Examples from these measurements are shown in Figs. 1and 2, and are compared to the model fit. In general, data for the longer-lived species predicts (within a half order of magnitude) the production for those shorter-lived ones.

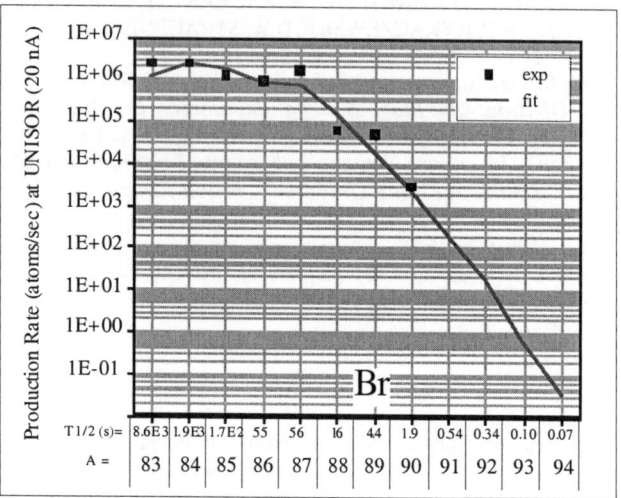

Figure 1. Br Production at U density of ~2.5 g/cm2 and a 20-nA beam

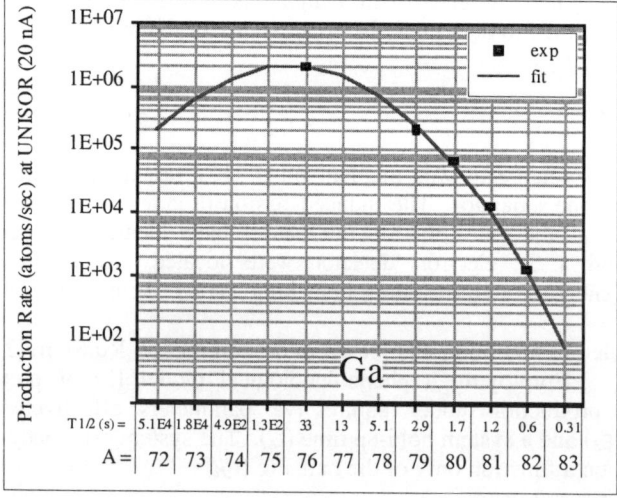

Figure 2. Ga Production at U density of ~2.5 g/cm2 and a 20-nA beam

A chart of the nuclides summarizing predicted production rates employing this model is shown in Fig. 3. It Includes this and previous data.

A design for another decay-spectroscopy facility within HRIBF is completed. At this location, the efficiency is approximately equal to that at UNISOR, and the target/ion-source operates for long periods with a 10-μA proton beam. Thus, production rates could be increased by almost three orders of magnitude.

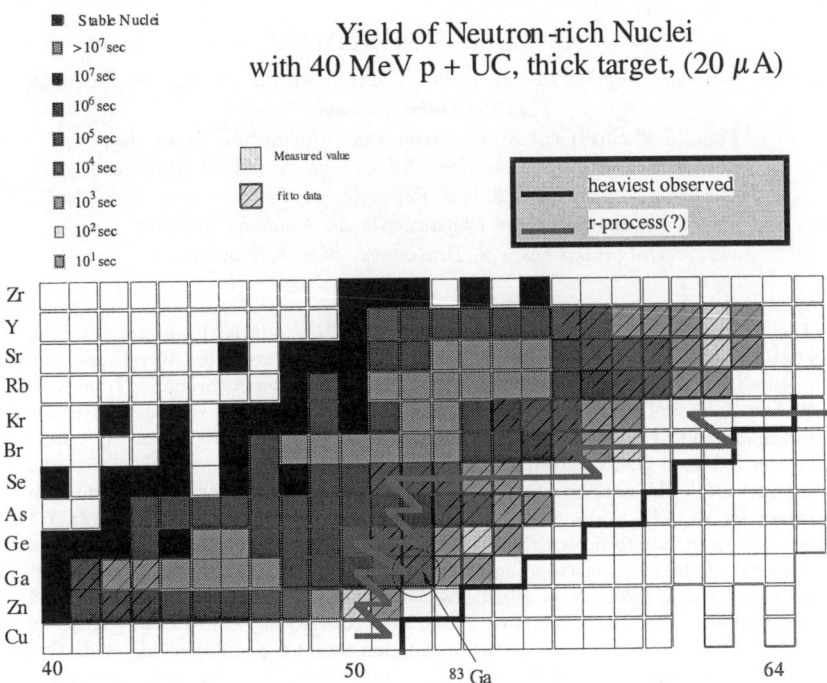

Figure 3. Summary of production rates measured and predicted.

Acknowledgments

This work has been supported by the U. S. Department of Energy under contracts DE-AC05-76OR00033, DE-AC05-96OR22464, DE-FG02-96ER40983, W-7405-ENG-48, and DE-FG02-96ER40978.

References

1. D.W. Stracener, in *Proceedings of the Sixteenth International Conference on the Application of Accelerators in Research and Industry,* edited by J.L. Duggan and I.L. Morgan, CP576, AIP Press, New York (2000) pp. 257-260.
2. G. Rudstam, **NIM A256** (1987) 465.
3. Yuliang Zhao, doctoral thesis, Tokyo Metropolitan University (1996).

THE USE OF THE PHOTOFISSION OF ^{238}U FOR A NEUTRON–RICH RADIOACTIVE ION BEAMS GENERATION

O. SZÖLLŐS[1], J. KLIMAN[2]

Flerov Laboratory of Nuclear Reaction, Joint Institute for Nuclear Research,
141980 Dubna, Russia.
[1] *Faculty of Electrical Engineering and Information Technology,*
Slovak University of Technology, Ilkovičova 3, 812 19 Bratislava,
Slovak Republic.
[2] *Department of Nuclear Physics, Slovak Academy of Sciences,*
Dúbravská cesta 9, Bratislava, Slovak Republic.

The fission fragments yield for photofission of ^{238}U, induced by bremsstrahlung photons with endpoint energies of 25 and $50 MeV$ was evaluated to estimate the possibility of producing the neutron–rich nuclei. The systematics coming from A.C. Wahl's Z_p model [1] for charge distribution of fission fragments were used. Results for xenon and krypton isotopes are compared with experimental data [2] obtained on the DRIBs [3] (**D**ubna **R**adioactive **I**on **B**eams) facility for neutron–rich nuclei production in Flerov Laboratory. The fission rate and fission density in production target for metallic uranium and UC_x compounds were simulated with Geant4 [4] simulation toolkit to design the target geometry. The fission rate dependence on material of the electron stopping target was examined. At nominal beam values on microtron MT-25 ($I_e = 20\mu A$, $E_e = 25 MeV$) up to 2.10^{11} fissions/s could be achieved. Then the production rate of neutron–rich isotopes reaching order of $10^9 s^{-1}$. The induced activity in the production target depending on an irradiation time was calculated for radiation protection purposes and target safety estimation. The cumulation of actinide nuclei was also calculated.

References

1. A. C. Wahl, *Phys. Rew.* **C32** 184 (1985).
2. Yu. P. Gangrsky *et al.*, *Part. and Nucl., Lett.* **6[103]** 5 (2000).
3. Yu. Ts. Oganessian *et al.*, *Proc. of the 4-th Int. Conf. on Dynamical Aspects of Nuclear Fission*, Častá Papiernička, Slovakia, 1998 (World Scientific, Singapore 2000) 1.
4. CERN/DRDC/94-29, DRDC/P58, "Geant4: an Object–Oriented toolkit for simulation in HEP", (1994).

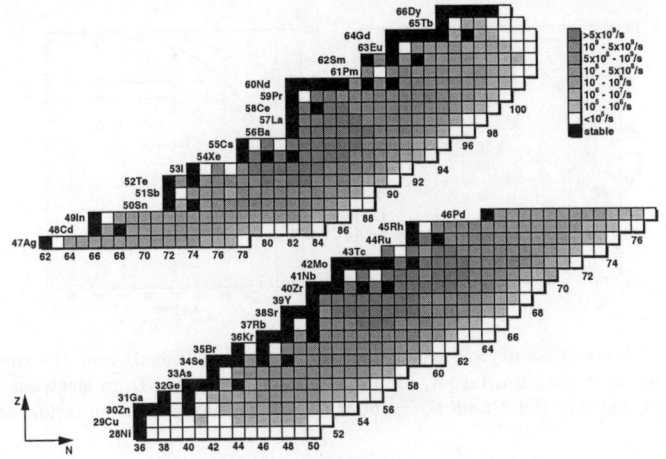

Figure 1. Cumulative fission fragments yield for ^{238}U calculated from Z_p model. Electron beam intensity $I_e = 20\mu A$, energy of electrons $E_e = 25 MeV$.

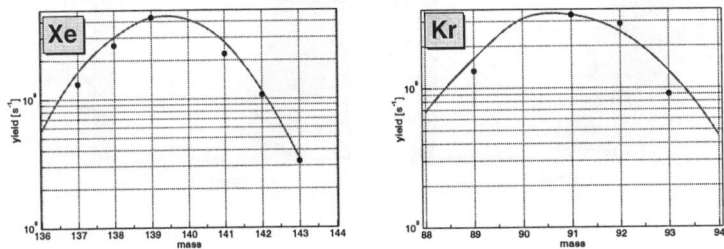

Figure 2. Independent yield of xenon and krypton isotopes calculated with Z_p model (lines) compared with experimental data (points) for photofission of ^{238}U induced by bremsstrahlung with endpoint energy $E_e = 25 MeV$.

Table 1. Target volume dimensions where 90% of fissions take place. W indicates the tungsten stopping target. $UC_x : 55\%_{mass}.U + 45\%_{mass}.C$, $\rho = 1.5 g/cm^3$.

Electron energy [MeV]	Target setup	Z [mm]	R [mm]	Fissions/e^-
25	U (metallic uranium)	20	9.6	0.00162
	W+U	20	20.2	0.00107
	W+UC$_x$	380	145	0.00086
50	U (metallic uranium)	23	8.3	0.00438
	W+U	21	14.3	0.00326
	W+UC$_x$	410	125	0.00252

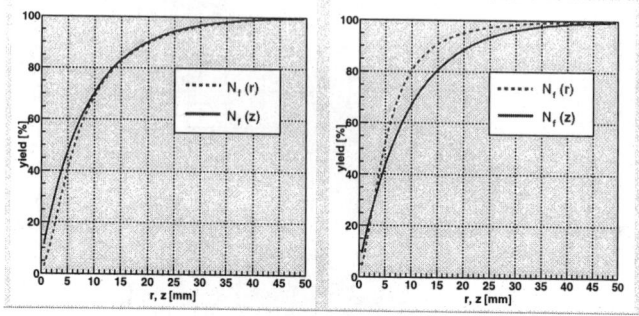

Figure 3. Integral fission yield as a function of radius (dashed) and thickness (solid curves) of metallic uranium target, induced by bremsstrahlung from electrons with energies $E_e = 25 MeV$ (left) and $E_e = 50 MeV$ (right) stopped in tungsten converter.

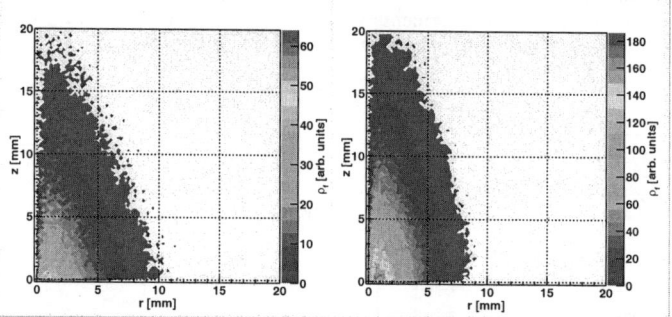

Figure 4. Volume density of fission in metallic uranium induced by bremsstrahlung from electrons with energies $E_e = 25 MeV$ (left) and $E_e = 50 MeV$ (right) stopped in tungsten converter.

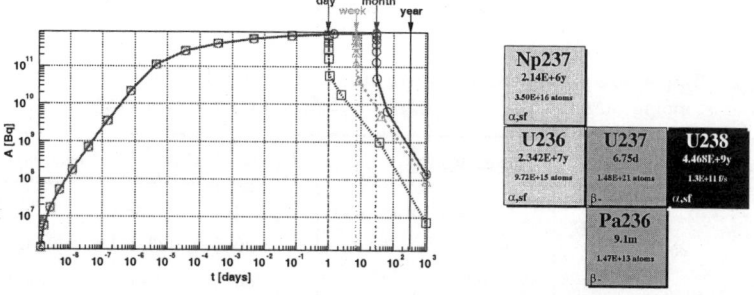

Figure 5. **left:** The ^{238}U target induced activity as a function of irradiation time. Beam–time: day ($\square$), week ($\triangle$) and month ($\bigcirc$). **right:** Cumulation of actinide nuclei after one day of target irradiation. Results are normalized to 1.3×10^{11} fissions/s.

Mass
Measurements
and
Mass Formula

PRECISION MASS MEASUREMENTS ON NEUTRON-RICH NUCLEI WITH THE PENNING TRAP MASS SPECTROMETER ISOLTRAP

K. BLAUM,* F. HERFURTH, A. KELLERBAUER[t]

CERN, Division EP-IS, 1211 Geneva 23, Switzerland

D. BECK, H.-J. KLUGE, C. SCHEIDENBERGER, D. RODRIGUEZ,
G. SIKLER, C. WEBER

GSI Darmstadt, Planckstraße 1, 64291 Darmstadt, Germany

G. BOLLEN, S. SCHWARZ

NSCL, Michigan State University, East Lansing MI 48824-1321, USA

R.B. MOORE

Dept. of Physics, McGill University, Montreal (Quebec) H3A 2T8, Canada

L. SCHWEIKHARD

Inst. f. Physik, Ernst-Moritz-Arndt-Universität, 17487 Greifswald, Germany

ISOLTRAP is a Penning trap mass spectrometer for on-line mass measurements of short-lived nuclides. A number of scientific highlights have already demonstrated the prominent role of ISOLTRAP in the precise determination of nuclear binding energies far from stability. The studies have included nuclides that are produced with yields as low as a few 100 ions/s and at half-lives well below 100 ms. The mass resolving power reaches 10^7 and the uncertainty of the resulting mass values has been pushed down to 10^{-8}. In this contribution, the current status of ISOLTRAP is presented, recent mass measurements on neutron-rich nuclei are summarized and planned measurements are outlined.

---

* *Corresponding author.* E-mail: klaus.blaum@cern.ch
[t] *Present address:* Sektion Physik, Ludwig-Maximilians-Universität, 85748 Garching, Germany

1. Introduction

Accurate mass measurements of short-lived nuclides are of high interest for a number of reasons: They allow direct observation of nuclear structure effects like the location of shell and sub-shell closures, pairing, or the onset of deformation. In combination with a precise study of super-allowed β emission they provide tests of the standard model. Additionally, masses of unstable nuclei are the most critical nuclear physics parameters for reliable nucleosynthesis calculations in astrophysics.

The tandem Penning trap mass spectrometer ISOLTRAP [1] at the on-line isotope separator ISOLDE [2] at CERN/Geneva plays a prominent role in this field [3]. The masses of more than 200 short-lived nuclides have been measured with a relative uncertainty of typically $\delta m/m \approx 1 \cdot 10^{-7}$ and even almost up to one order of magnitude better in some special cases [3-5]. Recently, the performance of the Penning trap mass spectrometer ISOLTRAP has been considerably enhanced. Major technical improvements were implemented to increase the range of accessible nuclei to those that are produced in minute quantities of only 100 ions/s and to nuclei with half-lives down to ≈ 50 ms as well as to decrease the typical relative uncertainty down to $\approx 1 \cdot 10^{-8}$ [6-8]. In particular, a linear radiofrequency quadrupole (RFQ) trap and a carbon-cluster source were added to the ISOLTRAP spectrometer. Since the unified atomic mass unit is defined as $1/12$ of the mass of ^{12}C the calibration of the magnetic field with carbon clusters allows absolute mass measurements.

Mass measurements on neutron-rich nuclei in the light, medium, and heavy mass range are of astrophysical importance as well as for understanding the shell structure in this region. The following is an overview of recent measurements of ISOLTRAP.

2. Experimental Setup

ISOLTRAP as shown in Fig. 1 consists of three functional parts: A radiofrequency quadrupole (RFQ) ion trap and two Penning traps [1,9]. A recent addition is a carbon-cluster source [6,11,12]. The 60-keV continuous ISOLDE beam is stopped in a linear gas-filled RFQ ion trap. To this end the ISOLDE ions are electrostatically retarded before they enter the RFQ, where they are cooled by energy loss due to helium buffer gas cooling. After a certain accumulation time (typically $10 - 20$ ms) the RFQ ion trap is brought to ground potential and the ions are transferred in a pulsed mode to the preparation Penning trap. Here, the ions are stored up to 1 s and

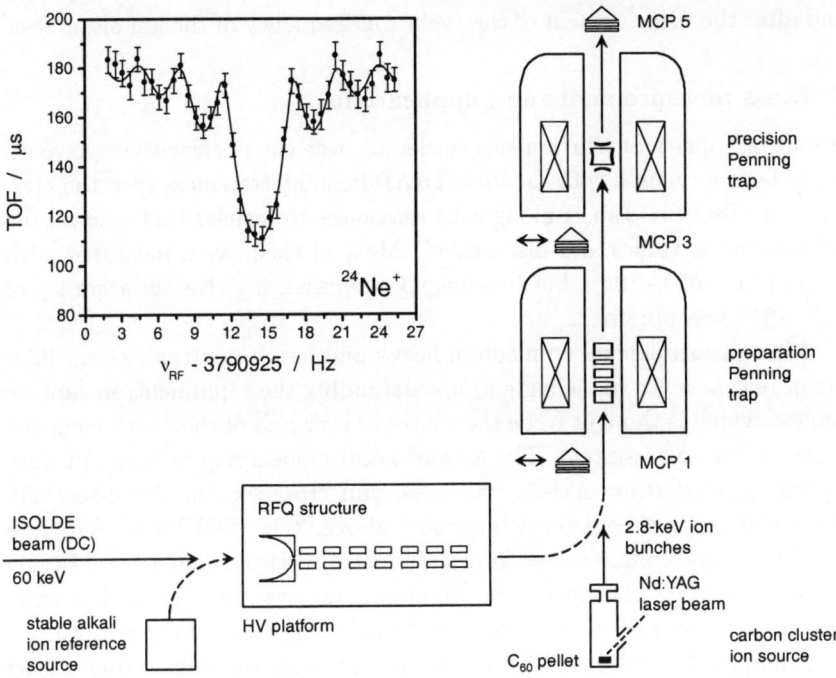

Figure 1. Schematic drawing of the mass spectrometer ISOLTRAP including the RFQ trap, the preparation Penning trap, the precision Penning trap, as well as the carbon cluster ion source. Micro-channel-plate (MCP) detectors are used to monitor the ion transfer as well as to record the time-of-flight resonance (MCP5) for the determination of the cyclotron frequency. The inset shows the cyclotron resonance of ^{24}Ne$^+$ with a fit of the theoretically expected curve [10] to the data points.

contaminant ions are removed by a mass-selective buffer gas cooling technique [13]. The ions are then transferred to the precision Penning trap for the actual mass measurement. The ions' cyclotron frequency is determined using a time-of-flight resonance technique [14]. To this end, the frequency of an azimuthal quadrupolar radiofrequency field is scanned. A characteristic cyclotron resonance curve for the neutron-rich ^{24}Ne$^+$ ion is shown in the inset of Fig. 1. A fit of the resonance curve to the theoretical function [10] yields the cyclotron frequency ν_c of an ion with mass m and charge q:

$$\nu_c = \frac{1}{2\pi} \cdot \frac{q}{m} \cdot B. \qquad (1)$$

The magnitude of the magnetic field B is measured by a determination of the cyclotron frequency of a reference ion with well-known mass both before and after the measurement of the cyclotron frequency of the ion of interest.

3. Mass measurements and applications

About 50 masses of neutron-rich nuclei all over the nuclear chart have already been measured with the ISOLTRAP Penning trap mass spectrometer. They are listed in Tab. 1 along with references to articles that contain the relevant mass values and discussions. Most of them were measured with an accuracy of $1 \cdot 10^{-7}$, but in some special cases, e.g. Ne, an accuracy of $\approx 5 \cdot 10^{-8}$ was obtained.

Mass measurements on medium heavy and heavy neutron-rich nuclides are of relevance for modeling and understanding the rapid neutron capture process which is thought to be the source of over half of the heavy elements found in the solar system. The astrophysical models require as input data nuclear ground state masses, half-lives, and cross sections for essentially the entire range of neutron-rich nuclides above ^{56}Fe. ISOLTRAP is able to provide accurate mass values for reliable astrophysical calculations. Beside the astrophysical relevance accurate atomic masses are in general mandatory for nuclear structure studies. In the following some of the motivations and the physical relevance for mass measurements on neutron-rich nuclei will be addressed in more detail.

3.1. *Masses of light nuclides near the drip line*

Nuclear binding energies provide key information on the isospin dependence of properties of nuclear matter. They tell us when a nuclear system becomes unbound, which defines the position of the drip lines. Bound/unbound transitions are critical phenomena, which depend on small changes in binding energies. High-quality input data as, for example, nuclear masses or, more precisely, nucleon separation energies are required. Concepts developed near stability such as the stabilization at magic nucleon numbers are sensitively tested in nuclei with extreme proton-to-neutron ratios. A well established phenomenon is the observation of disappearance of shell gaps at $N = 20$ and $N = 28$. Indications for new shell-closure like effects have been found at $N = 16$, $N = 26$, and $N = 34$.

A first step towards mass measurements in this mass region with ISOLTRAP was done by investigation neutron-rich Ne and Ar isotopes (see Tab. 1). Furthermore, the existence of halo structures is a phenomenon critically depending on very small neutron separation energies. In the near future it is planned to perform ISOLTRAP mass measurements on con-

Table 1. List of neutron-rich nuclei whose mass was measured with the ISOLTRAP Penning trap mass spectrometer.

Element	Mass number	Reference
Ne	23, 24	‡
Ar	42, 43	[15]
	44, 45, 46	†
Rb	88 ...94	[16]
Sr	91 ...95	[16]
Sn	128 ...132	[17]
Cs	134 ...142	[19]
	145, 147	‡
Ba	139 ...144	[19]
Bi	215, 216	‡
Fr	221, 222	[20]
	229	‡
Ra	226, 230	[19,20]
	229	‡

†measured in 2001, to be published
‡measured in 2002, to be published

firmed neutron halo nuclei and new candidates like ^{6,8}He, ^{11}Li, ^{11,12}Be, and $^{20-22}$N. In addition, ^{9}Li and $^{23-25}$Ne will be addressed because of their importance for laser spectroscopy experiments [21].

3.2. Masses in the vicinity of the doubly magic ^{132}Sn

The mass data in the vicinity of doubly closed shells for nucleons give highly valuable structure information. This has a particular significance for the regions far from stability where new phenomena as for instance shell quenching may arise due to low binding energies. Accurate binding energies, derived from the atomic masses, are also very important for the modelling of the astrophysical r-process. In addition to general trend studies in this mass region there exists a number of nuclides that are of special interest.

ISOLTRAP measurements resulted in new mass values e.g. for the nuclides $^{124,129-132}$Sn [17]. As can be seen from Fig. 2, important deviations from accepted table values [18] were found for the ground state Sn iso-

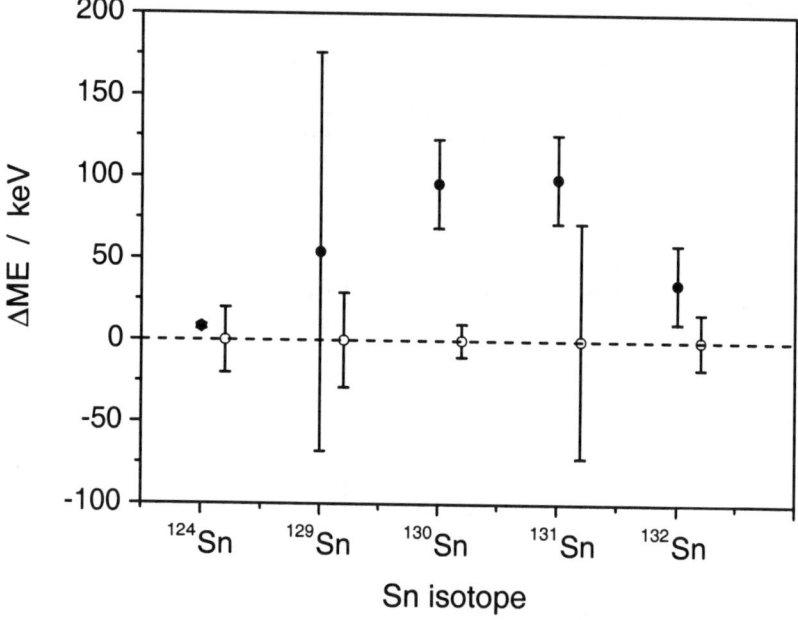

Figure 2. Mass excess (ME) for Sn isotopes in the ground state as given in the updated AME'95 [18] (filled circles) and as obtained by ISOLTRAP (open circles). The difference ΔME is plotted while the ISOLTRAP data define the zero line.

topes, which show the need for more accurate values in this region. The measured masses around the doubly magic ^{132}Sn are also inconsistent with shell model based predictions [22], whereas the value for ^{132}Sn is in agreement with recent results from improved Q_β measurements [23].

Recently we addressed neutron-rich nuclei in the Cs and Ba isotopic chains (see Tab. 1) as a continuation of successful earlier studies [19].

3.3. Masses of neutron-rich Tl, Pb, and Bi isotopes

Considerable interest exists in the determination of the properties of Tl, Pb, and Bi nuclei and the understanding of the nuclear structure as one proceeds away from the doubly magic ^{208}Pb. Despite their vicinity to the line of stable nuclides, information on single particle levels, decay properties, and ground state properties is extremely scarce for nuclides to the neutron-rich site of ^{208}Pb. Knowledge of binding energies is not only important for the understanding of nuclear structure in this region. An extension of experimentally known masses away from the valley of stability is decisive for constraints on nuclear mass models in the region where the r–process path may proceed towards the synthesis of the heaviest elements known in our solar system. First measurements in the close vicinity of ^{208}Pb are already performed (Tab. 1). Further mass measurements are planned.

4. Summary and Outlook

ISOLTRAP is able to perform mass measurements on radionuclides with an accuracy below 10^{-8}. In the near future ISOLTRAP aims to perform high-accuracy mass measurements especially in the light mass range (^{6,8}He, ^{9,11}Li, ^{11,12}Be, $^{17,25-27}$Ne, and $^{20-22}$N) where a new region should be explored. In addition masses in the region of the doubly-magic nuclei ^{78}Ni, ^{132}Sn, and ^{208}Pb will be addressed to provide data that improve the knowledge of masses in these important regions and may therefore contribute to a better understanding of the nuclear structure and of the nucleosynthesis via the r–process.

Acknowledgments

The support of the German Ministry for Education and Research (BMBF) as well as the support by the European Commission (EUROTRAPS network under contract number FMRX-CT-97-0144, RTD project EXO-TRAPS under contract number FMGC-ET-98-0099, and NIPNET network under contract number HPRI-CT-2001-50034) are gratefully acknowledged.

References

1. G. Bollen, et al., Nucl. Instrum. Methods **A368**, 675 (1996).
2. E. Kugler, Hyp. Int. **129**, 23 (2000).
3. G. Bollen, Nucl. Phys. **A693**, 3 (2001).
4. F. Herfurth, et al., Eur. Phys. J. **A15**, 17 (2002).
5. F. Herfurth, et al., J. Phys. B, (accepted for publication, 2002).
6. A. Kellerbauer, et al., Eur. Phys. J. **D**, (accepted for publication, 2002).
7. K. Blaum, et al., Nucl. Instrum. Methods **B**, (accepted for publication, 2002).
8. K. Blaum, et al., J. Phys. **B**, (accepted for publication, 2002).
9. F. Herfurth, et al., Nucl. Instrum. Methods **A469**, 254 (2001).
10. M. König, et al., Int. J. Mass Spec. Ion Processes **142**, 95 (1995).
11. C. Scheidenberger, et al., Nucl. Phys. **A701**, 574c (2002).
12. K. Blaum, et al., Eur. Phys. J. **A15**, 245 (2002).
13. G. Savard, et al., Phys. Lett. **A158**, 247 (1991).
14. G. Gräff, H. Kalinowsky, and J. Traut, Z. Phys. **A297**, 35 (1980).

15. F. Herfurth, *et al.*, *Phys. Rev. Lett.* **87**, 142501 (2001).
16. H. Raimbault-Hartmann, *et al.*, *Nucl. Phys.* **A706**, 3 (2002).
17. G. Sikler, *et al.*, *Eur. Phys. J.* **A15**, (accepted for publication, 2002).
18. G. Audi and A.H. Wapstra, *Nucl. Phys.* **A595**, 1 (1995); private communication (2001).
19. F. Ames, *et al.*, *Nucl. Phys.* **A651**, 3 (1999).
20. G. Bollen, *et al.*, *J. Mod. Optics* **39**, 257 (1992).
21. R. Neugart, *Eur. Phys. J.* **A15**, 35 (2002).
22. C.T. Zhang, *et al.*, *Phys. Rev. Lett.* **77**, 3743 (1996).
23. B. Fogelberg, *et al.*, *Phys. Rev. Lett.* **82**, 1823 (1999).

RESULTS FROM JYFLTRAP AT IGISOL[*]

J. SZERYPO

Sektion Physik, University of Munich (LMU)
Am Coulombwall 1, D-85748 Garching, Germany
E-mail: Jerzy.Szerypo@physik.uni-muenchen.de

V. S. KOLHINEN, T. ERONEN, J. HAKALA, A. JOKINEN[†], S. KOPECKY, A. NIEMINEN, S. RINTA-ANTILA AND J. ÄYSTÖ

Department of Physics, University of Jyväskylä
P.O. Box 35 (YFL), FIN-40351 UNIVERSITY OF JYVÄSKYLÄ, Finland
E-mail: Veli.Kolhinen@phys.jyu.fi

The IGISOL facility at the Department of Physics of the University of Jyväskylä (JYFL) is delivering radioactive beams of short-lived exotic nuclei, in particular the neutron-rich isotopes from the fission reaction. These nuclei are studied with the nuclear spectroscopy methods. In order to substantially increase the quality and sensitivity of such studies the beam should undergo beam handling: cooling, bunching and isobaric purification. The first two processes are performed with the use of an RFQ cooler/buncher. The isobaric purification is made by a Penning trap placed after the RF-cooler. This will yield a substantial background reduction in the nuclear decay spectroscopy experiments and enable studies of much more exotic nuclei, like the ones belonging to r-process path. This contribution describes the current status of the Penning trap project including the results of its first off-line and on-line tests.

1. Introduction

A project for improving the quality of radioactive ion beams produced at IGISOL[1], aiming at the enhancement of the quality of experiments, has been described already in several papers (e.g. [2,3]). The goal was to decrease substantially the beam energy spread ΔE (below 1 eV) as well as transverse emittance ξ, and increase the mass resolving power $R = M/\Delta M$ up to 10^5. The latter is opening a possibility of rejecting even isobaric contaminants and obtaining a pure monoisotopic beam. This improvement is achieved due to

---

[*] This work is supported by by the Academy of Finland under the Finnish Centre
of Excellence Program 2000-2005 (Project No. 44875) and by the EXOTRAPS project in the EU LSF-RTD program under contract no. ERBFMGECT980099.
[†] Present address: CERN, EP, CH-1211 Geneva 23, Switzerland.

radioactive beam handling, which consists of three steps: beam cooling (which improves both ΔE and ξ), bunching and purification (due to high R). The first two steps are done with the use of an RFQ cooler/buncher. The beam purification is performed applying a cylindrical Penning trap[4].

2. Project description

Main features of the project are described in ref. [3]. The RFQ cooler/buncher [2] exists at JYFL already. Its performance: $\Delta E = 0.6$ eV, ξ is about 4 πmmmrad (at 40 keV), transmission TR > 60% and cooling time around 1 ms. Bunching possibility was also demonstrated with bunches as short as 5 μs. It's application to the collinear laser spectroscopy experiments is described by another contribution to this conference[5].

Technical description of the Penning trap system is contained in ref. [6]. Here an outline of the ion movement excitations and the isobaric separation principle will be presented. In a Penning trap ions are confined in three dimensions by a superposition of static electric quadrupole and homogeneous magnetic fields (the latter usually created by means of a superconducting magnet). In such a field composition, an ion movement in the transversal plane containing the trap center is split into two components:

a) slow circular magnetron movement around the trap center with magnetron frequency ω_-,

b) fast circular reduced cyclotron movement with reduced cyclotron frequency ω_+, usually close to the classical cyclotron frequency $\omega_c = qB/m$. Basic dependence linking these quantities is: $\omega_c = \omega_- + \omega_+$.

It is possible to change the radii of the above movements by means of dipole RF excitations in the plane transversal to the trap axis. For that, the central ring electrode of the trap should be split into two halves to which the opposite RF signal phases should be connected. This excitation, if done at the magnetron frequency, is mass non-selective and increases the orbit radii of all ions in the trap. On the other hand, making such an excitation at the ω_+ frequency is mass-selective and excites only ions in a limited mass range (ultimately only one mass) depending on the excitation parameters.

To achieve an isobaric purification, mass-selective buffer gas cooling technique[7] can be used. In contrast to the dipole excitations described above, this excitation must be of quadrupole character and must be done at the ω_c frequency. To achieve that, one must split the ring electrode into four quadrants, then connect electrically opposite quadrants in pairs and connect each pair to the opposite phase of the RF signal. Such an excitation is decreasing the orbit radii of ions from a narrow mass range whose centroid corresponds to the mass in the $\omega_c = qB/m$ equation. This process can have a high mass resolving power, of the order of 10^5, which permits to reject even isobaric contaminants[4]. A necessary condition is that the trap ejection side is equipped with a diaphragm of a (small) radius r_d. Relevant experimental procedure is following:

- injecting the ionic bunch into the trap,
- de-centering all ions by means of magnetron excitation so that all ion orbits have the radii above r_d,
- centering back with a cyclotron excitation only the ions of interest (of a given mass) so that their final orbit radii are below r_d,
- ejecting the ions from the trap.

Resulting very high mass resolving power is particularly important for the experimental program at IGISOL, which in a future will be centered on exotic neutron-rich nuclei, produced in fission. The isobaric purification will allow for rejecting of all unwanted members of the isobaric chain, leaving only a species of interest. This will significantly improve signal-to-background ratio, sensitivity and precision of the experiments, and will extend the range of the isotopes investigated.

Summarizing, the task of the purification Penning trap at IGISOL is to:

1) accept cooled and bunched beams from the RFQ cooler/buncher,

2) perform the isobaric purification,

3) deliver clean, monoisotopic bunched beams for nuclear spectroscopy investigations, precise nuclear mass measurements and laser spectroscopy experiments.

The nuclear spectroscopy experiment will utilize a variety of gamma, X-ray and particle detectors. The bunched monoisotopic beam from the trap will be transferred to a measurement position and implanted in a collecting foil or a movable tape surrounded by the above-mentioned detector set-up.

The precise nuclear mass measurements of radioactive ions will be done using a second Penning trap. It will be placed directly after the purification trap in the same superconducting magnet. The measurement accuracy of up to 10^{-7} with a cylindrical trap and up to 10^{-8} with a hyperbolic one should be reachable[8]. It will enable mass measurements of many neutron-rich isotopes not reachable anywhere else (e.g. of refractory elements) and will significantly broaden experimental program at IGISOL.

As an extension of the conventional decay spectroscopy studies, nuclear spectroscopy in a Penning trap interior ("in-trap" spectroscopy) is foreseen[9]. This means placing the detectors of a needed type directly inside the trap and positioning the radioactive sample in front of them. Advantages of such type of measurements over conventional spectroscopy are mentioned in ref.[3].

3.   Off-line test results

3.1. *Low-voltage tests*

First tests of the Penning trap system were carried out at the voltages of the HV-platform[6] below 10 kV, with the use of the off-line ion source, delivering Xe ion beam. Transmission through the trap, measured for a continuous beam by means of Faraday cups, of about 25% was obtained. In the next step, continuous beam was captured inside the trap potential well by means of switching on and off the trap well injection and ejection walls sequentially. Ejected ionic bunch was detected by means of an MCP (microchannel-plate) detector and the TOF spectrum was measured with a Stanford SR430 multichannel scaler. In order to perform ion movement excitations inside the trap (see Section 2.) two HP33120A function generators were connected to the proper segments of the trap central ring electrode. This permitted to perform all three ion movement excitations in the Penning trap, in the azimuthal plane: dipole magnetron excitation at ω_-, dipole reduced cyclotron excitation at ω_+ and quadrupole cyclotron excitation at ω_c frequency. As an outcome, an optimal f_- frequency was determined as 1730 Hz (for the trap potential well depth of 100 V).

The ω_+-excitation was performed for the mass A=32, assumed as O_2, ions. Frequency scan around the theoretical value (Fig. 1) gave the experimental result of f_+=3357086 Hz. At this frequency, the peak corresponding to A=32 mass was totally removed from the TOF spectrum. Combining both frequencies permitted to determine the strength of the B-field in the trap and to find an experimental f_c-value (814591 Hz) for the ^{132}Xe mass. This value was used for the subsequent ω_c-excitation of the corresponding ions and their centering was demonstrated.

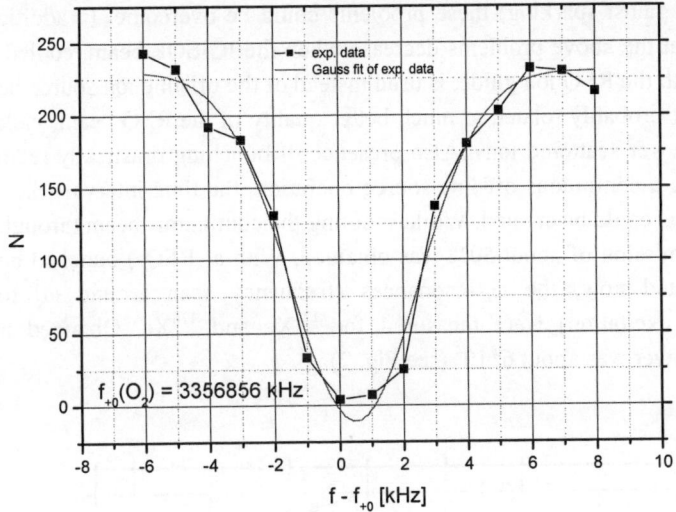

Figure 1. Frequency scan for the ω_+-excitation of A=32 ions. Centroid of the Gaussian fit is shifted with respect to f_{+0} by 230 Hz. The scan was done at the RF-amplitude of U=2 V, excitation time t=100 ms and pressure p=0.02 mbar in the He-gas inlet line (pressure in the trap center is below this value).

3.2. High-voltage tests

Lifting the HV-platform up to 30-40 kV brought at the beginning some undesirable effects:

- due to electron traps in the electrostatic potential at places where the magnetic field was high, substantial restgas ionization by the electrons circling in the field was taking place (Penning effect) resulting in a spurious current creation hindering original ion source current measurement (to the smaller extent, such problems were present also during the low-voltage tests),
- the above spurious current have hit some of the trap insulators, evoking their charge-up which, in turn, has influenced negatively ion trajectories creating big problems with the beam transmission,
- sparking inside the trap has caused damage of some electronic equipment.

After long and tedious work (liquidating the electron traps by the electrode structure changes and proper potential adjustment, securing electronic equipment against sparking) these problems could be overcome. In addition it occurred that the above problems decrease when the IGISOL beam, cooled and bunched with the RFQ ion guide, is used instead of the off-line ion source beam. This is most probably related to much better quality of the RFQ beam, yielding e.g. much fewer scattered ions. Also presence of bunching drastically reduced, compared to a continuous off-line source operation, the time interval when the scattered ions could be created. While shooting the continuous beam through the trap a transmission of about 50% was obtained. With a (RFQ-) bunched beam, mass-separated now, the ω_c-resonances (frequency scan around ω_c for a quadrupole excitation) were measured for ^{124}Xe and ^{126}Xe. Obtained mass resolving power was about $6*10^4$ (see Fig. 2).

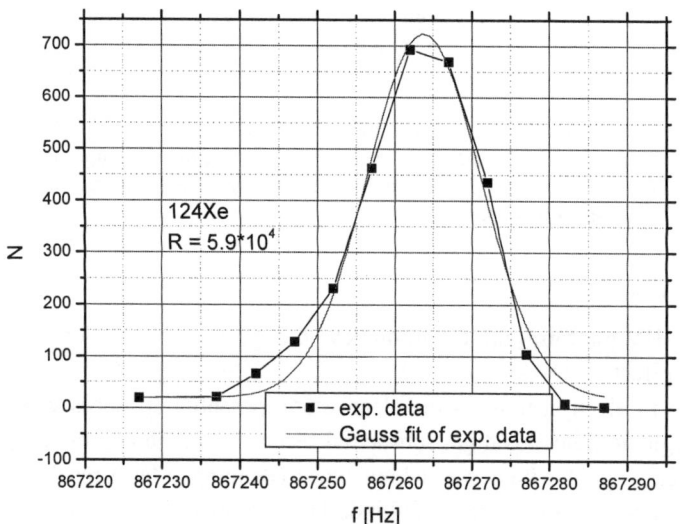

Figure 2. Frequency scan for the ω_c-excitation of ^{124}Xe, U=250 mV/t=70 ms. Before the ω_c-excitation the ions were de-centered with the ω_--excitation (see Section 2) at f= 1718Hz/U=140 mV/t=10 ms. The scan was done at p=0.04 mbar.

4. On-line test results

For the on-line Penning trap test with a radioactive beam, reaction ^{58}Ni(p,n) was used, delivering ^{58}Cu (3.2 s). This test was made at HV = 29 kV. For the ω_c-resonance measurement, the frequency scan covering both ^{58}Ni and ^{58}Cu masses was performed. Both resonances were very well separated. Mass resolving power obtained here amounted to $7*10^4$ (see ref. [10])

5. Outlook

First off-line and on-line tests of JYFLTRAP Penning trap system have shown that the apparatus is functioning very well. Mass resolving power obtained is matching the one delivered by the leading Penning trap system in this class, ISOLTRAP[11] at ISOLDE/CERN.

In the near future, further optimization of the beam transmission through the trap and of the ω_c-resonance measurement parameters (maximizing mass resolving power, minimizing excitation time) will be performed. Soon the trap system will be applied in a fission experiment and a final check of the trap performance in the nuclear spectroscopy experiment will be carried out.

Acknowledgments

One of authors (JS) gratefully acknowledges Conference's cost support from the Conference Organizers and travel support from NIPNET project under contract no. HPRI-CT-2001-50034.

References

1. J. Äystö, Nucl. Phys. A693, 477 (2001).
2. A. Nieminen, J. Huikari, A. Jokinen, J. Äystö, P. Campbell and E. C. A. Cochrane, Nucl. Instr. Meth. A469, 24 (2001).
3. J. Szerypo, A. Jokinen, V. S. Kolhinen, A. Nieminen, S. Rinta-Antila and J. Äystö, Acta Phys. Polonica B33 no. 1, 487 (2002).
4. H. Raimbault-Hartmann, D. Beck, G. Bollen, M. König, H.-J. Kluge, E. Schark, J. Stein, S. Schwarz and J. Szerypo, Nucl. Instr. Meth. B126, 378 (1997).
5. A. Jokinen, these proceedings
6. V. S. Kolhinen, T. Eronen, J. Hakala, A. Jokinen, S. Kopecky, S. Rinta-Antila, J. Szerypo and J. Äystö, Nucl. Instr. Meth. B (2003), to be published.
7. G. Savard, St. Becker, G. Bollen, H.-J. Kluge, R.B. Moore, T. Otto, L. Schweikhard, H. Stolzenberg and U. Weiss, Phys. Lett. A158, 247 (1991).
8. K. Blaum, these proceedings

9. L. Weissman, F. Ames, J. Äystö, O. Forstner, K. Reisinger and S. Rinta-Antila, Nucl. Instr. Meth. A, in press
10. V. S. Kolhinen, T. Eronen, J. Hakala, A. Jokinen, S. Kopecky, A. Nieminen, S. Rinta-Antila, J. Szerypo and J. Äystö, Nucl. Phys. A (2003), to be published
11. G. Bollen, S. Becker, H.-J. Kluge, M. König, R. B. Moore, T. Otto, H. Raimbault-Hartmann, G. Savard, L. Schweikhard, and H. Stolzenberg, Nucl. Instr. Meth. A368, 675 (1996).

MICROSCOPIC MASS FORMULAS AND THE FISSION BARRIERS OF NEUTRON-RICH NUCLEI

J. M. PEARSON*

Département de Physique,
Université de Montréal,
Montréal (Qc), H3C 3J7 Canada
E-mail: pearson@lps.umontreal.ca

M. SAMYN, S. GORIELY, A. MAMDOUH, AND M. RAYET

Institut d'Astronomie et d'Astrophysique,
CP 226 Université Libre de Bruxelles,
B-1050, Bruxelles, Belgium

We describe here our Hartree-Fock mass formulas, and also the ETFSI approximation to the HF method that made possible a calculation of the barriers of the nearly 2000 nuclei expected to be involved in the r-process, as well as of all measured barriers. We discuss the crucial impact of new mass data, which have radically modified the extrapolations to the neutron-drip line.

1. Introduction

The r-process of stellar nucleosynthesis depends crucially on the masses and fission barriers (among other quantities) of nuclei that are so neutron-rich that there is no hope of being able to measure them in the laboratory. It is thus of the greatest importance to be able to make reliable extrapolations of these quantities away from the known region of the nuclear chart, relatively close to the stability line, out towards the neutron-drip line. Until recently the masses and barriers used in all studies of the r-process were calculated via the macroscopic-microscopic approach, with the macroscopic part based on some form or other of the liquid-drop model, and shell corrections added by application of the Strutinsky theorem. The most sophisticated mass formula of this kind is that based on the "finite-range droplet model" (FRDM) [1].

---
*Work partially supported by NSERC (Canada).

However, in an attempt to put the extrapolations on as rigorous a footing as possible, the Brussels-Montreal group has for more than 15 years been developing a much more microscopic approach, based in general on the Hartree-Fock (HF) method with Skyrme forces and δ-function pairing forces. We have used the method to calculate both masses and barriers, the point being that any model that gives the nuclear binding energy as a function of deformation can be applied to the calculation of barriers.

Although nuclear HF calculations have been feasible for more than 30 years, the computer demands of our project are enormous. For the mass formula we fit the forces to essentially all of the some 2000 measured masses, which means that all these nuclei have to be computed hundreds of times, and *a priori* it must be supposed that all nuclei are deformed. Once the force is determined from the mass fit, the construction of a complete mass table requires the computation of the more than 8000 nuclei lying between the drip lines. As for barriers, there is no question of refitting the force, since in the interest of a unified treatment of the nuclear physics of the r-process we wish to use the same force for masses and barriers. Thus each nucleus has to be calculated once only, but the computer demands are even heavier than in the case of the mass formula, since the deformation space that has to be explored is very large indeed.

Thus we were at first obliged to adopt the ETFSI (extended Thomas-Fermi plus Strutinsky integral) approximation to the Skyrme-HF method [2,3,4,5,6]. This led to the ETFSI-1 mass formula [6], the first to be based entirely on microscopic forces: the Skyrme and pairing forces were fitted to essentially all the mass data available at the time (1992), and a complete mass table constructed. Later, very extensive barrier calculations were performed using the ETFSI method, with the same force that emerged from the ETFSI-1 mass fit; the nuclei considered included not only all those whose barriers have been measured [7], but also the nearly 2000 that could be relevant to the r-process [8].

By the beginning of the new millenium the computer power available to us had grown to the point where we could construct full HF mass formulas, without recourse to the ETFSI approximation (Section 2). HF calculations of barriers using the forces generated in these mass fits have likewise begun for selected nuclei (see Ref. [9]), but the ETFSI calculations of Refs. [7,8] still constitute the only available microscopic calculations covering all of the large numbers of nuclei required for the r-process (Section 3).

2. The HF Mass Formulas

So far we have developed three different HF mass formulas, all based on Skyrme-type forces of the conventional form

$$v_{ij} = t_0(1 + x_0 P_\sigma)\delta(\mathbf{r}_{ij}) + t_1(1 + x_1 P_\sigma)\frac{1}{2\hbar^2}\{p_{ij}^2\delta(\mathbf{r}_{ij}) + h.c.\}$$

$$+t_2(1 + x_2 P_\sigma)\frac{1}{\hbar^2}\mathbf{p}_{ij}.\delta(\mathbf{r}_{ij})\mathbf{p}_{ij} + \frac{1}{6}t_3(1 + x_3 P_\sigma)\rho^\gamma\delta(\mathbf{r}_{ij})$$

$$+\frac{i}{\hbar^2}W_0(\boldsymbol{\sigma}_i + \boldsymbol{\sigma}_j).\mathbf{p}_{ij} \times \delta(\mathbf{r}_{ij})\mathbf{p}_{ij} \quad , \tag{1}$$

where P_σ is the two-body spin-exchange operator. The HF formalism associated with this force is quite standard, and is summarized in Ref. [10]. Likewise, all these mass formulas involve a δ-function pairing force,

$$v_{pair}(\mathbf{r}_{ij}) = V_{\pi q}\ \delta(\mathbf{r}_{ij}) \quad . \tag{2}$$

There is also a Wigner term that acts only in light nuclei.

The first of our mass formulas [11], HFBCS-1, treats the pairing in the BCS approximation, and the parameters of the Skyrme and pairing forces, and of the Wigner term, are fitted to the 1768 measured masses of nuclei with $N, Z \geq 8$ given in the 1995 Audi-Wapstra compilation [12]. The final parameter fit (set MSk7) gave a rms error of $\sigma = 0.718$ MeV.

Because of the well known problems with the BCS approximation in highly neutron-rich nuclei we next developed a full HF-Bogolyubov (HFB) treatment [13]. With the original force, MSk7, there was a significant deterioration in the quality of the fit to the same data, but on refitting we achieved a rms error of $\sigma = 0.740$ MeV with our final parameter set (BSk1), only slightly worse than before. The new mass formula, HFB-1, was, in fact, quite similar to HFBCS-1, even in its extrapolations out to the neutron-drip line, and we conclude that as far as mass formulas are concerned the much simpler HFBCS method is just as reliable as the HFB method.

After the construction of these two mass formulas, measured masses for 382 new nuclei with $N, Z \geq 8$ were reported [14], and the predictions of both mass formulas were found to be badly overbound, the rms error for these new nuclei exceeding 1.1 MeV in both cases. It was shown that the problem lay mainly with the choice of cutoff prescription adopted for the δ-function pairing force. For both HFBCS-1 and HFB-1 (as for ETFSI-1) this had been chosen to lie at the value of $\hbar\omega = 41A^{-1/3}$ MeV. However, a great improvement was found if the cutoff energy was chosen to lie rather at $E_F + 15$ MeV for all nuclei, E_F denoting the Fermi energy for the nucleus in question (different, of course, for neutrons and protons). For the 2135 nuclei in the new data set we finally found a rms error of $\sigma = 0.674$ MeV; a part of this improvement came from a refinement to the Wigner term. (For the same data set the FRDM gives $\sigma = 0.676$ MeV, but it should be realized that it had been fitted to a much smaller data set.) This is the BSk2 force and the corresponding mass formula is labelled HFB-2. [15]

It now turns out that with the modification of the pairing cutoff imposed by the new mass data there are some drastic changes in the extrapolations out towards the neutron-drip line, even though the quality of the data fit is relatively unchanged. In particular, there is a considerable weakening of the neutron-shell gaps, as defined by $\Delta(N_0) = S_{2n}(Z, N_0) - S_{2n}(Z, N_0 + 2)$. Fig. 1 shows the situation for the magic neutron number $N_0 = 184$, the implications of which for fission barriers will be discussed below.

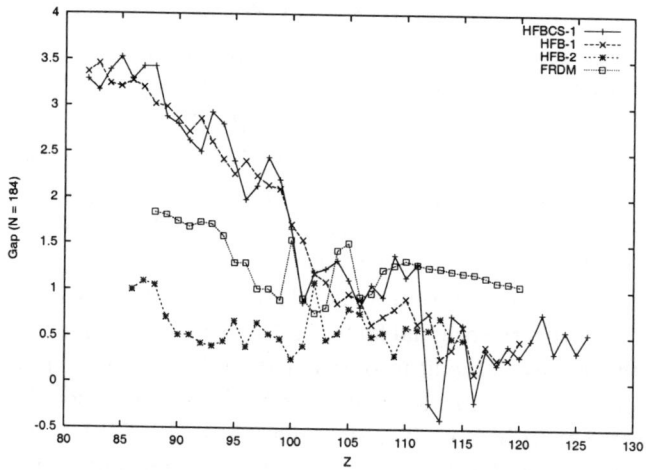

Figure 1. $N_0 = 184$ neutron-shell gap for HFBCS-1, HFB-1, HFB-2 and FRDM mass models.

3. The ETFSI Barriers

The ETFSI approximation to the Skyrme-HF method has been described in detail in Refs. [2,3,4,5,6], and here we recall only the main features. The starting point for a given Skyrme force is a semi-classical approximation to the HF method, the fourth-order extended Thomas-Fermi (ETF) method,

$$E[\rho_{HF}] \simeq E_{ETF}[\rho_{ETF}] \equiv \int \mathcal{E}_{ETF}\left(\rho_q(\mathbf{r}), \rho_q'(\mathbf{r}), \dots\right) d^3\mathbf{r} \quad , \qquad (3)$$

where the energy density $\mathcal{E}_{ETF}$ is, for the given Skyrme force, a function only of the nucleon densities $\rho_q(\mathbf{r})$ ($q = n$ or p) and their various derivatives. The HF condition that $E[\rho_{HF}]$ be stationary with respect to arbitrary variations of all the single-particle (s.p.) wave-functions carries over in Eq. (3) into the condition that E_{ETF} be minimized with respect to arbitrary variations in the densities $\rho_q(\mathbf{r})$. But E_{ETF} will then vary smoothly from one nucleus to another, the shell fluctuations that are an essential feature of the HF method being effectively lost in truncating the semi-classical expansion that characterizes the ETF approximation.

However, the ETF method itself serves as a basis in which the shell corrections are restored perturbatively. The central s.p. field $\tilde{U}_q(\mathbf{r})$ and the spin-orbit s.p. field $\tilde{\mathbf{W}}_q(\mathbf{r})$ are first generated by folding the given Skyrme force over the density distributions $\tilde{\rho}_q(\mathbf{r})$ that emerge from the minimization of E_{ETF}. The corresponding s.p. Schrödinger equation is then solved for the eigenvalues $\tilde{\epsilon}_{i,q}$. Denoting by $\tilde{\tau}_q$ and $\tilde{\mathbf{J}}_q$ the kinetic-energy and spin-current densities, respectively, determined in the ETF part of the calculation, the Strutinsky theorem now leads to the improved approximation

$$E[\rho_{HF}] \simeq E_{ETF}[\rho_{ETF}] + \sum_{i,q} \tilde{\epsilon}_{i,q}$$

$$- \sum_{q=n,p} \int d^3\mathbf{r} \left(\frac{\hbar^2}{2M_q} \tilde{\tau}_q + \tilde{\rho}_q \tilde{U}_q + \tilde{\mathbf{J}}_q . \tilde{\mathbf{W}}_q \right) \quad . \tag{4}$$

(We have not shown here the effect of the pairing, which was represented by a δ-function force, treated in BCS.) The first of the extra terms on the right-hand side of Eq. (4), the sum of the s.p. energies, clearly contains shell effects, while the integral term, which we call the "Strutinsky-integral" (SI), is just a smoothed version of the former, so that the two terms together constitute a shell correction. [a]

Instead of minimizing E_{ETF} with respect to arbitrary variations in the densities $\rho_q(\mathbf{r})$, we take for the latter a parametrization that reduces in the limit of spherical nuclei to the simple Fermi form

$$\tilde{\rho}_q(r) = \frac{\rho_{0q}}{1 + \exp(\frac{r - C_q}{a_q})} \quad . \tag{5}$$

---

[a] Note that the ETFSI method is applicable only when the Skyrme parameters are constrained in such a way that the effective and real nucleon masses are equal. Fortunately, this condition is compatible with a good agreement with experiment both for masses and fission barriers [4].

More generally, the ETFSI method takes account of deformations, although here we impose axial symmetry; deformed configurations are generated from the spherical distribution of Eq. (5) by means of the (c, h, α) prescription [16], as described in Refs. [3,7] (in the mass formula we also imposed left-right symmetry, whence $\alpha = 0$).

The ETFSI method described here is essentially equivalent to the HF method in the sense that when the underlying Skyrme-type force is fitted to the same data by one method or the other the two methods give very similar extrapolations out to the neutron-drip line: the discrepancy is less than 1 MeV for absolute masses, and less than 0.5 MeV for neutron-separation energies S_n and beta-decay energies Q_β, quantities of direct relevance to the r-process [2,3]. A similar agreement is found for fission barriers, provided these are less than 13 MeV high [9].

Although the energy (4) displays shell-model fluctuations, the ETFSI method, like any method based on the Strutinsky theorem, expresses these fluctuations in terms of the sum of s.p. energies that themselves vary smoothly with respect to N, Z, and (c, h, α), as do all other quantities appearing on the right-hand side of Eq. (4). We thus made extensive use of interpolation both in the construction of the ETFSI-1 mass table [6] and in the barrier calculations of Refs. [7,8], with the complete ETFSI calculation, as described above, being performed only for a restricted number of key nuclei and key deformations. We estimate that the complete ETFSI-1 mass table was constructed some 2000 times faster than would have been possible with a direct application of the HF method, even though the two methods are essentially equivalent in their extrapolations. While this speed advantage of the ETFSI method over the HF method is no longer crucial in the case of the mass formula, it is still the basis of the only extensive microscopic barrier calculations that have been performed [7,8], as we discuss below.

In any case, the results obtained with the ETFSI-1 mass formula are very similar to those of the HFBCS-1 and HFB-1 formulas, both with respect to the quality of the fit and to the extrapolations to the neutron-drip line, the same pairing-cutoff prescription having been adopted in all three cases. In particular, large gaps are again found at the neutron-drip line in the vicinity of $N = 184$.

As for the ETFSI-1 barriers, they were calculated with the same force that emerged from the ETFSI-1 mass fit, SkSC4. Full results have been presented in Refs. [7,8]. One of the most striking conclusions concerns the very high barriers found near the neutron-drip line in the vicinity of $N =$

184: see Fig. 2. Such high barriers imply that the r-process path may continue well beyond $A = 300$. However, it is reasonable to associate these high barriers with the large shell-model gaps found for these nuclei in ETFSI-1, and we have seen how the new mass data [14] imply, through the modified pairing-cutoff prescription that is required, much smaller gaps. It is thus conceivable that a new calculation (of necessity, HFB) with the force BSk2 of the HFB-2 mass formula, would show much lower barriers for such nuclei. We have checked just one case, that of ^{276}U, and find the barrier to be reduced by 4.6 MeV (see Ref. [9] for further details). As seen in Fig. 2, this barrier is still much higher than the value given by the drop-model calculations of Howard and Möller [17]. At least a large part of the remaining difference can be traced to the low symmetry coefficient a_{sym} of the Skyrme force BSk2 (see Ref. [8]).

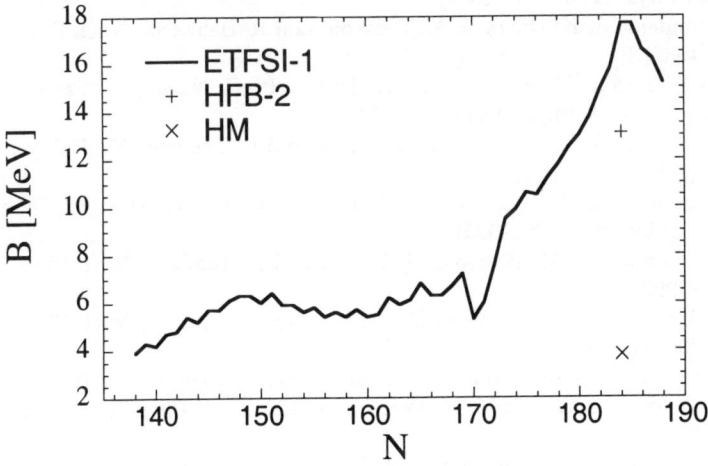

Figure 2. Primary fission barriers of the uranium isotopes ($Z = 92$) in three models: ETFSI-1 [8], HFB-2 [15], and Howard-Möller (HM) [17].

4. Concluding Remarks

We have described our various HF mass formulas, with particular emphasis on the implications of the new Audi-Wapstra mass data for extrapolations out to the neutron-drip line: the predicted shell-model gaps are considerably weaker than before. We also discussed the ETFSI-1 fission-barrier

678

calculations, which covered all nuclei whose barriers have been measured, as well as the nearly 2000 neutron-rich nuclei expected to play a role in the r-process. We showed that these barrier predictions for highly neutron-rich nuclei will have to be modified in the light of the new mass data (see Ref. [9]).

More generally, our experience with these new mass data serves to show how cautiously one should interpret all extrapolations out to the highly neutron-rich region of the nuclear chart: further accumulation of mass data might reveal more nasty surprises.

References

1. P. Möller, J. R. Nix, W. D. Myers, and W. J. Swiatecki, At. Data Nucl. Data Tables **59**, 185 (1995).
2. A. K. Dutta, J.-P. Arcoragi, J. M. Pearson, R. Behrman, and F. Tondeur, Nucl. Phys. **A458**, 77 (1986).
3. F. Tondeur, A. K. Dutta, J. M. Pearson, and R. Behrman, Nucl. Phys. **A470**, 93 (1987).
4. J. M. Pearson, Y. Aboussir, A. K. Dutta, R. C. Nayak, M. Farine, and F. Tondeur, Nucl. Phys. **A528**, 1 (1991).
5. Y. Aboussir, J. M. Pearson, A. K. Dutta, and F. Tondeur, Nucl. Phys. **A549**, 155 (1992).
6. Y. Aboussir, J. M. Pearson, A. K. Dutta, and F. Tondeur, At. Data Nucl. Data Tables **61**, 127 (1995).
7. A. Mamdouh, J. M. Pearson, M. Rayet, and F. Tondeur, Nucl. Phys. **A644**, 389 (1998).
8. A. Mamdouh, J. M. Pearson, M. Rayet, and F. Tondeur, Nucl. Phys. **A679**, 337 (2001).
9. M. Samyn, S. Goriely, and J. M. Pearson, next paper (this conference).
10. F. Tondeur, S. Goriely, J. M. Pearson, and M. Onsi, Phys. Rev. C **62**, 024308 (2000).
11. S. Goriely, F. Tondeur, and J. M. Pearson, At. Data Nucl. Data Tables **77**, 311 (2001).
12. G. Audi and A. H. Wapstra, Nucl. Phys. **A595**, 409 (1995).
13. M. Samyn, S. Goriely, P.-H. Heenen, J. M. Pearson, and F. Tondeur, Nucl. Phys. **A700**, 142 (2002).
14. G. Audi and A. H. Wapstra, private communication (2001).
15. S. Goriely, M. Samyn, P.-H. Heenen, J. M. Pearson, and F. Tondeur, Phys Rev. C **66**, 024326 (2002).
16. M. Brack, J. Damgaard, A. S. Jensen, H. C. Pauli, V. M. Strutinsky, and C. Y. Wong, Rev. Mod. Phys. **44**, 320 (1972).
17. W. M. Howard and P. Möller, At. Dat. Nucl. Data Tables **25**, 219 (1980).

REALISTIC FISSION BARRIER CALCULATIONS WITHIN THE SKYRME HARTREE-FOCK THEORY

M. SAMYN AND S. GORIELY

Institut d'Astronomie et d'Astrophysique, CP-226 ULB, Bvd du Triomphe,
B-1050 Brussels, Belgium
E-mail: msamyn@astro.ulb.ac.be

J.M. PEARSON

Dépt. de Physique, Université de Montréal, Montréal (Québec), H3C 3J7
Canada

We calculate fission barriers with the Hartree-Fock-BCS (HFBCS) method, using Skyrme forces fitted to essentially all the nuclear mass data with the same method. A first set of our results is compared to experimental data. We also briefly consider the calculation of barriers with the Hartree-Fock-Bogolyubov (HFB) method, discussing in particular the highly neutron-rich nucleus ^{276}U.

1. Introduction

As explained in the previous paper [1], the Brussels-Montreal group is developing a Hartree-Fock (HF) approach to the calculation of nuclear masses and fission barriers, the motivation being the need to be able to make reliable extrapolations from the data to the highly neutron-rich nuclei that are involved in the r-process. This same paper gives an overview of our efforts to obtain a reliable mass formula within the HFBCS and HFB approaches using Skyrme-type forces [2,3,4]. Here we will describe the present state of our efforts to extend the HFBCS and HFB calculations to the prediction of fission barriers. The emphasis will be on the former method, but we will also present some results for HFB calculations as well.

There have, of course, been several HFBCS and HFB calculations of the fission barriers of certain selected nuclei in the past, but for the elucidation of the r-process the barriers of nearly 2000 nuclei are required, and considerations of computer time become crucial. In fact, at present the only microscopic calculation of all the barriers required for the r-process was performed using the ETFSI (Extended Thomas-Fermi plus Strutinsky Integral) approximation to the HF method [5]: see Ref. [1] for a summary. However, with the much greater computer power at our disposal, we can now contemplate recalculating all these barriers with the full HFBCS method. The pilot calculations reported here are intended to show primarily how the HFBCS method leads to a satisfactory agreement with experiment.

Since we are aiming at a unified treatment of all the nuclear physics of the r-process, we use in these HFBCS barrier calculations the force that emerged from the HFBCS mass formula, MSk7.[2] Likewise in the HFB calculations we use the forces BSk1[3] and BSk2[4] given by the mass fits HFB-1 and HFB-2, respectively. (In the same way, the ETFSI barriers were calculated with the force given by the ETFSI-1 mass fit, SkSC4.) Many nuclei have more than one barrier, of course, but here we consider mainly the primary, i.e highest, barrier, since this is the most reliably measured, and the one most relevant to the r-process.

Our calculational procedures are described in Sect. 2, and are illustrated in Sect. 3 by the familiar case of the double-humped barrier of ^{240}Pu. In Sect. 4 we present our results for left-right symmetric HFBCS calculations, and compare them with experiment, while in Sect. 5 we consider the sensitivity of barriers to the pairing cutoff prescription, an issue raised in Ref. [1].

2. The constrained HFBCS model and the calculation of barrier heights

Our method of solving the HFBCS equations for Skyrme interactions has been presented earlier[2] and has now been extended to left-right asymmetric deformations. The HF states are expanded on an axially deformed oscillator basis. To closely follow the fission path, it is convenient to relate the deformation parameters of the basis to the so-called (c, h, α) parametrization used in the ETFSI calculations,[5] defined as the elongation, the necking and the left-right asymmetry parameters, respectively. It allows for the definition of a reference surface that more or less coincides with the actual surface of the fissioning nucleus.

All information needed to calculate the fission probability and the fission fragments distribution are obtained from the analysis of the multi-dimensional energy surface, where the dimensions are essentially the deformation parameters of the nucleus, which can be taken as the multipole moments. The only way to obtain such a surface is to constrain the HF calculation to *every* possible deformation of the nucleus. It is however not necessary to constrain on multipoles of higher order than four, as they are optimized by the self-consistency of HF and are believed not to play a significant role. The multipole moments that must be constrained (we use the method of the quadratic constraint) are thus the quadrupole, the octupole and the hexadecapole moments, all being calculated in the frame of the centre of mass. For extreme deformations where the reference surface is split, the HF calculations are not constrained to ensure numerical convergence. A proper description of a system at very large deformation would require a two-centre oscillator basis,[6] so that our model is not able to describe the fission process beyond scission.

The determination of barrier heights is in principle a simple matter if there are just two deformation parameters, e.g., (c, h): with the total energy E of the given nucleus calculated at a sufficient number of deformations one just makes a contour plot of E in the (c, h) plane. However, this procedure is not available in the present case, as we admit a left-right asymmetry, and thus have to determine the fission path in the 3-dimensional space spanned by the variables (c, h, α).

An ingenious solution to this problem is provided by the "flooding model" of Tondeur [7]. In two dimensions we imagine water being slowly poured into a model of the energy surface, and observe its depth, measured at the lowest point, i.e., at the ground state, as it spills over the various barriers. The virtue of this method is that its algorithm can be easily generalized to an arbitrary number of dimensions: see Ref. [5] for a detailed account. We stress that this model is applicable whether the energy E at each deformation is calculated in HF or ETFSI; indeed it has recently been used in liquid-drop calculations of fission barriers [8,9].

In all cases we begin with a first HFBCS (or HFB) calculation of the energy surface, assuming left-right symmetry, $\alpha = 0$. If the corresponding ETFSI calculation [5] indicates that a particular barrier in this surface is asymmetric, we calculate the HFBCS (or HFB) energy surface over the (c, h) plane in the vicinity of the concerned saddle point for each of the four values of $\alpha = n.\delta\alpha$, with $n = 0, 1, 2, 3$ and a suitable value of $\delta\alpha$ (note that α is never negative). The grid over the (c, h) plane corresponds to $\delta c = \delta h = 0.05$. Before applying the flooding model, every surface is interpolated in the c and h directions using the cubic spline method, and with respect to α using Lagrange interpolation.

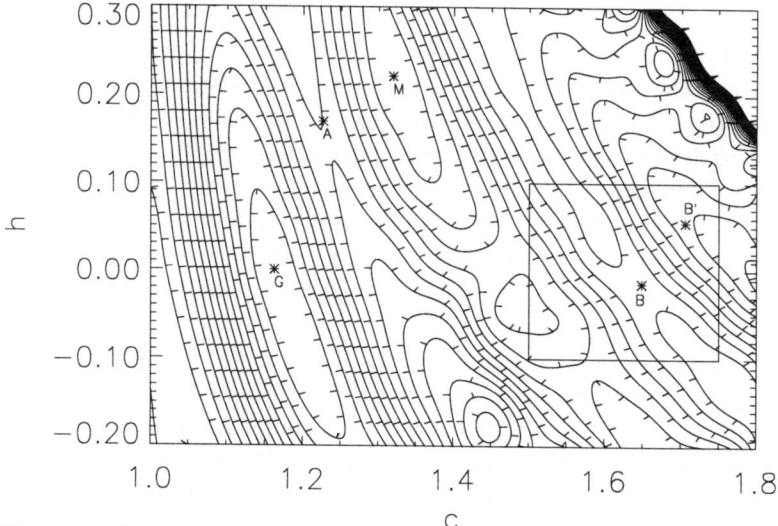

Figure 1. Contour plot of the left-right symmetric energy surface of ^{240}Pu in the $(c,h)_{\alpha=0}$ plane. The contour lines are spaced by .9 MeV to clarify the graph. Small tick marks along each contour point in the downhill direction. Letters G, A, M, B, B' refer to the ground-state ($Q = 2890\ fm^2$), the inner saddle-point ($Q = 5600\ fm^2$), the isomeric state ($Q = 8820\ fm^2$), the left-right symmetric and asymmetric outer saddle-point ($Q = 10200\ fm^2, O = -6250\ fm^3$), respectively. The square centred on B shows the limits in the (c,h) plane taken to calculate the local 3D energy surface with its saddle-point B'. The uper right zone of the panel describes the fissioned nucleus.

3. The case of ^{240}Pu

We consider this case in some detail in order to illustrate our general procedure, showing the energy surface for $\alpha = 0$ in Fig. 1. The calculated (measured) energy of the ground-state (G) is -1812.70 (-1812.67) MeV, which is in agreement with the quality of the fit of MSk7 to the measured masses. The inner barrier height, measured to be of 5.8 MeV, is predicted to be 5.5 MeV. The first shape isomeric state (M) is at 815 keV above the ground-state (experimentally, about 2.8 MeV). The outer symmetric barrier B (at 6.9 MeV) has been recalculated by the method explained before, including the third dimension α within a local variation of c and h as shown by the square centred on B. The resulting asymmetric saddle-point (B') gives an outer barrier height of 4.1 MeV, while it is measured to be 5.3 MeV. The effect of the left-right asymmetry on the outer barrier is to lower it by 2.8 MeV.

4. HFBCS symmetric fission barriers and comparison with data

It turns out that many of the experimentally measured primary barriers show left-right symmetry in the ETFSI calculations [5], and in the present pilot calculations we assume that these barriers remain so in HFBCS calculations, since this reduces the computer time by a factor of 4. Fig.2 compares our calculations with experimental data[10,11] for the primary barriers of all nuclei that were predicted to be symmetric in the ETFSI-1 calculations [5]. We here recall that the only input of the model, the Skyrme force MSk7, was determined entirely by the mass fit, with no subsequent adjustment to the fission data. The most notable success of the HFBCS calculations, as compared to the earlier ETFSI calculations, is for the 22 nuclei with $80 \leq Z \leq 86$, as seen in Fig. 2. Experimentally, the primary barriers of all these nuclei are higher than 13 MeV, but the ETFSI method invariably overestimates them, the mean error being around 3.4 MeV. This problem is now removed, the mean error in HFBCS being only 0.055 MeV for these 22 nuclei (rms error = 0.868 MeV). This suggests that the problem with the ETFSI method lies in its shape parametrization becoming inadequate at large deformations, rather than in any new dynamics, such as a dependence of the pairing strength on effective surface area, coming into play.

The results obtained for nuclei with $Z \geq 90$ are not yet well understood so that the overall agreement will only be expressed in terms of the rms and the mean errors: on these remaining 64 nuclei, the mean deviation to experiment is -0.81 (-0.003 MeV) MeV with an rms of 1.213 (0.788 MeV) MeV for HFBCS (ETFSI) calculations, accounting for a tendency of HFBCS values to underestimate the experimental data in specific regions of A ($A = 230-235$ and $A = 284-292$). We have not found the reason for such deviations, having tried other Skyrme forces with different properties, and even within the HFB scheme. The fission barrier of superheavies are known to be mainly due to the variation of the level density at the Fermi surface, which can be quantified in terms of a shell correction energy. Therefore, a significant decrease in the effective mass may result in an overall increase of the barrier heights because of a decrease of the level density at the barrier. Barrier heights have been calculated with the Skyrme force SLy6[12], which has a rather low effective mass (0.69 instead of 1.05 for MSk7), and found good agreement with experiment where MSk7 fails ($A \approx 235$ and $A \approx 286$), but SLy6 overestimates all other barriers by approximately 3 MeV. It is

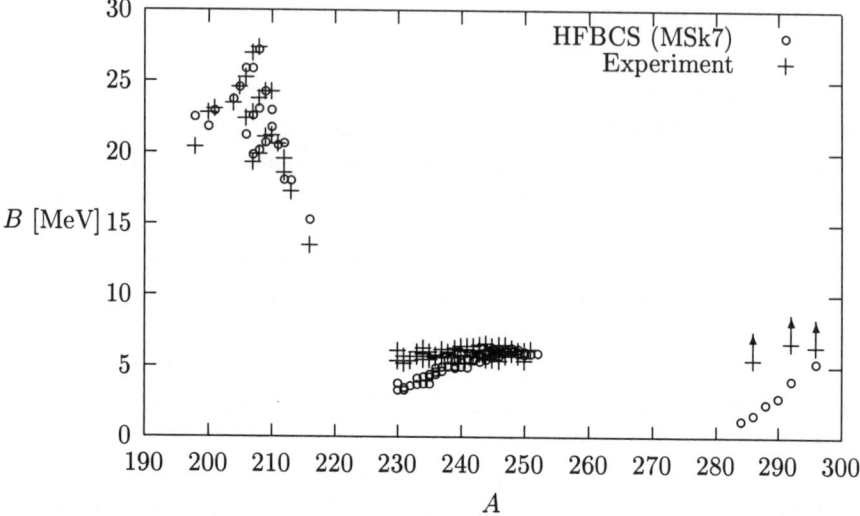

Figure 2. Comparison of HFBCS(MSk7) and experimental fission barrier heights [MeV] predicted to be left-right symmetric by the ETFSI model and measured to be primary (the highest barrier). Calculated barriers are upper limits, as the triaxiality has not been included. Experimental barriers of superheavy nuclei are lower limits. For $Z = 112$, results corresponding to $N = 172, 174, 176, 178$ are shown to underline the dependence of the barrier heights with N.

however difficult to assign the differences seen in Fig. 2 only to the effective mass. SLy6 has a density-dependent pairing, and was not determined by a fit to all nuclear masses. A detailed study of the dependence of the barrier heights with the Skyrme force used is currently being investigated and will be reported in a forthcoming paper. We note that a possible source of error could also come from the very indirect method used to extract the fission barriers from fusion and fission reactions through reaction rates analysis.[11]

We finally stress that the calculated barrier heights in Fig. 2 are upper limits, as left-right asymmetry and triaxiality are not included. Although we have considered here only those nuclei that have left-right symmetry in the ETFSI calculations, it could be that some of these nuclei would display asymmetry in HFBCS, if the α degree of freedom were released. As for triaxiality, Dutta et al.[13], working in the ETFSI framework, found that for a selected set of nuclei barriers could be lowered by 0.6 MeV on average, with a maximum effect of 1.3 MeV for ^{252}Cf. Concerning the superheavy nuclei, Ćwiok et al.[14] predicted the barrier of the neutron-rich isotope $^{288}112$ to be axially symmetric, while for $^{310}126$ triaxiality lowered the barrier by the enormous amount of 4 MeV.

5. The case of ^{276}U

As discussed in Ref. [1], it is now known that the pairing cutoff prescription used in the force MSk7 (one major shell above 0 MeV) is incompatible with the latest mass data, and that when the cutoff is modified to fit these data the shell-model gaps at the neutron-drip line become much smaller. It was therefore conjectured that the very high barriers found with the ETFSI calculations near the neutron-drip line in the vicinity of the magic number $N = 184$ would be considerably reduced when recalculated with a force incorporating the new cutoff prescription (15 MeV above the Fermi energy for all nuclei). So far the only force that has been completely fitted to all the mass data with this new pairing prescription is the force BSk2. Since this was fitted to the mass data in the HFB framework we are obliged to perform our new barrier calculation in this same framework, which is very time-consuming, and to compare it to a barrier calculation done with BSk1, the HFB force equivalent to MSk7 and SkSC4. We considered the single case of ^{276}U, and found that the primary barrier height with BSk2 fell to 13.1 MeV, as compared with 16.7 MeV if BSk1 is used. Thus the conjecture concerning the sensitivity of the barriers of drip-line nuclei to the cutoff prescription appears at first sight to be confirmed. However, the corresponding ETFSI and HFBCS barriers are 17.7 MeV and 14.0 MeV respectively, both obtained with the old pairing cutoff prescription. We are thus left with another dilemma, the discrepancy between MSk7 and BSk1 suggesting that HFBCS calculations themselves may not be sufficient, and that the very time-consuming HFB method may be necessary.

6. Conclusions

We have found in nearly all cases a good agreement with experiment for primary fission barriers calculated with the HFBCS method, using the Skyrme force MSk7. In particular, using the full HFBCS method seems to resolve the strong overestimate given by the ETFSI method in the case of high barriers. However some discrepancies for superheavy nuclei ($Z \approx 112$) are found.

We have confirmed, in the case of ^{276}U, that the strong shell quenching resulting from the modified pairing cutoff prescription that is implied by the new mass data does, as expected, lead to significantly lower barriers for highly neutron-rich nuclei in the vicinity of $N = 184$.

The present HF and former ETFSI calculation of fission barrier clearly shows that large-scale microscopic models can compete with more phenomenological highly parametrized models in the reproduction of experimental data. The large-scale HF calculation of the symmetric and asymmetric barriers for all the thousands nuclei of relevance for the r-process has now become feasible, although it still faces many technical difficulties. From the physics point of view, this problem is also open to many questions, one of the most important one concerning the sensitivity of the fission barriers to the different components of the effective force. A coherent and accurate determination of nuclear masses and fission barriers within one unique mean field approach definitely represents one of the most challenging issues for nuclear astrophysics applications in the future. Much work remains to be done.

Acknowledgments

M.S. and S.G. are FNRS Research Fellow and Associate, respectively. J.M.P. acknowledges financial support from NSERC (Canada). We thank M. Rayet and M. Bender for clarifying discussions.

References

1. J.M. Pearson, M. Samyn, S. Goriely, A. Mamdouh and M. Rayet, this conference.
2. F. Tondeur, S. Goriely, J.M. Pearson and M. Onsi, *Phys. Rev.* **C62**, 024308 (2000)
3. M. Samyn, S. Goriely, P.-H. Heenen, J. M. Pearson and F. Tondeur, *Nucl. Phys.* **A700**, 142 (2002).
4. S. Goriely, M. Samyn, P.-H. Heenen, J. M. Pearson and F. Tondeur, *Phys. Rev.* **C66**, 024326 (2002); (*www-astro.ulb.ac.be*).
5. A. Mamdouh, J. M. Pearson, M. Rayet and F. Tondeur, *Nucl. Phys.* **A644**, 389 (1998); *Nucl. Phys.* **A679**, 337 (2001); (*www-astro.ulb.ac.be*).
6. J.F. Berger and D. Gogny, *Nucl. Phys.* **A333**, 302 (1980).
7. F. Tondeur, private communication.
8. P. Möller and A. Iwamoto, Phys. Rev. **C61**, 047602 (2000).
9. P. Möller, D. G. Madland, A. J. Sierk, and A. Iwamoto, Nature **409**, 785 (2001).
10. G.N. Smirenkin, IAEA Report INDC(CCP)-359 (1993); (*iaeand.iaea.or.at*).
11. M.G. Itkis, Yu.Ts. Oganessian and V.I. Zagrebaev, *Phys. Rev.* **C65**, 044602 (2002).
12. E. Chabanat, P. Bonche, P. Haensel, J. Meyer and R. Schaeffer, *Nucl. Phys.* **A635**, 231 (1998).
13. A. K. Dutta, J. M. Pearson, and F. Tondeur, *Phys. Rev.* **C61**, 054303 (2000).
14. S. Ćwiok, J. Dobaczewski, P.-H. Heenen, P. Magierski and W. Nazarewicz, *Nucl. Phys.* **A611**, 211 (1996).

Superheavy Nuclei

82

EXOTIC CLUSTERS COMPOSED OF MATTER AND ANTIMATTER

W. GREINER

Institut für Theoretische Physik, J.W. Goethe-Universität, D-60054 Frankfurt,
Germany

T. BÜRVENICH

Theoretical Division, Los Alamos National Laboratory, Los Alamos NM 87545
USA

We discuss the possibility of producing a new kind of nuclear system by putting a few antibaryons inside ordinary nuclei. The structure of such systems is calculated within the relativistic mean–field model assuming that the nucleon and antinucleon potentials are related by the G–parity transformation. The presence of antinucleons leads to decreasing vector potential and increasing scalar potential for the nucleons. As a result, a strongly bound system of high density is formed. Due to the significant reduction of the available phase space the annihilation probability might be strongly suppressed in such systems.

1. Introduction

In this proceedings article we would like to report on some recent exciting results that have been obtained together with our friends and collaborators I. N. Mishustin, L. M. Satarov, J. A. Maruhn, and H. Stöcker [1]. Before embarking upon the physical discussion, we would like to introduce the ideas and the framework.

Presently it is widely accepted that the relativistic mean–field (RMF) model [2] gives a good description of nuclear matter and finite nuclei [3]. Within this approach the nucleons are supposed to obey the Dirac equation coupled to mean meson fields. Large scalar and vector potentials, of the order of 300 MeV, are necessary to explain the strong spin–orbit splitting in nuclei. The most debated aspect of this model is related to the negative–energy states of the Dirac equation. In most applications these states are simply ignored (no–sea approximation) or "taken into account" via the non–linear and derivative terms of the scalar potential. On the other

690

hand, explicit consideration of the Dirac sea combined with the G–parity arguments leads to such interesting conjectures as the existence of deeply–bound antinucleon states in nuclei [4] or even spontaneous production of nucleon–antinucleon pairs [5,6]. Unfortunately, the experimental information on the antinucleon effective potential in nuclei is obscured by the strong absorption caused by annihilation. The real part of the antiproton effective potential might be as large as 200–300 MeV, with the uncertainty reaching 100% in the deep interior of the nucleus.

Keeping in mind all possible limitations of the RMF approach, below we consider yet another interesting application of this model. Namely, we study properties of light nuclear systems containing a few real antibaryons. At first sight this may appear ridiculous because of the fast annihilation of antibaryons in the dense baryonic environment. But as our estimates show, due to a significant reduction of the available phase space for annihilation, the life time of such states might be long enough for their observation. In a certain sense, these states are analogous to the famous baryonium states in the $N\overline{N}$ system [7], although their existence has never been unambiguously confirmed. To our knowledge, up till now a self–consistent calculation of antinucleon states in nuclei has not been performed. Our calculations can be regarded as the first attempt to fill this gap. We consider first two nuclear systems, namely ^{16}O and ^{8}Be, and study the changes in their structure due to the presence of an antiproton. Then we discuss the influence of small antimatter clusters on heavy systems like ^{208}Pb.

2. Theoretical framework

Below we use the RMF model which previously has been successfully applied for describing ground–states of nuclei at and away from the β–stability line. For nucleons, the scalar and vector potentials contribute with opposite signs in the central potential, while their sum enters in the spin–orbit potential. Due to G–parity, for antiprotons the vector potential changes sign and therefore both the scalar and the vector mesons generate attractive potentials.

To estimate uncertainties of this approach we use three different parametrizations of the model, namely NL3 [8], NL–Z2 [9] and TM1 [10]. In this paper we assume that the antiproton interactions are fully determined by the G–parity transformation. We solve the effective Schrödinger equations for both the nucleons and the antiprotons. Although we neglect the Dirac sea polarization, we take into account explicitly the contribution of the an-

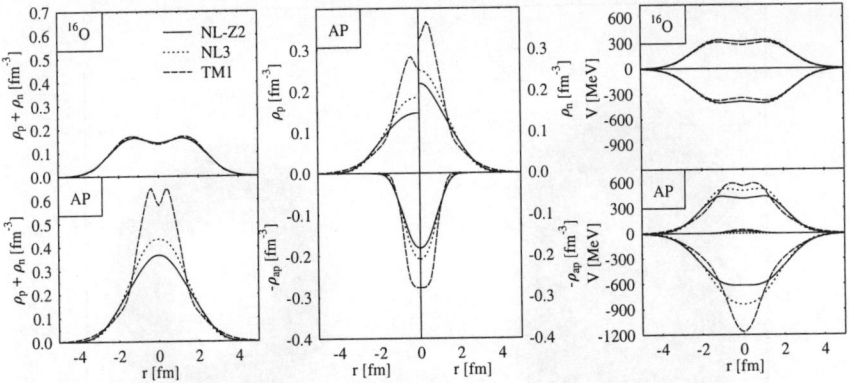

Figure 1. The left panel represents the sum of proton and neutron densities as function of nuclear radius for ^{16}O without (top) and with an antiproton (denoted by AP). The left and right parts of the upper middle panel show separately the proton and neutron densities, the lower part of this panel displays the antiproton density (with minus sign). The right panel shows the scalar (negative) and vector (positive) parts of the nucleon potential. Small contributions shown in the lower row correspond to the isovector (ρ–meson) part.

tibaryon into the scalar and vector densities. For protons and neutrons we include pairing correlations within the BCS model with a δ–force (volume pairing) [11]. Calculations are done within the blocking approximation [12] for the antiproton, and assuming the time–reversal invariance of the nuclear ground–state. The coupled set of equations for nucleons, antinucleons and meson mean fields is solved iteratively and self–consistently. The numerical code employs axial and reflection symmetry, allowing for axially symmetric deformations of the system.

3. Structure of light nuclei containing antiprotons

As an example, we consider the nucleus ^{16}O with one antiproton in the lowest bound state. This nucleus is the lightest nucleus for which the mean–field approximation is acceptable, and it is included into the fit of the effective forces NL3 and NL-Z2. The antiproton state is assumed to be in the $s1/2^+$ state. The antiproton contributes with the same sign as nucleons to the scalar density, but with opposite sign to the vector density. This

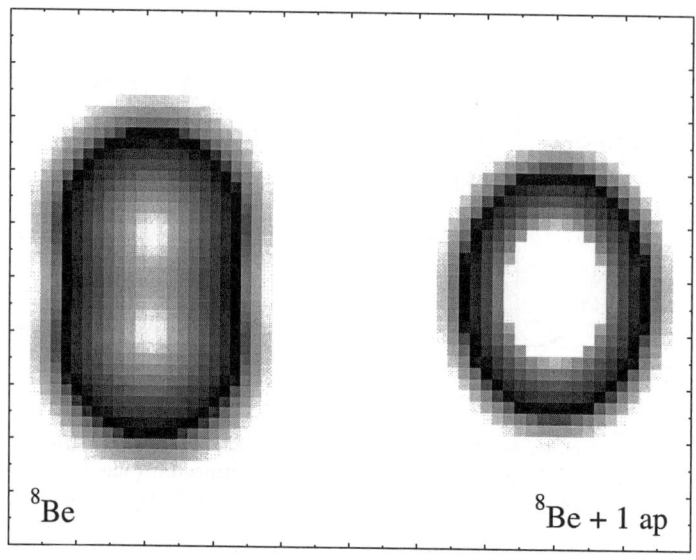

Figure 2. Contour plot of nucleon densities for ^{8}Be without (left) and with (right) antiproton calculated with the parametrization NL3. The maximum density of normal ^{8}Be is 0.20 fm^{-3}, while for the nucleus with the antiproton it is 0.61 fm^{-3}.

leads to an overall increase of attraction and decrease of repulsion for all nucleons. The antiproton becomes very deeply bound in the $s1/2^+$ state. To maximize attraction, protons and neutrons move to the center of the nucleus, where the antiproton has its largest occupation probability. This leads to a cold compression of the nucleus to a high density.

Figure 1 shows the densities and potentials for ^{16}O with and without the antiproton. For normal ^{16}O all RMF parametrizations considered produce very similar results. The presence of an antiproton dramatically changes the structure of the nucleus. The sum of proton and neutron densities reaches a maximum value of $(2-4)\,\rho_0$, where $\rho_0 \simeq 0.15\,\text{fm}^{-3}$ is the normal nuclear density, depending on the parametrization. The largest compression is predicted by the TM1 model. This follows from the fact that this parametrization gives the softest equation of state as compared to other forces considered here.

Since nucleons feel a deeper potential as compared to the nucleus without the antiproton, their binding energy increases too. The nucleon binding is largest within the NL3 parametrization. In the TM1 case, the $s1/2^+$ state is also deep, but higher levels are less bound as compared to the NL3 and NL–Z2 calculations. This is a consequence of the smaller spatial extension of the potential in this case. The highest $s1/2^-$ level is even less bound than for the system without an antiproton. The total binding energy of the system is predicted to be 828 MeV for NL–Z2, 1051 MeV for NL3, and 1159 MeV for TM1. For comparison, the binding energy of a normal ^{16}O nucleus is 127.8, 128.7 and 130.3 MeV in the case of NL–Z2, NL3, and TM1, respectively. Due to this anomalous binding we call these systems Super Bound Nuclei (SBN).

As a second example, we investigate the effect of a single antiproton inserted into the ^{8}Be nucleus. In this calculation only the NL3 parametrization was used (the effect is similar for all three forces). The normal ^{8}Be nucleus is not spherical, exhibiting a clearly visible 2 α structure with the deformation $\beta_2 \simeq 1.20$ in the ground–state. Inserting the antiproton gives rise to compression and change of nuclear shape. Its maximum density increases by a factor of three from 0.20 fm^{-3} to 0.61 fm^{-3}. The cluster structure of the ground state completely vanishes. A similar effect has been predicted in Ref. [13] for the case of the K^- bound state in the ^{8}Be nucleus. In our case the binding energy increases from 52.9 MeV (the experimental value is 56.5 MeV) to about 700 MeV!

4. Doubly-magic lead with antiproton and anti-alpha

We would like to discuss here the structural effect of an antiproton or an anti-alpha nucleus in the doubly magic lead nucleus. Contour plots of the sum of proton and neutron densities are shown in figures 3 (lead with an antiproton) and 4 (lead with an anti-alpha nucleus). In these cases we encounter a quite different scenario: again, the complete system is affected, but not in the sense that the whole nucleus shrinks and becomes very dense. Here, a small and localized region of high density develops within the heavy system. Additionally, the lead nucleus deforms itself. This effect is largest for the case of lead with $\bar{\alpha}$. The single-particle levels (Fig. 5) reflect this behaviour and indicate the cause for the deformation of lead: In a small region with a deep potential, only states with small angular momenta can be bound deeply. States with higher angular momenta do not have much overlap with the potential. This is exactly what can be concluded from Fig. 5. We see that basically only the lowest s- and p-states can be bound deeper than for lead without any antiparticles present. Higher lying states do not gain significantly binding or are even lesser bound.

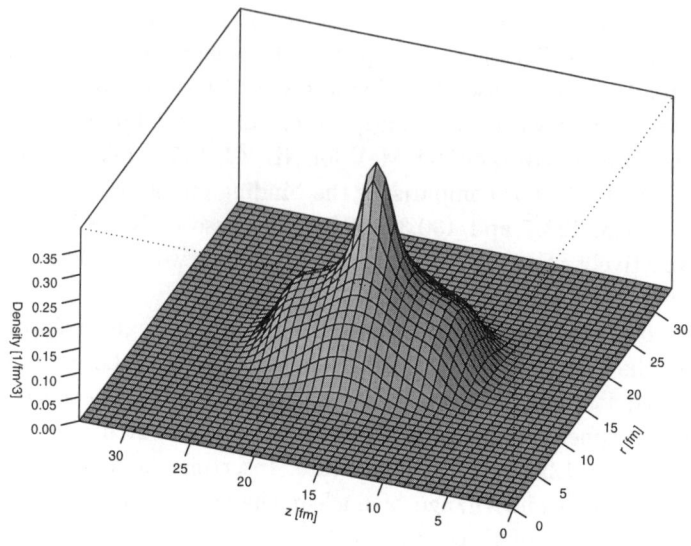

Figure 3. Surface plot of the sum of proton and neutron densities for the system ^{208}Pb with $\bar{p}$.

The deformation effect probably has two reasons: firstly, a deformation might be energetically favourable to gain some binding for the higher lying states. Secondly, the distortion of the system due to the presence of antiparticles destroys the magicitiy of the system.

5. Systems with total baryon number zero

It is interesting to consider finite systems having total baryon number zero, i.e. systems with the same amount of baryons and antibaryons. In the following we will present the cases of an α - anti-α and an ^{16}O - anti-^{16}O system. Fig. 6 shows the results for a system consisting of an α -$\bar{\alpha}$ system. The total system posseses a quite small radius and large baryon

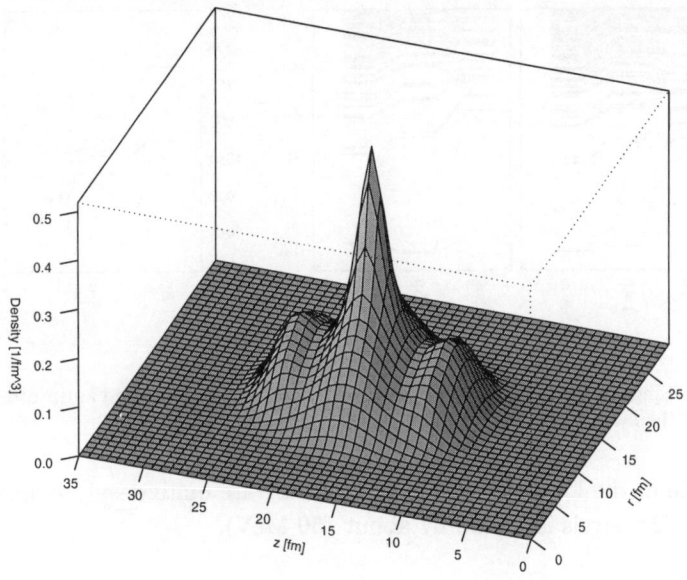

Figure 4. Surface plot of the sum of proton and neutron densities for the system ^{208}Pb with ā.

and antibaryon densities. The total baryon density is exactly zero. Due to the inverse coupling of antibaryons to the vector field and the negative charge of the antibaryons, all sources and potentials except for the scalar one vanish completely! All particles occupy a single particle state with extactly the same energy, and they all feel the same potential, namely the scalar potential, to which all particles present in the system couple alike. The binding energy of this nuclear system is huge: 2649 MeV with the force NL3 and 2235 MeV with NL-Z2. This is about 100 times larger than the binding energy of one α.

A similar structural effect occurs for the system of ^{16}O - anti-^{16}O. Again, all potentials except the scalar one vanish exactly. The binding energy of this system is 13227 MeV with NL3. All nucleons and antinucle-

696

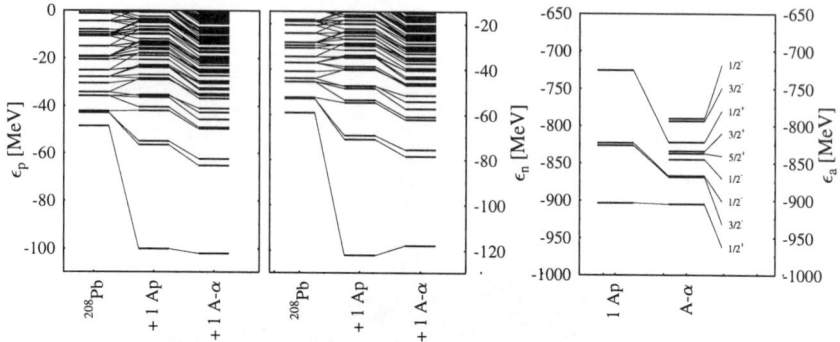

Figure 5. Single particle levels of protons (left), neutrons (middle) and antiproton levels (right) for the systems ^{208}Pb, ^{208}Pb + $\bar{p}$ and ^{208}Pb + $\bar{\alpha}$

ons reside in the lowest s- and p-levels. They are quite deeply bound (the lowest $s1/2+$ levels is bound by about 750 MeV).

6. Life time, formation probability and signatures of SBNs

The crucial question concerning a possible observation of the SBNs is their life time. The only decay channel for such states is the annihilation on surrounding nucleons. The mean life time of an antiproton in nucleonic matter of density ρ_B can be estimated as $\tau = < \sigma_A v_{\rm rel} \rho_B >^{-1}$, where angular brackets denote averaging over the wave function of the antiproton and v_{rel} is its relative velocity with respect to nucleons. In vacuum the $N\overline{N}$ annihilation cross section at low $v_{\rm rel}$ can be parametrized as [14] $\sigma_A = C + D/v_{\rm rel}$ with C=38 mb and D=35 mb. For $< \rho_B > \simeq 2\rho_0$ this would lead to a very short life time, $\tau \simeq 0.7$ fm/c (for $v_{\rm rel} \simeq 0.2$). However, one should bear in mind that the annihilation process is very sensitive to the phase space available for decay products. For a bound nucleon and antinucleon the available energy is $Q = 2m_N - B_N - B_{\overline{N}}$, where B_N and $B_{\overline{N}}$ are the corresponding binding energies. As follows from our calculations, this energy is strongly reduced compared to $2m_N$, namely, $Q \simeq 600 - 680$ MeV (TM1), 810–880 MeV (NL3) and 990–1050 MeV (NL–Z2) for the lowest antiproton states.

For such low values of Q many important annihilation channels involving two heavy mesons (ρ, ω, η, η', ...) are simply closed. Other two–body

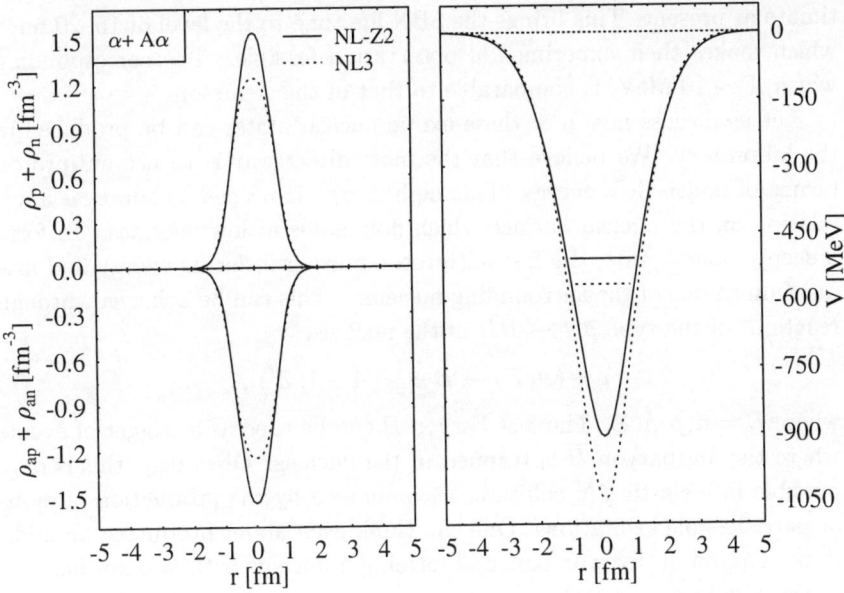

Figure 6. The left panel shows the sum of proton and neutron densities (top) as well as the corresponding sum for antibaryons for the α-$\bar{\alpha}$ system. The right panels shows the scalar potentials and the single particles levels of the nucleons and antinucleons.

channels such as $\pi\rho$, $\pi\omega$ are considerably suppressed due to the closeness to the threshold. As is well known, the two–pion final states contribute only about 0.4% of the annihilation cross section. Even in vacuum all above mentioned channels contribute to σ_A not more than 15% [15]. Therefore, we expect that only multi–pion final states contribute significantly to antiproton annihilation in the SBN. But these channels are strongly suppressed due to the reduction of the available phase space. Our calculations show that changing Q from 2 GeV to 1 GeV results in suppression factors 5, 40 and 1000 for the annihilation channels with 3, 4 and 5 pions in the final state, respectively. Applying these suppression factors to the experimental branching ratios [16] we come to the conclusion that in the SBNs the annihilation rates can be easily suppressed by factor of 20–30. There could be additional suppression factors of a structural origin which are difficult to estimate at present. This brings the SBN life time to the level of 15–20 fm/c which makes their experimental observation feasible. The corresponding width, $\Gamma \sim 10$ MeV, is comparable to that of the ω–meson.

Let us discuss now how these exotic nuclear states can be produced in the laboratory. We believe that the most direct way is to use antiproton beams of multi–GeV energy. This high energy is needed to suppress annihilation on the nuclear surface which dominates at low energies. To form a deeply bound state, the fast antiproton must transfer its energy and momentum to one of the surrounding nucleons. This can be achieved through reactions of the type $\bar{p}N \to B\bar{B}$ in the nucleus,

$$\bar{p} + (A, Z) \to B + \bar{B}(A - 1, Z') , \tag{1}$$

where $B = n, p, \Lambda, \Sigma$. The fast baryon B can be used as a trigger of events where the antibaryon $\bar{B}$ is trapped in the nucleus. Obviously, this is only possible in inelastic $\bar{p}N$ collisions accompanied by the production of pions or particle–hole excitations. One can think even about producing an additional baryon-antibaryon pair and forming a nucleus with two antibaryons in the deeply bound states. In this case two fast nucleons will be knocked out from the nucleus.

Without detailed transport calculations it is difficult to find the formation probability, W, of final nuclei with trapped antinucleons in these reactions. A rough estimate can be obtained by assuming that antiproton stopping is achieved in a single inelastic collision somewhere in the nuclear interior i.e. taking the penetration length of the order of the nuclear radius R. From the Poisson distribution in the number of collisions the probability of such an event is

$$w_1 = \frac{R}{\lambda_{\text{in}}} \exp\left(-\frac{R}{\lambda}\right) , \tag{2}$$

where $\lambda_{\text{in}}^{-1} = \sigma_{\text{in}}\rho_0$ and $\lambda^{-1} = (\sigma_{\text{in}} + \sigma_A)\rho_0$ (here σ_{in} and σ_A are the inelastic and annihilation parts of the $\bar{p}N$ cross section). The exponential factor in Eq. (2) includes the probability to avoid annihilation. For initial antiproton momenta $p_{\text{lab}} \simeq 10$ GeV we use $\sigma_{\text{in}} \simeq 25$ mb, $\sigma_A \simeq 15$ mb [16] and get $\lambda \simeq 1.6$ fm which is comparable with the radii of light nuclei. For an oxygen target, using $R \simeq 3$ fm leads to $w_1 \simeq 0.17$.

In fact we need relatively small final antiproton momenta to overlap significantly with the momentum distribution of a bound state, namely, $\Delta p \sim \pi/R_{\bar{p}}$, where $R_{\bar{p}} \simeq 1.5$ fm is the characteristic size of the antiproton spatial distribution (see Fig. 1). The probability of such a momentum loss can be estimated by the method of Refs. [17,18] which was previously used for calculating proton spectra in high–energy pA collisions. At relativistic bombarding energies the differential cross sections of the $\bar{p}p \to \bar{p}X$ and $pp \to pX$ reactions are similar. The inelastic parts of these cross sections drop rapidly with transverse momentum, but they are practically flat as a function of longitudinal momentum of secondary particles. Thus, the probability of the final antiproton momentum to fall in the interval Δp is simply $\Delta p/p_{lab}$. For $p_{lab} = 10\,\text{GeV}$ and $\Delta p = 0.4\,\text{GeV}$ this gives 0.04. Assuming the geometrical fraction of central events $\sim 20\%$ we get the final estimate $W \simeq 0.17 \times 0.04 \times 0.2 \simeq 1.4 \cdot 10^{-3}$. One should bear in mind that additional reduction factors may come from the matrix element between the bare massive antibaryon and the dressed almost massless antibaryon in a deeply bound state. But even with extra factors $\sim 10^{-1} - 10^{-2}$ which may come from the detailed calculations the detection of SBNs is well within the modern experimental possibilities.

Finally, we mention a few possible signatures of SBNs which can be used for their experimental detection. First of all, we remind the reader that according to the Dirac picture, any real antibaryon should be interpreted as a hole in the otherwise filled Dirac sea. Therefore, the nucleons from the positive-energy states of the Fermi sea can make direct transitions to the vacant negative-energy states of the Dirac sea. These super-transitions will be accompanied by the emission of a single pion or kaon depending on the nature of the trapped antibaryon. The energy of such a super-transition is fixed by the discrete levels of the initial and final baryons and according to our calculations should be of about 1 GeV. Obviously, this emission should be isotropic in the rest frame of the nucleus. The 1-pion or 1-kaon annihilation is a unique feature of finite nuclear systems. In vacuum such transitions are forbidden by the energy-momentum conservation. Therefore, the observation of a line in the pion or kaon spectrum at energies between 1 and 2 GeV would be a clear signal of the deep antibaryon states in nuclei. One can also look for narrow photon lines with energies in the range from 40 to 200 MeV corresponding to the transitions of nucleons and antibaryons between their respective levels. It is interesting to note that these signals will survive even if due to the lack of time the nucleus does not fully rearrange to a new structure.

Another strong signal may come from the collective response of the target nucleus to the presence of an antibaryon. Initially the nucleons will acquire radial acceleration due to the attractive interaction with the trapped antibaryon. This will lead to a collective motion similar to monopole oscillations around the compressed SBN state. Moreover, annihilation of the antibaryon will leave the nuclear remnant in a nonequilibrium state of high density. The nuclear system will expand and eventually break up into fragments. Therefore, the decay of the SBN state will result in nuclear multifragmentation with large collective flow of fragments. Both proposed signatures require rather ordinary measurements, which should be easy to perform with standard detectors.

7. Discussion and conclusions

Our main goal in this paper was to demonstrate that energetic antiproton beams can be used to study new interesting phenomena in nuclear physics. We discuss the possible existence of a completely new kind of strongly interacting systems where both the nucleons and the antinucleons coexist within the same volume and where annihilation is suppressed due to the reduction of the available phase space. Such systems are characterized by large binding energy and high nucleon density. Certainly, antinucleons can be replaced by antihyperons or even by antiquarks. We have presented the first self–consistent calculation of a finite nuclear system containing one antiproton in a deeply bound state. For this study we have used several versions of the RMF model which give excellent description of ordinary nuclei. The presence of an antiproton in a light nucleus like ^{8}Be or ^{16}O changes drastically the whole structure of the nucleus leading to a much more dense and bound state. In heavy systems the presence of a few antinucleons distorts and deforms the nuclear system leading to a localized central region of highly increased density. We find that that nuclear systems with total baryon number zero show extremely deep and symmetric states.

It is clear however that these structural changes can occur only if the life time of the antibaryons in the nuclear interior is long enough.

One should bear in mind that originally the RMF model was formulated within the Hartree and no–sea approximations. Implementing the Dirac sea may require serious revision of the model and inclusion of additional terms. Hartree calculations including the Dirac sea and Hartree–Fock calculations including exchange terms lead to smaller nucleon potentials in normal nuclei. Shallower potentials will produce smaller attraction for antinucleons, but the qualitative effect that the presence of antiprotons reduces repulsion and enhances attraction for nucleons will remain valid. We expect that the additional binding and compression of the nucleus will appear even for an antinucleon potential as low as 200 MeV.

In summary, on the basis of the RMF model we have studied the structure of nuclear systems containing a few real antibaryons. We have demonstrated that the antibaryons act as strong attractors for the nucleons leading to enhanced binding and compression of the recipient nucleus. As our estimates show the life times of antibaryons in the nuclear environment could be significantly enhanced due to the reduction of the phase space available for annihilation. Narrow peaks in the pion or kaon spectra at the energy around 1 GeV are proposed as the most clear signature of deeply-bound antibaryon states in nuclei.

References

1. T. Bürvenich, I. N. Mishustin, L. M. Satarov, J. A. Maruhn, H. Stöcker, and W. Greiner, Phys. Lett. B **542/3-4** (2002) 261
2. B.D. Serot and J.D. Walecka, Adv. Nucl. Phys., **16** (1985) 1.
3. P.G. Reinhard, Rep. Prog. Phys. **52** (1989) 439.
4. N. Auerbach, A.S. Goldhaber, M.B. Johnson, L.D. Miller, and A. Picklesimer, Phys. Lett. **B182** (1986) 221.
5. I.N. Mishustin, Sov. J. Nucl. Phys. **52** (1990) 722.
6. I.N. Mishustin, L.M. Satarov, J. Schaffner, H. Stöcker, and W. Greiner, J. Phys. **G19** (1993) 1303.
7. O.D. Dalkarov, V.B. Mandelzweig, and I.S. Shapiro, Nucl. Phys. **B21** (1970) 66.
8. G. Lalazissis, J. König, and P. Ring, Phys. Rev. **C55** (1997) 540.
9. M. Bender, K. Rutz, P.–G. Reinhard, J.A. Maruhn, and W. Greiner, Phys. Rev. **C60** (1999) 34304.
10. Y. Sugahara and H. Toki, Nucl. Phys. **A579** (1994) 557.
11. M. Bender, K. Rutz, P.–G. Reinhard, and J.A. Maruhn, Eur. Phys. J. **A8** (2000) 59.
12. K. Rutz, M. Bender, P.–G. Reinhard, J.A. Maruhn, and W. Greiner, Nucl. Phys. **A634** (1998) 67.
13. Y. Akaishi and T. Yamazaki, Phys. Rev. **C65** (2002) 044005.
14. C.B. Dover, T. Gutsche, M. Maruyama, and A. Faessler, Prog. Part. Nucl. Phys. **29** (1992) 87.
15. C. Amsler, Rev. Mod. Phys. **70** (1998) 1293.
16. J. Sedlák and V. Šimák, Sov. J. Part. Nucl. **19** (1988) 191.
17. R.C. Hwa, Phys. Rev. Lett. **52** (1984) 492.
18. L.P. Csernai and J.I. Kapusta, Phys. Rev. **D29** (1984) 2664.

THERMAL EMISSION OF INTERMEDIATE MASS FRAGMENTS AND THE LIQUID-VAPOR PHASE TRANSITION

L. G. MORETTO, J. B. ELLIOTT, L. PHAIR, AND G. J. WOZNIAK

*Nuclear Science Division, Lawrence Berkeley National Laboratory,
University of California, Berkeley, California, 94720*

We study the origin of anomalous caloric curves and negative heat capacities in the liquid-gas coexistence region. Coexistence is described in terms of clusterization in the vapor according to Fisher's formula. Multifragmentation data are used to determine the liquid-vapor coexistence line. The phase diagram is obtained for the finite system and an extrapolation is made to infinite nuclear matter.

The nuclear thin skin is the basis of the liquid drop model, which manages to reproduce the binding energies of nuclei to within 1%. A similar leptodermous treatment of nuclear systems at $T > 0$ should lead to an equivalently good reproduction of nuclear thermodynamical properties.

The appearance of a vapor phase at $T > 0$ opens two complementary perspectives for the characterization of phase coexistence: the liquid perspective and the vapor perspective. From the liquid perspective, one can determine the caloric curve in terms of vaporization enthalpy. From the vapor perspective one considers the extent to which nucleons are aggregated into clusters, as an indicator of incipient liquid condensation.

In the first part of this presentation we take the liquid perspective and derive analytically the caloric curve and the (negative) heat capacity for a drop undergoing an isobaric phase transition. In the second part we take the vapor perspective and show that clusterization in the 3d Ising model can be accounted for in terms of the leptodermous expansion.

Recently, first order phase transitions in small systems were associated with anomalous convex intruders in the entropy versus energy curves, resulting in back-bendings in the caloric curve, and in negative heat capacities[1]. In the context of nuclear physics, the claim has been made of an empirical observation of these anomalies, such as negative heat capacities in nuclear systems[2].

In this section we investigate the role of varying potential energies ("ground states") with system size on caloric curves and negative heat capacities. Our study applies to leptodermous (thin skinned) van der Waals-like fluids and to models such as Ising, Potts, and lattice gas.

Consider a macroscopic drop of a van der Waals fluid with A constituents in equilibrium with its vapor. The vapor pressure p is given by

$$p \simeq p_o \exp\left(-\frac{\Delta H_m}{T}\right), \tag{1}$$

where ΔH_m is the <u>molar</u> vaporization enthalpy and ΔV_m is the molar change in volume. Equation (1) represents the p-T univariant line in the phase diagram for a drop of finite size where ΔH_m must be corrected for the surface energy of the drop[3]

$$p = p_0 \exp\left(-\frac{\Delta H_m^0}{T} + \frac{a_s}{A^{1/3}T}\right) = p_{\text{bulk}} \exp\left(\frac{a_s}{A^{1/3}T}\right). \tag{2}$$

where a_s is the surface energy coefficient. At constant T the vapor pressure increases with decreasing size of the drop (see Fig. 1).

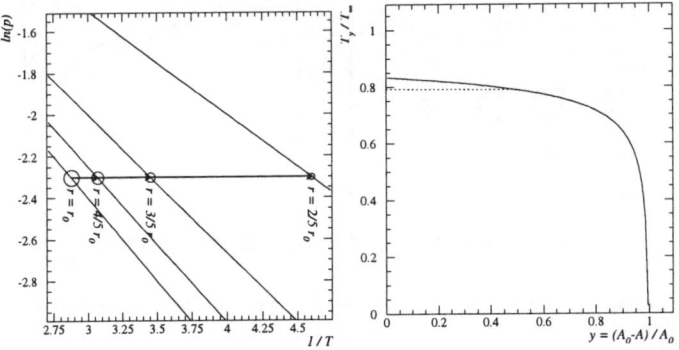

Figure 1. Left: the log of the saturated vapor pressure as a function of inverse temperature for different droplet radii. Arrows illustrate the path of evaporation at constant pressure. Right: The temperature as a function of droplet size for a drop evaporating at constant pressure (open boundary conditions). The solid line shows the case of a spherical drop, while the dotted line shows the case of a finite cubic lattice evolving as in Fig. 2 top.

Consider the case of isobaric evaporation of a drop starting from a drop with A_0 constituents and evaporating into a drop with $A < A_0$ constituents.

Let us define the drop size parameter $y = \frac{A_0 - A}{A_0}$. At constant pressure

$$p_0 \exp\left(-\frac{\Delta H_m^0}{T}\right) = p_0 \exp\left(-\frac{\Delta H_m(y)}{T_y}\right), \tag{3}$$

from which follows

$$\frac{T_y}{T_\infty} \simeq \frac{\Delta H_m(y)}{\Delta H_m^0} \simeq 1 - \frac{1}{A^{1/3}} \simeq 1 - \frac{1}{A_0^{1/3}(1-y)^{1/3}}. \tag{4}$$

A slight <u>decrease</u> in temperature is predicted as the drop evaporates isobarically, thus leading to a negative isobaric heat capacity in the coexistence region as illustrated in Fig. 1. As the drop is evaporating at constant pressure, the drop moves from one coexistence curve to another according to its decrease in radius, and thus to progressively lower temperatures. This slight effect is due <u>not</u> to an increase in surface since the drop surface of course <u>diminishes</u> as $A^{2/3}$, but to the slight increase of molar surface (see Fig. 2). Also, the formation of bubbles in the body of the drop is thermodynamically disfavored by the factor $f = \exp(-\gamma \Delta S/T)$ where ΔS is the surface of the bubble.

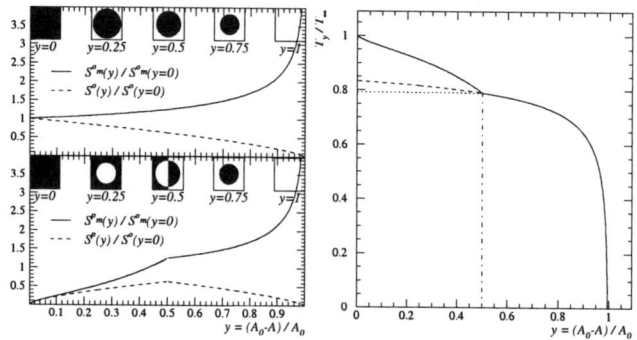

Figure 2.   Left Top: The surface S^o (dashed) and molar surface S_m^o (solid) area of a drop for open boundary conditions normalized to their values at $y = 0$. Left Bottom: The surface S^p (dashed) and molar surface S_m^p (solid) area of a drop for periodic boundary conditions normalized to their values at $y = 0$. In-sets show the configurations at various values of y. Right: the temperature as a function of droplet size for a drop evaporating at constant pressure in a system with periodic boundary conditions. The solid line shows the case of a finite cubic lattice with periodic boundary conditions evolving as in Fig. 2 bottom, while the dotted line and the dashed line are the same as in Fig. 1 and the vertical dash-dotted line indicates the case of 50% lattice occupation.

Let us now move to the cases of lattice gas, Ising, and Potts models. We consider first an evaporating finite system in three dimensions of size

$A_0 = L^3$, with open boundary conditions. For maximal density at $T = 0$ (the ground state) $y = 0$ and the entire cubic lattice is filled. For decreasing densities, always at $T = 0$ a single cluster of minimum surface is present, which evolves from a cube to a sphere. The associated change in surface is shown in Fig. 2. The caloric curve from $y = 0$ to $y = 1/2$ is essentially flat like in the infinite system, and the heat capacity is trivially infinite.

The introduction of periodic boundary conditions rids the system of "dangling bonds." At $y = 0$, the lattice is filled with particles so that $\Delta H_m(0) = \Delta H_m^0$ characteristic of the infinite system. As y increases at fixed lattice size, a bubble develops in the cube and surface is rapidly created (see Fig. 2). The bubble develops since the periodic boundary conditions prevent evaporation from the surface. The bubble grows with increasing y until it touches the sides of the lattice. This occurs for $y \approx 1/2$. At nearly $y = 1/2$ and beyond, the "stable" configuration is a drop that eventually vanishes at $y = 1$. The change in surface associated with the range $0 \le y \le 1$ as well as the molar surface are shown in the bottom left panel of Fig. 2.

The evaporation enthalpy thus becomes

$$\Delta H_m(y) \simeq a_v \left(1 - \frac{y^{2/3}}{A_0^{1/3}(1-y)}\right) \qquad (5)$$

from $y = 0$ to $y = 1/2$, and

$$\Delta H_m(y) \simeq a_v \left(1 - \frac{1}{A_0^{1/3}(1-y)^{1/3}}\right) \qquad (6)$$

from $y = 1/2$ to $y = 1$. As a consequence, for periodic boundary conditions

$$\frac{T_y}{T_\infty} \simeq 1 - \frac{y^{2/3}}{A_0^{1/3}(1-y)} \qquad (7)$$

from $y = 0$ to $y = 1/2$, while from $y = 1/2$ to $y = 1$ Eq. (4) holds.

The dramatic effect of periodic boundary conditions can be seen in Fig. 2. The temperature decreases substantially with increasing y, due to the fact that the molar enthalpy at $y = 0$ assumes its bulk value ΔH_m^0 and must meet the previous case of open boundary conditions for $y = 1/2$. This may well explain the calculated negative heat capacities reported in literature, as due to the unnatural choice of boundary conditions.

In the case of nuclei the quantity ΔH_m is determined by all the terms in the liquid drop model, which contribute to the mean binding energy per nucleon. One can immediately infer that when the binding energy per nucleon <u>decreases</u> with A, the heat capacity should be positive, and vice-versa. Thus, since the maximum binding energy per nucleon occurs at $A \sim 60$, negative heat capacities should be possible only for $A < 60$.

Explicitly,

$$C_p = \frac{\frac{(\Delta H_m(A))^2}{T}}{\frac{d\Delta H_m}{dA}}. \tag{8}$$

The derivative in the denominator can be evaluated approximately from the dependence on the binding energy per nucleon B upon the mass number $\frac{d\Delta H_m}{dA} = \frac{dB}{dA}$. The liquid drop model allows us to estimate such a derivative. From Fig. 3 it is apparent that the binding energy <u>decreases</u> with A for $A >\sim 60$. Consequently in all this region of A, positive specific heats should be expected. Only for $A <\sim 60$, negative specific heats are predicted.

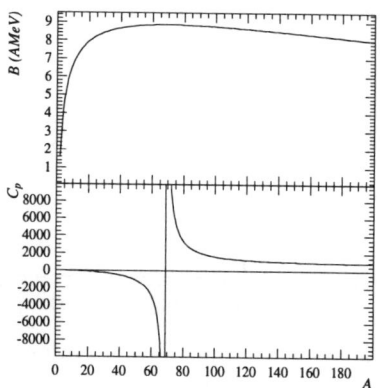

Figure 3.   The binding energy of atomic nuclei (top) and the associated heat capacity (bottom).

This straightforward result based on elementary thermodynamics and ground state binding energies raises serious questions as to the meaning of the negative heat capacities claimed in large nuclear systems[2].

In multifragmentation, reducibility is the property that the probability of observing n-fragments of a given size is expressible in terms of an elementary one-fragment probability. Both binomial, and its limiting form, Poissonian reducibilities have been extensively documented experimentally for nuclear multifragmentation[4,5].

Thermal scaling is the linear dependence of the logarithm of the one-

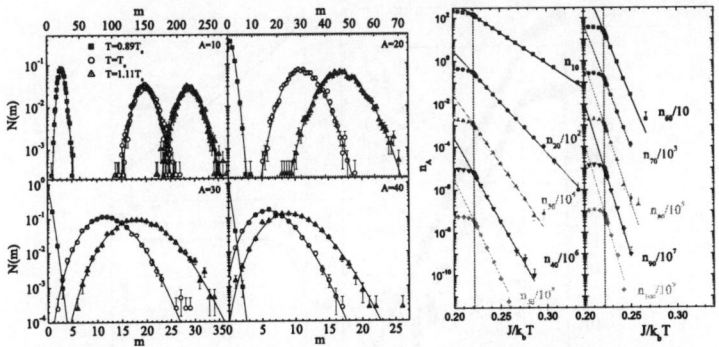

Figure 4. Left: The probability distributions for obtaining m fragments of size A at the three temperatures indicated. The solid lines are Poisson distributions with means given by the Monte Carlo data. Right: Arrhenius plots of the cluster distributions. The lines are fits of the form given in Eq. (9). The critical temperature is indicated by the dashed line.

fragment probability with $1/T$ (an Arrhenius plot) according to:

$$q_i = q_0 e^{-B_i/T} = q_0 e^{-c_0 A^\sigma /T} \tag{9}$$

where B_i is a "barrier" corresponding to the production process.

The combination of these two empirical features attests to a statistical mechanism of multifragmentation in general, and to liquid-vapor coexistence specifically[8].

In this light we analyze the Ising results in the same way as has been done with nuclear multifragmentation data[4,5]. Fig. 4 shows the multiplicity distributions for a sample of fragment sizes and temperatures. The solid lines represent Poisson distributions calculated from the corresponding mean multiplicities. The distributions are remarkably close to Poissonian for all masses and all temperatures below, at and above T_c.

If the fragment distributions exhibit thermal scaling, its Arrhenius plot should be linear. As shown in Fig. 4, this is the case over a wide range of temperatures ($0 < T < T_c$) and fragment sizes.

The features of reducibility and thermal scaling discussed above can be found united in Fisher's formula[6,7].

$$n_A(T) = q_0 A^{-\tau} \exp\left(\frac{c_0 A^\sigma}{T_c}\right) \exp\left(-\frac{c_0 A^\sigma}{T}\right) \tag{10}$$

where q_0 is a normalization constant, τ is a topological critical exponent, c_0 is the surface energy coefficient and $\varepsilon = (T_c - T)/T_c$. Therefore, a graph of the scaled cluster distributions $(n_A(T)A^\tau/q_0)$ as a function of $\varepsilon A^\sigma /T$

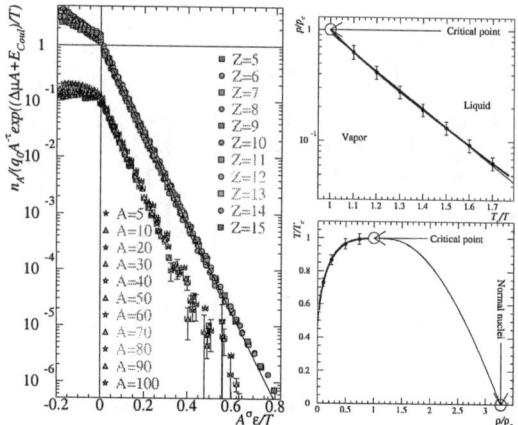

Figure 5. Left: the scaled yield distribution versus the scaled temperature for the ISiS data (upper) and $d = 3$ Ising model calculation (lower). Upper right: The reduced pressure-temperature phase diagram: Lower right: The reduced density-temperature phase diagram.

should collapse the distributions of all cluster sizes onto a single curve. This scaling behavior can clearly be seen in Fig. 5.

The ISiS data sets

The ISiS charge yields from AGS experiment of 8 GeV/c π + Au fragmentation data were fit to the following modified form of Eq. (10) which incorporates the Coulomb. The scaled data shown in Fig. 5 collapse to a single line over six orders of magnitude. This line is the liquid to vapor phase coexistence line in excited nuclei.

Fisher's model assumes that the non-ideal vapor can be approximated by an ideal gas of clusters. Accordingly, the total pressure is the sum of their partial pressures: $p/T = \sum n_A$. The resulting p, T diagram is given in Fig. 5.

Fitting the coexistence line $p/p_c = \exp(\Delta H/T_c(1 - T_c/T))$ which describes many fluids gives $\Delta H = 26 \pm 1$ MeV from which $\Delta E \approx 15$ AMeV, remarkably close to the nuclear bulk energy coefficient.

The system's density can be found from $\rho = \sum A n_A$ as shown in Fig. 5. Following Guggenheim[9]:

$$\frac{\rho_{l,v}}{\rho_c} = 1 + b_1(1 - \frac{T}{T_c}) \pm b_2(1 - \frac{T}{T_c})^{1/3}. \qquad (11)$$

Fitting the coexistence curve from the ISiS E900a data with Eq. (11) one obtains an estimate of the full ρ_v branch of the coexistence curve and changing the sign of b_2 gives the full ρ_l branch of the coexistence curve of finite neutral nuclear matter. The critical density is found to be $\rho_c \sim 0.3\rho_0$.

The EOS data sets

The EOS Collaboration has collected data for the reverse kinematics reactions 1.0 AGeV Au+C, 1.0 AGeV La+C and 1.0 AGeV Kr+C[10,11].

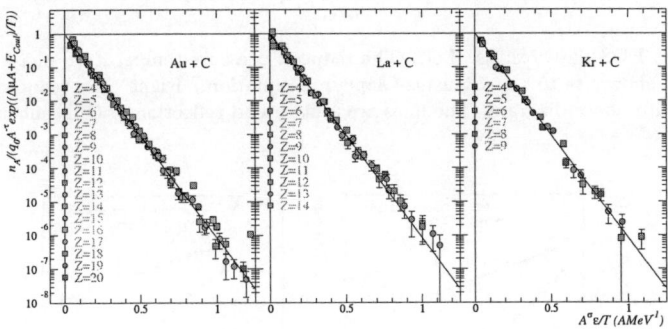

Figure 6. The scaled yield distribution versus the scaled temperature for the gold, lanthanum and krypton systems. The solid line has a slope of c_0.

Fig. 6 shows the Fisher plot of fragment mass yield distribution scaled by the power law pre-factor, the chemical potential and Coulomb terms. The scaled data for all three systems collapse onto a single line over several orders of magnitude. This collapse provides direct evidence for a liquid to vapor phase transition in excited nuclei.

The p-T and T-ρ coexistence curves can be determined from this analysis by transforming the information in Fig. 6 into the phase diagrams in Fig. 7. From these it is possible to make an estimate of the bulk binding energy of nuclear matter and the $\Delta E/A \approx 14$ MeV, close to the nuclear bulk energy coefficient of 15.5 MeV.

Finite size effects are paramount in nuclei. The binding energy per nucleon decreases from the ~ 15.5 AMeV of nuclear matter to about 8 AMeV for typical nuclei.

We can expect that such a drastic reduction affects the critical temperature as well. The Ising model can be used again as a simple testing ground. If a finite system is considered (no periodic boundary conditions) a surface is generated with the attendant surface energy. This allows us to write a "liquid drop" formula for the Ising model: $E = a_V A + a_S A^{2/3}$.

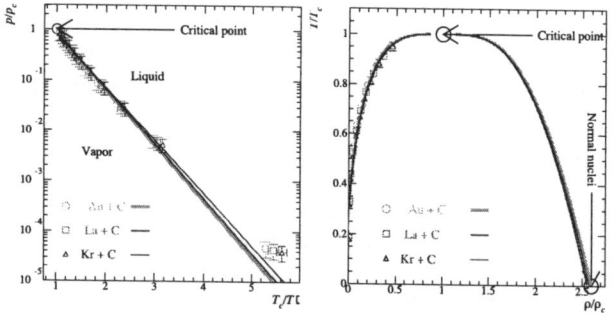

Figure 7. EOS data results. Left: The reduced pressure-temperature phase diagram. The lines show fits to the Clausius-Clapeyron equation. Right: the reduced density-temperature phase diagram. The lines are a fit to and reflection of Guggenheim's equation.

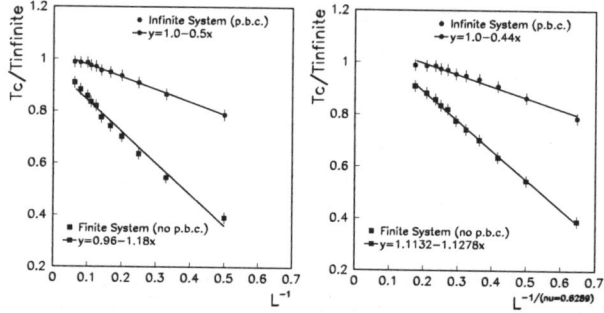

Figure 8. Left: the naive estimate of finite size scaling. Right: a more sophisticated estimate of finite size scaling. The data points and fits on the bottom of both figures show the results for lattices with open boundary conditions (no p.b.c.) and more closely represent the case of finite systems like nuclei.

We now determine the critical temperature for various sizes (lattices) and check its dependence on the lattice size (see Fig. 8). For a finite system all the quantities with the dimension of energy should scale with the binding energy per site corrected for the surface energy

$$\frac{T_c^{A_0}}{T_c^{\infty}} = \frac{a_V A_0 + a_S A_0^{2/3}}{a_V A_0} = 1 - \frac{1}{A_0^{1/3}} = 1 - \frac{1}{L} \qquad (12)$$

where A_0 is the number of sites in the lattice and L is the linear lattice side. As we can see in Fig. 8, this scaling works quite well.

The result of this exercise shows that the critical temperature of infinite nuclear matter can be obtained in a similar way. In each of the three EOS reactions, remnants of different sizes make a good range of A_0 accessible.

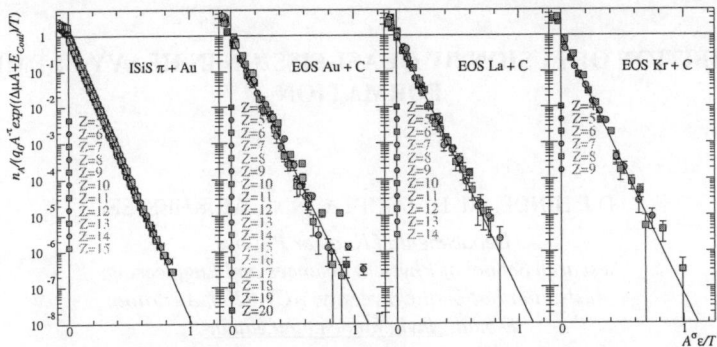

Figure 9. Fisher scaling and finite size scaling analysis of the ISiS and EOS data sets.

The preliminary results are shown in Fig. 9. The extracted values for the critical temperature of infinite nuclear matter are ~ 13.6 MeV from the ISiS data and ~ 12.9 MeV from the EOS data. These values agree with theoretical estimates of the critical temperature of bulk nuclear matter.

We thank Prof. C.M. Mader, Prof. R. Ghetti and Prof. J. Helgesson for their input and Ising model calculations and we acknowledge the experimental efforts of the ISiS and EOS collaborations. This work was supported by the US Department of Energy.

References

1. D. H. E. Gross, Phys. Rep. **279**, 119 (1997).
2. M. D'Agostino *et al.*, Phys. Lett. B, **473**, 219 (2000).
3. L. Rayleigh, Phil. Mag. **34**, 94 (1917).
4. L. G. Moretto, *et al.*, Phys. Rep. **287**, 249 (1997).
5. L. Beaulieu *et al.*, Phys. Rev. Lett. **81**, 770 (1998).
6. M. E. Fisher, Physics **3**, 255 (1967).
7. M. E. Fisher, Rep. Prog. Phys. **30**, 615 (1969).
8. J. B. Elliott *et al.*, Phys. Rev. Lett. **85**, 1194 (2000).
9. E.A. Guggenheim, J. Chem. Phys., **13**, 253 (1945).
10. J. A. Hauger *et al.*, Phys. Rev. C **57**, 764 (1998)
11. J. A. Hauger *et al.*, Phys. Rev. C **62**, 024626 (2000).

INHIBITION OF FUSION BY QUASI-FISSION IN HEAVY ELEMENT FORMATION

D.J. HINDE, M. DASGUPTA AND A. MUKHERJEE

Department of Nuclear Physics,
Research School of Physical Sciences and Engineering,
Australian National University, ACT 0200, Australia
E-mail: david.hinde@anu.edu.au

Comprehensive fission and evaporation residue measurements were made for three reactions, all of which lead to the ^{216}Ra compound nucleus, at energies from the Coulomb barriers upwards. The results show inhibition of fusion and the increasing presence of quasi-fission with decreasing mass-asymmetry of the projectile-target combination in the entrance channel. That work has been extended, by making measurements of evaporation residue cross-sections for the reaction of ^{16}O with ^{204}Pb, forming ^{220}Th, a nucleus that has previously been formed in more mass-symmetric reactions. The new data show that for the more symmetric reactions, evaporation residue cross-sections are inhibited by quasi-fission, typically by a factor of 10. This result overturns previous assumptions for such reactions.

1. Introduction

In collisions of heavy nuclei, capture inside the Coulomb barrier leads to formation of a dinucleus, which can evolve in two directions. The projectile may be absorbed, and a compact compound nucleus formed, which will decay by evaporation or fission; this is commonly called fusion. Alternatively, the dinucleus may remain elongated, and the lighter nucleus gains mass from the heavier one. The system may break into two fragments before mass equilibration has been achieved. This latter process, named quasi-fission, competes directly with the fusion process and suppresses the yield of fusion-evaporation residues (ERs), since a compact equilibrated compound nucleus is never formed. In fission measurements, experimental signatures of quasi-fission are mass distributions wider than those for fusion-fission, and larger fission fragment angular anisotropies.

Suppression of ER cross-sections has been observed in symmetric reactions only when Z_1Z_2 is larger than about 1600 [1]. However, in studies of fission, the presence of quasi-fission has been inferred for light projectiles ($A \geq 24$) interacting with ^{208}Pb ($Z_1Z_2 \geq 984$) [2] and for ^{32}S + ^{182}W ($Z_1Z_2 = 1184$) [3]. In reactions involving even lighter projectiles with deformed trans-actinide targets (such as $^{16}O + {}^{238}U$ with $Z_1Z_2 = 736$) at sub-barrier energies, evidence for quasi-fission has been claimed [4]. The apparent lower threshold for quasi-fission seen in the fission events, compared to that extracted from the ER cross-sections, may be due to the effects of angular momentum, or to the limited range of experimental ER data. To address this question, measurements were made for reactions leading to the compound nucleus ^{216}Ra, which showed evidence [5] both for suppression of fusion, and of the presence of quasi-fission near the Coulomb barrier in mass-asymmetric reactions with Z_1Z_2 as low as 711. This paper summarizes these results, before describing the new ER cross-section measurements for the $^{16}O + {}^{204}Pb$ reaction, and comparing them with existing data for more mass-symmetric reactions also forming ^{220}Th.

2. Reactions forming ^{216}Ra

The three reactions studied were $^{12}C + {}^{204}Pb$ ($Z_1Z_2 = 492$), $^{19}F + {}^{197}Au$ ($Z_1Z_2 = 711$) and $^{30}Si + {}^{186}W$ ($Z_1Z_2 = 1036$), fusion in each case forming ^{216}Ra. Fission and ER cross-sections and characteristics were measured from the Coulomb barrier upwards for each reaction. The ^{216}Ra nuclei were formed at the same excitation energy in the three reactions, ranging from 30-90 MeV.

Fission fragments were detected in coincidence, in two large area position-sensitive multi-wire proportional counters, allowing determination of the fission fragment mass-ratios and angular anisotropies. ERs were stopped in catcher foils at the target position. The individual xn, pxn and αxn ERs were identified by their ground-state decay α-particles, detected in an annular silicon surface-barrier detector between beam-bursts. Several $^{30}Si + {}^{186}W$ ER measurements were repeated with a 12% thicker catcher foil, and additional ER measurements were performed for the $^{12}C + {}^{204}Pb$ and $^{19}F + {}^{197}Au$ reactions, by detecting the decay α-particles after implantation in a silicon surface-barrier detector, which was positioned behind a velocity filter. There was good agreement between all measurements. Absolute cross-sections were determined through measuring elastic scattering in the monitor, fission and α-particle detectors, at low beam energies where the elastic scattering cross-sections follow the Rutherford scattering formula.

The measured ER cross-sections (triangles) and fission cross-sections (open circles) are shown for the three reactions as a function of centre-of-mass energy $E_{c.m.}$ in the top panels of Figure 1. The fission cross-sections may include quasi-fission, so the sum of ER and fission is defined as the capture cross-section. If there were no quasi-fission, the capture cross-sections would be the fusion cross-sections. Both capture cross-sections and barrier distributions [6]

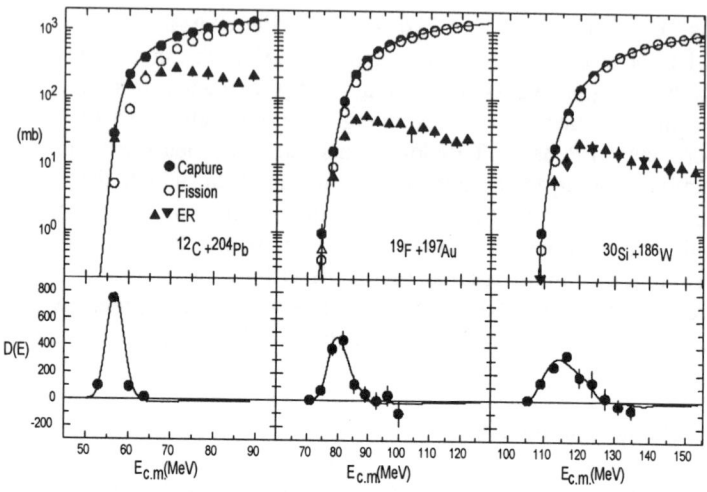

Figure 1. Cross-sections and capture barrier distributions for the three reactions. The upper frames show the sum (closed circles) of the measured fission (open circles) and ER (triangles) cross-sections. The lower frames show the experimental capture barrier distributions by the solid circles. Lines represent coupled--channels calculations.

were well-reproduced by quantum-mechanical coupled-channels calculations, using nuclear potential parameters consistent with other reactions, and coupling parameters from the known properties of the collective modes in each nucleus. The large prolate deformation of ^{186}W in particular broadens of the barrier distribution for the ^{30}Si + ^{186}W reaction.

From the coupled-channels calculations reproducing the experimental capture cross-sections, reliable angular momentum distributions were obtained for each reaction [6]. With increasing beam energy more angular momentum ($\lambda\eta$) is brought into the reaction. As the fission barrier reduces with increasing λ^2, there is a rapid rise in the fission probability P_f. This limits the production of ERs to low angular momentum values ($\lambda<30$). The cross-section for ERs at a given $E_{c.m.}$ and excitation energy E^* is given by

$$\sigma_{ER}(E_{c.m.}) = \pi\Delta^2 \sum_{l=0}^{\infty} (2l+1)T_l(E_{c.m.})(1 - P_f(l, E^*)),$$

where $T_\lambda(E_{cm})$ is the probability of capture for λ, and Δ is the reduced deBroglie wavelength, determined by the beam momentum. Division by $\pi\Delta^2$ gives the reduced cross-section, $\widetilde{\sigma}_{ER}(E)$. At beam energies sufficiently high above the

Coulomb barrier, T_λ $(E_{cm}) \approx 1$ for all λ which lead to ERs. Equation 1 then reduces to

$$\widetilde{\sigma}_{ER}(E_{c.m.}) = \sum_{l=0}^{\infty}(2l+1)(1 - P_f(l, E^*)).$$

Thus at the same E^*, the three reactions should give the same $\widetilde{\sigma}_{ER}$, as long as $P_f(\lambda, E)$ is independent of the reaction, in accordance with Bohr's hypothesis of the independence of compound nucleus formation and decay [8].

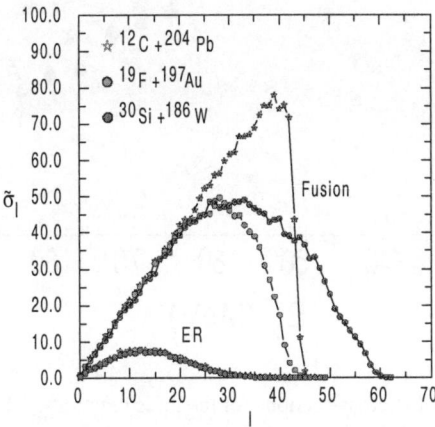

Figure 2. Calculated reduced capture and ER angular momentum distributions for the three reactions at an excitation energy of $E^* \approx 57$ MeV.

At an excitation energy of 57 MeV, Figure 2 shows that the capture probabilities are indeed close to unity for all angular momenta leading to ER production. Statistical model ER cross-sections are thus here independent of the capture angular momentum distributions for each projectile, and so at the same excitation energy the three reactions should give the same $\widetilde{\sigma}_{ER}$.

The experimental $\widetilde{\sigma}_{xn}$, the reduced xn ER cross-sections, are shown in Figure 3. Comparing the yields for Ra isotopes alone eliminates possible enhancement of the ER yields for ^{12}C due to incomplete fusion [5]. The data show a consistent dependence of the yield on projectile mass at energies where the capture λ-distributions fully populate the λ-values leading to ERs (at energies above 55 MeV). The results demonstrate clearly that there is significant inhibition of fusion for the heavier projectiles. Reduced xn cross-sections for ^{19}F and ^{30}Si in this region are respectively (0.64 ± 0.09) and (0.57 ± 0.08) of those for ^{12}C, whereas they would be expected to be equal in the absence of quasi-fission. Fusion with light projectiles like ^{19}F had previously

716

implicitly been assumed to proceed without any inhibition. If this inhibition is caused by the presence of quasi-fission, there should be evidence in the properties of the fission events.

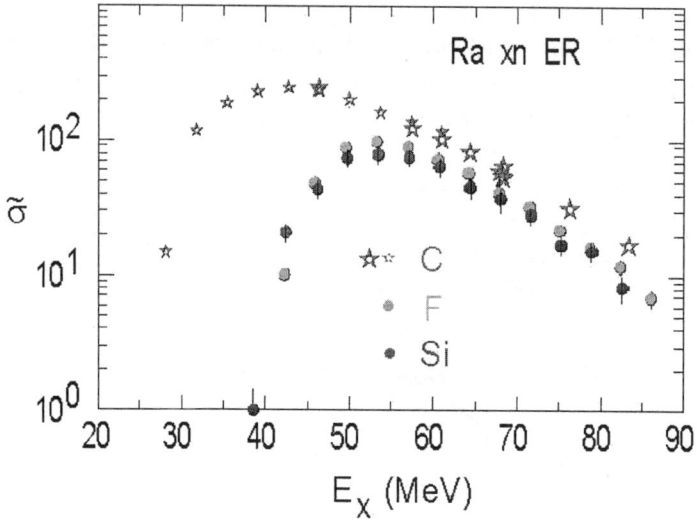

Figure 3 Reduced xn ER cross-sections for the three reactions. The larger stars are from a second experiment, from which only xn cross-sections were determined.

Fission fragment mass distributions for the three reactions were determined from the coincident fission fragment measurements. Normalised fission fragment mass distributions, defined as the mass of the fragment detected at backward angles divided by the summed mass, are shown in Figure 4(a) at an excitation energy of $E^* \approx 60$ MeV, over the angular range $125° \leq \theta_{c.m.} \leq 165°$. The widths of the distributions are not independent of the entrance channel, but rather become wider with increasing entrance-channel mass-symmetry.

The measured variances of the normalised fission mass-distributions, σ_M^2, are presented as a function of excitation energy in Figure 4(c). There is a systematic increase in width of the mass distributions with projectile mass for the same excitation energy. Although the mass distributions are expected to depend most strongly on excitation energy, they may depend on angular momentum. The average mean-squared angular momenta, $\langle\lambda^2\rangle$, obtained from the coupled-channels fits to the capture cross-sections, are presented in Figure 4(b) for the three reactions. Although the $\langle\lambda^2\rangle$ for ^{12}C matches that for the ^{30}Si reactions at one energy, and approaches that for the ^{19}F reaction at the highest energy, the mass distributions do not match at any energy, but show a systematic dependence on projectile mass as well as the expected increase with excitation energy. This result indicates an increasing contribution from quasi-fission for the heavier projectiles, consistent with the observed inhibition of fusion for the heavier projectiles.

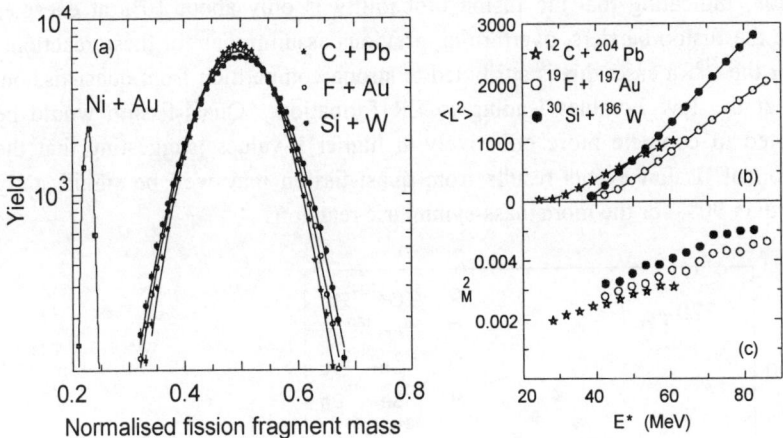

Figure 4. (a) Normalised fission fragment mass distributions (mass of observed fragment divided by summed mass) at an excitation energy of $E^* \approx 60$ MeV. The elastic scattering of ^{58}Ni from ^{197}Au shows the instrumental resolution. (b) Calculated mean-squared capture angular momenta introduced in the three reactions, and (c) measured variance of the normalised fission mass-distributions, as a function of excitation energy.

3. Reactions forming ^{220}Th

ER cross-sections for the more fissile compound nucleus ^{220}Th have been previously measured for reactions with the projectiles ^{40}Ar [9], ^{48}Ca [10], ^{82}Se [11] and ^{124}Sn [9]. In the interpretation of these reactions, it had seemed that quasi-fission does not inhibit the ER cross-sections at energies where the ER yields saturate. This result is in conflict with the above conclusions for ^{216}Ra. To investigate this difference, cross-sections for ERs have been measured [7] for the reaction ^{16}O+^{204}Pb, which also forms the compound nucleus ^{220}Th. The same decay alpha detection apparatus as used for ^{216}Ra was employed. As shown in Ref. [5], the yield of ERs can be enhanced by incomplete fusion with light projectiles, so to eliminate this contribution, the reduced cross-sections $\tilde{\sigma}_{xn}$ for those ERs which emitted only neutrons (i.e. Th ERs) are compared. In the absence of quasi-fission, at the same E*, all reactions should give the same $\tilde{\sigma}_{xn}$ where the λ-values leading to ERs are fully populated. The data (Figure 5) for the more mass-symmetric reactions show that the $\tilde{\sigma}_{xn}$ are typically 10 times lower than those for the $^{16}O+^{204}Pb$ reaction above the saturation energy for each reaction, indicating that the fusion probability is only about 10% at energies above the fusion barriers, overturning previous assumptions for these reactions. As for the ^{216}Ra case, this is attributed to strong competition from quasi-fission, even at the low λ-values leading to ER formation. Quasi-fission would be expected to compete more effectively at higher λ-values, suggesting that the fraction of fission events results from quasi-fission may well be significantly more than 90% for the more mass-symmetric reactions.

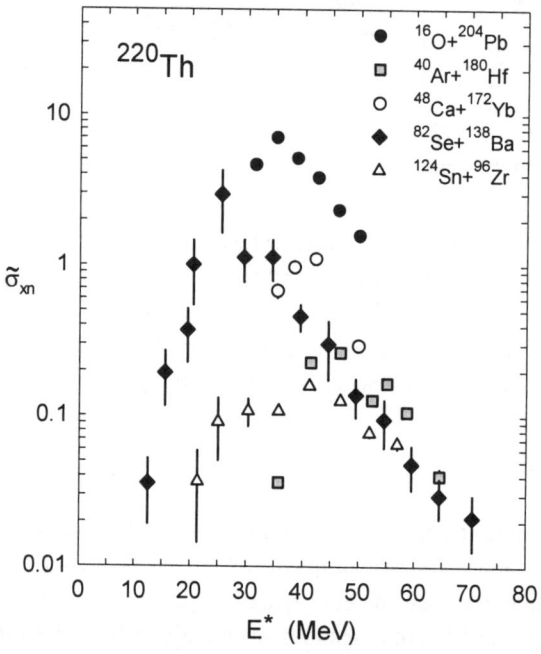

Figure 5. Reduced cross-sections for Th ERs in five reactions forming ^{220}Th.

The results can be most simply explained in terms of the mass-asymmetry dependence of the potential energy surface (PES) [5]. For collisions on the mass-asymmetric side of the Businaro-Gallone ridge point (e.g. $^{16}O+^{204}Pb$), the PES favours absorption of the lighter nucleus, leading directly to a compact shape. On the mass-symmetric side, the PES favours the lighter nucleus taking mass from the heavier partner, leading towards mass-symmetric elongated shapes more likely to result in quasi-fission. The measurement of ERs indicates that compact shapes can also evolve from mass-symmetric shapes.

The results emphasize the importance of dynamical aspects of fusion, dependent on mass-asymmetry, in controlling the fusion process, even when forming a nucleus as light as ^{220}Th. In reactions forming heavier nuclei, the inhibition of fusion by quasi-fission will be even more severe.

Significantly, the data show that most of the fusion inhibition occurs already for the ^{40}Ar-induced reaction, with evidence for only an additional small reduction for the more mass-symmetric reactions. Thus the difference in fusion inhibition due to mass-asymmetry alone for reactions with ^{48}Ca on trans-actinide nuclei (hot fusion) and of ^{64}Ni and heavier on ^{208}Pb (cold fusion) may be less than has been believed, as both are on the mass-symmetric side of the Businaro-Gallone ridge. This suggests that the role of nuclear structure in determining fusion may be crucial in both cold and hot fusion, as discussed in other contributions to these proceedings. Shell structure resulting in "cold valleys" seems likely to play a vital role in cold fusion. For hot fusion, the results reinforce the proposal [4] that it is the compact shapes resulting from contact with the equator of the heavy prolate target nuclei that enhance the fusion cross sections, counteracting the additional chances of fission from the more highly excited fused system. This may explain how hot fusion reactions can be competitive with cold fusion in forming very heavy nuclei.

Acknowledgments

MD acknowledges the support of the Australian Research Council.

References

1. Blocki, J.P., Feldmeier, H., and Swiatecki, W.J., Nucl. Phys. A459, 145 (1986).
2. Back, B.B. et al., Phys. Rev. C32, 195 (1985).
3. Back, B.B. et al., Phys. Rev. C53, 1734 (1996).
4. Hinde, D.J. et al., Phys. Rev. Lett. 74, 1295 (1995), Phys. Rev. C53, 1290 (1996).
5. Berriman, A.C. et al., Nature 413, 144 (2001).
6. Dasgupta, M. et al., Annu. Rev. Nucl. Part. Sci. 48, 401 (1998).
7. Hinde D.J. et al., Phys. Rev. Lett., accepted (2002).
8. Bohr, N., Nature 137, 344 (1936).
9. Vermeulen, D. et al., Z. Phys. A 318, 157 (1984).
10. Sahm, C.-C. et al., Nucl. Phys. A441, 316 (1985).
11. Satou, K. et al., Phys. Rev. C 65, 054602 (2002).

FUSION-FISSION FOR SUPERHEAVY NUCLEI *

G. G. ADAMIAN, N. V. ANTONENKO AND W. SCHEID

Joint Institute for Nuclear Research, 141980 Dubna, Russia,
Institute of Nuclear Physics, 702132 Tashkent, Uzbekistan,
Institut für Theoretische Physik der Justus–Liebig–Universität, D–35392
Giessen, Germany

The fusion and quasifission processes are considered within the dinuclear system concept.

1. Introduction

The superheavy nuclei with charge numbers $Z =110$, 111 and 112 were reached in cold fusion reactions with ^{208}Pb and ^{209}Bi targets during last years [1]. Recent successes in the synthesis of the nuclei $Z =114$ and 116 in hot fusion actinide-based reactions with ^{48}Ca beam assure for obtaining new elements in these reactions [2]. However, the cross sections for heavier elements are expected to be less than 1 pb.

As shown in [3,4,5], fusion and quasifission can be described as an evolution of a dinuclear system (DNS) which is formed in the entrance channel during the capture stage of the reaction after dissipation of the kinetic energy of the collision. In the DNS model [3,4] the compound nucleus is reached by a series of transfers of nucleons from the light nucleus to the heavy one. One can assume that the decay of the DNS which evolves in mass and charge asymmetry coordinates gives an adequate description of the charge, mass and kinetic energy distributions of the quasifission products.

2. Cross sections for superheavy nuclei

In accordance with the DNS concept the evaporation residue cross section is factorized as follows [3,4]

$$\sigma_{ER}(E_{\rm c.m.}) = \sigma_c(E_{\rm c.m.})P_{CN}(E_{\rm c.m.}, J = 0)W_{sur}(E_{\rm c.m.}, J = 0). \qquad (1)$$

*This work is supported by AvH- and VW-Stiftung, RFBR, DFG, STCU (Uzb-45).

The calculations of the evaporation residue cross sections demand an analysis of all three factors in (1). The value of $\sigma_c = \frac{\lambda^2}{4\pi}(J_{max}+1)^2 T(E_{c.m.}, J = 0)$ is the effective capture cross section for the transition of the colliding nuclei over the entrance (Coulomb) barrier with the probability T. The contributing angular momenta in the evaporation residue cross section are limited by the surviving probability $W_{sur}(E_{c.m.}, J)$ with $J_{max} \approx 10$ when highly fissile superheavy nuclei are produced for energies $E_{c.m.}$ near the Coulomb barrier. For cold fusion reactions, $T(E_{c.m.}, J = 0) = 0.5$ is chosen.

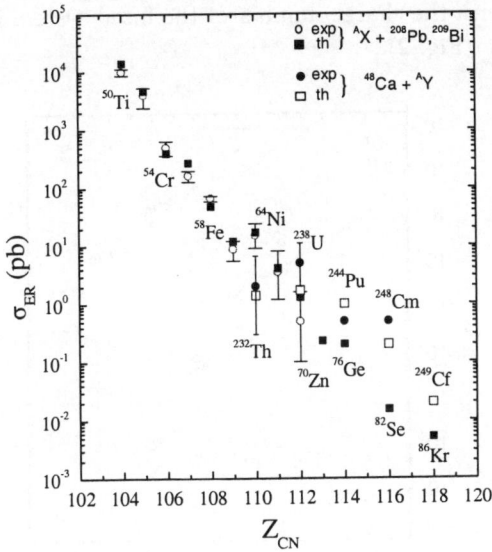

Figure 1. Calculated evaporation residue cross sections for the cold and hot fusion reactions are compared with available experimental data [1,2].

The probability of complete fusion P_{CN} in (1) depends on the competition between complete fusion in η mass asymmetry $\eta = (A_P - A_T)/(A_P + A_T)$ (A_P and A_T are the mass numbers of the DNS nuclei) and quasifission (decay of the DNS in the internuclear distance R after the capture stage). P_{CN} is expressed as follows

$$P_{CN} = \frac{\lambda_\eta^{Kr}}{\lambda_R^{Kr} + \lambda_\eta^{Kr}} - \frac{\lambda_\eta^{Kr}\lambda_R^{Kr}}{\lambda_R^{Kr} + \lambda_\eta^{Kr}} \frac{\tau_\eta - \tau_R}{1.72}. \tag{2}$$

For the quasi-stationary rates of fusion λ_η^{Kr} and quasifission λ_R^{Kr} through the fusion barrier ($B_\eta = B_{fus}^*$) in η and quasifission barrier ($B_R = B_{qf}$) in R, respectively, we used a two-dimensional Kramers-type expression

[4]. The second term in (2) is related to the transient times τ_R and τ_η to reach the quasi–stationary rates along the R and η coordinates ($\tau_R^{-1}, \tau_\eta^{-1} > \lambda_R^{Kr}, \lambda_\eta^{Kr}$). In the case that the fusion barrier is much higher than the quasifission barrier, $B_{\text{fus}}^* \gg B_{\text{qf}}$, i.e. if the transient time τ_η in η is larger (or equal) than the lifetime t_0 of the initial DNS, we obtain [4]

$$P_{CN} = \frac{\lambda_\eta^{Kr}}{1.72}[\tau_\eta(\exp[t_0/\tau_\eta] - 1) - t_0].$$

The available experimental data [1,2] on the evaporation residue cross section of ^{208}Pb- and ^{209}Bi-, and actinide-based reactions are well reproduced with the DNS concept [4] (Fig. 1). The strong decrease of the fusion cross sections with the charge number of the fused system is caused by the decrease of P_{CN} (Fig. 2).

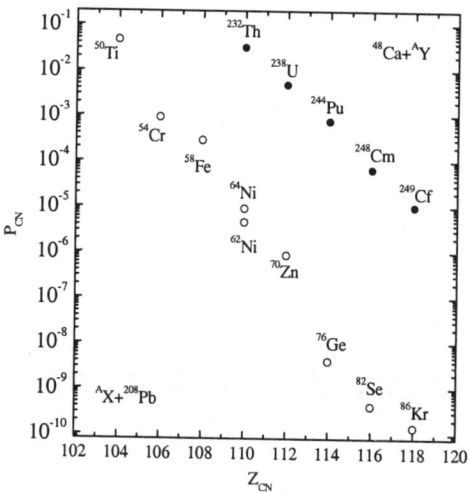

Figure 2. Calculated fusion probabilities in the indicated cold (open circles) and hot (closed circles) fusion reactions.

3. Quasifission

Quasifission means no formation of a compound nucleus but the decay of the DNS over the quasifission barrier B_{qf}. In the DNS concept the quasifission process is simultaneously described as the evolution of the DNS in the charge and mass asymmetry coordinates by nucleon transfer between the nuclei and the decay of the DNS into the direction of increasing internuclear distance. So, the quasifission process delivers products with different η. We use master equations in order to calculate the probability $P_{Z,A}(t)$ for finding the dinuclear system at time t in a configuration where the light fragment-nucleus has charge and mass numbers $Z_P = Z$ and $A_P = A$, respectively, and the heavy fragment-nucleus $Z_T = Z_{total} - Z$ and $A_T = A_{total} - A$:

$$\frac{dP_{Z,A}(t)}{dt} = \Delta_{Z+1,A+1}^{(-,0)}P_{Z+1,A+1}(t) + \Delta_{Z-1,A-1}^{(+,0)}P_{Z-1,A-1}(t)$$

$$+ \Delta_{Z,A+1}^{(0,-)}P_{Z,A+1}(t) + \Delta_{Z,A-1}^{(0,+)}P_{Z,A-1}(t)$$

$$- (\Delta_{Z,A}^{(-,0)} + \Delta_{Z,A}^{(+,0)} + \Delta_{Z,A}^{(0,-)} + \Delta_{Z,A}^{(0,+)} + \Lambda_{Z,A}^{qf})P_{Z,A}(t). \quad (3)$$

The microscopical transport coefficients $\Delta_{Z,A}^{(\pm,0)}$ and $\Delta_{Z,A}^{(0,\pm)}$ are probability rates for transfer of a proton and neutron [6], respectively, and $\Lambda_{Z,A}^{qf} = \lambda_R^{Kr}(Z,A)$ the decay rate in the coordinate R for quasifission. The measurable charge and mass yields for quasifission are obtained from

$$Y_{Z,A}(t_0) = \Lambda_{Z,A}^{qf} \int_0^{t_0} P_{Z,A}(t)dt, \quad (4)$$

where the reaction time t_0 is determined by solving the equation

$$\sum_{Z,A} \Lambda_{Z,A}^{qf} \int_0^{t_0} P_{Z,A}(t)dt = 1 - P_{CN}. \quad (5)$$

In order to calculate the average total kinetic energy (TKE) of the quasifission products and its dispersion, one has to regard the deformation of the fragments in addition. The distribution of the fragments in charge, mass and deformation can be written as

$$P = P(Z_P, A_P, \beta_P, Z_T, A_T, \beta_T) = Y_{Z_P,A_P}(t_0)P_{\beta_P}(Z_P, A_P)P_{\beta_T}(Z_T, A_T). \quad (6)$$

Here, $P_\beta(Z,A)$ is a distribution in deformation at fixed values of Z and A. It can be taken as

$$P_\beta(Z,A) = \frac{1}{\sqrt{2\pi\sigma_\beta^2}} \exp(-(\beta - <\beta>)^2/(2\sigma_\beta^2)) \quad (7)$$

with $\sigma_\beta^2 = \frac{\hbar\omega_{vib}}{2C_{vib}} \coth(\hbar\omega_{vib}/(2k\Theta))$, where $\omega_{vib}(Z,A)$ and $C_{vib}(Z,A)$ are the frequency and stiffness parameter of quadrupole vibrations, respectively, and $\Theta(Z,A)$ is the temperature of the dinuclear system. Applying the distribution (16) we calculate the average total kinetic energy as a function of the mass number A_P of light fragment:

$$<TKE(A_P)> = \int\int d\beta_P d\beta_T \sum_{Z_P} TKE \cdot P / (\int\int d\beta_P d\beta_T \sum_{Z_P} P)$$

$$\approx \sum_{Z_P} TKE|_{\beta_i = <\beta_i>} Y_{Z_P,A_P}(t_0) / \sum_{Z_P} Y_{Z_P,A_P}(t_0), \quad (8)$$

with $TKE = V_{nucl}(R_b) + V_{Coul}(R_b)$, where the radius $R_b = R_b(Z_P, A_P, \beta_P, Z_T, A_T, \beta_T)$ is the position of the Coulomb barrier. The variance of TKE can be written as a sum of the contributions of variances from the exchange of nucleons and from the deformations:

$$\sigma_{TKE}^2(A_P) \approx \sum_{Z_P} TKE^2|_{\beta_i = <\beta_i>} Y_{Z_P, A_P}(t_0) / \sum_{Z_P} Y_{Z_P, A_P}(t_0)$$

$$- < TKE(A_P) >^2 + (\sigma_{TKE}^{def}(A_P))^2{}_{\text{nucleus P}} + (\sigma_{TKE}^{def}(A_P))^2{}_{\text{nucleus T}}, \quad (9)$$

where

$$(\sigma_{TKE}^{def}(A_P))_j^2 = \sum_{Z_P} \left(\frac{\partial TKE}{\partial \beta_j} \right)^2 \Bigg|_{\substack{\beta_P = <\beta_P> \\ \beta_T = <\beta_T>}} \sigma_{\beta_j}^2 Y_{Z_P, A_P}(t_0) / \sum_{Z_P} Y_{Z_P, A_P}(t_0). \quad (10)$$

Calculated mean values of the TKE for the products of nearly symmetric splitting in all reactions considered follow the systematic which is closed to the one given in Ref. [7]. The experimental determination of the TKE has large systematic and statistical uncertainties. For the ^{48}Ca+^{248}Cm reaction, the calculated data on $Y(A_P)$ and $\sigma_{TKE}^2(A_P)$ are compared with the experimental data [8] in Fig. 3. Since the DNS has large moment of inertia, the data calculated at $J = 0$ and 70 are very similar. Considering the fluctuations of the DNS charge asymmetry (the fluctuations of the Coulomb interaction) at fixed mass asymmetry, the fluctuations of the quadrupole deformation parameters of the DNS nuclei, one can explain the large variance of the TKE distribution as a function of the mass numbers of the fragments. For $A_P > 100$, $(\sigma_{TKE}^{def}(A_P))^2$ mainly contributes to $\sigma_{TKE}^2(A_P)$. The same is for all reactions considered. The contribution to the variance of TKE due to the nucleon exchange is more important in the decay of quite asymmetric DNS. With increasing excitation energy the variance of the TKE of quasifission products with $A_P = A/2 \pm 20$ smoothly increases (Table 1) mainly due to the increase of $\sigma_{\beta_i}^2$ with Θ.

As is shown in Fig. 4 for the ^{48}Ca $+ ^{244}$Pu $\rightarrow ^{292}$114 fusion reaction, the nuclei up to Ne can be seen among the quasifission products with present experimental possibilities. The measurement of the products with $A_P \approx 20$ and 28 in the collisions near the Coulomb barrier would confirm the evolution of the DNS to the compound nucleus in mass asymmetry coordinate. The yield of light products is known to be larger for larger beam energies.

Since the quasifission barrier B_{qf} increases with decreasing Z and increasing number of neutrons in the system, the difference between the quasifission distributions in cold and hot fusion reactions is related to different

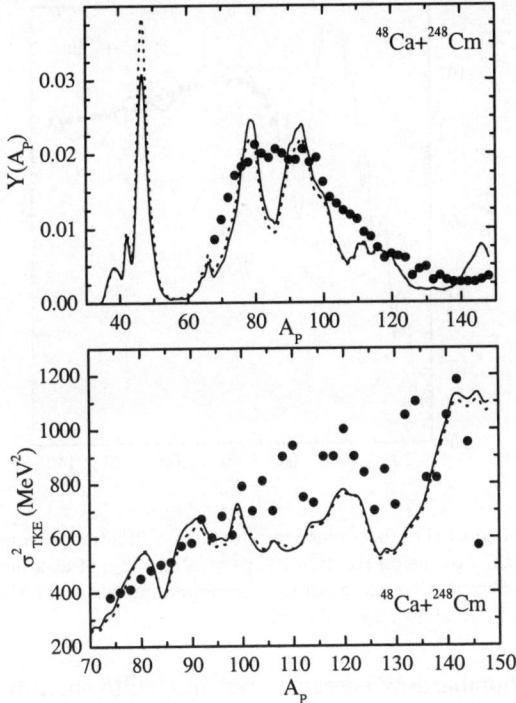

Figure 3.   The calculated mass yield (upper part) and variance of the TKE (lower part) of the quasifission products as a function of mass number of the light fragment for the hot fusion reaction ^{48}Ca+^{248}Cm→^{296}116 at the bombarding energy corresponding to an excitation energy of the compound nucleus of 37 MeV. The results calculated at $J = 0$ and 70 are presented by solid and dotted curves, respectively. The experimental data [8] are shown by solid points.

choices of colliding nuclei. We found for the reaction ^{86}Kr+^{208}Pb→^{294}118 that the quasifission products are practically associated with fragmentations near the initial DNS due to the small values of B_{qf}. In the Pb-based reactions the relative yield of nearly symmetric quasifission fragments decreases with increasing atomic number of projectile and is in general smaller than in the actinide-based reactions with ^{48}Ca. The yield of symmetric products of quasifission increases with number of neutrons in the system due to larger values of B_{qf}.

The average values of the variances of the TKE of quasifission products with $A/2 - 20 \leq A_P \leq A/2$ as well as relative contributions of the fusion-fission with respect to the quasifission to the yield of symmetric products are listed in Table 1 for various reactions. The contribution of fusion-fission is mainly determined by P_{CN} (Fig. 2). Although this contribution

726

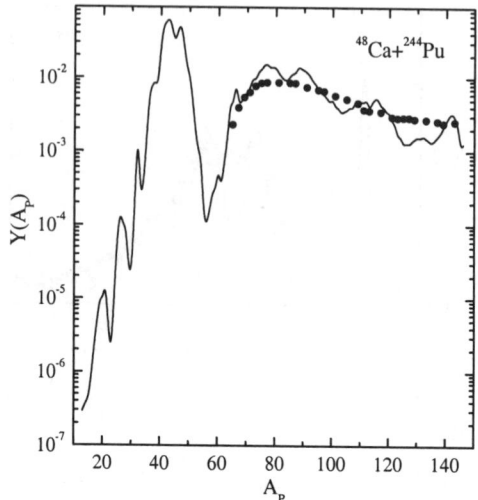

Figure 4. Mass yield of the quasifission products as a function of mass number of the light fragment for the hot fusion reaction $^{48}Ca+^{244}Pu \rightarrow ^{292}114$ at a bombarding energy corresponding to an excitation energy of the compound nucleus of 42 MeV. The available experimental data [8] are shown by solid points.

increases with bombarding energy, it remains quite small in the reactions considered. Therefore, in these reactions the quasifission mainly results the yield of nearly symmetric products. The ratio between the motions of the DNS to more asymmetric and more symmetric configurations decreases exponentially with an increasing charge number of the superheavy compound nucleus. The number of neutrons emitted from the DNS (pre-decay neutrons) is quite small in the reactions considered. The total number of neutrons, post-decay (Table 1), accompanying the quasifission is described well in our model.

4. Summary

1) If the compound nucleus is quite stable to be detected, the quasifission process is the main factor suppressing the complete fusion of heavy nuclei. In cold and hot fusion reactions the fusion-fission events are much smaller than the events of the production of certain quasifission products. The main contribution to symmetric and near symmetric fragmentations comes from quasifission process. 2) A comparison of theoretical and experimental data of the yield of asymmetric quasifission products constitutes a critical test for the dynamics of existing fusion models. The measurement of highly

asymmetric quasifission products would prove the motion of the DNS to the compound nucleus in mass asymmetry coordinate.

We thank Prof.V.V.Volkov and Mr.A.V.Andreev for fruitful discussions.

Table 1. The calculated average variances σ^2_{TKE} of the TKE for the nearly symmetric quasifission products with $A/2 - 20 \leq A_P \leq A/2$, fraction of the fusion-fission process with respect to the quasifission in the yield of nearly symmetric quasifission products, and the calculated total numbers of emitted neutrons for nearly symmetric quasifission splitting (M_n^{sym}) with $A/2-20 \leq A_P \leq A/2$ and for the quasifission splitting (M_n^{asym}) with $A_P < A/2 - 20$. The reactions and the energies E^*_{CN} of corresponding compound nuclei are indicated.

Reactions	E^*_{CN} (MeV)	σ^2_{TKE} (MeV^2)	$P_{CN}/\sum\limits_{A_P=A/2-20}^{A/2} Y(A_P)$	M_n^{sym}	M_n^{asym}
^{48}Ca+^{238}U	33.4	756	5.8×10^{-2}	7.0	5.4
	50	840	2.4×10^{-1}	8.1	6.4
^{48}Ca+^{244}Pu	34.8	805	1.4×10^{-2}	7.5	5.4
	50	893	1.1×10^{-1}	8.5	6.4
^{48}Ca+^{248}Cm	37	889	4.0×10^{-3}	8.2	5.9
	50	949	1.0×10^{-2}	9.2	6.9
^{48}Ca+^{249}Cf	30.6	808	1.3×10^{-3}	7.7	5.4
	50	949	3.2×10^{-2}	9.0	6.8
^{58}Fe+^{208}Pb	14.5	420	4.9×10^{-5}	3.5	2.1
	30	484	5.9×10^{-2}	4.8	3.2
^{64}Ni+^{208}Pb	12.5	499	1.2×10^{-5}	3.5	2.1
	30	609	8.5×10^{-3}	4.9	3.5
^{86}Kr+^{208}Pb	17	738	2.1×10^{-7}	4.8	2.8
	30	813	2.0×10^{-5}	7.0	4.7

References

1. S.Hofmann and G.Münzenberg, *Rev. Mod. Phys.*, **72** 733 (2000).
2. Yu.Ts.Oganessian et al., *Phys. Rev. Lett.*, **83** 3154 (1999).
3. V.V. Volkov, *Izv. AN SSSR ser.fiz.*, **50** 1879 (1986); N.V. Antonenko et al., *Phys. Lett. B*, **319** 425 (1993); *Phys. Rev. C*, **51** 2635 (1995); E.A. Cherepanov, preprint JINR, E7-99-27(1999); R.V.Jolos, A.K.Nasirov, A.I.Muminov, *Eur. Phys. J. A*, **4** 245 (1999).
4. G.G.Adamian, N.V.Antonenko, W.Scheid and V.V.Volkov, *Nucl. Phys. A*, **633** 409 (1998); **678** 24 (2000).
5. A.Diaz-Torres et al., *Phys. Rev. C*, **64** 024604 (2001).
6. G.G.Adamian et al., *Particles and Nuclei*, **25** 583 (1994).
7. J.P. Unik et al., *Int. Conf. on Physics and Chemistry of fission*, 1973, vol.2, IAEA, Vienna (1974), p.19.
8. M.G. Itkis et al., *Int. Conf. on Nuclear Physics at Border Lines*, eds. G.Fazio et al., World Scientific, Singapore, 2002, pp. 142, 157.

STUDY OF SUPERHEAVY ELEMENTS AT GSI - RECENT RESULTS AND PERSPECTIVES

S. HOFMANN

Gesellschaft für Schwerionenforschung, GSI,
Planckstrasse 1,
D-64291 Darmstadt, Germany
E-mail: s.hofmann@gsi.de

The nuclear shell model predicts that the next doubly magic shell-closure beyond ^{208}Pb is at a proton number between Z = 114 and 126 and at a neutron number N = 184. The outstanding aim of experimental investigations is the exploration of this region of spherical 'Superheavy Elements'. This article describes the experiments that were performed recently at the GSI SHIP. They resulted in an unambiguous identification of elements 110 to 112. They were negative so far in searching for elements 113, 116, and 118. The measured decay data are compared with theoretical predictions. Some aspects concerning the reaction mechanism are also presented.

1. Experimental Results

In this section, recent results are presented dealing with the confirmation of elements 110 to 112. Detailed presentations of the properties of elements 107 to 109 and of earlier results on elements 110 to 112 were given in previous review [1-3].

Element 110 was discovered in 1994 using the reaction ^{62}Ni + ^{208}Pb → ^{270}110* [4]. A total of three decay chains were measured (see also remarks at the end of this section). The main experiment was preceded by a thorough study of the excitation functions for the synthesis of ^{257}Rf and ^{265}Hs in order to determine the optimum beam energy for the production of element 110. The data revealed that the maximum cross-section for the synthesis of element 108 was shifted to a lower excitation energy, different from the predictions of reaction theories.

The heavier isotope ^{271}110 was synthesized with a beam of the more neutron-rich isotope ^{64}Ni [2]. The important result for the further production of elements beyond meitnerium was that the cross-section was enhanced from 2.6 pb to 15 pb by increasing the neutron number of the projectile by two, which gave hope that the cross-sections could decrease less steeply with more neutron-rich projectiles. However, this expectation was not proven in the case of element 112.

The even-even nucleus $^{270}110$ was synthesized using the reaction ^{64}Ni + ^{207}Pb [5]. A total of eight α-decay chains was measured during an irradiation time of seven days. Decay data were obtained for the ground-state and a high spin K isomer, for which calculations predict spin and parity 8^+, 9^- or 10^- [6]. The new nuclei ^{266}Hs and ^{262}Sg were identified as daughter products after α decay. Spontaneous fission of ^{262}Sg terminates the decay chain.

Element 111 was synthesized in 1994 using the reaction ^{64}Ni + ^{209}Bi → $^{273}111^*$. A total of three α chains of the isotope $^{272}111$ were observed [7]. Another three decay chains were measured in a confirmation experiment in October 2000 [8].

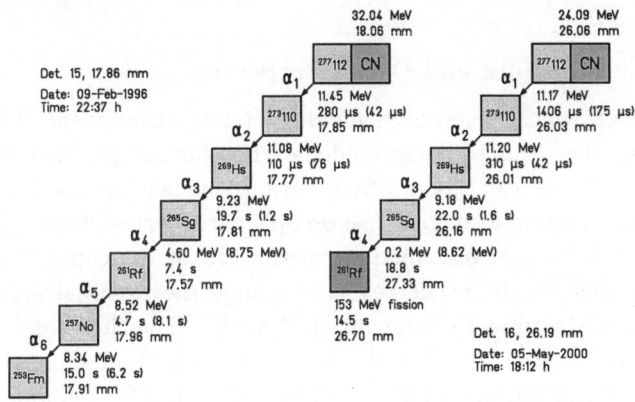

Figure 1. Two decay chains measured in experiments at SHIP in the cold fusion reaction ^{70}Zn + ^{208}Pb → $^{278}112^*$. The chains were assigned to the isotope $^{277}112$ produced by evaporation of one neutron from the compound nucleus. The lifetimes given in brackets were calculated using the measured α energies. In the case of escaped α particles the α energies were determined using the measured lifetimes.

Element 112 was investigated at SHIP using the reaction ^{70}Zn + ^{208}Pb → $^{278}112^*$ [9]. The irradiation was performed in January-February 1996. Over a period of 24 days, a total of 3.4×10^{18} projectiles were collected. One α-decay chain, shown in the left side of Fig. 1, was observed resulting in a cross-section of 0.5 pb. The chain was assigned to the one neutron-emission channel. The experiment was repeated in May 2000 aiming to confirm the synthesis of $^{277}112$ [8]. During a similar long measuring time, but using slightly higher beam energy, one more decay chain was observed, also shown in Fig. 1. The measured decay pattern of the first four α decays is in agreement with the one observed in the first experiment.

A new result was the occurrence of fission which ended the second decay chain at ^{261}Rf. A spontaneous-fission branch of this nucleus was not yet known, however, it was expected from theoretical calculations. The new results on ^{261}Rf were proven in a recent chemistry experiment [10,11], in which this isotope was measured as granddaughter in the decay chain of ^{269}Hs.

A reanalysis of all decay chains measured at SHIP since 1994, a total of 34 decay chains was analyzed, revealed that the previously published first decay chain of ^{277}112 [9] (not shown in Fig. 1) and the second of the originally published four chains of ^{269}110 [4] were spuriously created. Details of the results of the reanalysis are given in Ref. [8].

2. Nuclear Structure and Decay Properties

The basic step which is necessary for the determination of the stability of SHEs is the calculation of the ground-state binding energy. As a signature for shell effects, we can extract from various models the shell-correction energy by subtracting a smooth macroscopic part (derived from the liquid-drop model) from the total binding energy. In macroscopic-microscopic models the shell-correction energy is of course the essential input value which is calculated directly from the shell model. The shell-correction energy is plotted in Fig. 2a using the data from Ref. [12]. Two equally deep minima are obtained, one at $Z = 108$ and $N = 162$ for deformed nuclei with deformation parameters $\beta_2 \approx 0.22$, $\beta_4 \approx -0.07$ and the other at $Z = 114$ and $N = 184$ for spherical SHEs. Different results are obtained from self-consistent Hartree-Fock-Bogoliubov (HFB) calculations and relativistic mean-field models [13-16]. They predict for the spherical nuclei shells at $Z = 114, 120$ or 126 (indicated as dashed lines in Fig. 2) and $N = 184$ or 172, with shell strengths being also a function of the amount of nucleons of the other type.

For the calculation of partial spontaneous fission half-lives the knowledge of ground-state binding energies is not sufficient. It is necessary to determine the fission barrier over a wide range of deformation. The most accurate data were obtained for even-even nuclei using the macroscopic-microscopic model [12,17]. Partial spontaneous fission half-lives are plotted in Fig. 2b.

Partial α half-lives decrease almost monotonically from 10^{12} s down to

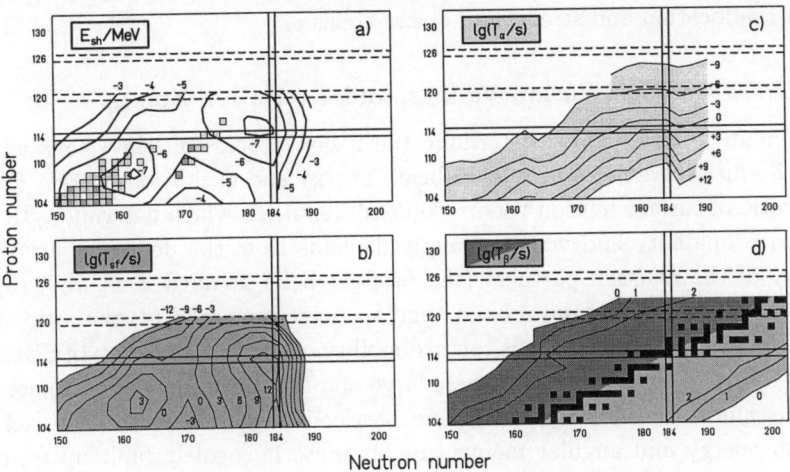

Figure 2. Shell-correction energy (a) and partial spontaneous fission, α and β half-lives (b-d). The calculated values in (a)−(d) are taken from Ref. [12,17] and in (d) from Ref. [18]. The squares in (a) mark the nuclei presently known or under investigation.

10^{-9} s near $Z = 126$ (Fig. 2c). The valley of β-stable nuclei (marked by black squares in Fig. 2d) passes through $Z = 114$, $N = 184$ [18]. At a distance from the bottom of the valley, the β half-lives decrease gradually down to values of one second.

The interesting question arises, if and how the uncertainty related with the location of the proton and neutron shell closures will change the half-lives of SHEs. Partial α and β half-lives are only insignificantly modified by shell effects, because the decay process occurs between neighboring nuclei. This is different for fission half-lives which are primarily determined by shell effects. However, the uncertainty related with the location of nuclei with the strongest shell-effects and thus longest partial fission half-life at $Z = 114$, 120 or 126 and $N = 172$ or 184, is inconsequential concerning the longest 'total' half-life of SHEs. The regions for SHEs in question are dominated by α decay. And α decay will be modified by only a factor of up to approximately 100, if the double shell closure will not be located at $Z = 114$ and $N = 184$.

The line of reasoning is, however, different concerning the production cross-section. The survival probability of the compound nucleus (CN) is determined among other factors significantly by the fission-barrier. Therefore all present calculations of cross-sections suffer from the uncertainty related

732

with the location and strength of closed shells.

3. Cross-sections, Fusion Valleys, and Excitation Energy

The main features which determine the fusion process of heavy ions are (1) the fusion barrier and related beam energy and excitation energy, (2) the ratio of surface tension versus Coulomb repulsion which determines the fusion probability and which strongly depends from the degree of asymmetry of the reaction partners (the product $Z_1 Z_2$ at fixed $Z_1 + Z_2$), (3) the impact parameter and related angular momentum, and (4) the ratio of neutron evaporation versus fission probability of the CN. In fusion of SHEs the product $Z_1 Z_2$ reaches extremely large and the fission barrier extremely small values. In addition, the fission barrier is fragile at increasing excitation energy and angular momentum, because it is solely built up from shell effects. For these reasons the fusion of SHEs is hampered, whereas the fusion of lighter elements is advanced through the contracting effect of surface tension.

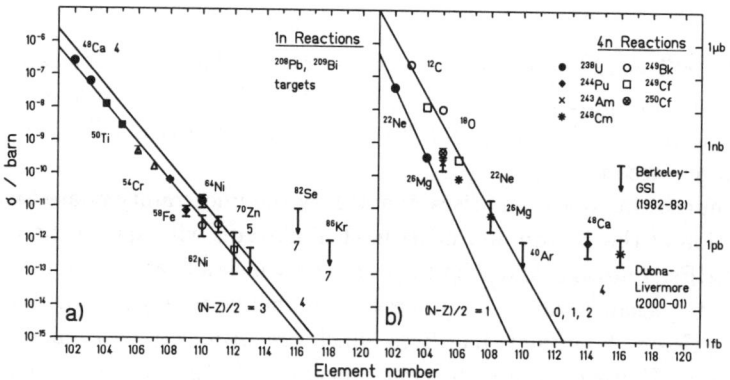

Figure 3. Measured cross-sections and cross-section limits for reactions using ^{208}Pb and ^{209}Bi targets and one neutron evaporation (a) and for reactions using actinide targets and four neutron evaporation (b).

The effect of Coulomb repulsion on the cross-section starts to act severely for fusion of elements beyond Fm. From there on a continuous decrease of cross-section was measured from microbarns for the synthesis of nobelium down to picobarns for the synthesis of element 112. The data obtained in reactions with ^{208}Pb and ^{209}Bi for the 1n evaporation channel at low excitation energies of about 10-15 MeV (therefore named *cold fusion*) and in reactions with actinide targets for the 4n channel at excitation energies of 35-45 MeV (*hot fusion*) are plotted in Fig. 3. Interesting for further investigation of SHEs are the relatively high cross-sections measured for the synthesis of elements 114 and 116 (4n channel) [19,20,21]. In both cases the obtained values of about 0.5 pb deviate considerably from the trend set by fusion of the lighter elements. An explanation could be a relatively high and wide fission barrier of the CN which is created by strong shell effects in the region of spherical SHEs. Note in this context that the experimental sensitivity increased by three orders of magnitude since the 1982-83 search experiments for element 116 using a hot fusion reaction [22].

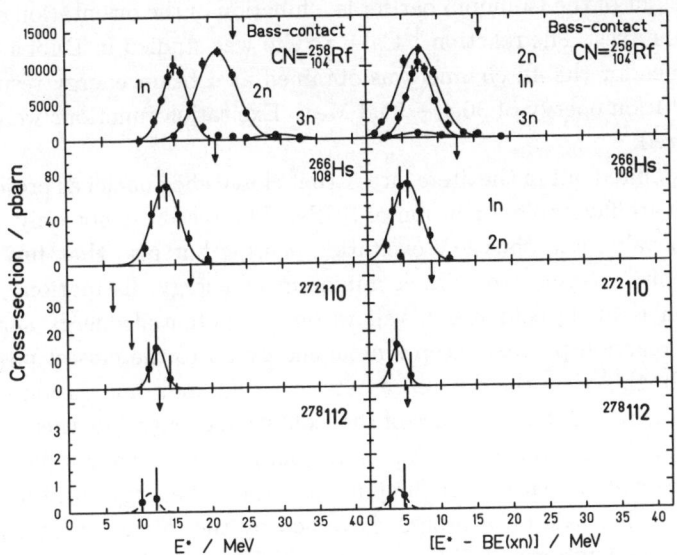

Figure 4.   Measured even element excitation functions.

A number of excitation functions was measured for the synthesis of elements from rutherfordium to 110 using Pb and Bi targets [3]. For the even elements these data are shown in Fig. 4. The maximum evaporation residue cross-section (1n channel) was measured at beam energies well below a fusion barrier calculated in one dimension [23]. At the optimum beam energy projectile and target are just reaching the contact configuration in a central collision. The relatively simple fusion barrier based on the Bass model [23] is too high and a tunnelling process through this barrier cannot explain for the measured cross-section.

Various processes are possible and are discussed in the literature which result in a lowering of the fusion barrier. Among these transfer of nucleons and excitation of vibrational degrees of freedom are the most important [24-33]. The theoretical studies are also aimed at reproducing the known cross-section data and further extrapolating the calculations into the region of spherical superheavy nuclei. The measured cross-sections for the formation of ^{257}Rf up to ^{277}112 are reproduced almost within about a factor of 2 by the various models. However, there are significant differences in the cross-section values for the synthesis of spherical SHEs beyond Z = 114.

In the case of actinide targets, the target nucleus is strongly deformed and the height of the Coulomb barrier is a function of the orientation of the deformation axes. The reaction ^{48}Ca + ^{248}Cm was studied in Dubna [19,20], and evidence for the 4n channel was obtained at a beam energy resulting in an excitation energy of 30.4 − 35.8 MeV. Excitation functions were not yet measured.

It was pointed out in the literature [34] that closed shell nuclei as projectile and target are favorable for fusion of SHEs. The reason is not only a low reaction Q value and thus low excitation energy, but also that fusion of such systems is connected with a minimum of energy dissipation. The fusion path is along cold fusion valleys on the potential energy surface, where the reaction partners keep kinetic energy up to the closest possible distance. In this view the difference between 'cold' and 'hot' fusion is not only a result from different values of the excitation energy, but there exists also a qualitative difference, which is on the one side based on a well ordered fusion process along paths of minimum dissipation of energy (cold fusion), and on the other side on a process governed by the formation of a more or less energy equilibrated CN (hot fusion). This qualitative explanation is well in agreement with the results from experimental studies of quasi-fission and compound-nucleus fission [35].

Acknowledgments

The recent experiments at SHIP were performed in collaboration with D. Ackermann, F.P. Heßberger, B. Kindler, J. Kojouharova, B. Lommel, R. Mann, G. Münzenberg, S. Reshitko, H.J. Schött (GSI Darmstadt); A. Popeko, A. Yeremin (JINR Dubna); S. Antalic, P. Cagarda, S. Saro (University Bratislava); M. Leino, J. Uusitalo (University Jyväskylä).

References

1. G. Münzenberg, *Rep. Prog. Phys.* A **51**, 57 (1988).
2. S. Hofmann, *Rep. Prog. Phys.* A **61**, 639 (1998).
3. S. Hofmann and G. Münzenberg, *Rev. Mod. Phys.* **72**, 733 (2000).
4. S. Hofmann *et al*, *Z. Phys.* A **350**, 277 (1995).
5. S. Hofmann *et al*, *Eur. Phys. J.* A **10**, 5 (2001).
6. S. Cwiok *et al*, *Phys. Rev. Lett.* **83**, 1108 (1999) and private communication.
7. S. Hofmann *et al*, *Z. Phys.* A **350**, 281 (1995).
8. S. Hofmann *et al*, *Eur. Phys. J.* A **14**, 147 (2002).
9. S. Hofmann *et al*, *Z. Phys.* A **354**, 229 (1996).
10. A. Türler *et al*, *Eur. Phys. J.* A , submitted (2002).
11. Ch.E. Düllmann *et al*, *Nature* **418**, 859 (2002).
12. R. Smolanczuk and A. Sobiczewski, Proc. XV. Nucl. Phys. Divisional Conf. on *Low Energy Nuclear Dynamics*, St.Petersburg, Russia, 1995, p.313 (World Scientific, Singapore, 1995).
13. S. Cwiok *et al*, *Nucl. Phys.* A **611**, 211 (1996).
14. G.A. Lalazissis *et al*, *Nucl. Phys.* A **608**, 202 (1996).
15. K. Rutz *et al*, *Phys. Rev.* C **56**, 238 (1997).
16. A.T. Kruppa *et al*, *Phys. Rev.* C **61**, 034313 (2000).
17. R. Smolanczuk *et al*, *Phys. Rev.* C **52**, 1871 (1995).
18. P. Möller *et al*, *Atomic Data and Nucl. Data Tables* **66**, 131 (1997).
19. Yu.Ts. Oganessian *et al*, *Phys. Rev.* C **63**, 011301 (2000).
20. Yu.Ts. Oganessian *et al*, *Phys. Atomic Nuclei* **64**, 1349 (2001).
21. Yu.Ts. Oganessian *et al*, *Nature* **400**, 242 (1999).
22. P. Armbruster *et al*, *Phys. Rev. Lett.* **54**, 406 (1985).
23. R. Bass, *Nucl. Phys.* A **231**, 45 (1974).
24. Y. Aritomo *et al*, *Phys. Rev.* C **59**, 796 (1999).
25. E.A. Cherepanov, *Pramana J. Phys.* **53**, 619 (1999).
26. R. Smolanczuk, *Phys. Rev.* C **59**, 2634 (1999).
27. R. Smolanczuk, *Phys. Rev. Lett.* **83**, 4705 (1999).
28. G. Giardina *et al*, *Eur. Phys. J.* A **8**, 205 (2000).
29. V.Yu. Denisov and S. Hofmann, *Phys. Rev.* C **61**, 034606 (2000).
30. G.G. Adamian *et al*, *Nucl. Phys.* A **678**, 24 (2000).
31. G.G. Adamian *et al*, *Phys. Rev.* C **62**, 064303 (2000).
32. R. Smolanczuk, *Phys. Rev.* C **63**, 044607 (2001).
33. V.I. Zagrebaev, in *Proc. Int. Workshop on Fusion Dynamics at the Extremes*, Dubna, Russia, 2000, p. 215 (World Scientific, 2001).
34. R.K. Gupta *et al*, *Z. Phys.* A **283**, 217 (1977).
35. M.G. Itkis *et al*, in *Proc. Int. Workshop on Fusion Dynamics at the Extremes*, Dubna, Russia, 2000, p. 93 (World Scientific, 2001).

SUPERHEAVY ELEMENT RESEARCH AT DGFRS

YU.TS. OGANESSIAN, V.K. UTYONKOV, YU.V. LOBANOV, F.SH. ABDULLIN,
A.N. POLYAKOV, I.V. SHIROKOVSKY, YU.S. TSYGANOV, G.G. GULBEKIAN,
S.L. BOGOMOLOV, B.N. GIKAL, A.N. MEZENTSEV, S. ILIEV, V.G. SUBBOTIN,
A.M. SUKHOV, O.V. IVANOV, G.V. BUKLANOV, A.A. VOINOV, K. SUBOTIC,
AND M.G. ITKIS

Joint Institute for Nuclear Research, 141980 Dubna, Russian Federation
E-mail: utyonkov@sungns.jinr.dubna.su

K.J. MOODY, J.F. WILD, N.J. STOYER, M.A. STOYER, R.W. LOUGHEED,
C.A. LAUE, J.B. PATIN AND D.A. SHAUGHNESSY

University of California, Lawrence Livermore National Laboratory, Livermore,
California 94551, USA
E-mail: moody3@llnl.gov

This paper presents results of experiments aimed at producing long-lived superheavy elements located near the spherical shell closures with $Z \geq 114$ and $N \geq 172$ in the reactions of neutron-rich isotopes ^{244}Pu, ^{248}Cm and ^{249}Cf with ^{48}Ca projectiles. The decay properties of the synthesized nuclei are consistent with the consecutive α-decays originating in the decays of parent nuclides ^{288,289}114, ^{292}116 and ^{294}118 produced in the $3n$ and $4n$-evaporation channels. The present observations can be considered to be experimental evidence of the existence of the "island of stability" of superheavy elements.

1. Introduction

Beyond the domain of the heaviest known nuclei the macroscopic-microscopic nuclear theory predicts the existence of a region of long-lived superheavy nuclei. Calculations performed over more than 30 years with different versions of the nuclear shell model predict a substantial enhancement of the stability of heavy nuclei when approaching the closed spherical shells at $Z=114$ and $N=184$.

The most neutron rich and, consequently, relatively stable isotopes of superheavy elements 114, 116 and 118 close to the predicted magic neutron shell $N=184$ are expected to be produced in fusion reactions of target and projectile nuclei with significant neutron excess, e.g., ^{244}Pu, ^{248}Cm and ^{249}Cf with the doubly magic ^{48}Ca projectiles. The resulting compound nuclei should have an excitation energy of 27-33 MeV at the Coulomb barrier and should de-excite with the greatest probability by the evaporation of three to four neutrons. The investigation of the heaviest even-even nuclei allows a better comparison with theoretical predictions due to their unhindered α-decays and spontaneous fission

(SF) decay modes. Our present experiments, involving the ^{244}Pu+^{48}Ca, ^{248}Cm+^{48}Ca and ^{249}Cf+^{48}Ca reactions were designed to attempt the production of elements 114, 116 and 118 at a cross section level of tenths of a picobarn, thus exceeding the sensitivity of the previous experiments for the synthesis of new elements in ^{48}Ca-induced reactions by about three orders of magnitude.

2. Experimental Technique

A beam of ^{48}Ca^{+5} ions was delivered by the U400 cyclotron at FLNR, JINR. The typical intensity of the continuous ion beam on the target was 0.7 pµA. The beam energy was determined with a precision of ~1 MeV by measuring the energies of scattered ions, and by a time-of-flight technique. The rotating targets contained the enriched isotopes ^{244}Pu (98.6%), ^{248}Cm (97.4%) and ^{249}Cf (97.3%) deposited onto 32-cm^2 of 1.5-µm Ti foils to a thickness of ~0.3 mg cm^{-2}.

The evaporation residues (EVRs) recoiling from the target were spatially separated in flight from the beam, scattered particles and transfer reaction products by the Dubna Gas-filled Recoil Separator (DGFRS) [1] consisting of a dipole magnet and two quadrupole lenses. The recoils passed through the hydrogen-filled volume of the separator (1 Torr), a Mylar window (~1 µm), then through a time-of-flight (TOF) system filled by pentane (~1.5 Torr), and were finally implanted in the focal-plane detector array. The transmission efficiency of the separator for Z=114-118 nuclei was estimated to be about 35-40% [2].

The focal-plane detector consisted of three 4×4 cm^2 silicon detectors, each with four strips having position sensitivity in the vertical direction. To increase the detection efficiency for full-energy α's, we arranged 8 detectors without position sensitivity in a box surrounding the focal-plane detector. Employing side detectors increased the α-particle detection efficiency from 53% to 87%. A set of 3 similar "veto" detectors was mounted behind the detector array in order to eliminate signals from low-ionizing light particles, which could pass through the focal-plane detector (300 µm) without being detected in the TOF system.

Alpha-energy calibrations were performed using the α peaks from nuclides produced in the bombardments of ^{nat}Yb and enriched $^{204,206-208}$Pb targets with ^{48}Ca ions [3]. The energy resolution for the detection of α-particles in the focal-plane detector was ≈45 keV at the beginning of the experiments and about 60 keV after a total beam dose of 5×10^{19} ^{48}Ca projectiles was delivered to the targets. For α's escaping from the focal-plane detector at different angles and absorbed in the side detectors, the energy resolution was ≈190 keV because of energy losses in the entrance windows and dead layers of both detectors and the pentane. The FWHM position resolutions of the signals of correlated decays of nuclei implanted in the detectors were 1.4 mm and 1.2 mm for EVR-α and EVR-SF signals, respectively, in the experiments of 1998. Values of 0.8 mm and 0.6 mm, respectively, were obtained in subsequent experiments. Fission fragments from ^{252}No implants produced in the ^{206}Pb+^{48}Ca reaction were used for a fission-energy calibration. The measured fragment energies were not corrected for the pulse-height defect of the detectors. The mean sum energy loss of fission fragments for ^{252}No was about 20 MeV.

3. Experimental Results

The ^{244}Pu+^{48}Ca bombardments were performed in Nov.-Dec., 1998, and June-Oct., 1999 [4]. A total of 1.5×10^{19} ^{48}Ca projectiles of energy ~236 MeV was delivered to the target. Taking into account the energy losses in the target and the overall beam energy and target thickness variations, we expected the resulting compound nuclei ^{292}114 to have excitation energy range E^{*}=31.5-39.0 MeV.

According to the concept of the "Island of Stability" of superheavy elements, as long as any α-decay chain leads to the edge of the stability region, it should be terminated by SF [5,6]. In the ^{244}Pu+^{48}Ca reaction we observed five SF events. Two such SF decays were detected within milliseconds following the implantation of corresponding position-correlated recoil nuclei and based on the lifetime were assigned to the SF of the 0.9-ms ^{244mf}Am isomer, a product of transfer reactions that are suppressed by the separator by more than five orders

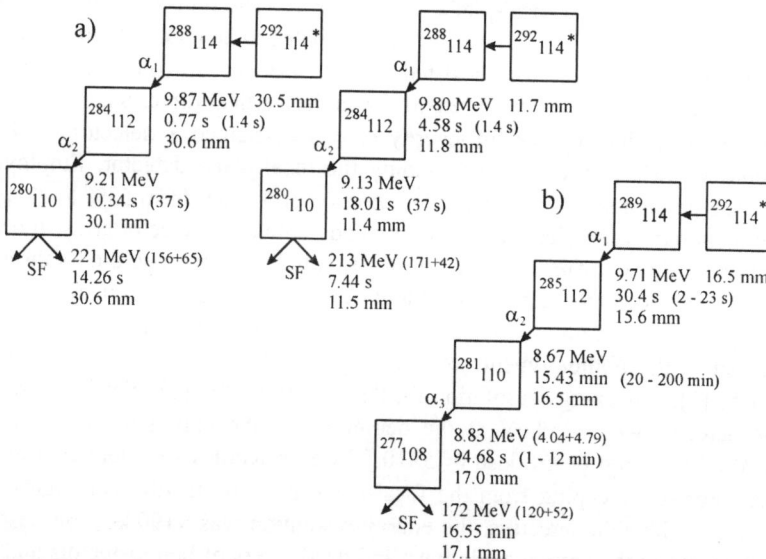

Figure 1. Time sequences in the decay chains observed in the ^{244}Pu+^{48}Ca reaction. The expected half-lives corresponding to the measured E_{α} values for the given isotopes are shown in parentheses following the measured lifetimes. Hindrance factors of 1 and 10 were assumed for α decay of nuclei with odd neutron numbers. Positions of the observed decay events are given with respect to the top of the strip.

of magnitude [1].

Three other SF events terminated the α-decay sequences of relatively long-lived nuclei. Two such SF events were observed in strips 2 and 8 [4]. The full decay chains including these two fission events are shown in Fig. 1a. The calculated probabilities that both these decay sequences were caused by the chance correlations of unrelated events at any position of the detector array and at the positions in which the candidate events occurred are less than 5×10^{-13}.

The formation of the nuclei which initiated the observed decays resulted from instantaneous ^{48}Ca beam energies of 237.6 and 237.0 MeV in the middle of the target, corresponding to excitation energy ranges E^*=33.6-39.7 and 33.2-39.1 MeV for the ^{292}114 compound nuclei, respectively. This would favor de-excitation of ^{292}114 by evaporation of 4 neutrons, which leads to the even-even nucleus ^{288}114. Indeed, the observed chains, including two α-decays and terminated by SF, match the decay scenario predicted for ^{288}114 [5]. The detected sequential decays have $T_{1/2}$ vs. E_α values that correspond well to the decays of the *even-even* isotopes of elements 114 and 112. To illustrate this, Fig. 1a presents the expected half-lives corresponding to the measured α-particle energies for the genetically related nuclides with the specified atomic numbers. For the calculation of half-lives with given Q_α values, the formula by Viola and Seaborg with parameters fitted to the T_α values of 65 known even-even nuclei with Z>82 and N>126 has been used. The measured total deposited energies for both fission events exceed the average value measured for ^{252}No by about 40 MeV, which also indicates the fission of a rather heavy granddaughter nucleus. From the above considerations, one can conclude that the detected decay chains originate from the parent even-even nuclide ^{288}114. We estimate the cross section for producing ^{288}114 in this reaction to be $0.5^{+0.8}_{-0.4}$ pb.

The next SF event was observed in strip 8. The entire position-correlated decay chain [4] is shown in Fig. 1b. The probability that this decay sequence was caused by the chance correlation of unrelated events is 6×10^{-3}. The best candidate for the parent nucleus is the even-odd isotope ^{289}114, produced in the $3n$-evaporation channel. Indeed, the α-decaying nuclides in this chain are characterized by lower decay energies than the corresponding members of the chain attributed to the decay of ^{288}114, and the decay sequence is terminated by SF at a later stage. The decay properties of the observed nuclei are also in agreement with calculations [7,8].

During June-July, Nov.-Dec., 2000, January and April-May, 2001, we carried out an experiment aimed at the synthesis of superheavy nuclei with Z=116 in the complete fusion reaction ^{248}Cm+^{48}Ca [9]. A beam dose of 2.3×10^{19} 240-MeV ^{48}Ca projectiles was collected in this experiment.

To improve background conditions for detecting long-time decay sequences, a special measurement mode was employed. The beam was switched off after a recoil signal was detected with parameters of implantation energy and

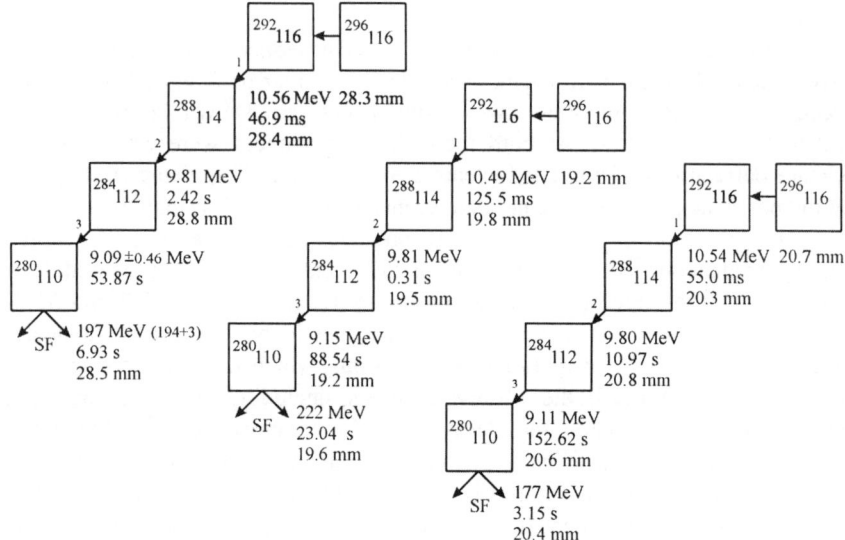

Figure 2. Time sequences in the decay chains observed in the ^{248}Cm+^{48}Ca reaction.

TOF expected for Z=116 evaporation residues, followed by an α-like signal with an energy of 10.05 MeV$\leq E_\alpha \leq$11.5 MeV, in the same strip, within a position window of 2 mm and time intervals of up to 5 s. Thus, all the expected sequential decays of the daughter nuclides with $Z\leq$114 could be observed in the absence of beam-associated background.

In this experiment three similar decay sequences were observed that are shown in Fig. 2. The implantation of EVRs in strips 4, 5, and 1 were followed by α-particles with E_α=10.53±0.06 MeV. These sequences switched the ion beam off, and further decays were detected under lower-background conditions. The probability of three observed event chains being totally of random origin is negligible. All the decays following the first 10.53-MeV α-particles agree well with the decay chains of ^{288}114, previously observed in the ^{244}Pu+^{48}Ca reaction (see Fig. 1a). Thus, it is reasonable to assign the observed decays to the

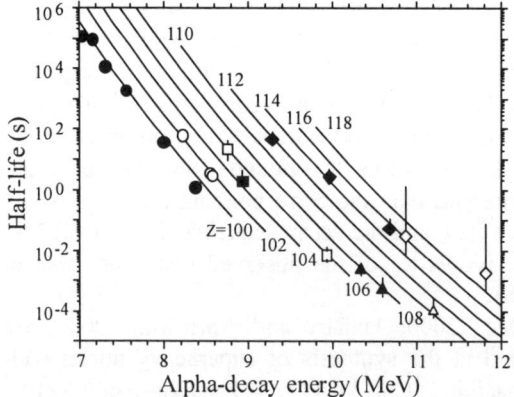

Figure 3. The dependence T_α vs. Q_α for known heavy even-even nuclei with Z=100 (●), 102 (○), 104 (■), 106 (□), 108 (▲), 110 (△) [10-12], and nuclei shown in Figs 1a and 2 (◆) and Fig. 4 (◇). The lines are drawn according to the formula by Viola and Seaborg.

nuclide $^{292}116$, produced via evaporation of four neutrons in the complete-fusion reaction $^{248}Cm+^{48}Ca$ with a cross section of $0.5^{+0.8}_{-0.3}$ pb.

The decay properties of $^{292}116$ ($E_\alpha=10.53$ MeV, $Q_\alpha=10.68\pm0.06$ MeV, $T_{1/2}=53^{+62}_{-19}$ ms), $^{288}114$ ($E_\alpha=9.82$ MeV, $Q_\alpha=9.96\pm0.06$ MeV, $T_{1/2}=2.6^{+2.0}_{-0.8}$ s), and $^{284}112$ ($E_\alpha=9.15$ MeV, $Q_\alpha=9.28\pm0.06$ MeV, $T_{1/2}=45^{+34}_{-14}$ s) measured in the $^{244}Pu+^{48}Ca$ and $^{248}Cm+^{48}Ca$ reactions are shown in Fig. 3 together with those of known lighter even-even nuclei. One can see that all the decay chain members follow the Geiger-Nuttall T_α vs. Q_α relationship for even-even nuclei.

During Feb.-June, 2002, we carried out an experiment aimed at the synthesis of element 118 in the reaction $^{249}Cf+^{48}Ca$. A beam dose of 2.5×10^{19} 245-MeV ^{48}Ca ions was collected in this experiment. The two fission events with high deposited energies $E_{tot}=207$ MeV and 223 MeV were preceded by EVRs detected with preceding time intervals of 0.56 s and 3.16 ms, respectively. The respective probability of detecting random recoil signals within given intervals is 9×10^{-3} and 6×10^{-6}. Only in one event (see Fig. 4) an EVR with energy and TOF signal close to that expected for a $Z=118$ nucleus was followed by an 11.65-MeV α-particle absorbed by the focal-plane detector and a 10.71-MeV α-particle detected both by the focal-plane and side detectors, which was followed by SF. The probability of observing a random sequence of 4 signals in strip 3 (even neglecting their vertical position correlation) is 1.5×10^{-6}. A 6-mm position deviation observed for the second α-particle may be due to the low energy (1.41 MeV) deposited in the focal-plane detector. For the second event (strip 8) no signals were detected in the short EVR-SF time interval. These events were observed at the instantaneous beam energy values of 245.6 MeV ($E^*=29.8\pm2.0$ MeV) and 246.1 MeV ($E^*=30.2\pm2.3$ MeV), respectively; these correspond to the expected maximum for the $3n$-evaporation channel, resulting in production of the even-even isotope $^{294}118$. The two α-decaying nuclei in the first decay chain follow the Geiger-Nuttall relationship for even-even nuclei (see Fig. 3). Measured Q_α values together with the known data for the even-even nuclides with $Z\geq100$ and theoretical values Q_α calculated in macroscopic-microscopic nuclear model [5,6] are shown in Fig. 5. The experimental Q_α values for $^{294}118$ and its descendants agree well

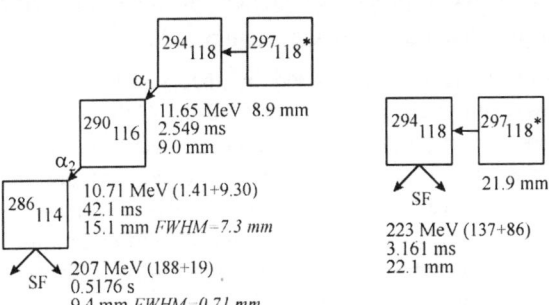

Figure 4. Time sequences in the decay chains observed in the $^{249}Cf+^{48}Ca$ reaction. FWHM position resolutions for escape α and SF with respect to EVR are shown in italics.

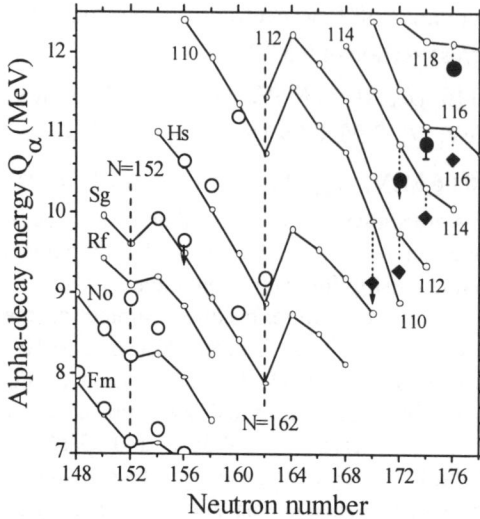

Figure 5. Alpha-decay energy vs. neutron number for even-even isotopes with $Z \geq 100$. Open circles - experimental data [10-12]. Solid diamonds and circles - values for isotopes shown in Fig. 1a, Fig. 2 and Fig. 4, respectively. Small circles connected with solid lines show theoretical Q_α values [5,6].

with the earlier data on heavier even-even isotopes with Z=114, 116 and also show a trend to deviate from the calculations to lower values. Note also that fission time of the final nuclide in the detected chain agrees with the calculated value of T_{SF}=1.5 s for [286]114 [5].

The second decay chain does not reproduce the first one. The decay time t_{SF}=3.16 ms is quite close to the first α-decay time (2.55 ms) in the previous chain. One can propose that the observed SF with rather high deposited energy E_{tot}=223 MeV (TKE~245 MeV) is related to fission of the EVR [294]118 itself produced in the reaction [249]Cf([48]Ca,3n).

4. Discussion

The lifetimes of the new isotopes, in particular [284]112 and [280]110, appear to be about 10^5 times longer than those of the known nuclei [277]112 and [270]110 [11].

The radioactive decay properties of the observed nuclei are in agreement with macroscopic-microscopic nuclear theory [5,6], which predicts both α-decay and SF properties of the heavy nuclei (see Fig. 5). A comparison of these calculations with the measured decay properties of new nuclides, including [280]110 ($T_{1/2}$=7.6$^{+5.8}_{-2.3}$ s), indicates that nuclei in the vicinity of spherical shell closures could be even more stable than predicted by theory. The α-decay energies of the heaviest new even-even nuclides are 0.35-0.5 MeV less than predicted values. Such a decrease in Q_α values leads to an increase of partial α-decay lifetimes by an order of magnitude. Calculations are far less definite regarding spontaneous fission; however, we note that the observed SF half-life of [280]110 exceeds the predicted value [5] by more than two orders of magnitude.

The α-decay properties of the synthesized nuclei agree also with self-consistent calculations [7,8]. All the above theoretical approaches predict the existence of the "island of stability" in the region of superheavy elements. The principal result of the present work is the observation of the considerable increase in lifetimes of superheavy nuclei with $Z \geq 110$ with increasing neutron number. In this respect, the decay properties of the new nuclides observed in present experiments confirm theoretical expectations and can be considered proof of the existence of enhanced stability in the region of superheavy elements.

Acknowledgments

This work has been performed with the support of INTAS under grant No. 991-1344 and of the RFBR under grants No. 01-02-16486 and 02-02-06190. Much of support was provided through a special investment of the Russian Ministry of Atomic Energy. The ^{244}Pu, ^{248}Cm and ^{249}Cf target materials were provided by the U.S. DOE through ORNL and by RIAR, Dimitrovgrad. Much of the support for the LLNL authors was provided through the U.S. DOE under Contract No. W-7405-Eng-48 with the University of California. These studies were performed in the framework of the Russian Federation/U.S. Joint Coordinating Committee for Research on Fundamental Properties of Matter.

References

1. Yu.Ts. Oganessian *et al.*, in *Proc. 4th Int. Conf. on Dynamical Aspects of Nuclear Fission*, Častá-Papiernička, Slovak Republic, 19-23 October 1998 (World Scientific, Singapore, 2000) p.334.

2. K. Subotic *et al.*, *Nucl. Instr. Meth. Phys. Res.* **A481**, 71 (2002).

3. Yu.Ts. Oganessian *et al.*, *Phys. Rev.* **C64**, 054606 (2001).

4. Yu.Ts.Oganessian *et al.*, *Phys. Rev. Lett.* **83**, 3154 (1999); *Phys. Rev.* **C62**, 041604(R) (2000); *Phys. At. Nucl.* **63**, 1679 (2000).

5. R. Smolańczuk and A. Sobiczewski, in *Proc. of XV Nuclear Physics Divisional Conference "Low Energy Nuclear Dynamics"*, St. Petersburg, Russia, 1995 (World Scientific, Singapore), p. 313; R. Smolańczuk, *Phys. Rev.* **C56**, 812 (1997); R. Smolańczuk, J. Skalski and A. Sobiczewski, *Phys. Rev.* **C52**, 1871 (1995).

6. I. Muntian, Z. Patyk and A. Sobiczewski, *Acta Phys. Pol.* **B32**, 691 (2001).

7. S. Ćwiok, W. Nazarewicz and P.H. Heenen, *Phys. Rev. Lett.* **83**, 1108 (1999).

8. M. Bender, *Phys. Rev.* **C61**, 031302 (2000); P.-G. Reinhard *et al.*, in *Proc. Tours Symposium on Nuclear Physics IV*, Tours, France, 2000 (AIP, New York, 2001), p.377.

9. Yu.Ts.Oganessian *et al.*, *Phys. Rev.* **C63**, 011301(R) (2001); *Phys. At. Nucl.* **64**, 1349 (2001).

10. *Table of Isotopes*, 8th ed., edited by R.B. Firestone and V.S. Shirley (Wiley, New York, 1996).

11. S. Hofmann *et al.*, *Eur. Phys. J.* **A10**, 5 (2001); S. Hofmann and G. Münzenberg, *Rev. Mod. Phys.* **72**, 733 (2000); G. Münzenberg G. *et al.*,

    Z. Phys. **A322**, 227 (1985); *Z. Phys.* **A324**, 489 (1986); F.P. Heβberger *et al.*, *Z. Phys.* **A359**, 415 (1997).

12. A. Türler *et al.*, *Phys. Rev.* **C57**, 1648 (1998); *GSI Scientific Report 2001*, GSI 2002-1, p. 1, GSI, 2002; Ch.E. Düllman *et al.*, *Nature* **418**, 859 (2002).

PRODUCTION OF ELEMENT 110 IN THE ^{208}Pb(^{64}Ni,n) REACTION

T. N. GINTER^a, K.E. GREGORICH^b, W. LOVELAND^c, D.M. LEE^b,
U.W. KIRBACH^b, R. SUDOWE^b, C.M. FOLDEN III^{b,d}, J.B. PATIN^{b,d},
N. SEWARD^e, P.A. WILK^b, P.M. ZIELINSKI^{b,d}, K. ALEKLETT^f,
R. EICHLER^g, H. NITSCHE^{b,d}, AND D.C. HOFFMAN^{b,d}

^{a} Michigan State University, East Lansing, MI 48824, USA
^{b} Lawrence Berkeley National Laboratory, Berkeley, CA 94720, USA
^{c} Oregon State University, Corvallis, OR 97331, USA
^{d} University of California, Berkeley, CA 94720, USA
^{e} University of Surrey, Surrey, England
^{f} Uppsala University, Uppsala, Sweden
^{g} Paul Scherrer Institute, Villigen PSI, Switzerland

In an experiment performed at the Berkeley Gas-filled Separator (BGS) of the Lawrence Berkeley National Laboratory, two chains of position- and time-correlated events were observed in the bombardment of a ^{208}Pb target with a 309 MeV ^{64}Ni beam. Each chain consisted of the implantation of an evaporation residue followed by the emission of α-particles. These chains are attributed to the decay of ^{271}110 produced with a cross-section of $8.3^{+11}_{-5.3}$ pb.

1. Introduction

The synthesis of element 110 has been reported at three laboratories: Lawrence Berkeley National Laboratory (LBNL) in the United States[1], Gesellschaft für Schwerionenforschung (GSI) in Germany[2,3,4,5,6], and the Joint Institute for Nuclear Research (JINR) in Russia[7,8,9]. Table 1 provides a summary of the isotopes observed, the production mechanisms, their observed decay modes, and references. None of these observations confirm the others since they all involved different isotopes. (The isotope ^{273}110 was reported both at JINR and GSI; although these results do not contradict each other[3], they also do not confirm each other because the α-decays observed did not have matching energies and lifetimes.) Futhermore, none of the work has been confirmed independently in an experimental setup not used in the original work; such verification is essential for establishing the credibility of the results because of the challenging nature of the experiments.

Table 1. Summary of previously observed element 110 isotopes.

Isotope	Production Mechanism	Decay Mode	Laboratory	Reference
^{267}110	^{209}Bi(^{59}Co,n)	α-decay	LBNL	1
^{269}110	^{208}Pb(^{62}Ni,n)	α-decay	GSI	2,3
^{270}110	^{207}Pb(^{64}Ni,n)	α-decay	GSI	4
^{271}110	^{208}Pb(^{64}Ni,n)	α-decay	GSI	5
^{273}110	^{277}112 α-decay	α-decay	GSI	3,6
	^{244}Pu(^{34}S,5n)	α-decay	JINR	7
^{280}110	^{284}112 α-decay	fission	JINR	8
^{281}110	^{285}112 α-decay	α-decay	JINR	9

In 1998, Hofmann, *et al.* produced ^{271}110 in the ^{208}Pb(^{64}Ni,n) reaction[5]. They observed a total of 9 events — 2 at a beam energy of 311.7 MeV, 6 at 313.0 MeV, and 1 at 315.5 MeV. The successful repetition of the synthesis of ^{271}110 using the same reaction is reported here.

2. Experimental Setup

This study was performed using the Berkeley Gas-filled Separator (BGS)[10] at the LBNL 88-Inch Cyclotron facility. The cyclotron delivered the ^{64}Ni^{14+} beam at an average current of $\sim$250 particle nA and at energies of 312.5, 315 ($\Delta E_{FWHM} = 3.9$ MeV [11]), and 317.5 MeV.

The ^{208}Pb target was located at the front of the BGS, about 5 mm downstream from the separator's 40-μg/cm^2 carbon entrance window. The target consisted of nine arc-shaped segments, each with a 500-μg/cm^2 thick lead layer sandwiched between layers of carbon with thicknesses of 40 μg/cm^2 (facing the beam) and 2 μg/cm^2. These segments were mounted around the periphery of a 14-inch diameter wheel which was rotated to minimize thermal stress on the target from beam heating. The beam energies at the center of the target were 306.7, 309.2 and 312.8 MeV [12]; the energy thickness of the target was 6 MeV for all three energies. Two silicon p-i-n detectors (mounted at $\pm$27 degrees with respect to the incident beam) monitored the product of beam intensity and target thickness by detecting beam particles that were elastically scattered from the target.

The BGS spatially separated the recoiling fusion-evaporation residues (EVR's, $E \sim 70$ MeV) in flight from both beam particles and transfer reaction products on the basis of their differing magnetic rigidities within the separator's 0.88-torr helium atmosphere. The magnetic rigidity ($B\rho$) of the ^{271}110 EVR's was estimated to be 2.1 T m [13]. The BGS magnetic field setting used for the $^{271}110$ reaction was extrapolated from the setting that centered the EVR distribution (with an estimated $B\rho$ of 1.47 T m) from the reaction of ^{64}Ni at 309 MeV on ^{120}Sn.

At the BGS focal plane the EVR's were deposited into a 300-μm thick passivated ion implanted silicon detector with an active area of 116 mm (horizontal) $\times$ 58 mm (vertical). This detector recorded the time, energy, and position of the implanted EVR's and of their subsequent α-decays. It had 32 independent vertical strips providing a 3.6-mm horizontal position accuracy. Resistive charge division provided vertical position resolution within each strip. The energy response of each strip of the detector was calibrated using the α-decay of implanted atoms; the strip detector had an average energy resolution of 70 keV for 5-9 MeV α-particles. The detector's geometrical efficiency for recording the full energy of an α-particle from the single decay of an implanted ion was 50%; this value results in an 81% efficiency for observing at least two full-energy events of a five-member α-decay chain.

A 10-cm $\times$ 10-cm parallel plate avalanche counter (PPAC) [14], placed 24 cm in front of the strip detector, recorded the time and position of recoiling ions before implantation. The presence or absence of signals from ions passing through the PPAC distinguished beam-related events in the strip detector and those from the α-decay of previously implanted ions. The average total counting rate ($E \geq 0.5$ MeV) over the entire strip detector (after applying the PPAC veto) was observed to be $\sim$1.3/s.

The primary difference in the LBNL measurement compared to the one at GSI was the use of a gas-filled separator to enhance the collection of EVR's produced in the reaction. The efficiency of the setup for transport and implantation of EVR's from the ^{64}Ni + ^{208}Pb reaction was estimated to be 70% based on a Monte Carlo simulation[15]. The assumptions in the simulation included: a Gaussian beam energy distribution; a Gaussian beam angle distribution defined by beamline limits; a linear beam energy loss in the target; and a 5-MeV (lab frame) FWHM Gaussian excitation function centered at the central thickness of the target. The simulation includes the effect of the EVR velocity spread from the evaporation of the neutron; scattering and energy loss of the EVR in the remaining target material; and charge exchange, scattering, and energy loss in the BGS helium.

3. Results

Three non-restrictive strategies were employed to search for possible chains of correlated events. The first search was for pairs of decay events with energies matching those previously observed for $^{271}110$, ^{267}Hs, ^{263}Sg, ^{259}Rf, and ^{255}No anywhere within a strip and within appropriate time windows. The second search was for recoil events which were followed by decay events with energies matching those previously observed for $^{271}110$, ^{267}Hs, or ^{263}Sg anywhere within the same strip and within appropriate time windows. The third search was for recoil events which were followed by any three decay events with energies above 0.5 MeV anywhere within the same strip and within a 0.5 second time window.

At the 315-MeV beam energy ($E_{center\ of\ target}$ = 309.2 MeV), two chains of events, correlated in position and time, were observed from the synthesis and decay of $^{271}110$. No correlated event sequences arising from $^{271}110$ were observed at the other two beam energies. A summary of the experiment is given in Table 2.

Table 2. Summary of experiment.

E_{lab} (^{64}Ni) at Center of Target (MeV)	Dose	Decay Chains Observed
306.7	2.7×10^{17}	0
309.2	2.7×10^{17}	2
312.8	1.1×10^{17}	0

Figure 1 details these decay sequences and shows them in comparison to the known decay data for $^{271}110$. In the first sequence, which occured in strip 27, all of the events clustered within a narrowly defined vertical position in the strip. In the second sequence, which took place in strip 19, the EVR event and the decay events at 9.88 MeV and 7.81 MeV also occurred within a narrowly defined vertical position. The fact that these strip-19 events took place at the other end of the strip from the one used to measure the position signals is consistent with the fact that no position data are available for the remaining decay events: the small signals from these low-energy escape events were below the threshold setting for the position ADC. No other strip 19 events occurred at times between those at 18.7, 0.90, 9.88, 0.65, and 2.13 MeV.

The final member of the decay chain, ^{255}No ($t_{1/2}$ = 3.1 m), decays either to ^{251}Fm ($t_{1/2}$ = 5.3 h) or to ^{255}Md ($t_{1/2}$ = 27 m, 92% EC decay to $t_{1/2}$ = 20.1 h ^{255}Fm); both of these branches are beyond the sensitivity of the experiment. The PPAC, which has a carbon-equivalent thickness

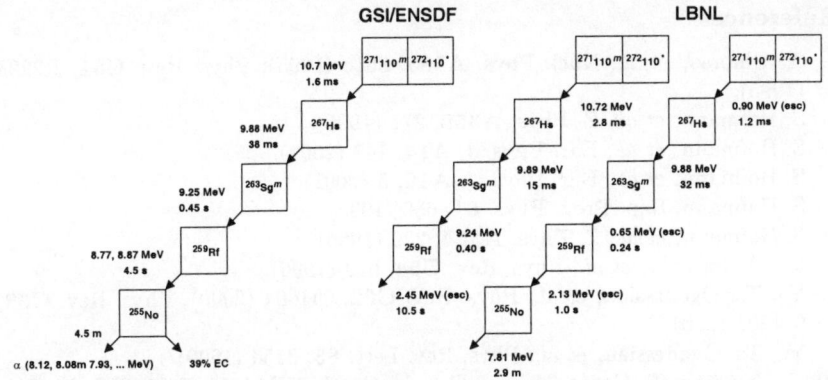

Figure 1.   Comparison of the known[5,16] decay sequence for $^{271}110$ with the two decay chains observed in this work. The notation "(esc)" marks decay events in which the α-particle presumably escaped from the beam-facing surface of the strip detector and, thus, deposited only part of its energy to be recorded.

of 0.6 mg/cm^2, lowers the energy of the EVR's from 70 MeV down to ~30 MeV; a pulse-height-defect of 50%, typical for very heavy ions[17], helps to explain the observed EVR energies of ~20 MeV.

The production cross section corresponding to the two $^{271}110$ decay sequences at 309.2 MeV is $8.3^{+11}_{-5.3}$ pb [18]. The "one event" upper limit cross section — *i.e.*, the cross section obtained by assuming one event is detected when none are present — for the bombardments at 306.7 and 312.8 MeV are 7.7 and 29 pb, respectively. The cross sections were calculated assuming their values to be constant for all beam energies throughout the target thickness. These results agree within the experimental uncertainties with the previous observations at GSI[5].

In conclusion, the production of element $^{271}110$ in the ^{208}Pb(^{64}Ni,n) reaction as reported in previous work[5] has been confirmed.

Acknowledgments

Victor Ninov was a participant in this work. The operations staff of the 88-Inch Cyclotron provided intense, steady beams of ^{64}Ni. M. Steiner, J. Yurkon, and D. J. Morrissey at Michigan State University provided the PPAC. Financial support was provided by the Office of High Energy and Nuclear Physics, Nuclear Physics Division of the U.S. Department of Energy, under contract DE-AC03-76SF00098 and grant DE-FG06-88ER40402.

750

References

1. A. Ghiorso, *et al.*, Nucl. Phys. **A583**, 861c (1995); Phys. Rev. **C51**, R2293 (1995).
2. S. Hofmann, *et al.*, Z. Phys. **A350**, 277 (1995).
3. S. Hofmann, *et al.*, Eur. Phys. J. **A14**, 147 (2002).
4. S. Hofmann, *et al.*, Eur. Phys. J. **A10**, 5 (2001).
5. S. Hofmann, Rep. Prog. Phys. **61**, 639 (1998).
6. S. Hofmann, *et al.*, Z. Phys. **A354**, 229 (1996).
7. Yu. A. Lazarev, *et al.*, Phys. Rev. **C54**, 620 (1996).
8. Yu. Ts. Oganessian, *et al.*, Phys. Rev. **C62**, 041604 (2000); Phys. Rev. **C63**, 011301 (2001).
9. Yu. Ts. Oganessian, *et al.*, Phys. Rev. Lett. **83**, 3154 (1999).
10. V. Ninov, K.E. Gregorich, and C.A. McGrath, ENAM98, B.M. Sherrill, D.J. Morrissey and C.N. Davids, ed., (AIP, Woodbury, 1999) p. 704.
11. C. M. Lyneis, private communication.
12. F. Hubert, R. Bimbot, and H. Gauvin, At. Data and Nucl. Data Tables **46**, 1 (1990).
13. A. Ghiorso, *et al.*, Nucl. Instru. Meth. **A269**, 192 (1988).
14. D. Swan, J. Yurkon, and D.J. Morrissey, Nucl. Instrum. Meth. **A348**, 314 (1994).
15. K.E. Gregorich, *et al.*, submitted to Phys. Rev. **C**.
16. S. Chu, L.P. Ekstrom, and R.B. Firestone, WWW Table of Radioactive Isotopes, database version 2.0, February 1999.
17. W. Loveland, R. Yanez, K. Aleklett, J.O. Liljenzin, and A. Ghiorso, Nuclear Chemistry Progress Report, Oregon State University 1993, p34.
18. K.-H. Schmidt, C.-C. Sahm, K. Pielenz, and H.-G. Clerc, Z. Phys. **A316**, 19 (1984).

COMPETITION BETWEEN FUSION-FISSION AND QUASIFISSION IN HEAVY-ION REACTIONS LEADING TO SUPERHEAVY ELEMENTS

L. KRUPA[5], A.A. BOGATCHEV, G. GIARDINA[3], F. HANAPPE[1], I.M. ITKIS,
M.G. ITKIS, M. JANDEL[5], J. KLIMAN[5], G.N. KNIAJEVA, N.A. KONDRATIEV,
I.V. KORZYUKOV, E.M. KOZULIN, T. MATERNA[1], YU.TS. OGANESSIAN,
I.V. POKROVSKY, V.A. PONOMARENKO, E.V. PROKHOROVA,
A.YA. RUSANOV[4], N. ROWLEY[2], L. STUTTGE[2], V.M. VOSKRESENSKI

Flerov Laboratory of Nuclear Reactions, JINR, 141980 Dubna, Russia
[1]Universite Libre de Bruxelles, Bruxelles, Belgique
[2]Institut de Recherches Subatomiques, CNRS-IN2P3, Strasbourg, France
[3]Department of Physics, University of Messina and INFN, Italy
[4]Institute of Nuclear Physics of the National Nuclear Center of Kazakhstan
480082 Almaty, Kazakhstan
[5]Institute of Physics SASc, 84228 Bratislava, Slovakia

The process of fusion-fission of heavy and superheavy nuclei (SHE) formed in the reactions with ^{48}Ca and ^{58}Fe ions at energies near and below the Coulomb barrier has been studied. The experiments were carried out at the U-400 accelerator of the Flerov Laboratory of Nuclear Reactions (JINR) using the time-of-flight spectrometer of fission fragments CORSET and the neutron multi-detector DEMON. The neutron multiplicities $<M_n>$ and the γ-ray multiplicities $<M_\gamma>$ in coincidence with fission fragments were investigated. A weak dependence of $<M_\gamma>$ as a function of the excitation energy E^* for quasi-fission in comparison with fusion-fission was found. The different dependence of $<M_\gamma>$ as a function of fission fragment mass or total kinetic energy (TKE) for fusion-fission and quasi-fission processes was observed. As well as for $<M_\gamma>$ noticeable differences have been observed in the values of neutron multiplicities $<M_n>$ for different mechanisms of superheavy compound nucleus decay (fusion-fission or quasi-fission). Thus, the fusion-fission and quasi-fission processes manifest itself in the γ-ray and neutron multiplicities as a function of mass, TKE, excitation energy of the compound system and entrance channel in significantly different ways.

1. Introduction

The interest in the study of the fission process of SHE is connected mainly with the possibility of obtaining information on the cross-section of compound nuclei production at excitation energies of $\approx$15-30 MeV (i.e. when the influence of shell effects on the fusion and the decay characteristics of the composite system is considerable). This is of basic importance concerning the synthesis of new heavy nuclides, because it helps to predict the probability of the SHE composite system to survive after evaporating 1, 2 or 3 neutrons, i.e. in "cold" or "warm" fusion reactions. However, to be able to do so, a much deeper insight into the mechanism of the fission process and a better knowledge of the fission–quasi-fission cross-section ratio as a function of the reaction entrance channel and excitation energy is necessary. As a result of the experiments, mass and energy distributions (MED) of fission fragments, fission and quasi-fission cross-sections, multiplicities of neutrons and γ-rays and their dependence on the mechanism of formation and decay of compound systems have been studied. Undoubtedly all these aspects are of great independent interest to nuclear fission physics.

This work presents the results of experiments aimed to study the fission process in the ^{48}Ca and ^{58}Fe - induced reactions. The choice of the underlined reactions was inspired by the recent experiments on producing the isotopes ^{283}112, ^{289}114 and ^{283}116 at Dubna [1,2] using the same reactions, while the ^{58}Fe projectile was chosen since the corresponding projectile-target combinations lead to the synthesis of even heavier elements.

2. ^{48}Ca – induced reactions

Figure 1 shows the data on MED of fission fragments of the ^{256}102, ^{286}112, ^{292}114 and ^{296}116 nuclei produced in the reactions with ^{48}Ca at one and the same excitation energy E*≈33 MeV. The main peculiarity of the data is the sharp transition from the predominant compound nucleus fission in the case of ^{256}102 to the quasi-fission mechanism of decay in the case of the ^{286}112 and heavier nuclei. It is very important to note that despite the dominating contribution of the quasi-fission process in the case of nuclei with Z = 112-116, in the symmetric region of fission fragment masses ($A_{CN}/2 \pm 20$) the process of fusion-fission of compound nuclei, in our opinion, prevails. It is demonstrated in the framings (see the right-hand panels of Fig. 1), where one can see that the mass distribution of fission fragments of compound nuclei is asymmetric in shape with the light fission fragment mass at about 132-134 a.m.u. The mass distribution first moment in the region of ~132 reflects the decisive role of shell effects, which is characteristic of SHE fission.

Figure 2 shows average total neutron $< M_n^{tot} >$ and average γ-ray multiplicity curves as a function of mass (left panels) and TKE (right panels) for above mentioned nuclei. The highest $<M_\gamma>$ is found for symmetric mass divisions. On the other hand in the case of asymmetric mass division, where the quasi-fission dominates $<M_\gamma>$ is much smaller. In addition for nucleus ^{256}102 local minimum is observed in $<M_\gamma>$ as a function of mass, suggesting the influence of nuclear structure of fission fragments. The minimum occurs for these cases when both fragments are near a closed shell, namely for A=128-132. As seen from the figures, the similar differences are observed in the total neutron multiplicities. The $< M_n^{tot} >$ is considerably lower for the region of fragment masses where the mechanism of quasi-fission predominates as compared with the region of fragment masses where, in our opinion, the process of fusion-fission prevails (in the symmetric region of fragment masses).

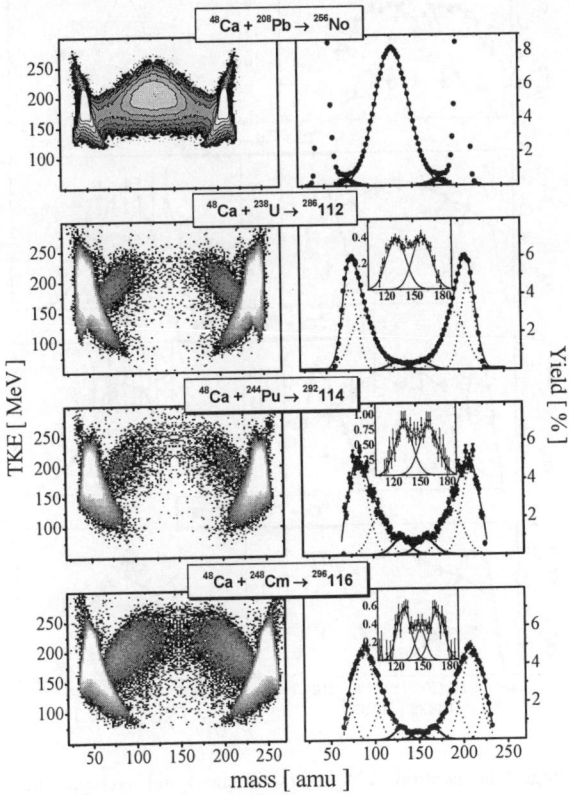

Figure 1. Double-differential cross-section $\partial^2 \Omega / \partial M \partial E_k$ [two-dimensional matrices $N(M, E_k)$] (left-hand side panels) and mass yields (right-hand side panels) of fission fragments of ^{256}No, ^{286}112, ^{292}114 and ^{296}116 nuclei produced in the reactions with ^{48}Ca at the excitation energy $E^* \approx 33$ MeV.

As regards the $<M_\gamma>$ as a function of TKE, one sees that $<M_\gamma>$ decreases with increasing TKE for symmetric mass splits ($A_{CN}/2 \pm 20$ amu), where the fusion-fission process dominates. On the other hand in the case of asymmetric mass division, where quasi-fission dominates $<M_\gamma>$ is almost constant as a function of

TKE, or a little decreases at low TKE. This trend is apparent in all reactions, where quasi-fission contributes to mass-energy distribution.

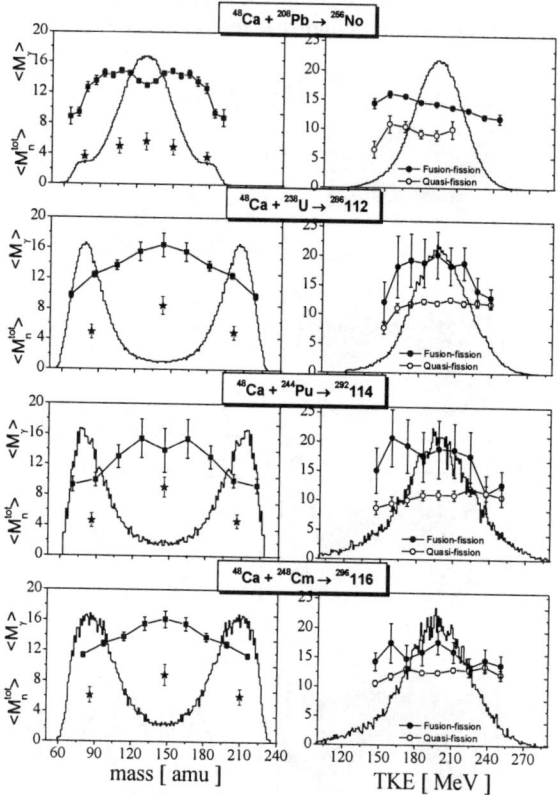

Figure 2. Average total neutron $< M_n^{tot} >$ (solid stars) and average γ-ray $<M_\gamma>$ (solid squares) multiplicities as a function of fragment mass (left-hand side panels) and as a function of TKE (right-hand side panels; the solid circles correspond to the fusion-fission region for a mass window of $A_{CN}/2\pm20$ amu and the open circles correspond to the quasi-fission region (see text)) for ^{48}Ca-induced reactions at the excitation energy $E^*\approx33$ MeV. Representative statistical bars and the mass and TKE distributions of fission fragments are also shown here.

3. ^{58}Fe – induced reactions

Figure 3 shows the data for the reactions of ^{58}Fe projectile on ^{232}Th, ^{244}Pu and ^{248}Cm targets, leading to the formation of the compound system ^{290}116 and the heaviest compound systems ^{302}120 and ^{306}122 (where N = 182-184). As seen from Figure 3, in these cases we observe an even stronger manifestation of the asymmetric mass distributions of ^{306}122 and ^{302}120 fission fragments with the light fragment mass at about 132. The corresponding structures are seen well in the TKE distribution as a function of fragment mass. Only for the reaction ^{58}Fe + ^{232}Th→ ^{290}116 (E*= 53 MeV) the valley in the region of M=A/2 disappears - this is seen from the mass distribution as well as from the average TKE dependence. This fact is connected with a reducing of the shell effects influence at so high excitation.

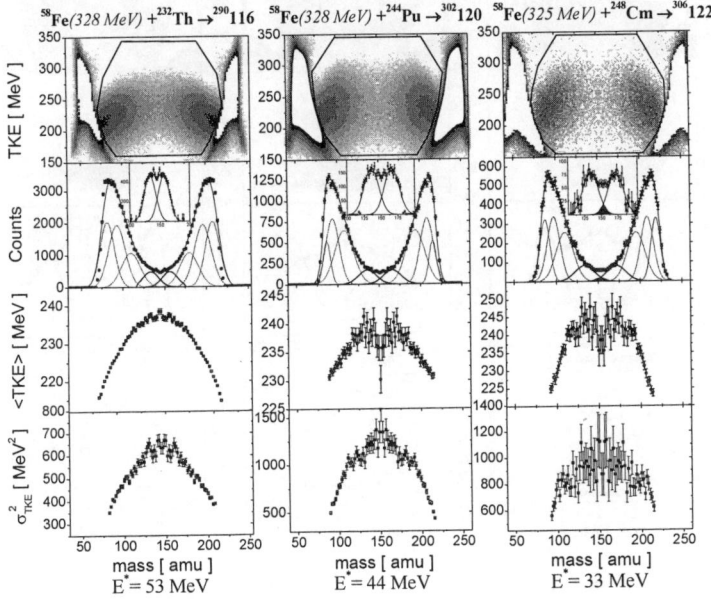

Figure 3. Two-dimensional matrices N(M, E$_k$), the mass yields, average TKE and the variances σ^2_{TKE} as a function of mass of fission fragments of ^{290}116, ^{302}120, ^{306}122, produced in the reactions with ^{58}Fe ions.

In Figure 4 the corresponding two-dimensional matrices $\langle M_\gamma \rangle (M, E_k)$ of average γ-ray multiplicity for above mentioned nuclei are shown. Only events in the region marked off by the thick polyline (upper panels) were used in the analysis of $\langle M_\gamma \rangle (M, E_k)$. In the following panels the $\langle M_\gamma \rangle$ and $< M_n^{tot} >$ as a function of fission fragment mass and TKE are shown. The behavior of $\langle M_\gamma \rangle$ and $< M_n^{tot} >$ is similar as in the case of ^{48}Ca - induced reactions. In the symmetric region of fragment masses the multiplicities are higher compared with asymmetric mass division. The $\langle M_\gamma \rangle$ curve decreases with increasing TKE for symmetric mass splits ($A_{CN}/2\pm20$ amu) and is almost constant for asymmetric ones.

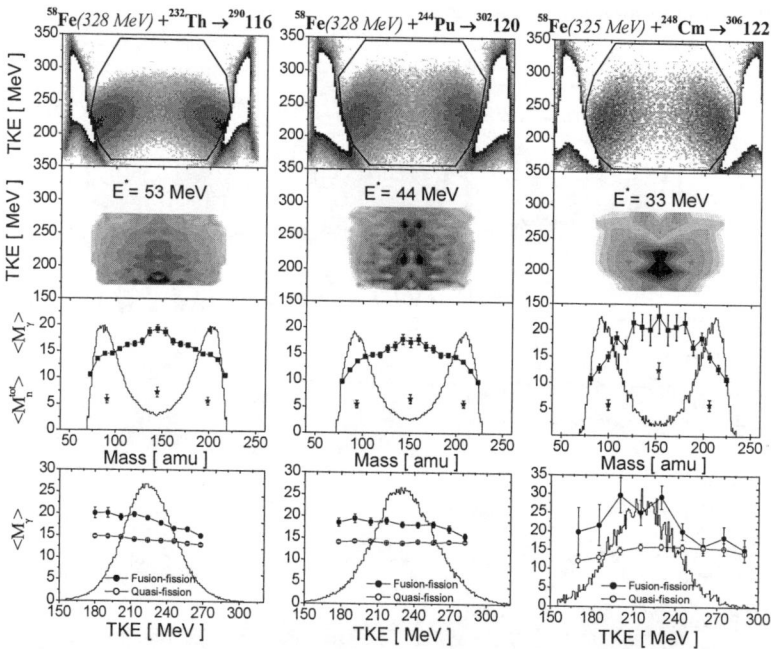

Figure 4. Two-dimensional matrices N(M, E_k), average γ-ray multiplicities as a function of fragment mass and TKE, average total neutron $< M_n^{tot} >$ (solid stars) and average γ-ray $\langle M_\gamma \rangle$ (solid squares) multiplicities as a function of fragment mass and as a function of TKE (the solid circles correspond to the fusion-fission region for a mass window of $A_{CN}/2\pm20$ amu and the open circles correspond to the quasi-fission region) produced in the reactions with ^{58}Fe ions. Representative statistical bars and the mass and TKE distributions of fission fragments are also shown here.

The differences between fusion-fission and quasi-fission processes are also manifested in $<M_\gamma>$ and $< M_n^{tot} >$ as a function of excitation energy and atomic number of compound nuclei. Figure 5 shows average γ–ray multiplicities for fusion-fission region ($A_{CN}/2\pm20$ amu) compared with the quasi-fission region as a function of the excitation energy of compound system measured in the reaction $^{58}Fe+^{208}Pb-> ^{266}108$. As one can see the $<M_\gamma>$ for fusion-fission region increases more steeply as in the case of quasi-fission region (the data from the work by Bock *et al* [3] are shown too). The similar behavior is seen for average total neutron multiplicities $< M_n^{tot} >$ as a function of atomic number (Figure 6).

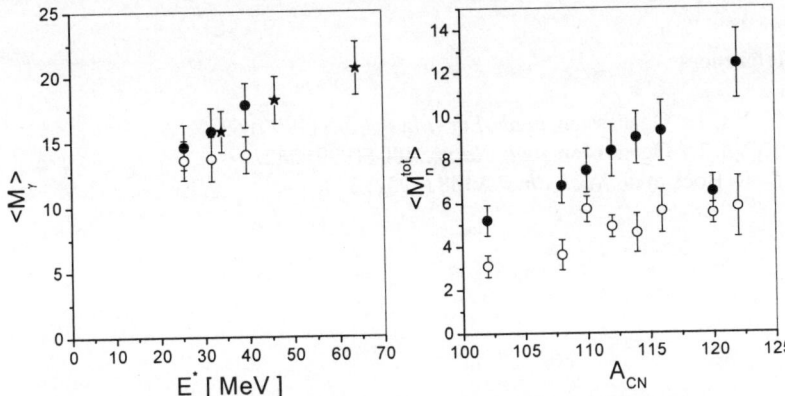

Figure 5. Average γ–ray multiplicities for fusion-fission region (solid circles) and quasi-fission region (open circles) as a function of the excitation energy in the reaction $^{58}Fe+^{208}Pb-> ^{266}108$. The data from the work by Bock *et al* [3] are shown as stars.

Figure 6. Total neutron multiplicities for fusion-fission region (solid circles) and quasi-fission region (open circles) as a function of atomic number of compound nuclei produced in the reaction with ^{48}Ca and ^{58}Fe at the excitation energy E^*=33-45 MeV.

4.   Conclusions

1.   In the case of the fusion-fission process as well as in the case of quasi-fission, the observed peculiarities of mass and energy distributions of the fragments, the ratio between the fission and quasi-fission cross sections, in dependence of the nucleon composition and other factors, are determined by the shell structure of the formed fragments.

2. It was established that TKE, neutron and γ-ray multiplicities for fusion-fission and quasi-fission of superheavy compound nuclei are significantly different. For nuclei with $Z > 100$ the TKE value is much smaller and neutron and γ-ray multiplicities are much higher (as a function of entrance channel, mass, TKE and excitation energy of compound system) in the case of fusion-fission as compared with the quasi-fission process. Thus, $<M_\gamma>$ as well as $<M_n>$ can serve as a criterion for distinguishing of above-mentioned processes.

This work was supported by the Russian Foundation for Basic Research under Grant № 99-02-17981 and by INTAS under Grant № 11929.

References

1. Yu. Ts. Oganessian *et al.*, *Eur. Phys. J*, *A***5** (1999) 63.
2. Yu. Ts. Oganessian *et al.*, *Nature* **400** (1999) 242.
3. R. Bock et al, *Nucl. Phys.* **A388** (1982) 334.

Radioactive Ion Beam Facilities

PROGRESS ON THE CONCEPTS AND DESIGN OF THE RARE ISOTOPE ACCELERATOR

D. F. GEESAMAN[†]

Physics Division.
Argonne National Laboratory
9700 S. Cass Avenue
Argonne, IL 60439
geesaman@anl.gov

The Rare Isotope Accelerator is a bold initiative for the U.S. nuclear science community that promises to revolutionize studies of the primary topics of this workshop: fission and neutron-rich nuclei. In this talk, a perspective on the role of RIA in enabling major progress in nuclear structure and nuclear reaction research and some important recent steps in technical progress for RIA are reviewed. The current status of RIA in the DOE project process and the actions needed by the scientific community are also discussed in the context of the worldwide mobilization for rare isotope beams and the goals of nuclear science for the next decade.

1. Introduction

At this conference there is no need to enter into an extensive scientific justification for the Rare Isotope Accelerator (RIA). The conference goals to understand neutron-rich nuclei and collective motions such as fission are primary examples why RIA is the ideal research tool in nuclear structure. Instead I will give a perspective from a millennium away. While hiking in Chaco canyon this summer, I came to the Supernova Pictograph (Figure 1) that is believed to be an anasazi artist's representation of the 1054 supernova in the Crab Nebula. I felt that the same sense of wonder and burning desire to understand natural phenomena that motivated this artist almost 1000 years ago shines bright in our quest to understand nuclei and supernova.

The time is right for RIA due to a confluence of events. First, experimental progress in understanding the QCD substructure of hadrons and nuclei has validated hadron-based models of the nucleus to quite short distance scales, confirming our picture of nuclear structure. At the same time advances in nuclear theory have made it clear that the solution to long-standing questions in nuclear structure lies in the many-body physics and has focused the physics discussions to issues such as: what is the isospin dependence of the nuclear

---

[†] This work is supported in part by the U. S. Department of Energy under contract No. W-31-109-ENG-38.

761

Figure 1: The Supernova Pictograph in Chaco Canyon.

three-body force[1]? What are the dependences on proton-neutron asymmetry of the mean-field spin-orbit force and of the nuclear equation of state? RIA is designed to answer these questions. At the same time, the astronomy and astrophysics communities are investing heavily in new generations of observatories and making amazing discoveries. Interpreting these results requires new insights from nuclear physics. What lets us be confident we can answer these questions is the rapid progress in accelerator and target technology and experimental technique that makes a bold leap forward possible at this time. I believe RIA will enable us to develop an overarching picture of the nature of nucleonic matter, understand the origin of the elements and test fundamental symmetries and search for physics beyond the standard model. This is why it received the highest priority for major new construction in the NSAC 2002 Long Range Plan.

In addition to vital advances in nuclear science, RIA will provide a copious bounty of isotopes for other basic research and societal applications. These major avenues include radioactive ion implantation for material science, isotope production for R&D by the biological and medical communities and stockpile stewardship applications. In my perspective as one of the organizers of the July DOE/NSF workshop on the Role of the Nuclear Physics Research Community in Combating Terrorism, the importance of the techniques and expertise of low energy nuclear physics to this high priority national need was clear. Developing the workforce and techniques of the future in this area will require the type of vigorous state-of-the-art research program that RIA makes possible.

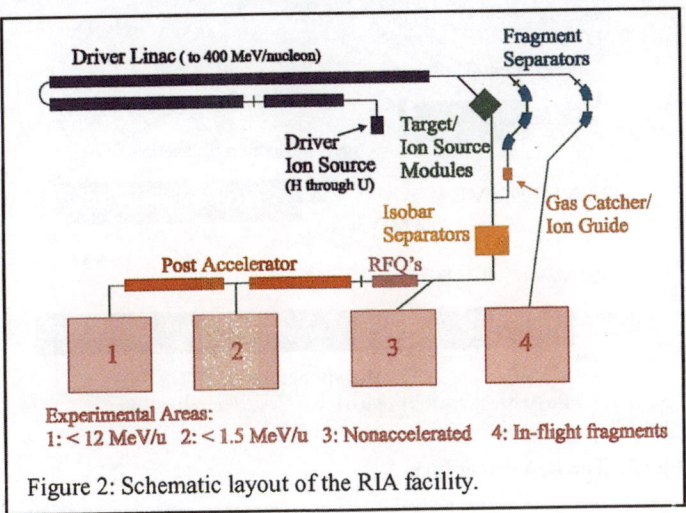

Figure 2: Schematic layout of the RIA facility.

2. The RIA Concept

The design of the Rare Isotope Accelerator evolved significantly in the late 1990's. Previously in, for example, the 1996 NSAC Long Range Plan, it was widely accepted that there were two complementary techniques to produce rare isotope beams: 1) break up a fast heavy beam on a light target and perform experiments with the fast fragments or 2) use a high power light ion beam to break up a heavy target, collect and ionize the isotopes desired and accelerate them for experiments, the classic isotope separation on-line (ISOL) technique. While planning at Argonne had included stopping fast ions for reacceleration for several years, in 1998 experiments carried out at Argonne to stop and inject radioactive ions into the Canadian Penning trap demonstrated that one could stop fast ions, independent of the specific ion chemistry, in helium gas and the resulting 1+ ions could be extracted rapidly (< 10 ms) and efficiently (up to 40%) by a combination of gas flow, DC and AC electric fields[2]. This created a new paradigm for the production of rare isotopes and made it clear that an accelerator was needed that could accelerate all stable species from protons to uranium with high (>100 kW) beam powers. We proposed that a superconducting linac provided the optimum solution to accelerate these beams[3]. A key feature of this proposal is that the large acceptance of such a linac allows multiple charge states of ions, which emerge following passage through a stripper foil, to be simultaneously accelerated to the same energy[4], resulting in an order of magnitude increase in beam power available for the heaviest beams even with current ion source technology. New concepts had to be developed for high power targets to deal with these high (100 kW) beam

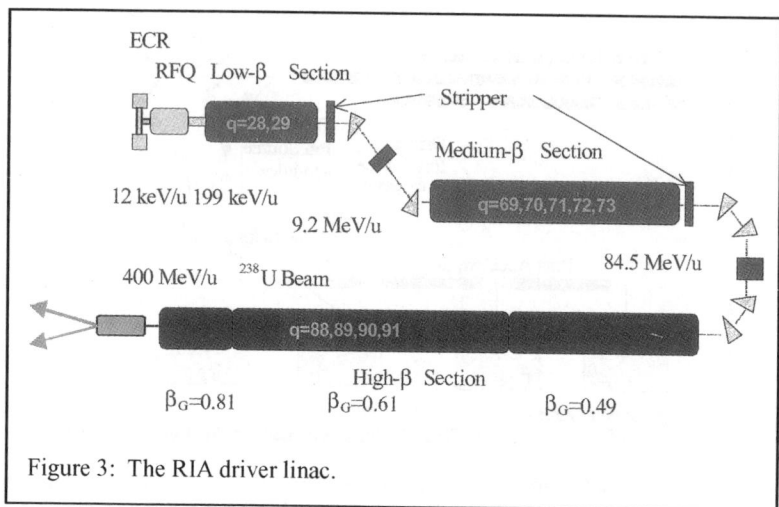

Figure 3: The RIA driver linac.

powers. Here several ideas proved useful, including liquid lithium technology adapted from the fusion reactor community and two-step targets that separate the energy deposited from the beam from the energy released in fission. The last idea has recently been incorporated at ISOLDE to reduce the backgrounds in neutron-rich radioactive beam experiments. Finally, having worked so hard to produce radioactive beams, it is critical to develop techniques to reaccelerate them with optimum efficiency. This required the development of a new generation of accelerating structures to efficiently accelerate very low energy high mass-to-charge-ratio 1+ ions.

In 1999, the NSAC ISOL Task Force adopted each of these ideas as the basis for the Rare Isotope Accelerator and concluded that the optimum performance for this facility required beams of all isotopes and energies of at least 400 MeV/u for uranium. Thus the RIA concept was defined as illustrated in the schematic layout of Figure 2. Each of the four proposed production mechanisms, fast-fragmentation, fast-in-flight-fission of uranium, two-step neutron-induced fission and standard ISOL production, is required to obtain the optimum yields[5]. For example, for high quality stopped and reaccelerated beams, ISOL and two-step fission provide orders of magnitude more beam intensity than fast beam techniques for the elements which diffuse rapidly through hot materials. In part, this is due to the expected limitation of the gas cell performance at high space charge densities. For Sn isotopes, for example, these light ion induced techniques provide the optimum reaccelerated beams from [107]Sn to [139]Sn. When available, such high yields are very important for many of the applications beyond nuclear physics that require harvesting or carefully controlled implantation of isotopes. Typically the fast, heavy-beam techniques are more efficient at the limit of proton and neutron rich isotopes, or for refractory elements that do not diffuse rapidly through materials.

Just as each of the production techniques is required to optimize the rare isotope production, each of the four experimental areas shown in Figure 2 has a special role to play in addressing the science case. Fast rare isotope beams provide the greatest reach at the limits of stability and in studies of the nuclear equation of state. The higher resolutions of reaccelerated beams offer major advantages in heavy nuclei with high densities of states and the lower velocities make them ideal for studying the interplay of collective and single particle degrees of freedom. One extremely important example in this regard is the reactions to produce the shell-stabilized superheavy elements. The range of predictions for the location of the center of the island (or peninsula) of superheavy elements is precisely a result of our lack of quantitative understanding of the interplay of collective and single particle degrees of freedom that RIA will resolve. While there are many indirect techniques for determining the reaction rates for nucleosynthesis in the rp-process and the breakout from the CNO cycle, as the current situation for the ^{7}Be(p,γ) reaction makes clear, presently there is still no substitute, when feasible, for the direct measurement of low energy reaction rates. Finally, the continued development of ion and atom traps offers great potential in the study of fundamental interactions and the high-precision determination of masses required for both astrophysics and fundamental interactions studies. Recent measurements using a Cf source, a gas-stopping cell and the Canadian Penning trap at ANL have determined the masses of 20 neutron-rich nuclei in the Ba-Pr region to typically better than 100 keV. These source measurements could provide ~100 new accurate mass measurements, but with RIA the number of isotopes which could be addressed with this technique is in the thousands. New experiments, for example a measurement of the electric dipole moment of ^{225}Ra and searches for scalar currents in the Argonne Advanced Penning Trap are also pioneering the technology for precision searches for physics beyond the standard model that RIA makes possible.

3. Technical Progress

Currently in the U.S. eight laboratories: ANL, JLab, LANL, LBNL, LLNL, NSCL/MSU, ORNL and Texas A&M are participating in the DOE-funded RIA R&D that received $2.8M in FY2002. Major progress is also being made internationally as new facilities are planned and constructed. Reference 6 provides a more extensive list of references to this work. To make sure that RIA is choosing the optimum technology, there is close cooperation on an international scale. In May 2002, the 2nd RIA Driver workshop brought experts

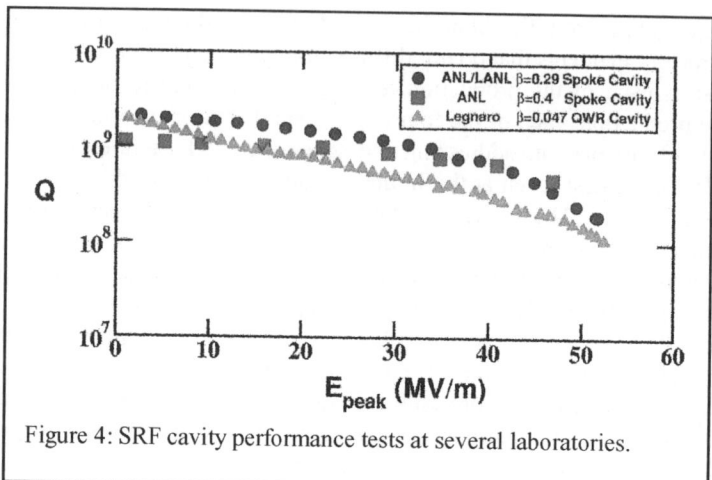

Figure 4: SRF cavity performance tests at several laboratories.

together to review the concepts for the superconducting driver linac (Figure 3). The conclusion of this workshop was that the RIA technology was sound; there are no hard failure modes. A number of areas were identified for optimization and cost minimization as the project planning progresses. I would like to review here some of the important technical developments of the past year.

The design of the accelerating cavities and the accelerator beam optics must proceed hand in hand to achieve a robust and reliable design. Considerable progress has been made in achieving higher accelerating field gradients in superconducting RF cavities at a number of labs over the past few years using electro-polishing and high-pressure rinse techniques. Examples of performance measurements of cavities at ANL and Legnaro are shown in Figure 4 surpassing the RIA design specifications. Potentially, this enhanced performance could result in significant cost savings for the project.

Decisions about how to implement this technology rest heavily on detailed calculations of the accelerator beam optics. It is especially important in such a high power machine to minimize beam loss to ensure one can properly perform accelerator maintenance. Extensive calculations show that one can maintain the high standards required by careful construction of the lattice and optimizing the phase space acceptance of the accelerator[7].

Targets that can handle the high beam power are a major challenge for RIA, both for the ISOL and fragmentation production mechanisms. Liquid lithium has the thermal properties and low Z to be ideal for this latter application. A joint ANL/MSU collaboration is constructing a hybrid Be-enclosed flowing liquid lithium target to be used as a fragmentation target at the NSCL. Figure 5 shows a prototype of a RIA windowless flowing liquid lithium jet that is operating at ANL. The next step is to establish that such cm diameter jets behave as predicted in a high power beam. At the other extreme, work is underway to establish if ~500 µgm/cm^2 thick windowless liquid lithium jets can provide a robust solution for stripping foils for the particle-µA heavy-ion beams at ~10 MeV/u.

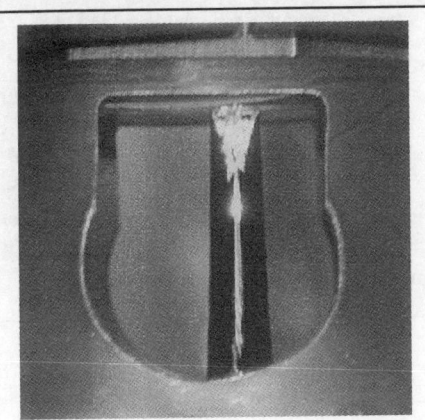

Figure 5: Windowless jet of flowing liquid lithium.

Full RIA-energy tests of the helium gas stopper cell are being performed by an international collaboration at GSI led by Guy Savard. In the first measurements, the energy loss and straggling of RIA energy beams were measured and it was demonstrated that the energy spread of the slowed beam, through the use of energy dispersion and precision graded absorbers, could be controlled to the level required. A full RIA scale 0.5 Atm-m gas cell, illustrated in Figure 6, has been constructed at Argonne where it will undergo detailed performance tests at ATLAS and then be transported to GSI for the full energy measurements. This sophisticated device contains over 7400 parts, over 4000 of which must be prepared to ultra-high vacuum specifications to maintain the clean environment required for the gas cell to operate properly.

As one final example of the progress, after working so hard to make rare isotopes, which most abundantly emerge from the sources as 1+ ions, it is important to efficiently reaccelerate these 1+ beams. This requires two new radiofrequency quadrupole accelerators. One is a CW split-coax RFQ recently developed at ANL that is capable of accelerating ions with a mass-to-charge ratio as high as 240[8]. By combining the acceleration of drift-tube cavities with the strong focusing of an RFQ, a new hybrid RFQ provides an efficient accelerating structure that is almost a factor of 2 improvement on traditional RFQs[9].

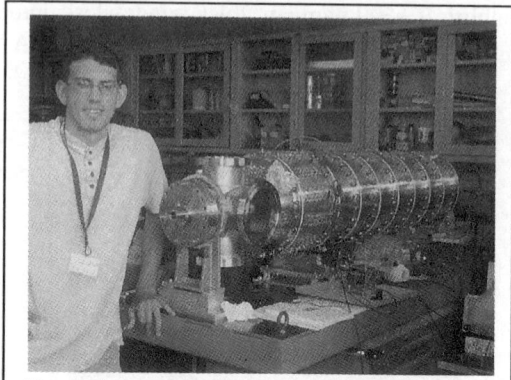

Figure 6: Prototype RIA gas stopper cell.

4. Summary and Call to Action

I have tried to summarize some of the history and the recent progress on RIA. The next critical step for RIA is for the Department of Energy to state that it wants the project by approving Critical Decision 0 (CD-0), a statement of Mission Need. What can you, the nuclear community, do to help RIA move forward? We must have your help now to make RIA a reality.

- Make it clear to your colleagues that this is tremendously exciting science.
- Make it clear that the cost of RIA is well understood and follows the estimates of the 1999 NSAC ISOL task force, only increasing with inflation.
- Join the RIA Users Group (www.orau.org/ria) and support its activities.
- Participate in the RIA experimental equipment workshops and RIA R&D activities.
- Encourage applications beyond nuclear science.

Common sense and the NSAC Long Range Plan make it clear that RIA cannot be built without increased funding for the DOE Office of Science. Illinois Congresswoman Judy Biggert is leading a resolution (HR-5270) calling for a doubling of the Office of Science budget in five years that is receiving broad based congressional support. Such political activity is essential. U.S. scientists who want RIA need to let their congressmen know that RIA is important to them.

In summary, the Rare Isotope Accelerator will open exciting new opportunities in nuclear science and a broad range of applications. RIA is ready to become a DOE project. We must continue to strive to make RIA a reality. The world-class scientific payoff will stimulate our field and many other areas of science for decades.

References

1. S. Pieper et al., *Phys. Rev. C* **64**, 014001 (2001).
2. G. Savard et al., *Proceedings of EMIS-14* to be published in *Nucl. Inst. Meth. B*, May 6-10, Victoria, BC Canada, 2002.
3. K. W. Shepard et al., "SC Driver Linac for a Rare Isotope Accelerator Facility" in the *Proceedings of the 9th International Workshop of RF Superconductivity,* Santa Fe, New Mexico, 1999 ed. B. Rusnak p. 345.
4. P. Ostroumov et al., *Phys. Rev. Lett.* **86**, 2798 (2000).
5. C.-L. Jiang et al., *Nucl. Intr. Meth. A* **492**, 57 (2002).
6. J. A. Nolen, "The U. S. Rare Isotope Acclerator Project" in the *Proceedings of the XXI International Linac Conference*, Gyeongju, Korea, 2002.
7. P. Ostroumov, "Design Features of High-Intensity Medium-Energy Superconducting Heavy-Ion LINAC" in the *Proceedings of the XXI International Linac Conference*, Gyeongju, Korea, 2002.
8. M. P. Kelly et al., *Proceedings of the 2001 Particle Accelerator Conf.*, June 18-22, Chicago, IL, 2001, p. 506.
9. P. N. Ostroumov et al., *Proceedings of the 2001 Particle Accelerator Conf.*, June 18-22, Chicago, IL, 2001, p. 4077.

RARE ISOTOPE RESEARCH CAPABILITIES AT THE NSCL TODAY AND AT RIA IN THE FUTURE[*]

C. KONRAD GELBKE

National Superconducting Cyclotron Laboratory
Michigan State University
East Lansing, MI 48824-1321, USA
E-mail: gelbke@nscl.msu.edu

The National Superconducting Cyclotron Laboratory (NSCL) at Michigan State University (MSU) has recently completed a coupled cyclotron upgrade capable of providing beams of rare isotopes with intensities higher by about 3 orders of magnitude than previously available in the U.S. Large additional gains, especially for medium-mass and heavy elements, will be possible with the Rare Isotope Accelerator (RIA), the highest priority recommendation for major new construction in the new (2002) Long Range Plan for Nuclear Science.

1. Overview

The National Superconducting Cyclotron Laboratory (NSCL) at Michigan State University (MSU) is a national user facility for rare isotope research and education, with over 250 employees (including 58 undergraduate and 45 graduate students) and serving a user community of over 500 researchers. Research at the NSCL is primarily in basic experimental and theoretical nuclear physics, nuclear astrophysics, accelerator physics, related instrumentation development, and applications to meet specific societal needs. Located on the campus of a major research university, the NSCL offers an ideal synergy of research and education and plays a major role in the training of the next generation of scientists.

The NSCL has recently completed an upgrade to a coupled cyclotron facility (CCF) capable of producing intense beams of primary heavy ions from hydrogen to uranium with maximum beam energies of 200 MeV/nucleon for lighter elements and 90 MeV/nucleon for uranium. A high-acceptance superconducting fragment separator immediately downstream of the coupled cyclotrons allows efficient production and separation-in-flight of a broad range of secondary rare isotope beams produced by projectile fragmentation or fission, opening new opportunities for basic nuclear physics and nuclear astrophysics research.

---

[*] This work is supported by the National Science Foundation under grant PHY-0110253

Insight and experience gained at MSU and other laboratories in producing radioactive beams via projectile fragmentation and in-flight separation has helped open new vistas for nuclear structure physics and nuclear astrophysics research, and NSCL faculty played a seminal role in developing the concept for the new Rare Isotope Accelerator (RIA). RIA will allow large additional gains in rare isotope intensities and is now the highest priority for new construction for the U.S. nuclear physics community.

In this talk I will briefly discuss the NSCL facility, the scientific reach of the CCF, and the additional gains made possible by RIA. Ongoing activities at the NSCL in support of RIA will be summarized, and a possible layout of the facility will be presented.

2. NSCL Facility

The floor plan of the NSCL experimental area is shown in Figure 1. The two cyclotrons, beam line and spectrometer magnets are based on superconducting technology and were designed and built in house.

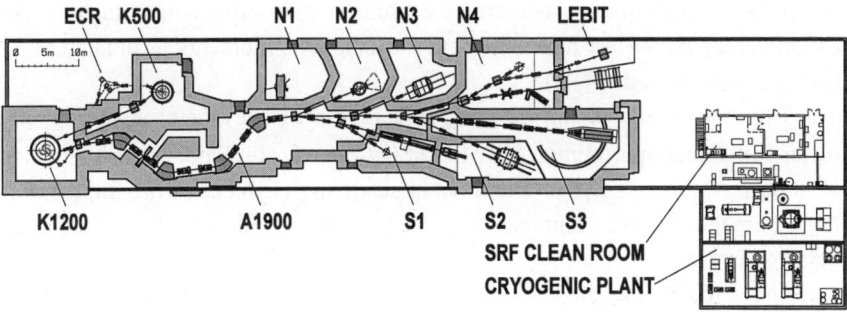

Figure 1. Floor plan of the NSCL high bay experimental area showing the two superconducting cyclotrons (K500 and K1200 on the left), the superconducting A1900 beam analysis system and subsequent beam lines, the various experimental vaults, the SRF R&D area, and the new cryoplant.

Two electron cyclotron resonance (ECR) ion sources produce the ions needed for acceleration. A beam switchyard allows injection from either ion source into either cyclotron allowing both coupled-cyclotron (CC) and K1200 stand-alone operations. For CC operations, intense beams of low-charge state ions accelerated in the K500 pass through a coupling line to the K1200 cyclotron where they are stripped and accelerated to full energy. Energetic primary beams can be used directly, or they can be converted into a broad range of nuclear species that are separated in-flight and delivered to one of the experimental vaults.

The new A1900 beam analysis system (momentum acceptance $\Delta p/p$ = 5.5%, maximum rigidity $B\rho_{max}$ = 6.0 Tm) can be used as a monochromator to define the energy and emittance of the primary beam, as a zero-degree magnetic spectrograph or, most importantly, as a fragment separator to select beams of rare isotopes produced by projectile fragmentation. Compared to the NSCL's previous fragment separator, the A1900 has about an order of magnitude higher acceptance and a 50% higher bending power.

Downstream from the A1900 is a beam switchyard that allows transportation of all radioactive ion beams to any experimental station at the NSCL.

The S1 vault contains the Reaction Product Mass Separator (RPMS) and a multi-purpose beam line primarily devoted to studies of Single-Electron-Events (SEE) in semiconductors. The RPMS achieves a mass resolution of 10^2 and a primary beam suppression factor of approximately 10^8 and used for low-background studies of nuclei far from stability, with its Wien filter providing additional purification particularly for studies with proton-rich nuclei. An experimental station for measuring nuclear magnetic moments is situated in S1 just upstream from the RPMS.

The "Superball" in the S2 vault is a high-efficiency neutron multiplicity meter containing approximately 17 m^3 of Gd-doped scintillator for the detection of neutrons in 4π geometry. The internal scattering chamber of the Superball is large enough to accommodate the Miniball/Miniwall 4π charged-particle-detector array.

The S800 in the S3 vault is a superconducting high-resolution magnetic spectrograph (energy resolution: 10^4, $B\rho_{max}$ = 4 Tm, $\Delta p/p$ = 5%, solid angle: 20 msr). The S800 beam line can be used to dispersion-match the beam to the S800 magnetic spectrograph or, alternatively, as a fragment separator ($\Delta p/p$ = 6%, $B\rho_{max}$ = 5.35 Tm, momentum resolution: 2000, solid angle: 6 msr). Both the S800 spectrograph and the S800 beam line deflect the ions vertically.

The 4π array in the N2 vault is a low-threshold "logarithmic" 4π detector consisting of successively layered shells of parallel plate multiwire detectors, segmented Bragg ionization chambers, and fast-slow plastic scintillator phoswich detectors. A number of forward arrays were built by outside user groups for experiments requiring higher resolution and/or granularity at small angles.

The N3 vault is the major general-purpose vault in the NSCL facility. It contains a large cylindrical multi-purpose scattering chamber (diameter: 231 cm, length: 271 cm). The chamber shell can be lifted from the vault to provide a large free space.

The N4 vault will be the normal location for a large-gap superconducting "sweeper" magnet (4 Tm), a high-acceptance magnetic spectrometer constructed by the High Magnetic Field Laboratory at Florida State University. The sweeper magnet can be combined with large area neutron detectors (neutron walls) for neutron time-of-flight (TOF) spectroscopy at very small angles.

The central beam line in N4 will feed the new LEBIT (Low Energy Beam and Ion Trap) facility [1]. The key element of the LEBIT facility is a high-pressure (up to 1 bar) helium gas cell for slowing down and collecting energetic rare isotopes from the A1900 fragment separator. The first experiments at the LEBIT facility will use a 9.4 T Penning trap for precision mass measurements of rare isotopes with half-lives as short as 10 ms and studying decays of free ions at rest, e.g., via low-energy conversion electron spectroscopy. The system is in the advanced stage of construction, and gas-stopping tests are underway. Experience gained with gas stopping and beam manipulation in LEBIT will provide valuable insight for the design and construction of a similar facility at the future Rare Isotope Accelerator (RIA).

In addition to the fixed major equipment described above, a number of special purpose detector arrays exist including the SEGA array (a set of 18 segmented germanium detectors) and a pair of position sensitive neutron time-of-flight walls (2m × 2m, liquid-scintillator filled). A collaboration of several universities and undergraduate colleges is currently building the more efficient modular neutron array, MONA (70% efficiency for $E_n > 50$ MeV).

3. Scientific Reach of the Coupled Cyclotron Facility (CCF)

To illustrate the scientific reach of the coupled cyclotron facility (CCF), Figure 2 depicts the predicted intensities after separation in flight with the A1900. For orientation, the approximate paths of nucleo-synthesis via the astrophysical rapid proton (rp) and rapid neutron (r) processes are indicated. With beams from the CCF it will be possible to study a large number of rp-process nuclei and r-process nuclei up to $A \approx 140$. However, investigation of the heavy r-process nuclei is beyond the reach of the CCF, and will require construction of a dedicated high-intensity facility such as RIA.

When discussing the scientific capabilities of rare isotope facilities, the beam intensity is not the only important figure of merit since other issues come

774

into play, such as back-ground suppression, efficient coincidence detection, thick target utilization, or short development times for beams of specific isotopes. In this context, it is noteworthy to point out that energetic beams of rare isotopes (E/A > 50 MeV) produced by projectile fragmentation (or fission) and in-flight separation offer a number of attractive features that allow studies of short-lived, neutron-rich nuclei even at rather low production rates, namely

- Economic production of medium-energy (E/A ≥ 50 MeV) beams of rare isotopes, without re-acceleration;
- Increased luminosity from the use of thick secondary targets (by up to a factor of 10,000);
- Reduced background from in-flight tracking and identification of individual isotopes in the beam on a particle-by-particle basis;
- Efficient particle detection from strong forward focusing;
- Short beam development times and low losses due to fast (sub-microsecond) and chemistry-independent separation and transport to the experiment.

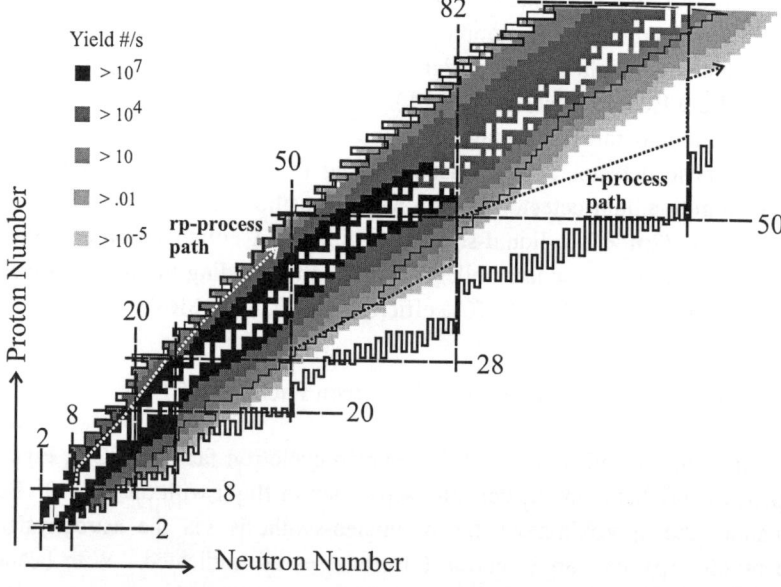

Figure 2. The scientific reach of the NSCL Coupled Cyclotron Facility is illustrated in terms of the projected intensities for fast beams of rare isotopes. Note of caution: Far from stability, the predicted intensities become increasingly uncertain due to a current lack of experimental data. (A color version of this plot can be found at the website: http://www.nscl.msu.edu/technology/ccf/reach/index.html.)

4.   RIA – A Vision for the Future

In developing plans for a next generation rare isotope research facility in the U.S., NSCL faculty proposed to replace the existing plans for a conventional ISOL facility and build instead a high-power heavy-ion facility capable of accelerating all elements up to energies per nucleon of at least 400 MeV. A high-acceptance fragment separator similar to the NSCL's A1900 would be used for in-flight separation of rare isotopes produced by projectile fragmentation or fission. After separation from the primary beam, the fragments could either be used directly for experiments or they could be stopped in a medium suitable for fast and efficient extraction and further manipulation, such as trapping or re-acceleration. After proving the viability of this approach with its inherent advantages, MSU faculty and colleagues from Argonne National Laboratory and several other laboratories in the U.S. refined these ideas into the Rare Isotope Accelerator (RIA) concept [2,3,4], schematically illustrated in Figure 3. RIA has been enthusiastically embraced by the U.S. nuclear science community [5] and is now the highest priority for new construction in the new (2002) long range plan for nuclear science of the DOE/NSF Nuclear Science Advisory Committee (NSAC).

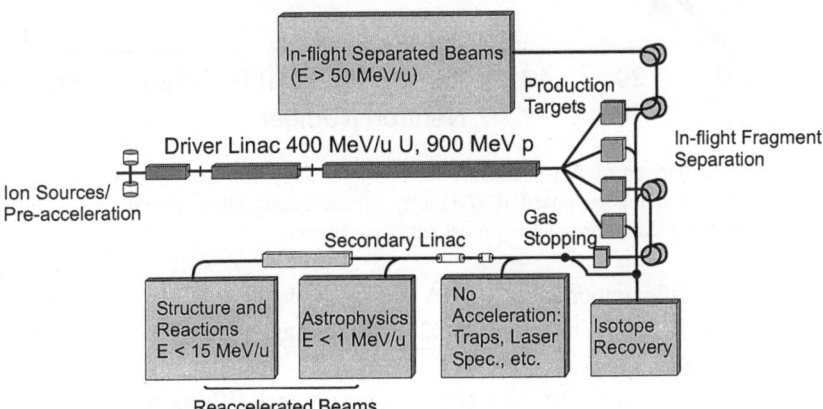

Figure 3. Simplified schematic of the RIA facility. Rare isotopes can be produced at rest via target fragmentation (or fission) or in flight via projectile fragmentation (or fission). Fast fragments, separated in flight, can be used directly for experiments, or they can be stopped in a gas cell from where they can be extracted for experiments at rest or for re-acceleration. Re-acceleration is also available for isotopes produced at rest. At least two experiments can be performed simultaneously.

776

Figure 4 gives a comparison of rare isotope intensities produced by projectile fragmentation or fission at RIA and the NSCL. For lighter masses RIA provides gains of one or two orders of magnitudes, but for heavier masses gains of four orders of magnitude and more can be realized. Nearly all r-process isotopes can be studied with fast beams at RIA, whereas research at the NSCL will be limited to the lighter r-process elements. This illustrates the paramount importance of RIA's fast beams capability and the ability to accelerate very heavy beams, including uranium.

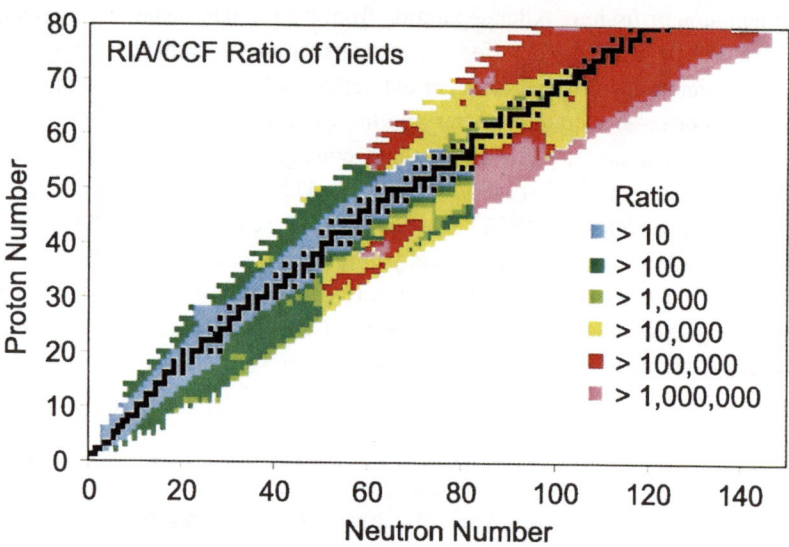

Figure 4. Comparison of fast beam rare isotope intensities at MSU's coupled cyclotron facility (CCF) and the proposed RIA facility. Fast beams allow achieving the large efficiency gains needed for reaching most r-process nuclei.

The scientific importance of RIA has been documented in a number of white papers [2,3,5-7], including the 2002 long-range plan for nuclear science [8]. In short, RIA is:

- Essential for maintaining cutting-edge research in basic nuclear science
- Essential for understanding how the elements that exist on earth were formed
- Essential for interpreting many of the high-quality astronomical observations made available from earth- and space-based observatories
- Essential for developing and testing theoretical models that can reliably predict rare isotope behavior, e.g., during stellar explosions

In addition, RIA will contribute to reliable nuclear weapons stockpile stewardship, and produce new radioisotopes for medical and materials sciences research and other applications.

RIA is the logical extension of the NSCL's ongoing rare isotope research program. Naturally, MSU strongly supports the construction of RIA and is offering to host RIA on the MSU campus. The MSU site is sufficiently large to allow a fully optimized layout that is unconstrained by existing legacy buildings and has ample space for future upgrades. A possible layout is shown in Figure 5. The example shown depicts a stretched driver linac with chicanes at the stripping stations. (Of course, a folded driver linac option is also possible if this should prove advantageous.) The layout provides for beam sharing between two experiments, but the site is sufficiently spacious to allow the incorporation of additional beam splits in the early design (at some additional cost) if this if this should be desirable.

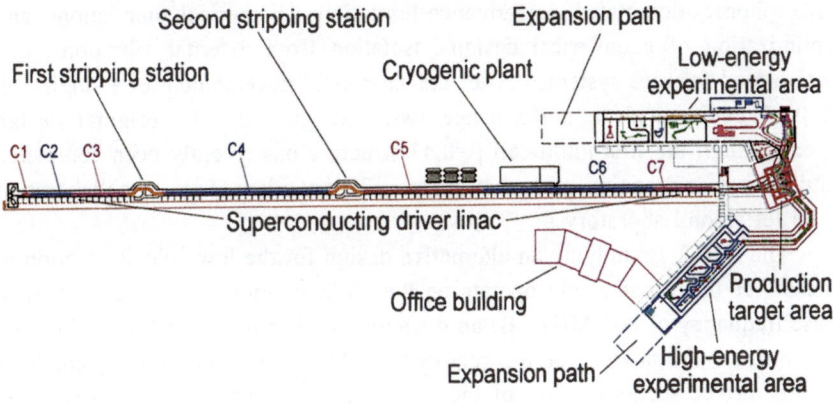

Figure 5. Optional layout of the RIA facility at MSU. Key elements of the facility are indicated in the figure. The site at MSU is sufficiently large to allow construction of a stretched driver linac (approximate length: 0.5 km), but a folded driver linac is also possible. The MSU site offers ample space for upgrades.

5. Accelerator Physics and Beam Dynamics

MSU is one of a few academic institutions with graduate programs in accelerator physics and beam dynamics. Traditionally, the hardware-related part of the accelerator physics program at MSU has been centered on cyclotron accelerators, beam line magnets, and magnetic spectrographs. An important new initiative in support of RIA is the development of superconducting radio-frequency (SRF) structures for use in linear accelerators.

Existing low-velocity (β = v/c $\approx$ 0.02 - 0.5) structures typically have lower performance and lower efficiency than state-of-the art high-velocity ($\beta \approx 1$) elliptical structures. One goal of the NSCL's SRF R&D program is to develop higher performance low-velocity structures by making use of existing know-how for building high-performance $\beta = 1$ structures. This strategy will include the use of high-purity niobium and maintenance of an ultra-clean environment during cavity fabrication and operation to reduce performance limits from field emission and thermal breakdown.

Most advanced high-velocity cavities ($\beta \approx 0.5$ - 1) consist of elliptically shaped multi-cell structures. When these cells are designed to accommodate lower velocity values, the cell-to-cell distance must be decreased and the walls of the cells become more planar, which makes the structures more susceptible to microphonics. A several strategies are being pursued to overcome or reduce microphonics-dominated performance-limitations, numerical simulations and optimization of geometrical designs, isolation from external vibrations, and active feed-forward systems. After the successful development of a single-cell β=0.47 cavity with a performance twice as good as the original design specification, the first multi-cell β=0.47 structure has recently been completed and is now undergoing extensive testing. (This work is done in collaboration with Jefferson Laboratory.)

The NSCL is studying an alternative design for the low velocity section of the driver linac that would operate on the 10-th harmonic of the agreed upon base frequency of 805 MHz. Beam dynamics studies performed so far indicate no intrinsic limitations for the required 80 MHz RFQ and subsequent low-energy driver sections. Use of the 10-th harmonic option explored at MSU would require only four different cavity types (instead of the six proposed by 14-th harmonic design pursued by ANL). Because of the higher frequency, these cavities would be smaller in size and promise improved microphonic characteristics. Mechanical fabrication of the first half-wave (spoke) resonator is complete. Chemical processing and testing of this prototype structure will be done during the first part of 2003. The design of the quarter wave resonators is in the advanced stages and fabrication of the first prototype is underway. (This work is done in collaboration with the University of Legnaro, Italy).

Currently, progress on development work for RIA is limited by insufficient federal funding and by the fact that the U.S. Department of Energy has not yet made the "Critical Decision Zero" (CD-0) that will be needed to allow work on the conceptual design report to proceed. As many excellent talks at this conference have demonstrated, several facilities around the world offer exciting opportunities for research with beams of rare isotopes and the rapid pace of progress and innovation in this field is truly impressive. The U.S. community is ready to work on the next step, RIA. It will likely take a decade before the science program at RIA will be underway, but in order to make that happen without further delays, the CD-0 milestone must be passed soon.

6. Acknowledgements

I would like to acknowledge the many excellent contributions of my colleagues, the faculty, students, and staff at the NSCL. Without their dedication and work there would be little to talk about. Operation of the NSCL as a national user facility is supported by the National Science Foundation under Grant No. PHY 01-10253.

References

1. S. Schwarz, M. Baird, G. Bollen, D. Lawton, P. Lofy, D. Morrissey, J. Ottarson, R. Ringle, P. Schury, T. Sun, *The LEBIT project at NSCL/MSU*, Proceedings 'Exotic Nuclei and Atomic Masses', July 2-7, 2001, Hämeenlinna, Finland, in print
2. ISOL Task Force Report, November 22, 1999, a copy can be downloaded from: http://srfsrv.jlab.org/isol/ISOLTaskForceReport.pdf
3. Scientific Opportunities with Fast Fragmentation Beams from RIA, March 2000, a copy can be downloaded from: http://www.nscl.msu.edu/future/ria/process/whitepapers/opportunitiesffbeam.pdf
4. See also the contribution by D. Geesaman to this conference.
5. OPPORTUNITIES IN NUCLEAR ASTROPHYSICS, Conclusions of a Town Meeting held at the University of Notre Dame, 7-8 June 1999, a copy can be downloaded from: http://www.nscl.msu.edu/future/ria/process/whitepapers/nuclearastrophysics.pdf
6. RIA Physics White Paper, summary of the town meeting held at the RIA 2000 Workshop in Durham, North Carolina, July 24-26, 2000, a copy can be downloaded from: http://www.nscl.msu.edu/future/ria/process/whitepapers/durham2000meeting.pdf
7. RIA Applications Workshop held at Los Alamos National Laboratory, October 30-31, 2000, a copy can be downloaded from: http://www.star.bnl.gov/STAR/nsac/papers/ApplicationWorkshopy.pdf
8. OPPORTUNITIES IN NUCLEAR SCIENCE -- A Long-Range Plan for the Next Decade, prepared by the DOE/NSF advisory committee, April 2002, a copy can be obtained from the U.S. Department of Energy, and can also be downloaded from:
http://www.nscl.msu.edu/future/ria/process/whitepapers/lrp_5547_final.pdf

THE GSI FUTURE FACILITY: BEAMS OF IONS AND ANTIPROTONS

WALTER F. HENNING

GSI.
Planckstr. 1,
64291 Darmstadt, Germany
E-mail: W.Henning@gsi.de

GSI, in close collaboration with the international science community, has developed the concept for a major new facility with intense primary beams of ions and, derived from that, intense secondary beams of short lived nuclei and of antiprotons. The scientific goals are the in-depth study and the comprehensive understanding of the structure of matter in the regime of the strong force.

1. Brief description of the science motivation

The research goals underlying the proposed facility span a fairly broad range. Within the context of this conference talk there is only time for a short summary description of the overall science, with the rest of the talk devoted to the description of the planned facility and to the specific research programs planned with beams of short-lived nuclei ('radioactive' beams).

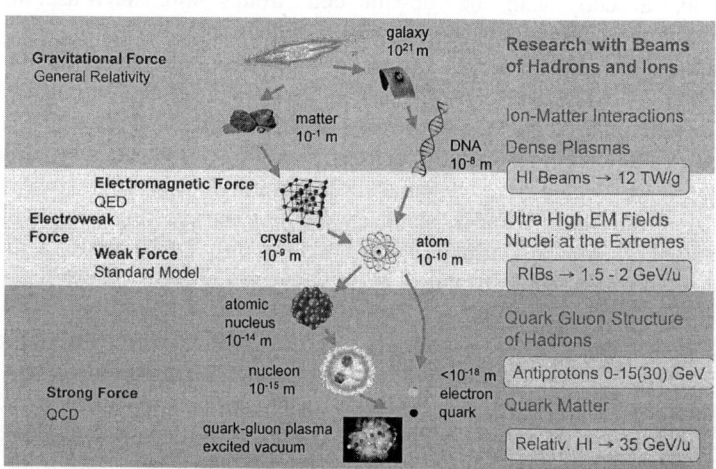

Figure 1. Schematic of the hierarchical structure of matter and the areas of research that will be addressed with beams from the proposed new GSI facility.

Figures 1 and 2 show cartoon-like overviews of the science areas. They include: i) the study of sub-nuclear degrees of freedom and of key aspects of the theory of the strong interaction, quantum chromo dynamics (QCD); this involves both, studies of hadron structure and the origin of the nuclear force as well as the quark-gluon structure of extended nuclear matter; ii) the exploration of the structure of nuclei, the nuclear many-body system, far from stability; and iii) ion-bulk matter interactions, including the physics of dense plasmas. The science case is worked out in detail in the Conceptual Design Report (CDR) for the facility [1]. Here we only mention one additional aspect: each level of the hierarchical structure of matter is, to a certain degree, directly connected to a stage in the evolution of the universe (Figure 2).

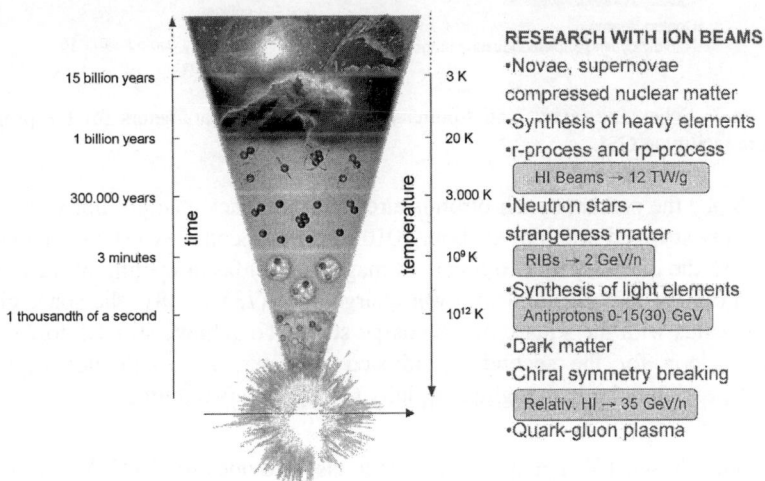

Figure 2. Schematic illustration of the potential interconnections between the evolutionary phases of the universe and the research areas to be pursued at the GSI facility with beams of ions and antiprotons.

2. Facility layout

The tools for this research program are intense primary and secondary ion beams, including beams of antimatter, i.e. antiprotons. These beams are generated by the proposed new facility in Figure 3. The existing Unilac-SIS18 system acts as injector. The core structure of the new facility is the SIS100/200 double-ring synchrotron system. Its main purpose is to provide a major step in primary (and thus secondary) ion beam intensities and, for certain uses, also in beam energy.

782

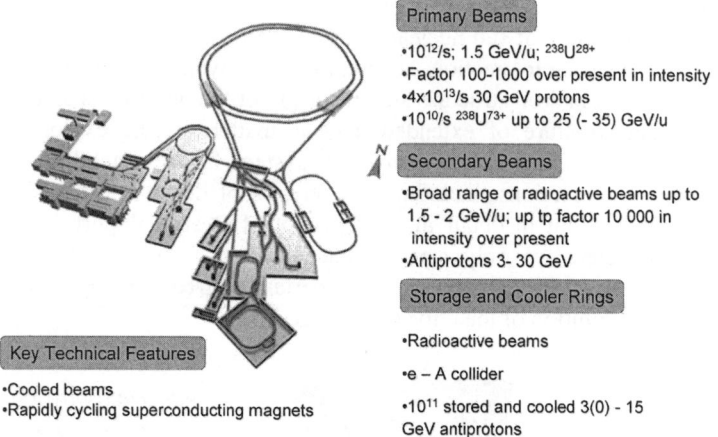

Primary Beams
- 10^{12}/s; 1.5 GeV/u; $^{238}U^{28+}$
- Factor 100-1000 over present in intensity
- 4×10^{13}/s 30 GeV protons
- 10^{10}/s $^{238}U^{73+}$ up to 25 (- 35) GeV/u

Secondary Beams
- Broad range of radioactive beams up to 1.5 - 2 GeV/u; up tp factor 10 000 in intensity over present
- Antiprotons 3- 30 GeV

Storage and Cooler Rings
- Radioactive beams
- e – A collider
- 10^{11} stored and cooled 3(0) - 15 GeV antiprotons

Key Technical Features
- Cooled beams
- Rapidly cycling superconducting magnets

Figure 3: Schematic layout and summary of performance parameters for the proposed future facility at GSI.

Since the present synchrotron is already at its space charge limit with beams of highly stripped uranium at about 1010 ions per second, two steps are made to increase the intensity by two orders of magnitude: i) faster cycling of the injector (from 0.3 Hz to 3 Hz) and ii) lower charge state (73+ to 28+; the space charge limit scales with the square of the charge state). To achieve still 1.5 to 2.0 GeV per nucleon for the secondary radioactive beams, an accelerator ring with correspondingly higher magnetic rigidity (~ 100 Tm) is required.

The chosen 100 Tm synchrotron then also provides for 30 GeV protons, the optimum energy for anti-proton production. The second ring - conservatively labeled SIS200 although a magnetic field level of six tesla, respectively 300 Tm bending power, is ultimately aimed for - serves, on the one hand, the important function of a stretcher and of slow, high duty-cycle extraction. Most importantly though it will, for high charge-state heavy ions, provide energies up to 25(35) GeV per nucleon at somewhat lower intensities, which however is appropriate for nucleus-nucleus collisions.

The key characteristics of the facility are summarized in Figure 3. For more details, we refer to the CDR [1].

3. Experimental facilities for beams of short-lived nuclei

Within the context of this conference, it seems appropriate to provide some details on the proposed research with beams of short-lived ('radioactive') nuclei. Figure 4 shows the layout, indicating as the central instrument a 2-stage fragment separator (the Super-FRS, where 'super' refers to the use of super-conducting magnets), plus three areas and systems for experimentation: for slowed (stopped), fast (relativistic), and stored and cooled beams. Figures 5 and 6 provide some details. Again, more information is found in the CDR [1].

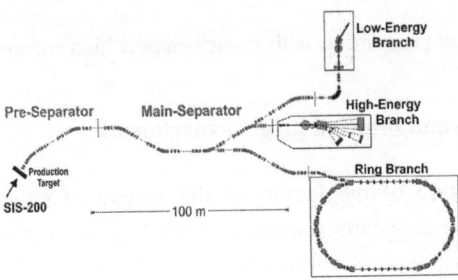

Figure 4: The Super-FRS at the new GSI facility: a large-acceptance high-resolution spectrometer for exotic nuclei, feeding three experimental areas.

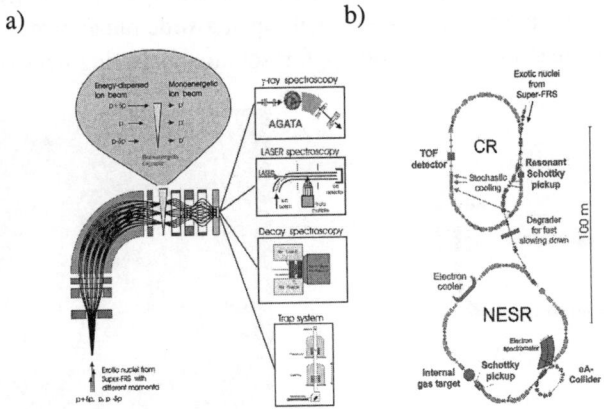

Figure 5: a) Schematic layout of precision experiments with exotic nuclei proposed for the low energy area at the new GSI facility. b) Layout of the ring section for experiments with stored and cooled exotic nuclei.

784

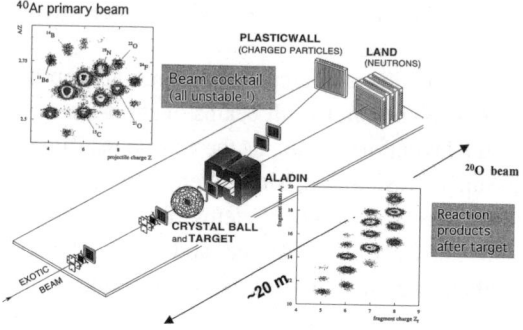

Figure 6: Schematic of experiments with exotic beams at high energies.

4. Beam cooling and internal target experiments

A second key feature of the facility is the system of collector, storage, and experimentation rings. Beam storage and cooling for high-energy heavy ion beams has evolved at the present facility as a technology with novel applications and research opportunities. The basic scheme of electron-beam cooling of ion beams, and the present cooler setup at the GSI-ESR applications are sketched in Figure 7. This technology is going to play an important role at the future facility, for radioactive ion beams as well as for high-energy anti-protons. The ion rings discussed in the previous chapter, and in particular also the high-energy storage and cooler ring for antiprotons (HESR), will open a wide range of new studies, in particular also reaction studies with high resolution and using internal targets [2].

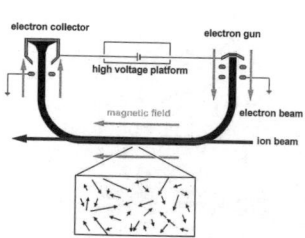 

Figure 7: Schematic of electron-beam cooling of ion beams (left) and the electron-beam cooler system at the present ESR at GSI (right).

5. Multiple use and synergies

The intrinsic nature of the ring systems allows for an important feature of the facility: a highly parallel use, that allows several programs to be carried out simultaneously. The beam manipulations in the rings require time spans, during which other beams can be accelerated or optimized and used in the other rings etc. This is a complex and intricate procedure that would require a broad discussion for which there is no room in the present context. Figure 8 illustrates with an example this important property. The CDR [1] provides for a more detailed analysis.

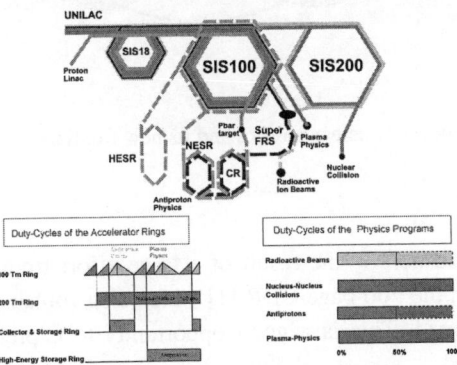

Figure 8: Schematic of the truly parallel operation that can be obtained for 3 to 4 independent physics programs at the future GSI facility.

The technical synergy which, by the way, has also important consequences for cost savings, is paralleled by a scientific synergy between the programs. For example, understanding QCD and the origin and structure of the nuclear force will allow a firmer underpinning of the relevant effective theories for the nuclear many-body system. Learning about confinement and chiral symmetry breaking in hadrons will hopefully help understand the confinement transition between hadronic and quark-gluon matter. The close linkage to the astrophysical settings and the evolution of the universe was already touched upon in the introduction. Figure 9 attempts to summarize these and other aspects and, at the same time, provides an outlook for the research areas that can hopefully be pursued with the realized facility.

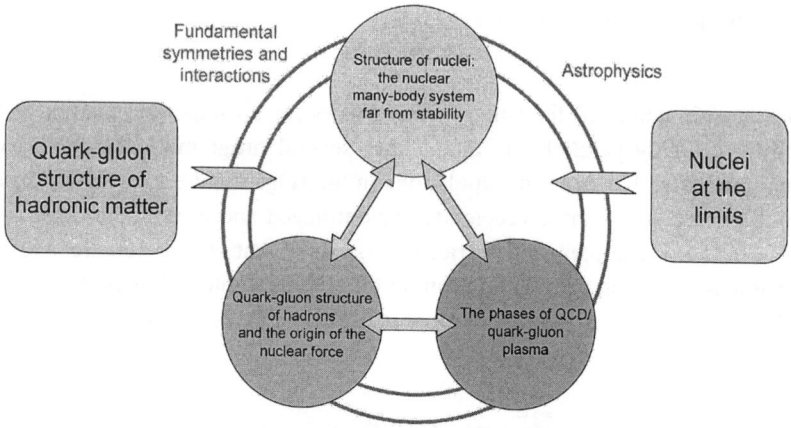

Figure 9: Summary view of nuclear physics and its new frontiers

Acknowledgments

The work described here is the result of a large effort from many individuals. This is reflected in the 700 page CDR [1] which also contains a list of names of those that contributed. This is a good opportunity to express GSI's thanks to everybody involved.

References

1. An International Accelerator Facility for Beams of Ions and Antiprotons; Conceptual Design Report, November 2001; http://www.gsi.de/GSI; Future/cdr/; this report is also available on CD. Mail to press@gsi.de to request a copy.
2. For a recent status report about the field of ion-beam cooling and storage rings, see for example: Proceedings of STORI02, International Conference on Nuclear Physics at Storage RingsUppsala, Sweden, June 16 - 20, 2002

THE SPIRAL II PROJECT AT GANIL

MAREK LEWITOWICZ

GANIL, BP 55027, 14076 Caen CEDEX France

Radioactive beams produced in-flight and as post-accelerated ISOL beams proved to be complementary with respect to their intensity, isotopic purity, available energy and optical quality. Both techniques currently in use at the GANIL/SPIRAL facility offer unique possibilities to perform experiments with light radioactive beams (A<80) in the energy range from 30 keV/nucl. to 80 MeV/nucl..

In order to extend the range of available ISOL radioactive beams at GANIL the SPIRAL II facility was proposed recently [1,2,3]. The facility, based on the high-intensity light and heavy-ion linear accelerator as a driver, should deliver high-intensity and high quality beams of neutron-rich fission fragments. To achieve the aim of at least 10^{13} fissions per second a 40 MeV deuteron beam with an intensity of up to 5 mA will be stopped in a graphite converter producing a high flux of fast neutrons, which in turn, will irradiate a low or high-density $^{238}UC_x$ target. The fission fragments extracted from the target and ionized in the $1^+/n^+$ charge breeding system will be accelerated in the existing CIME cyclotron in the energy range from about 1.7 up to about 6 MeV/nucl.. Simultaneously, the neutron-rich exotic nuclei might be available for studies at low energies right behind the high-resolution mass separator. The full accelerator complex of GANIL with SPIRAL and SPIRAL II should allow for a parallel operation delivering up to 5 different radioactive or stable beams.

The capability of the linac driver (LINAG Phase I) to deliver high-intensity (up to 1 pmA) heavy ions with a maximum energy of 14.5 MeV/nucl. (for the mass over charge ratio M/Q=3) opens also interesting possibilities in the production of exotic nuclei and radioactive beams in the fusion-evaporation reactions.

In the talk, recent results of experiments performed at GANIL on properties of the very neutron-rich nuclei will illustrate the physics case and the experimental opportunities of the SPIRAL II project coupled with the high efficiency spectrometers like EXOGAM and VAMOS.

Long range plans of GANIL, which include an extension of the LINAG accelerator up to energy of 100 MeV/nucl. for heavy ions as well as its use as a post-accelerator in the EURISOL project, will be presented.

788

References

1. SPIRAL Phase II, European RTT, Final Report, Report GANIL R 01 03, 2001; www.ganil.fr/research/sp/reports/
2. SPIRAL II: Preliminary Design Study, Report GANIL R 01 04, 2001
3. LINAG Phase I, A technical Report, GANIL, June 2002; www.ganil.fr/research/sp/reports/

THE HOLIFIELD RADIOACTIVE ION BEAM FACILITY, PRESENT AND FUTURE

JAMES R. BEENE

Physics, Division, Oak Ridge National Laboratory,[‡]
Oak Ridge, Tennessee, USA
E-mail:beenejr@ornl.gov

The Holifield Radioactive Ion Beam Facility was developed to take advantage of an existing accelerator complex, and to provide the U.S. with an ISOL (isotope separation on-line) radioactive ion beam facility. In spite of the very modest initial investment, the HRIBF is now producing a steady stream of important scientific results. This paper briefly reviews the operation of the facility, outlines some recent developments, and discusses how these have been applied to our scientific program. Finally, a brief discussion is provided of our vision of the development of the HIRBF during the roughly ten-year period that separates us from the availability of RIA.

1. Introduction

In the first four talks in this session you have heard about the ambitious and exciting plans for new radioactive ion beam (RIB) facilities in the U.S. [1,2], France [3], and Germany [4]. Early in the conference we heard about developments and plans at the ISAC facility at TRIUMF in Canada [5]. This paper concerns Holifield Radioactive Ion Beam Facility (HRIBF) at Oak Ridge National Laboratory, a facility much more modest in scale than the ones you have just heard about, but one which has nevertheless made impressive progress and produced results at the forefront of nuclear science and RIB production technology.

2. An Overview of HRIBF RIB Production Systems

HRIBF was developed in the mid-1990s at a very low initial cost (~$3M) to take advantage of an existing accelerator complex at ORNL and to provide the U.S. with an ISOL (isotope separator on line) radioactive beam capability. The HRIBF is not only making important contributions to ISOL RIB production, but also producing important scientific results with internationally acknowledged

---

[‡] Managed by UT-Battelle, LLC, for the U. S. Department of Energy under contract DE-AC05-00OR22725.

790

impact. It is the only facility in the U.S. dedicated to the production of high-quality post-accelerated beams of radioactive species for nuclear structure and nuclear astrophysics research, and it is the only facility in the world capable of delivering beams of medium mass neutron-rich nuclei post-accelerated to energies above the Coulomb barrier.

The basic layout (see Fig. 1) and operation of the HRIBF has been described in several publications (Refs 6 and 7 are recent examples), so only a brief discussion will be given here. The RIB production aspect of the facility consists of three major component systems, a production or driver accelerator, a post-accelerator, and the RIB injector system that is comprised of the ISOL production and beam preparation systems. Radioactive species are produced in hot targets, generally a few g/cm^2 thick, mounted in a target/ion source on the RIB injector platform. The beam used to produce the radioactive nuclei of interest is (at least at present) a beam of an isotope of hydrogen or helium provided by the K=100 Oak Ridge Isochronous Cyclotron. The RIB injector platform is operated up to -300 kV with respect to ground. The target/ion source assembly mounted on the RIB injector platform is operated up to ± 60 kV from platform ground. Radioactive atoms produced in the target are transported thermally by diffusion and effusion to the ionization region where they are ionized and formed into a beam. The target system is generally operated at high temperature (up to 2000°C) to speed up this transport. The

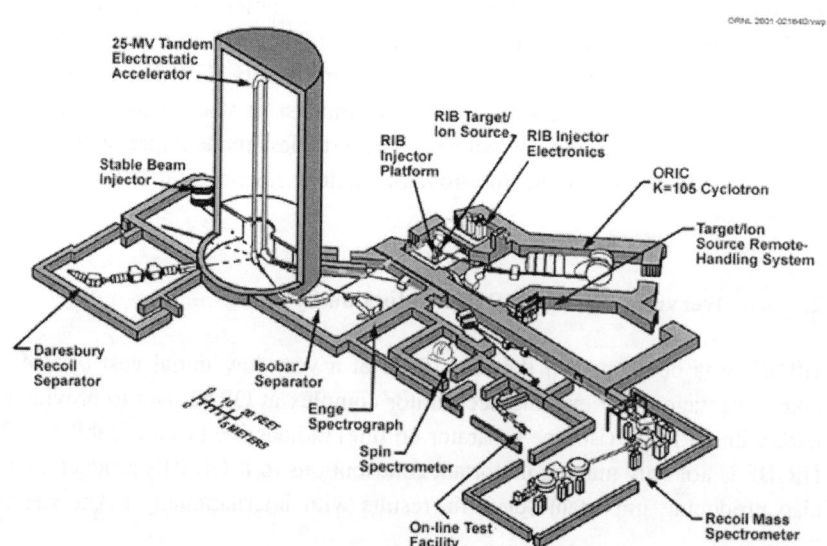

Figure 1. A schematic representation of the HRIBF

structure and format of the target as well as its temperature can dramatically influence the speed with which species of interest are transported for ionization [8]. Several varieties of ion sources are employed at HRIBF. Some are modeled closely on sources developed at other facilities but unique, new-concept sources have been developed at HRIBF to solve particular problems. An example is the kinetic-ejection negative-ion source (KENIS) [6]. The primary ion beam, which may be either positive or negative depending upon the ion source used, is passed through a first stage mass separation (m/Δm~1000) on the RIB injector platform to select the mass number of interest. Since our post-accelerator requires negative ions, positive ions are converted to negative ions by charge exchange in an alkali vapor cell. The negative-ion beam is accelerated off the RIB injector platform, mass analyzed again in a separator with resolving power m/Δm~20000 to suppress isobaric contamination, injected into the 25-MV tandem post-accelerator, and finally delivered to an experimental endstation. The tandem has been adapted to accelerate beams at terminal potentials ranging from 1 MV to over 25 MV. The resulting accelerated beam is of high quality, with an energy spread $\Delta E/E < 10^{-4}$ and a low transverse emittance. Beams with mass up to ~80 can be accelerated to energies of 5 MeV/nucleon with a transmission efficiency of ~10%. Heavier beams (up to A~150) can be accelerated to 5 MeV/nucleon at the cost of a factor of ~5 in intensity (transmission ~2%), since an extra stripping is required during acceleration.

More than 130 radioactive species with accelerated beam intensities $>10^3$ pps are now available from HRIBF. About eleven are proton-rich species, and the rest neutron-rich. All neutron-rich species are produced with proton induced fission of ^{238}U [6,9,10] as described in section 4.2. In contrast, only a few proton-rich species can be produced from a single production target system, so that each proton-rich beam carries a substantial overhead of development effort.

For many experiments RIB purity is even more important than total beam intensity. In many cases, we cannot take full advantage of our high-resolution final stage mass separation to provide isobarically pure beams. This is either because of mass differences smaller that the separator can deal with, or because the process of producing negative ions results in a sufficient loss of beam quality to make magnetic separation impractical. We have managed, however, to develop ad hoc beam purification techniques for particular species to make critical experiments possible.

3. Experimental Systems

Experiments with radioactive ion beams must meet many challenges, including low-beam intensities and high background rates. The careful design and optimization of experiments and experimental tools is critical to success. The HRIBF experimental equipment is centered on two major endstations. The nuclear structure endstation is based on the Recoil Mass Spectrometer and includes a set of highly optimized arrays for γ-ray, charged-particle, and neutron detection [11,12]. The nuclear astrophysics endstation is based on the Daresbury Recoil Separator (DRS), which has been adapted and optimized for study of proton and alpha capture reactions in inverse kinematics [13, 6]. A gas target and a number of detector arrays have been developed specifically for use at the DRS. In addition to these major endstations, the facility has two general purpose beam lines, one of which is equipped with a large scattering chamber, and an Enge spectrograph that has been adapted for use in gas-filled mode. An effort is underway to re-commission the Spin Spectrometer, a 4π NaI array [14]. The old UNISOR isotope separator has been adapted to serve as the on-line test facility (OLTF) [10] for developing and testing ion sources and production targets using low-intensity hydrogen and helium beams from the 25-MV Tandem. In addition to the OLTF, there are two off-line high-voltage platforms on which initial development and testing of new ion source concepts and measurements of ion source performance are carried out [6].

4. Beam Development and Experimental Results

The HRIBF has developed a suite of more than 130 beams that can be produced with intensities in excess of 10^3 pps. Thirty-eight of these have actually been used in experiments, including isotopes of F, As, Cu, Ge, Sr, Ag, Sn, and Te. Experiments have ranged from careful and precise measurements on resonances of astrophysical interest to measurements of masses in the A~80 region. Our heavy neutron-rich beams are unique worldwide and must be exploited, but excellent and exciting physics continues to be done at HRIBF with proton-rich beams.

4.1. *Proton-rich Beams*

The comparatively low energy of the ORIC production beams (p and d up to ~50 MeV, ^{4}He up to 100 MeV, and ^{3}He up to ~130 MeV) means that proton-rich radioactive species are produced at HRIBF primarily through fusion-evaporation reactions, such as (p,xn) and (α,xn). In general, a new target system must be developed for each new species. In the early stages of HIRBF development, beams of radioactive fluorine isotopes were chosen by our community of astrophysics users as the highest priority. This difficult problem was solved by development of a novel fiber target system and the KENIS [6,9,10]. These developments enabled us to deliver beams of pure [17]F and ^{18}F with intensities up to 10^7 pps to a variety of experiments. A series of high-impact experiments in astrophysics [15-20] and nuclear reactions [21,22] have been done using these beams. Current proton-rich beam development efforts are centered on [33,34]Cl, and [25]Al.

4.2. *Neutron Rich Beams*

In 1999, HRIBF received permission to use [238]U as a target for RIB production. This opened the possibility of producing neutron-rich radioactive species by proton induced fission of uranium. An intense effort was undertaken to develop high-release-efficiency uranium targets [6,8,10]. The targets now in use consist of thin (~10 μm) layers of uranium carbide deposited on an open-structured matrix of reticulated vitreous carbon fiber. Efforts to improve these targets continue and new target formats are being developed and tried, but the performance of current targets is very good.

The uranium target system has the great advantage that a single target can be used to produce several hundred radioactive species. However, it is almost always the case that several species are produced for each mass number, and that the species produced with the greatest intensity at a given mass are not the most physically interesting. Consider the case of A=132, which includes the particularly interesting double closed-shell nucleus [132]Sn. The relative yields of the A=132 species [132]Te, [132]Sb, and [132]Sn are ~85%, ~12%, and ~1% respectively. The mass differences between the isobars are too small to allow adequate purification with the Isobar Separator, especially after charge exchange. Fortunately, a chemical technique was developed [10] which has enabled us to make almost pure beams of radioactive Sn. Sulfur, in the form of H_2S is added to the target volume, and the desired beam extracted from the ion source as a molecular XS^+ ion. To obtain a [132]Sn beam, we extract and analyze [132]SnS^+ (mass 164). Almost no detectable yield of [132]TeS^+ or [132]SbS^+ is observed. After mass analysis, the molecular beam is passed through a Cs charge-exchange cell where the transformation [132]$SnS^+ \rightarrow$ [132]Sn^- is accomplished. The resulting beam is found to be >95% [132]Sn. This sulfide ion technique is found to result in significant enrichment of Ge beams as well. Figure 2 shows beams of A=132, 133, and 134 ions after sulfide ion purification, with Z-identification provided by a Bragg detector. As can be seen, the A=132 beam, which was initially ~1% Sn is now essentially pure Sn, while the A=134 beam in which the Sn component was much less that one part in one thousand prior to sulfide purification has been enriched to ~30% [134]Sn. The intensity of purified single-stripped Sn beam that has been achieved to date is 1×10^8 pps, 9×10^5 pps, 3×10^3 pps, and 3 pps for A= 130, 132, 134 and 136. The single-stripped Te beam intensity at A= 132, 134 and 136 is 3×10^7 pps, 2×10^6 pps and 3×10^5 pps. Other techniques are available at HRIBF for neutron-rich beam purification including the use of surface ionization sources. By applying these techniques, we can provide highly enriched beams of Ge, Sn, Rb, Cs, Br, and I isotopes.

A number of pioneering experiments with neutron-rich beams have been completed at HRIBF. The first systematic studies of Coulomb excitation with neutron-rich RIBs were carried out ^{132,134,136}Te and ^{126,128}Sn using the CLARION array of segmented clover Ge detectors and the HyBall array of charged particles detectors [23]. Interesting deviations from normal stable-nucleus systematics are expected as we push our studies to very neutron-rich species, however, it was surprising that deviations results turned up in ^{136}Te, only two neutrons beyond the N=82 closed shell. A smaller than expected B(E2) value was found for ^{136}Te–substantially smaller than that of the two-neutron hole counterpart ^{132}Te, which was also measured for the first time. This result has recently been explained [24] as a result of reduced neutron pairing strength beyond the N=82 closed shell.

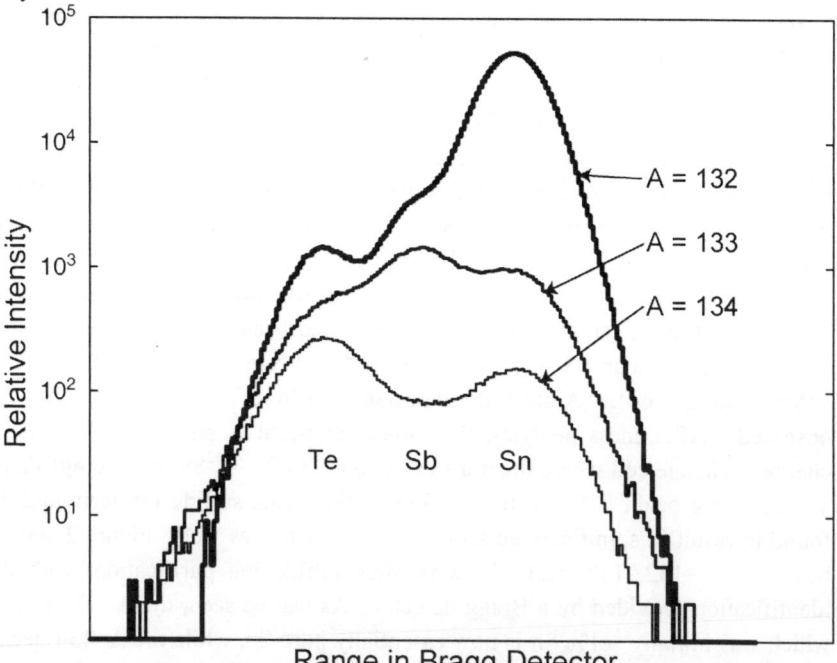

Figure 2. Bragg detector beam assays at A=132, 133 and 134. The mass 132 beam is essentially pure tin, while the 134 beam is about 25% Sn.

The ability to purify beams of Sn has enabled us to carry out a number of experiments that would otherwise have been untenable. These have included a measurement of the excitation function for fusion of ^{132}Sn+^{64}Ni and Coulomb excitation of the lowest 2^+ state in ^{128,130}Sn using the CLARION-HyBall setup [23] and for ^{132,134}Sn and ^{82}Ge using a setup optimized for low beam intensities, small cross sections and high gamma-ray energies.

The excitation function for evaporation residue in ^{132}Sn+^{64}N was measured [25] at six energies ranging from ~90% to ~115% of the fusion barrier. The measurements were made by tagging all beam particles with transmission timing detectors [26], and detecting and identifying evaporation residues in an ionization counter at zero degrees. Fusion cross sections were measured that are surprisingly strongly enhanced at sub-barrier energies compared to earlier measurements in ^{124}Sn+^{64}Ni.

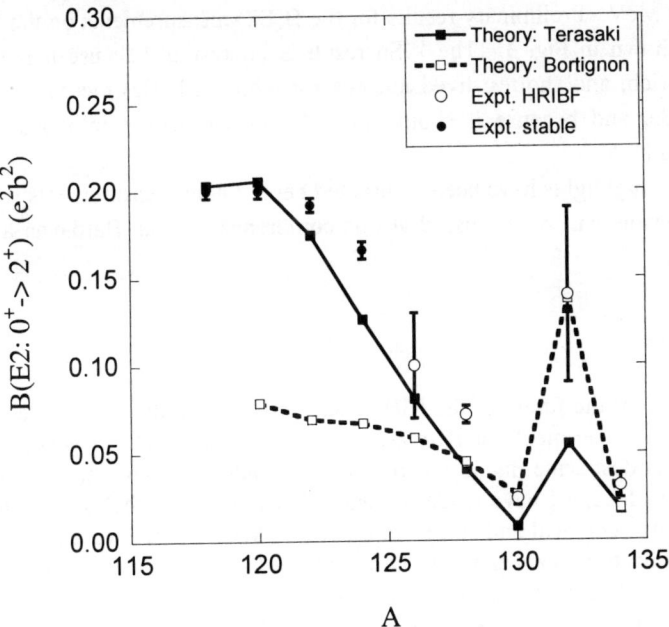

Figure 3. Experimental and theoretical B(E2) for the first 2^+ state in even Sn isotopes. The open data points were measured with radioactive beams at HRIBF. The data for ^{132,134}Sn are very preliminary.

A measurement of the B(E2) of the first 2^+ state in ^{132}Sn is an important piece of information relevant to the evolution of shell structure with increasing neutron number, and the structure of the N=82 Z=50 double-closed shell. Unfortunately, the state lies at 4.04 MeV, and consequently, has a very small excitation cross section. An experiment was designed to attack this problem. Gamma-rays were detected in the 150 crystal ORNL-NSCL-TAMU BaF$_2$ array with a full energy efficiency of almost 40% at 4 MeV. Scatter beam and target particles were detected in an annular double-sided Si detector with 48 radial and 16 azimuthal segments. A micro-channel-plate based beam tagging detector [26] and a Bragg detector were used to monitor beam. Measurements were carried out at ^{132}Sn beam energies of 470 MeV and 495 MeV, incident on a ^{48}Ti target. The double-stripped ^{132}Sn beam intensity averaged about 10^5 pps, and amounted to about >95% of the total A=132 beam. The ^{134}Sn Coulex measurement employed a ^{90}Zr target, and a beam of about 3000 ^{134}Sn pps at an energy of 400 MeV. Preliminary results for the B(E2) measurements on the Sn isotopes are shown in Fig. 3. The ^{134}Sn result is interesting because it is the most neutron-rich, and shortest-lived nucleus for which a B(E2) measurement has been made, and because it should provide further insight into neutron pairing beyond N=82.

Only a few highlights have been mentioned here. Other measurements with neutron-rich beams will be discussed at this conference by Dan Bardayan and Cyrus Baktash.

5. Future Plans

The main thrust for the future of the RIB science program in the U.S. is clearly RIA. However, experiments at RIA are at least a decade away. While the HRIBF is a modest-scale facility compared to some that you have heard discussed today, I hope I have made it clear that we are nevertheless carrying out an impressive technical and scientific program. We believe it essential that a vigorous radioactive beam program be maintained in the U.S. between now and the time RIA comes on line. If this is not done, an appropriately experienced community of users will not be available to use RIA. HRIBF can play a significant role as a test bed for ISOL techniques and a training ground for students as well as senior scientists. In order to fulfill this mission adequately, we must invest in improvement of HRIBF. Clearly a major investment of funds is not justified with RIA on the horizon, but modest investments can pay major dividends in the capability of the facility. Consequently, we have developed and presented to the Department of Energy, a modest, phased program of improvements which will enable us to fulfill our mission as a bridge to RIA. The upgrade program consists of development of a second radioactive beam production platform, and the purchase of a small commercial accelerator to serve as a supplement to ORIC for production of radioactive species. The second production platform will not only provide us with redundancy and hence improved reliability in this critical area, but it will also give us the opportunity to test and implement new and advanced target systems, ion sources and beam preparation techniques that are essentially impossible on our existing single platform. Likewise, provision of a supplementary driver accelerator, along with a companion program to add new capability to ORIC, can substantially improve both the reliability and performance of the HIRBF.

Acknowledgements

The author acknowledges that this paper reports on the work of the entire scientific and technical staff of the HRIBF. Their efforts are greatly appreciated.

References

1. K. Gelbke, these proceedings.
2. D. Geesaman, these proceedings.
3. M. Lewitowicz, these proceedings.
4. W. Henning, these proceedings.
5. A. Schotter, these proceedings.
6. G. D. Alton, et al., Indian J. Phys 76S, 9 (2002).
7. G. D. Alton and J. R. Beene, Journal of Physics G 24, 1347 (1998).
8. G. D. Alton, J. R. Beene and Y. Liu, NIM A 434, 190 (1999).
9. R. F. Welton, Nucl. Phys. A 701, 452c (2002).
10. D. W. Stracener, NIM A in press (2003), and Proceedings of the 16th International Conference on Application of Accelerators in Research and Industry, AIP vol. 576, pp257-260, American Institute of Physics, Woodbury NY, 2000.
11. C. J. Gross et al., NIM B 682, 363c (2000).
12. K. Rykaczewski et al., Nucl. Phys A 710, 179c (2002).
13. M. S. Smith, Proceedings of the 2nd Oak Ridge Symposium on Nuclear Astrophysics, Oak Ridge TN, 1997.
14. M. Jaaskelainen et al., NIM 204, 385 (1983).
15. D. W. Bardayan et al., Phys. Rev. Lett. 89, 262501 (2002).
16. D. W. Bardayan et al., Phys. Rev. C63, 065802 (2001).
17. J. C. Blackmon et al., Nucl. Phys. A688, 142c (2001).
18. D. W. Bardayan et al., Phys. Rev. C62, 055804 (2000).

798

19. D. W. Bardayan et al., Phys. Rev. C62, 042802 (2000).
20. D. W. Bardayan et al., Phys. Rev. 83, 45 (1999).
21. J. F. Liang et al., Phys. Rev. C65, 051603 (2002).
22. Gomez del Campo et al., Phys. Rev. Lett. 86, 43 (2001).
23. D. Radford et al., PRL 88, 222501 (2002) .
24. J. Terasaki., et al., Phys. Rev. C 054313 (2002).
25. F. Liang et al., submitted to Phys. Rev. Lett. March, (2003).
26. D. Shapira, T. A. Lewis and L. D. Hulet, NIM A 454, 409 (2000).

ISAC AT TRIUMF: STATUS AND FUTURE

A. C. SHOTTER

TRIUMF
4004 Wesbrook Mall,
Vancouver, BC, V6T 2A3, Canada

The ISAC-I Radioactive Beam Facility at TRIUMF is now successfully operating. The facility uses the ISOL method based on high-energy spallation reactions to produce the radioactive ions. Radioactive ions over a wide mass range can be used for a variety of studies, and in addition, ions below A=30 can be accelerated up to 1.5 MeV/A. ISAC-II, an extension of ISAC-I, is currently under construction and will accelerate all ions up to A =150 to 6.5 MeV/A. The facility will be described as well as aspects of the experimental program, with special emphasis on the current nuclear astrophysics program.

1. Introduction

The TRIUMF laboratory is Canada's national laboratory for particle and nuclear physics. The laboratory has facilities and interests than span various areas of subatomic physics. In the last few years, the laboratory has been developing a purpose-built facility (ISAC, Isotope Separation and Acceleration) for the production of intense radioactive ion beams (R.I.B.). The principle of production of R.I.B.s is based on the ISOL method, and uses the spallation reaction initiated by 500 MeV protons on various target materials. The R.I.B.s so produced are used for a variety of research problems; this paper will concern the use of these beams with special emphasis on nuclear astrophysics investigations.

2. The ISAC Facility

The TRIUMF laboratory facilities are based on a suite of five cyclotrons. These cyclotrons are used to service a range of activities ranging from pure particle and nuclear physics research to medical applications. The ISAC facility uses one of the beams from the main 500 MeV cyclotron. This cyclotron accelerates H$^-$ ions, so can simultaneously provide several beams of different intensities and energies to a variety of target stations. The ISAC facility is based on the ISOL method and uses a beam of up to 100 micro-amps, 500 MeV protons from this cyclotron. The beam is transported into a purpose-built target area where it is directed onto specially constructed targets. The resulting spallation reaction produces a variety of radioactive isotopes. The trick then is to extract these isotopes from the target, ionize them, mass separate them, select the appropriate isotope and then deliver this isotope beam to the experimenter. Since this is an online system, isotopes can be delivered to the experimenter with lifetimes as low as tens of milliseconds.

Due to the high radioactivity produced in the target, handling of the targets and its associated ancillary equipment has to be done remotely. The need to do this in a highly shielded area is one of the main costs associated with the ISAC facility, as indeed it will be for any high-powered ISOL facility. To increase the flexibility of the facility there are two target stations, one in use, and the other in waiting or in maintenance mode.

The ISAC facility has been operating up to the present time using a surface ion source. However an ECR source has now been installed, and there is rapid progress towards the development of a laser ion source. This will ensure a wide variety of unstable isotopes of different elements can be produced as pure isotope beams.

The most important factor in any ISOL facility is the composition and construction and mode of operation of the isotope production target. The ISAC target is designed to take up to 100 micro-amps of 500 MeV proton beam. The target material may come as a powder, pellets or compressed composite discs. The target is 1.8 cm in diameter and can be up to 19 cm long. Control of the target temperature is very important for efficient release of spallation produced isotopes. Generally for low intensity proton beams, the target must be externally heated, while for higher intensity proton beams, the target must be cooled. Different target materials are used depending on the particular isotope that needs to be produced. In this way a whole variety of R.I.B. isotopes ranging in mass from 8 to 160 have been produced for a range of experiments with an intensity over the range of 10^3 to 10^{11} particles per second. More details can be found on the TRIUMF web site: www.triumf.ca.

The present ISAC facility, designated ISAC-I, can produce a wide range of isotopes in the keV/A range. A post-accelerator system consisting of RFQ and DTL accelerators can now accelerate $A < 30$ ions to 1.5 MeV/A. To increase the range of ions that can be accelerated and to increase the acceleration energy, a new post-accelerator is under construction. The new accelerator complex, designated ISAC-II, will be capable of acceleration of R.I.B.s ions up to $A = 160$ and to an energy of 6.5 MeV/A. The type of post accelerator will be a superconducting linear accelerator. In addition to the accelerator, a charge state booster is being developed so that higher ion masses can be accelerated through the initial RFQ accelerator that has a restriction of $A/q < 30$. The current plans are to have first beams accelerated through this new accelerator in 2005.

One of the main problems with any ISOL facility is that generally only one isotope beam can be used at any time. Also, since the beam intensity is low, experiments can take a long time to complete. Expensive apparatus not currently being used with a particular R.I.B. is therefore remaining idle; methods should always be sought to increase this efficiency. For this reason, investigations are being undertaken at TRIUMF to increase the number of simultaneous R.I.B.s that can be used at any time. In this respect, the ISAC driver, the H⁻ cyclotron, can simultaneously produce two high intensity proton beams; each beam can then irradiate separate targets, therefore, producing two separate radioactive beams.

3. The ISAC Experimental Program

The current experimental program for ISAC-I covers a range of scientific fields that are linked to specific experimental apparatus. For radioactive beams in the KeV range, experimental facilities include the 8π γ-ray spectrometer used for structure studies of radioactive ions, LTNO apparatus to measure magnetic properties of isotopes and material science investigations, β-NMR apparatus for implantation of polarized (up to 70%) for material surface investigations, together with a γ-neutron decay facility and a γ-β decay facility. In addition work on an ion trap facility for mass measurements of radioactive ions is well under way. For current ISAC accelerated radioactive beams up to 1.5 MeV/A, the experimental program mainly relates to nuclear astrophysics problems. This is dealt with in the next section.

4. Experimental Nuclear Astrophysics

4.1 General comments

Main sequence stars have lifetimes of several millions to several billions of years depending on their mass. The energy emitted from stars during their lifetime mainly originates from nuclear reactions within the stellar interior. These reactions mostly involve stable nuclei. For example, reactions such as $^{14}N + p \rightarrow {}^{15}O + \gamma$, $3\alpha \rightarrow {}^{12}C + \gamma$, $^{12}C + {}^{12}C \rightarrow {}^{24}Mg + \gamma$, $\alpha + {}^{12}C \rightarrow {}^{16}O + \gamma$, etc are all important and involve reactions between stable nuclei.

Heavy stars come to the end of their life when material in the core of the star can no longer generate nuclear energy. This occurs when the material that was initially hydrogen and helium is transformed to iron. So a critical point is reached where the energy is no longer available to resist gravitational collapse. The rapid collapse of the core to nuclear densities, and the resulting rebound sends a massive shock wave through the star that results in a supernova explosion. The details of this explosive process are still not fully understood, but it is generally agreed that the explosive outburst process is associated with nuclear processing involving a wide range of isotopes; in fact, the majority of these reacting isotopes are unstable.

It is interesting to note that the majority of stars are closely linked to other stars either as binary stars or higher cluster systems. Binary stars provide a fascinating spectrum of different stellar environments dependent on the masses of the stellar partners, their evolutionary state, and the separation between the partners. Phenomena such as dwarf nova, nova, x-ray bursters, supernova of Type 1a, and even gamma-ray bursters have a natural explanation as being associated with a particular type of stellar binary system. So, for example, a nova explosion is thought to be associated with a binary system comprising of a white dwarf star and a main sequence star evolving towards the end of its life. As the main sequence star expands, material can pass over to the orbiting white dwarf forming a compressed layer of material on its surface. As the density and temperature of this layer increases, a point will be reached where a nuclear runaway situation can occur where the transferred hydrogen and helium explosively reacts with the heavier nuclei in the accumulated layers. Due to the explosive nature of this process, both stable and unstable nuclei are involved. Another similar situation occurs with a binary system comprising a neutron star and evolving main sequence star. For this situation the explosive runaway leads to an intense burst of x-rays – so called x-ray bursters. Here the rapid nuclear processing occurs at a higher temperature and density than for a nova explosion, so the hydrogen and helium material transferred from the main sequence star can be explosively converted to nuclei up to, maybe, mass 100.

From the above, in summary, stellar evolution is strongly associated with nuclear processing of isotopes. Reactions involving stable nuclei drive the evolutionary stages of main sequence stars, while reactions with unstable nuclei dominate the outcome of explosive events. For an understanding of stellar evolution in general, both types of reactions need to be studied in the laboratory. For reactions involving stable isotopes, in principle it is reasonably straightforward to study these reactions in the laboratory. However, the problem is the need to study the reactions at low energies corresponding to the interior temperatures of main sequence stars; this can present the experimenter with extreme challenges due to the low experimental yields associated with very low cross sections. For reactions involving unstable nuclei, due to much higher stellar temperatures, the cross sections needing to be measured are significantly higher, but the real challenge here is to produce in the laboratory the radioactive ions on which to measure the cross sections.

4.2 ISAC Nuclear Astrophysics Program

The radioactive isotopes of interest to nuclear astrophysics investigations are produced by the spallation of a target by 500 MeV protons, ionized in an appropriate ion source and then mass selected by a high resolution mass spectrometer. The ions leaving this spectrometer will have an energy of 2 keV per mass unit. These ions may then either be delivered to the experimenter as is, or be further accelerated. Generally, for nuclear astrophysics purposes, the ions are accelerated; as already noted this is undertaken by the use of an RFQ accelerator section, followed by electronic stripping before further acceleration through a DTL accelerator. The final ion energy can be between 0.15 to 1.5 MeV/A.

Many of the reactions of interest in explosive burning involve (p, γ), (α, γ), and (α, p) reactions. To study these reactions with unstable ions requires the use of inverse kinematics. Two experimental facilities have been constructed specially to study these types of reactions. The DRAGON recoil spectrometer is designed to measure radiative capture reactions. The radioactive beam of interest is directed onto a windowless gas target of either hydrogen or helium. The beam particles together with the few forward peaked radiative captive ions are then directed into the spectrometer, the main purpose of which is to cleanly separate the beam ions from the few reaction product ions.

For the DRAGON spectrometer this separation can be up to 1 in 10^{15} (i.e., for the passage of 10^{15} beam particles, only one of these particles will find its way in error to the end detector). This is achieved by a succession of ion charge and mass selection systems, together with detection of the radiative capture γ-rays by a BGO array surrounding the gas target. Recently this spectrometer has been used to measure the reaction ^{21}Na (p, γ)^{22}Mg at energies corresponding to nova astrophysical temperatures. Measurements such as this will be of crucial importance in interpretation of the new gamma ray satellite data as this becomes available.

The TUDA facility is a large general purpose scattering chamber specifically designed to accommodate a variety of charge particle solid state detector arrays. These arrays are capable of detecting emitted particles with a solid angle approaching 4π. So even with low intensity R.I.B.s, it is possible to study a variety of reactions. Currently the system has been used to complement reactions studied by the DRAGON spectrometer. In particular elastic scattering can be an excellent guide to identify study regions for investigating radiative capture reactions. As another example explosive stellar burning often involves (α, p) reactions. The TUDA facility is ideally suited to study such reactions.

The DRAGON and TUDA facilities are designed to study a wide range of reactions of astrophysical interest. However, the current configuration of the accelerators at ISAC restricts the acceleration of ions to $A < 30$. In this mass range there are various reactions of special interest involving the hot CNO cycle, the CNO cycle, Ne-Na cycle, and the Mg-Al cycle.

When ISAC-II is operational it can be used to address a range of astrophysical problems associated with heavy unstable isotopes. This will be particularly relevant for the astrophysical p and r process and also the higher reaches of the rp process. For these studies, a variety of well-established experimental techniques will be used, but they will need to be specially adapted for radioactive beam experiments. For ISAC-II there will be a need to build new particle and neutron detection arrays; high efficiency γ-ray detector arrays and magnetic spectrometers will also be needed.

5. In Summary

The need for the production of intense beams of radioactive nuclei is recognized within the world-wide scientific community. The ISAC facility at TRIUMF is one of the premier facilities in the world to produce such beams. The first stage of the project, ISAC-I, is working and producing a wide variety of radioactive beams and has the capability to accelerate beams up to 1.5 MeV/A. These accelerated beams are used mainly for nuclear astrophysics experiments, while the low energy beams (~ KeV/A) are used in a wide variety of experimental investigations. ISAC-II is currently under construction and will accelerate beams up to 6.5 MeV/A for all masses up to A =160; it will be completed in

2005. These beams will considerably extend the research capability of ISAC in the areas of nuclear physics and nuclear astrophysics.

All these facilities are open to the international community to use. Proposals for the use of these facilities should be submitted to the international Experimental Evaluation Committee (EEC); details can be found on TRIUMF's web site.

6. Acknowledgements

This paper gives an account of the current situation at ISAC. The design and construction of this facility has involved many people. I would like to pay a tribute to all who have made ISAC one of the leading R.I.B. facilities in the world.

PARTICIPANT LIST

Gurgen Adamian
Joint Institute for Nuclear Research, Dubna
Bogoluybov Laboratory of Theoretical
Physics
Dubna, Moscow Region 141982
RUSSIA
adamian@thsun1.jinr.ru

Duncan Appelbe
CLRC Daresbury Laboratory
Keckwick Lane
Warrington, Cheshire WA4 4AD
UNITED KINGDOM
d.appelbe@dl.ac.uk

Masato Asai
Japan Atomic Energy Research Institute
Advanced Science Research Center
Tokai, Ibaraki 319-1195
JAPAN
asai@tandem.tokai.jaeri.go.jp

Thomas Aumann
GSI Darmstadt
Planckstr. 1
Darmstadt D-64291
GERMANY
t.aumann@gsi.de

Cyrus Baktash
Oak Ridge National Laboratory
P. O. Box 2008
Bldg. 6000, MS 6371
Oak Ridge, TN 37831
USA
baktashc@ornl.gov

Dan Bardayan
Oak Ridge National Laboratory
P. O. Box 2008
Bldg. 6010, MS 6354
Oak Ridge, TN 37831
USA
bardayan@mail.phy.ornl.gov

Jon Batchelder
UNIRIB/ORAU
P. O. Box 2008
Bldg. 6008, MS 6374
Oak Ridge, TN 37831
USA
batcheld@mail.phy.ornl.gov

Cornelius Beausang
Yale University
Physics Department/WNSL
272 Whitney Avenue
New Haven, CT 06520-8124
USA
cornelius.beausang@yale.edu

James Beene
Oak Ridge National Laboratory
P. O. Box 2008
Bldg. 6000, MS 6368
Oak Ridge, TN 37831
USA
beenejr@ornl.gov

Jose Benlliure
University of Santiago de Compostela
Facultad de Fisica, Santiago de Compostela
A Corufla 15706
SPAIN
j.benlliure@usc.es

Fred Bertrand
Oak Ridge National Laboratory/UT
P. O. Box 2008
Bldg. 6000, MS 6369
Oak Ridge, TN 37831
USA
feb@ornl.gov

Carrol Bingham
University of Tennessee
Department of Physics
Knoxville, TN 37996-1200
USA
cbingham@utk.edu

808

Klaus Blaum
CERN EP-IS
Route de Meyrin
Geneva 23 1211
SWITZERLAND
klaus.blaum@cern.ch

Artur Blazkiewicz
Vanderbilt University
801 Hillview Hts. Apt 47
Nashville, TN   37204
USA
a.blazkiar@vanderbilt.edu

Piotr Borycki
University of Tennessee
3700 Sutherland Avenue, I6
Knoxville, TN 37919
USA
pborycki@utk.edu

William Brantley
Furman University
3300 Poinsett Avenue
Department of Phyiscs
Greenville, SC 29613
USA
bill.brantley@furman.edu

Malcolm Burns
University of Paisley
Paisley, Scotland PA1 2BE
UNITED KINGDOM
malcolm.burns@paisley.ac.uk

Ken Carter
UNIRIB/ORAU
P. O. Box 2008
Bldg. 6008, MS 6374
Oak Ridge, TN 37831
USA
carter@mail.phy.ornl.gov

Joakim Cederkall
CERN
Route de Meyrin
Geneve 23 1211

SWITZERLAND
joakim.cederkall@cern.ch

Robert Chapman
University of Paisley
Paisley, Scotland PA1 2BE
UNITED KINGDOM
chap-ph0@wpmail.paisley.ac.uk

Lakshman Chaturvedi
Banaras Hindu University
Deparment of Physics
Varanasi, 221005
INDIA
lakshman@banaras.ernet.in

Jolie Cizewski
Rutgers University
Department of Physics
136 Frelinghuysen Road
Piscataway, NJ 08854-8019
USA
cizewski@physics.rutgers.edu

Jerry Cole
INEEL
2725 Hallon Street
Idaho Falls, ID 83402
USA
jdc@inel.gov

Thomas Cornelius
Institut für Theoretische Physik
Robert-Mayer-Str. 8-10,
Frankfurt am Main, Hessen 60054
GERMANY
corneliu@th.physik.uni-frankfurt.de

Aldo Covello
Università di Napoli Federico II
Dipart. Scienze Fisiche
Complesso Univ. Monte S. Angelo
Via Cintia
Napoli 80126
ITALY
covello@na.infn.it

John Cowan
University of Oklahoma
Department of Physics & Astronomy
440 W. Brooks Street
Norman, OK 73069
USA
cowan@mail.nhn.ou.edu

David Dean
Oak Ridge National Laboratory
P. O. Box 2008
Bldg. 6011, MS 6373
Oak Ridge, TN 37831
USA
dean@mail.phy.ornl.gov

Zden k Dlouhý
Nuclear Physics Institute/ASCR
e, CZ-25068
CZECH REPUBLIC
dlouhy@ujf.cas.cz

Raul Donangelo
Instituto de Fisica - UFRJ
Cidade Universitaria - C.P. 68528
Rio de Janeiro 21941-972
BRAZIL
donangel@if.ufrj.br

Peter Egelhof
GSI Darmstadt
Planckstr. 1
Darmstadt, Hessen 64291
GERMANY
p.egelhof@gsi.de

Timo Enqvist
University of Jyväskylä
Department of Physics
P. O. Box 35
Jyväskylä FIN-40014
FINLAND
timo.enqvist@phys.jyu.fi

Thierry Ethvignot
Commissariat a l'Energie Atomique
Service de Physique Nucleaire, BP 12

91680 Bruyeres-le-Chatel
FRANCE
thierry.ethvignot@cea.fr

Herbert Faust
Institut Laue-Langevin
6, rue Jules Horowitz
Grenoble, Cedex 9 F-38042
FRANCE
faust@ill.fr

Hubert Flocard
Konrad Gelbke
CSNSM-CNRS
Bat. 104
Orsay, Campus 91405
FRANCE
flocard@csnsm.in2p3.fr

Dennis Fong
Vanderbilt University
Box 1807-B
Nashville, TN 37235
USA
d.fong@vanderbilt.edu

Nikolaos Fotiades
Los Alamos National Laboratory
MS H855, Lansce-3
Los Alamos, NM 87545
USA
fotia@lanl.gov

Stefan Frauendorf
University of Notre Dame
Nieuwland Science Building
Notre Dame, IN 46556
USA
sfrauend@nd.edu

Moshe Gai
University of Connecticut
Department of Physics
2152 Hillside Road, U3046
Storrs, CT 06269-3046
USA
gai@uconn.edu

Donald Geesaman
Argonne National Laboratory
9700 S. Cass Avenue
Argonne, IL 60439
USA
geesaman@anl.gov

40 **Konrad Gelbke**
Michigan State University/NSCL
East Lansing, MI 48824
USA
gelbke@nscl.msu.edu

Janine Genevey
Institut des Sciences Nucléaires
IN2P3-CNRS
53, Rue des Martyrs
Grenoble, Is re 38026
FRANCE
genevey@isn.in2p3.fr

Tom Ginter
Michigan State University/NSCL
164 S. Shaw Lane
East Lansing, MI 48824-1321
USA
ginter@nscl.msu.edu

Thomas Glasmacher
Michigan State University/NSCL
164 S. Shaw Lane
East Lansing, MI 48824-1321
USA
glasmacher@nscl.msu.edu

Philip Gore
Vanderbilt University
Box 1807-Station B
Nashville, TN 37235
USA
p.m.gore@vanderbilt.edu

Geoffrey Greene
University of Tennessee
Nielsen Physics 502
Knoxville, TN 37996

USA
greenegl@ornl.gov

Walter Greiner
Johann Wolfgang von-Göethe Universität
Robert-Mayer-Str. 8-10
P. O. Box 111932
Frankfurt am Main 11 60054
GERMANY
greiner@th.physik.uni-frankfurt.de

Konstantin Gridnev
St. Petersburg University
Institute of Physics
Uljanova 1
St. Petersburg, 198504
RUSSIA
gridnev@nuclpc1.phys.spbu.ru

Robert Grzywacz
ORNL/Warsaw University
P. O. Box 2008
Bldg. 6000, MS 6371
Oak Ridge, TN 37831
USA
grzywacz@mail.phy.ornl.gov

Vladimir Gudkov
University of South Carolina
Department of Physics & Astronomy
Columbia, SC 29208
USA
gudkov@sc.edu

J. B. Gupta
59 Saakshra Apts.
A-3 Paschim Vihar
New Delhi 110063
INDIA
j_b_gupta@hotmail.com

Franz-Josef Hambsch
EC-JRC-IRMM
Retieseweg, Geel 2440
BELGIUM
hambsch@irmm.jrc.be

Joseph Hamilton
Vanderbilt University
Box 1807-Station B
Nashville, TN 37235
USA
j.h.hamilton@vanderbilt.edu

Margareta Hellström
GSI Darmstadt
Planckstr. 1
Darmstadt, Hesse D-64291
GERMANY
m.hellstroem@gsi.de

Walter Henning
GSI Darmstadt
Planckstr. 1
Darmstadt D-64291
GERMANY
W.F.Henning@gsi.de

David Hinde
Australian National University
Department of Nuclear Physics
Building 57
Canberra, ACT 0200
AUSTRALIA
david.hinde@anu.edu.au

Sigurd Hofmann
GSI Darmstadt
Planckstr. 1
Darmstadt D-64291
GERMANY
s.hofmann@gsi.de

Hui Hua
University of Rochester
60 Crittenden Blvd.
Apt. 229
Rochester, NY 14620
USA
Hhua@exotic.pas.rochester.edu

Jae-Kwang Hwang
Vanderbilt University
Nashville, TN 37235

Box 1807-Station B
USA
jae-kwang.hwang@vanderbilt.edu

Takatoshi Ichikawa
Konan University
8-9-1 Okamoto
Kobe, Hyogo 658-8501
JAPAN
takichi@popsvr.tokai.jaeri.go.jp

Tetsuro Ishii
Japan Atomic Energy Research Institute
Shirakata-Shirane 2-4
Tokai, Ibaraki 319-1195
JAPAN
ishii@popsvr.tokai.jaeri.go.jp

Naoyuki Itagaki
University of Tokyo
Department of Physics
7-3-1 Hongo
Tokyo 113-0033
JAPAN
ctagaki@phys.s.u-tokyo.ac.jp

Cheng Lie Jiang
Argonne National Laboratory
9700 S. Cass Avenue
Argonne, IL 60439
USA
jiang@anlphy.phy.anl.gov

Ari Jokinen
University of Jyväskylä
Department of Physics
PB 35 (YFL)
Jyväskylä, FIN 40014
FINLAND
ari.s.jokinen@phys.jyu.fi\

Elizabeth Jones
Vanderbilt University
Box 1807-B
Nashville, TN 37235
USA
elizabeth.f.jones@vanderbilt.edu

812

Graham Jones
University of Liverpool
Oliver Lodge Laboratory
Brownlow Hill
Liverpool L69 7ZE
UNITED KINGDOM
jx@ns.ph.liv.ac.uk

Valery Kalinin
V.G. Khlopin Radium Institute
Second Murinsky av., 28
St. Petersburg 194021
RUSSIA
kalinin@atom.nw.ru

Jan Kormicki
Vanderbilt University
Box 1807-Station B
Nashville, TN 37235
USA
jan.kormicki@autocom.pl

Leonid Kravchuk
ORISE/SNS ORNL
701 Scarboro Road
Oak Ridge, TN   37830
USA
kravchuk@sns.gov

Vladimir Kravchuk
KVI
Zernikelaan 25
Groningen, Groningen 9747 AA
THE NETHERLANDS
kravchuk@kvi.nl

Lubos Krupa
Joint Institute for Nuclear Research, Dubna
Flerov Laboratory of Nuclear Reactions
Marie-Curie 6
Dubna 141980
RUSSIA
krupa@musum.jinr.ru

Alexander Laptev
Petersburg Nuclear Physics Institute
Gatchina, Leningrad Region 188300

RUSSIA
laptev@pnpi.spb.ru

Francois Le Blanc
Institut de Physqiue Nucléaire
IN2P3-CNRS
Orsay, Cedex 91406
FRANCE
leblanc@ipno.in2p3.fr

Marek Lewitowicz
GANIL
BP 55027
Caen CEDEX 14076
FRANCE
lewitowicz@ganil.fr

Ke Li
Vanderbilt University
Box 1807-Station B
Nashville, TN 37235
USA
ke.li@vanderbilt.edu

Yixiao Luo
Lawrence Berkeley National Laboratory
Nuclear Science Division
MS 50A-1148
Berkeley, CA 94720
USA
yluo@sseos.lbl.gov

John Lynn
Los Alamos National Laboratory
12 Calle Encanto
P. O. Box 631
Tesuque, NM 87574
USA
eric.lynn@worldnet.att.net

Wenchao Ma
Mississippi State University
Department of Physics
P. O. Box 5167
Mississippi State, MS 39762
USA
mawc@ra.msstate.edu

Paul Mantica
Michigan State University/NSCL
164 S. Shaw Lane
East Lansing, MI 48824-1321
USA
mantica@nscl.msu.edu

Richard Meyer
RAME', Inc & Clark University
129 Maravista Avenue
East Falmouth, MA 02536
USA
ramramruth@aol.com

George Miley
University of Illinois,Urbana-Champaign
103 S. Goodwin Avenue
Room 100
Urbana, IL 61801
USA
g-miley@uiuc.edu

Serban Misicu
Institut für Theoretische Physik
Robert-Mayer Strasse No. 8-10
Frankfurt Main, Hessen D-60054
GERMANY
misicu@th.physik.uni-frankfurt.de

Luciano Moretto
Lawrence Berkeley National Laboratory
Nuclear Science Division
MS 88
Berkeley, CA 94720
USA
lgmoretto@lbl.gov

Witold Nazarewicz
University of Tennessee
401 A. Nielsen Physics Building
Knoxville, TN 37996-1200
USA
witek@utk.edu

Volker Oberacker
Vanderbilt University

Department of Physics & Astronomy
VU Station B 351807
Nashville, TN 37235
USA
volker.e.oberacker@vanderbilt.edu

James Ollier
University of Paisley
Paisley, Scotland PA1 2BE
UNITED KINGDOM
olli-ph0@wpmail.paisley.ac.uk

Suresh Pancholi
University of Delhi
Department of Physics & Astrophysics
Delhi, 110 007
INDIA
pancholi@nde.vsnl.net.in

Joshua Patin
Lawrence Livermore National Laboratory
7000 East Avenue, L-231
Livermore, CA 94550
USA
jbpatin@llnl.gov

John Pearson
University of Montreal
Physics Department
Montreal Qc H3C 3J7
CANADA
pearson@lps.umontreal.ca

Heikki Penttila
University of Jyväskylä
P. O. Box 35 (YFL)
Jyväskylä FIN-40014
FINLAND
heikki.penttila@phys.jyu.fi

Marek Pfutzner
IEP, Warsaw University
Hoza 69
Warszawa PL-00-681
POLAND
pfutzner@mimuw.edu.pl

William Phillips
University of Manchester
Oxford Road
Manchester, England M13 9PL
UNITED KINGDOM
hf@mags.ph.man.ac.uk

Andreas Piechaczek
Louisiana State University
127 W. Farragut Road
Oak Ridge, TN 37830
USA
andreas@mail.phy.ornl.gov

Rodney Piercey
Embry-Riddle Aeronautical University
600 S Clyde Morris Blvd.
306 Lehman Bldg.
Daytona Beach, FL 32114-3900
USA
pierceyr@erau.edu

Jean Pinston
Institut des Sciences Nucleaires
53, Avenue des Martyrs
Grenoble, Isere 38026
FRANCE
pinston@isn.in2p3.fr

Dorin Poenaru
National Institute of Physics & Nuclear
Engineering
Atomistilor 407
P. O. Box MG-6
Bucharest-Magurele 76900
ROMANIA
poenaru@ifin.nipne.ro

A. V. Ramayya
Vanderbilt University
Box 1807-Station B
Nashville, TN 37235
USA
a.v.ramayya@vanderbilt.edu

John Rasmussen
Lawrence Berkeley National Laboratory
817 Oxford Street
Berkeley, CA 94707-2013
USA
jorasmussen@lbl.gov

Lee Riedinger
Oak Ridge National Laboratory
P. O. Box 2008
Bldg. 4500N
Oak Ridge, TN 37831-6263
USA
riedingerl@ornl.gov

Krzysztof Rykaczewski
Oak Ridge National Laboratory
P. O. Box 2008
Bldg. 6000, MS 6371
Oak Ridge, TN 37831
USA
rykaczew@mail.phy.ornl.gov

Mathieu Samyn
Universite Libre de Bruxelles
IAA, CP-226, ULB
Boulevard du Triomphe
Bruxelles, B-1050
BELGIUM
msamyn@astro.ulb.ac.be

Hendrik Schatz
Michigan State University/NSCL
164 South Shaw Lane
East Lansing, MI 48824-1321
USA
schatz@nscl.msu.edu

Paul Semmes
Tennessee Technological University
Department of Physics
Cookeville, TN 38505
USA
psemmes@tntech.edu

Olivier Serot
CEA-Cadarache,DEN/DER/SPRC/LEPh
Bat. 230
St. Paul Lez Durance, 13108
FRANCE
olivier.serot@cea.fr

Dawn Shaughnessy
Lawrence Livermore National Laboratory
P. O. Box 808, L-231
Livermore, CA 94551
USA
shaughnessy2@llnl.gov

Michihiro Shibata
Nagoya University
Furo-cho, Chikusa-Ku
Nagoya 464 8603
JAPAN
i45329a@nucc.cc.nagoya-u.ac.jp

Alan Shotter
TRIUMF
Vancouver, B.C.
CANADA

Gavin Smith
University of Manchester
Brunswick Street
Manchester, England M13 9PL
UNITED KINGDOM
gavin.smith@man.ac.uk

Eugene Spejewski
UNIRIB
444 Mariner Point Drive
Clinton, TN 37716-9481
USA
gene@mail.phy.ornl.gov

Mario Stoitsov
University of Tennessee/JIHIR
P. O. Box 2008
Bldg. 6011, MS 6373
Oak Ridge, TN 37831
USA
stoitsovmv@ornl.gov

Mark Stoyer
Lawrence Livermore National Laboratory
L-231
Livermore, CA 94550
USA
mastoyer@llnl.gov

Jerzy Szerypo
University of Munich
Am Coulombwall 1
D-85748 Garching
GERMANY
jerzy.szerypo@physik.uni-muenchen.de

Mohammed Tantawy
University of Tennessee
401 Nielsen Physics Bldg.
Knoxville, TN 37996-1200
USA
mtantawy@utk.edu

Jun Terasaki
University of Tennessee/JIHIR
P. O. Box 2008
Bldg. 6011, MS 6373
Oak Ridge, TN 37831
USA
terasaki@mail.phy.ornl.gov

Michael Thoennessen
Michigan State University/NSCL
164 S. Shaw Lane
East Lansing, MI 48824-1321
USA
thoennessen@nscl.msu.edu

Ronald Townsend
Oak Ridge Associated Universities
130 Badger Avenue
Oak Ridge, TN 37831-0117
USA
townsenr@orau.gov

Sait Umar
Vanderbilt University
Physics & Astronomy
6301 Stevenson Center

Nashville, TN 37235
USA
umar@compsci.cas.vanderbilt.edu

Anna Unzhakova
Joint Institute for Nuclear Research, Dubna
Frank Laboratory of Neutron Physics
Nuclear Physics Division
Dubna, Moscow Region 141980
RUSSIA
annu@thsun1.jinr.ru

Vladimir Utyonkov
Joint Institute for Nuclear Research, Dubna
Joliot-Curie 6
Dubna, Moscow Region 141980
RUSSIA
utyonkov@sungns.jinr.dubna.su

Alexander Vorobyev
Petersburg Nuclear Physics Institute
Orlova Roshcha
Gatchina, Leningrad District 188300
RUSSIA
alexander.vorobyev@pnpi.spb.ru

Takahiro Wada
Konan University
8-9-1 Okamoto
Kobe, Hyogo 658-8501
JAPAN
wada@konan-u.ac.jp

Michael Wiescher
University of Notre Dame
225 Nieuwland Science Hall
Department of Physics
Notre Dame, IN 46556
USA
wiescher.l@nd.edu

Andreas Woehr
Argonne National Laboratory
Department of Physics
9700 South Cass Avenue, Bldg. 203
Argonne, IL 60561
USA
andreas.woehr@anl.gov

Hans-Juergen Wollersheim
GSI Darmstadt
Plankstrasse 1
Darmstadt D-64291
GERMANY
h.j.wollersheim@gsi.de

Ching-Yen Wu
University of Rochester
NSRL, Department of Physics
University of Rochester
Rochester, NY 14627
USA
Wu@NSRL.rochester.edu

Edward Zganjar
Louisiana State University
Department of Physics & Astronomy
Baton Rouge, LA 70803
USA
zganjar@rouge.phys.lsu.edu

Author Index

Third International Conference on
Fission and Properties of Neutron-Rich Nuclei

Conference Photos

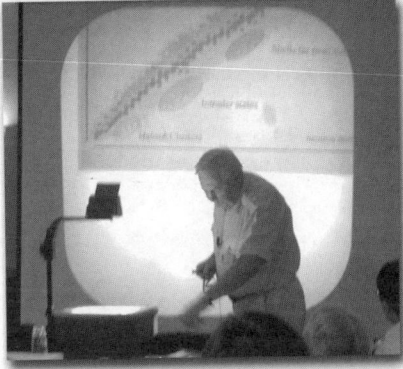

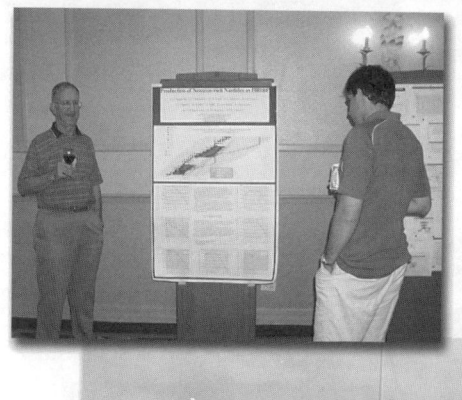